A GIANT STEP: FROM MILLI- TO MICRO-ARCSECOND ASTRONOMY

IAU SYMPOSIUM No. 248

COVER ILLUSTRATION: MICHAEL A.C. PERRYMAN

The second astrometric satellite Gaia was adopted within the scientific programme of the European Space Agency (ESA) in October 2000. It aims to measure the positions, parallaxes and proper motions of the 1 billion celestial bodies on the sky. Expected accuracies are in the 7–25 μ as range down to 15 mag and sub-mas accuracies at the faint limit (20 mag). The astrometric data are complemented by low-resolution spectrophotometric data in the 330–1000 nm wavelength range and, for the brighter stars, radial velocity measurements. As a result, the distances and motions of the stars in our Galaxy will be determined with extraordinary precision, allowing astronomers to determine our Galaxy's three-dimensional structure, space velocities of its constituent stars and, from these data, further our understanding of our Galaxy's origin and evolution.

The original figure is from http://sci.esa.int/science-e/www/object.

INTERNATIONAL ASTRONOMICAL UNION

UNION ASTRONOMIQUE INTERNATIONALE

A GIANT STEP: FROM MILLI- TO MICRO-ARCSECOND ASTROMETRY

PROCEEDINGS OF THE 248th SYMPOSIUM OF
THE INTERNATIONAL ASTRONOMICAL UNION
HELD IN SHANGHAI, CHINA
OCTOBER 15–19, 2007

Edited by

WENJING JIN
Shanghai Astronomical Observatoy, Shanghai, China

IMANTS PLATAIS
Johns Hopkins University, Baltimore, MD, USA

and

MICHAEL A.C. PERRYMAN
European Space Agency, Noordwijk, The Netherlands

CAMBRIDGE UNIVERSITY PRESS
The Edinburgh Building, Cambridge CB2 8RU, United Kingdom
32 Avenue of the Americas, New York, NY 10013-2473, USA
477 Williamstown Road, Port Melbourne, VIC 3207, Australia
Ruiz de Alarcón 13, 28014 Madrid, Spain
Dock House, The Waterfront, Cape Town 8001, South Africa

First published 2008

Printed in the United Kingdom at the University Press, Cambridge

Typeset in System LaTeX 2_ε

A catalogue record for this book is available from the British Library

Library of Congress Cataloguing in Publication data

ISBN 9780521874700 hardback
ISSN 1743-9213

Table of Contents

Session 1. Hipparcos Catalogue
Chair: Erik Høg

Session 2. Highlights of optical astrometry
Chair: Erik Høg **and** *François Mignard*

Session 3. Astrometry with radio interferometers
Chair: Edward B. Fomalont

Session 4. Space astrometry: status and the future

Chair: Alexandre H. Andrei

Session 5. Celestial reference frames at multi-wavelengths

Chair: Dafydd W. Evans

Session 6. Towards reference frame at the micro-arcsecond level

Chair: Dafydd W. Evans **and** *Toshio Fukushima*

Session 7. Stellar parameters and Galactic structure & evolution

Chair: Imants Platais **and** *Catherine Turon*

Session 8. Astrometric education and outreach
Chair: Catherine Turon

Session 9. Astrometry in the age of large surveys and virtual observatories
Chair: Norbert Zacharias

Closing remarks

Preface

IAU Symposium No. 248, *A Giant Step: From Milli- to Micro-Arcsecond Astrometry*, was held in Shanghai between 15–19 October 2007. Plans for the symposium had been initiated two years previously, and the meeting was the culmination of careful and lengthy preparation, both by the Scientific Organizing Committee (co-chaired by Michael Perryman of the European Space Agency and Shuhua Ye of the Shanghai Astronomical Observatory) and by the Local Organizing Committee (chaired by Huang Cheng of the Shanghai Astronomical Observatory). Nearly 200 participants attended the symposium, representing scientists from China and other Asian countries, Europe, North, Central and South America, and Australia.

The symposium intentionally addressed a very broad programme, in terms of wavelength coverage, instrumental techniques, and scientific objectives. This ensured a very wide scientific participation, a very high standard of invited and contributed presentations, many stimulating and productive discussions, and an excellent scientific atmosphere throughout.

The meeting was timed to coincide with the 10th anniversary of the Hipparcos and Tycho Catalogue release in 1997. It indeed marked a pivotal point in the remarkable and inexorable development of astrometry. It provided an opportunity for the scientific assessment of the notable successes of milliarcsec astrometry at both optical and radio wavelengths which have come to fruition over the past decade. The revolutionary technical successes of VLBI in the radio and the Hipparcos space astrometry mission in the optical have led to the International Celestial Reference Frame materialized almost simultaneously at the milliarcsec level at both radio and optical wavelengths. In the last few years, the Hubble Space Telescope Fine Guidance Sensor in the optical, and VLBA observations in the radio, have reached even higher astrometric precision. Many other projects, ranging from re-reductions within the Hipparcos reference frame of Schmidt plate measurements, UCAC and SDSS observations in the optical, and 2MASS in the infrared, have vastly extended the quantity of astrometric data at somewhat lower accuracy. Astrometric information is now available for around a billion objects, and growing rapidly in number and quality.

What has made these technical successes so exciting is the astrophysical panorama which opens up as accuracies have broken through the milliarcsec level: with Hipparcos, geometric distance determinations to large numbers of objects out to 100 pc and beyond have become possible, studies of Galactic kinematics and dynamics in the solar neighborhood and beyond have been brought into much sharper focus, and these various programmes have provided a rich source of empirical data for the verification and refinement of models of star formation, stellar structure and evolution, and of Galactic formation and dynamical evolution. Motions of stars orbiting the Galactic center using adaptive optics imaging in the infrared, and accurate direct distance determinations to nearby star-forming regions using VLBA, are other remarkable examples of the information contained in high-quality astrometric data.

As astrometry thus merges more firmly into astrophysics, the prospects for the future look even more remarkable. As prospects for parallaxes and proper motions for vast numbers of stars at the tens of microarcsec level appear on the astronomical horizon, the symposium participants addressed the attendant scientific harvest that will flow from these data: an astonishingly precise picture of Galactic space motions will provide robust empirical evidence for the origin, formation and evolution of our Galaxy, while models of stellar structure and evolution will be further expanded to embrace the effects of rotation, chemical diffusion, magnetic fields, convection and internal mixing, seismology, and mass transfer. Early and late stages of stellar evolution, dynamical and taxonomic studies of vast numbers of Solar System objects, the formation and evolution of exo-planets, unprecedented and detailed studies of the supermassive black hole at the Galactic center, precise studies of the applicability of General Relativity, and much more, will all become accessible. An even deeper understanding of the Milky Way's complex origin, evolution, and present-day structure will be directly tied to progress in observational astronomy at all wavelengths, and astrometry is now poised to take a leading role.

Technical approaches for reaching these accuracies were presented, spanning the optical, infrared, and radio wavelength regimes. By the curious parallel development of disciplines so often witnessed in science, it appears that accuracies of tens of microarcsec will become accessible almost simultaneously across this wavelength range, through developments of VLBI/VLBA, ALMA and SKA at radio frequencies, through the Gaia and SIM Planet Quest space missions in the optical, and through the accompanying development of theories of Earth rotation, of relativistic modelling, and through the supporting developments of data storage and data processing technologies at levels orders of magnitudes beyond the capabilities that existed only a decade or so ago. Astrometric (infrared) instrumentation at the ESO VLTI and Keck interferometers and elsewhere will also contribute to these microarcsec measurement capabilities. Massive ground-based surveys such as Pan-STARRS and LSST, where the data flows will be measured in Terabytes per night, should bear enormous astrometric harvest, although at a lower accuracy level. Many of these experiments and studies will be well under way within the coming decade.

The new and exciting opportunities that are opening up should also stimulate young scientists entering the field, and part of the symposium was devoted to highlighting the educational opportunities of modern astrometry based on the skills that will be required in the future.

The organizers express their gratitude for the active and constructive participation of the Scientific Organizing Committee throughout the preparation of the symposium. The symposium was sponsored and supported by IAU Divisions *I* (Fundamental Astronomy); and by the IAU Commissions No. *8* (Astrometry), No. *4* (Ephemerides), No. *19* (Rotation of the Earth), No. *26* (Double and Multiple Stars), No. *33* (Structure and Dynamics of the Galactic System), No. *35* (Stellar Constitution), No. *40* (Radio Astronomy), and by the following IAU Working Groups: General Relativity in Celestial Mechanics, Astrometry and Metrology; Natural Planetary Satellites, Nomenclature for Fundamental Astronomy; Optical/Infrared Interferometry; and Global VLBI.

All participants and their accompanying guests were pleased to express their collective gratitude to the Local Organizing Committee, comprised by Huang Cheng and his colleagues of the Shanghai Astronomical Observatory, who did so much to make the meeting run smoothly and efficiently, and to make it both enjoyable and highly memorable. Funding for the meeting, including essential travel sponsorship, was kindly and generously provided by the International Astronomical Union, the Chinese Astronomical Society, the National Natural Science Foundation of China, the National Science Foundation of the USA, the Shanghai International Culture Association, the Science and Technology Commission of Shanghai Municipality, the National Astronomical Observatories, the Shanghai Astronomical Observatory, and the Chinese Academy of Sciences.

Michael Perryman and Shuhua Ye (Co-Chairs of the Scientific Organising Committee)
Wenjing Jin, Imants Platais and Michael Perryman (Proceedings Editors)

THE ORGANIZING COMMITTEE

Scientific

A.H. Andrei (Brazil)
P.T. de Zeeuw (Netherlands)
D.W. Evans (UK)
W.J. Jin (China PR)
S.A. Klioner (Germany)
I.I. Kumkova (Russia)
R. Launhardt (Germany)
C.L. Lu (China PR)

F. Mignard (France)
M.A.C. Perryman (chair, Netherlands)
I. Platais (USA)
M.J. Reid (USA)
D.J. Schade (Canada)
M. Stavinschi (Romania)
C. Turon (France)
S.H. Ye (co-chair, China PR)

Local

C. Huang(chair)
L. Chen
G.X. Dong
C.L. Huang

J.L. Li
L.L. Shang
Z.H. Tang

Acknowledgements

The symposium is sponsored and supported by the IAU Divisions I (Fundamental Astronomy); and by the IAU Commissions No. 8 (Astrometry), No. 4 (Ephemerides), No. 19 (Rotation of the Earth), No. 26 (Double and Multiple Stars), No. 33 (Structure & Dynamics of the Galactic System), No. 35 (Stellar Constitution), No. 40 (Radio Astronomy); and by WG (Future Developments in Ground-Based Astrometry), WG (General Relativity in Celestial Mechanics, Astrometry and Metrology), WG (Natural Planetary Satellites), WG (Nomenclature for Fundamental Astronomy), WG (Optical/IR Interferometry), WG (Global VLBI).

The Local Organizing Committee operated under the auspices of the
Shanghai Astronomical Observatory, Chinese Academy of Sciences.

Funding by the International Astronomical Union,
Chinese Astronomical Society,
National Natural Science Foundation of China,
National Science Foundation of USA,
Shanghai International Culture Association,
Science and Technology Commission of Shanghai Municipality,
National Astronomical Observatories, Chinese Academy of Sciences,
and
Shanghai Astronomical Observatory, Chinese Academy of Sciences
is gratefully acknowledged.

CONFERENCE PHOTOGRAPH

Participants

Alexandre H. **Andrei**, (1) Observatorio Nacional C MCT; (2) Observatorio do Valongo - UFRJ, Brazil — oat1@on.br
Frédéric **Arenou**, GEPI, Observatoire de Paris, CNRS, Universite Paris Diderot, France — Frederic.Arenou@obspm.fr
Mochamad I. **Arifyanto**, Institut Teknologi Bandung, Indonesia — ikbal@as.itb.ac.id
Jean-Eudes **Arlot**, IMCCE/Paris obs./CNRS, France — arlot@imcce.fr
Carine **Babusiaux**, Observatoire de Paris GEPI, France — carine.babusiaux@obspm.fr
Octavian **Badescu** , Bucharest Faculty of Geodesy, Romania — octavian@aira.astro.ro
Norbert **Bartel**, York University, Canada — bartel@yorku.ca
Charles **Beichman**, Michelson Science Center, USA — chas@pop.jpl.nasa.gov
George F. **Benedict**, McDonald Observatory/University of Texas, USA — fritz@astro.as.utexas.edu
Andrew F. **Boden**, Michelson Science Center/Caltech, USA — bode@ipac.caltech.edu
Anthony **Brown**, Leiden Observatory, Netherlands — brown@strw.leidenuniv.nl
Andreas B. **Brunthaler**, Max-Planck-Institut fuer Radioastronomie, Germany — brunthal@mpifr-bonn.mpg.de
Beatrice **Bucciarelli**, INAF - Astronomical Observatory of Torino, Italy — bucc@oato.inaf.it
Alexey **Butkevich**, Lohrmann Observatory, Dresden Technical University, Germany — alexey.butkevich@tu-dresden.de
Nicole **Capitaine**, SYRTE/UMR8630 Observatoire de Paris France, France — n.capitaine@obspm.fr
Phillip A. **Cargile**, Vanderbilt University, USA — p.cargile@vanderbilt.edu
Jeffrey L. **Carlin**, University of Virginia, USA — jc4qn@mail.astro.virginia.edu
Brian **Chaboyer**, Department of Physics and Astronomy Dartmouth College, USA — Brian.Chaboyer@Dartmouth.edu
Patrick **Charlot**, Observatoire de Bordeaux, France — charlot@obs.u-bordeaux1.fr
Shami **Chatterjee**, The University of Sydney, NSW 2006 Australia, Australia — S.Chatterjee@physics.usyd.edu.au
Hao **Cheng**, Shanghai Astronomical Observatory, China — chz@shao.ac.cn
Li **Chen**, Shanghai Astronomical Observatory, China — chenli@shao.ac.cn
Yoonkyung **Choi**, University of Tokyo/NAOJ, Japan — yoonkyung.choi@nao.ac.jp
Yaoquan **Chu**, University of Science and Technology of China, China — yqchu@ustc.edu.cn
Maria-Teresa **Crosta**, INAF-Astronomical Observatory of Turin (OATo), Italy — crosta@oato.inaf.it
Chenzhou **Cui**, National Astronomical Observatories, CAS, China — ccz@bao.ac.cn
Jos **deBruijne**, ESA/ESTEC (SCI-SA), Netherlands — jdbruijn@rssd.esa.int
Xuemei **Deng**, Purple Mountain Observatory, China — dengxuemei403@163.com
Josselin **Desmars**, IMCCE, Paris Observatory, France — desmars@imcce.fr
Peng **Dong**, Purple Mountain Observatory, China — dongpeng@pmo.ac.cn
Christine **Ducourant**, Observatoire de Bordeaux LAB, France — ducourant@obs.u-bordeaux1.fr
Daniel **Egret**, Observatoire de Paris, France — Daniel.Egret@obspm.fr
Dafydd W. **Evans**, Institute of Astronomy, Cambridge, UK — dwe@ast.cam.ac.uk
Claus V. **Fabricius**, Institute for Space Studies of Catalonia and University of Barcelona, Spain — claus@am.ub.es
Jacqueline K. **Faherty**, Stony Brook University/American Museum of Natural History, USA — jfaherty@amnh.org
Edward B. **Fomalont**, National Radio Astronomy Observatory, USA — efomalon@nrao.edu
Mariagrazia **Franchini**, INAF-Osservatorio Astronomico di Trieste, Italy — franchini@oats.inaf.it
Toshio **Fukushima**, NAOJ, Japan — Toshio.Fukushima@nao.ac.jp
Laiwo M. **Fung**, Shanghai Astronomical Observatory, China/USA — lwfung@shao.ac.cn
Mario **Gai**, Istituto Nazionale di Astrofisica Osservatorio Astronomico di Torino, Italy — gai@oato.inaf.it
Ronny **Geisler**, Landessternwarte Heidelberg - ZAH, Germany — rgeisler@lsw.uni-heidelberg.de
Andrea M. **Ghez**, UCLA, USA — ghez@astro.ucla.edu
Stefan **Gillessen**, Max-Planck-Institute for extraterrestial physics, Garching, Germany — ste@mpe.mpg.de
Terrence **Girard**, Yale University, USA — girard@astro.yale.edu
Ana **Gomez**, Observatoire de Paris GEPI 92195 Meudon Cedex, France — ana.gomez@obspm.fr
Naoteru **Gouda**, JASMINE Project Office, NAOJ, Japan — naoteru.gouda@nao.ac.jp
Jean P. **Guibert**, Centre d'Analyse des Images,GEPI/Observatoire de paris, France — Jean.Guibert@obspm.fr
Jose C. **Guirado**, DAA, Universitat de Valencia, Spain — j.c.guirado@uv.es
Li **Guo**, Shanghai Astronomical Observatory, China — kent-gl@shao.ac.cn
Kazuya **Hachisuka**, Shanghai Astronomical Observatory, China/Japan — khachi@shao.ac.cn
Wenbiao **Han**, Shanghai Astromomical Observatory, CAS, China — wbhan@shao.ac.cn
Misha **Haywood**, Paris-Meudon Observatory, France — Misha.Haywood@obspm.fr
Daniel **Hestroffer**, IMCCE/Paris Observatory, France — hestro@imcce.fr
David A. **Hobbs**, Lund Observatory, Sweden — david@astro.lu.se
Erik **Hoeg**, Niels Bohr Institute Copenhagen University, Denmark — erik.hoeg@get2net.dk
Mareki **Honma**, NAOJ, Japan — mareki.honma@nao.ac.jp
Chengli **Huang**, Shanghai Astronomical Observatory, China — clhuang@shao.ac.cn
Cheng **Huang**, Shanghai Astronomical Observatory, China — hc@shao.ac.cn
Tianyi **Huang**, Department of Astronomy, Nanjing University, China — tyhuang@nju.edu.cn
Hiroshi **Imai**, Department of Physics, Faculty of Science, Kagoshima University, Japan — hiroimai@sci.kagoshima-u.ac.jp
Yago **Isasi**, Universitat de Barcelona (Spain), Spain — yagoisasi@gmail.com
Zeljko **Ivezic**, University of Washington, USA — ivezic@astro.washington.edu
David J. **James**, Vanderbilt University, USA — david.j.james@vanderbilt.edu
Weichun **Jao**, Georgia State University, USA — jao@chara.gsu.edu
Nianchuan **Jian**, Shanghai astronomical observatory, China — jnnccc@shao.ac.cn
Chengjin **Jin**, National Astronomical Observatories, China — cjjin@bao.ac.cn
Wenjing **Jin**, Shanghai Astronomical Observatory, China — jwj@shao.ac.cn
Carme **Jordi**, Dept. Astronomia i Meteorologia Universitat de Barcelona (ICC/IEEC), Spain — CARME@AM.UB.ES
Arkadiy **Kharin**, Main Astronomical Observatory of National Academy of Sciences of Ukraine, Ukraine — kharin@mao.kiev.ua
Sergei A. **Klioner**, Lohrmann Observatory, Germany — Sergei.Klioner@tu-dresden.de
Hideyuki **Kobayashi**, NAOJ, Japan — hideyuki.kobayashi@nao.ac.jp
Yukiyasu **Kobayashi**, NAOJ, Japan — yuki@merope.mtk.nao.ac.jp
Sergey **Kopeikin**, University of Missouri-Columbia, USA — kopeikins@missouri.edu
Yuri Y. **Kovalev**, Max Planck Institute for Radio Astronomy, Germany — ykovalev@mpifr-bonn.mpg.de
Vera **Kozhurina-Platais**, Space Telescope Science Institute, USA — verap@stsci.edu
Yevgen **Kozyryev**, RI Nikolaev Astronomical Observatory, Ukraine — tttt_na0@mail.ru
Irina I. **Kumkova**, Sobolev Astronomical Institute, Saint- Petersburg State University, Russia — kumkova@iperas.nw.ru
Tomoharu **Kurayama**, NAOJ, Japan — t.kurayama@nao.ac.jp
Uwe R. **Lammers**, European Space Astronomy Centre of European Space Agency, Spain — Uwe.Lammers@sciops.esa.int
Ralf **Launhardt**, Max Planck Institute for Astronomy, Germany — rl@mpia.de
Yveline **Lebreton**, Observatoire de Paris, France — Yveline.Lebreton@obspm.fr
Sebastien **Lepine**, Department of Astrophysics American Museum of Natural History, USA — lepine@amnh.org
Jean-Francois **Lestrade**, Observatoire de Paris CNRS, France — jean-francois.lestrade@obspm.fr
Lennart **Lindegren**, Lund Observatory, Sweden — lennart@astro.lu.se
Yu **Liu**, Shanghai Astronomical Observatory, China — yuliu@shao.ac.cn
Jinling **Li**, Shanghai Astronomical Observatory, China — jll@shao.ac.cn
Juan **Li**, Shanghai Astronomical Observatory, China — lijuan@shao.ac.cn
Zhengxin **Li**, Shanghai Astronomical Observatory, China — shaolzx@yahoo.com
Laurent R. **Loinard**, Centro de Radioastronomiay Astrofisica Universidad, Mexico — l.loinard@astrosmo.unam.mx
Chunli **Lu**, Purple Mountain Observatory, China — cllu@pmo.ac.cn
Eugene A. **Magnier**, Institute for Astronomy University of Hawaii, USA — eugene@ifa.hawaii.edu
Steven R. **Majewski**, Department of Astronomy, University of Virginia, USA — srm4n@virginia.edu
Yindun **Mao**, Shanghai Astronomical Observatory, China — dundun@shao.ac.cn
Jon M. **Marcaide**, University of Valencia, Spain — J.M.Marcaide@uv.es
Paolo D. **Marcantonio**, INAF Osservatorio Astronomico di Trieste, Italy — dimarcan@oats.inaf.it
Ivan **Marti-Vidal**, Dpt. Astronomia i Astrofisica Universitat de Valencia, Spain — i.marti-vidal@uv.es
Eduard **Masana**, Departament d'Astronomia i Meteorologia. Universitat de Barcelona, Spain — emasana@am.ub.es
Chopo **Ma**, NASA-Goddard Space Flight Center, USA — chopo.ma@nasa.gov
Francois **Mignard**, Observatoire de la Côte d'Azur, France — francois.mignard@obs-nice.fr
Masanori **Miyamoto**, National Astronomical Observatory, Japan, Japan — miyamo@nn.iij4u.or.jp
Makoto **Miyoshi**, NAOJ, Japan — makoto.miyoshi@nao.ac.jp
Carlo **Morossi**, INAF Osservatorio Astronomico di Trieste, Italy — morossi@oats.inaf.it
Serge **Mouret**, IMCCE - Paris observatory, France — mouret@imcce.fr
Jose L. **Muinos**, Real Instituto y Observatorio de la Armada, Spain — ppmu@roa.es
Akiharu **Nakagawa**, Kagoshima University, Japan — nakagawa@astro.sci.kagoshima-u.ac.jp

Yoshito **Niwa**, National Astronomical Observatory Japan / Kyoto University, Japan — kazin.niwa@nao.ac.jp
Weitou **Ni**, Purple Mountain Observatory, China — wtni@pmo.ac.cn
Karen **O'Flaherty**, European Space Agency, Netherlands — koflaher@rssd.esa.int
William J. **O'Mullane**, European Space Astronomy Centre, Spain — william.omullane@esa.int
Xiaopei **Pan**, Jet Propulsion Laboratory, USA — xiaopei.pan@jpl.nasa.gov
Petre V. **Paraschiv**, Astronomical Institute of Romanian Academy, Romania — paras@aira.astro.ro
Jean C. **Pecker**, Former General Secretary of the IAU — c.pecker@wanadoo.fr
Qingyu **Peng**, Jinan University, China — pengqy@pub.guangzhou.gd.cn
Michael A. C. **Perryman**, Astrophysics Missions Division (SCI-SA), ESTEC, Netherlands — mperryma@rssd.esa.int
Slawomir S. **Piatek**, New Jersey Institute of Technology & Rutgers University, USA — piatek@physics.rutgers.edu
Jinsong **Ping**, Shanghai Astronomical Observatory, CAS, China — pjs@shao.ac.cn
Gennadiy I. **Pinigin**, Research Institute "Nikolaev Astronomical Observatory", Ukraine — pinigin@mao.nikolaev.ua
Elena V. **Pitjeva**, Institute of Applied Astronomy of Russian Academy of Sciences, Russian — evp@ipa.nw.ru
Imants **Platais**, Johns Hopkins University, Depatment of Physics & Astronomy, USA — imants@pha.jhu.edu
Petre P. **Popescu**, Astronomical Institute of Romanian Academy, Romania — petre@aira.astro.ro
Dimitri **Pourbaix**, Intitute of Astronomy and Astrophysics Universite Libre de Bruxelles, Belgium — pourbaix@astro.ulb.ac.be
Yuri **Protsyuk**, Research Institute "Nikolaev Astronomical Observatory", Ukraine — yuri@mao.nikolaev.ua
Timo **Prusti**, European Space Agency, Netherlands — Timo.Prusti@rssd.esa.int
Shubo **Qiao**, Shanghai Astronomical Observatory, China — qsb@shao.ac.cn
Zhaoxiang **Qi**, Shanghai Astronomical Observatory, China — kevin@shao.ac.cn
Andreas **Quirrenbach**, Landessternwarte Heidelberg, Germany — A.Quirrenbach@lsw.uni-heidelberg.de
Tristan **Röll**, Friedrich-Schiller-Universitaet Jena, Germany — troell@astro.uni-jena.de
Matias C. **Radiszcz**, Universidad de Chile, Chile — mradis@das.uchile.cl
Mark J. **Reid**, Harvard-Smithsonian CfA, USA — reid@cfa.harvard.edu
Shulin **Ren**, Purple Mountain Observatory, China — rensl@pmo.ac.cn
Andrea **Richichi**, European Southern Observatory, Germany — arichich@eso.org
Noel **Robichon**, GEPI Observatoire de Paris, France — noel.robichon@obspm.fr
Annie C. **Robin**, Institut UTINAM BP1615 25010 Besancon cedex, France — annie.robin@obs-besancon.fr
Cyril **Ron**, Astronomical Institute AS CR, Czech Republic — ron@ig.cas.cz
Johannes **Sahlmann**, ESO, Geneva Observatory, Germany — jsahlman@eso.org
Tobias **Schmidt**, Astrophysical Institute, FSU Jena, Germany — tobi@astro.uni-jena.de
Ralf-Dieter **Scholz**, Astrophysikalisches Institut Potsdam, Germany — rdscholz@aip.de
Johny **Setiawan**, Max-Planck-Institut fr Astronomie, Germany — setiawan@mpia.de
Michael **Shao**, Jet Propulsion Laboratory, USA — michael.shao@jpl.nasa.gov
Allyson A. **Sheffield**, Vassar College/University of Virginia, USA — alsheffield@vassar.edu
Zhiqiang **Shen**, Shanghai Astronomical Observatory, China — zshen@shao.ac.cn
Kaixian **Shen**, National Time Service Center, China — shenkx@ntsc.ac.cn
Ming **Shen**, Nanjing University, China — shenming@nju.org.cn
Richard L. **Smart**, INAF OATo, Italy — smart@oato.inaf.it
Mitsuru **Soma**, NAOJ, Japan — Mitsuru.Soma@nao.ac.jp
Alessandro **Sozzetti**, INAF-OATo & Harvard Smithsonian Center for Astrophysics, Italy — sozzetti@oato.inaf.it
Magda **Štavinschi**, Astronomical Institute of the Romanian Academy, Romania — magda_stavinschi@yahoo.fr
Vojtoch **Štefka**, Astronomical Institute AS CR, Czech Republic — stefka@ig.cas.cz
Masahiro **Suganuma**, NAOJ, Japan — suganuma@merope.mtk.nao.ac.jp
Zhenghong **Tang**, Shanghai Astronomical Observatory, China — zhtang@shao.ac.cn
Andrew R. **Taylor**, University of Calgary Department of Physics and Astronomy, Canada — russ@ras.ucalgary.ca
Ramachrisna **Teixeira**, IAG C São Paulo University, Brazil — teixeira@astro.iag.usp.br
William **Thuillot**, IMCCE-Paris Observatory, FRANCE — thuillot@imcce.fr
Jordi **Torra**, Dept. Astronomia i Meteorologia Universitat de Barcelona (ICC/IEEC), Spain — jordi@am.ub.es
Rosa M. **Torres-Lopez**, Centro de Radioastronomia y Astrofisica - UNAM, Mexico — r.torres@astrosmo.unam.mx
Takuji **Tsujimoto**, National Astronomical Observatory, Japan — taku.tsujimoto@nao.ac.jp
Robert N. **Tubbs**, Max Planck Institute for Astronomy, Germany — tubbs@mpia.de
Catherine **Turon**, Observatoire de Paris, France — catherine.turon@obspm.fr
William F. **van Altena**, Yale University, USA — vanalten@astro.yale.edu
Floor **van Leeuwen**, Institute of Astronomy University of Cambridge, UK — fvl@ast.cam.ac.uk
Alberto **Vecchiato**, INAF - Astronomical Observatory of Torino, Italy — vecchiato@oato.inaf.it
Jan **Vondrák**, Astronomical Institute, Czech Republic — vondrak@ig.cas.cz
Nicholas A. **Walton**, Institute of Astronomy University of Cambridge, UK — naw@ast.cam.ac.uk
Shuhe **Wang**, Shanghai Astronomical Observatory, China — shwang@shao.ac.cn
Zhenyu **Wu**, National Astronomical Observatories, China — zywu@bao.ac.cn
Yi **Xie**, Astronomy Department of Nanjing University, China — yi.s.xie@gmail.com
Ye **Xu**, Max-Planck-Institut fuer Radioastronomie, Germany — xuye@mpifr-bonn.mpg.de
Yoshiyuki **Yamada**, Department of Physics, Kyoto University, Japan — yamada@amesh.org
Masahiro **Yamauchi**, University of Tokyo, NAOJ, Japan — yamauchi@merope.mtk.nao.ac.jp
Taihei **Yano**, NAOJ, Japan — yano.t@nao.ac.jp
Shuhua **Ye**, Shanghai Astronomical Observatory, China — ysh@shao.ac.cn
Yong **Yu**, Shanghai Astronomical Observatory, China — yuy@shao.ac.cn
Norbert **Zacharias**, U.S. Naval Observatory, USA — nz@usno.navy.mil
Marion I. **Zacharias**, 3450 Mass. Ave. NW Washington DC, 20392, USA — miz@usno.navy.mil
Alexander F. **Zakharov**, Institute of Theoretical and Experimental Physics, Russia — zakharov@itep.ru
Bo **Zhang**, Shanghai Astronomical Observatory, China — zb@shao.ac.cn
Mian **Zhang**, Shanghai Astronomical Observatory, China — jitai@shao.ac.cn
Haibin **Zhao**, Purple Mountain Observatory, China — meteorzh@pmo.ac.cn
Cheng **Zhao**, Purple Mountain Observatory, China — zhaocheng@pmo.ac.cn
Yongheng **Zhao**, National Astronomical Observatories, CAS, China — yzhao@lamost.org
Xingwu **Zheng**, Nanjing University, China — xwzheng@nju.edu.cn
Zi **Zhu**, Department of Astronomy, Nanjing University, China — zhuzi@nju.edu.cn

Opening address by the Local Organizing Committee

Dear colleagues,

On behalf of the Local Organizing Committee it is a pleasure to welcome all of you to the IAU Symposium No. 248 in our beautiful city of Shanghai.

Shanghai is a modern international metropolis, China's center of economy, science and technology, information and culture. This fast changing city has retained colorful and harmonious mixture of traditional and modern Chinese civilizations. It has many wonderful cultural opportunities for you to discover and to enjoy. Dear friends, please take your time to explore the present and the future of astrometry, to contribute your knowledge, to renew acquaintances, to make new friends, and to enjoy Chinese culture in this beautiful golden autumn.

The symposium is held at the conference hall of the Shanghai Astronomical Observatory (SHAO). It was formally established in 1962 following the merger of the former Xujiahui (originally spelt Zi-Ka-Wei) and Sheshan (Zô-Sè) observatories, which were founded by the French Mission Catholique in 1872 and 1900 respectively. It is an institute of the Chinese Academy of Sciences (CAS). The observatory's main research activities fall into four divisions: astro-geodynamics, astrophysics, very long baseline interferometry (VLBI), and the technical laboratories. Its observing facilities include a 25m radio telescope used for VLBI, a 1.56m optical telescope, and a 60cm satellite laser-ranging system (SLR), all of which were built by ourselves. There are also three technical laboratories researching hydrogen atomic clocks, VLBI techniques, and optical astronomy. SHAO is responsible for Chinese VLBI and SLR networks. It also hosts the central office of the Asia-Pacific Space Geodynamics Program (APSG), as well as a partner group of the Max Planck Institute for Astrophysics (MPA).

Today we clearly remember the last IAU Symposium No. 156 on **Developments in Astrometry and Their Impact on Astrophysics and Geodynamics** held also in Shanghai in September 1992. This Symposium achieved a satisfactory success under the efforts of all its participants. We sincerely hope that the IAU Symposium No. 248 will make a much greater success under your active participation and we will try our best to make you feel comfortable.

We are deeply grateful to several institutions for their financial or organizing support: the International Astronomical Union, Chinese Astronomical Society, National Natural Science Foundation of China, National Science Foundation of USA, Shanghai Association for Science & Technology, Science and Technology Commission of Shanghai Municipality, Shanghai International Culture Association, and National Astronomical Observatories of Chinese Academy of Sciences, especially SHAO.

We sincerely hope that you will enjoy the meeting.

Thank you very much and welcome!

Cheng HUANG, Chair of LOC
Shanghai China, September 12, 2007

A Giant Step: from Milli- to Micro-arcsecond Astrometry
Proceedings IAU Symposium No. 248, 2007
W. J. Jin, I. Platais & M. A. C. Perryman, eds.

The Hipparcos Catalogue:
10th anniversary and its legacy

C. Turon and F. Arenou

GEPI/UMR-CNRS 8111, Observatoire de Paris, 92190 Meudon, France
catherine.turon@obspm.fr, frederic.arenou@obspm.fr

Abstract. The European Space Agency decision to include the Hipparcos satellite into its Science Programme is placed in the context of the years 1965-1980 and in the historical perspective of the progress of astrometry. The motivation and ideas which lead to the Hipparcos design are reviewed as well as its characteristics and performance. The amount and variety of applications represent an impressive evolution from the original science case and opened the way to much more ambitious further space missions, especially Gaia, based on the same basic principles. A giant step in technology led to a giant step in science. Next steps are presented at this Symposium.

Keywords. astrometry, space vehicles, stars: distances, stars: kinematics, stars: fundamental parameters, Galaxy: fundamental parameters, Galaxy: solar neighbourhood

1. Introduction

It is quite difficult for us, in 2007, to transport ourselves in the context of the years 1965-1970 when Pierre Lacroute, at Strasbourg Observatory, first put forward the idea that the only way out to improve drastically the precision of astrometric observations was by observing stars from space. The best available positions and proper motions at that epoch were included in the fundamental catalogue FK4 (Fricke *et al.* 1963). The internal precision of position was from about 0.″04 at epoch 1950 for stars in the northern hemisphere to 0.″08 in the southern hemisphere. Moreover, systematic errors could reach the same order of magnitude in some areas of the sky. The expected improvements were very limited (for example, it was expected to lower the systematic errors to some 0.″03 by using a network of astrolabes for over 10 years...). The three main difficulties were the deterioration of stellar images caused by the atmosphere, the non-stability of the instruments and the inability of ground-based observatories to observe the whole celestial sphere.

Pierre Lacroute came to the conclusion that, after a long period of instrumental improvements, any new development would ineluctably come up against the main difficulty: the presence of the atmosphere itself, and that only slow and very limited progress could be expected from the ground. He then proposed a very daring solution that he himself qualified as 'weird' (farfelue) (Lacroute 1991): to make astrometric observations from a satellite (Lacroute 1966; Lacroute 1982; Kovalevsky 2005). His arguments were very simple: to build a reference system over the whole celestial sphere, you need to measure large angles. Superposing two fields of view at a fixed angular distance with a very stable complex mirror was possible on a satellite, without the atmosphere, thus allowing to measure small angles instead. This was the way to eliminate systematic errors. After a very first proposition to CNES in March 1966 (Lacroute 1966), the idea was presented to the IAU General Assembly in Prague (Lacroute 1967). It was received with much interest.

2. Why and how an astrometric satellite

2.1. *The first steps: 1966-1972*

After the very first steps described above, a new, more detailed version was presented to CNES in November 1967, and a feasibility study performed in 1969-1970 (Husson 1975). In addition to a grid made of regularly spaced opaque and transparent slits, a new crucial modification of the project was the introduction of a complex mirror, in place of the pentagonal prism of the previous version, used to project the two fields of view in the focal plane. This was the key to obtain a stable basic angle. It is noticeable that two of the basic characteristics of the future Hipparcos were already introduced in the project as early as 1967: the complex mirror and the grid. The stars of both fields of view were crossing the grid perpendicularly to the slits, their light was modulated, and measured by two photometers. It was possible to measure the angle between two stars of magnitude $V = 7$ to an accuracy of $0\overset{''}{.}01$.

The CNES feasibility study proposed a satellite of 140 kg to be launched by a Diamant rocket, able to observe at least 700 stars all over the sky with no systematic error and with a precision of about $0\overset{''}{.}01$. In addition, trigonometric parallaxes could be measured with an improvement of the accuracy with respect to ground-based observations by a factor 5 to 10. The scientific perspectives were underlined with enthusiasm by J. Kovalevsky in a note to CNES (Kovalevsky 1969) and by the referee Committee set up by CNES to examine the project of P. Lacroute (Kovalevsky 1970): '*Nearly all fields of astronomy will be concerned*'. In addition to the stellar reference system and its connection to the dynamic reference system, many other topics were quoted: stellar kinematics, Galactic dynamics, fundamental astronomical constants, motions of the Earth crust, stellar and cosmic distance scale, etc. The résumé was '*this experience, if successful, would completely revolutionize astrometry*'. However, there were a number a critical points, more noticeably the realisation of the complex mirror and the stabilisation of the platform, incompatible with a Diamant launcher.

In August 1970, during the IAU General Assembly in Brighton, P. Lacroute made presentations to Commissions 8 and 24 about the advantage of space observations for astrometry, drawing the attention of the astronomical community to the possibility of measuring absolute trigonometric parallaxes to better than $0\overset{''}{.}005$. Commission 8 unanimously adopted a resolution underlining the big interest of space for rapidly obtaining a major improvement of the astrometric accuracy, improvement which was very difficult, if not impossible, to obtain from the ground (Lacroute 1970).

2.2. *Two different options: SpaceLab and TD, 1973-1974*

In November 1973, P. Lacroute proposed to ESRO two possible options of a space astrometry satellite, further presented to the ESRO advisory structure by Jean Kovalevsky (Bacchus & Lacroute 1974). These two options were different in their scientific objectives and, as a consequence, in their principles. The TD option proposed to use a TD1 type satellite, systematically scanning the sky and observing all 150 000 stars brighter than a given magnitude. The SpaceLab option would on the contrary observe a pre-defined programme of up to 40 000 selected stars prepared in advance. This option permitted '*to include objects of special interest*' but required long and complex pointing. No less than 8 missions of 11 days over two years were planned to obtain the astrometric parameters of the programme stars (!).

In this context, ESRO was advised by the Launching Programme Advisory Committee to organize a symposium with a twofold motivation: the scientific significance and objectives of a space astrometry mission, and the support by the community. This Symposium was held in Frascati in October 1974 and was a big success. It marked the opening of a new era where space astrometry was considered not only for its astrometric applications, but also, and mainly, for its astrophysics science case. The conclusions of the Symposium recommended the SpaceLab option for its capability of pre-selection of programme stars.

2.3. *ESA studies: 1975-1979*

Following the success of the Frascati Symposium, demonstrating '*the fundamental importance of improving astrometric data for a large scientific community, and highlight*(ing) *the potential of* (astrometry) *space projects*', a 'Mission Definition Group' composed of five scientists and four ESA engineers was set up by ESA. Two configurations, intentionally very different, were studied in order to fully explore the range of capabilities of space astrometry (Mission definition study 1977): option A, with preselection of stars (under E. Høg's leadership), option B with the simplest observation procedure and systematic observation of all stars brighter than 10.5^{m} (under P. Lacroute's leadership). A major improvement in overall efficiency was obtained thanks to the introduction of an Image Dissector Tube (Høg 1975). This device allowed both a systematic scanning of the sky **and** the selection of the programme stars. It is noticeable that E. Høg also proposed simultaneous photometric observations (low dispersion spectra).

The Mission definition study was followed, in 1976-1979, by a feasibility study of a project to be launched by Ariane on a geostationary orbit (Phase A study 1978, 1979). The Hipparcos mission was aiming at the accurate measurement of trigonometric parallaxes and annual proper motions of about 100 000 selected stars with an expected average mean error of about $0''.002$.

2.4. *The inclusion of Hipparcos in the ESA Science Programme: 1979-1980*

Even though, in the 1970s, a '*new impetus* (was) *given to astrometry by the use of large reflecting telescopes, automatic measuring machines, radio astrometry and photoelectric astrometry, ... astrometry to many* (was looking) *dull and old-fashioned*' in comparison with the spectacular evolution of astrophysics (Høg 1979). Moreover, it is worth remembering that, at that epoch, unlike observations in X or infrared wavelengths or exploration of solar system objects, astrometry was not considered as a space science. However, the progress on many problems of Galactic and stellar astronomy was blocked because of the lack of reliable trigonometric parallaxes and proper motions. Hence, more and more astronomers (astrophysicists as well as astrometrists) were becoming very supportive of the necessity of finding ways to obtain these quantities with a much higher accuracy and for many more stars. For example, let us quote Hodge (1981): '*The determination of the extragalactic distance scale, like so many problems that occupy astronomers attention, is essentially an impossible task. The methods, the data, and the understanding are all too fragmentary at this time to allow a reliable result to be obtained. It would probably be a wise thing to stop trying for the time being and to concentrate on better establishing such things as the distance scale in our Galaxy*'.

The main reference for trigonometric parallaxes was the General Catalogue of Trigonometric Stellar Parallaxes (Jenkins 1963), including about 7000 stars, but mostly of quite poor quality. Only for about 5% of them, the closest to the Sun (mostly less than 10 pc), were the distances known with probable errors less than 10 %. For less than 300 stars were the trigonometric parallaxes accurate enough to use the derived absolute magnitude for calibration purposes of the spectral type-luminosity or colour-luminosity relations of main sequence stars. Moreover, the differences between individual measurements obtained at different observatories were not understood and the sample was strongly biased towards bright stars ($5 < V < 10$), towards common spectral types in the solar neighbourhood (main sequence G and K stars) and towards high proper motion stars (Gliese 1975, Turon 1975). For proper motions, the situation was better, but the same biases were present in the samples of stars with accurate proper motions. As a result, many studies in our Galaxy were using indirect methods, more or less reliable, to determine distances.

Between June 1978 and February 1980 a large promotional campaign was conducted throughout Europe in favour of Hipparcos. An early and informal call for proposals of observation, in an attempt to judge the scientific interest in the project, was released independently by E. Høg and C. Turon. They were able to collect about 170 research proposals submitted by 125 astronomers from 12 countries. The Hipparcos satellite was decided in this context, in March 1980. Later on, in March 1981, E. Høg proposed the addition of the two-colour star mapper channels. This led to the Tycho experiment.

3. Science case versus publications

It is enlightening to illustrate the evolution of the Hipparcos science case from the early proposal to the research proposals from which the Input Catalogue was built (Turon *et al.* 1992a) and, finally, to the actual use of Hipparcos.

The scientific programme as defined by Lacroute in March 1966 was planned in several steps, depending on the possible impact on telemetry link or power consumption of the satellite. By priority order he first defined a) as a minimum, the study of the FK4 errors, b) a precise reference sphere (up to 8200 stars with $V <7$), and he discussed at length the way to resolve the equation system, c) proper motions, repeating four times the operation over 15 yr, d) solar system for the Jupiter satellites and a few minor planets. Then Lacroute mentioned '*Other possible applications ?*' The first item was '*Trigonometric parallaxes ?*', with the comment '*Yes*', as if these were just a by-product. Instead, a factor 10 improvement could be obtained when compared to photographic observations with the same number of measurements. Finally, research for obscure companions (planets) to nearby stars was presented with a single sentence: '*the system appears well adapted to the study of small deviations of the proper motion on nearby bright stars*'. In the proposal of Nov. 1967, the very same programme was described, except that the question mark to '*parallaxes*' had disappeared.

The first conference devoted to the applications of Space Astrometry was organized by ESRO and held in Frascati on 22-23 October 1974 (ESRO 1975). As already mentioned above, the astronomical community at large was invited to present the consequences of high accuracy astrometric data on their respective fields of activity. Still, many of the communications at this Symposium were related to fundamental astronomy: improvement of the Earth-Moon system (Meyer & Froeschlé 1975), motion of Earth's pole (Proverbio & Uras 1975) extension of the stellar reference system by Walter (1975), or the definition of an absolute reference frame foreseen by Kovalevsky (1975). Besides, Gliese (1975) considered the trigonometric parallaxes as improving the absolute luminosities and thus the luminosity calibrations and luminosity function, the space velocities, and the Hyades distance as one first step in the cosmic distance scale ladder. This latter point was also the one developed by Turon (1975), with the knowledge of the galactic evolution and structure in mind, and by Murray (1975), though with the use of proper motions. Van de Kamp (1975) also mentioned the hunt of unseen companions. The first conclusion of the Frascati symposium, which could also be the one of the present symposium, 33 years later, was the following: '*Improvements in astrometric data have a fundamental scientific significance and would provide decisive progress in astronomy, astrophysics and geodynamics*'. This text, prepared by Jean Kovalevsky, insisted on the multiple improvements expected for astrophysics, depending on the duration of the mission and whether proper motions were obtained or not.

In 1976, the ESA Mission Definition study report (ESA 1977) discussed the current status of astrometry, noting first how the fundamental system of reference was affected by the then poor astrometric precision and spatial density. The report then recalled that '*the motions are the main data for deriving distances, and all theories of evolution of our Universe depend, therefore, on these motions*', with little improvement in parallax precision or number of stars to expect on ground in the future. Besides, due to the systematic errors (e.g. a north-south systematic difference amounting to 5 mas), little improvement could be obtained using groups of stars. Pointing out the role of the Hyades in the distance scale, the report insisted that if a '*space astrometry mission could determine the parallax of 30 or so*' Hyades stars, '*these observations alone would already warrant this mission*'! Obtaining a 1% precision on the Hyades distance would improve a) the age of open clusters, b) the distance scale, c) the Cepheid period-luminosity relation, d) stellar luminosities (stellar physics), e) the space velocities of subdwarfs. Apart from the Hyades, parallaxes combined to other observations would also improve the knowledge of the velocity dispersion, space distribution of stars, and reddening within 100 pc. Finally, and probably for the first time, the need for a ground-based parallel effort was also indicated, concerning radial velocities, colours and luminosity determination for astrophysics, and a higher density of relative position measurements for the reference system. Depending which satellite option was chosen (with or without input catalogue), the expected improvement of the HR diagram was discussed.

In addition, the Mission Definition Group '*strongly recommend*(ed), *as the most significant scientific improvements, that a Space Astrometry programme should be established, calling for the launching of a second spacecraft, identical or similar to the first one, after a time of some ten years*' which paved the way for Gaia more than two decades later. The main motivations were the improvement of proper motion for stellar kinematics and the detection of invisible companions, a small part indeed of the huge Gaia scientific case!

In 1976, the IAU Commission 8 urged '*that space astrometry should be developed and carried out as soon as possible*', but was also prudent: '*This however should not affect the planning of ground based programmes before the accuracy, reliability and long term continuity of space astrometry have been assured*'.

Three years later, the Hipparcos science case described in the report on the Phase A study (ESA 1979) was yet more mature, most of the future use being anticipated. Not less than 21 pages were devoted to the scientific objectives. The astrometric data were described as '*the very base of most astrophysical theories and of our conception of the Universe*'. Clearly, beside the improved reference system, space astrometry was now described as a tool for astrophysics: six open clusters would benefit from the parallax measurements, the fine structure of the H-R diagram would be determined, and a catalogue of astrometric parameters would be useful for '*an enormous variety of kinematical, dynamical and astrophysical investigations*', and for the detection and measurement of binaries. '*Indirect benefits on the distance scale, the chemical and dynamical evolution of the Galaxy, the testing of cosmological theories*' were also mentioned. An illustrative comparison between several ground-based parallax programs and that of Hipparcos showed that the annual rate of parallax information with Hipparcos would be 3000 times (in terms of statistical weight) that of the best ground based programs (USNO). It was foreseen that Hipparcos could allow to get at last an unbiased sample of the closest ($< 20\,\mathrm{pc}$) neighbourhood for stars intrinsically brighter than $M_V = 7.5$. While showing the performances of Hipparcos for the five nearby open clusters, it should be underlined that the Hipparcos potential weakness was already noticed in this phase A report, '*taking into account the reduction of statistical weight because of correlation between parallaxes of adjacent stars'.*

Capitalizing on this broad scientific interest, the observing program for Hipparcos was built from the 214 world-wide scientific proposals received in 1982. This work, performed by the 'Input Catalogue Consortium', established in 1981, had to take into account the scientific priorities allocated by the Hipparcos Scientific Selection Committee, but also the satellite observing capabilities (Turon *et al.* 1992b).

Now, ten years after the publication of the results, the scientific outcomes of the mission are considerable, and are described in this volume by M.A.C. Perryman. Up to 1996, the average number of scientific papers mentioning Hipparcos in their abstract (indicating an intensive use of Hipparcos results) was 50 papers per year on the average. Between 1997, year of the publication, and 2000, this increased to more than 400, dropping to 200 in 2004, but remaining constant since. In total, nearly 6000 publications quote either Hipparcos or Tycho in their abstract. In parallel, the number of publications citing the bibliographical reference of the Hipparcos Catalogue is constant at about 250-300 per year from 1998 onward. This number is probably, and will be more and more, underestimated, as one may quote the distance of a star as given by SIMBAD instead of quoting the source catalogue: the main compliment that can be paid to Hipparcos is that its results are now in the public domain.

The legacy of Hipparcos is not simply how the data has been used but also how it **will** be used, especially in a micro-arcsecond era. In this respect, one may note that the technological progresses are 'too fast': the initial suggestion done by the Hipparcos instigators of having another satellite launched to improve the proper motions is no more valid, as the recent gain in precision is now larger than the linear gain than can be obtained with time, so that the proper motion precision with Gaia (a few μas) is preferable to the 1 mas over 20 years = 50 μas. In spite of this, the outcome of the Hipparcos mission may still be significant 20 years later: Olling (2007) shows that Solar System Analogs having periods of several decades may be detected using SIM and Hipparcos as first epoch. The Hipparcos and Tycho data base will also be used for what concerns photometry: already the HD 209458 and HD 189733 stars have had their planetary transits unveiled in the Hipparcos data, and their transiting period found with exquisite precision.

4. Conclusion

Hipparcos has been the result of many convergent efforts: the vision and tenacity of its first proposer, P. Lacroute, his understanding that going to space would be a great advantage for astrometric measurements, the commitment of an increasing number of colleagues in the mission, and the decision of the European Space Agency to investigate further this original perspective, while staying open to new innovations. Once included in the ESA Science Programme, the close collaboration and constant exchanges between the Agency, the Scientific Consortia and Industry were crucial in leading to a satellite design which fully reflected the scientific requirements carefully established for the mission (Perryman *et al.* 1997). Finally, we should also be grateful to ESOC saving and operating a satellite on a wrong orbit.

A giant step in technology allowed to go from ground-based astrometry to space astrometry and reach milli-arcsecond accuracy for more than 100 000 target stars. A new giant step is now promising to lead to micro-arcsecond accuracy with second generation space astrometry missions: ESA's Gaia, direct heritage of Hipparcos, but with on-board systematic detection of targets objects, on-board observations of the third component of the space velocity, and on-board observations of astrophysical parameters; JAXA's

Jasmine, taking benefit of its observation in infrared to obtain direct distances in the galactic center; NASA's SIM the most accurate of the planned experiments. All these, and a few smaller projects, are described in this Symposium.

Acknowledgements

We gratefully thank Erik Høg, Jean Kovalevsky and Michael Perryman for their comments and the documentation they could recover. This text has also benefited from NASA's Astrophysics Data System.

References

Bacchus, P. & Lacroute, P. 1974, in: W. Gliese, C. A. Murray, and R. H. Tucker (eds), *New Problems in Astrometry*, IAU Symp. 61, Reidel, p. 277

ESA 1977, *Space Astrometry*, Report on the Mission definition study, ESA DP/PS(76)11, Rev. 1, 25 January 1977

ESA 1978, *Space Astrometry, Hipparcos*, Report on Phase A study, ESA DP/PS(78)13, 26 April 1978

ESA 1979, *Hipparcos, Space Astrometry*, Report on the Phase A study, ESA SCI(79)10, December 1979

ESRO 1975, *Space Astrometry*, ESRO SP-108, T. D. Nguyen & B. T. Battrick (eds)

Fricke, W., Kopff, A., Gliese, W. & al. 1963, *Veröff. des Astronomischen Rechen-Instituts Heidelberg*, Nr. 10

Gliese, W. 1975, in: ESRO 1975, p. 109

Hodge, P. W. 1981, *ARAA* 19, 357

Høg, E. 14 November 1975, Projects in optical astrometry

Høg, E. 1979, in: C. Barbieri & P. L. Bernacca (eds), *European Satellite Astrometry*, Université de Padoue, p. 7

Husson, J. C. 1975, in: ESRO 1975, p. 43

Jenkins, L. F. 1963, *General catalogue of trigonometric stellar parallaxes*, New Haven: Yale University Observatory

Kovalevsky, J. 28 July 1969, *Remarques sur le projet de Satellite Astrométrique de M. P. Lacroute*, Note to CNES

Kovalevsky, J. 13 May 1970, *Rapport sur l'intérêt scientifique du projet de Satellite Astrométrique de M. Lacroute*, Report to CNES on the proposal of M. Lacroute

Kovalevsky, J. 1975, in: ESRO 1975, p. 67

Kovalevsky, J. 2005, in: A. Heck (ed.), *The Multinational History of Strasbourg Astronomical Observatory*, ASSL, Springer, p. 215

Lacroute, P. 1966, *Projet d'observation spatiale en Astronomie Fondamentale*, March 1966, proposal to CNES

Lacroute, P. 1967, *Proceedings of the Thirteenth General Assembly, IAU Trans.*, XIIIB, 63

Lacroute, P. 1970, *Proceedings of the Fourteenth General Assembly, IAU Trans.*, XIVB, 95

Lacroute, P. 1982, in: M.A.C. Perryman & T. D. Guyenne (eds), *The Scientific Aspects of the Hipparcos Mission*, ESA SP-177, p. 3

Lacroute, P. 1991, Mémoires de l'Académie des sciences, arts et belles-lettres de Dijon, 132, 95

Meyer, C. & Froeschlé, M. 1975, in: ESRO 1975, p. 95

Murray, A. 1975, in: ESRO 1975, p. 129

Olling, R. P. 2007, ArXiv e-prints, 704, arXiv:0704.3059

Perryman, M. A. C., Lindegren, L., Kovalevsky, J., Høg, E., *et al.* 1997, A & A 323, L49

Proverbio, E. & Uras, S. 1975, in: ESRO 1975, p. 83

Turon, C. 1975, in: ESRO 1975, p. 115

Turon, C., Gòmez, A., Crifo, C., *et al.* 1992a, A & A 258, 74

Turon, C., Crézé, M., Egret, D., Gòmez, A. *et al.* 1992b, ESA SP-1136

Van de Kamp, P. 1975, in: ESRO 1975, p. 121

Walter, H. G. 1975, in: ESRO 1975, p. 73

A Giant Step: from Milli- to Micro-arcsecond Astrometry
Proceedings IAU Symposium No. 248, 2007
W. J. Jin, I. Platais & M. A. C. Perryman, eds.

© 2008 International Astronomical Union
doi:10.1017/S1743921308018528

Science highlights from Hipparcos

M. A. C. Perryman

European Space Agency, ESTEC, Noordwijk, The Netherlands

Abstract. The Hipparcos and Tycho Catalogues were published in 1997. In the intervening 10 years, several thousand papers making more-or-less direct use of the data have been published. I summarise a number of scientific applications which illustrate the variety of problems to which the data have been applied. This includes the re-reduction of old and contemporary astrometric observations, investigations of Galactic structure and dynamics, investigations of stellar structure and evolution, investigations of the Solar System and our Earth's environment, and various uses of the binary star and photometric data.

Keywords. astrometry, Galaxy: general, stars: general

---

1. Introduction

The fundamental task of measuring stellar positions, and the derived properties of distances and space motions, has preoccupied astronomers for centuries. As one of the oldest branches of astronomy, astrometry is concerned with measurement of the positions and motions of planets and other bodies within the Solar System, of stars within our Galaxy and, at least in principle, of galaxies and clusters of galaxies within the Universe as a whole. Accurate star positions provide a celestial reference frame for representing moving objects, and for relating phenomena at different wavelengths. Determining the systematic displacement of star positions with time gives access to their motions through space. Determining their apparent annual motion as the Earth moves in its orbit around the Sun gives access to their distances through measurement of parallax. All of these quantities, and others, are accessed from high-accuracy measurements of the relative angular separation of stars. Repeated measurements over a long-period of time provide the a stereoscopic map of the stars and their kinematic motions.

What follows, either directly from the observations or indirectly from modeling, are absolute physical stellar characteristics: stellar luminosities, radii, masses, and ages; and their dynamical signatures. The physical parameters are then used to understand their internal composition and structure, to disentangle their space motions and, eventually, to explain in a rigorous and consistent manner how the Galaxy was originally formed, and how it will evolve in the future. Crucially, stellar motions through space reflect dynamical perturbations of all other matter, visible or invisible.

Astrometry is in principle capable of measuring a whole host of higher-order phenomena: at the milliarcsec level, binary star signatures, General Relativistic light bending, and the dynamical consequences of dark matter are already evident; at the microarcsec level, targeted by the next generation of space astrometry missions currently under development, direct distance measurements will be extended across the Galaxy and to the Large Magellanic Cloud. Other effects will become routinely measurable at the same time: perspective acceleration and secular parallax evolution, more subtle metric effects, planetary perturbations of the photocentric motion, and astrometric micro-lensing; and at the nanoarcsec level, currently no more than an experimental concept, effects of optical interstellar scintillation, geometric cosmology, and ripples in space-time due to gravitational waves will become apparent. The bulk of this seething motion is largely below current observational capabilities, but it is there, waiting to be investigated.

The Hipparcos satellite was a space mission primarily targeting the uniform acquisition of milliarcsec-level astrometry (positions, parallaxes and annual proper motions) for some 120 000 stars. The satellite was launched in August 1989 and operated until 1993. The data processing was finalised in 1996, and the results published by ESA in June 1997 as a compilation of 17 hard-bound volumes, a celestial atlas, and 6 CDs, comprising the Hipparcos and Tycho Catalogues. The original Tycho Catalogue comprised some 1 million stars of lower accuracy than the Hipparcos Catalogue, while the Tycho 2 Catalogue, published in 2000, provided positions and accurate proper motions for some 2.5 million objects, through the combination of the satellite data with plate material from the early 1900s. Details of the satellite operation, and the successive steps in the data analysis, and in the validation and description of the detailed data products, are included in the published catalogue. The result have been in the scientific domain for 10 years.

2. The Hipparcos science

The Hipparcos results impact a very broad range of astronomical research, which I will classify into three major themes:

2.1. *Provision of an accurate reference frame*

This has allowed the consistent and rigorous re-reduction of a wide variety of historical and present-day astrometric measurements. The former category include Schmidt plate surveys, meridian circle observations, 150 years of Earth-orientation measurements, and re-analysis of the 100-year old Astrographic Catalogue (and the associated Carte du Ciel). The Astrographic Catalogue data, in particular, have yielded a dense reference framework reduced to the Hipparcos reference system propagated back to the early 1900s. Combined with the dense framework of 2.5 million star mapper measurements from the satellite, this has yielded the high-accuracy long-term proper motions of the Tycho 2 Catalogue.

The dense network of the Tycho 2 Catalogue has, in turn, provided the reference system for the reduction of current state-of-the-art ground-based survey data: thus the dense UCAC 2 and USNO B2 catalogues are now provided on the same reference system, and the same is true for recent surveys such as SDSS and 2MASS. Other observations specifically reduced to the Hipparcos system are the SuperCOSMOS Sky Survey, major historical photographic surveys such as the AGK2 and the CPC2, and the more recent proper motion programmes in the northern and southern hemispheres, NPM and SPM. Proper motion surveys have been rejuvenated by the availability of an accurate optical reference frame, and amongst them are the revised NLTT (Luyten Two-Tenths) survey, and the Lépine–Shara proper motion surveys (north and south). Many other proper motion compilations have been generated based on the Hipparcos reference system, in turn yielding large data sets valuable for open cluster surveys, common-proper motion surveys, etc..

The detection and characterisation of double and multiple stars has been revolutionised by Hipparcos: in addition to binaries detected by the satellite, many others have been revealed through the difference between the Hipparcos (short-term) proper motion, and the long-term photocentric motion of long-period binary stars (the $\Delta\mu$ binaries). New binary systems have been followed up through speckle and long-baseline optical interferometry from ground, through a re-analysis of the Hipparcos Intermediate Astrometric Data, or through a combined analysis of astrometric and ground-based radial velocity data. Other binaries have been discovered as common-proper motions systems in catalogues reaching fainter limiting magnitudes. I have compiled a list of some 25 papers together revising the analysis of more than 15 000 Hipparcos binary systems, providing new orbital solutions, or characterising systems which were classified by Hipparcos as suspected double, acceleration solutions, or stochastic ('failed') solutions.

I have also compiled a list of more than 30 papers together presenting radial velocities for more than 17 000 Hipparcos stars since the catalogue publication, not counting two papers presenting some 20 000 radial velocities from the CORAVEL data base, and some 25 000 RAVE measurements including some Tycho stars. These radial velocity measurements are of considerable importance for determining the three-dimensional space motions, as well as further detecting and characterising the properties of binary stars.

More astrophysically, studies have been made of wide binaries, and their use as tracers of the mass concentrations during their Galactic orbits, and according to population. Numerous papers deal with improved mass estimates from the spectroscopic eclipsing systems, important individual systems such as the Cepheid-binary Polaris, the enigmatic Arcturus, and favourable systems for detailed astrophysical investigations such as V1061 Cyg, HIP 50796, the mercury-manganese star ϕ Her, and the spectroscopic binary HR 6046. A number of papers have determined the statistical distributions of periods and eclipse depths for eclipsing binaries, and others have addressed their important application in determining the radiative flux and temperature scales. Studies of the distributions of detached and contact binaries (including W UMa and symbiotic systems) have also been undertaken.

The accurate reference frame has in turn provided results in topics as diverse as the measurement of General Relativistic light bending; Solar System science, including mass determinations of minor planets; applications of occultations and appulses; studies of Earth rotation and Chandler Wobble over the last 100 years based on a re-analysis of data acquired over that period within the framework of studies of the Earth orientation; and consideration of non-precessional motion of the equinox.

2.2. Constraints on stellar evolutionary models

The accurate distances and luminosities of 100 000 stars has provided the most comprehensive and accurate data set relevant to stellar evolutionary modeling to date, providing new constraints on internal rotation, element diffusion, convective motions, and asteroseismology. Combined with theoretical models it yields evolutionary masses, radii, and stellar ages of large numbers and wide varieties of stars.

A substantial number of papers have used the distance information to determine absolute magnitude as a function of spectral type, with new calibrations extending across the HR diagram: for example for OB stars, AFGK dwarfs, and GKM giants, with due attention given to the now more-quantifiable effects of Malmquist and Lutz–Kelker biases. Other luminosity calibrations have used spectral lines, including the Ca II-based Wilson–Bappu effect, the equivalent width of O I, and calibrations based on interstellar lines.

A considerable Hipparcos-based literature deals with all aspects of the basic 'standard candles' and their revised luminosity calibration. Studies have investigated the Population I distance indicators, notably the Mira variables, and the Cepheid variables (including the period–luminosity relation, and their luminosity calibration using trigonometric parallaxes, Baade–Wesselink pulsational method, main-sequence fitting, and the possible effects of binarity). The Cepheids are also targets for Galactic kinematic studies, tracing out the Galactic rotation, and also the motion perpendicular to the Galactic plane.

Hipparcos has revolutionised the use of red clump giants as distance indicators, by providing accurate luminosities of hundreds of nearby systems, in sufficient detail that metallicity and evolutionary effects can be disentangled, and the objects then used as single-step distance indicators to the LMC and SMC, and the Galactic bulge. Availability of these data has catalysed the parallel theoretical modeling of the clump giants.

For the Population II distance indicators, a rather consistent picture has emerged in recent years based on subdwarf main sequence fitting, and on the various estimates of the horizontal branch and RR Lyrae luminosities.

A number of different methods now provide distance estimates to the Large Magellanic Cloud, using both Population I and Population II tracers, including some not directly dependent on the Hipparcos results (such as the geometry of the SN 1987A light echo, orbital parallaxes of eclipsing binaries, globular cluster dynamics, and white dwarf cooling sequences). Together, a rather convincing consensus emerges, with a straight mean of several methods yielding a distance modulus of $(m - M)_0 = 18.49$ mag. Through the Cepheids, the Hipparcos data also provide good support for the value of the Hubble Constant $H_0 = 72 \pm 8 \, \mathrm{km \, s^{-1} \, Mpc^{-1}}$ as derived by the HST key project, and similar values derived by WMAP, gravitational lensing experiments, and Sunyaev–Zel'dovich effect.

A huge range of other studies has made use of the Hipparcos data to provide constraints on stellar structure and evolution. Improvements have followed in terms of effective temperatures, metallicities, and surface gravities. Bolometric corrections for the *Hp* photometric band have become available within the last few months through the work of Michael Bessell, opening the way for new and improved studies of the observational versus theoretical Hertzsprung–Russell diagram.

Many stellar evolutionary models have, of course, been developed and refined over the last few years, and Hipparcos provides an extensive testing ground for their validation and their astrophysical interpretation: these include specific models for pre- and post-main sequence phases, and models which have progressively introduced effects such as convective overshooting, gravitational settling, rotation effects, binary tidal evolution, radiative acceleration, and effects of α-element abundance variations. These models have been applied to the understanding of the HR diagram for nearby stars, the reality of the Böhm-Vitense gaps, the zero-age main sequence, the subdwarf main sequence, and the properties of later stages of evolution: the subgiant, first ascent and asymptotic giant branch, the horizontal branch, and the effects of dredge-up, and mass-loss. Studies of elemental abundance variations include the age–metallicity relation in the solar neighbourhood, and various questions related to particular elemental abundances such as lithium and helium. Other studies have characterised and interpreted effects of stellar rotation, surface magnetic fields, and observational consequences of asteroseismology, notably for solar-like objects, the high-amplitude δ Scuti radial pulsators, the β Cephei variables, and the rapidly-oscillating Ap stars.

Many studies have focused on the pre-main sequence stars, both the (lower-mass) T Tauri and the (higher-mass) Herbig Ae/Be stars, correlating their observational dependencies on rotation, X-ray emission, etc. The understanding of Be stars, chemically-peculiar stars, X-ray emitters, and Wolf–Rayet stars have all been substantially effected by the Hipparcos data. Kinematic studies of runaway stars, produced either by supernova explosions or dynamical cluster ejection, have revealed many interesting properties of runaway stars, also connected with the problem of (young) B stars found far from the Galactic plane.

Dynamical orbits within the Galaxy have been calculated for planetary nebulae and, perhaps surprisingly given their large distances, for globular clusters, where the provision of a reference frame at the 1 milliarcsec accuracy level has allowed determination of their space motions and, through the use of a suitable Galactic potential, their Galactic orbits, with some interesting implications for Galactic structure, cluster disruption, and Galaxy formation.

One of the most curious of the Hipparcos results in this area is probably the improved determination of the empirical mass–radius relation of white dwarfs. At least three such objects appear to be too dense to be explicable in terms of carbon or oxygen cores, while iron cores seem difficult to generate from evolutionary models. 'Strange matter' cores have been postulated, and studied by a number of theoretical groups.

2.3. *Galactic structure and dynamics*

The distances and uniform space motions have provided a substantial advance in understanding of the detailed kinematic and dynamical structure of the solar neighbourhood, ranging from the presence and evolution of clusters, associations and moving groups, the presence of resonance motions due to the Galaxy's central bar and spiral arms, the parameters describing Galactic rotation, the height of the Sun above the Galactic midplane, the discrimination of the motions of the thin disk, thick disk and halo populations, and the evidence for halo accretion.

Many attempts have been made to further understand and characterise the solar motion based on Hipparcos data, and to redefine the large-scale properties of Galactic rotation in the solar neighbourhood. The latter has been traditionally described in terms of the Oort constants, but it is now evident that such a formulation is quite unsatisfactory in terms of describing the detailed local stellar kinematics. Attempts have been made to re-cast the problem into the 9-component tensor treatment of the Ogorodnikov–Milne formulation, analogous to the treatment of a viscous and compressible fluid by Stokes more than 150 years ago. The results of several such investigations have proved perplexing. The most recent and innovative approach has been a kinematic analysis based on vectorial harmonics, in which the velocity field is described in terms of (some unexpected) 'electric' and 'magnetic' harmonics. They reveal the warp at the same time, but in an opposite sense to the vector field expected from the stationary warp model.

Kinematic analyses have tackled the issues of the mass density in the solar neighbourhood, and the associated force-law perpendicular to the plane, the K_z relation. Estimates of the resulting vertical oscillation frequency in the Galaxy of around 80 Myr have been linked to cratering periodicities in the Earth's geological records. Related topics include studies of nearby stars, the stellar escape velocity, the associated initial mass function, and the star formation rate over the history of the Galaxy. Dynamical studies of the bar, of the spiral arms, and of the stellar warp, have all benefited. Studies of the baryon halo of our Galaxy have refined its mass and extent, its rotation, shape, and velocity dispersion, and have provided compelling evidence for its formation in terms of halo substructure, some of which is considered to be in falling, accreting material, still ongoing today.

New techniques have been developed and refined to search for phase-space structure (i.e. structure in positional and velocity space): these include convergent-point analysis, the spaghetti method of Hoogerwerf & Aguilar, global convergence mapping, epicycle correction, and orbital back-tracking. An extensive literature has resulted on many aspects of the Hyades, the Pleiades, and other nearby open clusters, comprehensive searches for new clusters, and their application to problems as diverse as interstellar reddening determination, correlation with the nearby spiral arms, and the age dependence of their vertical distribution within the galaxy: one surprising result is that this can be used to place constraints on the degree of convective overshooting by matching stellar evolutionary ages with cluster distances from the Galactic plane.

In addition to studying and characterising open clusters, and young nearby associations of recent star formation, the Hipparcos data have revealed a wealth of structure in the nearby velocity distribution which is being variously interpreted in terms of open cluster evaporation, resonant motions due to the central Galactic bar, scattering from nearby spiral arms, and the effects of young nearby kinematic groups, with several having been discovered from the Hipparcos data in the last 5 years.

The Hipparcos stars have been used as important (distance) tracers, determining the extent of the local 'bubble', itself perhaps the result of one or more nearby supernova explosions in the last 5 Myr. The interstellar medium morphology, extinction and reddening, grey extinction, polarisation of star light, and the interstellar radiation field, have all been constrained by the Hipparcos stars.

2.4. *Other applications*

Superficially it may seem surprising that the Hipparcos Catalogue has been used for a number of studies related to the Earth's climate. Studies of the passage of nearby stars and their possible interaction with the Oort Cloud have identified stars which came close to the Sun in the geologically recent past, and others which will do so in the relatively near future. Analysis of the Sun's Galactic orbit, and its resulting passage through the spiral arms, favour a particular spiral arm pattern speed in order to place the Sun within these arms during extended deep glaciation epochs in the distant past. In this model, climatic variations are explained as resulting from an enhanced cosmic-ray flux in the Earth's atmosphere, leading to cloud condensation and a consequent lowering of temperature. A study of the Maunder Minimum, a period between 1645–1715 coinciding with the coldest excursion of the 'Little Ice Age', and a period of great hardship in Europe, was interpreted in the context of the number of solar-type stars out to 50–80 pc showing correspondingly decreased surface activity. Several studies have used the accurate distance data, accompanied by stellar evolutionary models, in an attempt to identify 'solar twins' (stars which most closely resemble the Sun in all its characteristics, and which may be the optimum targets for searches for life in the future), and 'solar analogues' (stars which will resemble the Sun at some past or future epoch, and which therefore offer the best prospects for studying the Sun at different evolutionary stages).

Many studies have used the accurate photometric data: as part of the construction of absolute or bolometric magnitudes, for their uniform colour indices, and for their extensive epoch photometry, which itself has been used for all sorts of variability analyses, including the rotation of minor planets, the study of eclipsing binaries, the complex pulsational properties of Cepheids, Mira variables, δ Scuti, slowly-pulsating B stars, and many others.

In addition to all of these, the Hipparcos and Tycho Catalogues are now routinely used to point ground-based telescopes, navigate space missions, drive public planetaria, and provide search lists for programmes such as exo-planet surveys; one study has even shown how positions of nearby stars at the milliarcsec level can be used to optimise search strategies for extraterrestrial intelligence.

3. The full review

This short synopsis provides an interim report of a full review, which will be published by Cambridge University Press in mid-2008. Titled *'Astronomical Applications of Astrometry: A Review Based on Ten Years of Exploitation of the Hipparcos Data'*, this book-length review covers the full range of the Hipparcos scientific findings, in an extensive summary and analysis of the scientific literature over the 10 years since the publication of the Hipparcos and Tycho Catalogues in 1997.

A Giant Step: from Milli- to Micro-arcsecond Astrometry
Proceedings IAU Symposium No. 248, 2007
W. J. Jin, I. Platais & M. A. C. Perryman, eds.

© 2008 International Astronomical Union
doi:10.1017/S174392130801853X

CCD survey with the CMASF from the Southern Hemisphere

M. Vallejo[1], J. L. Muiños[1], F. Belizón[1], F. J. Montojo[1]
C. C. Mallamaci[2], J. A. Pérez[2], L. F. Marmolejo[2],
J. L. Navarro[2], J. Sedeño[2]

[1] Real Instituto y Observatorio de la Armada, San Fernando, Spain. email: ppmu@roa.es

[2] Observatorio Astronómico Félix Aguilar, San Juan, Argentina. email: ccmalla@gmail.com

Abstract. A subcatalogue with positions and magnitudes of stars brighter than $V = 16$ and declinations between $-30°$ and $0°$ is presented. The observations have been carried out with the Círculo Meridiano Automático de San Fernando at the Carlos Ulrrico Cesco Observatory in San Juan (Argentina) in the framework of an agreement between the ROA and the OAFA. The final goal of this collaboration is to publish a survey of positions and magnitudes of stars brighter than $V = 16$ and with declinations between $-55°$ and $+30°$.

Keywords. astrometry, surveys, catalogs, stars: kinematics

1. Introduction

The Círculo Meridiano Automático de San Fernando (CMASF) is a fully automated meridian circle of the Real Instituto y Observatorio de la Armada (ROA) observing since 1996 at the Carlos Ulrrico Cesco Observatory (CUC) in San Juan (Argentina). In December 1999, a CCD camera with 1552×1032 pixels of 9 microns in size was installed on the CMASF. The main observing program for this instrument is a sky survey within the declination zone $-55° < \delta < +30°$ and the magnitude range $8 < V < 16$. The expected positional precision is ~ 50 mas in both coordinates and 0.04 mag in V. The instrument is managed jointly by the ROA and the Observatorio Astronómico Félix Aguilar (OAFA) of the University of San Juan (Argentina), also owner of the CUC.

In order to test the performance of CMASF, a subcatalogue of positions named the Hispano-Agentinian Meridian Catalogue No. 2 (HAMC2) has been derived for a smaller declination zone $-30° < \delta < 0°$. We expect to publish this subcatalogue in early 2008. In this paper we present preliminary analysis of the subcatalogue and provide comparisons with the UCAC2 and Tycho-2 catalogues, that we have made to investigate possible systematic errors.

2. Observations

The CMASF observations reported here have been carried out from December 1999 to May 2006. This automated meridian circle has a 176 mm aperture and a 2660 mm focal length (0″69/pixel). It is located at the CUC observatory on the eastern slopes of Andes with the following geographic coordinates: longitude 69° W, latitude of 31° S, altitude 2330 m above sea level.

The instrument operates in the drift-scan mode (Stone 1993) by observing each night several strips of the sky with a width of 18′ in declination each but having a variable length in right ascension ($20^m < RA < 3^h$). Then, a zone to be observed is divided into the sub-bands separated by $10'$ in declination, so that each sub-band has an $8'$ overlap in declination between the adjacent strips (see Evans, Irvin & Helmer 2002).

3. Reductions of observations

Initial reductions include detection of stellar images, their centering and deriving a number of parameters characterizing the images in each strip. Then, the Tycho-2 catalogue (Høg *et al.* 2000) is used as a reference frame to obtain preliminary ICRS coordinates in each strip at the epoch of observations. As soon as the observations in a zone of the sky are completed, the final reductions are carried out. For each strip, atmospheric fluctuations of the night are removed using a calibration function based on the overlapping bands between the current strip and the five adjacent strips above and another five adjacent strips below it (Evans, Irwin & Helmer 2002). Then a catalogue is formed by averaging right ascensions, declinations, and magnitudes of the stars. A more complete description of the method used to form the HAMC2 can be found in the documentation of the Carlsberg Meridian Catalogue 14 (2005).

4. Comparison with other catalogues

To investigate possible systematic errors, comparisons have been made between the HAMC2 positions and magnitudes and those in the UCAC2 (Zacharias *et al.* 2004) and Tycho-2 (Høg *et al.* 2000) to investigates possible systematic errors. No such error were found. The best achieved average formal precision of HAMC2 positions is 30–40 mas at $V\sim 12$.

5. Conclusions

The HAMC2 subcatalogue presented here shows that the precision in positions reaches the expectations for this survey. No systematic errors have been detected.

Acknowledgements

We would like to acknowledge Dafydd W. Evans who kindly provided software to construct the catalogue. Also, our gratitude is extended to the Estado Mayor de la Armada and the Ministerio de Educación y Ciencias (AYA2005-24974-E) of Spain and to the San Juan University of Argentina for financial support of this project.

References

Carlsberg Meridian Catalogue, La Palma 2005, No. 14, Copenhagen University Observatory, Institute of Astronomy of Cambridge, Real Instituto y Observatorio de la Armada en San Fernando

Evans D., Irwin M., & Helmer L. 2002, *A&A*, 395, 347

Høg E., Fabricius C., Makarov V. V., Urban S., Corbin T., Wycoff G., Bastian U., Schwekendiek, & Wicenec A. 2000, *A&A*, 355, L27

Stone R. C. 1993, in *Developments in Astrometry and Their Impact on Astrophysics and Geodynamics*, Kluwer Academic Publishers, p. 65

Zacharias N., Urban S. E., Zacharias M. I., Wycoff G. L., Hall D. M., Monet D. G., & Rafferty T. J. 2004, *AJ*, 127, 3043

A Giant Step: from Milli- to Micro-arcsecond Astrometry
Proceedings IAU Symposium No. 248, 2007
W. J. Jin, I. Platais & M. A. C. Perryman, eds.

© 2008 International Astronomical Union
doi:10.1017/S1743921308018541

The role of pre-Gaia positional data in determining binary orbits with Gaia data †

S. L. Ren and Y. N. Fu

Purple Mountain Observatory, Chinese Academy of Sciences, Nanjing, 210008, China
email: rensl@pmo.ac.cn or fyn@pmo.ac.cn

Abstract. Simulations show that incorporating the pre-Gaia positional data into the Gaia astrometric data can significantly improve the efficiency of determining binary orbits, especially for those binaries with periods from 8 to 25 years.

Keywords. stars: fundamental parameters, binaries: general, methods: data analysis

1. Introduction

In recent papers, simulations revealed that, even with their unprecedented precision, Gaia astrometric data can only be used to reliably measure the orbits of binaries with periods less than 6 years (Lattanzi *et al.*, 2004). But for those binaries with periods larger than 6 years, the reliability of orbital measurements becomes very low because the observational time span is shorter than the orbital period during Gaia mission. In other words, in order to obtain reliable orbit solutions of these binaries, one has to incorporate the other data based on long-term observations (pre-Gaia data). There are two kinds of pre-Gaia data, namely radial velocity data and positional data. The former one is believed to be useful as it can be inferred from similar situations of Hipparcos binaries (e.g., Jancart *et al.*, 2005). The role of latter one is discussed in this paper. Since 1990, the precision of the pre-Gaia positional data has already reached about several mill-arcseconds, but there are only a few data available for some binaries. Therefore, these data themselves aren't sufficient to be used to derive reliable orbits of binaries. This is the case for some binaries in the fourth catalog of interferometric measurements of binary stars (Hartkopf *et al.*, 2001). Compared to Gaia astrometric data, these data have the advantage of long time span. It is then interesting to know whether these data of the long observational time span can play a role in determining binary orbits after Gaia mission? In order to answer this question, we do the following simulations.

2. Simulations

First, one hundred simulated orbits are produced with periods randomly distributed in the range from 6 to 25 years, relative semi-major axis chosen as 50 mas (higher than resolution limits of many large telescopes), astrometric semi-major axis chosen as 1 mas (corresponds to Gaia data with signal-to-noise ratio about 100/1), and other parameters randomly distributed in their respective ranges.

Then, similar to Lattanzi *et al.*, (2000), Gaia astrometric data are simulated with precision 0.01 mas, and data number 60 on average. Pre-Gaia positional data are simulated

† Supported by the National Natural Science Foundation of China (Grant Nos. 10473025 and 10703014)

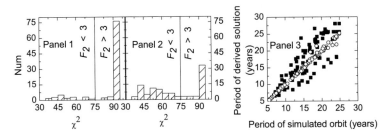

Figure 1. Panel 1: the distribution of χ^2 of the solutions derived from Gaia data only; panel 2: the distribution of χ^2 of the solutions from combining Gaia and pre-Gaia data; panel 3: periods of simulated orbits versus derived solutions. The filled squares indicate orbit solutions from Gaia data only, and the empty circles indicate the solutions from combining Gaia and pre-Gaia data.

with precision 5 mas, data number 5, and observational epochs randomly distributed in the range from 1990 to 2010 years.

By only fitting Gaia data and fitting both Gaia and pre-Gaia data, two kinds of orbit solutions can be derived for each binary.

3. Results

In order to decide the acceptability of a derived orbit solution, we use the F_2 indicator (e.g., Lattanzi *et al.*, 2004). If $|F_2| < 3$, the derived solution is acceptable. More than 60% of derived solutions from both Gaia and pre-Gaia data are acceptable, the percentage is reduced to 15% for the solutions derived from Gaia data only. The distributions of χ^2 are shown for the two different kinds of solutions in figure 1, respectively.

We compare the periods of derived solutions with that of simulation orbits in panel 3. If period < 8 years, solutions are all in accordance with that of simulation orbits, if period > 8 years, solutions derived from combining Gaia and pre-Gaia data are still in good agreement with that of simulation orbits, which is not the case for the solutions from Gaia data only.

4. Conclusion

Above simulations show that pre-Gaia positional data can significantly improve the efficiency of determining orbits of binaries, especially for those with periods > 8 years.

Acknowledgements

Thanks prof. W. J. Jin for her useful suggestions.

References

Hartkopf, W. I., McAlister, H. A. & Mason, B. D. 2001, *AJ* 122, 3480
Jancart, S., Jorissen, A., Babusiaux, C. & Pourbaix, D. 2005, *A&A* 442, 365
Lattanzi, M. G., Spagna, A., Sozzetti, A. & Casertano, S. 2000, *MNRAS* 317, 211
Lattanzi, M. G., Jancart, S., Morbidelli, *et al.* in C. Turon, K. S. O'Flaherty, M. A. C. Perryman eds., 2005 *Proceedings of the Gaia Symposium "Detection and Characterization of Extrasolar Planets with Gaia"*, P. 251

A Giant Step: from Milli- to Micro-arcsecond Astrometry
Proceedings IAU Symposium No. 248, 2007
W. J. Jin, I. Platais & M. A. C. Perryman, eds.
© 2008 International Astronomical Union
doi:10.1017/S1743921308018553

The understanding of the FK5 and Hipparcos proper-motion systems†

Z. Zhu

Department of Astronomy, Nanjing University, China
email: zhuzi@nju.edu.cn

Abstract. Comparing proper motions of the FK5 and Hipparcos, several authors declared that the two proper-motion systems are inconsistent with the value of the precession correction obtained from VLBI and LLR observations. Based on the proper-motion data from the PPM and ACRS catalogues which are constructed on the FK5 system, the inconsistent values of the precessional correction and of the time dependent term of equinox correction, derived from the different subsets of stellar samples, have been found. One of the reasons for those discrepancies should be mostly due to the internally biased proper-motion system of the FK5.

Keywords. astrometry, Galaxy: fundamental parameters, reference systems

1. Introduction

The Hipparcos system defines a quasi-inertial reference system on the ICRS. The pointing error of its coordinates with respect to ICRS in the mean observational epoch J1991.25 is ± 0.6 mas and the rotation error of the system is ± 0.25 mas yr^{-1} (Perryman, *et al.*1997). The FK5 system is a dynamical system and transfers to the inertial system through a precise determination of the precession constant and time-dependent term of equinox correction. From researches on VLBI and LLR it was found that the IAU 1976 precession constant was actually over-estimated, and this leads to the lunisolar precession correction of the FK5 system of $\Delta p \approx -3.0$ mas yr^{-1} (McCarthy & Captaine 2002).

From the comparison of the FK5 system and Hipparcos, one found that the speed of rotation ω, between two systems, can be expressed (ESA 1997)

$$\omega = (\omega_x, \omega_y, \omega_z) = (-0.10 \pm 0.10, 0.43 \pm 0.10, 0.88 \pm 0.10), \qquad (1.1)$$

in units of mas yr^{-1}. If both the FK5 and Hipparcos systems are rigid, then the speed vector ω should theoretically reflect the precession correction of the FK5 system, or we have $\omega_x = 0$, $\omega_y = -\Delta p \sin \epsilon$, and $\omega_z = +\Delta p \cos \epsilon - (\Delta \lambda + \Delta e)$. Here $\Delta \lambda$ and Δe are the planetary precession correction and time-dependent term of equinox correction, respectively. If the measurement results $\Delta p = -2.997 \pm 0.008$ mas yr^{-1} (McCarthy & Captaine 2002) and $(\Delta \lambda + \Delta e) = -1.16 \pm 0.26$ mas yr^{-1} (derived from ACRS proper motions, Miyamoto & Sôma 1993) are substituted into the above formulae, it is quite obvious that the three components are in contradiction with the speed vector ω in Eq.(1.1). Mignard & Froeschlé (2000) and Walter & Hering (2005) attempted to resolve the contradiction of the relation between the two systems with the precession constant corrections, but there were still a lot of questionable points awaiting clarification. In my previous work, I have more systematically analyzed the systematic errors of the PPM and ACRS proper motions with respect the Hipparcos, and pointed out that there are serious problems which exist in the past ground-based systems, such as the regional error, magnitude equation

† Supported by the National Natural Science Foundation of China(Grant Nos. 10333050 and 10673005)

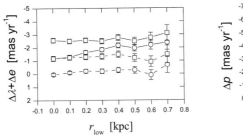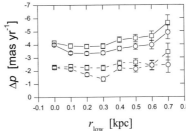

Figure 1. Precession correction and time-dependent term of equinox correction. The solid lines express results derived from PPM proper motions, while the dashed curves show those from ACRS data. Circles and squares indicate results determined from all type of stars and from K-M giants, respectively.

and color equation, etc. (Zhu 2000). In the present work we will go further to study the effects of the PPM and ACRS proper-motion systems on the determination of the precession constant correction in order to understand the FK5 and Hipparcos proper-motion Systems from another point of view.

2. Precession constant correction and discussion

For the processing, we again adopt the 3-D kinematics model, similar as Miyamoto & Sôma 1993) did. To consider problems such as arising from magnitude equations of PPM and ACRS proper motions and velocity dispersion of the nearby stars on the estimation of the parameters, we give separate results for the different subsamples in the heliocentric distance $r_{low} \leqslant r \leqslant 1.0$ kpc. Precession constant correction Δp and the correction $(\Delta\lambda + \Delta e)$ are illustrated in Figure 1, derived for all type of stars and for K-M giants, respectively.

From analysis of Δp and $(\Delta\lambda + \Delta e)$ given in Figure 1, we can cognize some features of the proper-motion systems and give a conclusive discussion as follows:

The results deduced from PPM and ACRS show obvious systematic differences, indicating an overall difference of ~ 1.5 mas yr^{-1} between the two proper-motion systems. Thus, neither PPM nor ACRS can completely represent the FK5 system. Furthermore, there are serious regional error, magnitude equation and colour equation in the FK5 system. In a certain sense, the FK5 system is a non-rigid reference system. Thus, the FK5 and Hipparcos proper motion systems cannot be connected simply by making use of the precession correction. Only when the FK5 is an ideal rigid system and the Hipparcos system is strictly established on the ICRS system, can this sort of simple relation hold.

References

ESA 1997, *The Hipparcos and Tycho Catalogues* SP-1200
McCarthy, D. D. & Captaine, N. 2002, *IERS Technical Note* 29, 9
Mignard, F. & Fraeschlé, M. 2000, *A&A* 354, 732
Miyamoto, M. & Sôma, M. 1993, *AJ* 105, 691
Perryman, M. A. C., Lindegren, L., Kovalevsky, J., *et al.* 1997, *A&A* 323, L49
Walter, H. G. & Hering, R. 2005, *A&A* 431, 721
Zhu, Z. 2000, *PASP* 112, 1103

A Giant Step: from Milli- to Micro-arcsecond Astrometry
Proceedings IAU Symposium No. 248, 2007
W. J. Jin, I. Platais & M. A. C. Perryman, eds.

Use of optical and radio astrometric observations of planets, satellites and spacecraft for ephemeris astronomy

E. V. Pitjeva

Institute of Applied Astronomy RAS, 10, Kutuzova emb., St.-Petersburg, 191187, Russia
email: evp@ipa.nw.ru

Abstract. Different types of radiometric observations of planets and spacecraft, as well as optical data used for constructing modern high-precision planet ephemerides are presented. New mass values for planets and asteroids are given. The IAA RAS EPM ephemerides (**E**phemerides of **P**lanets and the **M**oon) are the basis for the Russian "Astronomical Yearbook" and are used in the navigation program "GLONASS" and the cosmic program "Phobos-Grunt".

Keywords. astrometry, ephemerides, reference systems

Until the 1960's, classical analytical theories of planets were based entirely upon optical observations (accurate to $0''.5$). Detection of Venus radioechoes in 1961 has opened the era of astrometric radio observations. Progress in astrometric engineering and introduction of new astrometric methods resulted in a revolution in both astrometry and ephemeris astronomy.

Modern high-precision planet ephemerides are constructed by simultaneous numerical integrating the equations of motion of the nine major planets, the Sun, 300 or more biggest asteroids, the Moon, and the lunar physical libration taking into account perturbations due to the solar oblateness and the massive ring of small asteroids. The most recent planet ephemerides having about the same accuracy and being adequate to modern observations have been developed at JPL – DE414 (Standish, 2006), IAA RAS – EPM2006 (Pitjeva, 2007) and IMCCE – INPOP06 (Fienga, *et al.*, 2008). Different types of the observations of planets, their satellites and spacecraft used for improvement of planet ephemerides are presented in Table 1 (where N is a number of observations).

The significance of radar observations is conditioned by two factors. First, two new types of measurements have appeared in astrometry. They are time delay is associated with distance by the light velocity and doppler-shift which gives the radial velocity of an object. Second, radar measurements have high accuracy (on the order of one part in 10^{11}), and it surpasses the accuracy of classical optic observations by four orders. At present, radar observations of the inner planets can be carried out for any point of their orbits; however, the ranging data of a spacecraft near planets have the most accuracy that has reached 1 m. It was radar measurements that made it possible to produce ephemerides of the inner planets with the millisecond accuracy and to determine different astronomical constants.

The accuracy of the modern VLBI observations of a spacecraft near planets is 1 mas, which permits the orientation of the planet ephemerides to the International Celestial Reference Frame (ICRF), and the uncertainty of the inner planet system with respect to the ICRF to be no more than $0''.001$ or <1 km. The rotation angles between the EPM2006 ephemerides and the ICRF are (in mas): $\varepsilon_x = 1.9 \pm 0.1$, $\varepsilon_y = -0.5 \pm 0.2$, $\varepsilon_z = -1.5 \pm 0.1$.

The observations of satellites of outer planets are of great importance not only for determination of their orbits but also for determination of the orbit of the outer planets, as these are more accurate than the observations of their parent planets and practically free from the phase effect. The modern accuracy of CCD data has already reached $0''.06$, which have made it possible to significantly improve the orbital elements of the outer planets, their satellites, masses of these bodies and other parameters of the outer planet systems.

Table 1. Observations used for constructing planet ephemerides

Optical observations of outer planets and their satellites 1913–2006, N = 48690

USNO
Pulkovo
Nikolaev
Tokyo
Bordeaux
LaPalma
Flagstaff
TMO

Types	Years	*A priori* accuracy
optical transit	1913–1994	$1'' \to 0''.5$
photoelectric transit	1963–1998	$0''.8 \to 0''.25$
photographic	1913–1998	$1'' \to 0''.2$
CCD	1995–2006	$0''.2 \to 0''.06$

Radar observations of Mercury, Venus, Mars, N = 58112

Millstone
Haystack
Arecibo
Goldstone
Crimea

Types	Years	*A priori* accuracy
ranging	1961–1997	100 km $\to 150$ m

Spacecraft radiometric data obtained by DSN antennae 1971–2006, N = 305599

Mariner − 9	*Venus*
Pioneer − 10, 11	*Jupiter*
Voyager	*Jupiter*
Phobos	*Mars*
Ulysses	*Jupiter*
Magelan	*Venus*
Galileo	*Jupiter*
Viking − 1, 2	*Mars*
Pathfinder	*Mars*
MGS	*Mars*
Odyssey	*Mars*

Types	Years	*A priori* accuracy
ranging	1971–2006	6 km $\to 1$ m
differ.range	1976–1997	$1.3 \to 0.1$ mm/sec
radial veloc.	1992–1994	$0.1 \to 0.002$ mm/sec
Δ VLBI	1990–2003	12 mas $\to 0.3$ mas

Table 2. New mass values of planets and asteroids proposed to WG NSFA

Planet	Previous values	New values	Year	Authors
M_M/M_E	$1.23000345(5) \cdot 10^{-2}$	$1.23000371(4) \cdot 10^{-2}$	2006	Standish
$M_\odot/M_V$	$4.0852371(6) \cdot 10^5$	$4.08523719(8) \cdot 10^5$	1999	Konopliv, *et al.*
$M_\odot/M_{Ma}$	$3.098708(9) \cdot 10^6$	$3.09870359(2) \cdot 10^6$	2006	Konopliv, *et al.*
$M_\odot/M_J$	$1.0473486(8) \cdot 10^3$	$1.047348625(17) \cdot 10^3$	2003	Jacobson
$M_\odot/M_{Sa}$	$3.497898(18) \cdot 10^3$	$3.4979018(1) \cdot 10^3$	2006	Jacobson, *et al.*
$M_\odot/M_P$	$1.3521(15) \cdot 10^8$	$1.36564(28) \cdot 10^8$	2008	Tholen, *et al.*
$M_\odot/M_{Eris}$		$1.191(14) \cdot 10^8$	2007	Brown & Schaller
$M_{Ceres}/M_\odot$	$4.39(4) \cdot 10^{-10}$	$4.72(3) \cdot 10^{-10}$	2007	Pitjeva & Standish
$M_{Pallas}/M_\odot$	$1.59(5) \cdot 10^{-10}$	$1.03(2) \cdot 10^{-10}$	2007	Pitjeva & Standish
$M_{Vesta}/M_\odot$	$1.69(11) \cdot 10^{-10}$	$1.35(2) \cdot 10^{-10}$	2007	Pitjeva & Standish

Masses of bodies are the fundamental constants for ephemeris astronomy. Resent research have modified most of the planet masses due to high-precission observations of spacecraft orbiting or passing near the planets as well as data of planet satellites. As for masses of the largest asteroids, many new, significantly more accurate mass estimates have been obtained recently from close encounters with other asteroids and from their perturbations onto Mars. In collaboration with Dr. Standish, we have chosen the best estimates of masses for planets and asteroids (Table 2) and offered them to the IAU WG NSFA (see http://maia.usno.navy.mil/NSFA/CBE.html).

More than 400000 data-points presented in Table 1 were used for construction of the EPM2006 ephemerides, and about 230 parameters of these ephemerides were determined. The detailed description of all the data used, plots of the observersion residuals, a range of astronomical constants obtained from the adjustment of the EPM2006 (namely, the value of Astronomical Unit, the rotation parameters of Mars, masses of the bodies, parameters of topography of planet surfaces, the solar oblateness, PPN-parameters, the variability of the gravitational constant, and others) are given in papers (Pitjeva 2005a; Pitjeva 2005b; Pitjeva 2007). Formal standard deviations of orbital elements of planets of the EPM2006 ephemerides are shown in Table 3.

Table 3. The formal standard deviations of elements of the planet orbits

Planet	a [m]	$\sin i \cos \Omega$ [mas]	$\sin i \sin \Omega$ [mas]	$e \cos \pi$ [mas]	$e \sin \pi$ [mas]	λ [mas]
Mercury	0.333	1.392	1.347	0.105	0.084	0.343
Venus	0.219	0.056	0.030	0.004	0.003	0.031
Earth	0.138	—	—	0.001	0.001	—
Mars	0.267	0.003	0.002	0.001	0.001	0.002
Jupiter	615	2.419	2.166	0.313	0.362	1.467
Saturn	4256	3.061	4.117	3.900	2.959	3.501
Uranus	40294	4.440	6.276	5.057	3.635	7.509
Neptune	463307	4.411	8.520	13.115	18.740	24.484
Pluto	3412734	6.790	15.662	80.870	38.847	43.554

As seen from Table 3, the uncertainties obtained for the inner planets are significantly smaller than the uncertainties for Saturn, Uranus, Neptune, Pluto. Jupiter occupies the intermediate state between the inner and the outer planets due to the availability of a number of spacecraft Jupiter data.

The EPM ephemerides which have been the basis for the Russian "Astronomical Yearbook" since 2006 and have been used in the navigation program "GLONASS" and the cosmic program "Phobos-Grunt", are available to outside users via ftp://quasar.ipa.nw.ru/incoming/EPM2004 (anonymous).

References

Brown, M. E. & Schaller, E. L. 2007, *Science* 316, 1585, doi: 10.1126/science.1139415

Fienga, A., Manche, H., Laskar, J., Gastineau, M. 2008, *A&A* 477(1), 315

Jacobson, R. A. 2003, "JUP230 orbit solution"

Jacobson, R. A., Antreasian, P. G., Bordi, J. J., Criddle, K. E., Ionasescu, R., Jones, J. B., Mackenzie, R. A., Pelletier, F. J., Owen Jr., W. M., Roth, D. C., Stauch, J. R. 2006, *AJ* 132(6), 2520

Konopliv, A. S., Banerdt, W. B., Sjogren, W. L. 1999, *Icarus* 139, 3

Konopliv, A. S., Yoder, C. F., Standish, E. M., Yuan, D. N., Sjogren, W. L. 2006, *Icarus* 182, 23

Pitjeva, E. V. 2005a, *Solar System Research* 39, 176

Pitjeva, E. V. 2005b, in D. W. Kurtz (ed.), *Transit of Venus: new views of the solar system and galaxy*, Proc. IAU Colloquium No. 196 (Cambridge University Press), p. 230

Pitjeva, E. V. 2007, *Proceedings of IAA RAS* 17, 42

Standish, E. M. 2006, *Interoffice Memorandum* 343R-06-002

Tholen, D. J., Buie, M. W., Grundy, W. M., & Elliott, G. T. 2008, AJ 135(3), 777, arXiv: 0712.1261

A Giant Step: from Milli- to Micro-arcsecond Astrometry
Proceedings IAU Symposium No. 248, 2007
W. J. Jin, I. Platais & M. A. C. Perryman, eds.

© 2008 International Astronomical Union
doi:10.1017/S1743921308018577

HST FGS astrometry – the value of fractional millisecond of arc precision

G. F. Benedict[1], B. E. McArthur[1] & J. L. Bean[2]

[1]McDonald Observatory, University of Texas, Austin, TX USA 78712
email: `fritz or mca@astro.as.utexas.edu`
[2]Institute für Astrophysic, Georg-August-Universität, Göttingen, Germany
email: `bean@astro.physik.uni-goettingen.de`

Abstract. In a few years astrometry with the venerable combination of Hubble Space Telescope and Fine Guidance Sensor will be replaced by SIM, Gaia, and long-baseline interferometry. Until then we remain a resource of choice for sub-millisecond of arc precision optical astrometry. As examples we discuss 1) the uses which can be made of our parallaxes of Galactic Cepheids, and 2) the determination of perturbation orbital elements for several exoplanet host stars, yielding true companion masses.

Keywords. astrometry, Cepheids, distance scale, planetary systems, stars: low-mass, brown dwarfs

1. Introduction

In the early 1970's, Bob O'Dell, then director of Yerkes Observatory, twisted the arm of Bill van Altena, encouraging Bill to propose to study the feasibility of doing astrometry with what was then called the Space Telescope (ST). Bill was joined by Larry Frederick and Otto Franz. Their efforts insured that astrometry with a Fine Guidance Sensor (FGS) would be an integral science component for ST. In 1977 Bill Jefferys formed a team to respond to the ST Announcement of Opportunity. The first author was a member of that team, which then included Raynor Duncombe, Paul Hemenway, and Pete Shelus. We were eventually chosen as a component (along with van Altena, Franz, and Fredrick) of the final STAT (Space Telescope Astrometry Team). Barbara McArthur joined us in 1985.

ST was launched in 1990, becoming the Hubble Space Telescope (HST). Because of seismic disturbances caused by day/night cycle temperature effects on the original solar arrays and slow, but fortunately steady, changes in the FGS, we were unable to prove that precision astrometry could be carried out with the FGS until after the first refurbishment mission in late 1992. With new, re-designed solar arrays, HST and the two FGS units we have used over the years, originally predicted to produce single-measurement astrometry with 3 millisecond of arc (mas) precision, have both routinely produced 1 mas or better per-observation precision astrometry. Calibration issues, many of which are non-trivial, are discussed in McArthur *et al.* (2002, 2006), Benedict *et al.* (1994, 2002a), and for binary star fringe morphology analysis, Franz *et al.* (1998).

The many results from the original STAT include parallaxes of astrophysically interesting stars (Benedict *et al.* 2000, 2002a, 2002b, 2003, McArthur *et al.* 1999, 2001), a parallax for the Hyades (van Altena *et al.* 1997), a link between quasars and the HIPPARCOS reference frame (Hemenway *et al.* 1997), determination of low-mass binary star masses (Franz *et al.* 1998, Benedict *et al.* 2001), and searches for Jupiter-mass companions to Proxima Cen and Barnard's Star (Benedict *et al.* 1999).

Since 1999, Barbara and I have carried on as General Observers, contributing to the study of the lower main sequence mass-luminosity relationship (Henry *et al.* 1999), the intercomparison of dwarf novae (Harrison *et al.* 2004), and parallaxes of cataclysmic variables (Beuermann *et al.* 2003, 2004; Roelofs *et al.* 2007) and the Pleiades (Soderblom *et al.* 2005). Jacob Bean, now a postdoc, joined us as a graduate student in 2002.

Two major themes of our investigations over the last few years have included the cosmic distance scale and extrasolar planetary systems. The remainder of this paper will outline recent results, specifically drawing from our work (Benedict *et al.* 2007) on the Galactic Cepheid Period-Luminosity Relationship (PLR), and the determination of extrasolar planetary masses (Benedict *et al.* 2002c, McArthur *et al.* 2004, Benedict *et al.* 2006, Bean *et al.* 2007).

2. The Galactic PLR

Our goal was to determine trigonometric parallaxes for nearby fundamental mode Galactic Cepheid variable stars. Our target selection consisted in choosing the nearest Cepheids (using *HIPPARCOS* parallaxes, Perryman *et al.* 1997), covering as wide a period range as possible. These stars were the brightest known Cepheids at their respective periods. Our new parallaxes provided distances and ultimately absolute magnitudes, M, in several bandpasses. Additionally, our investigation of the astrometric reference stars provided an independent estimation of the line of sight extinction to each of these stars, a contributor to the uncertainty in the absolute magnitudes of our prime targets. These Cepheids, all with near solar metallicity, should be immune to variations in absolute magnitude due to metallicity variations, *e.g.* Groenewegen *et al.* (2004), Macri *et al.* (2006). Adding our previously determined absolute magnitude for δ Cep, Benedict *et al.* (2002b), we established V, I, K, and W_{VI} Period-Luminosity Relationships using ten Galactic Cepheids with average metallicity, $\langle$[Fe/H]$\rangle$=0.02, a calibration that can be directly applied to external galaxies whose Cepheids exhibit solar metallicity. Details are given in Benedict *et al.* (2007).

One of our interesting results is an independent determination of the distance modulus of the Large Magellanic Cloud (LMC), particularly given the recent exploration of possible sociological influences on such determinations (Schaefer, 2007). OGLE (Udalski *et al.* 1999) has produced the largest amount of LMC Cepheid photometry. We worked with an apparent W_{VI} PLR for 581 Cepheids in the LMC. These data were carefully preened, selecting only Cepheids with normal light curves and amplitudes from Ngeow *et al.* (2005). They provide a zero-point of 12.65 ± 0.01. Direct comparison of that W_{VI} zero-point with the zero-point of the left-hand $M_{W(VI)}$ PLR in Figure 1 yields an LMC distance modulus 18.51 ±0.03 with no metallicity corrections.

Macri *et al.* (2006) demonstrate that a metallicity correction is necessary by comparing metal-rich Cepheids with metal-poor Cepheids in NGC 4258. With a previously measured [O/H] metallicity gradient from Zaritsky *et al.*(1994), Macri *et al.* find a Cepheid metallicity correction in W_{VI}, $\gamma = -0.29 \pm 0.09_r \pm 0.05_s$ magnitude for 1 dex in metallicity, where r and s subscripts signify random and systematic. This value is similar to an earlier W_{VI} metallicity correction (Kennicutt *et al.* 1998) derived from Cepheids in M101 (−0.24± 0.16). Taking the weighted mean of the Kennicutt and Macri values and using the difference in metallicity of LMC and Galactic Cepheids (−0.36 dex from means of the data in Groenewegen *et al.* 2004 tables 3 and 4) we find a metallicity correction of −0.10± 0.03 magnitude with the Galactic Cepheids being brighter. With this estimated metallicity correction we obtain a corrected LMC modulus of 18.41 ± 0.05. One other recent determination is noteworthy for its lack of dependence on any metallicity corrections. Fitzpatrick *et al.*(2003) derive 18.42± 0.04, from eclipsing binaries, a modulus in close agreement with our new value.

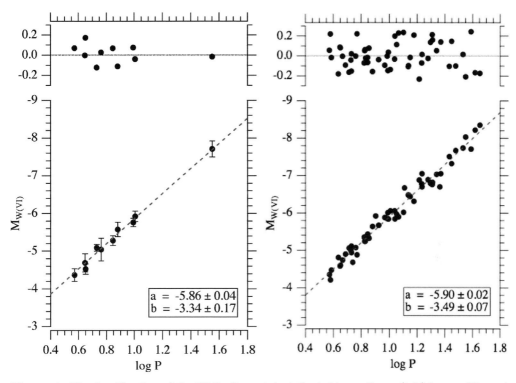

Figure 1. The densification of the PLR. Our original Cepheid parallaxes (left) have calibrated techniques applied to many other, more distant Cepheids (right, after Fouqué *et al.* 2007). Note the reduction in the error of the slope of the PLR.

As an additional check on our adopted LMC modulus, m–M= 18.41 ± 0.05, a Macri *et al.* (2006) differential modulus (LMC-NGC 4258) leads to a modulus of 29.29 ± 0.08 for NGC 4258, a value in excellent agreement with the maser-based distance for NGC 4258, m–M = 29.29 ± 0.15 from Herrnstein *et al.*(1999).

One of the first uses of our new Cepheid parallaxes was the calibration of various techniques used to provide distances for many more long period (and all far more distant) Cepheids as described in Fouqué *et al.* (2007). These techniques include infrared surface brightness and interferometric Baade-Wesselink, which compare measured physical diameters with radial velocities. This calibration results in a densification and extension of the Galactic Cepheid PLR, and is illustrated in Figure 1. The immediate value of the densification is a greater confidence in the absolute magnitudes of the longer-period, brighter Cepheids, the ones critical for extragalactic distance determination.

3. Extrasolar planet masses

Currently, fewer than 10% of the more than 200 candidate exoplanets (http://exoplanet.eu) orbiting nearby stars have precisely determined masses. Because the most successful technique for detecting candidate exoplanets, the radial velocity method, suffers from a degeneracy between the mass and orbital inclination for most of the known exoplanet candidates, only a minimum mass, $M sini$, is known. In principle, radial velocities alone can be used to determine the masses of exoplanets in multiplanet systems, if two or more planets are experiencing significant mutual gravitational interactions on short timescales. However, only one such system has been investigated, results from different groups vary significantly, and the effects of non-coplanarity remain to be considered.

Thus, establishing precise masses, rather than arguing statistically (using $M sini$, which everyone does, e.g. the otherwise excellent review of Udry & Santos 2007), for the majority of exoplanet candidates requires additional observations. The primary techniques that have been employed to break the mass-inclination degeneracy in radial velocity data are astrometry (Benedict *et al.* 2002c, McArthur *et al.* 2004, Benedict *et al.* 2006, Bean *et al.* 2007) and transit (e.g., Bakos *et al.* 2007) observations. Ideally, all candidates should have their masses determined. Obviously, not all systems are edge-on to us. And, unfortunately, the size of the perturbation due to the companion decreases linearly with distance, placing many of them beyond the reach of HST FGS astrometry.

Table 1. Masses from HST FGS Astrometry

Component	$M_*[M_\odot]$	[Fe/H]	d (pc)	ecc.	M_p (M_{Jup})	α (mas)	inc. (°)	P (d)
GJ 876 b[1]	0.32	-0.12^2	4.7	0.1	1.9±0.5	0.25	84±6	61
55 Cancri d[3]	1.21	0.32	12.5	0.33	4.9±1.1	1.9	53±7	4517
ϵ Eridani b[4]	0.83	−0.03	3.2	0.7	1.6±0.2	1.9	30±4	2502
HD 33636 B[5]	1.02	−0.13	28.1	0.48	142±11	14.2	4.1±0.1	2117

[1]Benedict *et al.* 2002c, [2]Bean *et al.* 2006a, 2006b, [3]McArthur *et al.*2004, [4]Benedict *et al.* 2006, [5]Bean *et al.* 2007

To partially rectify this lack of knowledge of the true masses, the HST allocation process awarded us over 100 orbits in Cycle 14 and 15 to be used to determine actual masses for HD 47536 b, HD 136118 b, HD 168443 c, HD 145675 b, HD 38529 c, and HD 33636 b. The object chosen for discussion in this contribution, HD 33636 (Bean *et al.* 2007) presents a cautionary tale. As we shall see, at least in some cases, $M sini$ is not a very good estimator of the actual companion mass.

We chose to work first on HD 33636 because we had finished collecting all the HST data for this object. Radial velocity data (Figure 2) were obtained from past investigators (Lick and Keck data from Butler *et al.* 2006 and Marcy (2007, private communication), Elodie data from Perrier *et al.* (2003)) and from a high-cadence campaign with the Hobby-Eberly Telescope High-Resolution Spectrograph (HRS; Tull 1998). The latter data were used to search for shorter-period companions, the unknown and unmodeled presence of which would have introduced noise into our velocity data. No such companions were found with detection limits detailed in Bean *et al.* (2007). Limits depend on period and assumed eccentricity of the tertiary. For example these velocity data would have permitted detection of a tertiary with $M > 0.5 M_{Jup}$ for $P < 300^d$ and eccentricity < 0.7.

Our techniques for combining radial velocity and HST FGS astrometry have been detailed in Benedict *et al.* (2001, 2006) and Bean *et al.* (2007). For HD 33636 we found an inclination, $i = 4.0° \pm 0.1°$ and, as shown in Figure 3, a perturbation semimajor axis, $\alpha = 14.2 \pm 0.2$ mas. Assuming a mass for the primary, HD 33636, $M = 1.02 \pm 0.03$ $M_\odot$ (Takeda *et al.* 2007), we obtain a mass for HD 33636 b, $M = 142 \pm 11 M_{Jup} = 0.14 \pm 0.01$ $M_\odot$. HD 33636 b is actually HD 33636 B, an M6V star. Including it in any

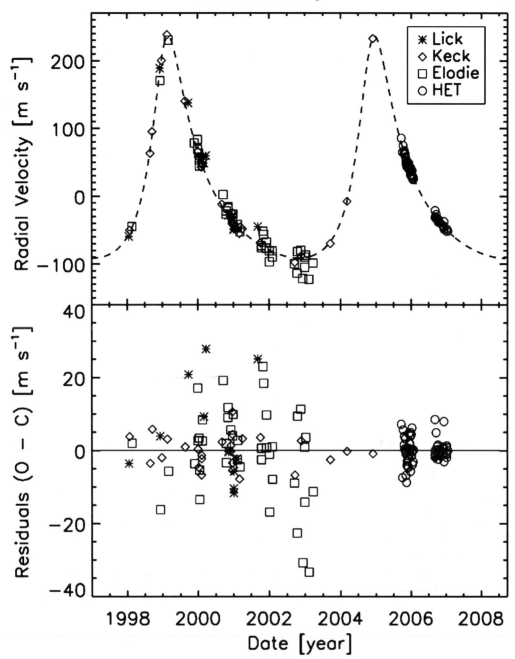

Figure 2. HD 33636: Radial velocities and residuals to our final orbit from combined velocities and astrometry. The HET residuals are slightly better than the Keck residuals.

statistical studies depending on $Msini$, such as metallicity, eccentricity, and semimajor axis, would corrupt those relationships.

We have made some progress. Table 1 collects the mass estimates that have already resulted from our combination of HST astrometry and high-precision radial velocities. Over the next year we will finish our analysis and determine actual masses for HD 47536 b, HD 136118 b, HD 168443 c, HD 145675 b, and HD 38529 c. But, before those results,

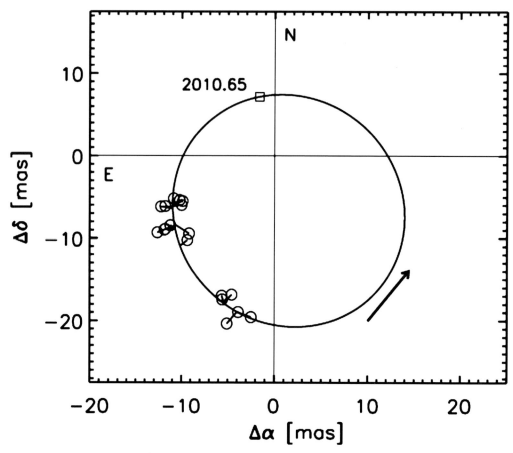

Figure 3. The astrometric perturbation due to the *stellar* companion to HD 33636. Based on radial velocities only this object had a minimum mass $Msini = 9.4M_{Jup}$. The inclination, $i = 4.1°$, and size of the semi-major axis of the perturbation, $\alpha = 14.2$ mas, identify the companion as an M dwarf star. The box marks the date of the next periastron passage.

we will have finished our analysis of the υ Andromedae system (McArthur *et al.* 2008). We have a preliminary indication that for the first time the degree of coplanarity of an extrasolar planetary system associated with a normal, main sequence star will be established. Beyond that, we are in the process of acquiring HST FGS data for similar coplanarity tests on the multiplanet systems associated with HD 128311, HD 202206, μ Ara, and γ Cep, paving the way for future studies of hundreds of such systems with ground-based long-baseline interferometry and the space-based projects, SIM and Gaia.

References

Bakos, G. A. *et al.* 2007, ApJ, 656, 552
Bean, J. L., *et al.* 2006a, ApJL, 653, L65
Bean, J. L. *et al.* 2006b, ApJ, 652, 1604
Bean, J. L. *et al.* 2007, AJ, 134, 749
Benedict, G. F. *et al.* 1998, AJ, 116, 429
Benedict, G. F. *et al.* 1999, AJ, 118, 1086
Benedict, G. F. *et al.* 2000, AJ, 119, 2382
Benedict, G. F. *et al.* 2001, AJ, 121, 1607

Benedict, G. F. *et al.* 2002a, AJ, 123, 473

Benedict, G. F., *et al.*. 2002b, AJ, 124, 1695

Benedict, G. F., *et al.* 2002c, ApJL, 581, L115

Benedict, G. F., et al. 2006, AJ, 132, 2206

Benedict, G. F., *et al.*. 2007, AJ, 133, 1810

Beuermann, K. *et al.* 2003, A&Ap, 412, 821

Beuermann, K. *et al.* 2004, A&Ap, 419, 291

Franz, O. G. *et al.* 1998, AJ, 116, 1432

Fitzpatrick, E. L. *et al.* 2003, ApJ, 587, 685

Fouqué, P., *et al.* 2007, ArXiv e-prints, 709, arXiv:0709.3255

Groenewegen, M. A. T. *et al.* 2004, A&Ap, 420, 655

Harrison, T. E. *et al.* 1999, ApJL, 515, L93

Harrison, T. E. *et al.* 2004, AJ, 127, 460

Herrnstein, J. R., *et al.* 1999, Nature, 400, 539

Kennicutt, R. C. *et al.* 1998, ApJ, 498, 181

Macri, L. M. *et al.* 2006, ApJ, 652, 1133

McArthur, B. *et al.* 1997. in Proc. 1997 HST Calibration Workshop, ed. S. Casertano, R. Jedrzejewski, T. Keyes, and M. Stevens, STScI Publication, Baltimore, MD

McArthur, B. E. *et al.* 1999, ApJL, 520, L59

McArthur, B. E. *et al.* 2001, ApJ, 560, 907

McArthur, B. *et al.* 2002, The 2002 HST Calibration Workshop, Edited by Santiago Arribas, Anton Koekemoer, and Brad Whitmore. Baltimore, MD: Space Telescope Science Institute, 2002, p.373

McArthur, B. E., *et al.* 2004, ApJL, 614, L81

McArthur, B. E. *et al.* 2006, The 2005 HST Calibration Workshop: Hubble After the Transition to Two-Gyro Mode, 396

McArthur, B. E., *et al.* 2008, in preparation

Ngeow, C.-C. *et al.* 2005, MNRAS, 363, 831

Perryman, M. A. C. *et al.* 1997, A&Ap, 323, L49

Roelofs, G. H. A. *et al.* 2007, ApJ, 666, 1174

Schaefer, B. E. 2007, ArXiv e-prints, 709, arXiv:0709.4531

Soderblom, D. R. *et al.* 2005, AJ, 129, 1616

Takeda, G.,*et al.* 2007, ApJS, 168, 297

Tull, R. G. 1998, Proc. SPIE, 3355, 387

Udalski, A. *et al.* 1999, Acta Astronomica, 49, 201

Udry, S. & Santos, N. C., 2007, Ann. Rev. Astron. & Astrophys., 45, 379.

van Altena, W. F., *et al.* 1997, ApJL, 486, L123

van Leeuwen, F. *et al.* 2007, MNRAS, 379, 723

Zaritsky, D. *et al.* 1994, ApJ, 420, 87

A Giant Step: from Milli- to Micro-arcsecond Astrometry
Proceedings IAU Symposium No. 248, 2007
W. J. Jin, I. Platais & M. A. C. Perryman, eds.

Astrometric detection and characterization of brown dwarfs

R.-D. Scholz[1], M. J. McCaughrean[2], S. Röser[3] and E. Schilbach[3]

[1]Astrophysikalisches Institut Potsdam, An der Sternwarte 16, D–14482 Potsdam, Germany
email: rdscholz@aip.de

[2]University of Exeter, School of Physics, Stocker Road, Exeter EX4 4QL, UK
email: mjm@astro.ex.ac.uk

[3]Astronomisches Rechen-Institut, Mönchhofstraße 12-14, D–69120 Heidelberg, Germany
email: roeser@ari.uni-heidelberg.de, elena@ari.uni-heidelberg.de

Abstract. As a result of failed star formation, brown dwarfs (BDs) do not reach the critical mass to ignite the fusion of hydrogen in their cores. Different from their low-mass stellar brothers, the red dwarfs, BDs cool down with their lifetime to very faint magnitudes. Therefore, it was only about 10 to 20 years ago that such ultracool objects began to be detected. Accurate astrometry can be used to detect them indirectly as companions to stars by the signature of the so-called astrometric wobble. Resolved faint BD companions of nearby stars can be identified by their common proper motion (CPM). A direct astrometric detection of the hidden isolated BDs in the Solar neighborhood is possible with deep high proper motion (HPM) surveys. This technique led to the discovery of the first free-floating BD, Kelu 1, and of the nearest BD, ϵ Indi B. Both were meanwhile found to be binary BDs. The astrometric orbital monitoring of ϵ Indi Ba+Bb, for which we know an accurate distance from the Hipparcos measurement of its primary, ϵ Indi A, will allow the determination of individual masses of two low-mass BDs. Hundreds of BDs have been identified for the last decade. Deep optical sky survey (SDSS) and near-infrared sky surveys (DENIS, 2MASS), played a major role in the search mainly based on colours, since BDs emit most of their light at longer wavelengths. However, alternative deep optical HPM surveys based on archival photographic data are not only sensitive enough to detect some of the nearest representatives, they do also uncover many of the rare class of ultracool halo objects crossing the Solar neighborhood at large velocities. SSSPM 1444, with the extremely large proper motion of 3.5 arcsec/yr, is one of the nearest among these subdwarfs with masses at the substellar boundary. We present preliminary parallax results for this and two other ultracool subdwarfs (USDs) from the Calar Alto Omega 2000 parallax program.

Keywords. astrometry, surveys, stars: low-mass, brown dwarfs, binaries: general, stars: distances, stars: kinematics, subdwarfs, solar neighborhood, Galaxy: halo

1. The pioneering role of astrometry in the search for brown dwarfs

After their initial deuterium burning, brown dwarfs (hereafter BDs) cool down throughout their lifetimes and become harder and harder to detect. Whereas young BDs appear as late-type M dwarfs, they fall into the L dwarf regime at an intermediate age between 100 Myr and 1 Gyr after their formation depending on their mass, and later transform into T dwarfs (Burrows *et al.* 2001). The Solar neighborhood preferentially consists of relatively old objects. Therefore, the majority of low-mass BDs near the Sun should be T-type ones, whereas young M-type BDs can probably only be found as members of young open clusters and associations, outside the 25 pc horizon typically defined as the Solar neighborhood "border".

The first BDs were discovered as companions to known nearby stars. Becklin & Zuckerman (1988) found GD 165B (later classified as L4 dwarf), a faint infrared companion of a white dwarf, the common proper motion (hereafter CPM) of which was measured by Zuckerman & Becklin (1992). Nakajima *et al.* (1995) detected Gl 229B (T7), a faint companion of an M2 dwarf providing also a first crude CPM measurement. In the same year, Rebolo *et al.* (1995) discovered the first BD in an open cluster, Teide 1 (M8), for which they confirmed astrometric membership by measuring the CPM with the Pleiades.

The first free-floating BDs in the Solar neighborhood were identified as result of a high proper motion (hereafter HPM) survey by Ruiz *et al.* (1997), and from the DEep Near-Infrared Survey (DENIS) without initial HPM measurements by Delfosse *et al.* (1997). However, there is one HPM object catalogued before all the above mentioned BD discoveries: LP 944-20 (Luyten 1979b), classified as an intermediate-age BD of spectral type M9 by Tinney (1998).

To our knowledge, there is only one successful BD detection by the measurement of the astrometric wobble: Pravdo *et al.* (2005) discovered a substellar-mass object around the nearby M5.5 dwarf GJ 802 within the STEPS (STEllar Planet Survey) program using the Palomar 5-m telescope.

2. Astrometric investigations of L and T dwarfs

In this section we consider only L and T dwarfs as the main reservoir of BDs, keeping in mind that (1) there are many M-type BDs in young open clusters and associations, (2) intermediate-age M-type BDs like LP 944-20 do also exist in the Solar neighborhood in very small numbers, (3) the early- to mid-L dwarfs population is a mixture of low-mass stars and BDs (Kirkpatrick 2005). The main reason for restricting our view to the new spectral types is the availability of a regularly updated complete database.

The L and T dwarf compendium housed at DwarfArchives.org and maintained by Gelino *et al.* (2007) currently contains 626 objects. As can be seen from Fig. 1, the Galactic plane is still a problematic region for searching BDs. Only 78 objects (12%) have trigonometric parallax measurements. These are mostly in the broader equatorial zone. Proper motions are available for 203 objects (32%). Whereas proper motions have been measured for almost all of the brighter ($J < 13.5$) objects, many bright objects are still lacking parallax measurements (Fig. 1).

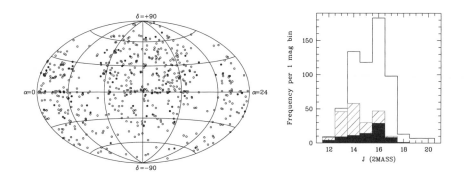

Figure 1. Left: Distribution of all known 626 L and T dwarfs over the sky (Gelino *et al.* 2007). Filled symbols show those 78 objects which have trigonometric parallax measurements. Right: J magnitude histogram of all objects in comparison with those having measured proper motions (hashed) and parallaxes (shaded), respectively.

Most of the existing parallaxes have been measured by Vrba *et al.* (2004) (28 objects) and Dahn *et al.* (2002) (19 objects). For 19 L and T dwarf companions the parallaxes come from Hipparcos (ESA 1997) measurements of their primary stars. The relatively low completeness concerning the proper motions is due to the fact that most of the L and T dwarfs were discovered in colour-based searches. These objects are generally fainter and more distant than objects discovered by their large proper motion. For 66 objects proper motions have been obtained/revised together with their parallaxes. Large numbers of proper motion measurements were also provided by Schmidt *et al.* (2007) (47 objects), Phan-Bao *et al.* (2007) (18 objects), and Hambly *et al.* (2001) (14 objects). The remaining proper motions have been published in many (discovery) papers dealing with only few L and T dwarfs.

3. The impact of new HPM surveys

HPM objects represent a mixture of *very nearby* neighbors of the Sun in the local population of the Galactic thin disk, and *very fast* representatives of the Galactic thick disk and halo, just passing through the neighborhood. By extending the existing HPM surveys to fainter magnitudes we expect to find the nearest isolated substellar objects with thin disk kinematics (section 4) as well as some of the lowest-mass visitors from the Galactic halo (section 5). Although halo stars are relatively rare when compared to the number density of disk stars, they are over-represented in HPM samples.

Table 1. Recent discoveries of objects with proper motions > 2 arcsec/yr

Name	proper motion [arcsec/yr]	Discovery paper	Distance (plx. ref.) [pc]	object type
SO 0253+1652	5.11	Teegarden+03	3.84 (1)	disk M6.5
ε Indi Ba,Bb	4.70	Scholz+03, McCaughrean+04	3.625 (2)	disk T1+T6
SSSPM 1444−2019	3.51	Scholz+04b	∼20	halo sdM9
2MASS 1114−2618	3.05	Tinney+05	∼7	disk T7.5
SCR 1845−6357 AB	2.66	Pokorny+03, Hambly+04, Biller+06	3.854 (1)	disk M8.5+T6
2MASS 0532+8246	2.60	Burgasser+03	26.7 (5)	halo sdL7
PM J13420−3415	2.55	Lépine, Rich & Shara 05	∼18	halo WD
LEHPM 3396	2.45	Pokorny+03, Phan-Bao+06	∼8	disk M9.0
LSR 1826+3014	2.38	Lépine+02	∼14	halo M8.5
F351-50	2.33	Ibata+00	35 (4)	halo cool WD
2MASS 0415−0935	2.26	Burgasser+02	5.74 (3)	disk T8.5
2MASS 0251−0352	2.17	Cruz+03, Schmidt+07	∼12	disk(?) L3.0
SCR 1138−7721	2.15	Hambly+04, Scholz+04a	8.18 (1)	disk M5.5

Trig. parallaxes: 1 - Henry+06, 2 - ESA97, 3 - Vrba+04, 4 - Ducourant+07, 5 - Burgasser+07

Table 1 lists 13 newly discovered objects (two of which were later resolved as binary systems). Compared to 73 previously known objects in the LHS catalogue (Luyten 1979a) with proper motions larger than 2 arcsec/yr, they represent a considerable increase in the number of the fastest stars on the sky. Except for two mid-M dwarfs and two white dwarfs all the new objects have masses at or below the substellar limit.

4. The nearest BDs to the Sun

A very red and extreme HPM ($\sim$4.7 arcsec/yr) object was discovered in an HPM survey of overlapping I band plates scanned in the SuperCOSMOS Sky Surveys (SSS). A very bright infrared counterpart was also detected at the expected position in the Two-Micron All Sky Survey (2MASS). The object was found to share its HPM with a well-known bright star separated by about 7 arcmin and consequently named ε Indi B (Scholz *et al.* 2003). With the given Hipparcos parallax of ε Indi A it was clear that the newly found object was the nearest BD to the Sun. Its proper motion was later improved by including additional epochs (Fig. 2). The final result (McCaughrean *et al.* 2004) agrees within 40 mas/yr with that of the primary star, which can be expected given the wide orbital motion.

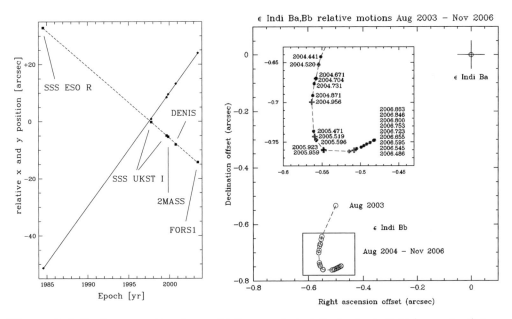

Figure 2. Left: Proper motion (x - solid line, y -dashed line) of ε Indi B (unresolved) from archival data, Right: Orbital motion of the resolved BD binary ε Indi Ba,Bb from the first 3 years of the adaptive optics monitoring program with NACO at the ESO VLT. The data are joined here with straight lines, not a fit.

McCaughrean *et al.* (2004) resolved ε Indi B as a close pair (sep. 0.7 arcsec) of two T-type BDs, ε Indi Ba,Bb, with an expected relative orbital period of only $\sim$15 years. This system provides the unique opportunity to measure the *individual* masses of both BDs by combining relative astrometry of the resolved pair (Fig. 2) with measurements of the absolute motion of the system barycenter against a network of field stars in wide-field CCD images. A corresponding long-term monitoring program (PI: M. McCaughrean) is currently being carried out.

Most of the known nearest BDs are T dwarfs. Among the 10 objects with measured parallaxes >100 mas in the compilation of Gelino *et al.* (2007), there are only two L dwarfs, discovered by Reid *et al.* (2000). Other bright L dwarfs discovered in HPM surveys (e.g. Scholz & Meusinger 2002; Salim *et al.* 2003) are still lacking parallaxes.

5. Ultracool subdwarfs in the Solar neighborhood

Ultracool subdwarfs (hereafter USDs) with spectral types later than sdM7 represent a new class of metal-deficient low-mass stars and BDs which have typically Galactic halo kinematics. Compared to 600 L and T dwarfs, only <20 USDs have been discovered so far (mostly as HPM objects). A proper classification scheme is still under development. For an overview we refer the reader to Burgasser *et al.* (2007a) and references therein.

Table 2. Preliminary parallaxes of ultracool subdwarfs

Name	Spectral Type	Ref.	$\mu_\alpha \cos\delta$ [mas/yr]	μ_δ [mas/yr]	π [mas]	Ref.
SSSPM 1444−2019	sdM9 (sdL:)	(1)	-2901.5 ± 2.5	-1977.9 ± 2.2	55.1 ± 2.8	(4)
2MASS 0532+8246	sdL7	(2)	$+2047.7 \pm 1.8$	-1658.6 ± 1.5	40.8 ± 1.7	(4)
			$+2041.9 \pm 1.8$	-1648.2 ± 1.8	37.5 ± 1.7	(6)
2MASS 0937+2931	sdT6	(3)	$+952.2 \pm 1.6$	-1308.3 ± 1.7	161.6 ± 1.8	(4)
			$+973.0 \pm 7.1$	-1297.8 ± 7.1	162.8 ± 3.9	(5)

References: 1 - Scholz *et al.* (2004b), 2 - Burgasser *et al.* (2003), 3 - Burgasser *et al.* (2002), 4 - Röser *et al.* in prep., 5 - Vrba *et al.* (2004), 6 - Burgasser *et al.* (2007b)

Trigonometric parallaxes of USDs are urgently needed. We (Röser, Schilbach, & Scholz) are currently running a dedicated parallax program using Omega 2000 at the 3.5m telescope of the Calar Alto Observatory, targetting 10 USDs. Omega 2000 with its 15 by 15 arcmin field and 450 mas per pixel provides a much larger number of astrometric reference stars which makes the astrometric solutions more accurate than with other instruments. In Table 2 we present first results for SSSPM 1444 and two other USDs, for which comparison measurements are available. Although the preliminary Calar Alto parallax results are based on only 2.5 years of observations (and small ∼1-2 mas corrections to absolute parallaxes are not yet included) they are in good agreement with those from the USNO infrared astrometry program (Vrba *et al.* 2004; Burgasser *et al.* 2007b).

6. Concluding remarks

9 of the 13 discoveries of extreme HPM objects (Table 1) including the nearest BDs and USDs were made by using digitized photographic Schmidt plates. This underlines that these archival data are still important and by far not yet fully exploited. However, large area multi-epoch near-infrared surveys (e.g. Kirkpatrick *et al.* 2007) will have the best chances to uncover many more of the hidden BDs in the Solar neighborhood.

Considerable efforts are still needed to complete the astrometric characterization of a sufficient number of benchmark BDs and of the poorly understood class of USDs. Ground-based observations will continue to play a major role for high-accuracy astrometry of brown dwarfs, since most of them are too faint to be seen by Gaia.

Acknowledgements

We would like to thank the SSS and 2MASS teams for providing excellent astrometric and photometric databases which served as the basis for many HPM discoveries. We also thank the European Southern Observatory and the Calar Alto Observatory for allocating time for the ε Indi Ba,Bb orbital monitoring and the USDs parallax programs, respectively.

References

Becklin, E. E. & Zuckerman, B. 1988, *Nature* 336, 656

Biller, B. A., Kasper, M., Close, L. M., Brandner, W., & Kellner, S. 2006, *ApJ* 641, L141

Burgasser, A. J., *et al.* 2002, *ApJ* 564, 421

Burgasser, A. J., *et al.* 2003, *ApJ* 592, 1186

Burgasser, A. J., Cruz, K. L., & Kirkpatrick, J. D. 2007a, *ApJ* 657, 494

Burgasser, A. J., *et al.* 2007b, *ApJ* accepted, arXiv:0709.1373

Burrows, A., Hubbard, W. B., Lunine, J. I., & Liebert, J. 2001, *Reviews of Modern Physics* 73, 719

Cruz, K. L., Reid, I. N., Liebert, J., Kirkpatrick, J. D., & Lowrance, P. J. 2003, *AJ* 126, 2421

Dahn, C. C., *et al.* 2002, *AJ* 124, 1170

Delfosse, X., *et al.* 1997, *A&A* 327, L25

Ducourant, C., Teixeira, R., Hambly, N. C., Oppenheimer, B. R., Hawkins, M. R. S., Rapaport, M., Modolo, J., & Lecampion, J. F. 2007, *A&A* 470, 387

ESA 1997, The HIPPARCOS and TYCHO catalogues, ESA SP 1200

Gelino, C. R., Kirkpatrick, J. D., & Burgasser, A. J. 2007, online database for 626 L and T dwarfs at Dwarfarchives.org (status: 1 October 2007)

Hambly, N. C., *et al.* 2001, *MNRAS* 326, 1279

Hambly, N. C., Henry, T. J., Subasavage, J. P., Brown, M. A., & Jao, W.-C. 2004, *AJ* 128, 437

Henry, T. J., Jao, W.-C., Subasavage, J. P., Beaulieu, T. D., Ianna, P. A., Costa, E., & Méndez, R. A. 2006, *AJ* 132, 2360

Ibata, R., Irwin, M., Bienaymé, O., Scholz, R., & Guibert, J. 2000, *ApJ* 532, L41

Kirkpatrick, J. D. 2005, *ARAA* 43, 195

Kirkpatrick, J. D., Looper, D. L., Burgasser, A. J., Cruz, K. L., Cushing, M. C., & Schurr, S. D. 2007, *AAS Meeting Abstracts* 210, #17.07

Lépine, S., Rich, R. M., & Shara, M. M. 2005, *ApJ* 633, L121

Lépine, S., Rich, R. M., Neill, J. D., Caulet, A., & Shara, M. M. 2002, *ApJ* 581, L47

Luyten, W. J. 1979a, LHS Catalogue: a catalogue of stars with proper motions exceeding 0.5" annually, Minneapolis, University of Minnesota

Luyten, W. J. 1979b, New Luyten Catalogue of stars with proper motions larger than two tenths of an arcsecond (NLTT), Minneapolis, University of Minnesota

McCaughrean, M. J., Close, L. M., Scholz, R.-D., Lenzen, R., Biller, B., Brandner, W., Hartung, M., & Lodieu, N. 2004, *A&A* 413, 1029

Nakajima, T., Oppenheimer, B. R., Kulkarni, S. R., Golimowski, D. A., Matthews, K., & Durrance, S. T. 1995, *Nature* 378, 463

Phan-Bao, N., *et al.* 2007, MNRAS submitted, arXiv:0708.4169

Phan-Bao, N., *et al.* 2006, *MNRAS* 366, L40

Pokorny, R. S., Jones, H. R. A., & Hambly, N. C. 2003, *A&A* 397, 575

Pravdo, S. H., Shaklan, S. B., & Lloyd, J. 2005, *ApJ* 630, 528

Rebolo, R., Zapatero-Osorio, M. R., & Martin, E. L. 1995, *Nature* 377, 129

Reid, I. N., Kirkpatrick, J. D., Gizis, J. E., Dahn, C. C., Monet, D. G., Williams, R. J., Liebert, J., & Burgasser, A. J. 2000, *AJ* 119, 369

Röser, S., Schilbach, E., & Scholz, R.-D. in preparation

Ruiz, M. T., Leggett, S. K., & Allard, F. 1997, *ApJ* 491, L107

Salim, S., Lépine, S., Rich, R. M., & Shara, M. M. 2003, *ApJ* 586, L149

Schmidt, S. J., Cruz, K. L., Bongiorno, B. J., Liebert, J., & Reid, I. N. 2007, *AJ* 133, 2258

Scholz, R.-D. & Meusinger, H. 2002, *MNRAS* 336, L49

Scholz, R.-D., McCaughrean, M. J., Lodieu, N., & Kuhlbrodt, B. 2003, *A&A* 398, L29

Scholz, R.-D., Lehmann, I., Matute, I., & Zinnecker, H. 2004a, *A&A* 425, 519

Scholz, R.-D., Lodieu, N., & McCaughrean, M. J. 2004b, *A&A* 428, L25

Teegarden, B. J., *et al.* 2003, *ApJ* 589, L51

Tinney, C. G. 1998, *MNRAS* 296, L42

Tinney, C. G., Burgasser, A. J., Kirkpatrick, J. D., & McElwain, M. W. 2005, *AJ* 130, 2326

Vrba, F. J., *et al.* 2004, *AJ* 127, 2948

Zuckerman, B. & Becklin, E. E. 1992, *ApJ* 386, 260

A Giant Step: from Milli to Micro-arcsecond Astrometry
Proceedings IAU Symposium No. 248, 2007
W. J. Jin, I. Platais & M. A. C. Perryman, eds.

Astrometry with ground-based interferometers

A. Boden[1] and A. Quirrenbach[2]

[1] Michelson Science Center, California Institute of Technolgy
Pasadena, CA, USA bode@astro.caltech.edu
[2] Zentrum fur Astronomie, University of Heidelberg
Heidelberg, Germany A.Quirrenbach@lsw.uni-heidelberg.de

Abstract. We review the status of ground-based interferometric astrometry. This will include a review of technology and results in differential techniques (e.g. relative orbit determination), as well as global astrometry techniques (globally-registered parallax and proper-motion estimates).

Keywords. instrumentation: interferometers, techniques: high angular resolution, techniques: interferometric, astrometry, binaries: general

1. Introduction

In this contribution we review the status on long-baseline (LB) optical/near-IR (OIR) interferometric astrometry. Operationally we define a long-baseline interferometer as a device that combines light from multiple independent telescopes to form and measure interference *fringes* – variations in detected power as a function of optical path difference between two telescopes. This definition excludes our consideration of interferometry using a single telescope aperture (e.g. "speckle" or aperture masking interferometry). And as the title implies we will discuss interferometric astrometry in the optical and near-IR; the many contributions and future potential of interferometric astrometry in radio and mm bands are discussed in other contributions to these proceedings (e.g. see manuscripts by Reid, Lestrade, and Loinard).

An astronomical interferometer is a device that measures interference (or attributes associated with the interference) in the incident radiation field from an astronomical source. This first such interferometer operating in OIR bands was constructed by Michelson at Mt. Wilson in 1920. The astrometric potential of LB OIR interferometers was appreciated by Michelson (1920) in their development; Anderson (1920) and Merrill (1922) observed a number of binary stars with the Michelson's Mt. Wilson interferometer, and published the first estimates of Capella's visual orbit. Since that time until today LB OIR interferometers have been used extensively in relative astrometry of stars over fields from milliarcseconds to tens of degrees, and when combined with external, global reference frames (e.g. FK5 or Hipparcos) such measurements have been used to compute differential corrections to stellar positions, with hopes of extending these analyses to globally-registered proper motion and parallax estimates.

In what follows we will survey astrometric methods used with LB OIR interferometers, and survey the scientific contributions made with these techniques.

2. Astrometric methods with interferometers

LB OIR interferometers perform astrometry of stellar fields by analyzing the *fringe patterns* of stars measured by the interferometer. (The reader unfamiliar with astronomical interferometry and interferometric observables is referred to Perley *et al.* 1989, Shao & Colavita 1992b, Thompson *et al.* 2001, Quirrenbach 2001, and Boden 1999a).

Depending on the parameters of the stellar field and the interferometer, the exact analysis methods can be different, but generally all methods fall into two broad classes:

• *Fringe amplitude Astrometry*: When the fringe patterns from multiple sources overlap (i.e. are within a fringe *coherence length*), the (power-normalized) fringe amplitude contains information on the relative separation and brightness of the sources. Fringe amplitude analysis necessarily applies when multiple sources are within a coherence length of each other, so this technique is applicable only over very small astrometric fields – field sizes of millarcseconds to on the order of 100 millarcseconds (depending on interferometer parameters). With such field sizes these amplitude techniques have typically been used in the resolution and orbital analysis of close binary stars. Notable in fringe amplitude analysis is the necessity to include the relative brightness between the sources in the astrometric analysis.

• *Separated fringe packet astrometry*: When fringe patterns from multiple sources are well separated (i.e. by multiple coherence lengths) in delay space, then the delay separation between fringe packets contains information on the relative separation of the sources. For fields larger than about 100 millarcseconds the separation between multiple fringe packets in delay space is the observable proxy for sky separation. Operationally various observation and measurement techniques are applied as a function of field size, but the common thread is the delay separation between objects in the field.

3. Visibility amplitude results

The first and largest body of scientific results from LB OIR interferometric astrometry follow from visibility amplitude analysis of close binary stars. As mentioned above, such binary star analysis was initially described by Michelson himself (1920), and using his interferometer Anderson (1920) and Merrill (1922) observed a number of binary stars, and published the first orbit determinations for the Capella system. The visibility amplitude effects created by the overlap of multiple fringe packets within a single coherence length are straightforward to compute (see Michelson 1920 and Boden 1999a). Astrometry using this method began with Capella, but has been broadly applied with contributions from most ground-based LB OIR interferometers (e.g. Table 1). Figure 1 illustrates a visual orbit on the nearby binary ι Pegasi (Boden *et al.* 1999b) reconstructed from interferometric visibility data from the Palomar Testbed Interferometer (PTI; Colavita *et al.* 1999). Interferometric visibility measurements serve as proxy for relative astrometry between the components, allowing an estimate of the component separation (and over time the orbit). Modern visibility analysis methods estimate the orbit directly from the visibility data without intermediate separation estimates (e.g. Herbison-Evans 1971 and Hummel 1993), and directly integrate radial-velocity (RV) data when available (e.g. Hummel 1998 and Boden 1999c).

The primary objective in most visibility amplitude/binary analyses is the determination of fundamental parameters (e.g. dynamical mass, radius, luminosity) for the components (a notable exception is work on Atlas by Pan *et al.* 2004 and Zwahlen *et al.* 2004 discussed elsewhere in these proceedings). Table 1 (reprinted from Cunha *et al.* 2007) lists the set of binary systems that have been analyzed using LB OIR interferometric data combined with double-lined RV data. There are a total of 34 systems listed spanning nearly the entire HR-diagram; all but three of these entries (discussed below) use interferometric visibility data as proxy for the component astrometry. Notable in this list are contributions on open cluster stars (e.g. θ^2 Tau; Armstrong *et al.* 2006), pre-main

Table 1. Interferometrically determined orbits and component masses for double-lined spectroscopic binaries. More details are available in the references listed below the table. The star κ Peg is a triple system; in this case the "wide" (A – Ba/Bb) and "narrow" (Ba – Bb) orbits are listed separately. Reprinted with permission from Cunha *et al.* (2007).

System	Spectral Types	a'' [mas]	M_1 [$M_\odot$]	M_2 [$M_\odot$]	Instr.	Ref.
HD 27483	F6V+F6V	1.3	1.38 ±0.13	1.39 ±0.13	PTI	K04
α Vir	B1III-IV+B3V:	1.5	10.9 ±0.9	6.8 ±0.7	Narrabri	H71
κ Peg B	F5IV+K0V:	2.5	1.662±0.064	0.814±0.046	PTI	M06
V773 Tau A	K2+?	2.8	1.54 ±0.14	1.332±0.097	KI	B07
θ Aql	B9.5III+B9.5III	3.2	3.6 ±0.8	2.9 ±0.6	Mk III	H95
β Aur	A2V+A2V	3.3	2.41 ±0.03	2.32 ±0.03	Mk III	H95
12 Boo	F9IV+F9IV	3.4	1.435±0.023	1.408±0.020	PTI	B00
		3.5	1.416±0.005	1.374±0.005	combined	B05b
σ Sco	B1III+B1V	3.6	18.4 ±5.4	11.9 ±3.1	SUSI	N07b
γ^2 Vel	O7.5II+WC8	3.6	28.5 ±1.1	9.0 ±0.6	SUSI	N07c
BY Dra	K4V+K7.5V	4.4	0.59 ±0.14	0.52 ±0.13	PTI	B01
o Leo	F9+A5m	4.5	2.12 ±0.01	1.87 ±0.01	combined	H01
HD 9939	K1IV+K0V	4.9	1.072±0.014	0.838±0.008	PTI	B06
σ Psc	B9.5V+B9.5V	5.6	2.65 ±0.27	2.36 ±0.24	PTI	K04
64 Psc	F8V+F8V	6.5	1.223±0.021	1.170±0.018	PTI	B99c
93 Leo	G5III+A7V	7.5	2.25 ±0.29	1.97 ±0.15	Mk III	H95
ζ^1 UMa	A2V+A2V	9.6	2.51 ±0.08	2.55 ±0.07	Mk III	H95
		9.8	2.43 ±0.07	2.50 ±0.07	NPOI	H98
ι Peg	F5V+G8V	10.3	1.326±0.016	0.819±0.009	PTI	B99b
η And	G8III+G8III	10.4	2.59 ±0.30	2.34 ±0.22	Mk III	H93
α Equ	G2III+A5V	12.0	2.13 ±0.29	1.86 ±0.21	Mk III	A92b
27 Tau	B8III+?	13.1	4.74 ±0.25	3.42 ±0.25	combined	Z04
HD 195987	G9V+?	15.4	0.844±0.018	0.665±0.008	PTI	T02
ζ Aur	K4Ib+B5V	16.2	5.8 ±0.2	4.8 ±0.2	Mk III	B96
θ^2 Tau	A7III+A:	18.6	2.1 ±0.3	1.6 ±0.2	Mk III	T95
		18.8	2.15 ±0.12	1.87 ±0.11	combined	A06
λ Vir	Am+Am	19.8	1.897±0.016	1.721±0.023	IOTA	Z07
HD 98800 B	K5V+?	23.3	0.699±0.064	0.582±0.051	KI	B05a
ϕ Cyg	K0III+K0III	23.7	2.536±0.086	2.437±0.082	Mk III	A92a
α And	B8IV+A:	25.2	5.5:	2.3:	Mk III	P92,T95
β Cen	B1III+B1III	25.3	11.2 ±0.7	9.8 ±0.7	SUSI	D05,Au06
β Ari	A5V+G0V:	36.1	2.34 ±0.10	1.34 ±0.07	Mk III	P90
λ Sco	B1.5IV+B2V	49.3	10.4 ±1.3	8.1 ±1.0	SUSI	T06
12 Per	F8V+G2V	53.2	1.382±0.019	1.240±0.017	CHARA	Ba06
α Aur	G1III+G8III	55.7	2.56 ±0.04	2.69 ±0.06	Mk III	H94
δ Equ	F7V+F7V	231.9	1.193±0.012	1.188±0.012	PTI	M05
κ Peg	F5IV+F5IV	235.0	1.549±0.050	composite	PTI	M06

References: A92a: Armstrong 1992a; A92b: Armstrong 1992b; A06: Armstrong 2006; Au06: Ausseloos 2006; Ba06: Bagnuolo 2006; B96: Bennett 1996; B99b: Boden 1999b; B99c: Boden 1999c; B00: Boden 2000; B01: Boden & Lane 2001; B05a: Boden 2005a; B05b: Boden 2005b; B06: Boden 2006; B07: Boden 2007; D05: Davis 2005; H71: Herbison-Evans 1971; H93: Hummel 1993; H94: Hummel 1994a; H95: Hummel 1995; H98: Hummel 1998; H01: Hummel 2001; K04: Konacki & Lane 2004; M05: Muterspaugh 2005; M06: Muterspaugh 2006; N07a: North 2007a; N07b: North 2007b; N07c: North 2007c; P90: Pan 1990; P92: Pan 1992; T95: Tomkin 1995; T02: Torres 2002; T06: Tango 2006; Z04: Zwahlen 2006; Z07: Zhao 2007

sequence stars (e.g. V773 Tau; Boden *et al.* 2007), low-abundance/Galactic thick-disk stars (e.g. HD 195987; Torres *et al.* 2002), subgiant (e.g. 12 Boo; Boden *et al.* 2005b) and giant (e.g. α Equ; Armstrong *et al.* 1992b) stars, and even triple systems (η Vir; Hummel *et al.* 2003). Prospects are excellent for continued important contributions in thcse areas.

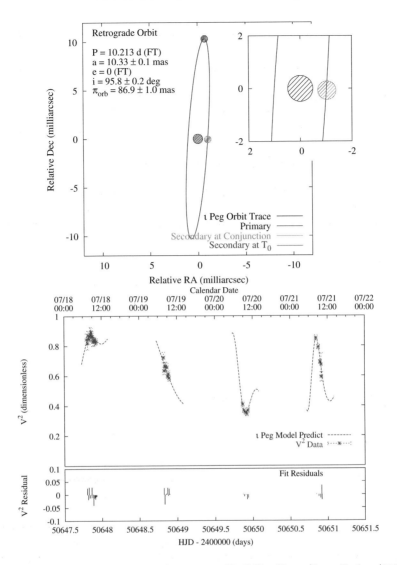

Figure 1. Visual Orbit of ι Pegasi and Supporting Visibility Data (from Boden (1999b)). Top: the relative visual orbit of ι Pegasi, with the primary rendered at the center, and the secondary position shown at both maximum elongation and conjunction. The ι Peg orbit is in fact circular but seen at a near-eclipsing orientation. Bottom: four consecutive nights of PTI calibrated visibility (V^2) data on ι Peg. Variations in the measured visibility can be used as proxy for the relative separation between the components.

4. Separated fringe-packet astrometry

Fringe amplitude astrometry techniques necessarily require the multiple sources to be within a single interferometric coherence length, typically on the order of tens to ≈ 100 mas projected on the sky. When astrometric fields increase beyond this 100 mas, multiple fringe packets are well-separated in interferometric delay space, and the delay difference itself becomes the observable proxy for relative astrometry. Above we have described such astrometry as separated fringe packet astrometry.

As a practical matter, separated fringe packet astrometry further subdivides into three different operating regimes that we will discuss in turn.

4.1. *Very narrow angle astrometry*

When the astrometric field is still within a interferometer sub-aperture (single telescope) diffraction pattern (e.g. $\approx$ 1 arcsec), then the multiple fringe packets can be observed simply by scanning the delay compensation device (typically a "delay line"; Colavita *et al.* 1999). By measuring the delay scan distance separating the multiple fringe packets (typically by laser metrology) one obtains an astrometrically-useful delay difference estimate. The first results of this kind are presented in Colavita (1994), and this method is described in detail in Lane & Muterspaugh 2004 (including performance improvements from synthetic phase stability allowed by the unique PTI dual-beam/fringe tracker design). When phase referencing is available typical relative astrometric precisions are on the order of 15 μarcseconds over a 0.5 arcsecond field (a fractional precision of 3 parts in 10^5). Astrophysical results from this method are included in the Table 1 compilation: δ Equ (Muterspaugh *et al.* 2005), κ Peg (Muterspaugh *et al.* 2006), and 12 Per (Bagnuolo et al 2006). Further, such astrometric precision enables planet-search programs, and the PTI PHASES astrometric planet search program is discussed in Muterspaugh *et al.* (2006) and Muterspaugh *et al.* (2007)

4.2. *Dual-beam narrow-angle astrometry*

While astrometric precisions at the arcsecond-scale are remarkable, the small working angles practically dictate that the methods are limited to arcsecond-scale binaries. A method to increase this field to expand sky coverage was presented by Shao & Colavita (1992a). In this narrow-angle astrometric scenario the fringes on multiple stars are tracked simultaneously to take advantage of significant correlation in the atmospheric phase noise over an *isoplanatic angle* ($\approx$ 30-40 arcsec in the near-IR; Quirrenbach 1999). The simultaneous fringe tracking requirement necessitates multiple beam combiners in a single facility, and guided the dual-fringe tracker design of PTI (Colavita *et al.* 1999). Hence this technique has been termed dual-beam narrow-angle astrometry.

This dual-beam astrometric technique has been demonstrated at PTI. Figure 2 shows the results of a three-months demonstration experiment performed in 1999 using the well-studied, nearby visual binary 61 Cygni. 61 Cyg is a nearby K-dwarf binary with an estimated 678-year orbital period and apparent separation at present epoch of roughly 31 arcseconds (Gorshanov *et al.* 2006). Using the PTI N-S baseline (so the principle measurement axis is in declination) PTI collected narrow-angle astrometric data on 61 Cyg over a three-month period in 1999. These astrometric measurements are made by tracking fringes on the two stars with two independent fringe trackers, and relating the relative delay between the two fringe trackers using a laser metrology system (details are given in Colavita *et al.* 1999). Projecting the derived 2-d separations into a space of declination vs time, the apparent night-to-night separation increase is clearly seen. Fitting a linear motion model to these separations, the ensemble RMS declination residual is 170 μarcseconds, with a particularly stable, contiguous 7-night run showing a 100 μarcseconds scatter around the same ensemble motion model. This scatter represents a fractional precision of 3 parts in 10^6. Application of this method to ground-based astrometric planet searches is planned for the VLTI PRIMA instrument, discussed in detail elsewhere in these proceedings (e.g. see contributions by Richichi and Launhardt).

4.3. *Wide-angle astrometry*

The narrow-angle method of Shao & Colavita (1992a) leverages the significant correlation in atmospheric turbulence over an isoplanatic angle (30-50 arcseconds at a typical site) to make 10-100 μarcseconds-class relative astrometric measurements. For interferometric measurements over a wider field this correlation is not applicable. Such wide-angle

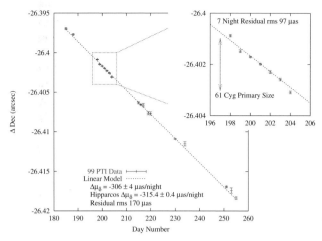

Figure 2. Principle Projection for 1999 PTI Dual-Beam Narrow-Angle Astrometry Demonstration. Roughly three months of relative astrometry data on the 30-arcsecond visual binary 61 Cygni are shown projected onto the principle (declination) measurement axis (all astrometric data was taken with the PTI N-S baseline) vs time. The apparent night-to-night relative motion due to the estimated 678-year orbit period is clear in these data. The apparent relative declination motion of 300 μarcsecond/day is very close to that measured by the Hipparcos satellite (the difference between the two estimates is consistent with acceleration from the binary orbit). The RMS scatter of these data around the best-fit linear motion model is 170 μarcsecond; data from a particularly stable 7-night run (inset) shows an RMS scatter of approximately 100 μarcsecond

measurements have been demonstrated by the Mark III interferometer (Shao *et al.* 1988) and Navy Prototype Optical Interferometer (NPOI; Armstrong *et al.* 1998).

Mark III wide-angle astrometry work is described in Mozurkewich *et al.* (1988), Shao *et al.* (1990), and Hummel *et al.* (1994b). This work focused on making serial delay measurements on a set of roughly 20 FK5 stars over a number of years, and using the data to compute differential corrections to the mean-epoch positions from the FK5 catalog. Shao *et al.* (1990) reports single-season RMS residuals of individual measurements on the order of 10 milliarcseconds, with Hummel *et al.* (1994b) reporting four-year ensemble RMS precisions on the order of 20 milliarcseconds. Over a wide-angle field on the order of 1 radian these measurements represent a fractional single-night precision of 1 part in 10^7.

Based on the success of Mark III wide-angle program, the NPOI was constructed to allow wide-angle astrometry over an extended period, with the objective of extending the proper-motion time baseline for Hipparcos stars. Benson *et al.* (2004) and Johnston *et al.* (2006) report on NPOI astrometry program status. At present NPOI astrometry is demonstrating 10-millarcsecond-class astrometric measurement precision in single-night runs, and the NPOI team anticipate multi-night ensemble astrometric solutions to be completed in the next year. NPOI wide-angle astrometry holds the promise to significantly improve the precision of Hipparcos proper motion and parallax estimates over the next few years, particularly in advance of Gaia (discussed elsewhere in these proceedings by Lindegren and Mignard).

5. Summary

We have summarized astrometric contributions by LB OIR interferometers. To date the largest body of results have focused on analysis of binary systems with an eye toward stellar fundamental parameter determinations, and interferometers will continue making important contributions in these areas (e.g. see Cunha *et al.* 2007 for a discussion). Further, the VLTI PRIMA facility is planning a broad astrometric exoplanet search program (see other contributions by Richichi and Launhardt in these proceedings); prospects are excellent for this (and other) astrometric survey(s) to contribute to understanding of the architecture and diversity of planetary systems.

Acknowledgments

We are happy the thank the organizers of IAUS 248, who graciously opened their city, country, and institution to all of us during the symposium. Further, we thank H. McAlister, D. Hutter, J. Benson, and K. Johnston for their advice and contributions in the preparation of this manuscript and presentation.

References

Anderson, J. 1920, *ApJ*, 51, 263.

Armstrong, J. *et al.* 1992a, *AJ*, 104, 2217.

Armstrong, J. *et al.* 1992b, *AJ*, 104, 241.

Armstrong, J. *et al.* 1998, *ApJ*, 496, 550.

Armstrong, J. *et al.* 2006, *AJ*, 131, 2643.

Ausseloos, M. *et al.* 2006, *A&A*, 455, 259.

Bagnuolo, W. *et al.* 2006, *ApJ*, 131, 2695.

Bennett, P. *et al.* 1996, *ApJ*, 471, 454.

Benson, J. *et al.* 2004, *SPIE*, 5491, 464.

Boden, A. 1999a, in *Proc. 1999 Michelson Summer School*, Lawson, P. editor (http://sim.jpl. nasa.gov/michelson/iss1999/index.html).

Boden, A. *et al.* 1999b, *ApJ*, 515, 356.

Boden, A. *et al.* 1999c. *ApJ*, 527, 360.

Boden, A. *et al.* 2000, *ApJ*, 536, 880.

Boden, A. & Lane, B. 2001 *ApJ*, 547, 1071.

Boden, A. *et al.* 2005a, *ApJ*, 635, 442.

Boden A., Torres G. & Hummel C. 2005b, *ApJ*, 627, 464.

Boden, A., Torres, G. & Latham, D. 2006. *ApJ*, 644, 1193.

Boden, A. *et al.* 2007, *ApJ*, 670, 1214 (`astro-ph/0706.2376`).

Colavita, M. 1994, *A&A*, 283, 1027.

Colavita, M. *et al.* 1999, *ApJ*, 510, 505.

Cunha *et al.* 2007, *Ast & Astrop Rev* in press (`astro-ph/0709.4613`).

Davis, J. *et al.* 2005, *MNRAS*, 356, 1362.

Gorshanov, D., Shakht, N., & Kisselev, A. 2006, *Ap*, 49, 386.

Herbison-Evans, D. *et al.* 1971, *MNRAS*, 151, 161.

Hummel, C. *et al.* 1993 *AJ*, 106, 2486.

Hummel, C. *et al.* 1994a, *AJ*, 107, 1859.

Hummel, C. *et al.* 1994b, *AJ*, 108, 326.

Hummel, C. *et al.* 1995, *AJ*, 110, 376.

Hummel, C. *et al.* 1998, *AJ*, 116, 2536.

Hummel, C. *et al.* 2001 *AJ*, 121, 1623.

Hummel, C. *et al.* 2003 *AJ*, 125, 2630.

Johnston, J. *et al.* 2006, *SPIE*, 6268, 6.

Konacki, M. & Lane, B. 2004, *ApJ*, 610, 443.

Lane, B. & Muterspaugh, M. 2004, *ApJ*, 601, 1129.

Merrill, P. 1922, *ApJ*, 56, 40.

Michelson, A. 1920, *ApJ*, 51, 257.

Mozurkewich, D. *et al.* 1988, *AJ*, 95, 1269.

Muterspaugh, M. *et al.* 2005, *AJ*, 130, 2866.

Muterspaugh, M. *et al.* 2006a, *ApJ*, 636, 1020.

Muterspaugh, M. *et al.* 2006b, *ApJ*, 653, 1469.

Muterspaugh *et al.* 2007, in "Planets in Binary Systems", N. Haghighipour Editor (Springer) (astro-ph/0705.3072).

North, J. *et al.* 2007a. *MNRAS*, 380, 1276.

North, J. *et al.* 2007b, *MNRAS*, 377, 415.

Pan, X. *et al.* 1990, *ApJ*, 356, 641.

Pan, X. *et al.* 1992, *ApJ*, 384, 624.

Pan, X., Shao, M., & Kulkarni, S. 2004, *Nature*, 427, 326.

Perley, R., Schwab, F., & Bridle, A. 1989, *Synthesis Imaging in Radio Astronomy, ASP Conf. Series 6* (Provo UT: Brigham Young Univ. Press).

Quirrenbach, A. 1999, in *Proc. 1999 Michelson Summer School*, Lawson, P. editor (http://sim.jpl.nasa.gov/michelson/iss1999/index.html).

Quirrenbach, A. 2001, *ARA&A*, 39, 353.

Shao, M. *et al.* 1988, *ApJ*, 327, 905.

Shao, M. *et al.* 1990, *AJ*, 100, 1701.

Shao, M. & Colavita, M. 1992a, *A&A*, 262, 353.

Shao, M. & Colavita, M. 1992b, *ARA&A*, 30, 457.

Tango, W. *et al.* 2006, *MNRAS*, 370, 884.

Thompson, R, Moran, J., & Swenson, G. 2001 *Interferometry and Synthesis on Radio Astronomy* 2nd Edition, (New York: John Wiley & Sons).

Tomkin, J., Pan, X., & McCarthy, J. 1995, *ApJ*, 109, 780.

Torres, G. *et al.* 2002, *AJ*, 124, 1716.

Zwahlen, N. *et al.* 2004, *A&A*, 425, L45.

Zhao, M. *et al.* 2007, *ApJ*, 659, 626.

A Giant Step: from Milli- to Micro-arcsecond Astrometry
Proceedings IAU Symposium No. 248, 2007
W. J. Jin, I. Platais & M. A. C. Perryman, eds.

The VLT Interferometer

A. Richichi

European Southern Observatory,
Karl-Schwarzschildstr. 2, 85748 Garching b. M., Germany
email: arichich@eso.org

Abstract. The ESO Very Large Telescope Interferometer (VLTI) is arguably the most powerful optical interferometric facility available at present. In addition to the wide choice of baselines and the light collecting power of its 8.2 m and 1.8 m telescopes, the VLTI also offers a smooth and user-friendly operation which makes interferometry accessible to any astronomer and covers a wide range of scientific applications. Behind the routine scientific operations, however, the VLTI is in constant evolution. I will present some of the technological and instrumental improvements which are planned for the near and mid-term future, and discuss their implications for astrometry in particular. Among them, the PRIMA facility and the proposed GRAVITY instrument are designed to reach the level of 10 microarcseconds in the near-infrared.

Keywords. instrumentation: high angular resolution, techniques: interferometric, astrometry

1. An introduction to the VLTI

Cerro Paranal in the chilean Atacama desert is a world-renowned astronomical site. It is home to four 8.2 m so-called Unit Telescopes (UTs), and to four 1.8 m Auxiliary Telescopes (ATs), as well as to other telescopes. The UTs and the ATs can be combined interferometrically by means of subterranean light ducts and delay lines. The maximum baselines that can be realized by combining UTs or ATs are about 130 m and 205 m, respectively, with a large number of different configurations in terms of baseline length and orientation (Fig. 1). In particular, the ATs can be moved over 30 stations, and their relocation can be accomplished in a few hours (Fig. 2).

The VLTI (Glindemann *et al.* 2003) had first fringes in March 2001. For about three years, the test instrument VINCI (Kervella *et al.* 2000) was used to commission the VLTI. The result is a public database of about 20,000 observations on several hundred of stars. A significant fraction of these data have been used, mainly by researchers outside ESO, to obtain impressive new scientific results, published in tens of refereed papers. By the end of 2002 and early 2004, respectively, the two facility instruments MIDI and AMBER arrived on Paranal. They were subsequently commissioned and made available to the community along the same guidelines as the single telescope instruments.

MIDI (Leinert 2004) is a 2-way beam combiner with spectral resolution up to $\lambda/\Delta\lambda = 230$ in the N band. It achieves a sensitivity of few tenths of Jy at the UTs, in self-fringe tracking mode. Better sensitivities are possible, mainly with the off-axis fringe-tracking capability of PRIMA discussed later. The combination of an mid-infrared constructive (i.e. not nulling) beam-combiner with the apertures and baselines of the VLTI is unique in the world, and a large number of the papers resulting from MIDI are indeed first time results in their own area.

AMBER (Petrov *et al.* 2007) can combine up to 3 beams at wavelengths between 1 and 2.5 μm, with spectral resolution up to $\lambda/\Delta\lambda = 10^4$. Although other near-IR beam combiners exist in the world with the capability of realizing three or more baselines, AMBER at the VLTI is unique in the simultaneous use of large telescopes and long

Figure 1. Aerial view of the Paranal site. In addition to the four 8.2 m Unit Telescopes, also two Auxiliary Telescopes (total of four by now) are visible. The ATs can be moved over a grid of 30 stations by rails. Observations are carried out from the Control Building on the left of the photo. All these telescopes can be combined interferometrically, with the exception of VST at the top right.

baselines. Results from AMBER are starting to be published at an encouraging rate, spanning a wide range of scientific topics.

Next to MIDI and AMBER, the VLTI includes a large number of complex subsystems, including adaptive optics on the UTs (MACAO, Arsenault al 2003), a visible tip-tilt corrector for the ATs (STRAP, Bonaccini al 1997), six delay lines (Hogenhuis *et al.* 2003), an H-band fringe tracker (FINITO, Gai *et al.* 2004), an IR tip-tilt corrector in the underground laboratory (IRIS, Gitton *et al.* 2004), a reference source (ARAL, Morel *et al.* 2004). The lab itself, now quite crowded, is schematically shown in Fig. 3.

ESO has conceived the VLTI to be not only a powerful interferometric facility, but also as one open and accessible to the general astronomical community. It is a fact that the majority of, if not all, interferometers are operated and used mostly by the same persons that designed and built them. Observations are often complex, the data demand local specialized software, and their analysis often requires an in-depth knowledge of the peculiarities of the interferometer used. At the VLTI however, most of these complications do not apply. Astronomers can request time, create the instructions for their observations and have them executed in service mode if so wished, and receive their data in a standard format as for any other single-telescope instrument at Paranal. Data are quality-checked and archived, and automated data reduction pipelines are run routinely.

Although the implementation of interferometry as a user-friendly technique has required a large effort at ESO, the outcome is certainly impressive. Each semester, about one hundred proposals are submitted and a large fraction of these are executed in service mode on MIDI and AMBER. Since the opening of the VLTI to the community, over 300 different programs have been or are about to be executed, with 116 different PIs from 15

Figure 2. A view of thee of the four Auxiliary Telescopes present on Paranal.

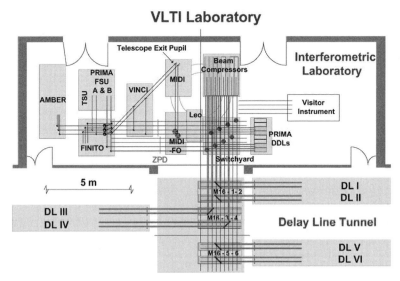

Figure 3. Scheme of the VLTI underground laboratory and delay lines, and of the main instruments and subsystems.

countries. It is important to stress that the VLTI is accessible not just to the researchers in the ESO member states, but to anyone in the world.

2. The future of astrometry at the VLTI

The VLTI is a fully functional facility with a complete set of instruments, but developments are under continuous evolution in order to maintain its leading status in interferometry throughout the next decade. Plans have started long ago to implement a dual-feed facility (PRIMA, Delplancke *et al.* 2006). This allows to perform interferometry, simultaneously and independently, on two targets separated by up to 1 arcminute. There are several uses of this mode: for example fringe tracking on a relatively bright nearby source, and long integrations (thus much fainter magnitudes) on a faint science target. Or, using the phase information between the two stars (one of which is supposedly with a known, fixed phase), to reconstruct images of the science target. Finally, and perhaps most interestingly for the purpose of this conference, there is the astrometric mode of PRIMA, in which minute changes of angular distance between the reference and the science target can be detected, and monitored in time. Accuracies of $50\,\mu$as are foreseen initially, up to $10\,\mu$as at a later stage when a precise models of the geometry of the telescope is built from many accumulated observations. This would allow, among other things, the direct detection of exoplanets. A consortium has been formed, to develop specific hardware, the so-called Differential Delay Lines, and software to this goal (see Launhardt in these proceedings).

PRIMA is largely built, and it will be deployed in Paranal in 2008. An intensive period of commissioning will follow, with the goal of first science observations later in the same year or early 2009.

Another step towards precise astrometry is GRAVITY (Gillessen *et al.* 2006), one of the second generation VLTI instruments which have recently completed their Phase A studies and have been recommended for further development. GRAVITY is designed specifically to achieve $10\,\mu$as astrometric accuracy on the faint infrared stars around the supermassive black hole at the center of our own Galaxy (see Gillessen in these proceedings). In addition to a direct measure of the black hole mass, it will be used to detect and investigate a number of strong gravitational effects, effectively serving as a unique physics laboratory. Of course, the high astrometric accuracy of GRAVITY can also be applied to a number of other astrophysical topics. The deployment of GRAVITY at the VLTI is foreseen around 2013.

References

Arsenault, R., Alonso, J., Bonnet, H. *et al.* 2003, in SPIE Proceedings, vol. 4839, p. 174

Bonaccini, D., Gallieni, D., Biasi, R. *et al.* 1997, in SPIE Proceedings, vol. 3126, p. 580

Glindemann, A., Algomedo, J., Amestica, R. *et al.* 2003, in SPIE Proceedings, vol. 4838, p. 89

Delplancke, F., Derie, F., Leveque, S. *et al.* 2006, in SPIE Proceedings, vol. 6268, p. 27

Gai, M., Menardi, S., Cesare, S. *et al.* 2004, in SPIE Proceedings, vol. 5491, p. 528

Gillessen, S., Perrin, G., Brandner, W. *et al.* 2006, in SPIE Proceedings, vol. 6268, p. 33

Gitton, P., Leveque, S., Avila, G., Phan Duc, T. 2004, in SPIE Proceedings, vol. 5491, p. 944

Hogenhuis, H., Visser, M., Derie, F. 2003, in SPIE Proceedings, vol. 4838, p. 1148

Kervella, P., Coude de Foresto, V., Glindemann, A. *et al.* 2000, in SPIE Proceedings, vol. 4006, p. 31

Leinert, C. 2004, in SPIE Proceedings, vol. 5491, p. 19

Morel, S., Vannier, M., Menardi, S. *et al.* 2004, in SPIE Proceedings, vol. 5491, p. 1079

Petrov, R., Malbet, F., Weigelt, G. *et al.* 2007 *A&A*, 464, 1

A Giant Step: from Milli- to Micro-arcsecond Astrometry
Proceedings IAU Symposium No. 248, 2007
W. J. Jin, I. Platais & M. A. C. Perryman, eds.

Micro-arcsecond relative astrometry by ground-based and single-aperture observations

T. Röll[1], A. Seifahrt[1,2] and R. Neuhäuser[1]

[1] Astrophysikalisches Institut und Universitäts-Sternwarte Jena,
email: troell@astro.uni-jena.de
email: rne@astro.uni-jena.de

[2] Institut für Astrophysik, Göttingen
email: seifahrt@astro.physik.uni-goettingen.de

Abstract. We present an observation method to obtain a relative astrometric precision of about $100 \ldots 150 \, \mu as$ with ground-based and single-aperture observations. By measuring the separation of double or triple stars we want to determine the astrometric signal of an unseen substellar companion as a periodic change in the separation between the stellar components. Using an adaptive optics system we correct for atmospheric turbulences and furthermore by using a narrow band filter in the near infrared we can suppress differential chromatic refraction effects. To reach a high precision we use a statistical approach. Using the new observation mode "cube-mode" (where the frames were directly saved in cubes with nearly no loss of time during the readout), we obtain several thousand frames within half an hour. After the verification of the Gaussian distributed behaviour of our measurements (done with a Kolmogorov-Smirnov-Test) the measurement precision can be calculated as the standard deviation of our measurements divided by the square root of the number of frames.

To monitor the stability of the pixel scale between our observations, we use the old globular cluster 47 Tuc as a calibration system.

Keywords. astrometry, methods: statistical, instrumentation: adaptive optics, binaries: general, planetary systems, globular clusters: individual (47 Tuc)

1. Introduction

Up to now, most of the extrasolar planets have been detected with the radial velocity technique. Due to the unknown inclination angle i this technique just yields the lower mass limit $M \sin i$ and not the true mass of the substellar companion. Therefore, all radial velocity planets should be regarded as planet candidates, until their true mass is determined. In contrast to the radial velocity technique, astrometry yields the inclination angle by measuring the astrometric signal of the substellar companion and hence its true mass. Recently, Bean *et al.* (2007) measured the astrometric signal of the radial velocity planet candidate HD 33636 b ($M \sin i = 9.3 \, M_J$) and obtained a value for the true mass of the companion of $M = 142 \pm 11 \, M_J$, thus it is a low mass star. This clarifies the importance of astrometric follow up observations to determine the true mass of radial velocity planet candidates. The mass is one of the most important stellar and substellar parameters and plays a key role in our understanding of the distribution, formation and evolution of substellar objects. Besides all other methods to determine the mass, which are using theoretical predictions (like evolutionary models), astrometry is a method, which is independent from theoretical assumptions (hence, from theoretical uncertainties) by measuring the dynamical mass of the objects. Up to now, three radial velocity planet candidates have been confirmed with absolute astrometry using the Fine Guiding Sensor (FGS) of the Hubble Space Telescope (HST), GJ 876 b by Benedict *et al.* (2002), 55 Cancri d by McArthur *et al.* (2004) and ϵ Eridani b by Benedict *et al.* (2006).

The idea, to search for extrasolar planets with astrometry is not a new one. Already van de Kamp (1969) observed Barnard's star and believed to find a planetary companion because of the measured non-linear movement of the star (absolute astrometry). Later, different groups like Gatewood *et al.* (1973) could not reproduce the detection of an astrometric signal of a possible substellar companion. The origin of the measured residuals in the position and the movement of Barnard's star in the data by van de Kamp (1969) were unknown systematic errors of the observation technique. This clarifies the complexity of doing astrometry, especially absolute astrometry. Until the beginning of the 1980s, about 50 stars (including Barnard's star and ϵ Eridani) with assumed unseen stellar and substellar companions were discussed. A summary of the astrometric search for unseen companions and the discussed stars at this time can be found in Lippincott (1978).

About 20 years later Pravdo and Shaklan (1996) measured the position of about 15 members of the open cluster NGC 2420 from the ground and reached an astrometric precision in the optical of about $150\mu as$. They identified the atmospheric noise and the differential chromatic refraction (DCR) as the limiting effects in the reached precision. Pravdo and Shaklan (1996) also mentioned the importance of a careful and long-term calibration to handle the systematic errors.

2. Observation method

To reach a precision comparable to the HST observation, we observe double and triple stars and measure the separation between all stellar components, thus using relative astrometry. In the case of an unseen substellar companion, we would measure the astrometric signal indirectly as a relative and periodic change in the separations.

The quest of measuring the astrometric signal of a substellar companion needs a careful handling of all noise sources, such as atmospheric noise, photon noise, background noise, readout-noise, DCR and others. Our observations on the southern hemisphere are done with the 8.2 meter telescope UT4 of the ESO Very Large Telescope (VLT) and the NACO S13 (NAOS-CONICA) infrared camera. Using the adaptive optics system NAOS (Nasmyth Adaptive Optics System) we correct for atmospheric turbulence and by using a narrow band filter centered in the near infrared ($\lambda_{cen} = 2.17\,\mu m$) we suppress DCR effects. Due to the use of the double-correlated readout mode, we suppress readout noise and by choosing a suitable exposure time (to reach a high signal to noise ratio) we can neglect photon and background noise.

The pixel scale of the detector (NACO S13) is about $13.25\,mas$, which means a Field of View (FoV) of about $14'' \times 14''$. A guide star for the AO system is always one of the stellar components. The separation of our observed multiple systems is typically four arcseconds. Hence, the angular separation is (with normal seeing conditions) always smaller than the isoplanatic angle.

Furthermore (besides the use of relative astrometry), we use a new observation mode, called "cube-mode". This mode saves frames directly into a cube and thus has nearly a zero loss of time during the readout. With the minimal exposure time of 0.35 seconds using the double-correlated readout mode it is possible to obtain 2500 frames in 15 minutes.

The following statistical principle is similar to the method of measuring the radial velocity with hundreds of spectral lines to reach a higher precision, which is used in the radial velocity technique. In our astrometric case we measure the separation between all stellar components in each frame and obtain several thousand measurements of the same separation. After a verification of the Gaussianity of measurements (done with a Kolmogorov-Smirnov-Test) and a two sigma clipping (to reject frames with low quality due to the non-constant performance of the AO system and the dynamical seeing behaviour), the measurement precision (Δ_{meas}) can be calculate as the standard deviation of the measurements (σ_{meas}) divided by the square root of the number of Gaussian distributed measurements (N), $\Delta_{meas} = \dfrac{\sigma_{meas}}{\sqrt{N}}$.

We have to keep in mind that the above value is just the measurement precision and describes only the statistically distributed sources of random errors. To determine the sources of systematic errors, which affect in the case of relative astrometry the pixel scale and the position angle, we need a special calibration reference system.

3. Calibration

Because we are dealing with relative astrometry we do not need an absolute astrometric calibration of our data, but we have to monitor the stability of our pixel scale to correct our measurements for possible variations of the pixel scale. The "normal way" of relative astrometric calibration is to use a Hipparcos binary system. This method results in the case of the NACO detector in a pixel scale of typically $13.25 \pm 0.05\, mas$ per pixel (see Neuhäuser et al. (2005)), which means a relative error of about 4/1000 of a pixel per pixel. Our measurement precision for our first observed binary system (HD 19994) are lower than 3/1000 of a pixel for a separation of about 175 pixel (see Fig. 1). This means, we need a calibration system where we can monitor the pixel scale down to a relative value of better than 2/100000 per pixel. Taking the $13.25\, mas$ as a given and fixed "reference pixel scale" for our first epoch, we have to use a calibration system, where we can detect changes in the pixel scale down to $0.2\, \mu as$ per pixel within one year.

The requirements for such a calibration system are a high intrinsic and known stability, a lot of "calibration stars" in the FoV of NACO and it should be bright enough for observations with the narrow band filter in the near infrared (due to DCR). We choose a core region of the old globular cluster 47 Tuc as a suitable calibration system for our targets on the southern hemisphere. The reasons are a relative large number of stars within the NACO FoV and a known intrinsic stability. McLaughlin et al. (2006) determined the transversal velocity dispersion of the 47 Tuc cluster members and obtained a value of about 630 μas yr^{-1}. To monitor the pixel scale we take hundreds of frames of 47 Tuc per epoch, measure the separation from each star to each star and compute the mean of all these separation measurements on every single frame. This mean of the separations represents the relative alignment of all observed cluster members and should have, within the errors (intrinsic instability and measurement errors), the same value every epoch. Using a Monte-Carlo-Simulation with our observed cluster members and a Gaussian distributed transversal velocity dispersion of $630\, mas$ yr^{-1}, we obtain an intrinsic stability of our used calibration cluster of about 3/100000 per pixel and year, which results for the given pixel scale of $13.25\, mas$ an intrinsic stability of about $0.4\, \mu as$ per pixel and year. This means, with 47 Tuc we are able to determine a change in the pixel scale down to $0.5 \ldots 1\, \mu as$ per pixel and year (including the typical measurement errors of the 47 Tuc cluster members). Due to the fact, that we are using a fixed and given "reference pixel scale" in our first epoch, we can not determine the absolute value of the pixel scale, but we are able to detect changes very precisely.

The difference between our intrinsic stability and the measurement precision of our first observed double star (HD 19994) shows, that we are limited by our used calibra-

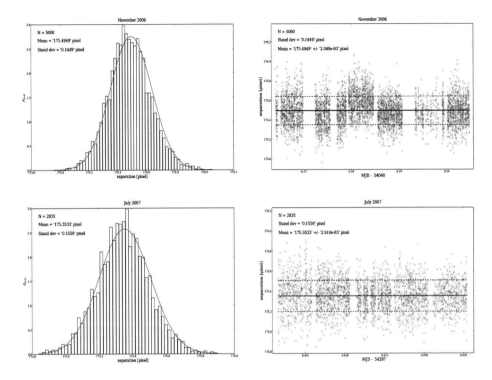

Figure 1. Separation measurements (right) and the Gaussian distribution of the measurements (left) of the stellar binary HD 19994 from 2006 (top) and 2007 (bottom)

tion system and not by noise sources like atmospheric noise, DCR or readout noise, which affect the measurement error. At the end, we achieve a total relative precision of $100 \ldots 150 \,\mu as$ per epoch for the double star HD 19994. Mayor *et al.* (2004) found a radial velocity planet candidate of $M \sin i = 1.68 \, M_J$ around HD 19994 A. The expected change in the separation (due to the astrometric signal of the planet candidate) depends on the orientation of both orbits (planetary and double star orbit) and is about $300 \,\mu as$, for $\Delta i = 35°$ ($M_{true} = 2.6 \, M_J$), about $450 \,\mu as$ for $\Delta i = 50°$ ($M_{true} = 4.9 \, M_J$, $2\,\sigma$ detection) and about $600 \,\mu as$ for $\Delta i = 80°$ ($M_{true} = 6.9 \, M_J$, $3\,\sigma$ detection). In the case of an astrometric non-detection we are able to exclude differences in the inclination of more than $\Delta i = 80°$ ($3\,\sigma$ detection) and thus to exclude true masses of more than $7\, M_J$.

References

Bean, J. L., McArthur, B. E., Benedict, G. F., Harrison, T. E. *et al.* 2007, *ApJ*, 134, 749
Benedict, G. F., McArthur, B. E., Forveille, T., Delfosse, X. *et al.* 2002, *ApJ*, 581, 115
Benedict, G. F., McArthur, B. E., Gatewood, G., Nelan, E. *et al.* 2006, *ApJ*, 132, 2206
Gatewood, G. & Eichhorn, H. 1973, *AJ*, 78, 769
Lippincott, S. L. 1978, *Space Sci. Revs*, 22, 153
Mayor, M., Udry, S., Naef, D. *et al.* 2004, *A&A*, 74, 238
McArthur, B. E., Endl, M., Cochran, W. D., Benedict, G. F. *et al.* 2004, *ApJ*, 614, 81
McLaughlin, D. E. & Anderson, J. & Meylan, G. *et al.* 2006, *ApJS*, 166, 249
Neuhäuser, R., Guenther, E. W., Wuchterl, G. *et al.* 2005, *A&A*, 435, 13
Pravdo, S. H. & Shaklan, S. B. 1996, *ApJ*, 456, 264
van de Kamp, P. 1969, *AJ*, 74, 238

A Giant Step: from Milli- to Micro-arcsecond Astrometry
Proceedings IAU Symposium No. 248, 2007
W. J. Jin, I. Platais & M. A. C. Perryman, eds.

Probing the properties of the Milky Way's central supermassive black hole with stellar orbits

A. M. Ghez[1], S. Salim[1,3], N. Weinberg[2,4], J. Lu[1], T. Do[1], J. K. Dunn[1], K. Matthews[3], M. Morris[1], S. Yelda[1], E. E. Becklin[1]

[1]UCLA Department of Physics and Astronomy
Los Angeles, CA 90095-1547 USA
email: ghez, jlu, tdo, jkdunn, morris, syelda, becklin@astro.ucla.edu

[2]California Institute of Technology, Pasadena, CA 91125 USA
email: kym@caltech.edu

[3]NOAO, 950 N Cherry Ave, Tucson, AZ 85719 USA
email: samir@noao.edu

[4]University of California Berkeley, Department of Astronomy
Berkeley, CA 94720-3411 USA
email: nnw@astron.berkeley.edu

Abstract. We report new precision measurements of the properties of our Galaxy's supermassive black hole. Based on astrometric (1995-2007) and radial velocity (2000-2007) measurements from the W. M. Keck 10 meter telescopes, the Keplerian orbital parameters for the short period star S0-2 imply a distance of 8.3 ± 0.3 kpc, an enclosed mass of $4.8 \pm 0.3 \times 10^6 M_\odot$, and a black hole position that is localized to within ± 1 mas and that is consistent with the position of SgrA*-IR. Astrometric bias from source confusion is identified as a significant source of systematic error and is accounted for in this study. Our black hole mass and distance are significantly higher than previous estimates. The higher mass estimate brings the Galaxy into better agreement with the relationship between the mass of the central black hole and the velocity dispersion of the host galaxy's bulge observed for nearby galaxies. It also raises the orbital period of the innermost stable orbit of a non-spinning black hole to 38 min and increases the Rauch-Tremaine resonant relaxation timescales for stars in the vicinity of the central black hole. Taking the black hole's distance as a measure of R_0, which is a fundamental scale for our Galaxy, and other measurements of galactic constants, we infer a value of the Galaxy's local rotation speed (θ_0) of 255 ± 13 km s^{-1}. With the precisions of the astrometric and radial velocity measurements that are now possible with Laser Guide Star Adaptive Optics, we expect to be able to measure Ro to an accuracy of $\sim 1\%$, within the next ten years, which could considerably reduce the uncertainty in the cosmological distance ladder.

Keywords. Galaxy: center, Galaxy: fundamental parameters, Galaxy: kinematics and dynamics, techniques: high angular resolution, stars: distances

1. Introduction

Ever since the discovery of fast moving (v > 1000 km s^{-1}) stars within 0.″3 (0.01 pc) of our Galaxy's central supermassive black hole (Eckart & Genzel 1997; Ghez *et al.* 1998), the prospect of using stellar orbits to make precision measurements of the black hole's mass (M_{bh}) and kinematics, the distance to the Galactic center (R_0) and, more ambitiously, to measure post-Newtonian effects has been anticipated (Jaroszynski 1998, 1999; Salim & Gould 1999; Fragile & Mathews 2000; Rubilar & Eckart 2001; Weinberg, Milosavlejic & Ghez 2005; Zucker & Alexander 2007). An accurate measurement of the

Galaxy's central black hole mass is useful for putting the Milky Way in context with other galaxies through the apparent relationship between the mass of the central black hole and the velocity dispersion, σ, of the host galaxy's bulge (e.g., Ferrarese & Merrit 2000; Gebhardt *et al.* 2000; Tremaine *et al.* 2002). It can also be used as a test of this scaling, as the Milky Way has the most convincing case for a supermassive black hole of any galaxy used to define this relationship. Accurate estimates of R_0 impact a wide range of issues associated with the mass and structure of the Milky Way, including possible constraints on the shape of the dark matter halo and, in comparison with future precision measurements of R_0 from tidal debris streams, the possibility that the Milky Way is a lopsided spiral (e.g., Reid 1993; Olling & Merrifield 2000; Majewski *et al.* 2006). Furthermore, if measured with sufficient accuracy ($\sim$1%), the distance to the Galactic center could influence the calibration of standard candles, such as RR Lyrae stars, Cepheid variables and giants, used in establishing the extragalactic distance scale. Measurements of deviations from a Keplerian orbit offer the exciting possibility of exploring the cluster of stellar remnants surrounding the central black hole, suggested by Morris (1993), Miralda-Escudé & Gould(2000), and Freitag *et al.* (2006). Estimates for the mass of the remnant cluster range from $10^4 - 10^5 M_\odot$. Its absence would be interesting in view of the hypothesis that the inspiral of intermediate-mass black holes by dynamical friction could deplete any centrally concentrated cluster of remnants. Likewise, measurements of post-newtonian effects would also provide a test general relativity, and, ultimately, could probe the spin of the central black hole.

Tremendous observational progress has been made over the last decade towards obtaining accurate estimates of the orbital parameters for the fast moving stars at the Galactic center. Patience alone permitted new proper motion measurements that yielded the first accelerations (Ghez *et al.* 2000; Eckart *et al.* 2002), which suggested that the orbital period of S0-2 could be as short as 15 years. The passage of more time then led to full astrometric orbital solutions (Schödel *et al.* 2002, 2003; Ghez *et al.* 2003, 2005a), which increased the implied dark mass densities by a factor of 10^4 compared to earlier velocity dispersion work and thereby solidified the case for a supermassive black hole. The advent of adaptive optics introduced radial velocity measurements of these stars (Ghez *et al.* 2003, Eisenhauer *et al.* 2003, 2005), which permitted the first estimates of the distance to the Galactic center from stellar orbits.

In this paper, we present new orbital models for S0-2, which provide the first estimates based on data collected with the W. M. Keck telescopes of the distance to the Galactic center. New astrometric and radial velocity measurements have been collected between 2004 and 2007, increasing the quantity of kinematic data available, and the majority of the new data was obtained with the laser guide star adaptive optics system at Keck, improving the quality of the measurements (Ghez *et al.* 2005b; Hornstein *et al.* 2007); Additionally, new data analysis has improved our ability to extract radial velocity estimates from past spectroscopic measurements, allowing us to extend the radial velocity curve back in time by two years. A full presentation of these results can be found in Ghez *et al.* (2008).

2. Results & Discussion

Orbit modeling of astrometric and radial velocity measurements of short period stars provides a direct estimate of the Milky Way's central black hole mass and distance. While it is possible to get very *precise* estimates of these quantities from existing data sets, this study shows that there are systematic uncertainties that must be accounted for to obtain *accuracy* in these estimates. Since a dominant source of systematic error in the data set

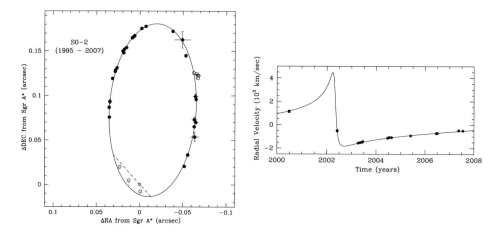

Figure 1. The best fit Keplerian orbital model with the astrometric and radial velocity data. The filled points were included in the formal fit, while the unfilled points are measurements that are excluded due to source confusion. Uncertainties are plotted on all points, except the unfilled/excluded points (here the uncertainties are comparable to the size of the points) for clarity. The data are well reproduced by a ten parameter model, which includes the black hole's mass, distance, and location in the plane of the sky as free parameters, and results in a reduced $\chi^2 \sim 1$ with the original estimates of the uncertainties. Dotted lines connect the measurements with their predicted location from the model, the dashed line shows the line of nodes, and the location of the black hole is marked by a circle, whose radius is approximately the black hole's positional uncertainty.

appears to be source confusion, we use only data from the brightest short orbital period star, S0-2, and only those measurements that are not confused with other known sources (see Figure 1). This results in a central black hole mass of $4.8 \pm 0.3 \times 10^6 \, M_\odot$ and distance of 8.3 ± 0.3 kpc (see Figure 2).

Our dynamical mass is larger than the $\sim 2 - 3 \times 10^6 \, M_\odot$ inferred from using projected mass estimators to derive the mass from measured velocity dispersions, even after accounting for the differences in distances (e.g, Eckart & Genzel 1997; Genzel et al 1997; Ghez *et al.* 1998; Genzel *et al.* 2000; see also Chakrabarty & Saha 2001). This discrepancy most likely arises from the assumptions intrinsic to the use of projected mass estimators. In particular, the projected mass estimators are based on the assumption that the entire stellar cluster is measured, which is not the case for the early proper motion studies as their fields of view were quite small (r $\sim$ 0.1 pc). Such pencil beam measurements can lead to significant biases (see discussions in Haller *et al.* 1996; Figer *et al.* 2003) An additional bias can arise if there is a central depression in the stellar distribution, such as that suggested by Figer *et al.* (2003). These biases can introduce factors of 2 uncertainties in the values of the enclosed mass obtained from projected mass estimates and thereby account for the difference between the indirect mass estimate from the velocity dispersions and the direct mass estimate from the orbital model fit to S0-2's kinematic data.

A higher mass for the central black hole brings our Galaxy into better agreement with the $M_{bh} - \sigma$ relation observed for nearby galaxies (e.g., Ferrarese & Merrit 2000; Gebhardt *et al.* 2000; Tremaine *et al.* 2002). For a bulge velocity dispersion that corresponds to that of the Milky Way ($\sim$100 km s^{-1}), the $M_{bh} - \sigma$ relationship from Tremaine *et al.* (2002) predicts a black hole mass of $9.4 \times 10^6 \, M_\odot$, which is a factor of 5 larger than

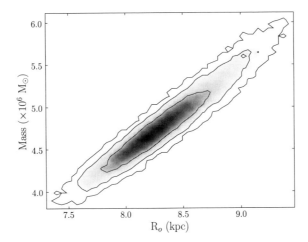

Figure 2. The probability distribution for central black hole's mass and distance from model fits to the orbit of S0-2. The best fit models imply a mass of $4.8 \pm 0.3 \times 10^6 M_\odot$ and distance of 8.3 ± 0.3 kpc. These quantities are not independent and the exact scaling depends on the relative impact of the astrometric and radial data on the model fits. Currently, the inferred mass scales with the inferred distance as $M \propto R_0^{1.8}$.

the value of the Milky Way's black hole mass used by these authors ($1.8 \times 10^6\ M_\odot$ from Chakrabarty & Saha 2001). The black hole mass presented here is a factor of of 2.7 larger than that assumed by Tremaine *et al.* (2002), bringing the Milky Way more in line with this relationship. With one of the most accurate and lowest central black hole masses, the Milky Way is, in principle, an important anchor for the $M_{bh} - \sigma$ relationship. However, the velocity dispersion of the Milky Way is much more uncertain than that of other nearby galaxies. Therefore our revised mass has only modest impact on the coefficients of the $M_{bh} - \sigma$ relation. More importantly, our revised mass estimate shows that factors of two changes in the black hole mass obtained from projected mass estimators are easily obtained and that much of the scatter in the $M_{bh} - \sigma$ relationship may easily arise from inaccuracies in the modeling of the black hole masses.

Revision of the central black hole's mass and distance can also, in principle, impact our understanding of the structure and timescales within our galaxy both on small and large scales. On the large scale, if we assume that the black hole is located at the center of our Galaxy, then its distance provides a measure of R_0. Its value from this study is consistent with the IAU recommended value of 8.5 kpc as well as the value of 8.0 ± 0.5 kpc suggested by Reid (1993), based on a "weighted average"† of all prior indirect measurements of R_0. Combining the value for R_0 from this study with the measured proper motion of Sgr A* along the direction of Galactic longitude (Reid & Brunthaler 2004; $\mu_{SgrA*,long} = -6.379 \pm 0.026$ mas yr^{-1}) and the Sun's deviation from a circular orbit (Cox 2000; 12 km s^{-1}) in the direction of Galactic rotation, we obtain an estimate of the local rotation speed, θ_0, of 255 ± 13 km s^{-1}, which is consistent with other recent measurements; these include a value of 222 ± 20 km s^{-1} from the review of Kerr & Lynden-Bell 1986 and 270 km s^{-1} derived by Méndez *et al.* 1999 from the absolute proper motions of $\sim$30,000 stars in the Southern Proper-Motion survey. As two of the fundamental Galactic constants, R_0 and θ_0 are critical parameters for axisymmetric models of the Milky Way. Under the assumption that the stellar and gas kinematics within our galaxy are well measured, the

† consensus value with consensus errors

values of R_0 and θ_0 determine the mass and shape of the Milky Way (Olling & Merrifield 2000; Olling & Merrifield 2001). Of particular interest is the value of the short-to-long axis ratio of the dark matter halo, q, as it offers a valuable opportunity to distinguish between different cosmological models. As Olling & Merrifield (2001) demonstrate, the uncertainty in q for the Milky Way is dominated by the large uncertainties in R_0 and θ_0. Based on this analysis, which is predicated on axisymmetric models of the Milky Way, a value of 8.3 kpc for R_0 and 255 km s^{-1} for θ_0 suggests that a highly flattened dark matter halo can be ruled out; this in turn disfavors two dark matter candidates (1) decaying massive neutrinos and (2) a disk of cold molecular hydrogen (Olling & Merrifield 2001).

Closer to the black hole, knowing its mass and distance from the Sun improves our ability to study the kinematics of stars within its sphere of influence. With the black hole's parameters in hand, much less kinematic information is needed to determine the orbital parameters for stars whose motion is dominated by the gravitational influence of the central black hole. This approach was used to estimate the possible range of orbital periods for the fast moving stars within 0″.5 of the central black hole, leading to the realization that these stars could have orbital periods as short as 15 years (Ghez *et al.* 2000; Eckart *et al.* 2002), which has indeed turned out to be the case (e.g., Schödel *et al.* 2002, 2003; Ghez *et al.* 2003, 2005; Eisenhauer *et al.* 2005). Further improvements in the constraints on the central black hole's properties and their degeneracies, as presented here, along with improved astrometry, has allowed us to derive orbital information for individual stars at much larger galacto-centric distances. With these measurements, in Lu *et al.* (2006), we test for the existence and properties of the young stellar disk(s), proposed by Levin & Beloborodov (2003) and Genzel *et al.* (2003b) from a statistical analysis of velocities alone. The direct use of individual stellar orbits out beyond a radius of 1″ reveals the existence of only one, relatively thin, disk of young stars (Lu *et al.* in prep).

On an even smaller scale, the mass and distance of the black hole set the magnitude and time-scale for various relativistic effects. Given estimated Keplerian orbital elements for stars at the Galactic center, we expect to able to measure their stellar orbits with sufficient precision in upcoming years to detect the Roemer time delay, the special relativistic transverse Doppler shift, the general relativistic gravitational red-shift, and the prograde motion of periapse (e.g., Weinberg *et al.* 2005; Zucker & Alexander 2007). The most likely star to be measured first is S0-2, as it has the shortest orbital period (P=15 yr), is quite eccentric (e=0.8830) and, as one of the brighter stars ($K_{S0-2} = 14.0$ mag), has more precise astrometric and spectroscopic measurements. The radial velocity signatures of the first three effects are expected to be comparable to each other and will impart a ~200 km/s deviation at closest approach (Zucker & Alexander 2007), when the star is predicted to have a line of sight velocity of -2600 km/s based on our updated Keplerian model. This effect is large compared to the radial velocity precision (~ 20 km/sec). Likewise, the expected apoapse center shift for S0-2, $\Delta s = \frac{6\pi G M_{bh}}{R_0 (1-e) c^2} = 0.9$ mas (see e.g., Weinberg 1972; Weinberg *et al.* 2005), is an order of magnitude larger than our current measurement precision ($\sigma_{pos} \sim 0.1$ mas). While stellar confusion in our present day adaptive optics measurements limits the accuracy of S0-2's positional estimates to only ~ 0.5 mas, improved adaptive optics systems on existing telescope and larger telescopes (see Weinberg *et al.* 2005) will improve the sensitivity to the predicted apocenter shift. To put this measurement into context with existing tests of general relativity, it is useful to note that one of the strongest constraints on general relativity to date comes from the Hulse-Taylor binary pulsar, PSR 1913+16, which has a relativistic parameter at periapse, $\Gamma = r_{sch}/r_{periapse}$, of only 5×10^{-6}, which is 3 order magnitude smaller than that of S0-2

(Taylor & Weisberg 1989; Zucker & Alexander 2007). The stars at the galactic center are therefore probing an unexplored regime of gravity, which, in its strong regime, is the least tested of the four fundamental forces of nature, and, with the larger black hole mass implied by the measurements presented here, the relativistic effects should be larger than previously anticipated.

Precession from general relativistic effects also influences the timescale for resonant relaxation processes close to the black hole (see, e.g, Rauch & Tremaine 1996; Hopman & Alexander 2006). When precession from general relativity dominates over that from the extended mass distribution, the resonant relaxation timescale is proportional to $M_{bh}^2 \times (J_{LSO}/J)^2 \times P$, where J and J_{LSO} are the orbital angular momenta for the orbit of interest and at the last stable orbit around the black hole, respectively, and P is the orbital period. For a given semi-major axis and accounting for the linear mass dependence of $(J_{LSO}/J)^2$, this results in a $M_{bh}^5/2$ dependency. Thus the higher black hole mass inferred from this study increases the timescale over which the black hole's loss cone would be emptied in the regime where general relativity dominates. For the regime where extended mass distribution dominates, the resonant relaxation timescale scales only as $M_{bh}^1/2$. A higher black hole mass also implies a longer period for the last stable orbit. If the central black hole is non-spinning, the innermost stable orbit has a period of 38.5 min. Periodicities on shorter timescales, such as the putative QPO at ~ 20 min (Genzel. et al. 2003a; Eckart *et al.* 2006; Bélanger *et al.* 2006) have been interpreted as arising from the inner most stable orbit of a spinning black hole. At the present mass, the spin would have to be 0.6 of its maximal rate to be consistent with the possible periodicity. However, it is important to caution that other mechanisms can give rise to such short periodicities, such as a standing wave pattern recently suggested by Tagger & Melia (2006). A further complication is that the temporal power spectrum is also statistically consistent with that observed from red-noise caused by disk instabilities (Do *et al.*, in prep).

3. Conclusions

The short orbital period star S0-2 has been intensively studied astrometrically (1995-2007) and spectroscopically (2000- 2007) with the W. M. Keck 10 meter telescopes. Fits of a Keplerian orbit model to these data sets, after removing data adversely affected by source confusion, result in estimates of the black hole's mass and distance of $4.8 \pm 0.3 \times 10^6 M_\odot$ and 8.3 ± 0.3 kpc, respectively. While the current analysis is dominated by 11 years of astrometric measurements that have ~ 1.2 mas uncertainties, the LGSAO data over the last 3 years have positional uncertainties that are an order of magnitude smaller. With higher strehl ratios and more sensitivity, LGSAO measurements are also less effected by source confusion; this is especially important for the closest approach measurements, which have to contend with source confusion from the variable source SgrA*-IR. Following S0-2 for another 10 years should result in the first measurement of the Sun's peculiar motion in the direction of the Galactic center from the orbit of S0-2 with a precision of a few km s^{-1} and 1% measurement of R_0. At this precision, the measurement of R_0 is of interest because it may be able constrain the cosmic distance ladder.

References

Bélanger, G., Terrier, R., de Jager, O. C., Goldwurm, A., & Melia, F. 2006, Journal of Physics Conference Series, 54, 420

Chakrabarty, D. & Saha, P. 2001, AJ, 122, 232

Cox, A. N. 2000, Allen's astrophysical quantities, 4th ed. Publisher: New York: AIP Press; Springer, 2000. Editedy by Arthur N. Cox. ISBN: 0387987460,

Eckart, A. & Genzel, R. 1997, *MNRAS*, 284, 576

Eckart, A., Genzel, R., Ott, T., & Schödel, R. 2002, *MNRAS*, 331, 917

Eckart, A., Schödel, R., Meyer, L., Trippe, S., Ott, T., & Genzel, R. 2006, *A&A*, 455, 1

Eisenhauer, F. *et al.* 2003, *ApJ*, 597, L121

Eisenhauer, F., *et al.* 2005, *ApJ*, 628, 246

Ferrarese, L. & Merritt, D. 2000, *ApJ*, 539, L9

Figer, D. F., *et al.* 2003, *ApJ*, 599, 1139

Fragile, P. C. & Mathews, G. J. 2000, *ApJ*, 542, 328

Freitag, M., Amaro-Seoane, P., & Kalogera, V. 2006, *ApJ*, 649, 91

Gebhardt, K., *et al.* 2000, *ApJ*, 539, L13

Genzel, R., Eckart, A., Ott, T., & Eisenhauer, F. 1997, *MNRAS*, 291, 219

Genzel, R., Schödel, R., Ott, T., Eckart, A., Alexander, T., Lacombe, F., Rouan, D., & Aschenbach, B. 2003a, *Nature*, 425, 934

Genzel, R., *et al.* 2003b, *ApJ*, 594, 812

Genzel, R., Pichon, C., Eckart, A., Gerhard, O. E., & Ott, T. 2000, *MNRAS*, 317, 348

Ghez, A. M. *et al.* 2003, *ApJ*, 586, L127

Ghez, A. M. *et al.* 2008, *ApJ*, to be submitted

Ghez, A. M., *et al.* 2005b, *ApJ*, 635, 1087

Ghez, A. M., Klein, B. L., Morris, M., & Becklin, E. E. 1998, *ApJ*, 509, 678

Ghez, A. M., Morris, M., Becklin, E. E., Tanner, A., & Kremenek, T. 2000, *Nature*, 407, 349

Ghez, A. M. *et al.* 2005a, *ApJ*, 620, 744

Ghez, A. M., *et al.* 2004, *ApJ*, 601, L159

Haller, J. W., & Melia, F. 1996, *ApJ*, 464, 774

Hopman, C., & Alexander, T. 2006, *ApJ*, 645, 1152

Hornstein, S. D., Matthews, K., Ghez, A. M., Lu, J. R., Morris, M., Becklin, E. E., Rafelski, M., & Baganoff, F. K. 2007, *ApJ*, 667, 900

Jaroszynski, M. 1998, Acta Astronomica, 48, 653

Jaroszyński, M. 1999, *ApJ*, 521, 591

Kerr, F. J., & Lynden-Bell, D. 1986, *MNRAS*, 221, 1023

Levin, Y., & Beloborodov, A. M. 2003, *ApJ*, 590, L33

Lu, J. R., Ghez, A. M., Hornstein, S. D., Morris, M., Matthews, K., Thompson, D. J., & Becklin, E. E. 2006, Journal of Physics Conference Series, 54, 279

Majewski, S. R., Law, D. R., Polak, A. A., & Patterson, R. J. 2006, *ApJ*, 637, L25

Méndez, R. A., Platais, I., Girard, T. M., Kozhurina-Platais, V., & van Altena, W. F. 1999, *ApJ*, 524, L39

Miralda-Escudé, J. & Gould, A. 2000, *ApJ*, 545, 847

Morris, M. 1993, *ApJ*, 408, 496

Olling, R. P. & Merrifield, M. R. 2000, *MNRAS*, 311, 361

Olling, R. P. & Merrifield, M. R. 2001, *MNRAS*, 326, 164

Rauch, K. P. & Tremaine, S. 1996, New Astronomy, 1, 149

Reid, M. J. 1993, *ARAA*, 31, 345

Reid, M. J. & Brunthaler, A. 2004, *ApJ*, 616, 872

Rubilar, G. F. & Eckart, A. 2001, *A&A*, 374, 95

Salim, S. & Gould, A. 1999, *ApJ*, 523, 633

Schödel, R. *et al.* 2002, *Nature*, 419, 694

Schödel, R. *et al.* 2003, *ApJ*, 596, 1015

Tagger, M. & Melia, F. 2006, *ApJ*, 636, L33

Tremaine, S., *et al.* 2002, *ApJ*, 574, 740

Weinberg, S. 1972 Gravitation and Cosmology: Principles and Applications of the General Theory of Relativity (New York: Wiley)

Weinberg, N. N., Milosavljević, M., & Ghez, A. M. 2005, *ApJ*, 622, 878

Taylor, J. H. & Weisberg, J. M. 1989, *ApJ*, 345, 434

Zucker, S. & Alexander, T. 2007, *ApJ*, 654, L83

A Giant Step: from Milli- to Micro-arcsecond Astrometry
Proceedings IAU Symposium No. 248, 2007
W. J. Jin, I. Platais & M. A. C. Perryman, eds.
© 2008 International Astronomical Union
doi:10.1017/S1743921308018632

Taming the binaries

D. Pourbaix†

Institut d'Astronomie et d'Astrophysique, Université Libre de Bruxelles, Brussels, Belgium
email:pourbaix@astro.ulb.ac.be

Abstract. Astrometric binaries are both a gold mine and a nightmare. They are a gold mine because they are sometimes the unique source of orbital inclination for spectroscopic binaries, thus making it possible for astrophysicists to get some clues about the mass of the often invisible secondary. However, this is an ideal situation in the sense that one benefits from the additional knowledge that it is a binary for which some orbital parameters are somehow secured (e.g. the orbital period). On the other hand, binaries are a nightmare, especially when their binary nature is not established yet. Indeed, in such cases, depending on the time interval covered by the observations compared to the orbital period, either the parallax or the proper motion can be severely biased if the successive positions of the binary are modelled assuming it is a single star. With large survey campaigns sometimes monitoring some stars for the first time ever, it is therefore crucial to design robust reduction pipelines in which such troublesome objects are quickly identified and either removed or processed accordingly. Finally, even if an object is known not to be a single star, the binary model might turn out not to be the most appropriate for describing the observations. These different situations will be covered.

Keywords. binaries: general, astrometry, methods: data analysis

1. Why worrying about binaries?

1.1. Binary effect

Before wondering what binaries can do for you, it is worth wondering what could happen if you do not care about them. The parallactic term in the equation of the stellar motion is a periodic term with a period of one year. Any additional signal with a period close to one year, e.g. a binary, therefore jeopardizes the derivation of the parallax. So, if the object is not known to be a binary and a single star model is assumed in the derivation of the parallax, the latter could be severely biased.

A similar bias can be noticed on the proper motion of long period binaries when the observations do not cover an integral multiple of the orbital period. In such a case, the orbital motion adds up to the proper motion and, when the single star model is assumed, the resulting proper motion is affected.

Those two situations are illustrated in Fig. 1 with two Hipparcos entries. In the left panel (R Hya=HIP 65835), the star was originally processed as a Variability Induced Mover (i.e. a case where the strong correlation between the total brightness of source and the position of the photocentre was explained with a binary model with one variable component, hereafter VIM). The resulting parallax was 1.62 ± 2.43 mas. Correcting for additional effects and adopting the basic single star model, Pourbaix *et al.* (2003) derived a parallax of 8.44 ± 1.0 mas, very consistent with other astrophysical results (Whitelock *et al.* 2000).

The second example is V815 Her (HIP 88848) taken from Fekel *et al.* (2005). The Hipparcos observations of that object could not be satisfactory fitted with any model and

† Research Associate, F. N. R. S., Belgium

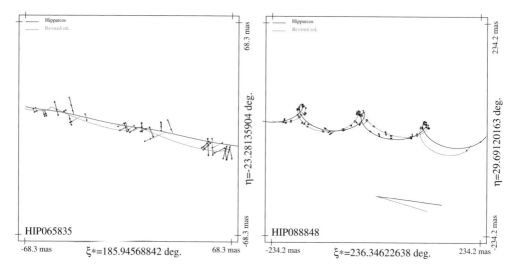

Figure 1. Incorrect binary assumption affecting the parallax (left panel) or the proper motion
(right panel).

a stochastic solution was therefore originally adopted. The resulting proper motion was
$(138.07, -18.58)$ mas/yr. Fekel and his collaborators noticed a third component in that
stellar system, with an orbital period of 2092 days. Once that orbit is accounted for, the
Hipparcos data yield an excellent fit leading to a revised proper motion of $(106.59, -30.84)$
mas/yr. This value is consistent with the long time-baseline proper motions(e.g. Tycho-2,
Høg *et al.* 2000).

1.2. *Binary fraction*

At this point, even if one agrees that binaries can bias the astrometric solution, one could
still argue that their number is so small that they will pop up in large sample as some
additional noise. Well, depending on the stellar population, the observed binary fraction
ranges from 15% in the lower end of the main sequence up to 80% in the upper end. Such
a high binary fraction is also observed out of the Galactic plane (18% of the halo stars
seem to be binaries, Carney *et al.* 2003).

Even though those fractions are pretty high, those binaries are not all troublesome for
the astrometrist. Indeed, unless they are very nearby, close binaries (with period of a few
days) are not and will not be resolved by any foreseen neither space nor ground based
program. Despite the high precision, even the photocentre might not be distinguished
from the barycentre of the system, thus making that system indistinguishable from a
genuine single star ... on the astrometric grounds.

Actually, despite the high fraction of binaries and the increasing observation precision,
astrometry is the least affected detection technique. Indeed, the period and distance act
against astrometric detection. The further the system, the smaller the angular disturbing
effect of the companion (unlike spectroscopic detection).

Mother nature also plays against astrometric binaries. Denoting a_r the semi-major
axis of the relative orbit of the secondary (the fainter component) around the primary
(the brighter star) and a_0 the semi-major axis of the orbit of the photocentre around the
centre of mass, the two are related through

$$a_0 = \left(\frac{M_2}{M_1 + M_2} - \frac{I_2}{I_1 + I_2}\right)a_r \tag{1.1}$$

where M denotes the mass and I the intensity of the related component. When the secondary is much fainter than the primary (falling below the detection limit), relation (1.1) reduces to $a_0 = a_1$, i.e. the size of the photocentric orbit is the same as the absolute orbit of the primary. As the secondary becomes brighter, a_0 decreases. In the extreme case of a perfect twin system, a_0 vanishes completely. Unfortunately, although nature does not really favour that kind of systems, it does not favour systems with $I_1 >> I_2$ either (Halbwachs *et al.* 2003).

1.3. *From astrometry to astrophysics*

So far, we have shown why astrometrists should worry about binaries. What about the other areas of astronomy? Well, anybody who uses the parallax of the object to change its apparent magnitude into its luminosity should worry about the binary nature. Since correcting for the binary nature might require the full orbital model to be known, being aware of the binary nature of an object already allows those scientists to discard it from their sample. So, for those researchers, getting a hint that a star is double is already enough (maybe to get rid of it). At the other end of the spectrum, there are scientists interested in getting more than just a detection. Among them, there are those who want to study the distribution of the orbital parameter (e.g. as a function of the spectral type, . . .). However, in order to carry out such an investigation, one needs as many well studied binaries as possible.

As already mentioned, spectroscopic detection of binaries is not affected by the distance to the systems. Spectroscopists are therefore pretty efficient in discovering binaries. However, the spectroscopic orbit lacks a key parameter which prevent them from being useful for astrophysics: the orbital inclination. Both the semi-major axis and the mass of the companion are entangled with the inclination. For instance, only the product of the mass with the sine of the inclination comes out of the equations.

$$\frac{(a_1 \sin i)^3}{P^2} = \frac{M_2^3 \sin^3 i}{(M_1 + M_2)^2}$$

Without the inclination, this product is essentially useless to the astrophysicist, even if M_1 is assumed from some stellar model.

That is where astrometry becomes important to the eyes of astrophysicists. They suddenly remember that if some orbital model can fit the astrometric wobble caused by the binary, the inclination can be obtained and the true mass of the secondary retrieved (Jancart *et al.* 2005). In the case of extrasolar planets, even when the astrometric orbit could not be fitted, the absence of orbit was sometimes enough to give an upper bound of the mass of the companion (Perryman *et al.* 1996). Some nevertheless tried to fit an orbit despite the low signal to noise ratio and ran into troubles (Zucker & Mazeh 2000, 2001).

2. Binary detection

2.1. *Ground-based observations*

In order to detect the binary (or non-single) nature of an object, one can rely upon several spectroscopic (yield spectroscopic binaries), photometric (eclipsing binaries), or optical (visual or astrometric binaries) observations. In the first and third case, one seeks any departure from the motion of a single star whereas, in the second case, one looks for a typical shape of the light curve. Whereas, for spectroscopic binaries, two independent observations are theoretically enough, more observations are required for the visual/astrometric cases and even more are needed for the eclipsing cases (as it

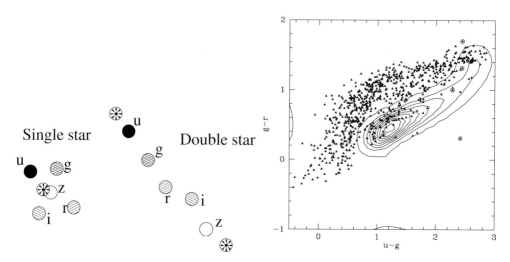

Figure 2. Positions of the photocentres according to the effective wavelength of the photometric filter (left panel) and binaries identified through that displacement of the photocentre in the SDSS DR5 (right panel).

requires a typical shape of the light curve rather than any departure from the constant radial velocity model for spectroscopic binaries).

Several teams are very active, sometime over a period exceeding the lifetime of their original instrument, and keep feeding the catalogues of optically resolved binaries (Hartkopf *et al.* 2001b; Mason *et al.* 2001) or even of astrometric orbits (Hartkopf *et al.* 2001a). Some of these teams presented their results during this meeting so a few others are listed here: McAlister and the CHARA Team continue with the Array (Bagnuolo *et al.* 2006), the work begun with speckle interferometry (McAlister 1997), Armstrong and Hummel who began with Mark III (Hummel *et al.* 1994) and then moved to NPOI (Armstrong *et al.* 1998; Hummel *et al.* 1998) and/or VLTI, and Davis and colleagues resolving binaries with the Sydney University Stellar Interferometer (Davis *et al.* 2005).

Besides those regular techniques (and the subsequent orbits), there are also less usual methods combining photometry and precise astrometry. The first approach consists in looking for a correlation between the position of the photocentre and the total brightness of the source (VIM). Such a method was suggested by Wielen (1996) and globally successfully applied to the Hipparcos observations (Pourbaix *et al.* 2003). Whereas a fixed configuration was assumed for Hipparcos, this assumption will be relaxed for Gaia (Halbwachs & Pourbaix 2005).

In case of VIM, one needs several observations of both the position of the photocentre and the total brightness of the source. An alternative/complementary approach consists in obtaining several positions at different wavelengths at just one epoch (Christy *et al.* 1983; Sorokin & Tokovinin 1985). In case of a single star, the scatter of those positions reflects the precision of the measurements (left panel of the left panel of Fig. 2). With a binary, those positions are scattered along the segment between the two stars. The presence of such an alignment is thus a hint for a double star (maybe an optical pair). That technique was applied (Pourbaix *et al.* 2004) to the SDSS data and yielded about 750 double stars (right panel of Fig. 2) when using the SDSS Data Release 5 (Adelman-McCarthy et al. 2007). This method extends the detection limits well within the stellar locus whereas a purely photometric approach would be limited to photometric outliers (Smolčić *et al.* 2004). The two methods are complementary.

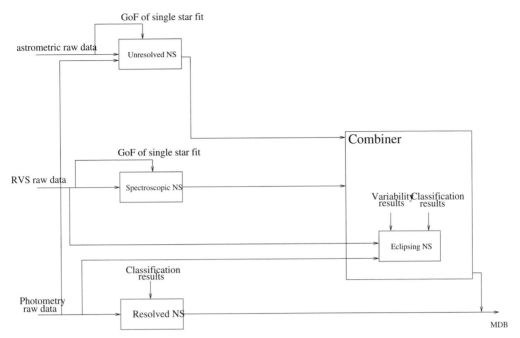

Figure 3. When entering the Object Processing coordination unit, each binary goes to a specific sub-pipeline depending on the nature of the observations yielding its detection.

2.2. *Hipparcos, a gold mine for binary seekers*

Even though the Hipparcos catalogue was released ten years ago (ESA 1997; Perryman 2008), there are still teams who are screening it for some binary signatures. Usually, those teams look for the astrometric signal of known binaries (e.g., Torres *et al.* 2006; Fekel *et al.* 2007). The goal is then to improve the understanding of those objects by obtaining the orbital inclination (the original Hipparcos astrometric solution can also be substantially revised).

Another use of the Hipparcos catalogue consists in comparing its proper motions (based on 3 years of observations) with the long time-baseline ones (e.g. Tycho-2 Høg *et al.* 2000). A discrepancy between the two proper motions is a strong hint towards a long period binary (Makarov & Kaplan 2005; Goldin & Makarov 2006; Frankowski *et al.* 2007)

3. Besides detection: the Gaia approach

In the Gaia data processing pipeline, binaries are identified as stellar objects with a poor fit in either of the different branches of the reduction which assumes a single star model (i.e. CU3, 5 and 6 Mignard *et al.* 2008). The Object Processing coordination unit #4 then picks up those outlying objects and dispatches them to sub-pipelines for further processing accordingly. The choice of a particular branch depends on the type of the observations in which the binary nature of the source was identified (Fig. 3).

The pipeline for astrometric binaries is depicted in Fig. 4. It consists in a cascade of models with increasing complexity. A quality indicator (generically called Goodness of Fit) is used to assess the solution. If it is good enough, the cascade stops and the solution is accepted. Otherwise, the data are transferred to the following box and the corresponding model fitted and evaluated.

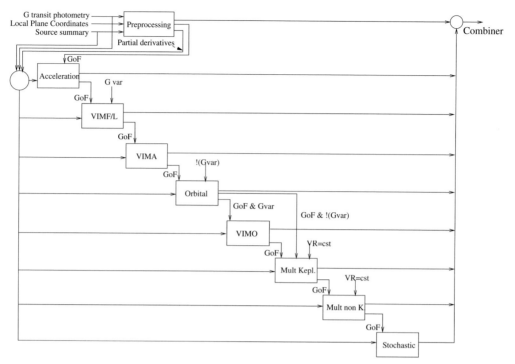

Figure 4. Specific pipeline for the binaries detected through their astrometric wobble

The cascade includes special considerations for binaries with a variable component, i.e. VIM, with either Fixed configuration, a Linear motion, an Accelerated motion or requiring a full Orbital solution. It also copes with extrasolar planetary systems with, potentially, a non-Keplerian description of the observed trajectory of the photocentre.

We are confident that although one will not obtain an orbital solution for the astrometric binaries detected by Gaia, the simplest model confidently describing the data will always be adopted.

4. Conclusion

If one does not want some astrometric results to be compromised, binaries have to be detected as early as possible in the data reduction pipeline and then processed with some dedicated models. Those two tasks are clearly identified and their implementation already quite advanced in Gaia DPAC (Mignard *et al.* 2008).

Acknowledgements

I thank J.-L. Halbwachs and A. Sozzetti for their help in the preparation of the Gaia data processing pipeline for the astrometric non-single stars. It is my pleasure to acknowledge the Belgian FNRS for its support as well the financial support of the PRODEX program through its grant 90078.

References

Adelman-McCarthy J. K., Agüeros M. A., Allam S. S., *et al.*, 2007, *ApJS*, 172, 634
Armstrong J. T., Mozurkewich D., & Rickard L. J., *et al.*, 1998, *ApJ*, 496, 550
Bagnuolo W. G. Jr., Taylor S. F., & McAlister H. A., *et al.*, 2006, *AJ*, 131, 2695

Carney B. W., Latham D. W., Stefanik R. P., Laird J. B., Morse J. A., Jan. 2003, *AJ*, 125, 293
Christy J. W., Wellnitz D. D., Currie D. G., 1983, Lowell Observatory Bulletin, 167, 28
Davis J., Mendez A., Seneta E. B., *et al.*, Feb. 2005, *MNRAS*, 356, 1362
ESA, 1997, The Hipparcos and Tycho Catalogues, ESA SP-1200
Fekel F. C., Barlow D. J., Scarfe C. D., Jancart S., Pourbaix D., 2005, *AJ*, 129, 1001
Fekel F. C., Williamson M., Pourbaix D., 2007, *AJ*, 133, 2431
Frankowski A., Jancart S., Jorissen A., 2007, *A&A*, 464, 377
Goldin A., Makarov V. V., 2006, *ApJS*, 166, 341
Halbwachs J. L., Pourbaix D., 2005, 575–578, ESA SP-576
Halbwachs J. L., Mayor M., Udry S., Arenou F., 2003, *A&A*, 397, 159
Hartkopf W. I., Mason B. D., Worley C. E., 2001a, *AJ*, 122, 3472
Hartkopf W. I., McAlister H. A., Mason B. D., 2001b, *AJ*, 122, 3480
Høg E., Fabricius C., Makarov V. V., *et al.*, 2000, *A&A*, 355, L27
Hummel C. A., Mozurkewich D., Elias N. M. II, *et al.*, 1994, *AJ*, 108, 326
Hummel C. A., Mozurkewich D., Armstrong J. T., *et al.*, 1998, *AJ*, 116, 2536
Jancart S., Jorissen A., Babusiaux C., Pourbaix D., 2005, *A&A*, 442, 365
Makarov V. V., Kaplan G. H., 2005, *AJ*, 129, 2420
Mason B. D., Wycoff G. L., Hartkopf W. I., Douglas G. G., Worley C. E., 2001, *AJ*, 122, 3466
McAlister H., 1997, In: Docobo J. A., Elipe A., McAlister H. (eds.) Visual double stars: formation, dynamics and evolutionary tracks, 3, Kluwer Academic Publishers
Mignard F., Bailer Jones C., Bastian U., *et al.*, 2008, In: W.J. Jin, I. Platais & M.A.C. Perryman (eds.), 2008, Proceedings of IAU Symposium 248, p.224
Perryman M. A. C., 2008, In: W. J. Jin, I. Platais & M. A. C. Perryman (eds.), 2008, Proceedings of IAU Symposium 248, p. 8
Perryman M. A. C., Lindegren L., Arenou F., *et al.*, 1996, *A&A*, 310, L21
Pourbaix D., Platais I., Detournay S., *et al.*, 2003, *A&A*, 399, 1167
Pourbaix D., Ivezić Ž., Knapp G. R., Gunn J. E., Lupton R. H., 2004, *A&A*, 423, 755
Smolčić V., Ivezić Ž., Knapp G.R., *et al.*, 2004, *ApJ*, 615, L141
Sorokin L. Y., Tokovinin A. A., 1985, SvAL, 11, 226
Torres G., Lacy C. H., Marschall L. A., Sheets H. A., Mader J. A., 2006, *ApJ*, 640, 1018
Whitelock P., Marang F., & Feast M., 2000, *MNRAS*, 319, 728
Wielen R., 1996, *A&A*, 314, 679
Zucker S., Mazeh T., 2000, *ApJ*, 531, L67
Zucker S., Mazeh T., 2001, *ApJ*, 562, 549

A Giant Step: from Milli- to Micro-arcsecond Astrometry
Proceedings IAU Symposium No. 248, 2007
W. J. Jin, I. Platais & M. A. C. Perryman, eds.

Astrometry of the solar system: the ground-based observations

J.-E. Arlot

IMCCE, UMR8028 du CNRS, Paris Observatory, UPMC, USTL
77 avenue Denfert-Rochereau, F-75014 paris, France
email: arlot@imcce.fr

Abstract. The main goal of the astrometry of solar system objects is to build dynamical models of their motions to understand their evolution, to determine physical parameters and to build accurate ephemerides for the preparation and the exploitation of space missions. For many objects, the ground-based observations are still very important because radar or observations from space probes are not available. More, the need of observations on a long period of time makes the ground-based observations necessary. The solar system objects have very different characteristics and the increase of the astrometric accuracy will depend on the objects and on their physical characteristics. The purpose of this communication is to show how to get the best astrometric accuracy.

Keywords. astrometry, solar system: general, ephemerides, occultations

1. Introduction

The astrometry of solar system objects is very specific since these bodies have a proper motion very fast and since they are, most of the time, not a point source but larger bodies with surface effects either depending on the structure of the object and/or on the phase angle. Note that we are interested in the astrometric position of the center of mass of the object and that we are observing, most of the time, its photocenter. Because of the large number of very different bodies in the Solar system, the observational techniques are also numerous in order to get the best astrometric accuracy. We will review the different techniques used by the observers and emphasize the ones providing the best accuracy and easily usable worldwide even with small telescopes.

2. Purpose of the solar system astrometry

Contrary to stars, the solar system bodies have proper motions very fast needing very complicated models for their motion. Celestial mechanics provide the tools to get the best dynamical models needing astrometric observations of the center of mass of the objects to fit the theoretical model to observations and to determine the parameters of the motion. Unfortunately, these motions, even very fast, have very long periodic terms depending on the motion of all the planets that require the observations to be spread regularly over a long period of time. So, the solar system astrometry should be made every day covering the orbits of all the objects. Theoretical models and astrometric observations will provide ephemerides of the solar system objects in order to help the preparation and the exploitation of the observations made either by space probes or by ground-based telescope. The ephemerides are also useful for the near-Earth objects on the orbits which could be a hazardous to the Earth. The determination of the scale of the solar system, its formation and evolution also need astrometric observations. The knowledge

Table 1. Observational techniques and their accuracy.

Technique	Accuracy	Objects	Comments
VLBI	2 to 10 mas	objects visited by space probes	all
Radar	10 to 100 m	Near-Earth objects	possible for Jupiter
LLR	1 to 3 cm	The Moon	
Transit circle	50 to 100 mas	magn. 6 to 15	except Mercury, venus and Mars
Scanning tel.	50 to 100 mas	until magn. 20	except the planets
Tangential focal plane	20 to 2000 mas	all	except the planet
Planets through sat.	20 to 50 mas	Giant planets	only Jupiter and Saturn
AO, IR	a few mas (relative)	inner satellites	objects close to primary
Photometric mutual events eclipses	1 to 30 km (1 to 10 mas) 500 km (150 mas)	main planetary satellites, asteroids Galilean satellites, Titan	occultations

of the dynamical reference system requires a good knowledge of the motions in the solar system. A better accuracy of the observations will allow us to make discoveries in the dynamics of the objects or in their physical structures. In fact, the non gravitational forces, not taken into account in theoretical models, can be detected from astrometric observations. Same, the tidal forces modify the orbits of the objects and astrometric observations will show the deviation of the real orbit relative to the theoretical one based upon a model not including these tidal forces. This shows the challenge to increase the accuracy of the astrometric observations of the solar system objects.

3. The different techniques of observation and their accuracy

Each technique of observations has its own precision and accuracy. Note that we must avoid to mix precision and accuracy. The precision is internal to the technique of observations and is calculated from the rms of observational residuals, made during a short period of time in which the residuals are not supposed to change. The accuracy is more difficult to determinate. It will be calculated from the residuals provided by a theoretical model, the precision of which being better than the one of the observations. We have to notice than the accuracy may be given either in angular units (arcsec or mas) or in distance unit (kilometers) depending on the technique itself.

- VLBI observations of space probes

The observations through differential VLBI are the most accurate but need to receive a radio signal from a space probe. The positions are relative to the nearby radio sources and linked to ICRF. The Deep Space network of the JPL is used for that purpose (Folkner *et al.* 1996, Standish 2000).

- Direct observations by space probes

Space probes take pictures of the objects that they are visiting with stars in the background: knowing the position of the probe, these pictures yield astrometric data. These observations (mainly the planetary satellites) are very rare but are added to the available series of data for dynamical purpose (see for example Jacobson 2004).

- Radar observations

Radar observations may be performed only for nearby objects and have been made for Near-Earth objects allowing to measure the distance to the Earth. Since the radar data have an accuracy until 10 times better than optical observations, this technique should be used more extensively for the calculation of the orbits of NEO (Ostro 2002).

- LLR (Lunar Laser Ranging)

The LLR is used only for the Moon thanks to the reflector put on the Moon first by Apollo 11. The ranging was made from several sites but only McDonald observatory, USA and Grasse Observatory, France, produced observations regularly and over decades (Dickey *et al.* 1994). The accuracy was about 20 cm at the beginning and now is a few centimeters. An improvement of this accuracy should be possible in the future thanks to the technical progress.

- Optical ground-based observations

In nearly all techniques described above, the optical ground-based observations are performed over centuries and are efficient since photographic plates allowed to record the data. The advantage of this technique is to be easily available at any time – important for solar system objects that need a continuous survey. The lower accuracy is replaced by the large number of observations and by their distribution over the time. Different techniques were developed in order to increase the astrometric accuracy as we will see below. Table 1 provides some comparisons between several optical ground-based observational techniques for the Galilean satellites and for the Uranian satellites. The photometric observations of occultations have the same accuracy for both systems but is better for the Uranian satellites since the accuracy is in kilometers at a very large distance.

Even if some non optical techniques provide a high level of accuracy, ground-based observations are necessary for several reasons:

-many objects are not visited by space probles making impossible VLBI or Doppler observations

-radar observation is a powerful technique but only the objects near the Earth can be observed even though the farther objects are reachable but with difficulties.

-objects for which a long period of observations is possible only with optical ground-based observations (objects beyond the main asteroid belt).

4. The meridian transit circle

The meridian transit circle observes the culmination of an object and measures directly its right ascension and declination. The automatic transit circles use a CCD target in TDI mode i.e. the CCD scans the sky rebuilding an image which may be very long. All the objects present in this strip are linked together. Each transit circle may be associated to its own star catalogue determining its absolute accuracy. Bright objects (magnitude from 6 to 15) have been extensively observed for many years and provide useful data (Stone 2001, Rapaport *et al.* 2002). These instruments have limits: small aperture, large pixels, small declination (at large declinations, the strip is not rectangular). A way to solve these problems is to use scanning telescopes along great circles of the celestial sphere such as LINEAR, working automatically, providing the majority of the asteroidal observations.

Table 2 provides the rms of the residuals of observations of Uranus and its satellite Oberon made at Flagstaff and at Bordeaux.

5. Tangent-plane astrometry

The most classical astrometric observations consist in making an image in the focal plane of the telescope. During most of the XXth century, photographic plates were made extensively either with short focus instruments or with long focus ones. Unfortunately, the star catalogues were poor and too few stars from these catalogues were present on the

Table 2. Rms of the residuals of transit circle observations in arcsec.

| | FASTT | | | | Bordeaux | | | |
| | Oberon | | Uranus | | Oberon | | Uranus | |
Year	R.A.	DEC	R.A.	DEC	R.A.	DEC	R.A.	DEC
1997					0.16	0.09	0.05	0.08
1998	0.15	0.14	0.12	0.15	0.11	0.18	0.09	0.09
2001	0.15	0.13	0.13	0.14	0.26	0.23	0.06	0.11
2002	0.13	0.11	0.16	0.11	0.11	0.31	0.08	0.09
2004	0.11	0.10	0.11	0.14	0.13	0.24	0.08	0.14
2005					0.22	0.28	0.08	0.25

Table 3. Characteristics of the main astrometric star catalogues.

Year	Name	Number of stars	Limit magnitude	Accuracy in mas	Accuracy in proper motion	Origin
1997	Hipparcos	120 000	12.4	< 0.78	< 0.88 mas/yr	obs. from space
2000	Tycho 2	2 500 000	16	25 to 100	< 2.5 mas/yr	from Tycho and 143 sources
1998	USNO A2	526 280 881				
2005	GSC II	1 billion	19.5	360		Schmidt plates
2003	USNO B1	1 billion	21	200		Schmidt plates
2004	UCAC2	48 000 000	7.5 to 16	20 to 70	1 to 7 mas/yr	CCD imaging
2004	Bright stars	430 000	< 7.5			Hipparcos + Tycho2
2005	Nomad	1 billion				compilation of best entries
2006	Bordeaux	2 970 674	15.4	50 to 70	1.5 to 6 mas/yr	$+11deg >$DEC$> +18deg$
2003	2MASS	470 000 000	16	60 to 100		Infrared K
2015	Gaia	1 billion	20	< 0.01		obs. from space

long-focus photographic plates because of the small number of stars in the catalogues at that time. Consequently, only relative astrometric positions were provided through reductions based upon the trail-scale method (Pascu 1996).

At the end of the XXth century, CCD technology appeared, allowing to capture images electronically which then are analyzed by computers. In fact, the early CCD chips were very small compared to photographic plates and, at the beginning, only relative astrometry was possible. However, the size of the CCD chips has increased along with the number of reference stars in catalogues, so that a sufficient number of astrometric stars are now available even in a small field. Table 3 provides a list of the main star catalogues and their characteristics, used in astrometric reductions.

5.1. Direct CCD imaging

CCD imaging provides, most of the time, small 3 to 12 arcmin fields. Thanks to the now available star catalogues, it is possible to find enough astrometric stars for the reduction (cf. Table 3). The accuracy of the measurement depends on several criteria:

- sampling of each image in pixels (the FWHM should be larger than 2.5, i.e. seeing and the pixel size must be adequate);
- signal/noise ratio must be high enough for centroiding of the image;
- atmospheric absorption must be taken into account for moving objects observed at

Table 4. Residuals and rms (in mas) from a re-reduced photographic plate made in 1994.

Date in JD	Io (O-C) in R.A.	(O-C) in DEC	Europa (O-C) in R.A.	(O-C) in DEC
2449521.577894	-41	-22	-43	+15
2449521.578356	-59	-24	-6	-23
2449521.578877	+15	-4	-6	-83
2449521.579456	-46	+44	+44	+22
mean (O-C)	-33	-2	-3	-2
rms (O-C)	43	28	31	45

small elevations above the horizon, far from the meridian transit with long exposures. The photocenter moves towards the zenith, contrary to the images of fixed stars (this effect is very different from refraction).

5.2. *Photographic plates*

Nowadays, photographic plates are no more in use in spite of the possibility of a very large field which are now made using mosaics of CCD chips. The low sensitivity of photographic plates has led to their abandonment for astronomical purpose. However, it appears to be important to scan the old photographic plates and to reduce them with the new star catalogues. This is equivalent to observing in the past with the benefit of getting "new" observations at the time when it is not possible to make them anymore. Several attempts were made showing the importance of this method (Pascu *et al.* 2005). Pluto, which has not been yet observed over a complete orbit, has a motion not very well understood. The reductions of old plates should add new data useful for such a purpose (see Table 4).

5.3. *Small field astrometry*

Some solar system objects, such as small satellites very close to their host planet, are difficult to image because of the brightness of a planet. Several techniques may be used in order to get measurable images. The goal is to make the image of a planet faint enough to minimise a halo of light overlapping with the target object. Two methods allow to get such astrometric images:

- adaptive optics: in fact, the brightness of a planet per square arcsec is similar to the one of the satellites and the large contrast between the planet and its satellite comes from the seeing effects. The light of a satellite is spread out over the detector while the light of a planet is concentrated on the same pixels. To solve this problem, it is useful either to observe in a site with a very good seeing or to use adaptive optics.

- infrared observations: we note that the giant planets are not that bright at specific infrared wavelengths where the light from the Sun is absorbed by the atmosphere of planet. Observations in the K-band allows us to get measurable images very close to the planet. Figure 1 shows an image of the Uranian satellites in the K-band: from left to right: Titania, Ariel, Miranda, Umbriel, Uranus, Oberon. The planet, usually very bright compared to its satellites, here is quite dark.

Figure 1. The system of Uranus in the K-band (c) ESO-NTT.

Table 5. Rms of the residuals of pseudo-observations of Saturn in arcsec.

Year	via Titan R.A.	DEC	via Hyperion R.A.	DEC
1999	0.05	0.10	0.17	0.25
2001	0.04	0.10	0.22	0.34
2003	0.05	0.09	0.26	0.19
2004	0.03	0.12	0.25	0.36
2006	0.08	0.08	0.17	0.28

5.4. *Close approaches of stars*

Another technique of observation possible even with a small field is the observations at close approach of a solar system object to a star, the position of which is known. Even without any other stars in the field, it is possible to use the motion of an object to calibrate the field, making a series of short exposures during the close approach. Even if the position of a solar system object is not known very well, its angular velocity can be predicted to a high accuracy. Relative position with respect to a star can be deduced with a high precision (Souchay *et al.* 2007).

5.5. *Pseudo-observations of planets*

Pseudo-observations of a planet consists in observing a satellite of the planet and following its positions through the ephemerides of the satellite. This method is used in some specific cases:

- the center of mass of a planet is not easy to observe because of a thick atmosphere (the case of Jupiter) or of the presence of a bright ring (Saturn);
- the ephemerides of satellites are more precise than the observations of planet ifself (the case of Galilean satellites and some Saturn's main satellites).

This method is interesting for Jupiter and Saturn if we have sufficient observations of absolute positions (right ascension and declination) of satellites. Table 5 shows the rms of the residuals of pseudo-observations of Saturn through Titan and Hyperion from transit circle observations. It is clear that the accuracy of this method depends on the satellite used for this purpose. It is known that Hyperion is more difficult to observe than Titan with a transit circle and that Hyperion has an ephemeris of less accuracy than Titan.

6. Photometric astrometry of mutual phenomena

During the years, astrometry consisted in the measurement of angles on the celestial sphere. Radar observations recently introduced the direct measurement of a distance between the observer and the object. However, other ground-based observations are able to provide distances between solar system objects: the occultations and eclipses.

The first phenomena observed for an astrometric purpose were the eclipses of a Galilean satellite by Jupiter. Observing the disappearance of a satellite in the shadow of Jupiter corresponds to a specific geometric configuration of the system Jupiter-satellite. Numerous observations are made when the configuration is changing with time, allowing to build a model of the motion of the satellites. The first dynamical model of the Galilean satellites was made thanks to observations of eclipses. Unfortunately, the astrometric accuracy (cf Table 1) may not be better than 100 mas because of the Jovian atmosphere. In contrast, the satellites have no atmosphere and their mutual occultations and eclipses can be accurately observed. The occurrence of such events corresponds to the equinox on the planet since all main satellites belong to the equatorial plane of a planet. The predictions of these events started in the 1970's when computers made possible such calculations. Predictions are now regularly made using the best dynamical models since the events are very sensitive to calculations (Arlot, 2002). Coordinated campaigns of observations allow to get numerous observations used for planetologic purpose (study of the surfaces) and for astrometry. The photometric observations provide often light curves which appear to be not symmetrical. This asymmetry is a source of information and allows to increase the accuracy of the relative positions deduced from the observations through the knowledge of the surface of objects. Note that the measures are in kilometers, more precise than in angle especially for remote objects. (Vasundhara *et al.* 2003, Emelianov and Gilbert 2006, Kaas *et al.* 1999).

Occultations of stars can provide accurate positions of the object relative to a star. However, these occultations are too rare to be extensively used for astrometric purpose. The main goal in this case, is to measure the size of an object.

7. Towards a higher astrometric accuracy

How to improve the accuracy of the Solar system object astrometry? We may ask for more space probes to be put around the Solar system objects but it is clearly a very limited option. Radar observations should be intensified for near-Earth objects since the distance provided by a radar are of a high accuracy, allowing to fit very accurately the dynamical models.

There are also other ways to improve the accuracy of observations: celestial mechanics predicts the position of the center of mass of an object and observations provide the position of a photocenter. An improvement is possible by studying the surface of the objects. First, the phase effect should be taken into account not only in its geometric effect but also in the reflected light, depending on the nature surface material. Several laws of diffusion and reflectance are helpful for this purpose, such as Hapke's laws (Hapke 1993). For some objects such as the main satellites of giant planets, the albedo maps may improve the astrometric accuracy of observations.

Another question is the arrival of Gaia. When the Hipparcos data arrived, a substantial progress was made and the interest in new data was expressed (Fienga *et al.* 1997). But would Gaia make useless all the old observations? A paper by Desmars (Desmars 2005) has shown that even very accurate observations made during a short period of time will not replace a set of numerous but less accurate observations made over a long time span period of time. Gaia will provide 50 observations on a period of 5 years for each solar system object with an accuracy of 0.1 to 1 mas, depending on the object. His conclusion was that Gaia will not bring more accuracy for objects extensively observed more than a century with a fairly good astrometric accuracy.

However, the interest in Gaia for the Solar system objects will be its catalogues of stars which will allow to re-reduce the old observations. We will be able to observe in the past with accuracies not reachable at the time of observations. In chapter 5.2 we have shown that re-reductions of photographic plates is very powerful even with the present-day catalogues. The arrival of Gaia catalogues will substantially increase the benefit of re-reduction of old observations.

8. Conclusion

The astrometry of the solar system may be improved by taking into account the physical characteristics of the objects. The arrival of new star catalogues such as Gaia will allow to re-reduce old observations providing much better data from the past. The observational effort must be continued in order to improve the quality of dynamical models for the motion of solar system objects.

References

Arlot, J. E. 2002, Predictions of the mutual events of the Galilean satellites of Jupiter occurring in 2002-2003 *Astron. & Astrophys.* 383, 719

Desmars, J. 2005, Etude prospective et dynamique des satellites planetaires dans la perspective des observations Gaia, *Rapport de Master 2 Recherche*, June 2005, Paris Observatory.

Dickey, J. O., Bender, P. L.,Faller, J. E.,Newhall, X. X., Ricklefs, R. L., Ries, J. G., Shelus, P. J., Veillet, C., Whipple, A. L., Wiant, J. R., Williams, J. G., & Yoder, C. F. 1994, Lunar laser ranging - A continuing legacy of the Apollo program *Science*, 265, 482

Emelianov, N. & Gilbert, R. 2006, Astrometric results of the observations of mutual occultations and eclipses of the Galilean satellites of Jupiter in 2003, *Astronomy and Astrophysics* 453, 1141.

Fienga, A., Arlot, J. E. & Pascu, D. 1997, Impact of HIPPARCOS Data on Astrometric Reduction of Solar System Bodies in *Proceedings of the ESA Symposium Hipparcos - Venice 97*, 13-16 May, Venice, Italy, ESA SP-402 (July 1997), p. 157-160

Folkner, W. M., McElrath, T. P., & Mannucci, A. J. 2005, Determination of Position of Jupiter from Very-Long Baseline Interferometry *Astron. J.*, 112, 1294

Hapke, B. 1993, *Theory of Reflectance and Emittance Spectroscopy*, Cambridge University Press.

Jacobson, R. A. 2004, The Orbits of the Major Saturnian Satellites and the Gravity Field of Saturn from Spacecraft and Earth-based Observations *Astron. J.*, 128, 492

Kaas, A. A., Aksnes, K., Franklin, F., & Lieske, J. H. 1999, Astrometry from the mutual phenomena of the Galilean satellites in 1990-1992, *Astron. J.* 117, 1933.

Ostro, S. J. 2002, Planetary radar astronomy, in *Encyclopedia of Physical Science and Technology*, vol. 12. Edited by R. A. Meyers, p. 295, Academic Press, New-York.

Pascu, D. 1996, Long-focus CCD astrometry of planetary satellites, in *Dynamics, ephemerides, and astrometry of the solar system: proceedings of the 172nd Symposium of the International Astronomical Union*, held in Paris, France, 38 July, 1995. Edited by Sylvio Ferraz-Mello, Bruno Morando, and Jean-Eudes Arlot, p. 373.

Pascu, D., Arlot, J. E., Lainey, V., & Robert, V. 2005, New observations of the Natural Planetary Satellites through the Natural Satellites USNO Plates Archive, *BAAS* 37, 753.

Rapaport, M., Teixeira, R., Le Campion, J. F., Ducourant, C., Camargo, J. B., & Benevides Soares, P. 2002, *Astronomy and Astrophysics* 383, 1054.

Stone, R. C. 2001, Positions of the outer planets and many of their satellites: FASTT observations taken in 2000-2001 *Astron. J.*, 122, 2723

Souchay, J., Le Poncin-Lafitte, C., & Andrei, A. H. 2007, Close approaches between Jupiter and quasars with possible application to the scheduled Gaia mission,*Astronomy and Astrophysics* 471, 335.

Standish, E. M. 2000, Dynamical reference frame - Current relevance and future prospects. In *Towards models and Constants for Sub-Microarsecond Astrometry, Proc. IAU Coll. 180*, K.J. Johnston, D. D. McCarthy, B. J. Luzum, and G. H. Kaplan, Eds), pp. 120-126. U.S. Naval Observatory, Washington D.C.

Vasundhara, R., Arlot, J. E., Lainey, V., & Thuillot, W. 2003, Astrometry from mutual events of the jovian satellites in 1997, *Astronomy and Astrophysics* 410, 337.

A Giant Step: from Milli- to Micro-arcsecond Astrometry
Proceedings IAU Symposium No. 248, 2007
W. J. Jin, I. Platais & M. A. C. Perryman, eds.
© 2008 International Astronomical Union
doi:10.1017/S1743921308018656

A new all-sky catalog of stars with large proper motions

S. Lépine[1], M. M. Shara[1], R.-M. Rich[2], A. Wittenberg[1], M. Halmo[1], B. Bongiorno[1]

[1] Dept. of Astrophysics, American Museum of Natural History, Central Park West at 79th street, New York, NY 10024, USA
email: lepine@amnh.org, mshara@amnh.org

[2] Dept. of Astrophysics, University of California at Los Angeles rmr@astro.ucla.edu

Abstract. A new all-sky catalog of stars with proper motions $\mu > 0.15''$ yr^{-1} is presented. The catalog is largely a product of the SUPERBLINK survey, a data-mining initiative in which the entire Digitized Sky Surveys are searched for moving stellar sources. Findings from earlier proper motions surveys are also incorporated. The new all-sky catalog supersedes the great historic proper motion catalogs assembled by W. J. Luyten (LHS, NLTT), and provides a virtually complete $> 98\%$ census of high proper motion stars down to magnitude $R = 19$.

Keywords. catalogs, solar neighborhood, stars: kinematics, stars: low-mass, brown dwarfs, stars: subdwarfs, stars: white dwarfs, Galaxy: stellar content

1. The SUPERBLINK survey

We have been conducting an all-sky survey for stars with large proper motions using data from the Digitized Sky Surveys (DSS). The scanned images in the DSS cover the entire sky in multiple bands and at various epochs, and the temporal baseline between the earliest and latest epoch is between 15 and 45 years for most areas on the sky.

The large motion displayed by high proper motion stars between the two epochs is detected directly from the scans by means of an image subtraction algorithm, described in detail in Lépine *et al.*(2002). Two-epochs finder charts are generated, which can be blinked on the computer screen. All objects detected in the survey are thus verified by eye, and spurious detections are excluded.

At the bright end, stars tend to become saturated on the DSS images, and are no longer properly detected by the code. The TYCHO-2 catalog is used to complete the census at the bright end. Also, all stars detected by SUPERBLINK are searched for a counterpart in the TYCHO-2. The positional and proper motion information from the TYCHO-2 catalog is used for all matching counterparts. The two-epoch charts are however examined to verify consistency; in some cases, it is found that the proper motion from TYCHO-2 must be in error, and the SUPERBLINK proper motion is used instead.

Counterparts in the the 2MASS All-Sky Catalog of Point Sources (Cutri *et al.*, 2003) are also identified for all SUPERBLINK detections. At the faint end, the positions are those extrapolated from the 2MASS catalog, and thus realized in the ICRS system and accurate to about 0.1". Counterpart are also identified in the USNO-B1 catalog (Monet *et al.*, 2003), and together with 2MASS, provide optical and infrared magnitudes for almost all the stars.

The systematic comparison with the 2MASS catalog allows us to systematically identify all common proper motion doubles which are resolved in the 2MASS images. The CCD observations from 2MASS have significantly higher resolution than the photographic images from the DSS. Numerous common proper motion doubles and multiple systems have being identified and their components will be listed as separate entries in the LSPM catalog. Many faint companions of Hipparcos stars are also being identified this way (e.g. Lépine & Bongiorno, 2007).

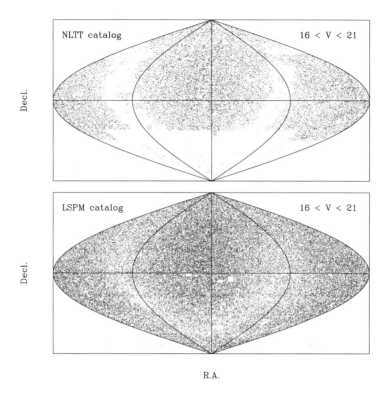

Figure 1. Comparison between the distribution of faint ($V > 16$) stars from the NLTT catalog (top) and the new LSPM catalog (bottom). The new catalog fills in all the gaps of the NLTT, particularly in the low Galactic latitude fields, and provides the most complete all-sky census of high proper motion stars to date.

2. Replacing the NLTT catalog

The NLTT catalog (Luyten, 1979) was notoriously incomplete in two main regions: the sky south of Decl.$=-30°$, and areas of high stellar density along the plane of the Milky Way. The incompleteness was most severe for stars fainter than magnitude V=16 (see Fig.1).

The northern part of the SUPERBLINK survey was completed first, and the results have already been published in Lépine & Shara(2005) as the LSPM-north catalog. The full LSPM catalog now fills in most of the remaining gaps in the south, and at last provides a true, all-sky census of faint stars with large proper motions (Fig.1). While there remains some level of incompleteness at low Galactic latitudes, especially toward the Galactic center, most of the variations in surface density observed in Fig.1 are due to selection effects from the high proper motion cutoff ($\mu > 0.15''$ yr^{-1}) of the survey. A combination of the Sun's motion through the local standard of rest and the asymmetric drift of the Galactic thick disk and halo stars results in more stars having large transverse motions at high Galactic latitudes, hence the larger density of stars detected there.

Compared with the 58,845 stars with proper motions $\mu > 0.18''$ yr^{-1} listed in the NLTT catalog, the LSPM catalog will list over 122,000 stars with proper motions $\mu > 0.18''$ yr^{-1}. With the increased sky coverage and completeness, the LSPM catalog makes the Luyten catalog obsolete, and from now on should be used as a replacement to the NLTT. For convenience, all NLTT stars will be identified in the LSPM both by their LHS designation and NLTT catalog number.

3. Stellar content and kinematics

A reduced proper motion diagram shows the stars in the LSPM to be composed of three main classes. Low-mass K and M red dwarfs from the disk population dominate, but a significant fraction are low-mass subdwarfs from the halo (sdK, sdM), and the catalog also contains thousands of white dwarfs. While we currently lack parallax distances for most of the stars in the LSPM catalog, photometric distances can be calculated for specific classes of objects. The red dwarfs, in particular, have a reasonably well calibrated $[M_v, V - J]$ color-magnitude relationship (Lépine, 2005).

With photometric distances and proper motions, it is possible to investigate the local kinematics of the dwarfs in the vicinity of the Sun ($d < 100$pc). By selecting stars in specific parts of the sky, one can obtain velocity-space projections in the UV, UW, and VW plane (Figure 2). Because of the high proper motion cutoff of the LSPM catalog ($\mu > 0.15''$ yr^{-1}), stars with low projected velocities are not represented in the census, which leaves a low-velocity "hole" in the maps of projected velocities. The hole increases for stars at larger distances. Despite this artifact, one can see that the velocity space projections of the nearby red dwarfs are not isotropic and show considerable structure. A comparison with the velocity space distribution of Hipparcos stars calculated by Nordström *et al.*(2004), shows very good agreement with our data. This shows how future astrometry, providing accurate parallaxes for all the LSPM stars, may have a major impact in uncovering fine structure in the kinematics of stars in the Solar vicinity.

4. Conclusions

The LSPM catalog now has all-sky coverage. The LSPM-south catalog will complement the already released LSPM-north, and yield a highly complete catalog of stars with proper motion $\mu > 0.15''$ yr^{-1}. The catalog is estimated to be $> 98\%$ complete for all H-burning stars and white dwarfs with proper motions in the range above, covering virtually all objects down to visual magnitude 19. The catalog is realized at the bright end by the Tycho-2 catalog, down to magnitude $\approx 10 - 12$. At the faint end, which encompass the vast majority of the stars, the proper motions are obtained from the SUPERBLINK software, while the positions are determined by the counterparts in the 2MASS catalog. Overall, the positional accuracy of the catalog is thus better than $0.12''$, while proper motions at the faint end have typical errors ≈ 10 mas yr^{-1} north of Decl.$=-30°$, and ≈ 20 mas yr^{-1} south of this. All double stars which are resolved in the 2MASS survey have been identified and are listed individually.

The SUPERBLINK survey is now being expanded to lower proper motion regimes, and future releases will expand the catalog to proper motions $\mu > 40$ mas yr^{-1}.

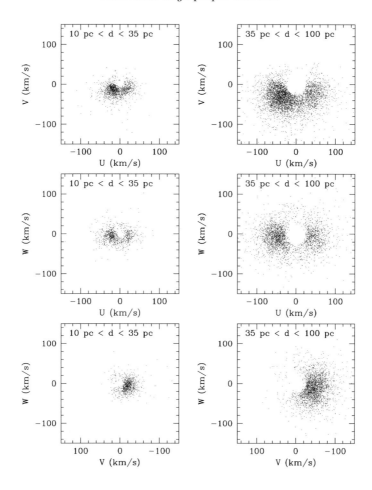

Figure 2. Projected motions in the UVW of red dwarf stars in the Solar neighborhood, based on LSPM catalog proper motions and photometric distances.The UW projection is from stars found at low Galactic latitude in the direction of the apex and antapex of the Sun's orbital motion around the Galaxy. The VW projection is obtained from stars in the direction of the Galactic center and anti-center.

References

Cutri, R. M., *et al.* 2003, The 2MASS All-Sky Catalog of Point Sources University of Massachusetts and Infrared Processing and Analysis Center (IPAC/California Institute of Technology.

Hog E., *et al.* 2000, *A&A*, 355, 27

Lépine, S., Shara, M. M., & Rich R.M.R. 2003, *AJ*, 124, 1190

Lépine, S. & Shara, M. M. 2005, *AJ*, 129, 1483

Lépine, S. 2005, *AJ*, 130, 1247

Lépine, S. & Bongiorno, B. 2007, *AJ*, 133, 889

Luyten W. J. 1979, New Luyten Catalogue of stars with proper motions larger than two tenths of an arcsecond (NLTT), University of Minnesota, Minneapolis

Monet, D. G., *et al.* 2003, *AJ*, 125, 984

Nordström, B. *et al.* 2004, *A$A*, 418, 989

A Giant Step: from Milli- to Micro-arcsecond Astrometry
Proceedings IAU Symposium No. 248, 2007
W. J. Jin, I. Platais & M. A. C. Perryman, eds.
© 2008 International Astronomical Union
doi:10.1017/S1743921308018668

Systematic biases and uncertainties of Hipparcos parallax

X. P. Pan

Jet Propulsion Laboratory, California Institute of Technology, Pasadena,CA 91007, USA
email: Xiaopei.Pan@jpl.nasa.gov

Abstract. Ground-based optical/IR interferometers have provided strong support to the space-based astrometric mission Hipparcos ever since the Hipparcos instrument was in operation in 1989. Interferometric observations also produced critical corrections of orbital motion to many targets, including radio stars, which link the Hipparcos system to the International Celestial Reference Frame (ICRF). In particular, orbital parallax from interferometers confirmed the 10% bias of the Pleiades distance from Hipparcos, and thus avoids revision of classical astronomy. Significant offsets and errors of Hipparcos parallax introduced by binary jitters are demonstrated in this work. By comparing the Hipparcos results with long baseline interferometry and other techniques including spectroscopy, multi-color photometry, Main-Sequence fitting, light curve measurements, Lunar occultation, Fine Guidance Sensor, etc., systematic biases and uncertainties of Hipparcos parallaxes are investigated and analyzed. We have established good models for major error sources of Hipparcos parallax, such as zonal bias, binary jitters, and luminosity-dependent errors. The lessons learned from the systematic biases of Hipparcos parallax are valuable to future space missions like SIM and Gaia.

Keywords. astrometry, techniques: interferometric, methods: data analysis

1. Introduction

Classical astrometry has limited accuracy and precision mainly for two reasons, i.e. the effects of diffraction and turbulence of the atmosphere. The Hipparcos spacecraft was launched into space to avoid atmospheric perturbations and achieve a milli-arcsecond (mas) precision. Ground-based optical/IR interferometers use long baselines ($5 * 10^7 \lambda$) to solve the difficulties of diffraction limits, and obtain unprecedented angular resolution of milli-arcsec. In addition, interferometers can freeze the atmosphere within seeing limits to obtain sub-mas measurement precision. Since 1987 the Mark III interferometer on Mt. Wilson has routinely provided scientific results in astrometry with high resolution and high accuracy (Shao, M. *et al.* 1988). For the first time in astronomy, spectroscopic binaries of milli-arcsec separation with large magnitude differences have been resolved with sub-mas precision (Pan *et al.* 1990). Combined with spectroscopic observations, the Mark III Interferometer provides orbital parallax with milliarcsec precision. A good example is the distance to the Hyades cluster, which was determined accurately via a binary star θ^2 Tau as 22.9 ± 0.9 mas in 1992 (Pan, Shao & Colavita 1992a), 22.0 ± 1.2 mas in 1995 (Tomkin, Pan & McCarthy 1995) and 21.0 ± 0.8 mas in 1997 (Torres, Stefanik & Latham 1997). The errors in the above distance are limited by radial velocity measurements, not by interferometry. In 1997 Hyades distance was confirmed by Hipparcos measurements as 21.89 ± 0.83 mas for that star (Perryman 1997) and 21.58 ± 0.13 mas for the Hyades cluster (Brown *et al.* 1997). In particular the radio star, Algol (HR 936), which is the prototype of eclipsing binary and a triple system, had its outer orbit determined (Pan, Shao & Colavita 1993). The orbit of Algol from the Mark III interferometer provided accurate corrections to link the Hipparcos reference frame to the International Celestial Reference

Frame(ICRF). Besides, the Mark III interferometer determined wide-angle star positions with ten mas precision in 1989 (Shao, M., Colavita, M. M, *et al.* 1990). Those wide-angle star positions provided critical references to confirm the high precision capabilities of Hipparcos preliminary measurements when the Hipparcos mission was troubled by its undesired orbit in 1989. The internal working note by the Hipparcos team admired valuable and speedy contributions from the Mark III interferometer (Perryman *et al.* 1990). Since the Hipparcos results are deeply troubled by binary stars, many accurate orbits of binary stars, which can be resolved only by interferometers, have been used in the Hipparcos catalogue in order to reduce the effects of binarity jitter in Hipparcos parallax (Pan *et al.* 1992b). Ever since the Hipparcos catalogue was published in 1997, the arguments on systematic errors in the Pleiades distance from Hipparcos have never stopped. Using ten years of observations from both the Mark III and the PTI interferometers the brightest binary star in the Pleiades cluster, Atlas, has its orbit determined without a help of the spectroscopic means, for the first time since development of interferometry (Pan, Shao & Kulkarni 2004). The orbital distance of the Pleiades, which has the big advantages of being without the influence of interstellar extinction, confirmed the 10% systematic bias of Hipparcos data for the Pleiades.

2. Comparisons between interferometry and Hipparcos

Binary stars not only play fundamental roles in astronomy and astrophysics, and also are extremely useful to check astronomical instrument performance. Binary stars have different separation, different color, different magnitude in 4-dimensional observations, and can be used to study resolution, accuracy and precision conveniently. For example, the Hipparcos catalogue used the interferometric orbit of α And (Pan *et al.* 1992b) and obtained its photocentric semi-major axis (a_0") to be 6.47 ± 1.16 mas (Perryman 1997). Recently a group of astronomers use Hipparcos and spectroscopic results to get the latest a_0" of α And equal to 7.26 ± 0.38 mas (Jancart, Jorissen, Babusiaux & Pourbaix). Those results, however, troubled them, even though the precision of 0.38 mas is much better than before. The problem is that value of a_0" produces $\Delta m = 3.9$ mag, which is inconsistent with the Mark III's measurements. The paper said that "The origin of this discrepancy is unknown". If the error analysis has not been understood properly, the small error bars cannot represent accurate measurements and precise analysis. Here we compare interferometric performance with Hipparcos measurements by calculating the same photocentric semi-major axis from two different techniques as shown in Table 1. It is obvious that interferometric precision (0.1 mas) is more than an order of magnitude better than the Hipparcos($\sigma = 1 - 5$ mas). Some authors combined the Hipparcos results with spectroscopy, and made the offset of α And much worse than before because of limited precision and number of visits for that target. The average of systematic offsets between Hipparcos and interferometry is about 1.3 mas, which is consistent with general error estimate of Hipparcos. We notice that many binaries, such as θ^2 Tau with a_0" = 4 mas, cannot be resolved by Hipparcos. Also, the minimum resolution of Hipparcos measurements is about 2 mas. It is impossible for Hipparcos to detect extra-solar planets like hot Jupiters (signature of hot Jupiters is about 0.1 mas or less).

3. Error models of Hipparcos parallax

Many papers have concluded that errors of Hipparcos parallax have random Gaussian distribution. By using an average of 100 stars parallax, the mean parallax of that cluster can have an error of $1/\sqrt{100} = 0.1$mas assuming 1 mas error for each star. That technique

Table 1. Comparison of Photocentric Orbits between Interferometry and Hipparcos

star (HIP)	a" (mas)	mass Ratio	Δmag (mag)	$cal - a_0$" (mas)	$Hip - a_0$" (mas)	Offsets (mas)
α And	24.15 ±0.13	0.33	2.63	4.66	6.47 ±1.16	1.81
					7.26 ±0.38	2.60
β Ari	36.10 ±0.30	0.64	1.99	20.00	11.32 ±1.33	8.68
					12.50 ±1.20	7.50
Atlas	12.94 ±0.11	0.42	1.68	3.15	4.23 ±0.97	1.08
HIP 91009	4.40 ±0.05	0.47	2.50	1.66	3.11 ±0.97	1.45
HIP 109176	10.33 ±0.10	0.38	1.70	2.16	4.07 ±0.27	1.91
HIP 101382	15.37 ±0.03	0.44	2.40	5.25	5.24 ±0.66	0.01
HIP 23453	16.20 ±0.10	0.45	2.20	5.48	4.18 ±0.90	1.30
HIP 24608	56.47 ±0.05	0.49	0.15	1.25	2.16 ±0.60	0.91
HIP 96683	23.70 ±0.04	0.49	0.30	1.39	1.60 ±0.61	0.21
HIP 28360	3.30 ±0.10	0.49	0.20	0.12	-0.54 ±0.76	0.66
HIP 99473	3.20 ±0.10	0.45	1.53	0.80	1.25 +1.04	0.45
HIP 2912	6.69 ±0.05	0.45	0.40	0.29	1.84 ±0.98	1.55
HIP 14576	94.61 ±0.22	0.27	2.92	19.88	19.00 ±0.57	0.88
θ^2 Tau	18.80 ±0.06	0.47	1.13	3.84		3.84
HIP 10280	2.10 ±0.90	0.46	4.00	0.91		0.91
HIP 69226	3.39 ±0.05	0.50	0.60	0.44		0.44
HIP 3810	6.52 ±0.06	0.49	0.11	0.09		0.09
HIP 47508	4.46 ±0.01	0.47	0.91	0.74		0.74
HIP 7564	4.94 ±0.02	0.56	2.06	2.13		2.13
HIP 69974	19.76 ±0.08	0.48	0.63	2.31		2.31
HIP 19762	2.78 ±0.06	0.46	0.56	0.25		0.25

brought many contradictory results for many clusters, and RR Lyr stars. In fact many error sources, such as instrument errors, binary jitter, zonal errors produced significant systematic errors in Hipparcos parallax(Pan & Makarov 2007). A typical example is the Hipparcos parallax of 23.41 ± 2.44 mas for HIP 32104 after correction of its orbital motion (a_0" $= 8.83 \pm 1.85$mas) (Perryman 1997). We used accurate orbital parameters and obtained a new parallax as 20.56 ± 0.71 mas, which demonstrates a binary jitter of 2.3 mas ($> 3\sigma$). We plot Hipparcos parallax error distributions as shown in Figure 1 for all 120000 stars, where the errors can be described as

$$\sigma_{Hip}^2 = \sigma_{sys}^2/f_N + \tau(mag)/f_N \qquad (3.1)$$

In the above model σ_{sys} is the systematic error; $\tau(mag)$ is the brightness dependent error; f_N is the number of visits. It is interesting to notice that the systematic error of 0.77 mas is the dominant error source for the majority of stars (f_N=25). When the stars are fainter than 8.7 mag, the brightness dependent errors are equal and larger than the systematic error. The obvious gap between 0.5 and 0.75 mas for bright stars (< 7 mag) is strong evidence for the effects of number of visits. Since it is impossible to have less than 10 visits for a meaningful fit, many stars above the green line ("+") with errors of $\tilde{1}.5$ mas are meaningless.

4. Conclusions

Long baseline optical/IR interferometers provide orbit determinations with precisions at a sub-mas level, which is critical to correct for binary jitter in Hipparcos observations. It has been demonstrated that the threshold of orbit detectability for Hipparcos is comparable to a single measurement precision of a few mas only. In addition, more

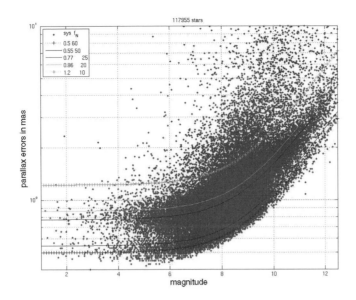

Figure 1. Error distributions of Hipparcos parallaxes

than 40% of binaries cannot be detected by Hipparcos. It is impossible for Hipparcos to detect the signature of exoplanets. A new model of Hipparcos parallax demonstrates a term of systematic errors at 0.8 mas. That model provides useful tools for identifying objects with large biases. The most important lessons we learned from the Hipparcos mission is that we have to do deeper investigations on systematic biases by using various techniques. Future space missions like Gaia and SIM will face the same challenging tasks but at the level of μas. It is important to use both Gaia and SIM to check and support each other.

References

Brown, A. G. A., Perryman, M. A. C., Kovalevsky, J., Robichon, N., Turon, C., & Mermilliod, J.-C. 1997, *Proceedings of the Hipparcos Venice ;97 symposium, ESA SP-402* 000, 681

Jancart, S., Jorissen, A., Babusiaux, C., & Pourbaix, D. 2002, *A&A*, 442, 365

Pan, Xiaopei, Shao, M. & Colavita, M. M., Mozurkewich, D., Simon, R. S. and Johnston, K. J. 1990, *ApJ* 356, 641

Pan, Xiaopei, Shao, M., & Colavita, M. M. 1992, in: H. McAlister & W. Hartkopf (eds.), *IAU Coll. 135*, A. S. P. Conf. ser. Vol. 32, p. 502

Pan, Xiaopei, Shao, M., Colovita, Armstrong, J. T., M. M., Mozurkewich, D., Vivekanand, M, Denison, C. S., Simon, R. S., & Johnston, K. J. 1992, *ApJ* 384, 624

Pan, Xiaopei, Shao, M., & Colavita, M. M. 1993, *ApJ (Letters)* 413, L129

Pan, Xiaopei, Shao, M., & Kulkarni, S. R. 2004, *Nature* 427, 326

Pan, Xiaopei & Makarov, V. 2007, *BAAS* 39, No.1, 12

Perryman, M. A. C. 1990, *Hipparcos Internal Working Note: Hipparcos/MK III report, 12/06*

Perryman, M. A. C. 1997, *The Hipparcos and Tycho Catologues, ESA SP-1200*

Shao, M., Colavita, M. M., Hines, B. E., Staelin, D. H., Hutter, D. J., Johnston, K. J., Mozurkewich, D., Simon, R. S., & Hershey, J. L., etc. 1988, *A&A* 193, 357

Shao, M., Colavita, M. M, Hines, B. E., Hershey, J. L., Hughes, J. A., Hutter, D. J., Kaplan, G. H., Johnston, K. J., Mozurkewich, D., Simon, R. S., & Pan, X. P. 1990, *AJ* 100, 1701

Tomkin, J., Pan, Xiaopei and McCarthy, K. 1995, *AJ* 109, 780

Torres, G., Stefanik, R. & Latham, D. W. 1997, *ApJ*, 485, 167

A Giant Step: from Milli- to Micro-arcsecond Astrometry
Proceedings IAU Symposium No. 248, 2007
W. J. Jin, I. Platais & M. A. C. Perryman, eds.

Towards a better understanding of Hipparcos

F. van Leeuwen

Institute of Astronomy, University of Cambridge, Madingley Road, Cambridge, CB3 0HA,
United Kingdom
email: fvl@ast.cam.ac.uk

Abstract. The formal errors on the astrometric data in the 1997 publication of the Hipparcos catalogue are, for stars brighter than about magnitude 9, largely the result of inaccuracies in the description of the along-scan attitude of the satellite. A detailed study of the dynamics of the Hipparcos satellite has led to a much improved understanding and modelling of the satellite attitude, taking into account peculiarities in the rotation of the payload (scan-phase jumps) as well as the detections of small hits. A new reduction of the Hipparcos data was initiated, in which the attitude modelling is a direct description of the dynamics of the satellite. In this so-called fully-dynamic attitude modelling the underlying torques acting on the satellite are reconstructed, and rates and error angles are obtained through integrations. Both the hits and the scan-phase jumps could be taken into account in the context of this model. The new model, including the provisions for discontinuities, led to a factor five reduction in the attitude noise. A new reduction, based on a global iterative solution like is also planned for Gaia, was started in 2004, and completed, after some 15 iterations, in 2007. In the process of this reduction, also the sensitivity of the solution to optimal connectivity between the data in the two fields of view was exposed and taken care of, as well as a couple of small-scale calibrations. The latter had not been possible to solve for in the original reductions. In the catalogue that resulted from this new reduction, the errors on the astrometric data for all but stars brighter than magnitude 4, are dominated by photon noise. Error-correlation levels in the underlying abscissa data are down by more than an order of magnitude, and play no longer any significant role. This feature very much simplifies the analysis of, for example, wide-binary stars and open cluster data.

Keywords. space vehicles, instrumentation: miscellaneous, surveys, astrometry

1. Introduction

The presence of a few unexpected results in the Hipparcos catalogue (ESA 1997) led to speculation concerning the overall reliability of the astrometric data contained in it, mainly based on an apparent discrepancy for the Pleiades cluster (see for example Pinsonneault *et al.* 1998; Pan *et al.* 2004; Soderblom *et al.* 2005). It has been verified, however, that for the individual measurements the formal errors as given in the catalogue are correct (Arenou *et al.* 1995; Lindegren 1995). The uncertainty in the published data is primarily caused by correlations in the underlying measurements, namely the abscissa residuals. These are derived from abscissae measurements, which were obtained from the modulated signal caused by a stellar image passing over a very regular grid in the focal plane of the telescope. This signal provided a transit time relative to the on-board clock. The along-scan attitude reconstruction provided the link between observation time and direction on the sky. This process used the same abscissa residuals that also formed input to the improvements of the astrometric parameters. The two processes use these data on very different time resolution, which allows them to be separated as part of an iterative process. This iterative process, referred to as the astrometric global iterative solution, is also planned for the Gaia data reductions (O'Mullane *et al.* 2006). These processes use as input the abscissa residuals, the differences between the measured and predicted abscissae as based on the results from the previous iteration step. Errors on the abscissa residuals become correlated for near-simultaneous measurements when there are modelling errors in the attitude reconstruction that are at least at the photon-noise level of the individual observations. Thus, bright stars are much more sensitive to such modelling errors than faint stars.

2. Attitude modelling

The attitude modelling for the 1997 catalogue was based on B-splines applied to the abscissa residuals as a function of time. The attitude solution was done as part of a process referred to as the great-circle reduction (van der Marel & Petersen 1992), in which the data collected over a period of about 6 to 8 hours was projected on a reference great circle, and the local corrections for the abscissa residuals, per object, and the along-scan attitude were solved for simultaneously. One of the consequences of this procedure is that local problems in the attitude reconstruction tend to replicate with the basic angle between the two fields of view. The reason for this is that such errors affect the estimate of the abscissa residuals per object, which are in this solution at the same time incorporated in the attitude fitting at other parts of the great circle, where in turn they affect the data of yet other objects. The correlations between errors thus created are most visible for the brighter stars, and can, at separations of a few tenths of a degree, quite easily reach levels of around 0.3 and higher.

In the new reduction of the Hipparcos astrometric data (van Leeuwen 2005; van Leeuwen & Fantino 2005; van Leeuwen 2007a,b) the attitude is modelled separately from the abscissa corrections, as a fully dynamic model (van Leeuwen & Fantino 2003). The component modelled is the underlying torque, from which the inertial rates are obtained through integrating over the Euler equation for a free-floating rigid body in space, and the error angles through integrating over the rates. The observables, however, are not the accelerations but instead the abscissa residuals. The fitting of these residuals therefore is done as an iteration. The residuals are fitted with a fifth-order spline function, of which the second derivative provides an update to the torques model in the form of a third-order spline function. It also provides corrections to starting points for the rate and error-angle integrations. These are then used to re-calculate the abscissa residuals, and the process is repeated. The dynamical model provides a further, physical, constraint to the attitude.

3. Peculiarities in the satellite dynamics

Two kinds of peculiarities have been taken into account in the attitude modelling for the new reduction, hits by micro-meteorites (or space debris), and non-rigidity events. The first manifest themselves as discontinuities in the rotation rates, the second as discontinuities in the rotation phase. Both were understood to exist in the original reductions, but the possibilities to detect these events from the data were very limited. The frequency with which the scan-phase discontinuities occur was in addition much underestimated (see section 11.4 in Volume 2 of ESA 1997). In the preparations for the new reduction, some 80 hits and 1600 scan-phase jumps were identified. Once identified, they were easy to incorporate in the attitude modelling without causing significant data loss.

Scan-phase jumps happen due to small, but discrete, thermal movements of one of the three solar panels of the satellite. They can range up to 120 mas, while below 20 mas they become difficult to distinguish with certainty. To make the payload rotate by that amount, a solar panel needs to shift by a few microns. These shifts occurred as a result of thermal changes, mostly due to eclipses, but also due to the perigee-passage conditions. From an examination of the latter it is clear that one panel in particular was causing the problems, as most of these scan-phase jumps are concentrated around a specific rotation phase of the satellite. The rotation phase, in the way it is defined here, is equivalent to the direction of the sun as seen from the satellite. Once the jumps and hits had been identified and incorporated in the reductions, the noise-level of the along-scan attitude reconstruction dropped dramatically, by a factor four to five. Such a reduction significantly reduces he noise on the astrometric parameters for all stars brighter than about magnitude 8 to 9.

4. Data correlations

The reduction in attitude noise also shows clearly in the abscissae error-correlation levels, which are reduced by more than an order of magnitude compared to the 1997 reduction, down to a level that is insignificant. This is important for the recovery of open cluster parallaxes and proper motions, as it simplifies the process and removes a major uncertainty (the actual abscissa-error-correlation levels for stars of different magnitudes) from the analysis.

To assess the accumulation of correlated errors, the scan-coincidence fraction needs to be determined. This shows what fraction of scans for a given star is shared with another star. Significant values are only found for stars at small, less than about 2 degrees, separation on the sky. Combining the error-correlation statistics with the coincidence statistics for the new reduction clearly shows that no significant level of correlated errors should be expected for the astrometric parameters of neighbouring stars. This was not the case for the 1997 catalogue. In fact, an examination of, for example, the parallax differences between the new and the old reduction shows a very significant correlation level for stars with separations less than one degree on the sky. It also shows, however, that these correlations disappear completely beyond a separation of 3 degrees. That observation is important for the Pleiades parallax determination as based on the old catalogue. Given that the cluster extends over a field of about 9 degrees diameter, the errors on the astrometric parameters of the 54 single cluster members spread over this area could only in the centre of the cluster have been significantly correlated, leaving more than half the number of stars with effectively uncorrelated errors.

5. Connectivity

In the construction of the new catalogue, the importance of fine-tuning the connectivity between the two fields of view became apparent. If left uncorrected, the data weight difference between the two fields of view can be considerable, more than a factor ten in some cases. The highest weights are naturally found there where bright stars are concentrated in relatively small areas on the sky, such as open cluster cores and OB associations. This issue is important for the along-scan attitude reconstruction, which provides, together with the basic-angle calibration, the linking between the two fields of view. This linking, in turn, is what makes it possible to determine absolute parallaxes from the Hipparcos data. If an area of the sky generally asserts more to much more weight than any other area it is connected to through the satellite scanning, then what may happen is that the astrometric data in that field do not get properly connected to the catalogue, and define their own local reference. This became clear during the first iteration tests for the construction of the new catalogue. Leaving the weights of the two fields uncorrected effectively meant that the new reduction was, for bright stars, mostly modelling the errors of the 1997 reduction. By adjusting the weight ratio between the two FoVs to a maximum value (2.7 was finally chosen), it was possible to observe how gradually in the iterations the "memory" of the old solution was lost. Thus, the new reduction should not have a connectivity problem for the astrometric data of concentrations of bright stars in small fields, such as the core of the Pleiades cluster.

6. Transit data

The astrometric parameters for the Hipparcos catalogue are derived from accumulated transit data. In the 1997 catalogue these transit data represented mean abscissa residuals for all field-of-view transits within the 6 to 8 hour period of observations used in the great circle reduction. All data were projected onto a reference great circle, which represents a kind of average for the instantaneous scan directions of the individual measurements. There are a couple of disadvantages in this representation, which were resolved in the new reduction.

In the new reduction the transit data are accumulated per individual field-of-view (FoV) transit, preserving the instantaneous scan direction of the transit as a basic parameter. An individual FoV transit can accidentally be affected by a superimposed image from the other FoV, but this situation will never repeat itself. Such disturbance is more difficult to eliminate after combining the transits as was done in the 1997 reductions. Individual FoV transits each have a characteristic transit ordinate, which is not exactly constant during the transit, but changes in general by no more than a few arcseconds. This allows the resolution of small-scale grid distortions as a function of transit position. Systematic residuals as a function of ordinate, and at a level of just under 1 mas, were shown to exist. These consist of two main contributions that reflect the way the grid was manufactured. The 168 by 46 scan-field prints forming the grid showed a common non-linearity and per row an average tilt. There were also small formal error drifts as a function of transit ordinate, different for the two FoVs.

Finally, the resolution of the data in the new reduction also allowed for an analysis of residual chromaticity corrections. The new reduction was able to use the epoch-colour information for very red variable stars, provided by Dimitri Pourbaix (Platais *et al.* 2003; Knapp *et al.* 2001, 2003). Thus, in general the data for red to very red stars should be much improved. Accumulation of residuals left after the astrometric-parameter determinations as a function of the colour index $V - I$ exposed the need for residual corrections as well as small, but systematic, adjustments for the formal errors. These adjustments could easily be explained as corrections to the simplified model with which the formal errors had initially been derived. In comparison, the 1997 reductions contained a much simpler model for the chromaticity terms. All these corrections affect mainly the brighter stars, for which they improve the general reliability of both the data and the associated formal errors.

One very noticeable effect of resolving the data in FoV transits concerns double stars with separations above about 10 arcsec. In the examination of the fainter components in these systems, the exact relative position of the primary with respect to the modulating grid is essential. The scan-orientation angle was given by the local orientation of the reference great circle in the 1997 reductions, which is not accurate enough. It is given as the instantaneous scan direction per FoV transit in the new reduction. These fainter components in double systems with separations between 10 and 30 arcsec have generally very poorly determined astrometric parameters in the 1997 catalogue, but have been well resolved in the new reductions, often exposing the double system as a real physical binary

through the similarity in proper motion and parallax with the primary component. Some
200 stars were affected this way.

7. Stochastic solutions

A so-called stochastic solution was implemented for objects that failed all other approaches of fitting astrometric parameters to the abscissa residuals. It was generally
considered that these stochastic solutions would ultimately be resolved as due to orbital
motion of double stars, but investigations by Jancart *et al.* (2005); Goldin & Makarov
(2007) could resolve only few as actual binaries. The 1997 catalogue contained around
1500 such cases, and of these nearly 1000 were resolved in the new reduction with simple
5-parameter astrometric solutions. It appears that for about two thirds of these stochastic
solutions the reason was an accidental accumulation of local attitude-fitting problems.

In the new solution some 500 of the old stochastic solutions were confirmed, and some
700 new ones detected. The latter is simply the consequence of the increased accuracy
for the brighter stars, which also increases the sensitivity to positional disturbances.

8. Variability-induced movers

A variability-induced mover or VIM solution applied to variable stars with a binary
component or a parasitic star positioned very close by, and of compatible brightness.
Two situations can be distinguished, the companion is either the brighter or the fainter
star in the system. In the first case, the duplicity effects become more noticeable when
the variable star reaches maximum light, in the second case, when the variable star
reaches minimum light. Correction coefficients can be introduced in the astrometric solution. These coefficients do not determine the system parameters completely, but only
a combination of those: separation and a reference magnitude difference.

In trying to process these VIM detections for the 1997 catalogue, it was noticed by
Pourbaix *et al.* (2003) that many are in fact spurious. This was confirmed in the new
reduction, which showed a very low confirmation rate of VIM solutions. In particular
detection for low-amplitude variables appeared to be vulnerable to anomalous detections.
This was not entirely unexpected, as the detection limit for these solutions had been set
to a level such that even the slightest indication from the data would be sufficient to
trigger this type of solution. For the new reduction the amplitude of the variations below
which no attempt was made for a VIM solution was set at 0.4 magnitudes. In addition,
the detection criterion was set such that, if the distribution of the statistic is Gaussian,
only about 2 per cent of the detections could be spurious. The number of detections thus
obtained has dropped dramatically, to only 44. Some of these were already known as
binaries.

9. Binary stars

The processing of binary stars in the Hipparcos data is a very labour-intensive job,
which for the new reduction has only been partly completed. Tests were carried out on
the differential parameters (separation, magnitude difference and orientation) for close
binaries, which showed some small-scale systematic differences that appear to be dependent on the analysis method. A small systematic difference was noted for the magnitude-
difference determinations, the origin of which is not yet clear. The astrometric parameters
for binary stars have been determined assuming the differential parameters as presented
in the 1997 catalogue, and some further optimization is therefore still possible. The data

required for this are included in the publication of the new reduction. The electronic publication of the data through AstroGrid will include provisions to update the catalogue with improved solutions.

10. Formal errors and accuracy assessment

The formal errors on the astrometric data in the 1997 catalogue were determined based on the errors assigned to the abscissa residuals, which could contain a significant contribution from the attitude reconstruction (see for details van Leeuwen 2005). Errors were furthermore adjusted in a final merging of the two independent reductions of the data, where also correlation levels between the results were established. Final corrections for error distributions lead to various adjustments of the formal errors (see further Chapter 17 of Volume 3 in ESA 1997).

In the new reduction the formal errors on the measurements were effectively fixed based on the integrated photon counts of the measurements and the relative amplitude of the first harmonic in the signal modulation. The errors are followed through the entire reduction process. The final check is obtained from the astrometric parameter fitting. Showing the residuals as a function of signal intensity still shows the same relation with the signal intensity, with only small corrections, no larger than a few per cent in most cases, of the formal errors as a function of transit ordinate and star colour (see above). The internal system of formal errors is therefore fully understood and internally consistent. It is in addition based on a single reduction, eliminating uncertainties associated with combining two sets of partly-correlated results.

The accuracy assessment of the catalogue is more difficult. A full discussion of such an assessment for the new reduction is presented by van Leeuwen (2007a). The most important, as it is relatively independent of assumptions, element in the accuracy assessment is the presence and distribution of negative parallaxes. These are the direct result of having measurement errors larger than the actual parallaxes, as will be the case for some very distant objects. Here the new reduction has created a problem. The smallest errors found on the parallaxes are just over 0.09 mas, but there are no negative parallaxes found for any of the 1000 stars with parallax errors below 0.22 mas, and for errors below 0.38 mas, only 10 negative values are found. This by itself is a clear indication that the formal errors in the new catalogue must at least be approximately right. The number of negative parallaxes is too small, and the values spread over too wide a range of actual parallax errors, to subject them to any sensible statistical test. The limited tests that could be done show that there is no proof of any additional noise on the parallaxes. Such noise could represent small distortions in the parallax-reference system, and may reflect some of the inherited problems of the 1997 catalogue. A contribution at a level of about 0.2 mas could, however, not be excluded either.

11. Publication of the new catalogue and some preliminary results

Background information for the construction of the new catalogue can be found in a series of papers, starting with two papers on torque analysis and attitude modelling (Fantino & van Leeuwen 2003; van Leeuwen & Fantino 2003). In the same volume also the orbit of the satellite was analyzed in detail (Dalla Torre & van Leeuwen 2003), and an overview was presented of the many peculiarities that affected the Hipparcos mission (van Leeuwen & Penston 2003).These papers were followed by a paper on the various peculiarities in the construction and data contents of the 1997 catalogue (van Leeuwen 2005) and a preview of the construction of the new catalogue (van Leeuwen & Fantino

2005). The catalogue has been presented on a DVD, containing also the intermediate astrometric data and various calibration files, as part of a book (van Leeuwen 2007b) describing not only how the new catalogue was derived, but also how these data can be used in astrophysical applications, such as luminosity calibrations, cluster-parameter determinations, and studies of galactic dynamics. These studies are mostly based on near-final iteration results and may therefore give slightly different values when repeated with the final catalogue. One of these studies, on the Cepheids, has already been presented (van Leeuwen et al. 2007). Several other studies are in progress. Of those still in preparation, the result on the Pleiades cluster is possibly at this moment the most interesting, as it confirms the earlier estimates that the cluster is closer to the Sun than seems to be indicated by other measurements. A partial confirmation of the earlier result was not entirely unexpected, as the cluster covers a 9 degree diameter field, and correlations in the 1997 catalogue are only effective over areas with a diameter of about two to three degrees. In addition, the position of the main sequence as defined by the new parallax measurement appears to be in good agreement with those of clusters of similar age, IC 2602 and IC 2391. Similarly, it is found that the Hyades and Preasepe isochrones coincide very well. One approach for the further interpretation of these data could be to examine the characteristics of the empirical cluster isochrones in the context of the theoretical isochrones, and see where features differ, rather than to try fitting each individual cluster to the theoretical isochrones.

References

Arenou F., Lindegren L., Frœschlé M., et al., 1995, A&A, 304, 52

Dalla Torre A. & van Leeuwen F., 2003, Space Sci. Rev., 108, 451

ESA (ed.) 1997, The Hipparcos and Tycho Catalogues, no. 1200 in SP, ESA

Fantino E. & van Leeuwen F., 2003, Space Sci.Rev., 108, 499

Goldin A. & Makarov V.V., Nov. 2007, ApJS, 173, 137

Jancart S., Jorissen A., Babusiaux C., & Pourbaix D., Oct. 2005, A&A, 442, 365

Knapp G., Pourbaix D., Jorissen A., 2001, A&A, 371, 222

Knapp G. R., Pourbaix D., Platais I., & Jorissen A., 2003, A&A, 403, 993

van Leeuwen F., Fantino E., 2003, Space Sci. Rev., 108, 537

van Leeuwen F. & Penston M. J., 2003, Space Sci. Rev., 108, 471

Lindegren L., 1995, A&A, 304, 61

van der Marel H. & Petersen C. S., 1992, A&A, 258, 60

O'Mullane W., Lammers U., Bailer-Jones C., et al., 2006, ArXiv Astrophysics e-prints

Pan X., Shao M., & Kulkarni S. R., 2004, Nature, 427, 326

Pinsonneault M. H., Staufer J., Soderblom D. R., King J. R., & Hanson R. B., 1998, ApJ, 504, 170

Platais I., Pourbaix D., Jorissen A., et al., Jan. 2003, A&A, 397, 997

Pourbaix D., Platais I., Detournay S., et al., Mar. 2003, A&A, 399, 1167

Soderblom D.R., Nelan E., Benedict G.F., et al., Mar. 2005, AJ, 129, 1616

van Leeuwen F., Aug. 2005, A&A, 439, 805

van Leeuwen F., Nov. 2007a, A&A, 474, 653

van Leeuwen F., Sep. 2007b, Hipparcos, the New Reduction of the Raw Data, Springer, Dordrecht

van Leeuwen F. & Fantino E., Aug. 2005, A&A, 439, 791

van Leeuwen F., Feast M. W., Whitelock P. A., & Laney C. D., Aug. 2007, MNRAS, 379, 723

A Giant Step: from Milli- to Micro-arcsecond Astrometry
Proceedings IAU Symposium No. 248, 2007
W. J. Jin, I. Platais & M. A. C. Perryman, eds.
© 2008 International Astronomical Union
doi:10.1017/S1743921308018681

Solution of Earth orientation parameters in 20th century based on optical astrometry and new catalog EOC-3

J. Vondrák, C. Ron and V. Štefka

Astronomical Institute,
Boční II, CZ-14131, Prague, Czech Republic
email: vondrak@ig.cas.cz

Abstract. A new star catalog has been derived recently, which is based on combination of Hipparcos/Tycho Catalogues with long lasting ground-based astrometric observations. In addition to 'classical' mean positions and proper motions of the stars, it contains also information on periodic motions of many stars that are due to double or multiple stellar systems. The catalog, called EOC-3, contains 4418 different objects, out of which 585 have significant periodic motions. This improved catalog is used as a reference frame in optical wavelength to derive a new series of Earth Orientation Parameters. To this end, almost five million observations of latitude/universal time variations made at 33 observatories are used. The new EOP series covers almost the entire 20th century (namely the interval 1899.7-1992.0).

Keywords. astrometry, catalogs, reference systems

1. Introduction

Some years ago, we collected the astrometric observations of latitude/universal time variations made worldwide at 33 observatories. These observations, re-analyzed with the Hipparcos Catalogue, were then used to determine the Earth Orientation Parameters (EOP) at 5-day intervals, covering the interval 1899.7–1992.0 (Vondrák *et al.* 1998, Vondrák *et al.* 2000, Ron *et al.* 2005). The observations, accumulated during almost a century of monitoring the Earth orientation, contain a valuable and rich astrometric material that was only seldom used to construct astrometric catalogs. Later on, new astrometric catalogues such as ARIHIP (Wielen *et al.* 2001) or TYCHO-2 (Høg *et al.* 2000) appeared as combination of Hipparcos/Tycho positions with ground-based catalogues. These catalogues yield more accurate proper motions than the original Hipparcos Catalogue.

Many of the objects observed in the programmes of monitoring Earth orientation from the ground are double or multiple systems, having non-negligible periodic motions. We tried to obtain a star catalogue with improved proper motions and quasi-periodic terms reflecting orbital motions of the stars observed in these programmes. To this end, we used about 4.5 million observations of latitude/universal time variations, and combined them with the catalogues ARIHIP, TYCHO-2, to obtain a new astrometric catalog, 'tailored' for long-term Earth orientation studies. The third version of the Earth Orientation Catalogue (EOC-3) contains 4418 different objects (i.e., stars, components of double stars, photocenters), out of which 585 have statistically significant periodic motions (Vondrák & Štefka 2007). Namely this catalog is used in the solution whose description follows.

2. The solution

Observations of variations of latitude ($\Delta\varphi$), universal time (UT0-UTC) and equal altitude differences δh with 47 different instruments, located at 33 observatories, were first recalculated with the new astrometric catalogue EOC-3, nutation IAU2000 (Mathews *et al.* 2002) and precession IAU2006 (Capitaine *et al.* 2003), and then used to determine Earth Orientation Parameters (EOP) at 5-day intervals: Coordinates of the pole in terrestrial frame x, y, and Universal time differences UT1-TAI (only after 1956).

For each instrument, we determine constant, linear, annual and semi-annual deviation in latitude / longitude dev_φ, dev_λ, and rheological parameter (combination of Love numbers) $\Lambda = 1 + k - l$ for the tidal variations of the vertical.

For the whole interval, we determine celestial pole offsets with respect to IAU2000/2006 precession-nutation model dX, dY, represented as quadratic functions of time.

Data from the following instruments were used:

• 10 PZT's (measuring $\Delta\varphi$, UT0–UTC): 3 at Washington; 2 at Richmond and Mizusawa; 1 at Mount Stromlo, Punta Indio & Ondřejov;

• 7 photoelectric transit instruments (measuring only UT0–UTC): 3 at Pulkovo; 1 at Irkutsk, Kharkov, Nikolaev & Wuhang;

• 16 visual zenith-telescopes & similar instruments (measuring $\Delta\varphi$ only): 7 ZT at ILS stations; 2 ZT at Poltava, 1 ZT at Belgrade, Blagoveschtchensk, Irkutsk, Jósefoslaw & Pulkovo; FZT at Mizusawa; VZT at Tuorla-Turku;

• 14 instruments for equal altitude observations - AST, PAST, CZ (measuring δh): 1 AST at Paris, Santiago de Chile, Shanghai, Simeiz & Wuhang; 2 PAST at Shaanxi, 1 PAST at Beijing, Grasse, Shanghai & Yunnan; 1 CZ at Bratislava, Prague & Pecný.

The geographic distribution of all 33 participating observatories is depicted in Figure 1.

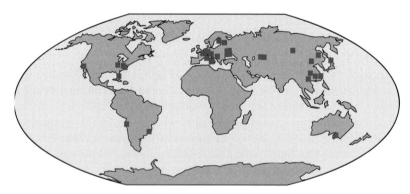

Figure 1. Geographic distribution of participating observatories.

To derive the EOP's, we use a procedure very similar to the one used earlier (Vondrák *et al.* 1998), i.e., we used the observation equations, which in a simplified form read

$$\Delta\varphi = x\cos\lambda - y\sin\lambda - dX\cos\alpha - dY\sin\alpha + dev_\varphi + \Lambda D_\varphi,$$
$$15\cos\varphi(\text{UT0–UTC}) = 15\cos\varphi(\text{UT1–UTC}) + \sin\varphi(x\sin\lambda + y\cos\lambda) +$$
$$+ \cos\varphi\tan\delta(dY\cos\alpha - dX\sin\alpha) + dev_\lambda + 15\Lambda D_\lambda\cos\varphi, \quad (2.1)$$
$$\delta h = 15\cos\varphi\sin a(\text{UT1–UTC}) + x(\cos\lambda\cos a + \sin\varphi\sin\lambda\sin a) -$$
$$- y(\sin\lambda\cos a - \sin\varphi\cos\lambda\sin a) + dY(\sin q\sin\delta\cos\alpha - \cos q\sin\alpha) -$$
$$- dX(\sin q\sin\delta\sin\alpha + \cos q\cos\alpha) + dev_\varphi\cos a + dev_\lambda\sin a +$$
$$+ \Lambda(D_\varphi\cos a + 15D_\lambda\cos\varphi\sin a),$$

where φ, λ are the observatory's geographic coordinates, α, δ, a and q are right ascension, declination, azimuth and parallactic angle of the star, respectively, and D_φ, D_λ are tidal variations of the local vertical computed for rigid Earth.

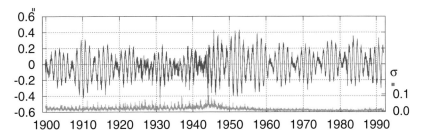

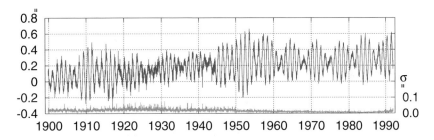

Figure 2. Polar motion x (upper plot), y (lower plot), and its uncertainties.

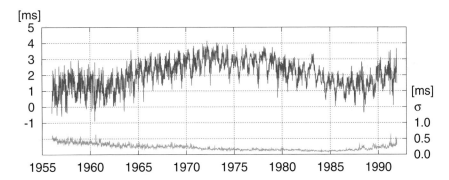

Figure 3. Length-of-day without short-period tidal variations, and its uncertainties.

The results are displayed in Figures 2–3, where both components of polar motion x, y and length-of day changes, calculated as the rate of UT1–UTC, are given. In the latter, the tidal variations after Yoder *et al.* (1981), with periods shorter than 35 days, were removed. The rheological parameters Λ, calculated for all participating observatories, are shown in Figure 4. The observatories are ranged by their increasing longitudes.

The celestial pole offsets, expressed with respect to IAU2000 model of nutation and IAU2006 model of precession, are very small so that it is impossible to detect their quasi-periodic variations from optical astrometry. Therefore they were estimated only as quadratic functions of time (in milliarcseconds):

$$dX = -7.4 + 28.6T + 30.0T^2 \tag{2.2}$$
$$dY = -5.9 + 9.4T + 1.2T^2,$$

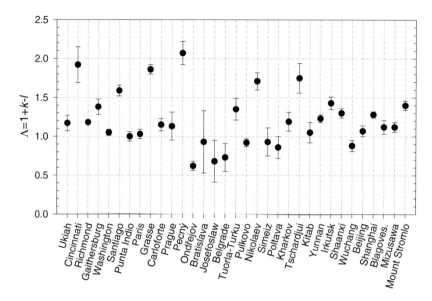

Figure 4. Rheological parameter $\Lambda = 1 + k - l$.

where T counts in centuries from 1956.0. The uncertainties of constant, linear and quadratic terms are ± 0.4 mas, 1.1 mas/cy and 3.5 mas/cy^2, respectively.

3. Conclusions

The new solution, based on catalog EOC-3, used 4 541 706 individual observations. It yields slightly better results than the one based on its previous version, EOC-2 (Vondrák 2004); the average standard error of one observation is $\sigma_\circ = 0.190''$ (in contrast to its previous value $0.191''$). The results are quite similar, especially the celestial pole offsets, given by Eqs. (2.2), still contain rather large linear and quadratic terms compatible with those found in (Ron et al. 2005). The new solution will serve for further long-term Earth rotation studies.

References

Capitaine, N., Wallace, P. T., & Chapront, J. 2003, *A&A*, 412, 567

Høg, E., Fabricius, C., Makarov, V. V. *et al.* 2000, *A&A*, 355, L27

Mathews, P. M., Herring, T. A., & Buffet, B. A. 2002, *J. Geophys. Res.*, 107 (B4), doi: 10.1029/2001JB000390.

Ron, C., Capitaine, N., & Vondrák, J. 2005, in: N. Capitaine (ed.) *Journées 2004 Systèmes de référence spatio-temporels*, (Observatoire de Paris), p. 110

Vondrák, J. 2004, *Serb. Astron. J.* 168, 1

Vondrák, J. & Štefka, V. 2007, *A&A*, 463, 783

Vondrák, J., Pešek, I., Ron, C., & Čepek, A. 1998, *Publ. Astron. Inst. Acad. Sci. Czech R.* 87

Vondrák, J., Ron, C., & Pešek, I. 2000, in: S. Dick, D. D. McCarthy, & B. Luzum (eds.), *Polar motion: Historical and Scientific Problems, Proc. IAU Coll. 178*, ASP Conf. Series 208, p. 206

Wielen, R., Schwan, H., Dettbarn, C. *et al.* 2001, *Veröff. Astron. Rechen-Inst. Heidelberg*, 40, Kommissions-Verlag G. Braun, Karlsruhe.

Yoder, C. F., Williams, J. G., & Parke, M. E. 1981, *J. Geophys. Res.* 86, 881

A Giant Step: from Milli- to Micro-arcsecond Astrometry
Proceedings IAU Symposium No. 248, 2007
W. J. Jin, I. Platais & M. A. C. Perryman, eds.

© 2008 International Astronomical Union
doi:10.1017/S1743921308018693

CCD astrometric observations of faint satellites and update of their orbits†

K. X. Shen[1], Z. H. Tang[2], R. C. Qiao[1], S. H. Wang[2], Y. R. Yan[2] and X. Cheng[1,3]

[1] National Time Service Center (NTSC), The Chinese Academy of Science, P. O. Box 18, Lintong, Shaanxi 710600, China. email: `shenkx@ntsc.ac.cn`
[2] Shanghai Astronomical Observatory, Chinese Academy of Sciences, Shanghai 200030, China
[3] Graduate School of the Chinese Academy of Sciences, Beijing 100039, China

Abstract. This paper reports on our observing campaign of faint satellites performed at the National Time Service Center and Sheshan station of SHAO from 1994 up to today. In the past few years due to benefit from using a large size CCD and the publication of the modern catalogues (UCAC2), a series of observations of faint satellites were obtained by us. Moreover the work of improving the orbit of Phoebe via numerical fit to the observations over a century is also presented.

Keywords. astrometry, solar system: general, planets and satellites: general, ephemerides

1. Introduction

Motivated by the space achievement, in recent years many astronomers have focused their attention on the faint satellites. As part of our ongoing program observing the planetary satellites in successful implement since 1994, we has been carrying out an observing campaign for some faint satellites in last decade. These satellites having magnitude more than 12 are quite faint, the astrometric observations of which are very difficult to make. In our observing program included are Mimas and Phoebe of Saturn, Miranda of Uranus, Triton and Nereid of Neptune.

For these satellites only few astrometric observations were obtained over nearly one century in the past since their discovery mainly due to that the objects are too faint to observe. It results that the accuracy of the existing theoretical models and the ephemeredes are decreasing, the need to improve the ephemeredes urges continually developing new astrometric observations for them.

2. Astrometric calibration

Before 2002 our observations were made using a 1.56m reflector with long focal distance (15.6 m) and a CCD detector of small size. From the instrumental characteristics one gets very small fields. Furthermore the previous star catalogs have a too low density, in the small fields of view no many catalogue stars can be found, so it is incapable to define precisely a known reference system. In this case calibration of CCD is more problematic.

The classical procedure in overcoming this difficulty consists in the construction of a secondary astrometric catalogue of faint stars in the neighborhood of the satellites, however it is obviously inconvenient. Accordingly, the calibration of CCD device was

† Supported by the National Natural Science Foundation of China (Grant Nos. 10673026, 10333050 and 10573018).

made usually using some special methods, for example, in reducing Saturnian satellite system the so-called 'bright moon method' was employed by us. This method relies on positions predicted from pre-existing ephemeredes, in which positions of the bright satellites were used to define a known reference system in every frame.

However, after 2002, we had the opportunity to benefit from two advances in technique: a new large size CCD chip (2048×2048) was used to replace the original one; another important advance should be attributed to publishing the high dense and accurate catalogue UCAC2. Generally about 15-20 UCAC2 reference stars are available for each CCD frame, thus it is enough to allow us to make the classical astrometric reduction directly. For more accurate measurement the typical 6 constants model of plate reduction was selected according to the procedure previously described by Tang (2002).

3. Observation of faint inner satellites

In observing two types of faint satellite, the different difficult might be encountered, for the inner satellites closing to a bright primary, the satellites are embedded in the primary's halo light. Thus in the frames the measured center of image shifts towards the planet's center. The use of CCD provides us possibility to perform the observing faint satellites, which was not possible earlier.

Many authors have presented their methods to minimize the systematic effects of gradients in the background near the planet on the satellite's measured position; all of them made the attempts of using a polynomials surface of degree one or two to fit the inclined sky background. Yan (2007) got a judgment that third degree polynomial is a most judicious choice and is enough to reduce the residuals after a detailed analysis via a numerical simulation as the real data.

In our program observing faint inner satellites included are Saturnian Mimas (mag 12.9) and Uranian Miranda(mag 16.3). Because of the unavoidable effect from halo light in the proximity of the primary, centering of the satellite image is generally difficult to measure. The 44 measurements of the positions of Mimas during 1997–2000 and the 83 measurements for Miranda during 1995–1997 were obtained.

In reduction of the satellite we have successfully applied the 'brighter moon method' to the calibration of the CCD chip (Shen et al. 2002; Qiao *et al.*, 2004), four better-known satellites were used as calibration satellites. The analytical theories TASS1.7 and GUST86 were used for theoretical computation. The calculations exhibit that for Mimas and Miranda the poor residuals are as large up to $0\rlap{.}''1$.

4. Observation of faint outer satellites

For the outer satellites in lager distance to the primary, the main difficulty consists in that no planet or major satellites with the satellite are present in the same field of view even when using CCD of large size; the satellites appear as an isolated object. Thus, the differential measurement relative to the major satellites of Saturn or reference satellite is very unlikely. As a consequence, the previous methods of astrometric reduction used for the planet's inner satellites cannot be performed. The following faint outer satellites have been studied.

(1) **Phoebe**. Phoebe is a most distant known outer satellite of Saturn (more than 12 million km). In each CCD frame this satellite appears as an isolated object with extremely faint visual magnitude (mag 16.45).

At 6 nights in December 2003 and at 3 nights in March 2004 a total of 115 frames were obtained. In this work, because of the use of a CCD detector with a wider field, the object to be measured and a much lager number of background stars are available in the field of view. The 'CENTER' command was used in batch mode to determine accurately the positions of all the targets, furthermore the Gaussian method, which is a 2-dimensional Gaussian function, including a term to represent the background level was used; the rough sky background bias was estimated and removed from each image. For the detailed procedure the reader can be referred to the paper Qiao *et al.* (2006).

In 2005, we completed a re-determined Phoebe's orbits (Shen *et al.*, 2005). In order to best determine Phoebe's orbit, it is necessary that the observations cover the longest time span as possible, in this reduction we used the 686 Earth-based astrometric observations available from 1905 to 2004, including the 101 new CCD observations from Qiao and Tang (2006) and 57 observations from Peng (2004).

The numerical integration was calculated using the 12th-order Runge-Kutta-Nystrom formula of Brankin. For Phoebe, the overwhelming perturbation is due to the Sun. The other perturbation induced from Jupiter and Uranus were computed using their positions derived from the JPL planetary ephemeris DE406. The perturbation from Titan is included as sole perturbing satellite.

(2) **Triton**. Before our observing Triton (mag 13.47), only less than 400 positions have an accuracy of better than 0.″15. Qiao reported that he and his colleagues took the 943 astrometric observations of Triton in the period of 1996–2006 (Qiao *et al.*, 2007).

A two-dimensional Gaussian to each image and a third-degree polynomial were considered by Yan (2007) for simulating sky background. The observed positions of Triton are compared with theoretical positions generated by the JPL and IMCCE ephemeredes respectively. The residuals of observations are about 0.″04.

(3) **Nereid**. Nereid is a very faint satellite of Neptune (mag 18.7), very poorly observed since its discovery. In 2006 we started observing campaign of Nereid using the 2.16m telescopes in National Astronomical Observatories (including original Beijing Astronomical Observatory, Yunan Astronomical Observatory and institute of astronomical instruments), with which the 71 observations have been taken. At present the observations have been in reduction. For permitting to record a faint image of this satellite long exposure time (more than 20 min) is usually needed.

5. Summary

We have presented a report on our astrometric campaign on faint satellites spanning the period from 1994 to 2007 as part of our ongoing observing program of the planetary satellites, which was initiated at Sheshan station in 1994. In the past 15 years a large number of highly accurate observations have been obtained. We have shown that the observations are highly accurate and significant for orbit determination of faint, poorly observed satellites.

References

Qiao, R. C., Shen, K. X., Harper, D., & Liu, J. R. 2004, *A&A*, 422, 377
Qiao, R. C., Tang, Z. H, Shen, K. X., Dourneau, G. *et al.* 2006, *A&A*, 454, 379
Qiao, R. C., Yan, Y. R., Shen, K. X. *et al.* 2007, *MNRAS*, 376, 1707
Peng, Q. Y., Vienne, A., Han, Y. B. *et al.* 2004, *A&A*, 424, 339
Shen, K. X., Qiao, R. C., Harper, D., Hadjifotinou, K. G. *et al.* 2002, *A&A*, 391, 775
Shen, K. X., Harper, D., Qiao, R. C., Dourneau, G.. *et al.* 2005, *A&A*, 437, 1109
Tang, Z. H, Wang, S. H., & Jin, W. J. 2002, *AJ*, 123, 125
Yan 2007, *Master Degree Thesis, Shanghai Astronomical Observatory*

A Giant Step: from Milli- to Micro-arcsecond Astrometry
Proceedings IAU Symposium No. 248, 2007
W. J. Jin, I. Platais & M. A. C. Perryman, eds.

Influence of the astrometric accuracy of observation on the extrapolated ephemerides of natural satellites

J. Desmars[1], J.-E. Arlot[1], A. Vienne[1,2]

[1]Institut de Mécanique Céleste et de Calcul des Éphémérides - Observatoire de Paris,UMR 8028 CNRS, 77 avenue Denfert-Rochereau, 75014 Paris, France
email: desmars@imcce.fr

[2]Université de Lille, 59000 Lille, France

Abstract. The accuracy of planetary satellites ephemerides is determined not only by the accuracy of dynamical model (internal accuracy) but also by the accuracy of the observations (external accuracy) used to fit the initial parameters of a model. This external accuracy extrapolated in the future is unknown most of the time and tends to degrade the global accuracy of ephemerides. Even if we can estimate the quality of the ephemerides by comparison with observations, we do not know how to determinate the evolution of the accuracy outside the period of observations. We will present a statistical method, resampling of observations, which allows a better estimation of the extrapolated accuracy in the future.

Keywords. planets and satellites: individual (Saturn), astrometry, ephemerides

1. Introduction

The accuracy of the ephemerides is determined by the accuracy of the model (internal accuracy) which is often well estimated and quite good, and by the accuracy of the observations (external accuracy) which depends on the quality and the distribution of the observations, and degrades the global quality of ephemerides. During the observational periods, the accuracy can be determined by the difference between observed positions and computed positions (O-C) and remains quite good. Outside the observational periods, the accuracy deteriorates and its estimation remains difficult. So the question is to know how to estimate the real accuracy of ephemerides outside the observational periods. The problem has already been studied for asteroids. Muinonen & Bowell (1994) , Virtanen *et al.* (2001) use statistical methods to determine the orbital uncertainties of asteroids. They succeeded in rediscovering asteroids observed in past but lost. But our problem is quite different because asteroids have a slow motion and few observations exist, whereas the natural satellites have a fast motion and are frequently observed. Statistical methods, however, remain a good way of study.

2. Statistical methods

One of the statistical methods is the bootstrap resampling (Efron & Tibshirani, 1993). Bootstrap samples are generated from the original data set (with N elements). Each bootstrap sample has N elements generated by sampling with replacement N times from the original set. Thus, a great number of new samples is allowed.

Here we use TASS model of main Saturnian satellites (Vienne & Duriez, 1995). The parameters of the model have been fitted to each sample, positions have been computed during 1850-2050 period and compared with positions computed with initial parameters. A catalogue of real observations of Saturnian satellites from 1874 to 2007 has been generated. Two observational periods can be separated. The first one, old observations, from 1874 to 1947, with a priori lower quality and the second one, recent observations from 1961 to 2007 with a priori better quality. The differences between positions in (α, δ) coordinates for Mimas are represented in Fig1. Fifty bootstrap samples were used. The differences are not very important during the observational period but becomes important outside the period. For the fit of old observations, the differences can reach 1.5" in α. For the fit to recent epochs, the differences are even more important (~ 10"). It is amazing since recent observations are a priori better than the old ones. This difference can be explained because the old observatinal period stretches from 1874 to 1947 (73 years) whereas recent observations stretch from 1961 to 2007 (46 years). So, a long period of average observations seems to be better than a short period of accurate observations, for a better accuracy outside the observational period.

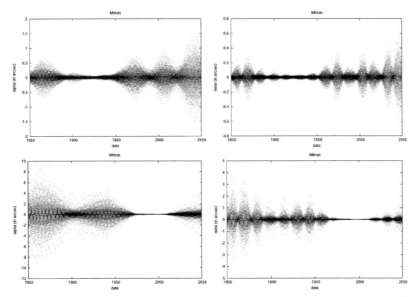

Figure 1. Difference for 50 bootstrap samples of old and recent observations in (α, δ) for Mimas

3. Conclusion and perspectives

Bootstrap resampling allows us to estimate the accuracy of ephemerides. The positions of satellites are not so accurate outside the observational period, especially if the model has been fitted on a short period of observations. Other statistical methods also enable to estimate the accuracy. For example, jackknife consists in resampling the original set by leaving out one observation at a time. Such a method leads us to define "one" observation and more precisely when two observations are independent. A next study on this problem of independence of observations will be undertaken.

References

Efron, B. & Tibshirani, R. J. 1993, in: Monographs on Statistics and Applied Probability *An Introduction to the Bootstrap*
Muinonen, K. & Bowell, E., 1993, *Icarus* 104, 255
Vienne, A. & Duriez, L., 1995, *A&A* 297, 588
Virtanen, J., Muinonen, K & Bowell, E., 2001, *Icarus* 154, 412

A Giant Step: from Milli- to Micro-arcsecond Astrometry
Proceedings IAU Symposium No. 248, 2007
W. J. Jin, I. Platais & M. A. C. Perryman, eds.

Parallax programs at the Bordeaux Observatory at sub-mas level accuracy

C. Ducourant[1], R. Teixeira[2], J.-F. le Campion[1] and G. Chauvin[3]

[1]Observatoire Aquitain des Sciences de l'Univers, CNRS-UMR 5804, BP 89, 33270 Floirac,
France. email: ducourant@obs.u-bordeaux1.fr, email: lecampion@obs.u-bordeaux1.fr

[2]Instituto de Astronomia, Geofísica e Ciências Atmosféricas, Universidade de São Paulo, Rua
do Matão, 1226 - Cidade Universitária, 05508-900 São Paulo - SP, Brasil. email:
teixeira@astro.iag.usp.br

[3]Laboratoire d'Astrophysique Observatoire de Grenoble BP 53 F-38041 GRENOBLE Cdex 9,
France. email: Gael.Chauvin@obs.ujf-grenoble.fr

Abstract. We present ongoing parallax programs developed at the Bordeaux Observatory several years ago. We describe the necessary steps leading to sub-milliarcsecond accuracy of calculated parallax. We show the importance of global methods for accurate parallax determination.

Keywords. astrometry, stars: distances, solar neighborhood, methods: data analysis

1. Introduction

With the increasing interest in extra-solar planets, parallax measurements appear to be crucial since most of known stars hosting a planet are situated close to the Sun (d⩽100pc) and therefore accessible to ground-based astrometry.

The use of 4m-class telescopes allows to derive parallaxes with sub-milli-arcsecond accuracy after 2 years of regular observations. In such a small time-base programs with so high accuracy, it is important to minimize all potential sources of errors and biases and to realistically evaluate the errors, and their origins.

We present here the Bordeaux parallax programs, describing several steps leading to high accuracy parallax determination. We investigate through Monte-Carlo simulations the achievable precision on parallaxes for ground-based observations using global iterative methods (central overlap methods Eichhorn (1997)). We present the dependency of errors on the reference catalogue used to calibrate the CCD and on the star density of the field (number of objects participating in the global fit).

2. The Bordeaux Parallax Programs

In the last years, we developed in Bordeaux several programs of parallax measurements aiming at the characterisation of the target through distance determination :

• Systematic search of nearby stars in wide field survey : Ducourant *et al.* (1998).
• Free Floating Brown dwarfs characterisation : Hawkins *et al.* (1998).
• Halo High velocity white dwarf candidates segregation : Ducourant *et al.* (2007a).
• An accurate distance and mass determination for the first imaged planetary system 2M1207Ab : Ducourant *et al.* (2007b).
• Distance determination to the TW Hydrae association, under realisation (see Teixeira *et al.* in this volume for details).

For these programs we developed a code of global iterative treatment (Eichhorn 1997) of the observations which take advantage of any well-measured star in each image to participate in the global fit, allowing to define a re-inforced reference system with respect to which the target parallax can be defined.

3. Methodology

To reach sub-milliarcsecond accuracies, it is important to minimize each source of bias and error. The major steps are :

• The observation strategy (S/N of objects, frequency, number of repeated frames, distance to zenith, filter, length of program). It appears that a crucial point concerns the zenith distance which has to be minimized in order to minimize differential colour refraction effects between target and field stars. A large number of frames taken at each epoch appears important for a correct evaluation of the errors.

• The selection of reference stars used to calibrate pixel scale and orientation of CCD (number, repartition in the field, catalogue astrometry) is an important step. It appears that a catalogue as 2MASS is suited for that purpose although it is not a proper motion catalogue and that its central epoch is 2000.0. We observe that it introduces less bias in the astrometry than the use of UCAC2, more accurate and with proper motion but so poorly dense (See also I. Platais, same volume, for needs of Deep Astrometric Standards).

• The global iterative resolution with a large number of objects participating to the global solution (fit stars) (also important to average differential colour refraction effects) to correctly evaluate the correction from relative to absolute parallax.

4. Results

We show through numerous Monte-Carlo simulations that at the sub-mas level the target's measured parallax is strongly dependent on the number of reference stars and that methods (conventional or global) that rely on low number of fit stars may result in a biased solution.

References

Ducourant, C., Teixeira R., Hambly N. *et al.*, 2007a, *A&A*, in press.
Ducourant, C., Teixeira R., Chauvin, G., Daigne, G., Le Campion, J. F., Song, I., & Zuckerman, B., 2007b, *A&A*, submitted.
Ducourant, C., Dauphole, B., Rapaport, M., Colin, J., & Geffert, M., 1998, *A&A*, 333, 882.
Eichhorn, H., 1997, *Astron. Astrophys.*, 327, 404.
Hawkins, M. R. S., Ducourant, C., Rapaport, M., & Jones, H. R. A., 1998, *MNRAS*, 294, 505.

A Giant Step: from Milli- to Micro-arcsecond Astrometry
Proceedings IAU Symposium No. 248, 2007
W. J. Jin, I. Platais & M. A. C. Perryman, eds.

GRAVITY: microarcsecond astrometry and deep interferometric imaging with the VLTI

F. Eisenhauer[1], G. Perrin[2], C. Straubmeier[3], W. Brandner[4], A. Boehm[4], F. Cassaing[5], Y. Clenet[2], K. Dodds-Eden[1], A. Eckart[3], P. Fedou[2], E. Gendron[2], R. Genzel[1], S. Gillessen[1], A. Graeter[1], C. Gueriau[2], N. Hamaus[1], X. Haubois[2], M. Haug[1], T. Henning[4], S. Hippler[4], R. Hofmann[1], F. Hormuth[4], K. Houairi[5], S. Kellner[1], P. Kervella[2], R. Klein[4], J. Kolmeder[1], W. Laun[4], P. Lena[2], R. Lenzen[4], M. Marteaud[2], D. Meschke[4], V. Naranjo[4], U. Neumann[4], T. Paumard[2], M. Perger[3], D. Perret[2], S. Rabien[1], J. R. Ramos[4], J. M. Reess[2], R. R. Rohloff[4], D. Rouan[2], G. Rousset[2], B. Ruyet[2], M. Schropp[1], B. Talureau[2], M. Thiel[1], J. Ziegleder[1] and D. Ziegler[2]

[1] Max-Planck-Intitut für extraterrestrische Physik, 85748 Garching, Germany
mail: `eisenhau@mpe.mpg.de`
[2] Observatoire de Paris, LESIA, 92190 Meudon, France
[3] I. Physikalisches Institut, Universität zu Köln, 50937 Köln, Germany
[4] Max-Planck-Institut für Astronomie, 69117 Heidelberg, Germany
[5] ONERA, 92322 Chatillon Cedex, France

Abstract. We present the adaptive optics assisted, near-infrared VLTI instrument GRAVITY for precision narrow-angle astrometry and interferometric phase referenced imaging of faint objects. With its two fibers per telescope beam, its internal wavefront sensors and fringe tracker, and a novel metrology concept, GRAVITY will not only push the sensitivity far beyond what is offered today, but will also advance the astrometric accuracy for UTs to 10μas. GRAVITY is designed to work with four telescopes, thus providing phase referenced imaging and astrometry for 6 baselines simultaneously. Its unique capabilities and sensitivity will open a new window for the observation of a wide range of objects, and — amongst others — will allow the study of motion within a few times the event horizon size of the Galactic Center black hole.

Keywords. instrumentation: interferometers, instrumentation: adaptive optics, Galaxy: center

1. Fundamental measurements over a wide range of astrophysics

GRAVITY, an interferometric imager with 10μas astrometric capability, coupled with spectroscopic and polarization modes and optimized to exploit the exquisite sensitivity of the 4x4 VLTI system, will revolutionize dynamical measurements of celestial sources interacting through gravity. It will carry out the ultimate test of determining whether or not the Galactic Centre harbours a $4 \times 10^6 \, M_\odot$ black hole. It has the potential to directly measure the space-time metric around this black hole, and thus may be able to test General Relativity in the presently unexplored strong curvature limit. GRAVITY will also be able to unambiguously detect and measure the mass of black holes in massive star clusters throughout the Milky Way and in many AGN. It will make unique measurements on gas jets in YSOs and AGN. It will explore binary stars, exoplanet systems and young stellar disks. Because of its superb sensitivity GRAVITY will excel in milli-arcsecond phase-referenced imaging of faint objects of any kind. Because of its outstanding astrometric capabilities, it will detect motions throughout the local Universe and perhaps beyond.

Because of its spectroscopic and polarimetric capabilities it is capable of detecting gas motions and magnetic field structures on sub-milliarcsecond scales.

2. Instrument design and perfomance

The VLTI, with its four 8m telescopes and a collecting area of 200m^2, is the only interferometer to allow direct imaging at high sensitivity and image quality. GRAVITY will for the first time utilize the unique 2" field of view of the VLTI, providing simultaneous interferometry of two objects with four telescopes. This permits narrow angle astrometry with a precision of 10μas. The application of phase referenced imaging — instead of closure phases — is a major advantage in terms of model-independence and fiducial quality of interferometric maps with a sparse array such as the VLTI. The second major new element of GRAVITY is the use of infrared wavefront sensors to open a new window for interferometry. In addition to broad band (K) imaging and astrometry, GRAVITY also features modest resolution spectroscopy and polarization analysis capabilities. The following Table 1 gives an overview of the expected performance of GRAVITY.

Table 1. Expected performance of GRAVITY.

Adaptive optics on K=7 star	36 % Strehl
Fringe tracking on K=10 star	270 nm rms OPD on science channel
Astrometry on K=10 primary and K=15 secondary star	10 μas in 5 minutes
Interferometric imaging on K=16 in 100 s	S/N Visibility = 10
Size and position measurements	K $\geqslant$ 19 in 6 hours

The baseline for the infrared wavefront sensor is a Shack Hartmann system. It will be located in the VLTI laboratory, thus also correcting for tunnel seeing in the VLTI optical train. The wavefront correction will be applied to the MACAO deformable mirrors located at the UT Coude focus. The interferometric beam combiner is based on fiber-fed integrated optics. The instrument is equipped with polarization-control, differential delay lines, and fast tip/tilt and fringe tracking actuators. GRAVITY will have two beam combiners for two objects. The first is optimized for phase referencing/fringe tracking at a high frame rate. The second, the science beam combiner, is optimized for long integrations. Its spectrometer provides moderate spectral resolution (R $\approx$ 500), and a Wollaston prism for polarization analysis. GRAVITY will have all its components enclosed in a vacuum cryostat for optimum stability and background suppression. The GRAVITY metrology is optimized for astrometric accuracy. Laser light is back-propagated from the GRAVITY beam combiners up to the telescope secondary mirrors, producing fringe patterns, which carry the differential optical path information. These fringe patterns are observed in scattered light through cameras mounted on the telescopes. Other than classical laser metrologies, the GRAVITY metrology measures the full beam and covers the entire optics train except the primary mirror, therefore reducing all systematic errors from non-common optical paths to a minimum.

References

Eisenhauer, F. *et al.* 2008, in *The Power of Optical/IR Interferometry*, eds. A. Richichi, F. Delplancke, A. Chelli, F. Paresce, ESO Astrophysics Symposia, Springer, 431

Paumard, T. *et al.* 2008, in *The Power of Optical/IR Interferometry*, eds. A. Richichi, F. Delplancke, A. Chelli, F. Paresce, ESO Astrophysics Symposia, Springer, 313

A Giant Step: from Milli- to Micro-arcsecond Astrometry
Proceedings IAU Symposium No. 248, 2007
W. J. Jin, I. Platais & M. A. C. Perryman, eds.

© 2008 International Astronomical Union
doi:10.1017/S1743921308018735

Brown Dwarf Kinematics Project (BDKP)

J. Faherty[1], K. Cruz[2], A. Burgasser[3], F. Walter[1] and M. Shara[4]

[1]Stony Brook University, Stony Brook, NY, 11794 email: jfaherty@amnh.org

[2]Caltech, MC 105-24 Pasadena, CA 91125 email: kelle@astro.caltech.edu

[3]MIT,77 Massachusetts Avenue Cambridge, MA 02139 email:ajb@mit.edu

[4]American Museum of Natural History, 79th Street and CPW, New York, NY 10023
email:mshara@amnh.org

Abstract. We report on the progress of the Brown Dwarf Kinematics Project (BDKP), which aims to measure the 6D positions and velocities of all known brown dwarfs within 20 pc of the Sun and select sources of scientific interest. In this paper we report on the status of the 33 targets on our parallax list as well as the results of our proper motion survey where we have measured over 400 new proper motions for known late M, L and T dwarfs.

Keywords. stars: low-mass, brown dwarfs, stars: kinematics, stars: fundamental parameters

1. Introduction

Brown dwarfs are faint, substellar mass objects that share properties of both stars and planets. They encompass three spectral classes; late type M, L, and T with over 600 members in the lowest temperature spectral classes (L and T). Brown Dwarfs offer a link between planet and star formation, challenges for atmospheric models, and clues into the early history of our galaxy. Only a small percentage of all identified objects have full 6D positions: 13% of known L and T dwarfs have reported parallaxes, less than half have proper motion measurements, and a sparse 8% have measured radial velocities. Despite these limited statistics the kinematics of late-type dwarfs have already provided insight into the origins and long-term evolution of this low mass population. From the analysis of space velocities of a sample of 21 L and T dwarfs within 20 pc of the Sun, Osorio, et al. (2007) have concluded that as a group these objects are relatively young (mean age ~1Gyr), a finding inconsistent with population synthesis simulations (Burgasser 2004). Furthermore, nearly half of their sample appears to have kinematics consistent with membership in the Hyades moving group (~625Myr). This implies that many of the nearby brown dwarfs may be very low mass members of young recently uncovered moving groups (Zuckerman & Song 2004). We currently have 33 confirmed brown dwarfs on a parallax list, most of which show potential membership in young associations. We also have a large observing program to measure proper motions for all known L and T dwarfs so we can begin a kinematic analysis of the entire population of substellar mass objects.

2. Overview of parallax program

We are using the ANDICAM 1024×1024 IR Hawaii detector on the 1.3m telescope at CTIO to measure brown dwarf parallaxes in the J band. Our instrument has a plate scale of 0.137"/pixel and a 2.4 arcminute field of view. Our target list (see table 1.) contains 3 L-subdwarfs, 11 late M to early L's that show gravity sensitive features indicative of youth,14 early to late L's that have photometric distances within 20pc and 4 Calibrators (not shown) with previously measured parallaxes. All of our parallax frames contain at

Table 1. PARALLAX TARGETS

ID	P.M.	SpT.	J	H	K	N	ID	P.M.	SpT.	J	H	K	N
6	—	Lsd	12.91	12.30	12.02	8	20	878	L1.5	12.45	11.69	11.15	4
7	—	Lsd	14.54	14.12	13.88	4	21	769	L2	12.81	12.04	11.46	9
8	—	Lsd	12.55	12.14	11.93	7	22	574	L0	13.58	11.69	11.15	4
9	—	M6	15.10	14.22	13.45	5	23	487	L1	13.65	13.03	12.57	4
10	—	M9.5	13.60	12.90	12.39	5	24	447	L2	13.41	12.57	11.98	8
11	3600	M9	11.38	10.82	10.39	7	25	426	L5	14.49	13.63	13.05	4
12	210	L3	14.98	13.71	12.96	5	26	375	L5	14.48	13.34	12.60	6
13	150	L0	15.07	14.00	13.42	7	27	329	L7	15.51	14.44	13.73	5
14	89	M9	12.69	12.00	11.50	7	28	236	L2	13.52	12.63	12.05	5
15	68	M8.5	13.97	13.28	12.81	6	29	194	L6.5	14.41	13.37	12.81	8
16	49	M8.5	13.03	12.36	11.89	5	30	—	L1.5	13.12	12.27	11.75	4
17	41	M8	13.63	12.97	12.45	4	31	—	L6	15.48	14.35	13.56	4
18	—	L0	14.78	13.86	13.27	5	32	—	L6	14.55	13.53	13.01	4
19	—	M6	12.76	11.77	11.28	5	33	—	L7	15.19	14.08	13.34	4

Table 1: Proper Motions(P. M.) are in milli-arcsec/year. N is the number of parallax frames.

least 7 astrometric reference stars. The target objects are carefully placed on the same X,Y pixel position at each observation so as to minimize differential distortions and ensure the same reference stars in each frame. Our initial transformations demonstrate an accuracy of 7–10 mas or better. The majority of objects are expected to be within 20pc therefore we expect to measure distances with 20% accuracy or better. We started observing our targets in January 2007 and have attempted to obtain at least one parallax frame per target per month while it is observable. Targets must be observed at least once every six months near times of maximum parallactic shift to accurately measure a parallax. Our observations will continue through January 2009 and we plan on reporting our preliminary measurements after 2 years of collected data.

3. Overview of proper motion program

We are using telescopes in the southern and northern hemisphere to measure proper motions for all known L and T dwarfs. As stated less than half of all known L and T dwarfs have measured proper motions. This is mainly due to the youthfulness of IR technology and the lack of a second epoch to follow-up on the first epochs of 8–10 years ago. We are using CPAPIR, a near-IR camera on the 1.5m at CTIO, TIFKAM a near-IR imager on the 1.3m at KPNO, and the I band imager on the 0.9m at CTIO to target known brown dwarfs. To date we have measured proper motions for over 400 known late-type M,L, and T dwarfs. We calibrate our second epoch to the 2MASS catalogue and with an average baseline of 7.3 years obtain errors between 10 to 30 mas/yr for our targets. In our sample, 25% are moving faster than 0.5"/year, and 5% are moving faster then 1.0"/year. We are currently analyzing our sample for kinematic trends and will have a detailed paper out soon.

References

Burgasser, A. J. 2004, *ApJS*, 155, 191-207

Osorio, M. R. Z., Martin, E. L., Bejar, V. J. S., Bouy, H., Deshpande, R., & Wainscoat, R. J. 2007, *ApJ*, 666, 1205-1218

Zuckerman, B. & Song, I.. 2004, *A&AR*, 42, 685-721

A Giant Step: from Milli- to Micro-arcsecond Astrometry
Proceedings IAU Symposium No. 248, 2007
W. J. Jin, I. Platais & M. A. C. Perryman, eds.

© 2008 International Astronomical Union
doi:10.1017/S1743921308018747

Astrometric detection of faint companions – the Pluto/Charon case study

A. H. Andrei[1,2], V. Antunes Filho[2], R. Vieira Martins[1], M. Assafin[2], D.N. da Silva Neto[2,3] and J. I. B. Camargo[2]

[1]Observatório Nacional/MCT
R. Gal. Jos Cristino 77, RJ, Brasil
email: oat1@on.br

[2]Observatório do Valongo/UFRJ, Brasil

[3]Universidade Estadual da Zona Oeste/RJ, Brasil

Abstract. The resolution of pairs of objects closer than the scale of seeing, and of difference of magnitude larger than ten percent is unreliable by direct imaging. The resulting image FWHM differs from a true PSF by no more than four percent. Yet, the peak of the associated Gaussian is shifted to a larger proportion.

The main results are the description of the FWHM and peak location shifts as function of the seeing scale, the centers separation, and of the magnitudes difference. Analytically, the estimators of variation were the resulting Gaussian amplitude, mean value, and standard deviation. The later is shown to be the most reliable estimator.

Keywords. methods: data analysis, techniques: image processing, astrometry, planets and satellites: general, binaries: general

1. Results

Figure 1 presents, for the Pluto/Charon case, the composition of the profiles of the brighter and dimmer companions. It is seen that the resulting compound curve is well represented by a Gaussian which differs from a single object PSF. The analysis of the ensemble of PSFs in the field permits to indicate the one revealing the signature of a compound system.

Figure 2 presents the nearly helicoidal wandering of the photocentre of the compound image of the Pluto/Charon system (i.e., without intervention of the dynamical pulling), from the survey campaign at the 0.6m B&C telescope at the LNA, Itajubá, Brasil. On the right, the same effect for a single Charon's orbit.

Once correcting the observed image from the photocentre wandering effect, the true Pluto's observed positions can be compared with its ephemeris. Systematics trends are seen both in RA (left) and DEC (right), the former with amplitude about twice of the latter. The trends cannot be of observational nature, due to the time span (6 years). Instead they can be reconciled to RA and DEC ephemeris corrections for the Pluto's orbit, matching recent determinations (Sicardy *et al.*, 2006).

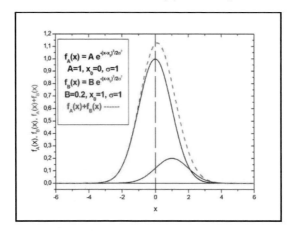

Figure 1. Gaussian composition (dashed curve) of the brighter (continuos line, taller curve) and dimmer companions (continuos line, shorter curve)

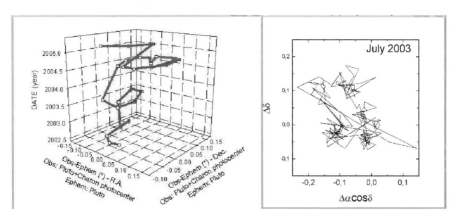

Figure 2. Charon's signature upon the astrometric variation of the compound photocentre of the Pluto/Charon system.

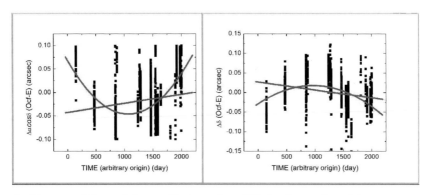

Figure 3. Linear and Second degree fits to series of Pluto observed positions, once corrected from the photocentre wandering, minus the ephemeris.

References

Sicardy, B., Bellucci, A., Gendron, E. *et al.* 2006, *Nature* 439, 52

A Giant Step: from Milli- to Micro-arcsecond Astrometry
Proceedings IAU Symposium No. 248, 2007
W. J. Jin, I. Platais & M. A. C. Perryman, eds.

© 2008 International Astronomical Union
doi:10.1017/S1743921308018759

Fringe Tracker for the VLTI Spectro-Imager

M. Gai[1], D. F. Buscher[2], L. Corcione[1], S. Ligori[1] and J. S. Young[2]

[1]Istituto Nazionale di Astrofisica – Osservatorio Astronomico di Torino, Italy
email: gai@oato.inaf.it

[2]Astrophysics Group, Cavendish Laboratory, University of Cambridge, UK
email: dfb@mrao.cam.ac.uk

Abstract. Interferometry in the near IR aims at providing imaging resolution on the *mas* scale, and astrometry at the few *μas* level, from ground based infrastructures using current telescope technology. To take advantage of simultaneous combination of four to eight telescopes, an international consortium is proposing to ESO the development of the VLTI Spectro-Imager. One of the key sub-systems, to measure and correct the atmospheric perturbations relative to the beam phase, is the fringe tracker, aimed at providing the science combiner with long, stable observing conditions. The fringe tracker function in interferometer is equivalent to adaptive optics for conventional telescopes. The fringe tracker concept under study, using minimum redundancy combination and bulk optics, is described.

Keywords. instrumentation: interferometers, techniques: interferometric

1. Introduction

Several major national and international astronomical facilities for interferometry have been, and are being, built around the world. The goal is to improve on angular resolution and/or measurement precision with respect to individual large telescopes: the diffraction limit of a single aperture diameter $D \sim \lambda/D$, drops to λ/B, defined by the baseline B separating the telescopes. Current baselines are on order of 100 to 200 m, e.g. at the Very Large Telescope Interferometer (VLTI) operated by the European Southern Organisation (ESO), providing a potential improvement of one order of magnitude with respect to the resolution of an individual 8 m class telescope. The imaging resolution in the near infrared (NIR, 1 to 2.5 μm) becomes of order of a few milli-arcseconds. The potential astrometric performance, e.g. the measurement precision of the separation of a binary, is provided by the same geometric factor, scaled by the visibility and photometric signal to noise ratio (SNR), and can therefore reach the micro-arcsecond range, as is the goal of the PRIMA facility at VLTI. Some characteristics of its Fringe Sensor Unit, which contributed to setting up the concept described herein, are described in Gai *et al.* (2004).

In modern telescopes, the ideal flat wavefront from a remote star is partially recovered thanks to active and adaptive optics. The differential piston between telescopes, however, has to be directly measured at interferometric level; this sets a physical limitation to the duration of an elementary scientific exposure, and to the brightness of a source which could be measured with acceptable SNR before coherence degradation. The Fringe Tracker (FT) has the function of measuring in real time the optical path difference (OPD) perturbations on a bright source for on-line (hardware) or off-line (software) correction, thus allowing long scientific exposures. Moreover, it may correct additional disturbances introduced by the realistic limitations of the interferometer and of the environment. Several critical aspects are discussed in Basden & Buscher (2005).

Current VLTI instruments provide interferometric measurements related to combination of either two or three telescopes at a time; this is less efficient than simultaneous combination of a larger number N of apertures, also because of the operation overheads. The number of baselines grows more than linearly with N, thus providing significant advantages for new, multiple beam combiners. A promising concept for combination of four, six or eight beams, able to cope also with significant upgrades of VLTI, is proposed by an European team to ESO: the VLTI Spectro-Imager (VSI), described in Malbet *et al.* (2006). VSI aims at NIR measurements with good spectral and spatial resolution. Our study of a suitable Fringe Tracker, identified as a crucial contribution, has been endeavoured in parallel to development of the scientific beam combiner.

2. The VSI Fringe Tracker Concept

A key difference between a scientific N beam combiner and a fringe tracker is that the latter does not require measurement on all baselines involved. Since any telescope is affected by an independent piston contribution, the problem's complexity (a number of unknowns) grows linearly with N. Using pair-wise combination, each beam is combined with only two other beams, thus providing N complementary interferometric outputs which allow for the N simultaneous phase estimates. This is actually a redundant measurement, since the interferometer is not usually sensitive to a common piston contribution, as in the case of a single telescope. The redundancy is minimal, as only one additional measurement is performed with respect to the minimum $N - 1$ value. However, this improves the robustness at a system level, in particular with respect to infrastructure and environmental disturbances introducing occasional flux dropouts or large piston fluctuations in any of the beams. Such events could be identified on the appropriate beam, but the remaining telescopes may remain linked and stabilised with respect to each other, providing a graceful degradation during the disturbance and easier interferometer recovery when the nominal conditions are restored. Minimum redundancy thus provides better rejection of system noise.

The FT concept is based on few bulk optical components, simultaneously serving all (or most) of the N beams; this ensures mutual alignment of the beams, with respect to both angular and longitudinal degrees of freedom, to within a fraction of wavelength, thanks to normal optical engineering tolerances. Such approach was used e.g. in laboratory demonstrators described in Ribak *et al.* (2007), with quite satisfactory results in easing the alignment and overall resiliency to disturbances. Bulk optics is a proven technique which can be extended efficiently to a combination of up to eight telescopes.

3. Conclusions

The VSI and FT concepts have been submitted to ESO and to national agencies for evaluation of the concept maturity and approval of the development plan.

Acknowledgements

We acknowledge the financial support of INAF - ref. 1478 (2005) for contributions to the FT lab prototype.

References

Malbet, F. *et al.* 2006, *Proc. SPIE* 6268, 62680Y
Basden, A. G. & Buscher, D. F. 2005, *MNRAS* 357, 656
Gai, M. *et al.* 2004, *Proc. SPIE* 5491, 528
Ribak, E. N., Gai, M,, Loreggia, D., & Lipson, S. G. 2007, *Optics Letters* 32, 1075

A Giant Step: from Milli- to Micro-arcsecond Astrometry
Proceedings IAU Symposium No. 248, 2007
W. J. Jin, I. Platais & M. A. C. Perryman, eds.
© 2008 International Astronomical Union
doi:10.1017/S1743921308018760

Non-tidal vertical variations and the past star catalogs

Z. X. Li

Shanghai Astronomical Observatory, Shanghai 200030, China
email: shaolzx@yahoo.com

Abstract. Tidal vertical variations have been removed in astrometric observations, but the non-tidal ones remain unknown. From the repeated observations of gravimetric networks in China, the non-tidal vertical variations at Beijing-Tangshan and West Yunnan are determined, which are of the order of 0″.2. Astrometric results, including past star catalogs, based on these observations should be corrected.

Keywords. astrometry, catalogs

Are there non-tidal vertical variations on Earth? It is an important problem in astronomy since ground optical observations are related to the local vertical (IAU, 1991). Until now only tidal vertical variations are removed in the observations while the non-tidal ones remain unknown. Past star catalogs are still suffering from the uncorrected non-tidal vertical variations. It is one of the reasons why the problem of non-tidal vertical variations could not be solved by astrometric techniques while these contaminated star catalogs were being used. Gravimetric technique is able to determine it but repeated gravity measurements performed on a network are needed. In China two gravimetric networks, located in Beijing-Tangshan and West Yunnan, were established in the 1980s, and gravity measurements with high precision are being performed 2 to 4 times per year ever since. Our study has demonstrated the possibility of using these gravimetric observations to determine non-tidal vertical variations (Plumb Line Variations: PLV) there (Li *et al.*, 2005). Tangshan (118.6°E, 39.7°N) and Midu (100.4°E, 25.3°N) have been chosen as the two sample sites where the PLV are determined. Figures 1 and 2 show the results.

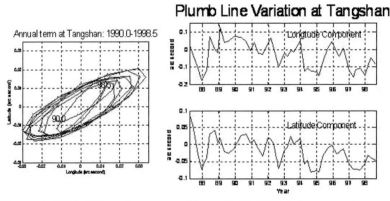

Figure 1. PLV at Tangshan: Annual term (left); PLV (after removing annual term)

It can be seen that the annual term of the PLV at these two sites is of the order of 0″.02–0″.03, while the PLV (after removing annual terms) is of the order of 0″.2. In Fig.3, it appears that the PLV at Tangshan is related to the earthquakes around ("PLV main

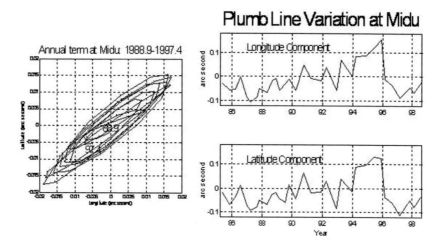

Figure 2. PLV at Midu: Annual term (left); PLV (after removing annual term)

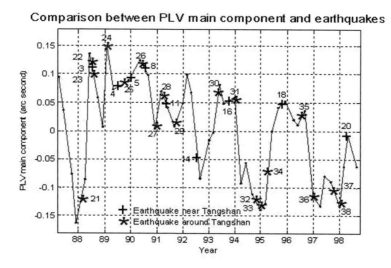

Figure 3. Comparison between the PLV at Tangshan and the earthquakes during 1987-1998

component" is the PLV component in the direction of azimuth of $66.5°$, while larger PLV components are shown). Similar phenomenon has also been found in the case of Midu PLV at West Yunnan.

Beijing and Yunnan observatories are located in or near these two regions, but in the past the non-tidal vertical variations (PLV) have never been considered. They exist in the past observations and in the star catalogs based on these uncorrected observations.

References

IAU 1991, *Trans Int Astronomical Union*, XXIB, 209

Li, Z. X., Li, H., & Han, Y. B. 2005, *Journal of Geodesy*, 78, 593

A Giant Step: from Milli- to Micro-arcsecond Astrometry
Proceedings IAU Symposium No. 248, 2007
W. J. Jin, I. Platais & M. A. C. Perryman, eds.

© 2008 International Astronomical Union
doi:10.1017/S1743921308018772

Application of MGC method to centering saturated and stretched images†

Y. D. Mao[1,2], Y. Li[1,2] and Z. H. Tang[1]

[1]Shanghai Astronomical Observatory, Chinese Academy of Sciences, Shanghai 200030, China
email: dundun@shao.ac.cn

[2]Graduate School of the Chinese Academy of Sciences, Beijing 100039, China

Abstract. Sometimes saturated and stretched star images with bad data in their central parts may result in astronomical CCD observations. Considering the special characteristics of the saturated and stretched star images, a method known as MGC(Mean-Geometric-Centering) is put forward for determining the centers of these images. The basic principles and processing pipelines of MGC are introduced, and several noteworthy items in the realization of MGC are discussed. Results of applying MGC to simulated and real data are also presented.

Keywords. techniques: image processing, methods: data analysis, astrometry

1. Introduction

Due to the restriction of linear response range, long-time exposure usually leads to saturated images, especially when using photographic plates we did in the early days. Because artificial satellites and stars have different motions, stretched images always exist whether we are tracking satellites or stars. None of the existing methods(including modified moment, Gaussian function fit, median, and derivative search, etc. Stone(1989)) is available for determining the centers of the saturated and stretched images. Based on the principle of marginal extraction from images smeared with noise, we come up with a new method known as MGC for centering these images. According to the results of applying MGC to simulated as well as real data obtained with 1.56m telescope at Shanghai, MGC was found to be better than the other methods for centering the images.

2. Pipelines and processing results of MGC

The basic principles and pipelines of MGC are as follows:
• Extract a series of edges, starting from the outside and moving to the inside of the saturated or stretched images with growing thresholds that are increased step by step.
• Link the extracted edges and get a series of closed curves that are just like contours of the images, corresponding to different thresholds.
• Calculate the geometric centers of these closed curves. According to an analysis of empirical data, when SNR is greater than certain threshold SNR0, the geometric centers (xc1, yc1), (xc2, yc2), (xc3, yc3), ... (xcn, ycn) are found to be very stable for thresholds in some particular range [TL, TH].
• Average the geometric centers with thresholds in the range [TL,TH] to get new coordinates (xc, yc) ,which we take as the center coordinate of the saturated or stretched star images.

† Supported by the National Natural Science Foundation of China (Grant Nos. 10673026, 10333050, and 10573018) and Science Technology Commission of Shanghai Municipality(06DZ22101).

Table 1 Results with for centering method for saturated star images

Methods	σ_x	σ_y	δ_x	δ_y
Moment			1.8017	1.3749
2D Gaussian	0.1058	0.1082	0.6396	-0.1294
1D Gaussian	0.3153	0.1439	-0.3323	-0.3926
MGC	0.1294	0.0619	-0.1017	-0.1500

Table 2 Results with two centering method for stretched images

FITS	Moment	MGC
1013i13	0.98713±0.00368	0.99231±0.00079
	-0.12031±0.00368	-0.12018±0.00079
	0.17384±0.04564	0.26934±0.01667
	-0.04317±0.04564	0.02710±0.01667
Nref=4	$\sigma_\alpha cos\delta$=1.220	$\sigma_\alpha cos\delta$=0.262
	σ_δ=0.512	σ_δ=0.187
0925i100	0.98732±0.00301	0.99350±0.00040
	0.12031±0.00301	-0.12024±0.00040
	0.02380±0.05775	0.02586±0.00526
	0.11815±0.05775	-0.01775±0.00526
Nref=4	$\sigma_\alpha cos\delta$=1.292	$\sigma_\alpha cos\delta$=0.174
	σ_δ=1.215	σ_δ=0.111

In comparison with other centering methods, we applied MGC to centering a large number of saturated and stretched images. Table 1 and Table 2 contrast the results obtained with different methods for the saturated and stretched star images respectively. According to our experience, a linear model with four parameters is usually good enough. Tycho-2 is used as a reference star catalogue (Høg 1999).

3. Conclusion

• The SNR of star images is a major factor for the centering accuracy. MGC is found to be much better than the other conventional algorithms for centering saturated and, especially, stretched star images.

• Usually star images obtained through marginal extraction are seriously infected with imperfections due to cosmic rays, hot pixels and bad pixels, etc., which may directly affect the processing results of MGC. Hence it is very important to preprocess the images with an appropriate filtering method. To avoid distorting the original images, we suggest using a median filter.

• Although MGC has many advantages as mentioned above for centering the stretched images, there are still some shortcomings. For example, too many man-induced factors have been included in determining SNR0 and [TL,TH]. In order to be more objective, we need to first experiment with simulated data that were generated with identical observing conditions, and then determine the SNR0 and [TL,TH] values through an analysis of the experiment results.

References

Stone R. C., *et al.* 1989, *AJ*, 97, 1227
Høg E., *et al.* 1999, *A&A*, 357, 367

A Giant Step: from Milli- to Micro-arcsecond Astrometry
Proceedings IAU Symposium No. 248, 2007
W. J. Jin, I. Platais & M. A. C. Perryman, eds.

© 2008 International Astronomical Union
doi:10.1017/S1743921308018784

Antarctic Project for astrometric observations

P. P. Popescu, P. V. Paraschiv, D. A. Nedelcu and O. Badescu

Astronomical Institute of the Romanian Academy
Str. Cutitul de Argint 5 Bucharest 040557, Romania
email: [petre,paras,nedelcu,octavian]@aira.astro.ro

Abstract. Astronomical Institute of Romanian Academy initiated this project in 2005. The results of researches related to the construction of mechanic systems and the proposed astrometric tasks will be finished in 2008. The module PROTEL is a result of several research groups and it will perform astrometric control observations remotely.

Keywords. astrometry, reference systems, surveys

1. Objectives

The aim of this project is to achieve a robotic telescope adapted to polar environment activity. The autonomous module for Antarctic astronomical surveys PROTEL (Polar Robotic Telescope)- represents an automatic telescope designed to work without human intervention in conditions of low temperature, imposed by polar environment. Astroclimate studies have indicated an annual average temperature falling down to $-25°C$, in area where the instrument is intended to be sited. That implies the use of special kinds of materials, optical components, electronic devices, and operating systems adapted to work in extreme cold. The lack of human participation imposes a completely automated, easy to use instrument. The astroclimate conditions there allow a highly precise detection of images. Thus it is possible to: observe near-Earth celestial bodies, improve the stellar reference frame in the neighborhood of extragalactic radio stars, study the mutual phenomena of the bodies in the Solar System, study stellar systems visible only from the Southern hemisphere, study the vertical deviation by astro-geodetic methods. Carrying out astronomical-geodetic determinations in the Southern hemisphere will allow to enlarge of the database which contains: local reference points of the extra-galactic radio-sources, positions, proper motions and rotation periods of the NEOs, mutual phenomena of the bodies in the solar system.

2. Scientific and technical aspects of the project

The first step performed is the technological study which contains the main parameters and attaining goals. Second step is the elaboration of the astroclimate study using the already obtained data-base and in site determination at Romanian-Australian scientific polar station Law-Racovita. The module will be built based on the following studies, projects and testing of materials and components in Romania and in polar region. After performing the final test in Romania, the module shall be installed in site by our scientific associates, after a preliminary training. Observational results are intended to be stored in data-bases, for analyzing and processing. The astro-geodetic observational programs will be performed through satellite communications. The meteorological conditions factor we mean the astroclimatic conditions during the testing and then during the achievement of the results is the major risk which the working team will incur. The materials and the technologies used can be greatly influenced by the environment conditions, there may be differences between the tests made at home and the real conditions in the work area. We do not exclude the possibility of contrary conclusions in some researches, which, however, cannot affect the general background of the problem, conclusions which will necessitate supplementary efforts and project modifications. The optical tube is a OGS - Ritchey Chretien 14".25 f/8 telescope, adapted to hard work at low temperatures ($-40°$C). The optical field of such a telescope in conjunction with a corresponding CCD camera is estimated to be at least 15×15 arcminutes, having an angular pixel resolution of 0.3 arcseconds/pixel. The positioning system adopted for the optical tube is a completely automated Fork mount. By means of a software downloadable in the memory of main command and control system, the mount enables the pointing of telescope toward the target - stellar areas. A GPS used as a temporal reference system synchronizes the inner time of instrument with the universal time.

3. Conclusion

We identify some new elements in the design of a Romanian astronomical instrument: a completely autonomous running module conceived as a specific data capturing robot, successful operating in the extreme conditions of environment, maintenance by data flow via satellite communications.

References

Popescu, P. *et al.* 2007, in: Fifty Years of Romanian Astrophysics *AIP Conference Proceedings* *895*, 195

Popescu, P. *et al.* 2006, in: Astronomy in Antarctica *IAU GA Prague, SPS7*, 487

A Giant Step: from Milli- to Micro-arcsecond Astrometry
Proceedings IAU Symposium No. 248, 2007
W. J. Jin, I. Platais & M. A. C. Perryman, eds.

© 2008 International Astronomical Union
doi:10.1017/S1743921308018796

Astrometry from mutual event and small-separation CCD imaging †

Q. Y. Peng[1,3], N. V. Emelyanov[2], L. Zhou[1] and W. R. Gu[1]

[1]Department of Computer Science, Jinan University, Guangzhou 510632, China
email: pengqy@pub.guangzhou.gd.cn

[2]Sternberg Astronomical Institute, 13, Universitetskij prospect, 119992 Moscow, Russia
email: emelia@sai.msu.ru

[3]Joint Laboratory for optical astronomy, CAS, Kunming 650011, China

Abstract. In order to determine precisely positions of the Galilean satellites of Jupiter, it is useful that shortly before and/or after photometric observations of a mutual event, normal CCD imaging observation is also performed for the two small-separation satellites. Experimental observations showed that the two observational types, on the whole, could derive mean (O-C) s with very good internal and external agreement (about 15–20 mas).

Keywords. planets and satellites: general, astrometry, eclipses, occultations

1. Introduction

At present, the newest theories of the Galilean satellites are the L1 in the IMCCE and JUP230 in NASA/JPL. Among modern ground-based CCD observations, normal CCD imaging and mutual event are two commonly used types for precise measurement of the Galilean satellites. Normal CCD imaging refers to the measurement of relative coordinates in the focal plane of the used telescope, and mutual event to the measurement of the epoch of an event (an eclipse or occultation of a satellite due to another satellite). Normal CCD imaging needs usually a long focal length telescope but it is not easy to calibrate its small field of view of the CCD due to the shortage of the suitable reference stars. However, for a small-separation satellite-pair, the calibration could be good enough since we can deliver calibration parameters (scale factor and orientation) by comparing the pixel positions of greater-separation satellites with their positions from theoretical ephemerides. These parameters would have enough precision for the measurement of small-separation satellite-pair. Therefore, it is very useful that before and/or after photometric observation of a mutual event, normal CCD imaging is also performed for the two small-separation satellites. In this paper, we describe our two types of observations taken at the Yunnan Observatory by 1-m telescope and their reductions, and show their results.

2. Observations and their reduction

In 2003, some mutual events of the Galilean satellites of Jupiter predicted by Arlot (2002) were observed at the Yunnan Observatory by the 1-m telescope with 1024 × 1024 CCD. Seven effective light curves for the mutual events were obtained. For four of these mutual events, we also performed normal CCD imaging observations before these events. In order to measure the pixel position and photometry of the Galilean satellites, we used a

† supported by the National Science Foundation of China (Grant Nos 10573008 and 10778617)

Table 1. (O-C) results (unit: arcsec)derived from mutual events and normal CCD imaging, respectively. J1, J2, J3 and J4 represent Io, Europa, Ganymede and Callisto, respectively.

Obs date	Event	Mutual	event	CCD	imaging
		(O-C)x	(O-C)y	<(O-C)x>	<(O-C)y>
Feb 18	J4OJ3	-0.053	0.117	0.034	0.025
Feb 20	J1OJ2	-0.014	0.021	-0.028	0.028
Feb 27	J1OJ2	-0.030	0.015	-0.036	0.034
Feb 28	J1EJ4	0.052	0.176	-0.012	-0.004

Notes:
[1].for the event J4OJ3 on Feb 18, observers suspected some time-recording error in the observations due to the lack of experience
[2].the event J1EJ4 on Feb 28 was interrupted during observing soon after the maximum drop since the eclipsed satellite entered the shadow of Jupiter

new image-processing technique (Peng *et al.* 2007). To reduce our normal CCD imaging observations, the newest theory of L1 (Lainey 2004a, 2004b) in the IMCCE for the Galilean satellites and JPL DE405 for Jupiter was adopted to compute their theoretical positions. The reduction of mutual events is the same as that in Emelyanov & Gilbert (2006). Specifically, we do not model the apparent relative motion of one satellite with respect to the other but the deflection of the observed relative motion from the predicted motion provided by L1 ephemeris instead. We take the reflectance properties of satellites from Hapke theory with Hapke parameters. In addition, we use the dependence of the magnitudes of Galilean satellites on angle of rotation and take the solar limb darkening into account.

3. Results and conclusions

After the reduction of the observations of mutual events and normal CCD imaging, we can derive the differences of observed minus computed positions (O–C) in right ascension and declination, respectively. Table 1 shows the detailed results. Despite our lack of experience for the first mutual event observation (J4OJ3) and the unexpected situation took place (J1EJ4) in our observations, it is shown that there is very good agreement for the event J1O J2 taken on Feb 20 and Feb 27, 2003. This agreement exists not only for internal observations (within ~15 mas) of the same observational type but also in the external ones (within ~20 mas) for two different types of observations. However, more observations are needed to test that. It is well known that the principal drawback of mutual event observations is the rarity of such phenomena. But for small-separation satellite-pair observations, they have obvious significance that the 50 arcsec separations occur more frequently than the mutual events—averaging about 50 per month (Pascu 1994). Therefore, we might draw a conclusion that normal CCD imaging observations for a small-separation satellite-pair may have a good precision comparable with that from the mutual event observations.

References

Arlot, J. -E. 2002, *A&A*, 383, 719-723
Emelyanov, N. V. & Gilbert, R. 2006, *A&A*, 453, 1141-1149
Lainey, V., Duriez, L., & Vienne, A. 2004a, *A&A*, 420, 1171-1183
Lainey, V., Arlot, J. -E., & Vienne, A. 2004b, *A&A*, 427, 371-376
Pascu, D., 1994, in: L. V. Morrison & G. F. Gilmore (eds.), *Galactic and Solar System Optical Astrometry*(Cambridge),p. 304-311
Peng, Q. Y., Vienne, A., Lainey, V., & Noyelles, B. 2007, *P&SS* (in press)

A Giant Step: from Milli- to Micro arcsecond Astrometry
Proceedings IAU Symposium No. 248, 2007
W. J. Jin, I. Platais & M. A. C. Perryman, eds.

Searching for sub-stellar companions among nearby white dwarfs, through common proper motion

M. Radiszcz[1,2] and R. A. Méndez[1,3]

[1]Departamento de Astronomia, Universidad de Chile, Casilla 36-D, Santiago, Chile
[2]email: madis@das.uchile.cl
[3]email: rmendez@das.uchile.cl

Abstract. The purpose of this work is to do a systematic, deep search for sub-stellar objects orbiting nearby white dwarfs (WDs). The scientific interest spans testing specific predictions of models of common envelope phase, as well as providing constraints to planetary system evolution in advanced stages of its parent star. Additionally, we seek to explore the hypothesis about the origin of metal lines in hydrogen WDs, produced by the accretion of tidal disturbed asteroidal or cometary material. This could be linked to the presence of a planetary object. Here, we show preliminary results of our near-infrared astrometric project.

Keywords. astrometry, binaries: close, stars: low-mass, brown dwarfs, stars: white dwarfs

1. Sample selection and observations

We select our targets from the most complete known sample of WDs within 20 pc of the Sun, compiled by Holberg *et al.* (2002). Revisions and additional objects have been added (Kawka *et al.* 2004;2006). From WDs with $-80^o < \delta < 20^o$, we select those WDs with young cooling + main sequence ages (<3Gyr), metal lines detected at their atmospheres, and with a possible near infrared excess. Targets in this sample have high proper motion (>0.1 arcsec) , so it is possible to confirm the sub-stellar companion candidates by looking for common proper motion pairs. We have obtained first epoch observations using VLT+NACO in Paranal Observatory in the J-band for 28 WDs. The frames obtained have a high-resolution (FWHM~0.1 arc-sec) go very deep ($J_{limit} \sim$23.5–24 mag). We take advantage of the adaptive optical system by using the WD, which is brighter in the V-band, to correct the observations in the near infrared. We obtained high contrast and resolution diffraction-limited images at J-band, where cooler objects are more prominent. Nevertheless, the small FoV ($\sim$28×28 arcsec^2) does not allow us to explore larger projected separations. Therefore, to achieve a more extended FoV (2×2 arcmin^2) , we have carried out additional observations with PANIC+Baade at Las Campanas Observatory. We obtained first epoch observation for 38 WDs, reaching a 5 σ detection limit at 21< J <22 mag ($\sim$5–10 M_{Jup} orbiting a WD).

2. Data analysis and results

For PANIC data reduction, we use the IRAF pipeline *gopanic*, with very good results. We check for any signs of optical distortion that may have been left uncorrected. For NACO data reduction, we use the pipeline jitter provided by ECLIPSE. We find a very little *"pincushion"* distortion, but it is not significant enough (<0.05 pix for a typical 7 arc-sec of offset) to introduce problems in the reduced images and astrometric solutions.

From 38 first epoch objects observed with PANIC, only 11 have at least a 1 year baseline for second epoch observations. These objects do not show common proper motion companions. For most of the objects, a 1 year of baseline is enough to detect common proper motion over 3σ dispersion from the fainter objects detected on the frames (fainter objects exhibit larger residual RMS).

For the NACO frames, we have second epoch observations for 9 promising targets of which the data for 8 objects have been reduced. Some of these objects have faint candidates around 1 arc-sec from the WD. To properly study objects closer than 1 arc-sec, it is necessary to execute a good PSF subtraction. IRAF DAOPHOT PSF models was not completely satisfactory. The best result could be obtained using a look-up table PSF model, based on a very similar WD (ideally the same WD without close companion) observed with the same instrumental configuration. Residuals after the PSF subtraction could be either real objects, optical effects or simply a PSF fit which does not exactly reproduce the central object. To improve the PSF subtraction, we observed the second epoch objects on two different rotation angles for the FoV. In this way we were able to discard some false detections produced by the NACO optical system (see a PSF subtracted image in Fig. 1-right). Despite the poor PSF subtraction within 1 arc-sec from the WD, it was possible to detect objects with a contrast of $\triangle J \sim 10$ mag, located just ~ 1 arc-sec off the target. Figure 1-left show a possible common proper motion companion, but it is not clear yet whether this object is real or not. Careful analysis of PSF substraction is necessary to confirm this detection. Until now, we have not had a confirmed detection of massive extra-giant planets among WDs. But a more detailed and complete analysis is needed to get a definitive detection.

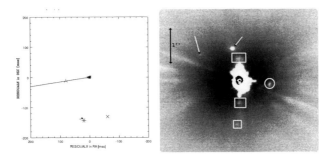

Figure 1. Left-Residuals of the position of stars between two frames from NACO WD residuals are plotted with the filled circle and consistent with the prediction (asterisk symbol) from proper motion (cross) and parallax (symbol over the line). A close object to the WD with a similar residual is marked with a open crossed square. To confirm common proper motion, a better analysis is necessary to discard a ghost image from the WD. Right-Residuals from one of our targets after of the PSF subtraction. Some residuals correspond to real objects (arrows); while others are due to optical effects or imperfect fitting of the PSF model (circles, squares).

References

Debes, J. H. & Sigurdsson, S. 2002, *ApJ*, 572, 556

Livio, M. & Soker, N. 1984, *MNRAS* 208, 763

Holberg, J. B. , Oswalt, T. D., & Sion, E. M. 2002, *ApJ* 571, 512

Kawka, A. & Vennes, S. 2006, *ApJ* 643, 402

Kawka, A., Vennes, S., & Thorstensen, J. R. 2004, *AJ* 127, 1711

Willes, A. J. & Wu, K., 2005, *A&A* 432, 1091

A Giant Step: from Milli- to Micro-arcsecond Astrometry
Proceedings IAU Symposium No. 248, 2007
W. J. Jin, I. Platais & M. A. C. Perryman, eds.

© 2008 International Astronomical Union
doi:10.1017/S1743921308018814

The VLTI as a tool to study eclipsing binaries for an improved distance scale

K. Shabun[1] A. Richichi[1] U. Munari[2] A. Siviero[2] and B. Paczynski[†]

[1]ESO, European Southern Observatory
email: kshabun@eso.org

[2]INAF, Osservatorio Astronomico di Padova

Abstract. Long-baseline interferometry with facilities such as the ESO VLTI is beginning to have the capability to measure directly in the range of milliarcsecond and less the angular separation and the angular diameter of some selected eclipsing binary systems. We have begun to carry out such observations with the AMBER instrument. In the special case of double-lined eclipsing binaries with well-detached components, from radial velocity and light curves it is possible to obtain a full solution of all orbital and stellar parameters, with the exception of the effective temperature of one star, which is normally estimated from spectral type or derived from atmospheric analysis of the spectrum or reddening-corrected photometric colors. In particular, we aim at deriving directly the effective temperature at least of one component in the proposed system, thereby avoiding any assumptions in the global solution through the Wilson-Devinney method. We have obtained an independent check of the results of this method concerning the distance to the system. This represents the first step toward a global calibration of eclipsing binaries as distance indicators. Our results will also contribute to the effective temperature scale for hot stars. The extension of this approach to a wider sample of eclipsing binaries could provide an independent method to assess the distance to the LMC.

Keywords. techniques: interferometric, binaries: eclipsing, stars: distances, stars: fundamental parameters, Magellanic Clouds

1. Introduction

We are using AMBER at the VLTI, currently offered on UTs in JHK bands, in the combination that provides the highest angular resolution, i.e. triplets which include the UT1-UT4 baseline and the low-resolution LR-HK mode. This mode provides dispersed visibilities over the $1.5 - 2.4$ micron range.

Our goal is to measure a restricted number of EB systems with a (semi-major axis) in the range $1 - 5$ mas, and angular diameters θ_1, θ_2 to $0.4 - 1.2$ mas. This will be accomplished by observing each system at least four times, at a precise separated phase of the orbital period P corresponding to the maximum separation. Time has been granted to observe δ Ori, $a = 1.4$, $\theta_1 = 0.98$, $\theta_2 = 0.55$, $P = 5.73$d (distance 280 pc), in the ESO Period 78. Unfortunately only two observational runs have been successfully accomplished in December 2006 and March 2007, the rest two runs were deferred to Period 80. Also four additional observations of δ Ori has been granted for the Period 80, as is the time for observations of η Ori, $a = 1.7$, $\theta_1 = 0.84$, $\theta_2 = 0.7$, $P = 7.98$d and R CMa, $a = 1.1$, $\theta_1 = 0.65$, $\theta_2 = 0.5$, $P = 1.13$d.

2. Overview

Each observation gives us 3 visibilities and one closure phase for the system. These values allow us to determine the separation and the orientation of the system. We observe

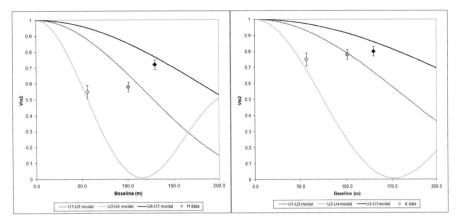

Figure 1. Averaged H- and K-band squared visibilities of δ Ori, observed on 31.12.2006 with AMBER UT1-UT3-UT4 (dots). The model fit (curves): separation 1.9 mas, position angle 75 degrees, diameters of 0.9 and 0.6 mas respectively. Each dot/curve color corresponds to a different baseline.

the systems at the time of maximum separation, at the opposite orbital phases. The figure 1 shows the averaged squared visibilities of δ Ori observed by AMBER at 31.12.06. The symbols are the visibilities averaged over the wavelengths, H-band 1.72 and K-band 2.26 microns, in reality we have 16 channels in the K and 11 channels in H band. The solid curves in both figures represent the (preliminary) best fit to the data of the separation and position angle of the binary, with respect to the length and position angle of the baselines. The figures show a good match between the model and observed visibilities. Not all parameters used in the fit are available from the literature, and in addition there is a wide scatter of values. However, we note a general agreement with the results of our preliminary fit.

The high quality radial velocity and photometric light-curves for the proposed targets are currently being collected at Asiago Observatory. A total of 49 high resolution Asiago Echelle spectra of δ Ori have already been collected in 10 different nights during the time interval of ESO Period 78. The spectra cover the range from 3800 to 7300 Å at a resolving power 30 000. We have achieved a S/N per pixel on the extracted spectrum in excess of 150. In addition, accurate B,V photoelectric photometry has been collected in 12 separate nights to map out the lightcurve of an eclipse.

We stress that this observational breakthrough is made possible for the first time by a combination of the long VLTI baselines with the accuracy, sensitivity and wavelength range of the UT+AMBER. Full solutions (masses, radii, effective temperatures, distances) will be obtained for the proposed EBs after the completion of the full observational program. If our initial observations are successful, we plan to expand them to a wider sample of binaries with diverse spectral types. One of the possible applications of the completed calibration is the precise determination of the distance to the Large Magellanic Cloud.

References

Paczynski, B. 1997, *Space Tel. Sc. Inst.* p. 273

Munari, U., Siviero, A. *et al.* 2004, *A&A* 418, L31

Siviero A., Munari U. *et al.* 2004, *A&A* 417, 1083

Richichi, A. *et al.* 2005, *A&A* 431, 773

A Giant Step: from Milli- to Micro-arcsecond Astrometry
Proceedings IAU Symposium No. 248, 2007
W. J. Jin, I. Platais & M. A. C. Perryman, eds.

© 2008 International Astronomical Union
doi:10.1017/S1743921308018826

The new CCD Zenith Tube

C. Ron, V. Štefka and J. Vondrák

Astronomical Institute of the Academy of Sciences of the Czech Republic,
Boční II, 141 31 Prague, Czech Republic
e-mail: ron@ig.cas.cz, stefka@ig.cas.cz, vondrak@ig.cas.cz

Abstract. The reconstruction of the Photographic Zenith Tube (PZT) at Ondřejov has been finished recently. The main improvement of the instrument consists in the replacement of the photographic plate with the CCD chip, and a completely new electronic control system. In addition to the astrometric use of the observations we intend to derive the variations of the local vertical and compare them with gravimetric observations of the nearby station Pecný.

Keywords. astrometry, telescopes

1. Introduction

In 2004, after more than 30 years of the permanent observations of PZT at Ondřejov Observatory we were confronted with the question either to close the observations and dismount the instrument or to attempt to raise money for a radical reconstruction. The project was financially supported by the Academy of Sciences of the Czech Republic.

2. The original PZT

The Photographic Zenith Tube (PZT) manufactured by Carl Zeiss, Jena in 1969 was installed at the Ondřejov Observatory in 1972. The permanent observations started in 1973 and interrupted at the end of 2005. The instrument, described in detail by Weber (1977), has an objective 250mm in diameter and 3780mm in focal length. The photographic plate was placed in a carriage which followed the star during the exposure. The synchronization with UTC was done by a crystal-controlled phonic motor. The controlling quartz clock was synchronized with the time signal OMA50 till 1993, and with DCF77 afterwards. During the whole period of 33 years 306 stars of 4.6–11.0 mag were observed. Several readjustments of the working catalogue were done by Weberová & Weber (1976), Vondrák (1980), Ron & Vondrák (1985), and finally Ron & Vondrák (2003). Due to precession, stars are leaving the visual field of the instrument and less than 200 stars from the catalogue were observable in 2005. The photographic plates (40×40mm) were measured at the measuring machine ASCORECORD and the reduction of the observation after Vondrák (1978) was used. The average rms of about 7μm was reached in the last few years.

3. The new CCD Zenith Tube

The reconstruction started in 2006 and was finished in July 2007. The principle of the PZT has not been changed. The photographic plate was replaced by the CCD chip, Kodak KAF 6303, with the size 20×30 mm and 2048×3072 pixels. The CCD chip is smaller than the original photographic plate so the visual field is narrower by about 3′ in declination. The mechanics of the instrument has been simplified substantially using the stepper motor to move the carriage with the chip. The exposure time is optional

Figure 1. The binary HIP 106038 at the photographic plate (2 figures on the left) and at the CCD (on the right). The distance of the components is $9''$ ($1''$ corresponds to $18\mu m$ or 2 pixels on the chip). The components (9.5 and 10.6 mag) were not separable on photographic plates.

from 2s for the bright stars (4–5mag) to 16s for the faintest stars. To speed up the readout time of a chip, only a half of image with the observed star is saved after each exposure. This enables to simplify requirements for the cooling of the chip. The cooling is done by two connected Peltier coolers located outside the instrument. The time is synchronized with GPG24A, the precise time and frequency generator synchronized with the UTC time scale via GPS satellite system. The device generates second time marks with the accuracy better than $0.1\mu s$, which is used to start the exposures directly. The working catalogue Ron & Vondrák (2003) has been supplied with new stars taken from the catalogue ARIHIP (Wielen *et al.* 2001) and comprises now 325 stars up to 12 mag. The observations started in August 2007 and 15 observing nights have been performed till October 2007. The coordinates of the optical center of the star image are determined using program SExtractor (Bertin 2007). The reductions of observations after Vondrák (1978) is used again with a few modifications and the average rms reaching 3-4μm.

The CCD Zenith Tube in Ondřejov will be used for astrometry and to determine the variation of the local vertical, both in automatic operational mode. The variations of the vertical will be compared with gravimetric observations at the nearby station Pecný equipped with the absolute and superconducting gravimeters. The substantial improvement of the resolution and sensitivity (up to 15 mag) is achieved, see Fig. 1 and the accuracy is better about 2–3 times.

Acknowledgements

The contribution of M. Wudia († in July, 2007) to the successful reconstruction of PZT is indubitable. This study was supported through the Research plan AV0Z10030501 of the Academy of Sciences of the Czech Republic and through the grant LC506 of the Ministry of Education, Youth and Sports of the Czech Republic. The catalogue update done by M. Jovanović and I. Milić from the University of Belgrade during their stay at Ondřejov observatory is appreciated.

References

Bertin, E. 2007, `http://terrapix.iap.fr/soft.sextractor`
Ron, C., & Vondrák, J. 1985, *Bull. Astron. Inst. Czechosl.*, 36, 289
Ron, C., & Vondrák, J. 2003, in *Proc. Journées 2002 SRST*, Bucharest, 191
Vondrák, J. 1978, *Bull. Astron. Inst. Czechosl.*, 29, 97
Vondrák, J. 1980, *Bull. Astron. Inst. Czechosl.*, 31, 89
Weber, R. 1977, *Jenaer Rundschau*, 22, 92
Weberová, L. & Weber, R. 1976, Wiss. Z. Tech. Univers. Dresden, 25, 919.
Wielen, R., Schwan, H., Dettbarn, C., Lenhardt, H., Jahreiß, H., Jährling, R., & Khalisi, E. 2001, *Veröff. Astron. Rechen-Inst. Heidelberg*, 40, Kommissions-Verlag G. Braun, Karlsruhe.

A Giant Step: from Milli- to Micro-arcsecond Astrometry
Proceedings IAU Symposium No. 248, 2007
W. J. Jin, I. Platais & M. A. C. Perryman, eds.

A compiled catalogue of reference stars around the selected ERS in the northern sky

V. P. Ryl'kov[1], N. Narizhnaja[1], A. Dement'eva[1], N. Maigurova[2], G. I. Pinigin[2] and Y. Protsyuk[2]

[1]Main Astronomical Observatory of RuAS, MAO RuAS, Russia
email: vryl@gao.spb.ru;

[2]Nikolaev Astronomical Observatory, NAO, Ukraine
email: pinigin@mao.nikolaev.ua

Abstract. A compiled catalogue consisting of more than 22000 reference stars in the range of $10^m - 16^m$ was obtained with 235 fields of δ from $-17°$ to $+80°$ around extragalactic radio sources (ERS) selected from the ICRF list. Initial catalogues were obtained from photographic (PG) and CCD observations.

Keywords. astrometry, reference systems

1. Observational data

One of the important problems of ground-based positional astrometry is the refinement of link between the radio and optical coordinate systems (de Vegt *et al.* 2001, Ryl'kov *et al.* 2005, Ryl'kov *et al.* 2005). A compiled catalogue consisting more than 22000 reference stars (RS) in the range of $10^m - 16^m$ with 235 fields of δ from $-17°$ to $+80°$, centered in the optical CCD-image of faint astrometric ERS (ICRF ERS) was obtained for a link to VLBI-observations (ICRS system). For this purpose 4 original star catalogues were used. Three of them were obtained by PG method at the astrograph (AG) telescopes, the rest at the Axial meridian circle (AMC) with a CCD camera.

Pul ERS. PG observations were obtained at the Normal AG (330/3464) in Pulkovo observatory in the 1990s. There are 74 ERS fields (35 are processed by now) in ~ 300 plates. The number of stars in the catalogue is ~ 4500. They were obtained in a field of $20'$ radius around ERS. The limiting magnitude is 17^m.

PIRS-K. Kiev (PG Intermediate Reference Stars) Catalogue. The observations were carried out for 115 ERS fields (2875 stars) with the $200/4126mm$ AG of Kiev University. In one field there are 25 RS in average within the field of $1°$ diameter centered in ERS.

PIRS-B. Bucharest. The observations were made at the $380/6000mm$ double AG in Romanian National Observatory in Bucharest using the field of $60'$ and 188 ERS fields (~ 4700 stars) were photographed in the 1990s.

AMC1B. The CCD observations were made in the near of ERS at the AMC $180/2480mm$ of NAO. The field ($\alpha \times \delta$) is $60' \times 24'$. 208 ERS fields (~ 17000 stars) were obtained. The AMC catalogue was compiled using the UCAC2 catalogue as reference one.

It was decided to sum up these catalogues, made approximately in the same epoch, for increasing density and accuracy of positions of stars around ERS. It was necessary to bring their positions to the same epoch and to study differences of coordinates of the stars for decreasing the random and systematic deviations because these catalogues were obtained with different telescopes and even by different methods of image registration.

The stars, without proper motions in initial catalogues (IC), were processed in the following way. We believe that the positions from the IC may be averaged, because the faint stars have small proper motions and the difference of epoch of observations does not exceed 10 years. Proper motions of stars were taken from the UCAC2 (N.Zacharies *et al.* 2000) to study the systematic deviations of the significant part of catalogue stars.

2. Forming and research of compiled catalogue

Comparison of positions for common stars in these IC was made. The stars with large differences in positions were excluded from the IC. About 30000 of star measurements were processed in total from four catalogues. More than 10000 stars in 151 fields around ERS with proper motions of the UCAC2 were found by averaging; ~ 11000 stars with no proper motion data were averaged and included in the compiled catalogue. All stars which have proper motions (in δ up to $+45°$) were reduced to the epoch and equinox J2000, for the rest positions have the mean epoch of observations and equinox J2000.

Thus, we obtained the catalogue of positions for 21355 basic stars in 235 fields of $20' - 30'$ radius with ERS. Almost 10650 stars have proper motions from the UCAC2. At present the compiled catalogue is not homogenous because many fields contain few RS, less than ten. They are mainly fields which were not included into programs of observation in Pulkovo and Nikolaev. They will be extended with observations in future.

The individual catalogues of each field compiled by us should be checked up for errors by comparison, at least, with 2 independently compiled catalogues. We chose these:
 – the UCAC2, which has stars brighter than 16^m in δ up to $+40° - +50°$;
 – the CMC13 (Carlsberg Meridian Circle, N13) for stars in δ from $-3°2$ to $+30°2$.

At first we chose ten fields in declinations from $0°$ to $+30°$, where the number of common stars is great for studying the accuracy of the compiled catalogue. It is evident that the systematic deviations in α and δ exist. It should be noted, that they are less than $100mas$ in α, but reach $200mas$ in δ in several fields. If you take into account that the comparison catalogues (CMC13 and UCAC2) have both internal and external errors of $50 - 100mas$ for $12^m - 16^m$ stars, then the compiled catalogue is in good agreement with the accuracy limits. The majority of the stars have the photometry of $12^m - 15^m$ from the UCAC2 magnitude and accuracy scatter in the limits $\pm 0''5$ for both coordinates.

If you take into account that stars without identification with the UCAC2 are mainly faint, it is possible to consider that the centre value of distribution from magnitude in range of $14^m - 15^m$ is based to the fainter stars. We found mean values of deviations for the coordinates of RS from their positions in the UCAC2. There were reduced all common stars for 200 fields with ERS. Several fields were compared with CMC13.

One can make a conclusion from the comparison, that several fields have significant deviations in α and δ. These circumstances indicate requirements of regular observations and improvement of stellar positions of the reference catalogues for obtaining the positions of ERS.

The research results of the compiled catalogue for 235 fields with ERS is available on magnetic disks in laboratory of astrometry of MAO RuAS and NAO, Ukraine.

References

de Vegt, C., Hindsley, R., Zacharias, N., & Winter, L. 2001, *A.J.*, 121, 2815-2818

Ryl'kov, V., Dement'eva, A., Narizhnaya, N., Pinigin, G., Maigurova, N., Protsyuk, Yu., Bocsa, G., Popescu, P., & Kleschenok, V. 2005, *Kinematics and Physics of Celestial Bodies, Suppl.*, 5, 328-332

Ryl'kov, V., Dement'eva, A., Narizhnaya, N., Maigurova, N., Pinigin, G., Protsyuk, Yu., Bosca, G., & Popescu, P., 2005, *Romanian Astronomical Journal, Suppl.*, 15, 131-136

A Giant Step: from Milli- to Micro-arcsecond Astrometry
Proceedings IAU Symposium No. 248, 2007
W. J. Jin, I. Platais & M. A. C. Perryman, eds.

Astrometry with the VLTI: calibration of the Fringe Sensor Unit for the PRIMA astrometric camera

J. Sahlmann[1,2†], R. Abuter[1], S. Ménardi[1] and G. Vasisht[1,3]

[1]ESO, Karl-Schwarzschild-Str. 2, 85748 Garching bei München, Germany
email: jsahlman@eso.org

[2]Observatoire de Genève, 51 Ch. des Maillettes, 1290 Sauverny, Switzerland

[3]JPL-Caltech, 4800 Oak Grove Dr., Pasadena CA 91109, USA

Abstract. The future PRIMA facility at the Very Large Telescope Interferometer (VLTI) in astrometric mode offers the possibility to perform relative narrow-angle astrometry with 10 micro-arcsecond accuracy. This is achieved with a dual-beam interferometer concept, where a reference star and the scientific target, confined in a 60 arcsecond field, are observed simultaneously. The angular separation of the two stellar objects gives rise to an optical delay in the interferometer, which is measured by the Fringe Sensor Unit (FSU) and an internal laser metrology. PRIMA is using two FSU fringe detectors, each observing the interference of stellar beams coming from one of the two objects and measuring the corresponding phase and group delay. The astrometric observable, yielding the angular separation, is deduced from the group delay difference observed between the two objects. In addition, the FSU phase delay estimate is used as error signal for the fringe stabilisation loop of the VLTI. Both functions of the FSU require high precision fringe phase measurements with a goal of 1 nm rms (corresponding to $\lambda/2000$). These can only be achieved by applying a calibration procedure prior to the observing run. We discuss the FSU measurement principle and the applied algorithms. The calibration strategy and the methods used to derive the calibration parameters are presented. Special attention is given to the achieved measurement linearity and repeatability. The quality of the FSU calibration is crucial in order to achieve the ultimate accuracy and to fulfill the primary objective of PRIMA astrometry: the detection and characterisation of extrasolar planetary systems.

Keywords. instrumentation: interferometers, astrometry, techniques: interferometric

1. Introduction and measurement principle

The Fringe Sensor Unit is the central element of the PRIMA (Phase Referenced Imaging and Micro-arcsecond Astrometry) dual-feed facility (Delplancke *et al.* 2006, Mottini *et al.* 2005). Two identical FSU fringe detectors deliver real-time estimates of optical path difference (OPD), group delay (GD) and fringe visibility for the two observed targets. Periodic variations of the relative target position can reveal the presence of an orbiting planet (Launhardt *et al.* 2008). The FSU operates in the infrared K-band and produces four ABCD-signals with a phase spacing of 90° in three spectral bands, respectively. Phase delay (ϕ_i), OPD and GD are computed from the fluxes I_A, I_B, I_C, I_D in the three bands with a modified ABCD algorithm (Eq. 1.1).

$$\tan \phi_2 = \frac{(I_A - I_C)\,\gamma - (I_B - I_D)\,\alpha}{(I_B - I_D)\,\beta - (I_A - I_C)\,\delta}, \quad \text{OPD} = \frac{\lambda_2}{2\pi}\,\phi_2, \quad \text{GD} = \frac{1}{2\pi}\frac{\lambda_3 \cdot \lambda_1}{\lambda_3 - \lambda_1}\cdot(\phi_3 - \phi_1), \quad (1.1)$$

† Present address: ESO, Karl-Schwarzschild-Str. 2, 85748 Garching bei München, Germany.

where λ_i are the effective wavelengths and $\alpha = \sin \phi_C$, $\beta = 1 + \cos \phi_C$, $\gamma = \cos \phi_B + \cos \phi_D$ and $\delta = -\sin \phi_B - \sin \phi_D$ take the phase shift errors ϕ_B, ϕ_C and ϕ_D into account. λ, ϕ_B, ϕ_C and ϕ_D in three bands are the parameters to be derived from the calibration.

2. Calibration procedure

We use Fourier Transform Spectroscopy to derive the effective wavelengths. Several OPD scans over the white light fringe packet are performed, while the FSU fluxes and the internal OPD are recorded. For each channel, we combine the consecutive scans and compute the effective wavelength from the Fourier Transform modulus. The relative phase shifts of the ABCD channels are derived by crosscorrelating their fringe packets. Offsets of the crosscorrelation-functions with respect to the autocorrelation of channel A are converted into phase on the basis of the effective wavelengths.

3. Results

Two criteria are used to assess the calibration quality: first the OPD and GD measurement linearity and second the repeatability over consecutive calibrations. We define the linearity parameter as the deviation of the measurement slope from unity. We find that the linearity of the OPD measurement over the central fringe (Figure 1) is within the specifications at ± 10 %. In contrast, the GD estimates exhibit large cyclic errors and the linearity is poor with 45 % RMS over the central 12μm range (± 20 % is specified). For 14 calibrations within 35 min, we find the central effective wavelength $\lambda_2 = (2.312\pm0.003)\mu$m. The central band phase shifts are measured to 74° (nominal 90°), 155° (180°) and 233° (270°) for channel B, C and D, respectively, with a 1° uncertainty. The large deviation from the nominal value is due to imprecisions of the FSU optics.

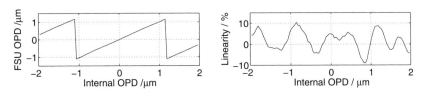

Figure 1. OPD measured by the FSU (left) and the corresponding linearity (right).

4. Conclusions

A calibration procedure for the FSU is established and methods to derive the calibration parameters are presented. The OPD measurement capability of the calibrated FSU is found to be satisfactory. In contrast, the GD linearity is insufficient. Corrective measures on the FSU in terms of hard- and software are currently under implementation. The wavelength determination accuracy has to be improved in order to reach the accuracy goal stated above, whereas the accuracy of the deduced phase shifts is satisfactory.

References

Launhardt, R., Henning, T., Quirrenbach, A., Delplancke, F., Elias II, N. M., Pepe, F., Reffert, S., Ségransan,D., Setiawan, J., Tubbs, R., **and** the ESPRI consortium 2008, *Proceedings of IAUS 248*, in this volume, p.417

Delplancke, F. *et al.* 2006, *SPIE* 6268

Mottini, S., Cesare, S., & Nicolini, G. 2005, *SPIE* 5962, p. 631

A Giant Step: from Milli- to Micro arcsecond Astrometry
Proceedings IAU Symposium No. 248, 2007
W. J. Jin, I. Platais & M. A. C. Perryman, eds.
© 2008 International Astronomical Union
doi:10.1017/S1743921308018851

Finding orbital motion of sub-stellar companions - the case of TWA 5B

T. Schmidt, R. Neuhäuser and M. Mugrauer

Astrophysikalisches Institut und Universitäts-Sternwarte, Universität Jena,
Schillergäßchen 2-3, 07745 Jena, Germany
email: tobi@astro.uni-jena.de

Abstract. TWA 5B is a brown dwarf companion of H= 12 mag, 2" off the ∼5 mag brighter triple star CoD-33° 7795 (=TWA 5), a member of the TW Hydrae association of T Tauri stars at ∼ 55 pc. This object is the first brown dwarf around a pre-main-sequence star (confirmed by common proper motion) ever found. In the last year we have newly reduced VLT NaCo data originally taken in 2003 and combined it with all the available astrometric data of the system to investigate possibly detectable orbital motion of the system. Indeed we were able to find linear orbital motion of the system combining data from HST, VLT and Gemini-North.

Keywords. stars: low-mass, brown dwarfs, stars: pre-main-sequence, stars: imaging, binaries: close, stars: individual (TWA 5A, TWA 5B)

1. Introduction

In 1999, at the time when only two brown dwarfs were confirmed to be companions to normal stars by both spectroscopy and proper motion [Gl 229B (Nakajima *et al.* 1995), G 196-3 B (Rebolo et al. 1998)], Lowrance *et al.* (1999) (here L99) and Webb *et al.* (1999) (W99) suggested independently a sub-stellar companion of TWA 5 in the ∼ 8–10 Myr young TW Hydrae association (Fig. 1). The companion TWA 5B is ∼5 mag fainter than the primary star in the infrared, and its IHJK colors are consistent with spectral type M8 to M8.5 (L99, W99). Weintraub *et al.* (2000) presented additional HST NICMOS narrow-band filter photometry, also consistent with a young late M-type brown dwarf. Neuhäuser *et al.* (2000) presented for the first time infrared spectra and proper motion of the H=12 mag object 2" off the brighter spectroscopic binary star TWA 5 finding the object to be co-moving with TWA 5A from observations with FORS and ISAAC and hence to be the 4th brown dwarf companion around a normal star confirmed by spectrum and proper motion. They derived the mass of TWA 5B to be between ∼15 and 40 M_{Jup} assuming a distance of 55 ± 16 pc estimated from the observation of four other members of the TW Hydrae association by Hipparcos and taking into account the age of TWA. After TWA 5A was resolved as binary (Macintosh *et al.* 2001) with ∼ 55 mas separation, recently Torres *et al.* (2003) and Mohanty *et al.* (2003) reported that TWA 5A is a triple, with one of the resolved stars being a spectroscopic binary. Here, we present first evidence for orbital motion of the young brown dwarf companion TWA 5B.

2. Methods and instruments

We gathered all the available archival observation results from Lowrance *et al.* (1999), Weintraub *et al.* (2000), Lowrance *et al.* (2001), Macintosh *et al.* (2001), Neuhäuser *et al.* (2000), Brandeker *et al.* (2003) and Neuhäuser *et al.* (2001). From these publications we took separation and position angle (measured from north over east to south) data and combined it with a new data point from data taken at VLT-UT4 with NaCo by Masciadri *et al.* (2005) and rereduced by us to look for first indications of orbital motion of the brown dwarf TWA 5B around its primary.

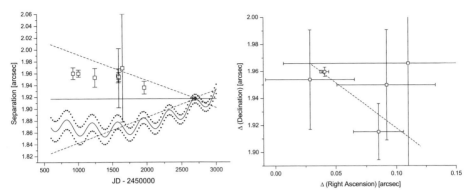

Figure 1. Left: Separations from the literature and our reduction of the 2003 data. The constant straight line indicates the case of bound motion (common proper motion). The wavy continuous line is the change expected if the sub-stellar B component is a non-moving background star. The opening cone enclosed by the dotted lines are its estimated errors. The waves of this cone show the differential parallactic motion which has to be taken into account if one of the components is a non-moving background star. The opening short dashed cone indicates the amplitude of possible orbital motion in the case of circular edge-on orbit. Right: Position of TWA 5B relative to TWA 5A.

3. Conclusions

From separation measurements the background hypothesis can be rejected by $6\,\sigma$ and from position angle by $25\,\sigma$. The deviations from the co-moving case of constant separation and position angle give more than $7\,\sigma$ & $10\,\sigma$ significance respectively for the presence of orbital motion. A linear fit to the nine separation measurements found give a reduced χ^2 of 0.2833 for a declining separation of 9.71 mas/yr corresponding to a probability of 96.1 %. In the near future curvature may be observable in the orbital motion of TWA 5B as final proof of companionship.

References

Brandeker, A.; Jayawardhana, R.; Najita, J. 2003, *AJ*, 126, 2009
Lowrance, P. J.; McCarthy, C.; Becklin, E. E. *et al.* 1999, *ApJ*, 512, 69 (L99)
Lowrance, P.; Becklin, E. E.; Schneider, G. *et al.* 2001, *ASPC*, 244, 289
Macintosh, B.; Max, C.; Zuckerman, B. *et al.* 2001, *ASPC*, 244, 309
Masciadri, E.; Mundt, R.; Henning, T. *et al.* 2005, *ApJ*, 625, 1004
Mohanty, S.; Jayawardhana, R.; Barrado y Navascus, D. 2003, *ApJ*, 593, 109
Nakajima, T.; Oppenheimer, B. R.; Kulkarni, S. R. *et al.* 1995, *Nature*, 378, 463
Neuhäuser, R.; Guenther, E. W.; Petr, M. G. *et al.* 2000, *A&A*, 360, 39
Neuhäuser, R.; Potter, D.; Brandner, W. 2001, *astro.ph*, (010)6304
Rebolo, R.; Zapatero Osorio, M. R.; Madruga, S. *et al.* 1998, *Science*, 282, 1309
Torres, G.; Guenther, E. W.; Marschall, L. A. *et al.* 2003, *AJ*, 125, 825
Webb, R. A.; Zuckerman, B.; Platais, I. *et al.* 1999, *ApJ*, 512, 63 (W99)
Weintraub, D. A.; Saumon, D.; Kastner, J. H.; Forveille, T. 2000, *ApJ*, 530, 867

A Giant Step: from Milli- to Micro-arcsecond Astrometry
Proceedings IAU Symposium No. 248, 2007
W. J. Jin, I. Platais & M. A. C. Perryman, eds.

© 2008 International Astronomical Union
doi:10.1017/S1743921308018863

Observation of the fast NEO objects with prolonged exposure

O. Shulga, Y. Kozyryev and Y. Sibiryakova

Research Institute "Nikolaev Astronomical observatory" (RI NAO), Ukraine.
Email: tttt_nao@mail.ru

Abstract. An original way of application of drift-scan imaging – electronic tracing technique for observation of the fast moving asteroids is presented in this contribution.

Keywords. astrometry, solar system: general, techniques: miscellaneous

1. Introduction

There is a difficulty of observating the fast NEO objects with a speed $V > 10$" sec^{-1}. That kind of objects cannot be observed with prolonged exposure as in classical method which is used for observing stars. Thus, the limiting magnitude of telescope cannot be reached. In order to increase the magnitude of observing objects a tracking mode is necessary. It is possible to have the following tracking modes:

- Digital – stacking of shifted images obtained with short exposure time;
- Mechanical – moving telescope around its two axes precisely;
- Electronic – tracing the electronic charge across the CCD chip.

The electronic tracking technique was designed in RI NAO. Observations are carried out when object is passing through the field of view of an unmovable telescope. Then, the exposure time depends on object's speed and a field size. This method can be used for observations of objects, ranging from stars to satellites in low Earth orbits.

2. The electronic tracing technique

The electronic tracing technique, designed in RI NAO, is based on Drift-Scan Imaging or Time Delayed Integration (TDI). The Drift-Scan Imaging technique is usually used to image long continuous strips of the sky. To make a scan it is necessary to read the lines of the CCD in perfect synchronization with the movement of the stars on the focal plane. It is also important to perfectly align the lines in the east-west direction. In RI NAO this technique has been used on Meridian Circle telescope since 1995. The field of application of the Drift-Scan Imaging technique can be extended considerably with possibility of rotation of the CCD camera to the certain angle. In that case it can be used for observation of any object which is uniformly moving through the field of view. For that purpose special rotation system was designed in RI NAO. The rotation system is a device in which a stepping motor and a angle encoder were equipped and which makes the rotation of CCD camera around optical axis of telescope.

3. Test observations of the fast NEO object

The electronic tracing technique was tested on very small telescope so only relative advantage can be showed. The test of observing asteroid 2007dt103 was realize with this method. On 2007-08-03 the asteroid speed is -3".7/min in right accession, -12".4/min

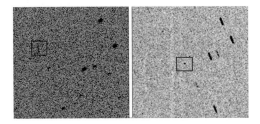

Figure 1. Images from different drift-scan observations. Left: Classical drift-scan imaging, right: Drift-scan with rotation system

in declination, and magnitude 14.1. The asteroid was observed on a unmovable telescope with the classical drift-scan imaging technique and with the rotation system (Figure 1). If exposure time is 170 sec, during the exposure time the asteroid passed $36\overset{''}{.}6$ or 17 pixels on the focal plane.

4. Drift-scan imaging with variable exposure time

The other method to extend field of application of the Drift-Scan Imaging technique is to use variable exposure time. The exposure time of classical drift-scan method depends only on the speed of a target and field size which can't be changed (in full exposure time). But the way to use variable exposure time exists. During the specified exposure time CCD camera works in drift-scan mode (speed of reading out matches with the speed of a target and pixel scale). When the specified exposure time is over frame readout with maximum speed is started without CCD flushing. Variable exposure time is efficiently to use in the case if full exposure time is superfluous and good result can be obtained with much less exposure. The other advantage of variable exposure time is that several frames of the object can be obtained when it is passing through the CCD matrix.

5. Additional reference stars frames

Images obtained with object tracking mode (mechanical or electronic) contains round image of object but stretched images of stars which are difficult for data reduction. To avoid this problem the special additional frames with different mode are made before and after object frame. Telescope is unmovable so it is not a problem to connect the object coordinates and the reference stars coordinates obtained on different frames at different time. The easiest way to obtain reference stars frame on stare telescope is to use very short exposure time (0.05–0.5 sec). When the fast NEO objects are observed with an unmovable telescope it is possible to use drift-scan imaging with variable short exposure time to obtain reference stars frame.

6. Conclusion

The electronic tracking technique can provide results comparable to that obtained by means of a precise tracking telescope for observing various types of objects, such as fast NEO, fast satellites in low Earth orbit and others. The majority of telescopes in Ukraine and Russia which are used for asteroid observations are old and cannot track such objects. The easiest and inexpensive way of modernization for tracking capability is to install the rotating system and CCD camera which supports the drift-scan mode. RI NAO in collaboration with Shanghai Astronomical Observatory carries out a joint project to use the electronic tracking technique.

A Giant Step: from Milli- to Micro-arcsecond Astrometry
Proceedings IAU Symposium No. 248, 2007
W. J. Jin, I. Platais & M. A. C. Perryman, eds.

Occultation by (22) Kalliope and its satellite Linus

M. Sôma[1], T. Hayamizu[2], K. Miyashita[3], T. Setoguchi[4] and T. Hirose[5]

[1] National Astronomical Observatory of Japan, Mitaka, Tokyo 181-8588, Japan
email: mitsuru.soma@nao.ac.jp

[2] Sendai Space Hall, Satsumasendai City, Kagoshima Pref. 895-0005, Japan
email: uchukan@bronze.ocn.ne.jp

[3] Chikuhoku Junior High School, Omi Village, Higashi-Chikuma Gun, Nagano Pref. 399-7701, Japan
email: k_miyash@js5.so-net.ne.jp

[4] Japanese Occultation Information Network, c/o Kagoshima Hi-tech College, Kagoshima City, Kagoshima Pref. 891-0141, Japan
email: set@bronze.ocn.ne.jp

[5] International Occultation Timing Association, 1-13, Shimomaruko 1-chome, Ota Ku, Tokyo 146-0092, Japan
email: thirose@cam.hi-ho.ne.jp

Abstract. The occultation of a 9.1 magnitude star by asteroid (22) Kalliope and its satellite Linus was successfully observed in Japan in 2006 November 7.826 UT. This was the first definite observation of an occultation of a satellite of an asteroid that was discovered previously by other means. As a result the position of the satellite relative to Kalliope was obtained to be $d = 0.246 \pm 0.011$ (arcsec), and $P = 313.8 \pm 2.7$ (deg), where d is the angular distance and P is the position angle. The derived size for Kalliope is (209 ± 40)km $\times (136 \pm 26)$km (with the major axis in position angle of (8 ± 17) deg), and that for Linus is (33 ± 3) km. From the observations, the occulted star is also found to be a close double star whose separation is about 0.7 mas in position angle of about 300 deg, and the magnitudes of the components are found to be almost the same (~ 9.9 mag).

Keywords. occultations, minor planets, asteroids, binaries: close

1. Observations

The occultation of the 9.1 magnitude star TYC 1886-01206-1 (SAO 78190) by asteroid (22) Kalliope and its satellite Linus was successfully observed in Japan at around 2006 November 7.826 UT. The occultation by Kalliope was predicted by the International Occultation Timing Association (IOTA) and the one by Linus was by Jerome Berthier, IMCCE (Institut de Mécanique Céleste et de Calcul des Éphémérides), France. Berthier's prediction was forwarded to us by Jean Lecacheux some 19 hours before the event. Those predictions were publicized to Japanese amateur observers through Japanese Occultation Information Network (JOIN), and observation results were also collected through JOIN.

The occultation by Kalliope was observed at eight stations and that by Linus was observed at other eight stations. Unfortunately the observations could not be made in the northern Japan due to bad weather. Five observers among the eight for Kalliope and three among eight for Linus used video equipment with precise time stamp for the observations so that their times were precise within about 0.03 sec. There was another observer for Linus who made the observation by video, but he failed to get precise times.

2. Results of the analysis

From the observations the diameters of Kalliope and Linus, and the relative position of Linus to Kalliope at the time of the occultation were obtained by a weighted least squares method. The apparent figure of Kalliope is assumed to be an ellipse, and the lengths of its major and minor axes were obtained together with the position angle of the major axis, but for Linus only the western part was observed and furthermore its visual observations were not reliable (the event occurred by about one minute earlier than they expected because their observations were made according to the prediction of the occultation by Kalliope) and therefore for Linus we assumed the apparent figure to be a circle.

The analysis of the observations gives the following results:
the angular separation d and position angle P of Linus relative to Kalliope for the epoch 2006 November 7.826 UT of the occultation are:
$d = 0.246 \pm 0.011$ (arcsec) $P = 313.8 \pm 2.7$ (deg).
The derived diameters for Kalliope are:
(209 ± 40)km $\times$ (136 ± 26)km
and the position angle of the major axis is: (8 ± 17) deg.
The derived diameter of Linus is: (33 ± 3) km.

3. Duplicity of the star

Among the nine video observations eight video tapes of the observations were provided to us. While analyzing the video tapes we found that all of the videos indicated that the occulted star is a close double star. From the analyses of the light curves we have found that their separation is about 0.7 mas at a position angle of about 300 deg, and the magnitudes of the components are almost identical (~ 9.9 mag).

4. Conclusion

An occultation of a 9.1 magnitude star by (22) Kalliope and its satellite Linus was observed in total at sixteen stations in Japan at around 2006 November 7.826 UT. As a result the angular separation of 0.246 ± 0.011 (arcsec) of Linus relative to Kalliope in position angle of 313.8 ± 2.7 (deg) was obtained for the epoch of the occultation. The derived size for Kalliope is (209 ± 40)km $\times$ (136 ± 26)km (with the major axis in position angle of (8 ± 17) deg), and that for Linus is (33 ± 3) km. From the observation the occulted star is also found to be a close double star whose separation is about 0.7 mas at a position angle of about 300 deg, and the magnitudes of the components are found to be almost the same (~ 9.9 mag). Details of the observations and analyses will be published elsewhere.

A Giant Step: from Milli- to Micro-arcsecond Astrometry
Proceedings IAU Symposium No. 248, 2007
W. J. Jin, I. Platais & M. A. C. Perryman, eds.

ESPRI data-reduction strategy and error budget for PRIMA

R. Tubbs[1], N.M. Elias II[1,2], R. Launhardt[1], S. Reffert[2], F. Delplancke[3], A. Quirrenbach[2], T. Henning[1], D. Queloz[4] and ESPRI Consortium[1-4]

[1] Max Planck Institute for Astronomy (MPIA),
Königstuhl 17, D-69117 Heidelberg, Germany
email: tubbs@mpia.de

[2] Landessternwarte (LSW), Heidelberg University Centre for Astronomy (ZAH),
Königstuhl 12, D-69117 Heidelberg, Germany

[3] European Southern Observatory (ESO),
Karl-Schwarzschild-Straße 2, D-85748 Garching bei München, Germany

[4] Observatoire Astronomique de l'Université de Genève
1290 Sauverny, Switzerland

Abstract. The Exoplanet Search with PRIma (ESPRI) will use the PRIMA dual-feed astrometric capability on the Very Large Telescope Interferometer (VLTI) to perform astrometric detections of extra-solar planets. We present an overview of our data-reduction strategy for achieving 10-μarcsecond accuracy narrow-angle astrometry using the PRIMA instrument. We discuss the error budget for astrometric measurements, and those aspects of our strategy which are designed to minimise the astrometric measurement errors.

Keywords. astrometry, planetary systems, instrumentation: interferometers, techniques: interferometric, methods: data analysis, techniques: high angular resolution, infrared: stars

1. Introduction

The Phase-Referenced Imaging and Microarcsecond Astrometry (PRIMA) instrument will provide a dual-beam capability for the ESO VLTI. It is designed to provide 10-μarcsecond-accuracy relative astrometry using the VLTI 1.8-m diameter Auxiliary Telescopes (ATs), as well as phase-referenced imaging with both the ATs and the 8-m Unit Telescopes. Observations will typically measure the separation vector between unresolved targets less than 20 arcseconds apart, with at least one target brighter than K = 11.

2. Error budget for astrometry with VLTI-PRIMA

Fig. 1 shows a summary of the principle sources of error in a 30-minute measurement of the separation of two stars within 10 arcseconds of each other, where one star has K = 8, and the other has K = 13.5 (the dependence of astrometric accuracy on the stellar brightnesses is shown graphically in Launhardt *et al.* (2008)).

The two error sources which remain a particular concern are associated with the VLTI ATs. The ATs were designed before the PRIMA instrument, so they were not built with an interferometric dual-feed capability (for observing two stars simultaneously), and the ATs were not specified to provide a stable narrow-angle baseline for differential astrometry. ESO is currently fitting two ATs with *Star Separator* units to allow dual-feed interferometry. The narrow-angle baseline is defined by the separation vector between the

132

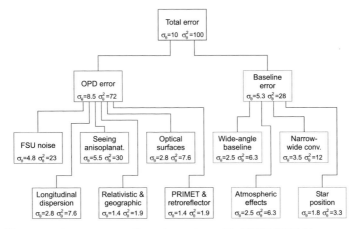

Figure 1. Error requirements tree for astrometry with VLTI-PRIMA. σ_θ represents the allowed RMS contribution towards the error in the measured separation, in μarcseconds.

images of the metrology retroreflector reference points in the entrance pupils of the two ATs. The ESPRI consortium has undertaken an assessment of the narrow-angle baseline stability, and has reached the following conclusions about the current AT design:

(*a*) The motion of the narrow-angle baseline end-points due to typical wind-loading on the ATs is at the limit of that is acceptable for narrow-angle astrometry, when the AT autoguider correctly compensates the induced pointing errors.

(*b*) With the current AT design, the baseline variation due to motion of the AT derotator axis is too large for the stringent astrometric requirements described in Fig. 1. The consortium is in the process of determining the most cost-effective approach for monitoring and/or stabilising the narrow-angle baseline end-points on the ATs. Possible solutions include careful monitoring of the derotator, or relocation of the metrology retroreflector to a point before the AT derotator (probably on M2).

3. Data-reduction strategy

The astrometric data-reduction software (ADRS) processes fringe, delay, environmental, and calibration data. The on-line pipeline averages, fits, or interpolates all raw data onto a fixed $\sim$ one-second time grid. A number of intermediate diagnostic quantities are also generated. The off-line processing corrects delays for instrumental, environmental, and astrometric effects. After searching very large datasets, long-term instrumental trends will be found and fit every $\approx$ six months (see Elias *et al.* (2008)).

The primary observable, after corrections, is the differential delay between the target and reference stars, which is ultimately converted to a separation angle. An ensemble of such data is used to fit planetary orbits with no ambiguity due to the unknown inclination. Other data include squared visibilities and Fourier-transform spectra.

References

Elias II, N. M., Tubbs, R., Köhler, R., Reffert, S., Stilz, I., Launhardt, R., de Jong, J., Quirrenbach, A., Delplancke, F., Henning, T., Queloz, D. & ESPRI Consortium 2008. Astrometric Data Reduction Software and error budget for PRIMA. *In: Proc. IAUS 249 (in press)*.

Launhardt, R., Henning, Th., Queloz, D. *et al.*, 2008, this volume p. 417.

A Giant Step: from Milli- to Micro arcsecond Astrometry
Proceedings IAU Symposium No. 248, 2007
W. J. Jin, I. Platais & M. A. C. Perryman, eds.

Astrometric performance of the Schmidt telescope at the Xuyi station of the Purple Mountain Observatory†

Y. Yu[1], Z. H. Tang[1], Z. X. Qi[1,2] and J. F. Wu[3]

[1]Shanghai Astronomical Observatory, Chinese Academy of Sciences, Shanghai 200030, China
email: yuy@shao.ac.cn

[2]Graduate School of the Chinese Academy of Sciences, Beijing 100039, China

[3]Surveying and Mapping Institute, Zhengzhou 450052, China

Abstract. On January 26, 2006, an area of the sky of 80 deg^2 at $\alpha = 9h$, $\delta = 20^o$ (J2000.0) was observed with the 100/120cm Schmidt telescope at the Xuyi station of the Purple Mountain Observatory. Astrometric performance of the telescope is analyzed by this group of CCD observations. The results show that: (1) the CCD images suffer from complicated distortions, and a third order plate model is recommended for reductions; (2) there is no obvious magnitude equation for objects brighter than 16.5 mag, and (3) for this group of CCD images, the astrometric precision for objects brighter than 16.5 mag is better than 70 mas per coordinate when reduced against the UCAC2 catalog.

Keywords. astrometry, methods: data analysis

1. Introduction

The 100/120 cm Schmidt telescope at the Xuyi station of the Purple Mountain Observatory is mainly used to detect and study the Near-Earth asteroids. The focal length of the telescope is 1.8 m, and the CCD field of view (FoV) is 2 × 2 deg^2 with 4096 × 4096 pixels (1 pixel = 15 microns).

On January 26, 2006, an area of the sky of 80 deg^2 at $\alpha = 9h$, $\delta = 20^o$ (J2000.0) was observed. Altogether 46 CCD images were obtained. The exposure time was set at 40 seconds, and the limiting magnitude is about 18.5 mag. The FWHM of the objects is about 2–3 pixels. The UCAC2 catalog (Zacharias *et al.* 2004) is used as a reference. There are about 1800 reference stars on each CCD image. The measured coordinates of objects are obtained by the method of Gaussian contour fitting.

2. Investigation of astrometric performance

Classical CCD adjustment is performed with the 1–5 order model against the UCAC2 catalog. The potentially poor stars are eliminated if the post-fit residuals are larger than 3 σ. In the case of a 2nd-order model, the standard deviation is about 160 mas, but for the 3rd-order model, it is improved to about 75 mas and that is equivalent to those with higher order models. This precision is also close to the combined error of the reference catalog and measurements. The residual vector distribution, when using a 3rd-order model, is shown in Fig.1(a). The residuals are distributed randomly, indicating that the

† Supported by the National Natural Science Foundation of China(Grant Nos. 10673026, 10333050, and 10573018) and Science & Technology Commission of Shanghai Municipality(06DZ22101).

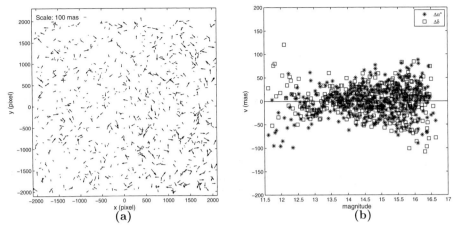

Figure 1. (a) Residual vectors distribution on images reduced with the 3 order plate model; (b)Relationship between the residuals and magnitudes. One point represents 5 stars.

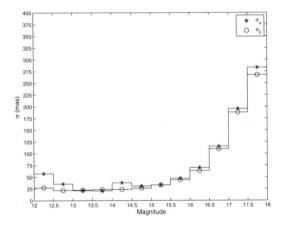

Figure 2. Precision per coordinate as a function of magnitude.

3rd-order model is suitable for reductions. The relationship between the residuals and magnitudes is plotted in Fig.1(b), where one point represents 5 reference stars. Apart from brighter stars (< 13 mag), there is no obvious magnitude equation with respect to the reference stars.

Based on repeated observations, a histogram of the precision as a function of magnitude is shown in Fig. 2. For this group of CCD images, the astrometric precision for objects brighten than 16.5 mag is better than 70 mas per coordinate.

References

Zacharias, N., *et al.* 2004, *AJ*, 127, 3043

A Giant Step: from Milli- to Micro-arcsecond Astrometry
Proceedings IAU Symposium No. 248, 2007
W. J. Jin, I. Platais & M. A. C. Perryman, eds.

© 2008 International Astronomical Union
doi:10.1017/S1743921308018905

A method of accurate guiding for LAMOST†

Y. Yu[1], Z. X. Qi[1,2], Y. D. Mao[1], Z. H. Tang[1] and M. Zhao[1]

[1]Shanghai Astronomical Observatory, Chinese Academy of Sciences, Shanghai 200030, China
email: yuy@shao.ac.cn

[2]Graduate School of the Chinese Academy of Sciences, Beijing 100039, China

Abstract. The main optical instruments of the Large Sky Area Multi-Object Fiber Spectro-scopic Telescope (LAMOST) are installed separately. Guiding is realized by adjusting the normal direction of the mirror MA. In order to keep the central star imaged on the rotation center of the fiber plug plate, the link between celestial and fiber plug plate coordinate systems has to be established. Some local parameters contain errors so that the normal direction of the MA calculated beforehand may not be accurate enough. The execution error of the telescope also drives the image of the central star to deviate from the rotation center. Therefore, in the process of observations the correction has to be calculated and applied periodically. Here we present the method of calculating the guiding parameters as well as the correction parameters for the LAMOST.

Keywords. astrometry, techniques: miscellaneous

1. Principle of calculating the guiding parameter

LAMOST is a special reflecting Schmidt telescope (Wang *et al.* 1996). Its focal length is 20m and the field of view is 5 degrees. 4000 optical fibers are going to be installed in a fiber plug plate laid on the focal plane of the telescope to perform multi-object fiber spectroscopic observations A total of 4 CCDs are installed in the plate to perform the guiding. Different from general purpose telescopes, the Schmidt correction mirror MA, main mirror MB and fiber plug plate are all fixed with respect to ground. Guiding is realized by adjusting the normal direction of the MA. This determines the novelty of realizing the guiding. During observations the central star is required to be continuously imaged onto the rotation center of the fiber plug plate. So the normal direction of the MA is given by

$$\vec{n} = < \vec{S}_{00}^* + \vec{S}_{00} >, \qquad (1.1)$$

where $\vec{S}_{00}^*$ and $\vec{S}_{00}$ represent the direction of the central star and the rotating center of the fiber plug plate respectively. In order to perform calculation, both of them should be expressed in horizontal coordinate system, i.e.

$$\vec{S}_{00}^* = [\mathbf{N}] \begin{pmatrix} cos\alpha_{00} cos\delta_{00} \\ sin\alpha_{00} cos\delta_{00} \\ sin\delta_{00} \end{pmatrix} = [\mathbf{Z}]\overline{R}_2(90^0 - \phi)\overline{R}_3(S_l) \begin{pmatrix} cos\alpha_{00} cos\delta_{00} \\ sin\alpha_{00} cos\delta_{00} \\ sin\delta_{00} \end{pmatrix}, \qquad (1.2)$$

$$\vec{S}_{00} = [\mathbf{F}] \begin{pmatrix} 0 \\ 0 \\ 1 \end{pmatrix} = [\mathbf{Z}]\overline{R}_3(-A_{00})\overline{R}_2(-90^o + H_{00}) \begin{pmatrix} 0 \\ 0 \\ 1 \end{pmatrix}, \qquad (1.3)$$

† Supported by the National Natural Science Foundation of China(Grant Nos. 10673026, 10333050, and 10573018) and Science & Technology Commission of Shanghai Municipality(06DZ22101).

where $[\mathbf{N}]$, $[\mathbf{F}]$ and $[\mathbf{Z}]$ represent equatorial, fiber plug plate, and horizontal coordinate system; A_{00} and H_{00} are horizontal coordinates of the rotation axis of the fiber plug plate, and ϕ and S_l are the astronomical latitude of the MA center and the local sidereal time.

On the other hand, the rotation angle of the fiber plug plate should be adjusted to compensate for the diurnal rotation of the image field and differential rotation caused by atmospheric refraction and aberration (Li *et al.* 2006). Based on the reflection law, the direction vector of the i-th guide stars is calculated and expressed in the fiber plug plate plane coordinate system $[\mathbf{F}]$ by

$$\vec{S}_i = < 2(\vec{S}_i^{*'} \vec{n}) - \vec{S}_i^* > = [\mathbf{F}] \begin{pmatrix} cos\theta_i sin\rho_i \\ sin\theta_i sin\rho_i \\ cos\rho_i \end{pmatrix}. \tag{1.4}$$

Then equation (1.5) is adopted to determine the rotation angle, assuring that each guide star is located near the center of its corresponding CCD. Hereafter λi is the coordinate parameter of the i-th CCD center in the fiber plug plate.

$$\sigma = \frac{\sum_{i=1}^{4}(\theta_i - \lambda_i)}{4} \tag{1.5}$$

2. Detecting deviations of the guiding parameters

In the whole process of observation, correction has to be calculated and applied periodically by observing guide stars on surrounding CCDs.

The theoretical coordinates of the guide stars in the fiber plug plate system $[\mathbf{F}]$ can be calculated based on their catalog positions, i.e.

$$\vec{S}_{ic} = < 2(\vec{S}_i^{*'} \vec{n}) - \vec{S}_i^* > = [\mathbf{F}] \begin{pmatrix} cos\lambda_{ic} sin\gamma_{ic} \\ sin\lambda_{ic} sin\gamma_{ic} \\ cos\gamma_{ic} \end{pmatrix}. \tag{2.1}$$

The corresponding observational coordinates can be obtained from CCD observations and expressed as

$$\vec{S}_{io} = [\mathbf{F}] \begin{pmatrix} cos\lambda_{io} sin\gamma_{io} \\ sin\lambda_{io} sin\gamma_{io} \\ cos\gamma_{io} \end{pmatrix}. \tag{2.2}$$

Then the observation equation is

$$\vec{S}_{io} - \vec{S}_{ic} = \Delta\vec{S}_{00} + \Delta\sigma sin\gamma_{ic} < \vec{S}_{00} \times \vec{S}_{ic} > . \tag{2.3}$$

Using 4 guide stars, offset of the central star relative to the rotating center and rotation angle can be calculated together. Consequently, correction of the normal direction of the MA can be calculated by

$$\Delta\vec{n} = -\frac{1}{|\vec{S}_{00}^* + \vec{S}_{00}|}(< \vec{S}_{00}^* + \vec{S}_{00} > \times(< \vec{S}_{00}^* + \vec{S}_{00} > \times\Delta\vec{S}_{00})). \tag{2.4}$$

References

Li, J. L., Zhang, B., Yu, Y., Qi. Z. X. & Zhao, M. 2006, *Chin. J. Aston. Astrophys.* 6, 495
Wang, S. G., Su, D. Q., Chu, Y. Q., Cui, X. Q. & Wang, Y. N. 1996, *Applied Optics* 35, 5155

A Giant Step: from Milli- to Micro-arcsecond Astrometry
Proceedings IAU Symposium No. 248, 2007
W. J. Jin, I. Platais & M. A. C. Perryman, eds.

© 2008 International Astronomical Union
doi:10.1017/S1743921308018917

Study of centering CCD image of faint satellites near a bright primary object†

Z. H. Tang[1], Y. Li[1,2], Y. R. Yan[1,2], S. H. Wang[1], R. C. Qiao[3] and K. X. Shen[3]

[1] Shanghai Astronomical Observatory, Chinese Academy of Sciences, Shanghai 200030, China.

[2] Graduate School of the Chinese Academy of Sciences, Beijing 100039, China

[3] National Time Service Center (NTSC), Chinese Academy of Science, Shaanxi 710600, China.

Abstract. The polynomial-fit method is applied to remove the uneven background of a satellite when it is near a bright primary object. Detailed analysis of this method is given. Some useful conclusions are drawn from the results of simulated data.

Keywords. astrometry, solar system: general, techniques: image processing, planets and satellites: general

1. Introduction

When observing a faint satellite close to a bright primary object, the satellite is embedded in the primary's halo light, which has a gradient across the satellite's image. Usual methods, such as 2-dimension Gaussian with a constant background fitting, cannot provide a correct center of the satellite.

Many authors presented various methods to remove the systematic effects on centering the position of a satellite's image. But each method is only suitable for a special case. Here we attempt to find a general approach to tackle this problem. The "polynomial-fit' method has been mentioned and employed previously for removing uneven background without any detailed discussions or argumented conclusions given. The goal of this work is to analyse which order of the polynomial is suitable for most cases.

2. Results from simulated data

First, the simulated images of a satellite near a bright primary were created under various assumptions, such as different distances between the satellite and the primary, different intensity ratios between the satellite and the primary, and so on. Second, different orders of the polynomial-fit method are used to fit the background when calculating the center of a satellite. Third, comparisons between the calculated and simulated centers are made.

Normally the images of both the satellite and the primary object can be represented by a two-dimension Gaussian model, with respectively different parameters of peak intensities I_s and I_p, center positions (x_s, y_s) and (x_p, y_p), and Gaussian radius parameters R_s and R_p. The combined intensity of a satellite and a primary can be simulated as:

† Supported by the National Natural Science Foundation of China(Grant Nos. 10673026, 10333050, and 10573018) and Science & Technology Commission of Shanghai Municipality(06DZ22101).

$$I(x,y) = I_p EXP(-\frac{(x-x_p)^2 + (y-y_p)^2}{2R_p^2}) + I_s EXP(-\frac{(x-x_s)^2 + (y-y_s)^2}{2R_s^2}) + N \quad (2.1)$$

where N is the random error. To calculate the center of the satellite from the simulated data, a square area around the center of the faint satellite image is selected first. Then the following model is used to fit the center of the satellite.

$$I(x,y) = I_0 EXP(-\frac{(x-x_0)^2 + (y-y_0)^2}{2R_0^2}) + \sum_{i=0}^{n}\sum_{j=0}^{n} a_{ij}(x-x_0)^i (y-y_0)^j \quad (2.2)$$

where I_0 is the fitted intensity of the image of the satellite, (x_0, y_0) is fitted center of the satellite, and n is the order of the polynomial (n=1,...5). Here I_0, R_0, (x_0, y_0) and a_{ij} are fitted by least squares.

Since the correct positions of the satellite are known when simulated under different orders of polynomial $n(n = 1, 2, 3, 4, 5)$, the mean difference between the simulated and calculated center positions (under different random errors) can be used to judge the validity of different orders of the polynomial. Table 1, 2 and 3 respectively show the simulation results for various values of I_p/I_s, r_{ps} and R_p.

Table 1. Results of five models with different (I_p/I_s) under different r_{ps}/R_p when R_p=32 pixel. Note: " $\times$ " means the result is not convergent with this model, " $\triangle$ " means the residual is very large with this model, and " $\bigcirc$ " means good results for this model.

model	I_p/I_s (r_{ps}/R_p=2)			I_p/I_s (r_{ps}/R_p=3)			I_p/I_s (r_{ps}/R_p=4)
	(12.5~25)	(25~37.5)	(37.5~250)	(12.5~25)	(25~100)	(100~250)	(1.0~250)
1-deg	$\times$	$\times$	$\times$	$\triangle$	$\triangle$	$\times$	$\bigcirc$
2-deg	$\triangle$	$\triangle$	$\triangle$	$\bigcirc$	$\bigcirc$	$\times$	$\bigcirc$
3-deg	$\bigcirc$	$\triangle$	$\triangle$	$\bigcirc$	$\triangle$	$\triangle$	$\bigcirc$
4-deg	$\bigcirc$	$\bigcirc$	$\triangle$	$\bigcirc$	$\bigcirc$	$\bigcirc$	$\bigcirc$
5-deg	$\bigcirc$	$\bigcirc$	$\bigcirc$	$\bigcirc$	$\bigcirc$	$\bigcirc$	$\bigcirc$

Table 2. Results of five models with different r_{ps}/R_p under different R_p (unit: pixel).

model	r_{ps}/R_p (R_p=8)			r_{ps}/R_p (R_p=16)			r_{ps}/R_p (R_p=32)		
	(1~4)	(4~5)	(>5)	(1~3)	(3~4)	(>4)	(1~2)	(2~3)	(>3)
1-deg	$\times$	$\times$	$\bigcirc$	$\times$	$\triangle$	$\bigcirc$	$\times$	$\times$	$\bigcirc$
2-deg	$\times$	$\times$	$\bigcirc$	$\times$	$\triangle$	$\bigcirc$	$\triangle$	$\bigcirc$	$\bigcirc$
3-deg	$\triangle$	$\triangle$	$\bigcirc$	$\bigcirc$	$\bigcirc$	$\bigcirc$	$\bigcirc$	$\bigcirc$	$\bigcirc$
4-deg	$\triangle$	$\bigcirc$	$\bigcirc$	$\bigcirc$	$\bigcirc$	$\bigcirc$	$\bigcirc$	$\bigcirc$	$\bigcirc$
5-deg	$\triangle$	$\bigcirc$	$\bigcirc$	$\bigcirc$	$\bigcirc$	$\bigcirc$	$\bigcirc$	$\bigcirc$	$\bigcirc$

3. Discussion

Synthetically produced star images have been used to investigate the properties of the polynomial-fit method. Five polynomial models have been tested under a wide range of

Table 3. Results of five models with different R_p (unit:pixel) under different r_{ps} (unit: pixel). Here $I_p/I_s = 25/8$

model	R_p (when $r_{ps}=30$)			R_p (when $r_{ps}=50$)			R_p (when $r_{ps}=80$)
	(10~12)	(12~16)	(16~34)	(10>12)	(12~14)	(14~34)	(10~34)
1-deg	×	×	×	○	○	×	○
2-deg	×	×	×	○	△	△	○
3-deg	○	×	×	○	○	○	○
4-deg	△	△	○	○	○	○	○
5-deg	○	○	○	○	○	○	○

observational conditions. The numerical calculations show that the polynomial-fit method is useful for correcting the effects of the strong halo light while centering the image of a faint satellite. In addition, the applicable range of the 1st-order polynomial model is limited, while the 5th-order polynomial model produces the highest precision. Usually, the 2nd, 3rd and 4th-degree polynomials can also satisfy the requirements of accuracy level for actual observations. After obtaining the approximate values of I_p/I_s, r_{ps}/R_p, and R_p, we can determine which model is suitable with the help of Tables 1-3.

References

Beurle, K., Harper, D., Jones, D. H. P., *et al.* 1993, *A&A*, 269, 564
Colas, F. & Arlot, J. E. 1991, *A&A*, 252, 402
Pascu, D., Seidelman, P. K., Schmidt, R. E., *et al.* 1987, *AJ*, 93, 963
Peng, Q. Y., Vienne, A., & Shen, K. X. 2002, *A&A*, 383, 296
Stone, R. C. & Harris, F. H. 2000, *AJ*, 119, 1985

A Giant Step: from Milli- to Micro-arcsecond Astrometry
Proceedings IAU Symposium No. 248, 2007
W. J. Jin, I. Platais & M. A. C. Perryman, eds.

Micro-arcsecond astrometry with the VLBA

M. J. Reid

Harvard-Smithsonian CfA, Cambridge, MA, 02138, USA
email: reid@cfa.harvard.edu

Abstract. The VLBA is now achieving parallaxes and proper motions with accuracies approaching the micro-arcsecond domain. The apparent proper motion of Sgr A*, which reflects the orbit of the Sun around the Galactic center, has been measured with high accuracy. This measurement strongly constrains Θ_0/R_0 and offers a dynamical definition of the Galactic plane with Sgr A* at its origin. The intrinsic motion of Sgr A* is very small and comparable to that expected for a supermassive black hole. Trigonometric parallaxes and proper motions for a number of massive star forming regions (MSFRs) have now been measured. For almost all cases, kinematic distances exceed the true distances, suggesting that the Galactic parameters, R_0 and Θ_0, are inaccurate. Solutions for the Solar Motion are in general agreement with those obtained from Hipparcos data, except that MSFRs appear to be rotating slower than the Galaxy. Finally, the VLBA has been used to measure extragalactic proper motions and to map masers in distant AGN accretion disks, which will yield direct estimates of H_0.

Keywords. Galactic: structure, stars: kinematics, black hole physics, masers, astrometry, instrumentation: interferometers, stars: fundamental parameters, cosmological parameters

1. The proper motion of Sgr A*

Since 1995, the VLBA has been used to measure the proper motion of the compact radio source Sgr A* relative to background quasars. The apparent motion comes predominantly from observing in the solar system, which orbits the Galactic center. With single epoch positional accuracies now better than about 0.1 mas, the VLBA is able to detect the effects of this 225 My orbit in less than one month's observing!

Were Sgr A* a stellar-mass object, for example an X-ray binary including a neutron star or black hole, it would be moving at thousands of km s^{-1}, as stars close to the position of Sgr A* are observed to do. However, for a supermassive black hole (SMBH) at the Galactic center, one expects a very small motion of ~ 0.3 km s^{-1} owing to small random perturbations from the $\sim 10^6$ stars (and perhaps upwards of 10^4 stellar remnants) within Sgr A*'s gravitational sphere of influence. With 8 years of VLBA data, Reid & Brunthaler (2004) find that the true motion of Sgr A* in the direction perpendicular to the Galactic plane is indeed very small: -0.4 ± 0.9 km s^{-1}. Preliminary analysis of data taken in 2007, shown in Fig. 1, extends the time baseline to 11 years and should reduce the uncertainty in Sgr A*'s motion below ± 0.7 km s^{-1}.

Infrared observations of stars in elliptical orbits about the position of Sgr A* by Schödel et al. (2002) and Ghez et al. (2005) demonstrate conclusively that there is an unseen mass of $\sim 4 \times 10^6$ M$_\odot$ near the position of Sgr A*. The existence of this dark mass, coupled with the upper limit for the intrinsic motion of Sgr A*, implies a lower limit for Sgr A*'s mass of $\sim 4 \times 10^5$ M$_\odot$. Combining this mass lower limit with an upper limit of ≈ 0.5 AU for the intrinsic (non-scatter broadened) size of Sgr A* (see Bower et al. (2004) and references therein), results in a mass density $> 7 \times 10^{21}$ M$_\odot$ pc^{-3}. This lower limit is within only about two orders of magnitude of a SMBH within the inner-most stable orbit

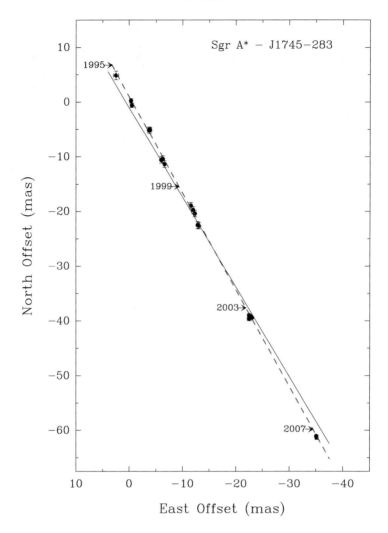

Figure 1. Apparent proper motion of Sgr A* relative to the extragalactic source J1745-283 from Reid & Brunthaler (in preparation). Time is indicated at the beginning of every fourth year. The apparent motion is caused predominantly by our moving vantage point in the solar system as it orbits about the Galactic center every ≈ 225 My. The *red dashed line* is the best fit proper motion and the *blue solid line* is the orientation of the IAU Galactic plane. The difference between these lines can be accounted for by the motion of the Sun toward the north Galactic pole, implying that Sgr A* is motionless to within about 1 km s^{-1} in this direction.

for a Schwarzschild black hole. This provides overwhelming evidence that Sgr A* is a SMBH.

Recognizing that Sgr A* is a SMBH that anchors the dynamical center of the Galaxy and that its apparent proper motion is dominated by the orbit of the Sun about the Galactic center, one could use this to provide a dynamical definition of the plane of the Galaxy. The current IAU definition of the Galactic plane is based primarily on the distribution of HI, with the Sun arbitrarily defining zero latitude. The Sun is now known to be ∼ 10 − 20 pc above the IAU plane and Sgr A* is ≈ 6 pc below the plane. A better definition of the Galactic plane could come from the direction of the apparent proper motion of Sgr A* (corrected for the small, well measured, peculiar motion of the

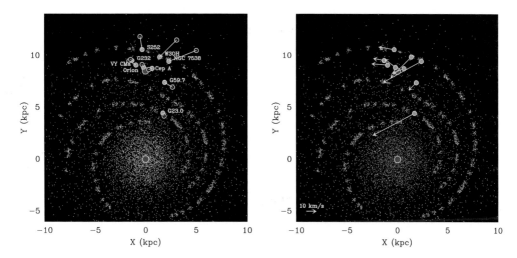

Figure 2. *Left panel:* Comparison of trigonometric parallaxes (*light cyan circles*) with kinematic distances (*dark blue circles*), assuming $R_0 = 8.5$ kpc and a flat rotation curve with $\Theta_0 = 220$ km s^{-1}. For all sources, except Cep A, the kinematic distances are larger than the trigonometric parallaxes. The background is a schematic model of the Milky Way with spiral arms from Taylor & Cordes (1993). The position of the Sun (0,8) and Sgr A* (0,0) kpc are also indicated. *Right panel:* Peculiar motions of MSFRs after removing a standard model of Galactic rotation ($R_0 = 8.5$ kpc and a flat rotation curve with $\Theta_0 = 220$ km s^{-1}) and Solar Motion parameters from Dehnen & Binney (1998).

Sun perpendicular to the plane), with Sgr A*'s position defining the zero of Galactic longitude and latitude. Of course, as the Sun orbits the Galactic center, the position of Sgr A* changes by about 6 mas y^{-1}; thus the zero of Galactic longitude would become a function of time.

2. Parallax and proper motions of star forming regions

An international team, including A. Brunthaler, K. Menten, L. Moscadelli, M. Reid, Y. Xu and X.-W. Zheng are conducting a large program with the VLBA to map the spiral structure of the Milky Way and to define its rotational kinematics and mass profile. This involves measuring trigonometric parallaxes and proper motions to methanol (and some other) masers in massive star forming regions. We are observing primarily in the 12.2 GHz methanol transition, since it falls in one of the VLBA bands.

Observations for the first group of masers are complete, and preliminary parallaxes for many sources are shown in Fig. 2. We have accurate distances to massive star forming regions in several spiral arms. All sources, with the possible exception of G59.7+0.1, lie close to these spiral arms. With more measurements we plan to map the locations of the major spiral arms of the Galaxy. In Fig. 2, we also show kinematic distances. As Xu *et al.* (2006) found for W3OH, the Perseus arm sources S252 and NGC 7538 also have a large kinematic anomalies, with kinematic distances roughly double their trigonometric parallax distances.

Almost all sources appear to have kinematic distances that are greater than trigonometric parallax distances. Decreasing R_0 and/or increasing Θ_0, from the adopted IAU values, would reduce the kinematic distances and bring them into better agreement with the trigonometric parallaxes. We note that the proper motion of Sgr A* by Reid & Brunthaler (2004), after a small correction for Solar Motion, implies $\Theta_0/R_0 = 29.45 \pm 0.15$ km

s^{-1} kpc^{-1}. If $R_0 = 8.0$ kpc (Reid 1993), this requires $\Theta_0 = 236$ km s^{-1}; using these Galactic parameters would reduce most kinematic distances by roughly 10-15%.

Parallax observations require solving for proper motions and, with current VLBA accuracies, proper motions with ~ 1 km s^{-1} are often achieved. Proper motions, coupled with radial velocities obtained from Doppler shifts of maser or thermally excited lines, yield full space velocities. The space velocities of the MSFRs generally indicate small motions out of the plane of the Galaxy, as expected for MSFRs. However, the components of the space velocities in the plane of the Galaxy reveal a surprising result.

After removing the effects of the rotation of the Galaxy, we plot the residual motions of the MSFRs in Fig. 2 superposed on a "plan view" of the Milky Way. Nearly all sources show large residual motions, notably counter to Galactic rotation. One possibility explanation for these large residuals is that the Galactic rotation model, based on IAU standard parameters for R_0 and Θ_0, needs modification. Indeed, decreasing R_0 and increasing Θ_0, can reduce the residual motions somewhat. However, the residual motions can only be reduced to ≈ 5 km s^{-1} in magnitude by adopting a Solar Motion correction that is markedly different from those determined from Hipparcos data.

Figure 3 shows the Hipparcos data used by Dehnen & Binney (1998) to determine the three components of the Solar Motion. Also plotted are the Solar Motion components required to remove the large systematic residuals for the MSFRs with VLBA parallaxes and proper motions. The components of the Solar Motion in the direction of the Galactic center (U) and toward the north Galactic pole (W) are conceptually simple to measure and there is excellent agreement between the Hipparcos and VLBA results. Also, the proper motion of Sgr A*, assuming it is nearly motionless at the Galactic center, provides a direct estimate of W, which is in agreement with other methods.

However, the VLBA results require the component of Solar Motion in the direction of Galactic rotation (V) to be about 18 km s^{-1}. This should be compared with the Hipparcos value of about 5 km s^{-1}, when the "asymmetrical drift" is extrapolated to zero stellar dispersion (ie, the dynamical Solar Motion; see Dehnen & Binney (1998) for a more complete discussion). The VLBA determined value for V is about 13 km s^{-1} greater than the dynamical Solar Motion. This suggests that MSFRs as a group rotate 13 km s^{-1} slower than the Galaxy spins. This result is fairly insensitive to the values adopted for R_0 and Θ_0. A possible explanation for this finding is that MSFRs are born near apocenter in slightly elliptical Galactic orbits, perhaps owing to the effects of spiral density wave shocks. Such shocks could compress gas clouds and induce star formation and remove angular momentum from originally circular Galactic orbits.

3. Extra-galactic VLBA astrometry

Even with near micro-arcsecond astrometry, one cannot measure parallaxes to other galaxies with high accuracy. However, proper motion accuracy increases with the spanned observing time, T, as $T^{-3/2}$. An international team including A. Brunthaler, H. Falcke and M. Reid have succeeded in measuring both the internal angular rotation and the proper motion of M33 with the VLBA. These observations constrain the dark matter in the Local Group. For more details, see the paper by A. Brunthaler in these proceedings.

One of the most interesting problems in contemporary cosmology is to determine the equation of state (w) of dark energy. Observations of primordial fluctuations in the Cosmic Microwave Background, supplemented with data from supernovae, gravitational lensing and galaxy distributions, suggest w near unity. However, this assumes either a flat universe or a value for the Hubble Constant (H_0). An independent and accurate determination of H_0 is crucial for a more accurate measure of w.

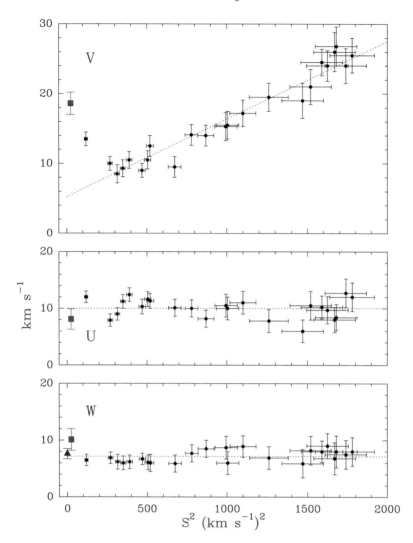

Figure 3. Solar Motion estimates. Plotted are the three components of Solar Motion (U toward the Galactic center, V in the direction of Galactic rotation; W toward the north Galactic pole) versus the dispersion of groups of stars used to determine the Solar Motion. *Black points* are from Hipparcos data by Dehnen & Binney (1998); *red squares* are from MSFR parallaxes and proper motions; the *blue triangle* is from the proper motion of Sgr A*. The well-known "asymmetrical drift" (*dotted line*) is seen in the V plot of the Hipparcos data; extrapolating to zero stellar dispersion defines the dynamical Solar Motion.

The Seyfert 2 galaxy NGC 4258 has a sub-pc scale accretion disk in its nucleus that can be traced by VLBI imaging of its H_2O maser emission. Herrnstein *et al.* (1999) determined a highly accurate geometric distance of 7.2 Mpc to this galaxy. Since the recessional velocity of NGC 4258 is only 470 km s^{-1}, the possibility of non-Hubble flow motions of several hundred km s^{-1} precludes a direct estimate of H_0. However, NGC 4258 has been used to re-calibrate the Cepheid-based extragalactic distance scale.

If galaxies like NGC 4258, but more distant and into the "Hubble flow" can be imaged with VLBI observations, then direct determinations of H_0 are possible. The Water Maser Cosmology Project (WMCP) is an international collaboration, including J. Braatz,

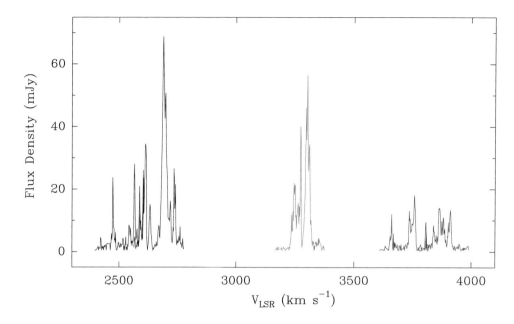

Figure 4. Interferometric spectrum of the H_2O masers towards UGC 3789 from a map made with VLBA plus GBT data. Note the similarities with the spectrum of NGC 4258: blue- and red-shifted high-velocity emission complexes that straddle the systemic emission complex centered near 3300 km s^{-1}.

J. Condon, L. Greenhill, C. Henkel, K.-Y. Lo, and M. Reid, that seeks to do this. Its goal is to measure H_0 directly with an accuracy of better than 3%. The WMCP uses the VLBA, along with large aperture telescopes such as the GBT and the Effelsberg 100-m telescope, to both monitor and map the H_2O maser emission from NGC 4258-like accretion disks in AGN.

One of the most promising candidates for distance measurement is UGC 3789, which with a recessional velocity of near 3300 km s^{-1} is into the Hubble flow. This source was discovered by J. Braatz using the GBT. The H_2O masers have been imaged with the VLBA/GBT and an interferometer spectrum is presented in Fig. 4. This spectrum displays the characteristics of a sub-pc scale accretion disk surrounding a SMBH: systemic velocity components flanked by high-velocity components separated by up to ± 750 km s^{-1}.

A map of the H_2O masers in UGC 3789 with positional accuracies approaching $\sim 5\ \mu$as is shown in the left panel of Fig. 5. As for NGC 4258, the maser spots fall in a linear pattern with the blue- and red-shifted high velocity components straddling the systemic velocity components, indicative of an edge-on disk. The angular extent of the detected spots is about 1.5 mas, about a factor of 10 smaller than for NGC 4258 and consistent with UGC 3789's much greater distance.

The right panel of Fig. 5 shows a position–velocity plot constructed along the position angle of the linear pattern of spots. One can clearly see a Keplerian rotation curve for the high velocity components, indicating a rotation speed of ≈ 600 km s^{-1} at an angular radius of ≈ 0.5 mas. This implies a central SMBH of $10^7\ (D/50\ \text{Mpc})\ M_\odot$, where D is the distance.

The distance to such a source with an edge-on rotating disk is given by $D = V^2/A\theta$, where V and A are the rotational velocity and centripetal acceleration at an angular

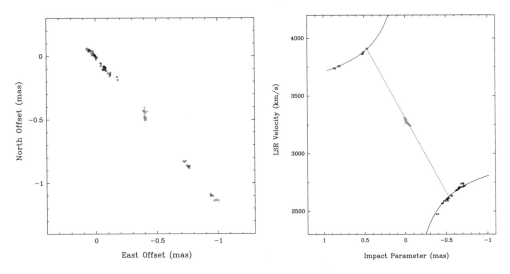

Figure 5. Map *(left panel)* and position-velocity plot *(right panel)* of H_2O masers toward UGC 3789. In the map the blue-shifted (red-shifted) high-velocity components are offset to the East (West) of the systemic masers near $(-0.4, -0.5)$ mas. In the position-velocity plot, the blue- and red-shifted components indicate a Keplerian rotation.

radius θ. Centripetal accelerations can be measured from the drift over time of Doppler shifts of maser features. Preliminary measurement of A, based on one year's monitoring with the GBT, gives distances consistent with H_0 near 70 km s^{-1} Mpc^{-1} and an anticipated uncertainty of about $\pm 10\%$. If achieved, this would give an independent value of H_0 with an accuracy comparable to that of the Hubble Key Project (Freedman *et al.*2001).

References

Bower, G. C. Falcke, H. Herrnstein, R. M, Zhao, J.-H., Goss, W. M., & Backer, D. C. 2004, *Science* 304, 704

Dehnen, W. & Binney, J. J. 1998, *MNRAS*, 298, 387

Freedman *et al.* 2001, *ApJ*, 553, 47

Ghez *et al.* 2005, *ApJ*, 620, 744

Herrnstein *et al.* 1999, *Nature*, 400, 539

Reid, M. J. & Brunthaler, A. 2004, *ApJ* 616, 872

Schödel *et al.* 2002, *Nature*, 419, 694

Taylor, J. H. & Cordes, J. M. 1993,*ApJ* 411, 674

Xu, Y., Reid, M. J., Zheng, X. W., & Menten, K. M. 2006, *Science*, 311, 54

A Giant Step: from Milli- to Micro-arcsecond Astrometry
Proceedings IAU Symposium No. 248, 2007
W. J. Jin, I. Platais & M. A. C. Perryman, eds.

© 2008 International Astronomical Union
doi:10.1017/S1743921308018930

Phase referencing VLBI astrometry observation system: VERA

H. Kobayashi[1], N. Kawaguchi[1], S. Manabe[1], K. M. Shibata[1], M. Honma[1], Y. Tamura[1] O. Kameya[1], T. Hirota[1], T. Jike[1], H. Imai[2], and T. Omodaka[2]

[1] Mizusawa VERA Observatory, National Astronomical Observatory of Japan

[2] Division of Astronomy, Department of Science, Kagoshima University

email: hideyuki.kobayashi@nao.ac.jp

Abstract. VERA aims at astrometric observations using phase referencing VLBI techniques, whose goal is a 10 micro arc-second accuracy for annual parallax measurements. VERA has four 20-m diameter VLBI radio telescopes in Japanese archipelago with the maximum baseline length of 2,300 km. They have the two-beam observing system, which makes simultaneous observations of two objects possible. This leads to very accurate phase referencing VLBI observations. An important science goal is to make a 3-dimensional map of the Galaxy and reveal its dynamics. In order to achieve this, VERA has the 22GHz and 43GHz bands for H2O and SiO maser objects, respectively. Maser objects are compact and suitable for astrometry observations. VERA's construction was started in 2000 and the array became operational in 2004. We have already measured annual parallaxes and proper motions of some galactic objects. In the future, VERA will collaborate with Korean and Chinese VLBI stations.

Keywords. instrumentation: interferometers, masers

1. Introduction

VLBI (Very Long Baseline Interferometer) technique provides the highest angular resolution in astronomical observation toolkit. Its typical angular resolution can be a few milli-arc-seconds or better. However, it was very difficult to determine the absolute position of the objects because of the uncertainty of absolute phases. Recently phase referencing technique has been developed, which usually makes observations using nodding two objects with a short time interval. However, phase errors still remain even after phase-referencing, which is caused by the short-term atmospheric fluctuations and instrumental instabilities. Thus, we have developed VERA (VLBI Exploration of Radio Astrometry) with 2-beam observing system, which allows us to perform simultaneous observation of two objects for accurate astrometric observations. The construction of VERA started in 2000 and completed in 2003. After the completion, verification and check-out were carried out for two years, the scientific observations started in 2004. In 2006, we have succeeded in measuring trigonometric parallaxes for some objects.

2. System Overview of VERA

VERA consists of four VLBI stations in Japanese archipelago (Fig. 1) with the maximum baseline length of 2,300 km and the minimum of 1,000 km. Observing frequency bands are 2, 8, 22, and 43 GHz. The 2 and 8 GHz bands are used for geodesy observations and 22 and 43 GHz bands are used for H_2O and SiO maser observations, respectively. Each station has the same 20-m diameter radio telescope with a 2-beam observational

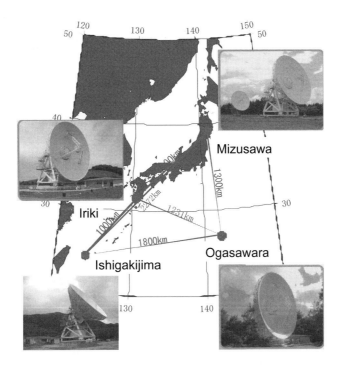

Figure 1. Distribution of VERA stations with baseline lengths of the array.

system. The 2-beam observational system makes it possible to observe two celestial objects simultaneously within a 2.2-deg separation angle. Usually one object is selected as a target source (e.g., a maser) and the other is an extra-galactic AGN object that serves as a phase calibrator. Observed phase variations of a calibrator is applied to the visibility data of the target object, and then a relative position of the target is determined with respect to the reference calibrator. Since the calibrator is extra-galactic and has no parallax, trigonometric parallax can be measured based on the relative astrometry between the target and the reference. In Fig. 2, a schematic view of the VERA 2-beam system is shown. Receivers can move over the focal plane of a telescope, which has a field curvature of its field-of-view at 2.2 deg.

In Fig. 3, an example of phase fluctuation measurements with two beams is shown (Honma et. al, 2003). In this observation, two strong maser sources were observed at the same time, and phase fluctuations of both sources showed a similar trend. In fact, the difference between the visibility phases is nearly flat, showing that phase-referencing with the 2-beam system works well to cancel out the fluctuations of troposphere.

For correlation process of the VERA array, we use tape-base recording system. The data are correlated with the Mitaka FX correlator. The correlator was originally developed for the VSOP (VLBI Space Observatory Program) project, and modified to VERA's high speed recording system, which provides a data rate of 1 Gpbs. The Mitaka FX correlator can process 5 stations at maximum data rate of 1 Gbps.

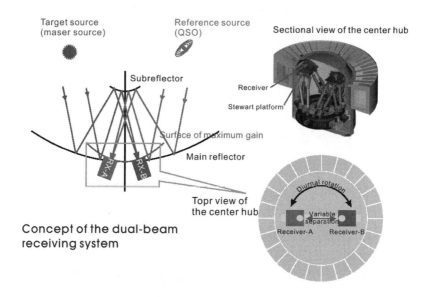

Figure 2. A schematic view of the 2-beam observing system with moving receivers.

3. Scientific goal

VERA is an astrometric tool for the galactic maser sources. H_2O and SiO maser sources are thought to be compact and suitable for the astrometric observations. The scientific goal of VERA is the construction of the Galaxy's map based on trigonometric measurements of these masers. Also the dynamics of the Galaxy will be revealed by the measurements of the distances and proper motions. In order to make the Galaxy's map with an accuracy better than 20%, one requires a parallax accuracy of 10 micro-arc-second, which is the goal of VERA's measurement accuracy. Among thousands of known maser sources, we expect that we can measure the trigonometric parallax and proper motions of more than 500 sources. Possible distribution of candidate sources (based on estimated distances) is shown in Fig. 4. Since H_2O and SiO maser sources are mostly star-forming regions and late type stars (AGB stars), VERA will also provide new information for studies of these objects, in particular precise distances as well as internal gas motions near proto-stars as well as those in the circumstellar envelopes of AGB stars.

4. Sensitivity

The aperture efficiency of VERA 20-m telescope is around 40% and 50% at 43 GHz and 22 GHz, respectively. Typical system noise temperatures are 200 K and 400 K at 22 GHz and 43 GHz, respectively. Usual fringe detection sensitivity for AGN calibrators is 100 mJy at 22 GHz, and is 250 mJy at 43 GHz, both at 100 sec of integration time. If the fringe of calibrators is detected, target sources are possible to integrate for a long time, e.g., 6 hours. Thus, the sensitivity of observations is limited by the flux intensity of a calibrator. In the case of a bright maser and a faint calibrator, one may use a maser source

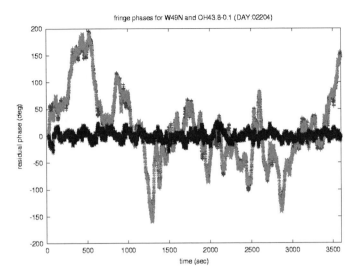

Figure 3. Phase variations of W49N and OH43.8-0.1 (rapidly-varying curves, nearly identical to each other), simultaneously observed with the 2-beam system. The difference of the two is also shown in the curve around 0 (Honma *et al.* 2003)

as a 'reference' to solve for the tropospheric fluctuations and obtain a phase-referenced map of that calibrator. This method can be used for masers with intensity of ∼10 Jy or brighter. In both cases, a target and a reference should be located within 2.2 deg from each other, which is the mechanical limit of the 2-beam system.

5. Geodesy

The station position should be accurate for precise astrometric observations. For the astrometric accuracy of 10 micro-arc-second and a 2,300 km baseline length, ∼2 mm and ∼5 mm accuracies are required in horizontal and vertical directions, respectively. To achieve this, VERA performs VLBI geodetic observations every two weeks as well as continuous station-position measurements using GPS. VLBI observations for geodesy are carried out at 2/8 GHz and also at 22GHz. Currently the horizontal accuracy of 2 mm and vertical of 4 mm (1-σ) are achieved. Moreover, we plan to measure vertical displacements by using a gravity meter to improve the accuracy. Figure 5 shows comparison of baseline accuracies with various geodesy experiments. VERA has already achieved the top accuracy.

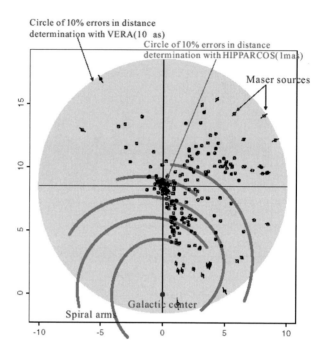

Figure 4. Possible distribution of H$_2$O maser sources in the Galaxy. The filled circle corresponds to a area of 10% distance error with a parallax accuracy of 10 μas.

6. Recent Results

We have succeeded in measuring trigonometric parallax for some objects. For instance, toward Orion KL, which is one of the most famous massive star forming regions, we have measured the trigonometric parallax of H$_2$O maser sources, providing the distance of 437±19 pc (Hirota, et. al, 2007). Our distance is slightly smaller than the distance estimate of 480 pc based on statistical parallax (Genzel et. al, 1981). Recently, parallax measurements of non-thermal stars in Orion region were published by other investigators, providing a distance of 414±7 pc (Menten, 2007, Sandstrom, et. al, 2007). We are conducting astrometry of the SiO maser in Orion KL and its preliminary distance based on half a year worth of data is 419±6 pc. When the observations of SiO maser are completed, it will hopefully provide the most accurate distance measurement to this Orion region.

In addition to Orion, around 20 sources are selected for regular astrometric monitoring to measure the distance and proper motions. For S269, we have successfully measured a parallax beyond 5 kpc with its distance at 5.28±0.24 kpc (Honma et al. 2007). S269 is one of the most distant sources for which a trigonometric parallax has been measured. Proper motions as well as parallax distance of S269 were used to constrain the Galactic rotation velocity at the position of S269, which is 13 kpc away from the Galaxy center, providing a direct confirmation of flat rotation curve out to 13 kpc. For molecular cloud NGC 281, which is associated with an HI super-bubble and is located 300 pc away from the Galactic plane, Sato et al.(2007) found that the NGC 281 system has a systematic

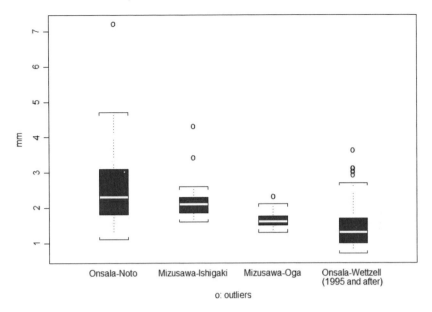

Figure 5. Comparison of baseline errors among geodesy experiments. VERA has achieved 2–3 mm accuracy in horizontal direction.

motion of ~ 20 km/s, moving away from the Galactic plane. This is the first detection of proper motions for sources associated with Galactic super-bubble, and provides a direct evidence that these super-bubble were blown out from the Galaxy plane, most likely by multiple SNe (Sato *et al.* 2007).

Also, some of the nearby star-forming regions have been observed with VERA and the distances were already obtained for two molecular clouds. Hirota *et al.*(2008) obtained the parallax distance of NGC 1333 to be 235±18 pc based on the H_2O maser associated with SVS 13. The distance of ρ-Oph cloud was measured to be 178^{+18}_{-37} pc by Imai *et al.*(2007), which is consistent with the results recently obtained by Loinard *et al.* (2008) based on VLBA parallax measurements of radio-emitting young stars in this region. In addition to star forming regions, there are some late-type stars for which astrometric observations were successfully made. A distance measurement of nearby semi-regular variable star S Crt is reported in these proceeding (Nakagawa 2008), providing a parallax distance of 430±25 pc. Choi *et al.*(2008) also measured a parallax distance of super-giant star VY CMa as 1.1±0.1 kpc, which is slightly smaller than the previous estimates.

7. Future works

Currently we have been carrying out 4 astrometric programs: 1) structure of nearby star-forming region, 2) galactic structure within 5 kpc of the Sun, 3) period-luminosity relation for Mira variables and 4) distance measurement of the galactic center. More than 20 objects are programmed for the measurements of trigonometric parallaxes and

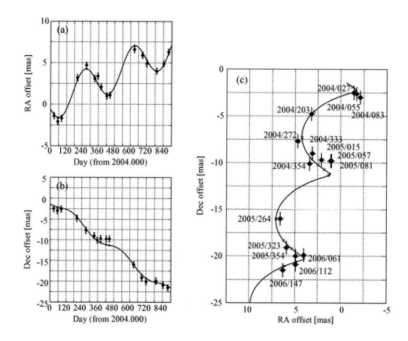

Figure 6. VERA measurement result for the trigonometric parallax and proper motions of
Orion-KL H$_2$O maser (Hirota *et al.* 2007).

proper motions. In addition to an operation as VERA, we have a plan to collaborate
with Korean VLBI Network (KVN), which will have multi-frequency phase referencing
capability. KVN will have 3 telescopes in Korea with 21-m diameter, with operating
frequency up to 129 GHz. The combined Korean-Japanese VLBI array will have a dense
UV coverage, providing shorter baselines than those of VERA itself and, hence, yielding a
better imaging. Also, China has a 4-station VLBI network, called the CVN (Chinese VLBI
Network). We have started discussions about an East Asia VLBI network which would
include China, Japan, and Korea. Figure 7 shows the distribution of VLBI stations in this
area. Including VLBI stations operated by Japanese universities and research intitutes
other than NAOJ, there will be nearly 20 VLBI stations in the East Asian region. Thus,
when these stations are combined to form a bigger array, the array will provide a better
imaging capability and a higher resolution, and thus will significantly contribute to the
better accuracy in VLBI astrometry.

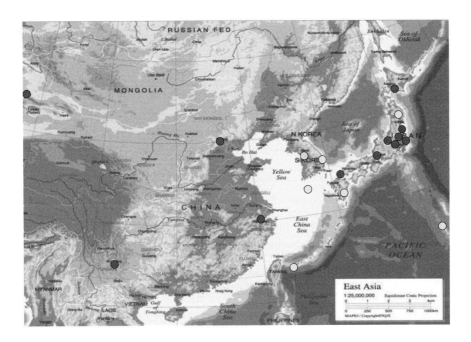

Figure 7. Distribution of VLBI stations in the East Asia region. The East Asia VLBI Network is under consideration.

References

Choi, Y. K., *et al.*, 2008, in this volume p.192

Genzel, R., *et al.*, 1981, *ApJ*, 244, 884

Loinard, L., *et al.*, 2008, in this volume p.186

Hirota, T., *et al.*, 2007, *PASJ*, 59, 897

Honma, M., *et al.*, 2003, *PASJ*, 55, L57

Honma, M., *et al.*, 2007, *PASJ*, 59, 889

Menten, K., *et al.*, 2007, *A&A*, 474, 515

Nakagawa, A., *et al.*, 2008, this volume p.206

Sandstrom, M., *et al.*, 2007, *ApJ*, 667, 1161

Sato, M., *et al.*, 2007, *PASJ*, 59, 743

A Giant Step: from Milli- to Micro-arcsecond Astrometry
Proceedings IAU Symposium No. 248, 2007
W. J. Jin, I. Platais & M. A. C. Perryman, eds.

© 2008 International Astronomical Union
doi:10.1017/S1743921308018942

Astrometric observations of neutron stars

S. Chatterjee

School of Physics, The University of Sydney,
NSW 2006, Australia
email: S.Chatterjee@physics.usyd.edu.au

Abstract. Precision astrometry can yield model-independent distances and velocities for neutron stars. Such measurements can be exploited, for example, to locate neutron star birth sites, establish reference frame ties, model the Galactic electron density distribution, and constrain the astrophysics of supernova explosions. As a case study, I discuss recent some parallax and proper motion measurements, and their scientific implications for supernova core collapse and the velocities of ordinary pulsars versus magnetars. I also outline the calibration techniques that are enabling sub-milliarcsecond astrometry of neutron stars with VLBI. In the short term, systematic surveys and high sensitivity on very long baselines will produce ongoing science dividends from precision astrometry at radio wavelengths. In the longer term, new technology such as focal plane arrays, new telescopes such as the Square Kilometre Array, and synergy with new instruments such as Gaia, LSST, and GLAST, all hold great promise in an upcoming era of microarcsecond astrometry.

Keywords. astrometry, techniques: interferometric, techniques: high angular resolution, stars: neutron, pulsars: general

1. Introduction

Neutron stars are exotic laboratories for some of the most extreme physics in the Universe. Astrometric observations of neutron stars (NS) can be exploited to constrain their origins, evolution, and environments, as well as to establish meaningful constraints on nuclear physics, particle physics, and theories of gravitation.

The basic observable, as in most other astrometric applications, is the position of the NS. For individual objects, rather than an ensemble of stars over the entire sky, such measurements are necessarily relative in nature, but absolute positions can be inferred from measurements relative to sources that define the ICRF (Ma *et al.* 1998). Over time, repeated measurements of the position $\vec{\theta}$ allow a proper motion $\vec{\mu}$ to be derived. For a precise proper motion, the primary consideration is a long time baseline, limited by the stability of the reference frame and the variability of the frame-defining sources. Finally, with enough astrometric precision, a trigonometric parallax π may also be measurable for NS. The primary consideration for such measurements is appropriate sampling over the Earth's orbital phase, not just a long time baseline.

Neutron stars emit over a broad range of wavelengths, and astrometric observations of NS have been conducted at wavelengths from radio to X-rays. For example, Kaplan *et al.* (2007) have used optical observations with the Hubble Space Telescope to measure a proper motion $\mu = 107.8 \pm 1.2$ mas yr^{-1} and a parallax $\pi = 2.8 \pm 0.9$ mas for RX J0720.4−3125. At X-ray wavelengths, the resolution that can be achieved with the current generation of telescopes is a limiting factor, but Winkler & Petre (2007) have used the Chandra X-ray Observatory to measure a proper motion of 165 ± 25 mas yr^{-1} for the RX J0822−4300, in the center of the Puppis A supernova remnant.

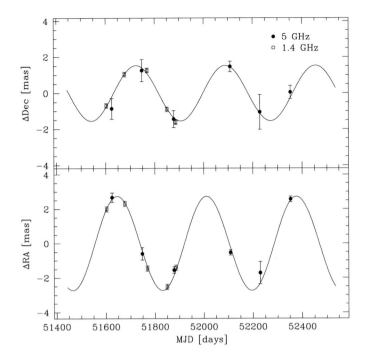

Figure 1. The parallax signature of PSR B1929+10 in right ascension and declination, after subtracting the best-fit proper motion from the astrometric positions measured with the VLBA at 5 GHz (closed circles; Chatterjee *et al.* 2004) and 1.4 GHz (open squares; Brisken *et al.* 2002). Sinusoids corresponding to the best fit parallax $\pi = 2.77$ mas are overplotted. Figure adapted from Chatterjee *et al.* (2004).

The majority of known NS are radio pulsars, and pulse timing is routinely used to refine their positions and proper motions. A subset of recycled ("millisecond") pulsars have rotation rates that are stable enough to permit sub-milliarcsecond astrometry based on pulse time of arrival. For example, van Straten *et al.* (2001) measure the general relativistic Shapiro delay for the binary pulsar J0437−4715, as well as a precise proper motion and parallax, based on pulse timing at Parkes. However, most pulsars do not have such stable rotation, particularly when they are young, and Very Long Baseline interferometry (VLBI) has usually been utilized to determine their astrometric parameters. Such efforts have a long history (e.g., Gwinn *et al.* 1986), but have become much more feasible with the Very Long Baseline Array (VLBA), which provides full-time, dedicated VLBI capabilities with identical antennas, allowing good control of systematic errors and leading to many recent parallax measurements (e.g., Fomalont *et al.* 1999; Brisken *et al.* 2000; Chatterjee *et al.* 2001; Brisken *et al.* 2002; Chatterjee *et al.* 2004, see Figure 1).

2. Scientific applications of neutron star astrometry

Precise and accurate measurements of the position, proper motion, and parallax of NS can be exploited to address a variety of scientific questions.

Position: Measuring the position of an object in two different coordinate frames permits the very fundamental operation of tying the two reference frames together. Since they are compact sources and are accessible to astrometry with different techniques and at different wavelengths, NS are particularly well suited to such reference frame ties.

Specifically, pulse timing provides radio pulsar positions in the Solar system reference
frame, while VLBI measurements are tied to the distant quasars. Thus, simply measuring
precise positions for recycled pulsars enables fundamental reference frame ties between
the Solar system and the extragalactic ICRF (e.g. Bartel *et al.* 1996). A few pulsars also
have optical counterparts, and PSR J0437−4715, for example, may provide a frame-tie
between the optical reference frame and the ICRF for the Space Interferometry Mission
(SIM).

Proper Motion: Precise proper motions for NS allow them to be traced back to their
birth sites in massive stellar clusters, and to associations with runaway stars (Hoogerwerf
et al. 2000; Vlemmings *et al.* 2004; Chatterjee *et al.* 2005, see Figure 2). For very young
objects, associations with their progenitor supernova remnants can be verified or refuted,
leading to independent age estimates for both the NS and the supernova remnant itself
(e.g., Gaensler & Frail 2000; Migliazzo *et al.* 2002; Blazek *et al.* 2006). The rotation of
such young pulsars is typically not stable enough for astrometry via pulse timing, and
interferometry is thus required. Combined with estimates for their distances, the proper
motions of NS also lead to velocity estimates. The high velocity tail of the distribution
implies that large kicks are imparted to proto-neutron stars during core collapse, as
discussed further below.

Parallax: When a parallax measurement is possible, it provides a *model independent*
estimate for the distance and velocity of the neutron star. Each such measurement cal-
ibrates global models of the Galactic electron density (Taylor & Cordes 1993; Cordes
& Lazio 2002), thus improving distance estimates from pulse dispersion measure for the
rest of the radio pulsar population, as well as probing the distribution of electron density
in the local interstellar medium (e.g., Toscano *et al.* 1999).

Observed thermal radiation from the NS surface can be used, in combination with
a precise distance, to constrain the 'size' of the photosphere, the NS radius, and thus
the Equation of State of matter at extreme pressures and densities (Yakovlev & Pethick
2004; Lattimer & Prakash 2004). For radio pulsars, uncertainties in the magnetospheric
emission restricts such an exercise to very young and hot objects (e.g., PSR B0656+14,
Brisken *et al.* 2003), while isolated NS which are X-ray bright and radio quiet pose a
challenge for optical astrometry (e.g., RX J1856.5−3754, Kaplan *et al.* 2002; Walter &
Lattimer 2002).

3. Case study: the astrophysics of pulsar kicks

Since their discovery, radio pulsars have been known to form a high velocity population. Statistical studies of the velocity distribution of ordinary young radio pulsars (Lyne
& Lorimer 1994; Arzoumanian *et al.* 2002; Hobbs *et al.* 2005; Faucher-Giguère & Kaspi
2006) yield mean three-dimensional population velocities of 300–500 km s^{-1}, while the
existence of objects such as PSR B2224+65 (the Guitar Nebula pulsar, with a transverse velocity ∼800—1600 km s^{-1}) demonstrates the presence of a long high-velocity tail
(Cordes *et al.* 1993; Chatterjee & Cordes 2004).

A number of physical mechanisms have been suggested to account for the high speeds,
including the disruption of binaries through mass loss in supernovae (Blaauw 1961; Iben
& Tutukov 1996) and the electromagnetic rocket effect (Harrison & Tademaru 1975), but
it now appears likely that NS gain a large part of their observed velocity from a "birth
kick" that involves asymmetries in the natal supernova explosion (Lai *et al.* 2001). The
nature of the asymmetry remains unclear: while hydrodynamic or convective instabilities
are plausible (Burrows & Hayes 1996; Janka & Mueller 1996), more exotic mechanisms

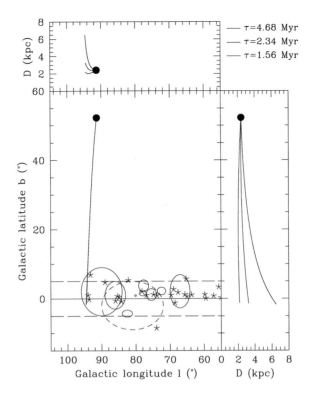

Figure 2. Possible orbits for PSR B1508+55, traced back in the Galactic potential. The solid dot denotes the current pulsar position and the thick solid line the path it has followed for an age $\tau = 2.34$ Myr (the spindown age, with braking index $n = 3$) and a radial velocity $v_r = 200$ km s^{-1}. Other possible orbits are shown (thinner lines) for a range of ages and radial velocities which lead to birth sites in the Galactic plane ($\tau = 4.69$ Myr for $n = 4$ and $v_r = -300$ km s^{-1}, $\tau = 1.56$ Myr for $n = 2$ with $v_r = 700$ km s^{-1}). Also indicated are the Cygnus superbubble (2 kpc away, dashed ellipse), the Cygnus OB associations (solid ellipses), and Galactic supernova remnants identified in this region. The solid horizontal line is the Galactic plane and the horizontal dashed lines indicate the pulsar birth scale height from Arzoumanian *et al.* (2002) at the distance of the Cygnus superbubble. At its characteristic age, the pulsar can thus be traced back to a birth site in the Galactic plane and near massive stars in OB associations which are plausible NS progenitors. Figure adapted from Chatterjee *et al.* (2005).

such as asymmetric neutrino emission in the presence of strong magnetic fields cannot be ruled out (Arras & Lai 1999).

The high velocity tail of the pulsar birth velocity distribution is of particular interest, since these kicks impose the most stringent constraints on mechanisms for supernova core collapse. Currently, two-dimensional hydrodynamic models can provide a supernova detonation, a NS kick, and an asymmetric supernova remnant by invoking neutrino energy deposition (Scheck *et al.* 2004, 2006). However, the first full three-dimensional simulations of supernovae (Fryer 2004) have difficulty in producing the required magnitude of birth kicks, possibly implying the presence of extremely high magnetic fields $> 10^{15}$ G (Arras & Lai 1999) or some combination of the above effects (Socrates *et al.* 2005) in driving kicks.

A magnetic field-driven kick mechanism immediately suggests the possibility that the population of magnetars (which have magnetic field decay as their primary energy source and usually emit no detectable radio pulsations) should have much larger kick velocities than the pulsar population. Indeed, in proposing the existence of magnetars, Duncan & Thompson (1992) suggested that an initial short birth period might be responsible for the generation of the high magnetar fields through an efficient large scale dynamo, and that these high magnetic fields could mediate asymmetric neutrino emission at birth, resulting in extreme space velocities (Thompson & Duncan 1993).

Astrometric observations have established significant constraints on these questions. Specifically, a VLBA proper motion and parallax for PSR B1508+55 (Chatterjee *et al.* 2005) provides a model independent transverse velocity estimate $V_\perp = 1083^{+103}_{-90}$ km s^{-1}, posing a significant challenge to those supernova core collapse simulations which do not provide large kicks. The extremely ordinary spin and spindown characteristics of the pulsar also challenge exotic kick scenarios that require extreme magnetic fields for high birth velocities.

Measuring the velocity of a population of magnetars and comparing them to the radio pulsar population would allow a clear test of magnetic field-driven kick mechanisms, and efforts to make such measurements are underway using X-ray and infra-red observations. In the meantime, the discovery of radio pulses from the magnetar XTE J1810−197 (Camilo *et al.* 2006), while posing a variety of theoretical challenges to ideas about magnetars, also allowed a rapid proper motion determination (Helfand *et al.* 2007) using the much higher angular resolutions offered by the VLBA. Helfand *et al.* (2007) find $V_\perp \sim 212 \pm 35$ km s^{-1} for a distance of 3.5 ± 0.5 kpc. Only the transverse velocity is measured, and for only a single magnetar, but with those caveats, the estimate is completely consistent with the population velocity of ordinary pulsars.

Finally, if birth kicks also influence the initial spin rate of the NS core (Spruit & Phinney 1998), then the spin and kinematic histories of each NS would be closely linked, but it is unknown whether or not such a link exists (Deshpande *et al.* 1999; Romani & Ng 2003). The Crab pulsar has a well defined symmetry axis, traced by the X-ray jet—torus structure and the optical equatorial wisps (Hester *et al.* 2002; Ng & Romani 2004), and the axis is naturally associated with the NS spin, since every other vector is rotation averaged. Kaplan *et al.* (2008) use archival HST observations to determine a precise proper motion for the Crab pulsar, and find a proper motion $(\mu_\alpha, \mu_\delta) = (-11.9 \pm 0.4, +3.9 \pm 0.4)$ mas yr^{-1}. However, in comparing the proper motion vector to the position angle of the spin axis, they find that the unknown motion of the progenitor star contributes a systematic uncertainty that dominates the statistical errors, leading to an estimate of the misalignment of $14° \pm 2° \pm 9°$, consistent with a broad range of values, including zero. Thus, the alignment of NS spin axes with their kick vectors is better tested over a statistical ensemble rather than with individual objects.

4. Calibration techniques and future directions

As illustrated above, astrometry can bring powerful constraints to bear on a variety of scientific questions about neutron stars, their origins, astrophysics, evolution, and environments. The primary obstacle is the difficulty of such astrometric observations. At optical, infra-red, and X-ray wavelengths, only a few NS can be observed and the required resolution for high-precision astrometry is difficult to attain.

At radio wavelengths, many pulsars (and a couple of magnetars) can be observed with VLBI. However, not only are most pulsars faint, but the most interesting categories, namely the youngest pulsars and the recycled ones, appear to be disproportionately faint. Pulsar gating (which accumulates signal only during the predicted on-pulse periods for pulsars) is now routinely employed to boost the signal to noise ratios for VLBA

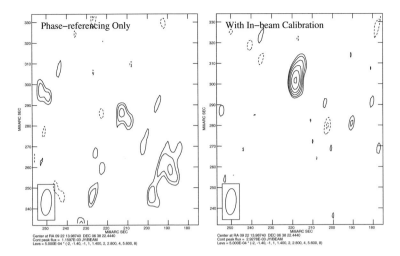

Figure 3. VLBA images of PSR B0919+06 at 1.4 GHz (left) after phase referencing to a calibrator $\sim 4°$ away, and (right) after in-beam calibration with a calibrator $12'$ away, within the primary beam of the antennas. On this occasion, a disturbed ionosphere does not allow useful astrometry with phase referencing alone, but with in-beam calibration, precise astrometric information can be effectively recovered.

astrometric observations. To make matters worse, however, it is not just a lack of sensitivity but systematic errors contributed by the ionosphere and the troposphere which are the primary impediments to sub-milliarcsecond astrometry. Progress has been made, for example, with GPS-based ionospheric calibration schemes. In-beam calibration (Fomalont *et al.* 1999), using a faint source in the same primary beam as the target source, has also proved to be effective (see Figure 3).

Currently, large dishes (like the Arecibo and Green Bank telescopes) are being employed in concert with the VLBA to boost sensitivity for pulsar astrometry, though such high-sensitivity arrays pose calibration challenges. In the Southern hemisphere, the Long Baseline Array is conducting astrometric observations, though the available baseline lengths are currently limited. Further in the future, focal plane feed arrays have been identified as a key technology for instruments now on the drawing board, including the Square Kilometer Array (SKA). By effectively offering much larger fields of view, a focal plane array allows the use of many faint reference sources, thus opening up a large fraction of the pulsar population to high-precision astrometry. It is also noteworthy that telescopes like the SKA will require a significant long baseline component in order to avoid being hopelessly confusion limited by faint sources, and such a long baseline component is ideal for high-sensitivity NS astrometry.

At optical and infra-red wavelengths, adaptive optics, high-sensitivity detectors, and new space-based missions like SIM and Gaia provide significant opportunities for astrometry of faint NS. Meanwhile, LSST, GLAST, and other survey instruments at a wide range of wavelengths will depend on precise astrometric observations to analyze and interpret their results.

Acknowledgements

I acknowledge a generous travel grant from the organizing committee for this conference, and support from the University of Sydney Postdoctoral Fellowship program. I also acknowledge my collaborators on a variety of projects, including Walter Brisken,

Jim Cordes, Ed Fomalont, Bryan Gaensler, Miller Goss, David Helfand, David Kaplan, Michael Kramer, Joe Lazio, Andrew Lyne, and Wouter Vlemmings.

References

Arras, P. & Lai, D. 1999, *ApJ*, 519, 745

Arzoumanian, Z., Chernoff, D. F., & Cordes, J. M. 2002, *ApJ*, 568, 289

Bartel, N., Chandler, J. F., Ratner, M. I., Shapiro, I. L., Pan, R., & Cappallo, R. J. 1996, *AJ*, 112, 1690

Blaauw, A. 1961, *Bull. Astron. Inst. Netherlands*, 15, 265

Blazek, J. A., Gaensler, B. M., Chatterjee, S., van der Swaluw, E., Camilo, F., & Stappers, B. W. 2006, *ApJ*, 652, 1523

Brisken, W. F., Benson, J. M., Beasley, A. J., Fomalont, E. B., Goss, W. M., & Thorsett, S. E. 2000, *ApJ*, 541, 959

Brisken, W. F., Benson, J. M., Goss, W. M., & Thorsett, S. E. 2002, *ApJ*, 571, 906

Brisken, W. F., Thorsett, S. E., Golden, A., & Goss, W. M. 2003, *ApJ Lett.*, 593, L89

Burrows, A. & Hayes, J. 1996, *Physical Review Letters*, 76, 352

Camilo, F., Ransom, S. M., Halpern, J. P., Reynolds, J., Helfand, D. J., Zimmerman, N., & Sarkissian, J. 2006, *Nature*, 442, 892

Chatterjee, S. & Cordes, J. M. 2004, *ApJ Lett.*, 600, L51

Chatterjee, S., Cordes, J. M., Lazio, T. J. W., Goss, W. M., Fomalont, E. B., & Benson, J. M. 2001, *ApJ*, 550, 287

Chatterjee, S., Cordes, J. M., Vlemmings, W. H. T., Arzoumanian, Z., Goss, W. M., & Lazio, T. J. W. 2004, *ApJ*, 604, 339

Chatterjee, S., Vlemmings, W. H. T., Brisken, W. F., Lazio, T. J. W., Cordes, J. M., Goss, W. M., Thorsett, S. E., Fomalont, E. B., Lyne, A. G., & Kramer, M. 2005, *ApJ Lett.*, 630, L61

Cordes, J. M. & Lazio, T. J. W. 2002, ArXiv e-print, astro-ph/0207156

Cordes, J. M., Romani, R. W., & Lundgren, S. C. 1993, *Nature*, 362, 133

Deshpande, A. A., Ramachandran, R., & Radhakrishnan, V. 1999, *A&A.*, 351, 195

Duncan, R. C. & Thompson, C. 1992, *ApJ Lett.*, 392, L9

Faucher-Giguère, C.-A. & Kaspi, V. M. 2006, *ApJ*, 643, 332

Fomalont, E. B., Goss, W. M., Beasley, A. J., & Chatterjee, S. 1999, *AJ*, 117, 3025

Fryer, C. L. 2004, *ApJ Lett.*, 601, L175

Gaensler, B. M. & Frail, D. A. 2000, *Nature*, 406, 158

Gwinn, C. R., Taylor, J. H., Weisberg, J. M., & Rawley, L. A. 1986, *AJ*, 91, 338

Harrison, E. R. & Tademaru, E. 1975, *ApJ*, 201, 447

Helfand, D. J., Chatterjee, S., Brisken, W. F., Camilo, F., Reynolds, J., van Kerkwijk, M. H., Halpern, J. P., & Ransom, S. M. 2007, *ApJ*, 662, 1198

Hester, J. J., Mori, K., Burrows, D., Gallagher, J. S., Graham, J. R., Halverson, M., Kader, A., Michel, F. C., & Scowen, P. 2002, *ApJ Lett.*, 577, L49

Hobbs, G., Lorimer, D. R., Lyne, A. G., & Kramer, M. 2005, *MNRAS*, 360, 974

Hoogerwerf, R., de Bruijne, J. H. J., & de Zeeuw, P. T. 2000, *ApJ Lett.*, 544, L133

Iben, I. J. & Tutukov, A. V. 1996, *ApJ*, 456, 738

Janka, H.-T. & Mueller, E. 1996, *A&A.*, 306, 167

Kaplan, D. L., Chatterjee, S., Gaensler, B. M., & Anderson, J. 2008, *ApJ*, accepted, ArXiv e-print 0801.1142

Kaplan, D. L., van Kerkwijk, M. H., & Anderson, J. 2002, *ApJ*, 571, 447

—. 2007, *ApJ*, 660, 1428

Lai, D., Chernoff, D. F., & Cordes, J. M. 2001, *ApJ*, 549, 1111

Lattimer, J. M. & Prakash, M. 2004, Science, 304, 536

Lyne, A. G. & Lorimer, D. R. 1994, *Nature*, 369, 127

Ma, C., Arias, E. F., Eubanks, T. M., Fey, A. L., Gontier, A.-M., Jacobs, C. S., Sovers, O. J., Archinal, B. A., & Charlot, P. 1998, *AJ*, 116, 516

Migliazzo, J. M., Gaensler, B. M., Backer, D. C., Stappers, B. W., van der Swaluw, E., & Strom, R. G. 2002, *ApJ Lett.*, 567, L141

Ng, C.-Y. & Romani, R. W. 2004, *ApJ*, 601, 479

Romani, R. W. & Ng, C.-Y. 2003, *ApJ Lett.*, 585, L41

Scheck, L., Kifonidis, K., Janka, H.-T., & Müller, E. 2006, *A&A.*, 457, 963

Scheck, L., Plewa, T., Janka, H.-T., Kifonidis, K., & Müller, E. 2004, *Physical Review Letters*, 92, 011103

Socrates, A., Blaes, O., Hungerford, A., & Fryer, C. L. 2005, *ApJ*, 632, 531

Spruit, H. C. & Phinney, E. S. 1998, *Nature*, 393, 139

Taylor, J. H. & Cordes, J. M. 1993, *ApJ*, 411, 674

Thompson, C. & Duncan, R. C. 1993, *ApJ*, 408, 194

Toscano, M., Britton, M. C., Manchester, R. N., Bailes, M., Sandhu, J. S., Kulkarni, S. R., & Anderson, S. B. 1999, *ApJ Lett.*, 523, L171

van Straten, W., Bailes, M., Britton, M., Kulkarni, S. R., Anderson, S. B., Manchester, R. N., & Sarkissian, J. 2001, *Nature*, 412, 158

Vlemmings, W. H. T., Cordes, J. M., & Chatterjee, S. 2004, *ApJ*, 610, 402

Walter, F. M. & Lattimer, J. M. 2002, *ApJ Lett.*, 576, L145

Winkler, P. F. & Petre, R. 2007, *ApJ*, 670, 635

Yakovlev, D. G. & Pethick, C. J. 2004, *Ann. Rev. Astron. Astrophys.*, 42, 169

A Giant Step: from Milli- to Micro-arcsecond Astrometry
Proceedings IAU Symposium No. 248, 2007
W. J. Jin, I. Platais & M. A. C. Perryman, eds.

The Square Kilometre Array

A. R. Taylor

Centre for Radio Astronomy
University of Calgary
2500 University Dr. N.W.,
Calgary, Alberta, Canada, T2N 1N4
email: russ@ras.ucalgary.ca

Abstract. The SKA is a global project to plan and construct the next-generation international radio telescope operating at metre to cm wavelengths. More than 50 institutes in 19 countries are involved in its development. The SKA will be an interferometric array with a collecting area of up to one million square metres and maximum baseline of at least 3000 km. The SKA reference design includes field-of-view expansion technology that will allow instantaneous imaging of up to several tens of degrees. The SKA is being designed to address fundamental questions in cosmology, physics and astronomy. The key science goals range from the epoch or re-ionization, dark energy, the formation and evolution of galaxies and large-scale structure, the origin and evolution of cosmic magnetism, strong-field tests of gravity and gravity wave detection, the cradle of life, and the search for extraterrestrial intelligence. The sensitivity, field-of-view and angular resolution of the SKA will make possible a program to create a multi-epoch data base of wide-angle relative astrometry to a few μas precision for $\sim$10,000,000 radio sources with $S > 10$ μJy.

Keywords. instrumentation: interferometers, techniques: high angular resolution, surveys

1. Introduction

The Square Kilometre Array (SKA) is a next-generation radio telescope being planned and developed by a consortium of institutions in 19 countries, including Argentina, Australia, Brazil, Canada, China, France, Germany, India, Italy, The Netherlands, New Zealand, Poland, Portugal, Russia, South Africa, Sweden, United Kingdom and the United States. It will be an interferometric array with collecting area of order one million square metres, providing a sensitivity about 50 times higher than the largest currently existing radio telescopes. Taking advantage of technology developments in radio frequency devices and digital processing it will achieve an sky imaging capacity 10,000 times faster than our best imaging radio telescopes.

The science case for the SKA has been under development for over a decade, e.g. Taylor (1999), Carilli & Rawlings (2004). The major leap in our ability to observe the universe enabled by the SKA will advance a broad range of modern astrophysics. The SKA science community has identified five key science areas where the SKA is targeted to make transformational advances in questions of fundamental importance in physics and astrophysics.

- *Strong-field tests of gravity using pulsars and black holes:* Surveys will detect tens of thousands of new pulsars including binary systems and potentially black hole companions. Thousands of milli-second pulsars will form a pulsar timing array for detection of gravity waves.
- *The origin and evolution of cosmic magnetism.* Surveys of polarization properties of the sky will yield measures of polarized synchrotron radiation arising from relativistic

particles interacting with magnetic fields. Faraday rotation for $\sim 10^8$ polarized extragalactic radio sources to cosmological distances.

• *Galaxy evolution and cosmology.* Atomic hydrogen emission will be detectable in normal galaxies to high redshift, providing measure of the cosmic evolution of HI and star formation. The large-scale distribution of galaxies out to high z will allow precise studies and determination of the equation of state of of dark energy.

• Probing the Dark Ages: At its lowest frequencies the SKA will probe the structure of the neutral IGM before and during the epoch of reionization.

• *The cradle of life:* Sub-AU imaging of thermal emission will trace the process of terrestrial planet formation. The raw sensitivity of the SKA will allow "leakage" radiation to be detected from potential civilizations in planetary systems around millions of solar-type stars.

Complete and current information in the SKA can be found at the project web site http://www.skatelescope.org.

2. The Square Kilometre Array

2.1. *SKA technology*

The technical specifications of the SKA are listed in Table 1, and Fig. 1 shows an artists conception of the SKA. The telescope will consist of an inner core region about 5 km across. Outside of the inner core arrays of "stations" will be distributed out to the

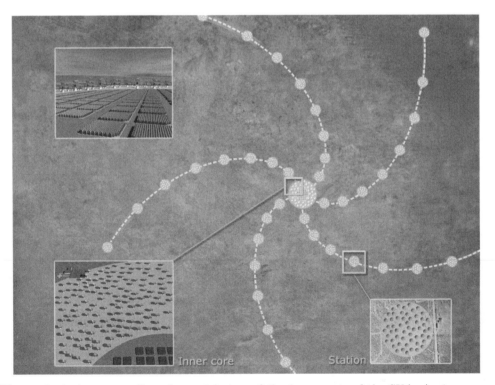

Figure 1. Artists conception of an arial view of the inner parts of the SKA. An inner core of about 5 km in diamter will contain about 50% of the collection area (see Table 3). Outside this area the SKA will be comprised of "stations", each made up of many antenna elements, extending to baselines of at least 3000 km.

maximum baseline. Both the inner core and the stations will be comprised of many antenna elements.

The lowest frequencies (below about 500 MHz) will be detected with arrays of tiles of aperture-plane phased-array feed systems, providing instantaneous field-of-view of hundreds of square degrees. At higher frequencies, large numbers of 10-15m class parabolic antennas will be equipped with new technology feed systems: focal-plane, phased-array feeds at GHz frequencies, and very wide-band, single-pixel feeds at higher frequencies. Focal-plane phased array feeds are critical technology for field-of-view expansion in the GHz regime. This technology is under rapid development. Figure 2 shows a prototype phase-array feed with 180 antenna element that has been constructed at the Dominion Radio Astrophysical Observatory of the Herzberg Institute of Astrophysics. This system will be installed and tested on a new-technology parabolic antenna at DRAO.

The SKA antenna technology thus separates in three frequency regimes, low, mid and high. This frequency-dependence of the SKA technology is summarized in Table 2. In the low- and mid-range, field-of-view expansion technology based on aperture-or focal-plane phased array feeds combined with the large leap in sensitivity makes the SKA a prodigious synoptic survey imaging telescope.

Table 1. SKA Specifications

Parameter	Specification
Frequency Range	70 MHz – 25 GHz
Sensitivity	$\frac{A_{\mathrm{eff}}}{T_{\mathrm{sys}}}$ = 5,000 to 10,0000, depending on frequency
Field-of-view	200 to 1 square degree, depending on frequency
Angular Resolution	0.1 arcsecond at 1.4 GHz
Instantaneous bandwidth	25% of band centre, maximum 4 GHz
Calibrated polarization purity	10,000 : 1
Imaging dynamic range	1,000,000 : 1 at 1.4 GHz
Output data rate	1 Terrabyte per minute

Figure 2. A focal-plane phase array demonstrator system (PHAD) at the Herzberg Institute of Astrophysics, Dominion Radio Astrophysical Observatory. This system will be tested on a new-technology 10-metre parabolic antenna as part of the technology development for the SKA and the Australian SKA Pathfinder.

Table 2. SKA Technology Dependent Frequency Ranges

SKA	Frequency Range	$\dfrac{A_{\mathrm{eff}}}{T_{\mathrm{sys}}}$	Field of View
low	$200 - 500$ MHz	$4{,}000 - 10{,}000$	200 sq degrees
mid	$0.5 - x$ GHz	10,000	10's of square degrees
high	$x - 25$ GHz	5,000	$\left(\frac{1.4}{\nu}\right)^2$ square degrees

Notes:
[1] The value of x, which is the frequency where the sensor technology changes from mid to high, will be determined by studies and the technology demonstrators planned for the 2008-2012 time frame (see discussion in text). Initial studies suggest that x will be $\sim$10 GHz

2.2. *SKA development*

The SKA will be built in a staged program. The development timeline is shown in Fig. 3. Construction of SKA technology pathfinder telescopes have begun at radio-quiet sites in Western Australia and South Africa, and will be completed by 2012. These pathfinder will have $\sim$1% of the collecting area of the complete SKA and will demonstrate the low-mid band SKA technologies. The demonstration of field-of-view expansion technology on the pathfinders will provide them with significant scientific potential as synoptic survey instruments. Early, pre-SKA, survey science with the pathfinders will be carried out between 2012 and 2015.

Following further technical studies of the properties of the proposed sites, a selection will be made in 2011. Construction of the low-mid technology of the SKA will begin around 2015 at the selected sites, and the low-mid SKA will be rolled-out over five years leading to completion of the low-mid SKA by around 2020.

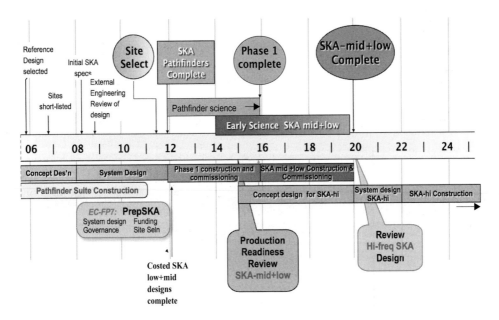

Figure 3. Timeline for SKA Development and Science. Pathfinder instruments at 1% scale of the the SKA collecting will be completed in Western Australia and South Africa by 2012. The rollout of the SKA will begin a few years later with construction start for SKA mid-low around 2015.

3. Astrometry with the SKA

With a maximum baseline of 3000 km, the highest angular resolution of the SKA will be similar to that currently achieved with global VLBI arrays. However, the main limitation to astrometric precision with current VLBI comes not from the angular resolution but from the need to have nearby, strong compact sources as phase references for differential astrometry. This is limited by the sensitivity of existing telescopes and their limited field-of-view. The SKA will have a significant fraction of its aperture out to the largest baselines (see Table 3), ensuring good sensitivity at the highest resolution. The combination of wide field-of-view and sensitivity of the SKA will instantaneously provide a large number of reference sources.

Table 3. Distribution of the SKA Aperture

Region	Distance	Cumulative Fraction of Aperture
core	< 1 km	20%
inner	< 5 km	50%
mid	< 150 km	75%
outer	at least 3000 km	100%

For a 3000 km maximum baseline the angular resolution is

$$\theta = \frac{20}{\nu_{\mathrm{GHz}}} \text{ mas} = 2 \text{ mas at 10 GHz.} \tag{3.1}$$

At 10 GHz the SKA will achieve a minimum detectable flux density of 0.3 μJy (5σ) in one hour. The SKA will be designed for high dynamic range imaging (see Table 1). At this frequency the imaging dynamic range will be less challenging than lower frequency so noise-limited images over the full field-of-view will be routine. The angular precision of phase-referenced astrometry is given approximately by

$$\Delta\theta = \frac{\theta}{\mathrm{SNR}}, \tag{3.2}$$

where SNR is the signal-to-noise ratio on the radio source. In a deep VLA integration at 8.4 GHz by Fomalont et. al. (2002), 0.64 sources per arcmin^2 were detected above 7.5 μJy. From fits to the number of sources versus flux density they derive an expression for the integral source counts,

$$N(> S) = 0.099 \left(\frac{S}{40}\right)^{-1.11}. \tag{3.3}$$

The SKA will achieve SNR > 2000 in one hour at 10 GHz on sources with flux density above 200 μJy. Within one square degree there will be on average 105 sources above this flux limit. The astrometric precision for these source will be $\sim$1 μas. In the same area there will be more than 1000 source above 12 μJy which will have differential astrometry to $\sim$10 μas. The SKA could therefore very quickly establish a dense grid of more than 10,000,000 sources with astrometric accuracy of a few μas that can be re-observed at multiple epochs.

In combination with the next-generation space VLBI mission, VSOP2, even higher precision astrometry will be possible. Assuming Earth-space baselines of about 30,000 km, angular resolution of 2 μas will be achieved at 8 GHz and 5-σ detection limit of 1.0 μJy. Sub-μas astrometry will be possible to sources with flux densities down to 0.4. mJy.

The astrometric capabilities of the SKA will be used for a range of science goals. In the most recent SKA science case Fomalont & Reid (2004) list a number of areas where SKA astrometry at the μarc-second level will enable significant advance

- distance, proper motions and orbital motions of many thousands of radio stars
- planetary detections
- masses of degenerate stars
- calibration of the universal distance scale from 10 to 10^7 pc,
- AGN core-jet interactions
- fundamental frame tie to <10 μarc-second
- terrestrial dynamics
- fundamental physics via solar and Jovian deflections
- pulsar array geometry for gravity wave detection

4. Conclusion

The combination of sensitivity and imaging field-of-view of the SKA will allow μas precision astrometry to 10's of millions of compact radio sources, including AGN and thousands or radio stars. In combination with VSOP2, astrometry on mJy sources will be possible to well below a μas. This capability will both provide a dense, μas astrometric grid of radio sources and will underpin a range of fundamental astronomy based on ultra-precise astrometry.

References

Carilli, C. L. & Rawlings, S., 2004, *New Astron. Revs*, 48, 979.

Fomalont, E. B., Kellermann, K. I., Partridge, R. B., Windhorst, R. A., & Richards, E. A., 2002, *AJ*, 123, 2402.

Fomalont, E. & Reid, M., 2004, *New Astron. Revs*, 48, 1473

Schillizi, R. T., 2004 in J. M. Oschmann Jr. (ed.), *Ground-based Telescopes*, Proceedings of the SPIE, 4569, p. 62

Taylor, A. R. 1999, in M. P. van Haarlam (ed.), *Perspectives on Radio Astronomy: Science with Large Antenna Arrays*, ASTRON, p. 1

A Giant Step: from Milli- to Micro-arcsecond Astrometry
Proceedings IAU Symposium No. 248, 2007
W. J. Jin, I. Platais & M. A. C. Perryman, eds.

© 2008 International Astronomical Union
doi:10.1017/S1743921308018966

Astrometry with ALMA:
a giant step from 0.1 arcsecond to
0.1 milliarcsecond in the sub-millimeter

J.-F. Lestrade

Observatoire de Paris-CNRS
77 av. Denfert Rochereau, F75014, Paris, France
email: `jean-francois.lestrade@obspm.fr`

Abstract. We discuss astrometric capabilities of the future interferometer ALMA that will be located at a high altitude site (5000m) in Northern Chile to operate in the sub-millimeter range. In this paper, we estimate the astrometric precision of ALMA to be ~ 0.18 milliarcsecond at the optimum observing frequency of 345 GHz from an error budget including the thermal noise and the systematic errors caused by uncertainties in antenna coordinates, reference source coordinates, Earth orientation parameters, dry atmosphere parameter and by phase fluctuations due to moisture above the site. We briefly discuss three applications: first, astrometric search of exoplanets around 446 nearby stars detectable by ALMA; second, proper motions and parallaxes of pre-stellar cores and protostars; third, the rotation rate of the debris disk around ϵ Eri to test the theory of dust trapping in mean motion resonances with unseen planets.

Keywords. instrumentation: high angular resolution, instrumentation: interferometers, astrometry, submillimeter, circumstellar matter, stars: formation

1. Introduction

ALMA (Atacama Large Millimeter Array) is an international project to build and operate a large millimeter/sub-millimeter interferometer array (up to 80 antennae) at a high altitude site (5000m) in Northern Chile in the Atacama desert. Partners of the project are ESO, US/Canada, Japan and Taiwan. The first concept of the array was the MMA (MilliMeter Array) described in a NRAO memo by Owen (1982). The project began in 2002. Site construction, hardware production lines and software development are now well underway, the first antenna is being installed as of this writing. Early science with a limited operational array is planned for 2010, and full science operations are expected to start in 2012.

ALMA will provide high-angular resolution and high sensitivity images of the molecular Universe and of the cold dust Universe in order to study the origins of planets and stars, and stellar formation in primordial galaxies at high z.

The present array is planned with 54×12-m antennae and 12×7-m antennae placed in different configurations to realise baselines from 150 m, for the most compact array, to 20 km, for the most extended. Ten frequency bands will be covered by the receivers between 31 GHz and 950 GHz. The bandwidth will be 8 GHz with dual polarization. For continuum observations, relevant to astrometry, 1σ flux sensibility will be 0.2mJy in 1 minute of integration time at the optimum frequency of 345 GHz in median atmospheric conditions at Atacama. This sensitivity is unprecedented in the sub-millimeter range.

Several major projects are planned for astrometry in the optical (*e.g.*, Gaia) that will dominate the field in the two decades to come. However, there are intrinsically cold celestial objects that "shine" only at submillimeter wavelengths, for which ALMA will be the instrument of choice for precise astrometry. Here we present the error analysis to estimate the expected astrometric precision of ALMA and discuss a few astrophysical applications.

2. Astrometric precision of ALMA

Astrometric accuracy depends both on statistical and systematic errors. If there were no systematic errors, *e.g.*, no errors in the station coordinates or in atmospheric pressure, then the theoretical precision of interferometer would be limited by thermal noise only:

$$\sigma_{\alpha,\delta} = \frac{1}{2\pi} \times \frac{1}{SNR} \times \frac{\lambda}{B}. \tag{2.1}$$

Assuming $\lambda = 850\mu$m (345 GHz), SNR=30, B=10 km, $\sigma_{\alpha,\delta}$ would be 0.1 mas. This is unheard of in the sub-millimeter range for astrometry since the existing arrays (IRAM, SMA and CARMA) provide a precision of $\sim$ 0.1 arcsecond. We shall discuss whether or not systematic errors can be controlled to the level that allows us to approach this theoretical precision. To the usual systematic errors encountered in radio interferometry at cm-λ (Thompson, Moran & Swenson (1986)) — uncertainties in the station coordinates, coordinates of reference source, Earth Orientation Parameters, and dry atmosphere zenith delay — rapid phase fluctuations caused by the moisture in the atmosphere – all become critical at sub-millimeter wavelengths. We assume that the ALMA astrometric observations are relative to a reference source that removes long term phase fluctuations from the atmosphere and electronics allowing to expand the integration time to tens of minutes or more as done in VLBI (Lestrade *et al.* (1990)). A typical angular separation between the target and a phase reference source is $\Delta\theta = 6°$, based on the density of strong sources at sub-millimeters (Holdaway *et al.* (2005)).

In our analysis, we use the simplified interferometric formula for the phase: $\phi = f\frac{B \sin\theta}{c}$, where θ is the angle between the direction of the source and the normal to the baseline vector, B is the baseline length, f is the effective frequency of observation, c is the speed of light. The differential phase between target and reference sources separated by $\Delta\theta$ is $\Delta\phi \sim f\frac{B \cos\theta_0}{c} \times \Delta\theta$, where $\Delta\theta$ is small ($6°/57° \sim 0.1$ radian). Differentiating this equation again — but with respect to baseline B and to reference source position θ_0 this time — yields the error $\delta\Delta\theta$ in the angular separation between these sources that we want to estimate.

Uncertainties in the station coordinates, i.e. baseline error δB, cause the error $\delta\Delta\theta \sim \frac{\delta B}{B} \times \Delta\theta$. The ALMA station coordinates can be determined at a level of 50μm after proper calibration of the array, making $\delta B \sim 100\mu$m (Wright (2002) and Conway (2004)). By using $B = 10$ km and $\Delta\theta = 6°$, one gets $\delta\Delta\theta = 0.2$ mas over one baseline. Hence, an array consisting of 66 antennae with uncorrelated coordinate errors yields an overall $\delta\Delta\theta = 0.2/\sqrt{66} = 0.03$ mas for the angular separation between target and reference sources.

Uncertainties of the reference source coordinates $\delta\theta_0$ cause the error $\delta\Delta\theta \sim \delta\theta_0 \times \Delta\theta$. Assuming the precision of absolute positions for the ALMA reference sources to be $\delta\theta_0 = 1$ milliarcsecond – a plausible estimate for ALMA – one gets $\delta\Delta\theta = 0.1$ mas.

Uncertainties in the Earth Orientation Parameters: pole x,y and UT1-UTC are known to a ~ 0.2 mas accuracy as routinely determined by VLBI on daily basis. This uncertainty causes an error similar to $\delta\theta_0$ just mentioned. Thus, $\delta\Delta\theta = 0.02$ mas.

Dry atmosphere : uncertainty of surface pressure. At the elevation E, the path delay through dry atmosphere is given by $\tau = \tau_z/\sin E$, with the zenith dry delay τ_z. The elevation E_1 and E_2 at stations 1 and 2 form a small difference $\Delta E = E_2 - E_1 \sim 0.1°$ over the longest baseline (20 km) for ALMA (unlike to VLBI where $E_1 - E_2$ is usually large). Consequently, the dry atmospheric interferometer phase is $\phi_D = f \times \tau_z \times cosE/sinE^2 \times \Delta E$, where ΔE is small. Thus, the differential dry phase between the target and reference sources is $\Delta\phi_D \sim f \times c\tau_z/c \times \Delta E \times \Delta\theta$, where all angles are in radians and small, and any error $\delta\tau_z$ will produce an error $\delta\Delta\phi_D \sim f \times c\delta\tau_z/c \times \Delta E \times \Delta\theta$. The zenith dry delay τ_z is proportional to the atmospheric pressure. High-quality barometers located at all antennae must be used to calibrate the effect. At Atacama, pressure is about half the pressure at sea level, i.e. $c\tau_z$ is $\sim$ 1.15m. Assuming the barometers to be accurate to ± 2 millibars, $c\delta\tau_z$ is 5mm and that yields $\delta\Delta\theta \sim 0.02$ mas.

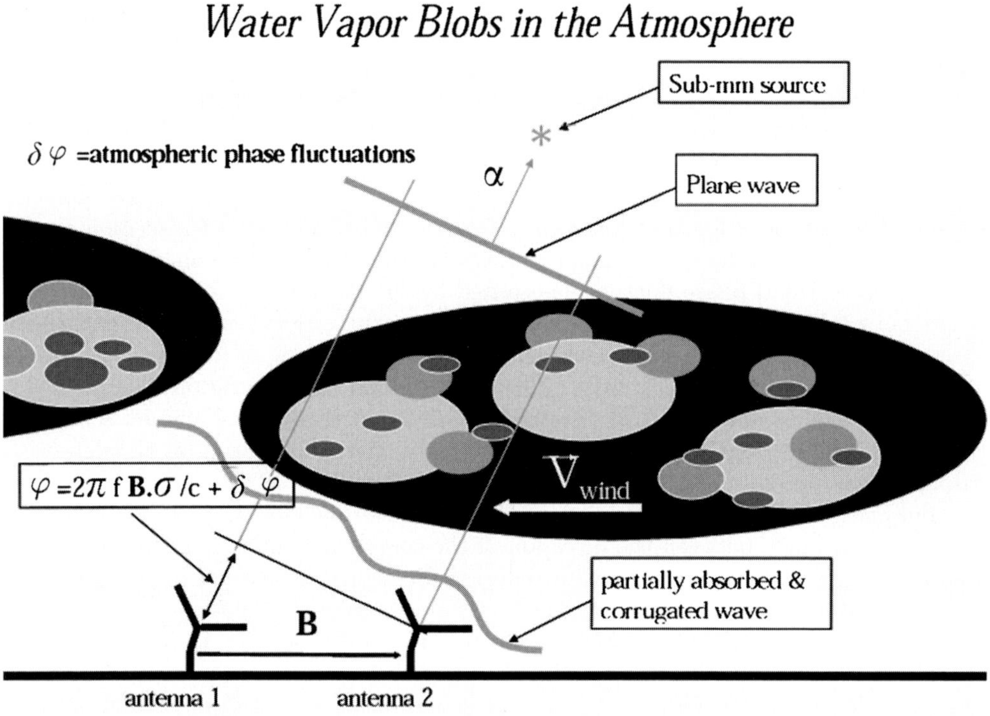

Figure 1. Sketch of the water vapor blobs blown by the wind above ALMA.

Moisture in the atmosphere and phase fluctuations . Fig. 1 shows a sketch of the blobs of water vapor blowing above ALMA. Moisture absorbs millimeter waves and adds fluctuating phase delay to the static dry atmosphere delay. A site as dry as the Atacama desert alleviates this effect but does not make it negligible. In astrometry, absorption impacts only the SNR, and thus the thermal noise of observations (Eq. 2.1). Phase delay fluctuations are more important and should be discussed in some details. These fluctuations have several time scales that depend on the sizes of the water vapor blobs and wind speed over the site. Fluctuations that come from the small blobs have small amplitudes and are rapid ($\sim$1 Hz) for a typical wind speed (10 m/s). Fluctuations that come from the largest blobs (outer scale of the atmosphere) have large amplitudes and are slow ($\sim 10^{-3}$ Hz). The standard approach to quantify fully the atmospheric phase fluctuations is to measure the phase structure function $D_\phi(\rho)$, *i.e.* the phase variance as a function of baseline length ρ. This function is difficult to measure in practice. The best phase structure function above an astronomical site is the one measured for the

VLA at 22 GHz by Carilli & Holdaway (1999) and sketched in Fig 2. Three regimes can be recognized ; for baselines $B = 0$ to 1.2 km, the phase rms ($\sqrt{D_\phi(\rho)}$) grows as the power-law $B^{0.85\pm0.03}$, characteristic of a turbulent Thick Screen (3-D) ; for baselines B = 1.2 to 6 km, the phase rms grows as $B^{0.41\pm0.03}$, characteristic of a turbulent Thin Screen (2-D) ; and finally for $B > 6$ km, theoretically the phase rms saturates, which is consistent with practice (see Fig. 7 in Carilli & Holdaway (1999)). Saturation of the phase rms occurs when baseline B is long enough to sample the full outer scale of the atmosphere. The two transition baselines between the 3-D and 2-D turbulent regimes (1.2 km on figure) and between the 2-D and saturation regimes (6 km on figure) depend on the site, the frequency, and weather conditions. Needless to say that characterizing a site is demanding. Theory predicts that the outer scale should be the scale height of the atmosphere which is 1-2 km (Pérez-Beaupuits *et al.* (2005)), unlike 6 km found for the VLA.

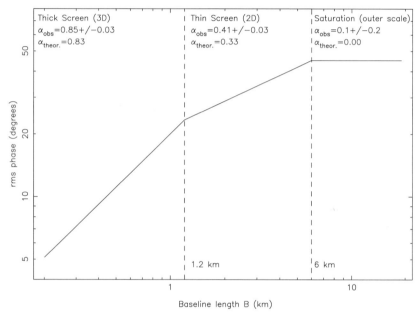

Figure 2. Sketch of the phase structure function measured above the VLA at 22 GHz (in reality, $\phi_{rms} = \sqrt{D_\phi}$ is plotted). Consult Carilli & Holdaway (1999) for the original Fig. 7 inluding data points that allowed to determin α_{obs}.

At the Atacama site, a programme is set up to monitor atmospheric phase fluctuations with an interferometer made of two telescopes, separated by 300 m, to track the carrier of a geostationary satellite at 11.2 GHz since 1995 (Radford *et al.* (1996)). This has allowed to model the phase rms by a power-law : $a \times (\rho/300)^\alpha$, where baseline length ρ is in meters and coefficients a and α are in Table 1 reprinted from Holdaway & Pardo (2001)). The baseline length for which saturation of this function occurs has not been determined yet for Atacama.

Based on this information, the largest phase rms is $\sigma_\phi = 1.86 \times (2000 \text{ m}/300 \text{ m})^{0.50} \times$ (345 GHz/11.2 GHz) = 148° at 345 GHz, in median weather conditions (a=1.86 in Table 1), and assuming an outer scale of 2000 m for the Atacama site. Such phase fluctuations have a time scale of ~ 3 minutes with a wind speed of 10 m s^{-1} (2000 m/10 m s^{-1}). The phase fluctuations can be removed from the data either by monitoring water vapor in the direction of the source by radiometers or by alternating observations between

Table 1. Parameters of the phase structure function $a \times (\rho/300)^{\alpha}$ at the Atacama site from Holdaway & Pardo (2001).

Weather conditions (percentile)	**a** zenith 11.2 GHz (degrees)	α
5% (best)	0.69	0.50
10%	0.91	0.54
20%	1.35	0.56
30%	1.86	0.57
40%	2.46	0.57

the target and reference sources faster than this time scale. A radiometer inverts sky brightness temperature into precipitable water vapor enabling phase fluctuations to be determined. An example of such monitoring between the CSO and JCMT radio telescopes ($B = 160$ m) at 345 GHz during a normal night at Mauna Kea (precipitable water vapor $= 2.6$ mm) is provided by Wiedner $et\ al.$ (2001). In their paper, the phase rms is $60°$ before radiometer corrections, and drops to $26°$ after corrections (see their Fig. 4).

It is expected that the phase rms can be corrected at least to this level (rms $= 26°$) on the longest baselines of ALMA because Atacama is drier than Mauna Kea. In the Wiedner $et\ al.$ (2001) phase plot, the uncertainty of the mean phase μ_ϕ is $\sigma_{\mu_\phi} = 26°/\sqrt{180} = 2°$, assuming Gaussian noise for all 180 data points of the corrected phases. This is relevant to astrometry because in the presence of phase fluctuations, a point-like celestial source can be located within the interferometer fringe spacing with a precision of $(\sigma_{\mu_\phi}/2\pi) \times \lambda/B$. Thus, with $\lambda = 850\mu$m (345 GHz) and $B = 10$ km, the mean phase uncertainty σ_{μ_ϕ} of $2°$ induces an astrometric uncertainty of 0.1 mas. In relative astrometry, this increases to $\delta\Delta\theta = \sqrt{2} \times 0.1$ mas due to the separation uncertainty. Note that this uncertainty is for a single baseline and will scale down as a function of the number of antennae. This functions will depend on the phase fluctuations at all scales ; small scale fluctuations are uncorrelated between antennae, while large scale fluctuations are uncorrelated only if the outer scale of the atmosphere is smaller than the array size. Conservatively, we retain the maximum value $\sqrt{2} \times 0.1$ milliarcsecond for our final error budget.

Table 2. Error budget for ALMA astrometry with a phase reference source separated from the target source by $6°$

Error sources	plausible systematic errors	Resulting errors in ALMA α, δ (milliarcsecond)
Ref. Sour. coord.	1 mas	0.1
Station coord.	50μm	0.03
EOP	0.2 mas	0.02
Dry atm. (pressure)	2 millibars	0.02
Moisture (Radiometer)	26° (post-correction rms)	0.14
Thermal noise (SNR)	SNR=30	0.1
Total	all	0.18

Final error budget : Table 2 summarizes the result of our analysis. Combining in quadrature all 6 sources of errors, we estimate the total astrometric precision of ALMA to be ~ 0.18 milliarcsecond, at the optimum observing frequency of 345 GHz, with the extended configuration, with a target-reference source separation of 6°, and with water vapor monitoring by radiometry. The analytical approach adopted to derive the present error budget for ALMA has been used previously for determining VLBI astrometric errors (Shapiro *et al.* (1979)) that compare satisfactorily to observations (Lestrade *et al.* (1990)) and to numerical simulations (Pradel *et al.* (2006)).

3. Astrometric applications

We explore briefly three possible astrometric applications of ALMA.

3.1. *Search for extra-solar planets*

We have computed the thermal flux densities at 345 GHz for all stars from the CNS3 (Catalogue of Nearby Stars 3rd Edition 1991, by Gliese and Jahreiss), and have found that 446 nearby stars are expected to have flux densities $\geqslant 0.1$ mJy (Table 3) and thus are detectable by ALMA with SNR=30 at reasonable integration time (right-hand plot of Fig. 3). The photospheres of these stars will not be angularly resolved even by the most extended configuration of the array. These are stable sources of emission and can accurately track the reflex motion of unseen planets.

Table 3. Distribution of the CNS3 stars detectable by ALMA above 0.1 milliJansky at 345 GHz. For some stars, the catalogue provides only approximate spectral types that have been interpreted in our analysis as the following : a-f=F0, f=F5, f-g=G0, g=G5, g-k=K0, k=K5, k-m=M0, m=M2, m+=M6.

Spectral Type	Number of stars in the whole CNS3 (mostly d $\leqslant$ 25 pc)	N of stars detectable by ALMA	Fraction of the CNS3
O	0	0	0 %
B	0	0	0 %
A	69	54	78 %
F	266	158	59 %
G	495	125	35 %
K	824	71	9 %
M	1804	36	2 %
Total	3461	446	13 %
Miscellaneous	341		

We have computed also the minimum planetary masses measurable by ALMA over a 10 year long programme and with an astrometric precision of 0.1 mas. The amplitude of the wobble of a central star due to the gravitational pull by planet in a circular orbit is :

$$\theta = \frac{m}{M}\frac{a}{d} = \frac{m}{d}\left(\frac{P}{M}\right)^{2/3}$$ where θ is in arcseconds, a in AU, P in years, d in pc, m and M,

the masses of the unseen planet and of the central star, in $M_\odot$. We plot the distribution of planetary masses that ALMA can probe by setting $\theta = 0.1$ mas, $P = 10$ yr, M and d to their values in the CNS3 catalogue, and by solving for m (left-hand plot of Fig. 3). See further details in Lestrade (2003).

3.2. *Pre-stellar cores and Protostars*

Tracing stellar formation phases in star forming regions (e.g. ρ Oph cloud at 160 pc) is one of the major current topics in astrophysics. Pre-stellar cores (age from -1 Myr

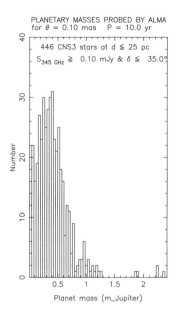
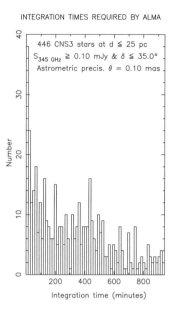

Figure 3. Left: *minimum masses detectable astrometrically by ALMA for possible unseen planets orbiting 446 nearby stars. The corresponding astrometric programme must last at least 10 years and sustain the astrometric precision of 0.1 mas.* **Right:** *the integration time required to reach the theoretical astrometric precision of 0.1 mas by ALMA for the selected 446 stars.*

to 0, stellar birth line), Class 0 protostars (age $\sim 10^4$ yr) and Class I protostars (age $\sim 10^5$ yr) "shine" only, or mainly, in the sub-millimeter range (André *et al.* (1993)). As suggested by P. André, measurements of proper motions can verify whether or not the rms velocity of pre-stellar cores is <1 km s^{-1} ($<$ 1 mas yr^{-1} at the distance of ρ Oph cloud) while rms velocity of Class 0 and Class I protostars is >1 km s^{-1} as predicted by theory. This is because pre-stellar cores are still coupled to their original molecular cloud and its magnetic field, while Class 0 and Class I protostars have decoupled, and stellar motions are in virial equilibrium in the gas-mass dominated gravitational potential of the cloud. ALMA with its astrometric precision of proper motions at ~ 0.1 mas yr^{-1} will be able to survey a large sample of these objects to validate this theory. Along with proper motions, trigonometric parallaxes can be produced from such a survey. They will provide the depth of star forming regions.

3.3. *Debris disks*

Stars develop **debris disks** at the end of the planet formation phase when the protoplanetary disk of primordial gas and dust has dissipated. A debris disk is composed of left-over planetesimals (asteroids, comets, Kuiper Belt Objects, Trojans) that could not agglomerate into larger planets during this early phase ($<$ 10 Myrs for solar-mass stars). They are the exosolar analogues of the Kuiper Belt and of the asteroid belt in the Solar System. Collisions between these left-over planetesimals generate dust grains whose large emitting surface make them observable from Mid-IR to sub-mm. The best example of this kind is the disk around a nearby K2V star ϵ Eri imaged by JCMT at 345 GHz (Greaves *et al.* (2005). Their image shows a disk seen almost face-on, and hosting a central cavity. The mean radius of the disk is 60 AU, similar to the Kuiper Belt. The image exhibits a clumpy annular structure that is interpreted as emitting dust grains trapped into mean motion resonances with an unseen planet orbiting within the central cavity. Simulation of this mechanism permits to find the putative planet by adjusting its orbital parameters in order to achieve a best fit to the image (Wyatt (2003)). If the clumps of the disk structure are compact enough for ALMA, *i.e.* angularly unresolved by the array, astrometric observations can reveal the rotation of the disk which should make all clumps move by ~ 0.5"/yr along directions tangential to a circle (Poulton *et al.* (2006)). With the astrometric precision that we have estimated for ALMA, this can be done with observations between two epochs separated by much less than a year ! The strategy of observations may take advantage of the fact that the photosphere of ϵ Eri can be detected easily by ALMA and used as the reference source. In this case, the proper motion and parallax are common to both the star and its disk, so that they do not bias the measurement of the rotation rate.

References

André, P., Ward-Thompson, D., & Barsony, M., 1993, *ApJ*, 406, 122

Carilli, C. L. & Holdaway, M. A., 1999, *Radio Sc.*, 34, 817

Conway, J., 2004, *Alma Memo Series*, memo#503

Greaves, J. S., *et al.*, 2005, *ApJ*, 619, L187

Holdaway, M. A. & Pardo, J. R., 2001, *Alma Memo Series*, memo#404

Holdaway, M. A., Carilli, C., Weiss, A., & Bertoldi, F., 2005, *Alma Memo Series*, memo#543

Lestrade, J.-F., Rogers, A. E. E., Withney, A. R., *et al.* 1990, *AJ*, 99, 1663

Lestrade, J.-F., 2003, *ASP Conf. Ser.*, 294, 587

Owen, F. N. 1982, *Alma Memo Series*, memo#1

Pérez-Beaupuits, J. P., Rivera, R. C., & Nyman, L.A, 2005, *Alma Memo Series*, memo#542

Poulton, C. J., Greaves, J. S., & Cameron, A. C., 2006, *MNRAS*, 372, 53.

Pradel, N., Charlot, P., & Lestrade, J.-F., 2006, *A&A*, 452, 1099.

Radford, S., Reiland, G., & Shillus, B., 1996, *PASP*, 108, 441

Shapiro, I. I., Wittels, J. J., & Counselman, C. C., 1979, *AJ*, 84, 1459

Thompson, A. R., Moran, J. M., & Swenson, G., 1986, *Interferometry ad Synthesis in Radioastronomy*, publishers : John Wiley & Sons

Wiedner, M. C., Hills, R. E., Carlstrom, J. E., & Lay, O. P., 2001, *ApJ*, 553, 1036.

Wright, M., 2002, *Alma Memo Series*, memo#427

Wyatt, M. C., 2003, *ApJ*, 598, 1321.

A Giant Step: from Milli- to Micro-arcsecond Astrometry
Proceedings IAU Symposium No. 248, 2007
W. J. Jin, I. Platais & M. A. C. Perryman, eds.

© 2008 International Astronomical Union
doi:10.1017/S1743921308018978

The FAST telescope and its possible contribution to high precision astrometry

C. J. Jin, R. D. Nan and H. Q. Gan

National Astronomical Observatories, Chinese Academy of Sciences
Chaoyang District, Datun Road, A.20, Beijing 100012, China
email: cjjin@bao.ac.cn

Abstract. In this report we give a brief introduction to the Five hundred meter Apeture Spherical Telescope (FAST). Some possible contributions of FAST to high precision astrometry are discussed. The illuminated aperture of FAST in normal operation mode is 300m in diameter. With special feeding mechanism, the whole 500m aperture could be used. FAST will cover frequencies from 70MHz to 3GHz, and observe at zenith angle of up to 40 degrees without a notable gain loss. As the most sensitive single dish radio telescope, FAST would be able to discover more mega-masers and measure the radial velocities of masers with higher precision. This may yield more delicate dynamics of their maser spots. FAST will increase the precision of time of arrival (ToA) measurements for pulsars. This will help in detecting the stochastic gravitational wave background and in establishing an independent timing standard based on the long-term stability of the rotations of a group of millisecond pulsars. FAST might also work as a very powerful ground station for the future space missions. In a three-way communication mode, FAST should be able to provide precise ranging and Doppler measurements. Moreover, by joining the international VLBI network, FAST would help to improve the precision of the VLBI astrometry measurements.

Keywords. telescopes, masers, pulsars: general, gravitational waves, reference systems

1. Introduction

FAST, the Five hundred meter Aperture Spherical Telescope, is one of the mega science facilities for fundamental research in China (Nan 2006). After more than 13 years of studies of its science case and key technologies, funding for the FAST project was approved by the National Development and Reform Commission (NDRC) in July, 2007. The construction of FAST is expected to be completed around 2014. FAST will be about 10 times more sensitive than the fully steerable 100m telescope Effelsberg and GBT. Compared with the 305m Arecibo telescope, FAST will be about 3 times more sensitive, and has a much wider sky coverage. As the largest single dish radio telescope ever built, FAST would help to carry out astrometric observations with unprecedented sensitivity.

In the following section we describe the general specifications and key technologies of FAST. Some possible contributions of FAST to high precision astrometry are discussed in the third section. A concluding remark is drawn in the last section.

2. The general specifications and key technologies

Figure. 1 shows a 3-D image of the FAST telescope and its optics. FAST will be built in a Karst depression. The terrain shown in Fig. 1 is taken from the real depression where FAST will be built. The shape of the main reflector is spherical with a radius of 300m. The overall diameter is 500m. The focus cabin is suspended about 140m above the reflector. During observations, part of the main reflector will be deformed instantaneously into a

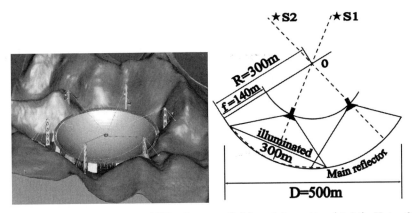

Figure 1. A 3-D image of the FAST telescope (left) and its optics (right). Note that the terrain in the left image is taken from the real depression at Dawodang.

Table 1. The main specifications of FAST telescope.

Item	Specification
Spherical reflector	Radius 300m, Aperture 500m, Opening angle $110 \sim 120^\circ$
Illuminated aperture	$D_{illu} = 300\text{m}$
Focal ratio	f/D=0.4665
Sky coverage	zenith angle 40°
Frequency	$70 \sim 3000 MHz$
Multi-beam(L-band)	19, beam number of future FPA > 100
Slewing	< 10min
Pointing accuracy	$8''$

paraboloid through active control. This illuminated part has a diameter of 300m. The optics of FAST allows the use of standard feed design. With special feeding mechanism, the whole 500m aperture can be illuminated. FAST will cover frequencies from 70MHz to 3GHz, and it will observe at a zenith angle up to 40 deg without a notable gain loss. The main specifications of FAST is listed in Table 1

The technical studies of FAST have been focused on 5 key technologies: the site, active main reflector, focus cabin suspension, measurement & control, and receivers.

Site. An extensive site survey was carried out in the Karst region in Guizhou province (Nan *et al.* 2002). After evaluation on the various aspects of depressions, including the shape, distribution of surrounding peaks, natural sink holes for drainage and RFI situations, Dawodang depression ($N25.647222^\circ$, $E106.85583^\circ$) is finally selected as the site for FAST.

Active main reflector. The central part of sphere is close to paraboloid as the focal length is properly chosen. In the case of FAST, the radial deviation between the sphere and the paraboloid is less than 0.67m across the illuminated area (Qiu 1998). The back structure of the main reflector is made from steel cables. The elasticity of the cable is used to implement the required deformations (Nan *et al.* 2003).

Focus cabin suspension. During observations, the feed needs to be located at the focus of the instantaneous paraboloid with high accuracy. A coarse drive by cables and a secondary fine adjusting stewart platform are utilized to achieve the desired positional accuracy. Extensive theoretical study and model experiments were carried out on the focus cabin suspension system (Zhu *et al.* 2004).

Measurement & control. The shape of the main reflector and the position of the feed need to be monitored and controlled in real time. This involves monitoring the position of the feed at long distance with high sampling rate and surveying large number of nodes on the main reflector with relatively low sampling rate. Fusion of various measuring techniques is adopted for FAST (Zhu *et al.* 2004).

Receiver. The optics of FAST allows the use of standard feed design. In 2000, a layout design of receivers for FAST was made jointly by the National Astronomical Observatories, Chinese Academy of Sciences (NAOC) and Jodrell Bank Observatory (JBO). The layout was based on the existing proven design at the various frequency bands. In 2006, this layout was upgraded. An optimization of the band division and receiver design is still ongoing, to make the best match to the science drivers.

3. The possible contributions to high precision astrometry

As the largest single dish radio telescope ever built, FAST would observe the radio sky with unprecedented sensitivity.

Masers. Astronomical masers associated with stars and active galactic nuclei provide useful probes to study the dynamics of them (e.g., Reid & Moran 1981; Lo 2005). In some cases, maser observations could yield distance measurement by trigonometric technique. The shells of OH 1612 MHz emission surrounding OH-IR sources offer such a possibility. The light-travel diameter of the shell could be inferred by measuring phase-lags in the OH spectrum, but the angular diameter is measured by radio interferometry. The distance is then obtained geometrically (Shepherd *et al.* 1992). Such IR-OH sources would then provide an independent calibration to the period-luminosity of the RR Lyrae that belongs to the same system.

Pulsars. Pulsars are cosmic clocks that provide ultra stable periodic pulses. The stability of some of the millisecond pulsars are competitive to or even better than the most stable atomic clocks. A group of millisecond pulsars well-distributed on the sky may provide an independent timing standard that could complement the atomic standard. Long-term monitoring of the ToA of a group of millisecond pulsars would also help in detecting the stochastic gravitational wave background. The high sensitivity and large sky coverage make FAST a powerful tool for discovering more pulsars and measuring the ToA of pulsars with much better precision. A more detailed discussion on observing pulsars using FAST could be found in Nan *et al.* (2006).

Ground station for space mission. FAST might also work as a very powerful ground station for the future space missions. The large collecting area will enable the downlink data rate increase by orders of magnitude. In three-way communication mode, FAST would be able to provide more precise ranging and Doppler measurements. In the case of a flyby event, FAST would then allow a better mass determination of the flyby object.

VLBI element. Due to its large collecting area and geographical location, FAST could greatly improve the capability of the current international VLBI network. FAST lies at the edge of most of the VLBI networks. Figure 2 shows the simulated uv-coverage of some of EVN antennas plus FAST. It is clearly seen that FAST contributes mostly to long baselines. With an aperture of 300m, FAST would increase the baseline detection sensitivity by an order of magnitude. This may yield more precise position measurement and weaker sources to be detected. Although in current design the highest frequency of FAST is 3GHz, it is feasible to extend it to X band when more delicate measurement and

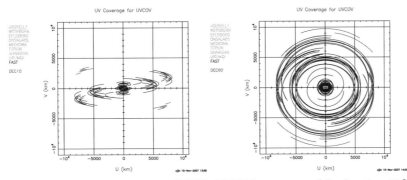

Figure 2. The simulated uv-coverage of EVN+FAST for sources of Declination 10 ° and 60 °

control mechanisms are employed in the future. This will enable FAST to participate in the S/X astrometric and geodetic VLBI measurements.

4. Concluding remarks

FAST is going to be the largest and the most sensitive single dish radio telescope in the world. Because of its large collecting area, FAST will be able to carry out astronomical observations with unprecedented sensitivity. Although FAST is not a dedicated astrometric telescope, it will be able to contribute to high precision astrometry through single dish observations of masers and pulsars and by joining the S/X astrometric and geodetic VLBI network.

Acknowledgements

The authors thank the SOC of IAUS248 for the acceptance of this presentation. We also thank Mark Reid, Chopo Ma and Xingwu Zheng for the valuable discussions on the FAST-related topics during the symposium. This work is supported by the Chinese Academy of Sciences and a key project grant 10433020 from the National Natural Science Foundation in China.

References

Lo, K. Y. 2005, *ARAA*, 43, 625
Nan, R. D., Peng, B., Su, Y., Zhu, W. B., Zhu, L. C., Qiu, Y. H., Zhu, B. Q., & Nie, Y. P. 2002, in: J. Vernin, Z. Benkhaldoun & C. Muñoz-Tuñón(eds.), *Astronomical Site Evaluation in the Visible and Radio Range*, Proc. ASP, Vol. 266, p. 408
Nan, R. D., Ren G. X., Zhu, W. B., & Lu, Y. J. 2003, *Radio Studies of Galactic Objects, Galaxies and AGNs* Proc. Sino-German Radio Astronomy Conference Radio Studies of Galactic Objects, Acta Astronomica Sinica, Supplement Issue, Vol. 44, p. 13
Nan, R. D. 2006, *Science in China Series G Physics, Mechanics & Astronomy*, 49(2), 129
Nan, R. D., Wang, Q. M., Zhu, L. C., Zhu, W. B., Jin, C. J., & Gan, H. Q. 2006, in: N. Wang, R. N. Manchester, B. J. Richett & A. Esamdin (eds.), *ChJAAS* Vol. 6, Issue S2, Proc. 2005 Lake Hanas International Pulsar Symposium, p. 304
Qiu, Y. H. 1998, *MNRAS*, 301, 827
Reid, M. J. & Moran, J. M. 1981, *ARAA*, 19, 231
Shepherd, M. C., Cohen, R. J., Gaylard, M. J., & West, M. E. 1992, in: C. de Jager, & H. Nieuwenhuijzen (eds.), Proc. International Colloquium, Amsterdam, p. 78
Zhu, W. B., Nan, R. D., & Ren, G. X. 2004, *Experimental Astronomy*, Vol. 17, Issue 1-3, p. 177
Zhu, L. C., Nan, R. D., Wu, S. B., Zhu, W. B., & Wang, Q. M. 2004, in: H. Lewis & G. Raffi (eds.), *Advanced Software, Control, and Communication System for Astronomy*, Proc. SPIE, Vol. 5496, p. 335

A Giant Step: from Milli- to Micro-arcsecond Astrometry
Proceedings IAU Symposium No. 248, 2007
W. J. Jin, I. Platais & M. A. C. Perryman, eds.

© 2008 International Astronomical Union
doi:10.1017/S174392130801898X

The Chinese VLBI network and its astrometric role

J. L. Li, L. Guo and B. Zhang

Shanghai Astronomical Observatory, Shanghai 200030, China
email: jll@shao.ac.cn

Abstract. In this report we present the current status of the Chinese VLBI network (CVN), and the results of satellite tracking experiments in the past few years related to realtime processing of the CVN dataflow and the reliability and precision of the CVN measurements. We briefly outline the prospective astrometric roles of CVN in the studies of long-term monitoring of extragalactic radio sources, densification of radio celestial reference frame, determination of Earth Orientation Parameters and linkage parameters of reference frames as well as the observation of pulsars for deep space autonomous navigation.

Keywords. astrometry, ephemerides, reference systems, Earth, Moon, planets and satellites: general

1. Current status of the Chinese VLBI network

The two 25m radio telescopes at Sheshan of Shanghai and Nanshan of Urumqi began astrometric observations since 1987 and 1993, respectively. The Chinese mobile VLBI consists of a 3m antenna and an S2 system. Observations could also be recorded by a hard disk system, which is compatible with the Mark5. In May 2006, two new antennas with diameters of 50m at Miyun near Beijing and 40m at Fenghuangshan of Kunming participated in tracking experiment of the Smart-1, a lunar satellite of the European Space Agency (ESA). The correlation center of the Shanghai Astronomical Observatory (SHAO) has successfully developed a five-station FX correlator and a system with PC-based hard disk recorder and playback. The dataflow rate per station is 256 Mbps. Up to now, in the background of the Chinese lunar exploration project, Chang'E-1 (CE-1), the Chinese VLBI Network (CVN) has been developed into four permanent antennas, one mobile station and a correlation center.

As shown in Fig. 1, the CVN extends geographically by over 34° east-west and 18° south-north, with baselines ranging from 1115km to 3249km. Table 1 lists some technical specifications of the four CVN permanent antennas, which are all equipped with the Mark5 recording system.

2. CVN and CE-1

Since the 1960s VLBI techniques have been demonstrating high precision in astrometry and geodesy studies, especially in establishing the celestial and terrestrial reference frames as well as determination of the Earth Orientation Parameters (EOP). In the Chinese lunar exploration project CE-1, VLBI is expected to contribute to real-time monitoring of the satellite orbit especially during the lunar capture stage. CVN is mandated to fulfil a 10-minute task, that is, to provide the spherical coordinates of the satellite within a delay of 10 minutes, with a precision of 0.2 arc-seconds near perilune. This presents a great challenge to the ordinary processing of the VLBI dataflow, including the preparation

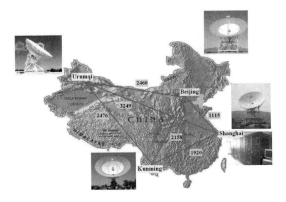

Figure 1. Geographical distribution of CVN

Table 1. Technical specifications of CVN antennas

Specifications	Shanghai	Beijing	Urumqi	Kunming
Begins operation	1987	2006	1993	2006
Structure	BWGC[1]	Prime focus	BWGC	BWGC
Diameter[2]	25m	50m	25m	40m
Pointing precision (arcsec)	20	19	15	30
Slewing rate (AZ/EL, d/s)	1.0/0.6	1.0/0.5	1.0/0.5	1.0/0.5
Receiver	L, C, S/X, K	S/X	UHF, L, C, S/X, K	S/X
Efficiency at S/X band	38%/40%	60%/68%	54%/52%	64%/47%
Recording terminal	Mark5A, VLBA, S2	Mark5A	Mark5A	Mark5A

[1]BWGC — Beam wave guide Cassegrain
[2]Diameter of the main reflection plane

of schedule, antenna tracking, data recording and transfer, correlation, extraction of VLBI observables, correction of systematic behavior and atmosphere delays, as well as deduction of the target spherical coordinates from observations.

Figure 2 shows a realtime processing of the CVN dataflow. The schedule is sent via internet to the antennas. The tracking data are also transferred via internet to the correlation center. The data are processed simultaneously by the hard and software correlators and then observables are extracted. As the antennas are tracking the satellite, observations from collocated GPS receivers are sent to the correlation center too and atmospheric delay information is extracted. This information is combined with the VLBI observables and the systematic behavior corrections deduced from observations of extragalactic radio sources in order to deduce the angular position of the satellite. All these steps are processed with a delay of less than ten minutes. At the end of the tracking pass the orbit is also determined.

In the middle of 2006 ESA offered a great opportunity to the CVN people to test their hardware and software with a goal of completing the 10-minute task by tracking the Smart-1, an ESA lunar satellite. Figure 3 shows the difference between the angular positions of the Smart-1 deduced from CVN realtime measurements and the ESA reconstructed orbit, which was post-stage determined with high precision. After removing some outliers, the standard deviation of the difference is about 0.1arcsec, which demonstrates CVN measurements are reliable and with sufficient precision for identifying the lunar capture of CE-1 satellite.

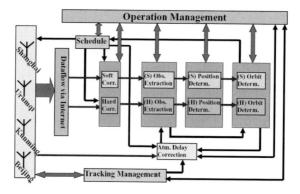

Figure 2. The realtime processing of CVN dataflow

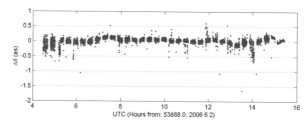

Figure 3. Comparison of the CVN realtime measurements and the ESA reconstructed orbit

3. CVN and the prospective high precision astrometry

CVN has undergone tremendous development by contributing to Chinese national projects and will definitely continue to do so. For example, in subsequent stages of the Chinese lunar exploration Chang'E-2 and Chang'E-3, VLBI will be applied to track the orbiter and lander relative to radio sources. In the Chinese Martian exploration project Yinghuo-1, VLBI will be the main facility for determining satellite's position and orbit. At the same time, CVN is also capable of doing some high-precision astrometric studies, especially in the framework of international cooperation. If CVN is combined with the Russian antennas, the geographical coverage will be about 3500km north-south and 6000km east-west, that would be very promising in studies of space geodesy, astrometry and spacecraft monitoring. The coordinates of the new antennas must be precisely determined, and a local survey at collocation sites with other space geodesy techniques should be done.

3.1. Spacecraft tracking and reference frames tie

As a strong support for ranging and Doppler observations, VLBI is very useful in improving angular position precision of spacecrafts. The CVN geometry is very competitive in high precision positioning of spacecrafts, significantly contributing to deep space exploration. By tracking an orbiter, the dynamical origin of longitude (equinox) can be determined. With phase referencing techniques, the tie parameters between the radio and dynamical reference frames can be determined. Analyzing VLBI observations of the orbiter, lander and extragalactic radio sources helps to reach a comprehensive understanding of the probing data and to refer them to a unified reference system.

3.2. *Monitoring extragalactic radio sources*

The extragalactic radio sources (ERS) are fiducial objects of the International Celestial Reference Frame (ICRF) (Ma, *et al.* (1998)), which represents one of the most outstanding scientific contributions of VLBI. However, these sources may have structures and be subjected to changes. Historical observations of the defining ICRF sources are very limited, and the number of observations per source is not even. For instance, among all observed sources by the end of 2005, about 70% were observed only in one to three sessions. It has been encouraged internationally to conduct regular monitoring of these sources. With the current CVN geometry and equipment infrastructure, several dozens of defining sources could be monitored on a regular basis. With the participation of large antennas (70m antenna in Ukraine and the Chinese FAST (Jin *et al.*(2008))) some weak sources, especially those near the ecliptic, could be observed, which would densify ICRF and serve to deep space exploration.

3.3. *Radio stars and pulsars*

It is believed that many radio stars may have planetary systems. By referring to ERS, a precise time series of the proper motion of a radio star can be accumulated (Boboltz, *et al.* (2007)), and small variation in the proper motion would reveal the existence of an extrasolar planetary system. Based on the rotation and orbital motion of the Earth, pulsar timing observations could reveal information of the dynamical equinox, which serves as the origin of longitude. By referring to ERS the proper motion and annual parallax of pulsars can be precisely determined, which is useful to the studies of stellar evolution and to deep space autonomous navigation.

3.4. *Quick EOP service*

By making the EOP results quickly available to the community, would then enable some important scientific and practical applications. Currently IVS has two sessions per week, R1 and R4, and are processed rapidly. CVN can regularly conduct the EOP determination. At the start of this program, the CVN geometry requires careful checks to evaluate the EOP precision and accumulate experience.

3.5. *Monitoring the baseline length*

VLBI is capable of determining continental distances of several thousand kilometers with the precision to a centimeter and even down to a few millimeters. Simultaneous observations of CVN and international antennas could be used in the studies of the Earth's crust motion and deformation. Variations in the baseline length is also related to tides, loading, and atmosphere modelling.

Acknowledgements

This work is supported by NSFC (*No. 10778635, No. 10173019, No. 10473019*), Chinese lunar exploration project CE-1 and STC of Shanghai Municipality (06DZ22101).

References

Boboltz, D. A., Fey, A. L., Puatua, W. K., Zacharias, N., Claussen, M. J., Johnston, K. J., & Gaume, R. A. 2007, *AJ* 133, 906
Jin, C. J., Nan, R. D., & Gan, H. Q. 2008, *this volume*, p.178
Ma, C., Arias, E. F., Eubanks, T. M., Fey, A. L., Gontier, A.-M., Jacobs, C. S., Sovers, O. J., Archinal, B. A., & Charlot, P. 1998, *AJ* 116, 516

A Giant Step: from Milli- to Micro-arcsecond Astrometry
Proceedings IAU Symposium No. 248, 2007
W. J. Jin, I. Platais & M. A. C. Perryman, eds.

VLBA determinations of the distances to nearby star-forming regions

L. Loinard[1], R. M. Torres[1], A. J. Mioduszewski[2] and L.F. Rodríguez[1]

[1]Centro de Radiastronomía y Astrofísica, Universidad Nacional Autónoma de México
Apartado Postal 72-3 (Xangari), 58089 Morelia, Michoacán, México
email: l.loinard,r.torres,l.rodriguez@astrosmo.unam.mx

[2]National Radio Astronomy Observatory, Array Operations Center
1003 Lopezville Road, Socorro, NM 87801, USA
email: amiodusz@aoc.nrao.edu

Abstract. Using phase-referenced multi-epoch *Very Long Baseline Array* observations, we have measured the trigonometric parallax of several young stars in the Taurus and Ophiuchus star-forming regions with unprecedented accuracy. The mean distance to the Taurus complex was found to be about 140 pc, and its depth around 20 pc, comparable to the linear extent of Taurus on the sky. In Ophiuchus, 4 sources have been observed so far. Two of them were found to be at about 160 pc (the distance traditionally attributed to Ophiuchus), while the other 2 are at about 120 pc. Since the entire Ophiuchus complex is only a few parsecs across, this difference is unlikely to reflect the depth of the region. Instead, we argue that two physically unrelated sites of star-formation are located along the line of sight toward Ophiuchus.

Keywords. astrometry, stars: distances, stars: formation, radio continuum: stars, stars: magnetic fields, radiation mechanisms: nonthermal, techniques: interferometric, binaries: general, ISM: clouds

---

1. Introduction

To provide accurate observational constraints for pre-main sequence evolutionary models, and thereby improve our understanding of star-formation, it is crucial to measure as accurately as possible the properties (e.g., age, mass, luminosity) of individual young stars. The determination of most of these parameters, however, depends critically on the often poorly-known distance to the object. While the average distance to nearby low-mass star-forming regions (e.g. Taurus or ρ Oph) has been estimated to about 20% precision using indirect methods (Elias 1978a,b; Kenyon et al. 1994; Knude & Hog 1998; Bertout & Genova 2006), the line-of-sight depth of these regions is largely unknown, and accurate distances to individual objects are still missing. Even the highly successful Hipparcos mission (Perryman *et al.* 1997) did little to improve the situation (Bertout *et al.* 1999) because young stars are heavily embedded in their parental clouds and are, therefore, faint in the optical band observed by Hipparcos.

Low-mass young stars often generate non-thermal continuum emission produced by the interaction of free electrons with the intense magnetic fields that tend to exist near their surfaces (e.g. Feigelson & Montmerle 1999). If the magnetic field intensity and the electron energy are sufficient, the resulting compact radio emission can be detected with Very Long Baseline Interferometers (VLBI –e.g. André *et al.* 1992). The relatively recent possibility of accurately calibrating the phase of VLBI observations of faint, compact radio sources using nearby quasars makes it possible to measure the absolute position of these objects (or, more precisely, the angular distance between them and the calibrating

Table 1. Distances to the sources in Taurus and Ophiuchus

Unit: pc

Complex	Taurus			Ophiuchus			
Source	T Tau	Hubble 4	HDE 283572	S1	DoAr 21	VSSG 14	WL 5
Distance	147.6 ± 0.6	132.8 ± 0.5	128.5 ± 0.6	$116.9 {}^{+7.2}_{-6.4}$	$121.9 {}^{+5.8}_{-5.3}$	$165.6 {}^{+6.2}_{-5.8}$	$168.3 {}^{+8.2}_{-9.3}$

quasar) to better than a tenth of a milli-arcsecond (Brisken *et al.* 2000, 2002; Loinard *et al.* 2005, 2007; Torres *et al.* 2007; Xu *et al.* 2006; Hachisuka *et al.* 2006; Hirota *et al.* 2007; Sandstrom *et al.* 2007). This level of precision is sufficient to constrain the trigonometric parallax of sources within a few hundred parsecs of the Sun (in particular of nearby young stars) with a precision better than a few percent using multi-epoch VLBI observations. With this goal in mind, we have initiated a large project aimed at accurately measuring the trigonometric parallax of a sample of magnetically active young stars in the most prominent and often-studied northern star-forming regions within 1 kpc of the Sun (Taurus, ρ−Ophiuchus, Perseus, Serpens). We use the 10-element Very Long Baseline Array (VLBA) of the National Radio Astronomy Observatory (NRAO). Here, we will summarize the results obtained so far in Taurus and Ophiuchus.

2. Observations and parallax determination

All the observations used here were obtained in the continuum at 3.6 cm (8.42 GHz) with the VLBA of the *National Radio Astronomy Observatory* (NRAO). A total of 7 sources were studied so far: three in Taurus (T Tau, Hubble 4, and HDE 283572), and four in Ophiuchus (S1, DoAr 21, VSSG 14, and WL 5). In all cases, between 6 and 12 observations spread over 1.5 to 2 years were obtained. The data were edited and calibrated following the standard VLBA procedures for phase-referenced observations (see Loinard *et al.* 2007, and Torres *et al.* 2007 for details).

The displacement of the sources on the celestial sphere is the combination of their trigonometric parallax (π) and proper motion (μ). For isolated sources (such as Hubble 4, and HDE 283572 in our case), it is common to consider linear and uniform proper motions, so the right ascension and declination vary as a function of time t as:

$$\alpha(t) = \alpha_0 + (\mu_\alpha \cos \delta)t + \pi f_\alpha(t) \qquad (2.1)$$
$$\delta(t) = \delta_0 + \mu_\delta t + \pi f_\delta(t), \qquad (2.2)$$

where α_0 and δ_0 are the coordinates of the source at a given reference epoch, μ_α and μ_δ are the components of proper motion, and f_α and f_δ are the projections over α and δ, respectively, of the parallactic ellipse.

For sources in multiple systems, however, the proper motions are affected by the gravitational influence of the other members of the system. As a consequence, the motions are curved and accelerated, rather than linear and uniform. If the orbital period is long compared with the timespan covered by the observations (as it is for T Tau –see Loinard *et al.* 2003), it is sufficient to include a uniform acceleration in the fit. This leads to the functions:

$$\alpha(t) = \alpha_0 + (\mu_{\alpha 0} \cos \delta)t + \frac{1}{2}(a_\alpha \cos \delta)t^2 + \pi f_\alpha(t) \qquad (2.3)$$

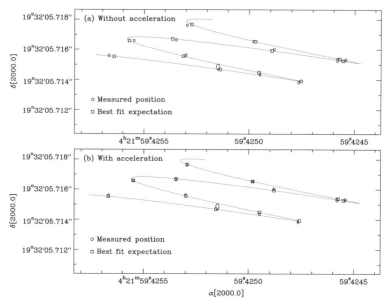

Figure 1. Measured positions of T Tau Sb and best fit without (a) and with (b) acceleration terms. The observed positions are shown as ellipses, the size of which represents the magnitude of the errors. Note the very significant improvement when acceleration terms are included.

$$\delta(t) = \delta_0 + \mu_{\delta 0} t + \frac{1}{2} a_\delta t^2 + \pi f_\delta(t), \tag{2.4}$$

where $\mu_{\alpha 0}$ and $\mu_{\delta 0}$ are the proper motions at a reference epoch, and a_α and a_δ are the projections of the uniform acceleration (see Fig. 1 for a comparison between fits with and without acceleration terms).

Finally, if a source is a member of a multiple system whose orbital period is shorter than, or comparable with the timespan covered by the observations (as will be the case for all Ophiuchus sources), then a full Keplerian fit is needed, but additional observations are required to properly constrain that fit. These observations are currently being obtained for all Ophiuchus sources, but are not yet analyzed. As a consequence, the fits presented below for the Ophiuchus sources are based on Eqs. 2.1 and 2.2. The resulting uncertainties will be much larger than for the sources in Taurus because the orbital motions generate an unmodelled scatter around the mean positions of sources. These increased errors should disappear once we include new observations, and perform a full Keplerian fit.

3. Results and discussion

The distance to the seven sources studied here are given in Table 1. For the Taurus sources, Hipparcos parallaxes (Bertout *et al.* 1999), and/or estimates based on a modified convergent point method (Bertout & Genova 2006) are available. Our measurements are always consistent with these values, but ours are one to two orders of magnitude more accurate. Only one source in Taurus (V773 Tau; Lestrade *et al.* 1999) has a VLBI-based parallax determination. Taking the mean of that and our 3 measurements, we estimate the mean distance to the Taurus cluster to be $\bar{d} = 137$ pc. The dispersion about that mean leads to a full width at half maximum depth of about 20 pc, comparable to the linear extent of Taurus on the sky.

Traditionally assumed to be at a distance of 165 pc (Chini 1981), Ophiuchus has recently been proposed to be somewhat nearer, at 120 pc (Knude & Hog 1998). Interestingly, two of our sources are consistent with the traditional value, but the other two are consistent with the distance at 120 pc. Since Ophiuchus is only a few parsecs across on the plane of the sky, it is very unlikely to have a depth of 40 pc. It is noteworthy that the two sources at about 120 pc (S1 and DoAr 21) are associated with the sub-condensation Oph A, whereas the two sources at $\sim$ 165 pc (VSSG 14 and WL 5) are associated with the condensation Oph B (See Motte *et al.* 1998 for an overview of the Ophiuchus complex). Thus, a plausible explanation of our results is that two physically unrelated star-forming regions are located along the line-of-sight toward Ophiuchus. Interestingly, Knude & Hog (1998) noticed an effect that would be consistent with this possibility in their analysis of extinction towards Ophiuchus. While a clear extinction step was visible at 120 pc, extinction was apparently extending up to about 160 pc. Additional VLBI observations will be needed to confirm the existence of two regions of star-formation towards Ophiuchus, and to investigate at which distance the other condensations in Ophiuchus are located.

4. Conclusions and perspectives

The present results show that VLBA observations of non-thermal sources associated with young stars have the potential to improve very significantly our knowledge of the spatial distribution for star-forming regions in the Solar neighborhood. Indeed, the precision obtained by these measurements is sufficient even to probe a 3D structure of star-forming regions. Coupled with pre-main sequence evolutionary models, such information could be used to reconstruct the history of star-formation in individual regions.

References

Andre, P., Deeney, B. D., Phillips, R. B., & Lestrade, J.-F. 1992, ApJ, 401, 667
Bertout, C., Robichon, N., & Arenou, F., 1999, A&A, 352, 574
Bertout, C. & Genova, F. 2006, A&A, 460, 499
Brisken, W. F., Benson, J. M., Beasley, A. J., Fomalont, E. B., Goss, W. M., & Thorsett, S. E. 2000, ApJ, 541, 959
Brisken, W. F., Benson, J. M., Goss, W. M., & Thorsett, S. E. 2002, ApJ, 571, 906
Chini, R. 1981, A&A, 99, 346
Elias, J. H. 1978a, ApJ, 224, 857
Elias, J. H. 1978b, ApJ, 224, 453
Feigelson, E. D. & Montmerle, T., 1999, ARAA, 37, 363
Hachisuka, K., *et al.* 2006, ApJ, 645, 337
Hirota, T., *et al.* PASJ, in press (http://arxiv.org/abs/0705.3792)
Kenyon, S. J., Dobrzycka, D., & Hartmann L., 1994, AJ, 108, 1872
Knude, J. & Hog, E. 1998, A&A, 338, 897
Lestrade, J.-F., Preston, R. A., Jones, D. L., *et al.*, 1999, A&A, 344, 1014
Loinard, L., Torres, R. M., Mioduszewski, A. J., Rodríguez, L. F., González, R. A., Lachaume, R., Vázquez, V., & González, E. 2007, ApJ, 671, 546
Loinard, L., Mioduszewski, A. J., Rodríguez, L. F., González, R. A., Rodríguez, M. I., & Torres, R. M. 2005, ApJL, 619, L179
Loinard L., Rodríguez, L. F., & Rodríguez, M. I., 2003, ApJ, 587, L47
Motte, F., Andre, P., & Neri, R. 1998, A&A, 336, 150
Perryman, M. A. C., Lindegren, L., Kovalevsky, J., *et al.*, 1997, A&A, 323, L49
Sandstrom K. M., *et al.*, ApJ, in press (http://arxiv.org/abs/0706.2361)
Torres, R. M., Loinard, L., Mioduszewski, A. J., & Rodríguez, L. F. 2007, ApJ, in press
Xu, Y., Reid, M. J., Zheng, X. W., & Menten, K. M. 2006, Science, 311, 54

A Giant Step: from Milli- to Micro-arcsecond Astrometry
Proceedings IAU Symposium No. 248, 2007
W. J. Jin, I. Platais & M. A. C. Perryman, eds.

VLBI astrometry for the NASA/Stanford gyroscope relativity mission Gravity Probe B

N. Bartel[1], R. R. Ransom[1]*, M. F. Bietenholz[1], D. E. Lebach[2], M. I. Ratner[2], I. I. Shapiro[2] and J.-F. Lestrade[3]

[1] York University, Toronto, ONT, Canada,

[2] Harvard-Smithsonian Center for Astrophysics, Cambridge, MA., U. S. A.

[3] Observatoire de Paris, Paris, France

*now at DRAO, Penticton, Canada

Abstract. We used VLBI observations at 8.4 GHz between 1991 and 2005 to determine the motion of the RS CVn binary IM Pegasi (HR 8703), the guide star for the NASA/Stanford gyroscope relativity mission, Gravity Probe B (GP-B). The motion was determined relative to our primary reference, the core of the quasar 3C 454.3. The stability of this core was checked relative to two other extragalactic sources, B2250+194 and B2252+172, the former of which was tied to the ICRF. The core of 3C 454.3 is stationary relative to these two sources to within 30 μas yr^{-1} in each coordinate. IM Pegasi's radio morphology varies, but appears to be on average centered on the primary. We estimate the proper motion of IM Pegasi with a statistical standard error (sse) of 30 μas yr^{-1} in each coordinate. We also estimate the parallax with a statistical standard error of 75 μas and parameters of the orbit with sse's corresponding to 110 μas on the sky. Coupled with our upper limit of three times the sse on any systematic errors in each parameter estimate, these results ensure that the uncertainty of IM Pegasi's proper motion makes only a small contribution to the uncertainty of GP-B's tests of general relativity.

Keywords. relativity, techniques: interferometric, reference systems, stars: imaging, stars: distances, radio continuum: stars

1. Introduction

Gravity Probe B (GP-B) is the spaceborne relativity experiment developed by NASA and Stanford University to test two predictions of general relativity (GR) with gyroscopes in low-earth orbit. GP-B sought to measure the precession of the gyroscopes relative to distant inertial space, represented by an extragalactic reference frame. Because of technical restrictions, the spacecraft could not measure the precession directly relative to that frame but only to an optically bright star, the guide star chosen to be the RS CVn binary IM Pegasi (HR 8703). We determined IM Pegasi's motion independently relative to the extragalactic reference frame. In particular, we used VLBI to determine this motion relative to the core of the quasar 3C 454.3 which we tied to two other extragalactic sources. The motion of one of them was measured directly within the International Celestial Reference Frame (ICRF).

2. Observations and results

Between 1997 and 2005, we carried out 35 sessions of VLBI observations at 8.4 GHz of IM Pegasi and two extragalactic sources, 3C 454.3 and B2250+194, supplemented by another (likely) extragalactic source, B2252+172 for the last third of the sessions. We used a worldwide network of radio telescopes including the 10 dishes of the Very Long Baseline Array (VLBA) and the Very Large Array of the National Radio Astronomy Observatory in the USA, the Effelsberg telescope of the Max-Planck-Institut für Radiostronomie in Germany, and the 70-m dishes of the NASA Deep Space Network in the USA, Spain, and Australia. Each session used typically 15 telescopes and was of about 15 hours in length. The data were correlated using the VLBA processor in Socorro, NM, USA. We also used results from 1991-1994 observations of IM Pegasi and 3C 454.3.

We located three major components in 3C 454.3 by fitting Gaussians to the brightness distributions. The most easterly of these components is the most stable, and stationary relative to the two other extragalactic sources to within 30 μas yr^{-1} in each coordinate, corresponding to a limit of 0.75 c in each coordinate for a flat universe with $H_0 = 70$ km s^{-1} Mpc^{-1}, $\Omega_M = 0.3$, and $\Omega_\lambda = 0.7$. This component is most likely closely related to the core and the gravitational center of the quasar and is used by us as the primary reference for IM Pegasi. The radio morphology of IM Pegasi varies, having sometimes one and sometimes two components (Figure 1), and at one epoch an even more complex structure. The radio emission is thought to originate in flares and magnetic loops above the surface of the star. Our astrometry shows that the radio emission is centered close to the primary on the plane of the sky (Figure 1).

We fit a 9-parameter model to the position of the centroid of the radio images of IM Pegasi at all 39 epochs; our estimate of the proper motion has a statistical standard error (sse) of 30 μas yr^{-1} in each coordinate. In addition we estimate the parallax with an sse of 75 μas and the parameters of the orbit with sse's corresponding to 110 μas on the sky. The systematic errors for each of these estimates are almost surely no more than threefold higher than the sse's. The release of our other astrometric results will be coordinated with the NASA/Stanford GP-B team. Our VLBI astrometry program surpassed its accuracy goal, thereby ensuring that the uncertainty of IM Pegasi's proper motion will make only a small contribution to the uncertainty of the GP-B tests of general relativity.

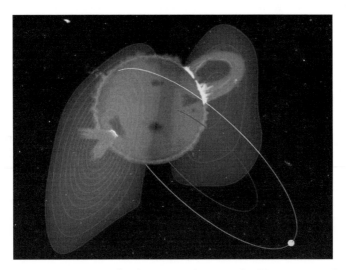

Figure 1. An artist's impression of IM Pegasi, showing the K2 primary, the much smaller secondary in the lower right, and their orbits. The radio emission seems to originate near the primary. Two flares represent the possible source of the radio emission. Superposed are contours representing the radio emission for epoch 1999 September 18/19. Two radio components appeared at this epoch. Contours are at 5, 10, 20, 30, ..., 80, and 90% of the peak brightness of 11.5 mJy/beam. The scale and orientation of this figure will be released later (see text).

A Giant Step: from Milli- to Micro-arcsecond Astrometry
Proceedings IAU Symposium No. 248, 2007
W. J. Jin, I. Platais & M. A. C. Perryman, eds.

© 2008 International Astronomical Union
doi:10.1017/S1743921308019017

Astrometry of red supergiant VY Canis Majoris with VERA

Y. K. Choi[1,2], T. Hirota[2], M. Honma[2], and H. Kobayashi[2]

[1]Department of Astronomy, Graduate School of Science, The University of Tokyo,
7-3-1 Hongo, Bunkyo-ku, Tokyo 113-0033, Japan

[2]Mizusawa VERA Observatory, National Astronomical Observatory of Japan,
2-21-1 Osawa, Mitaka, Tokyo 181-8588, Japan
email: yoonkyung.choi@nao.ac.jp

Abstract. We present observational results on the red supergiant VY Canis Majoris with VERA. We have observed 22 GHz H_2O masers and 43 GHz SiO masers (v=1 and 2 J=1-0) around VY CMa for 13 months. We succesfully detected a parallax of 0.87 ± 0.08 mas, corresponding to the distance of $1.15\,^{+0.10}_{-0.09}$ kpc using H_2O masers. As the result of phase-referencing analyses, we have measured absolute positions for both H_2O masers and SiO masers. The H_2O maser features show rapid expansion off the central star.

Keywords. astrometry, masers, stars: distances, supergiants, stars: individual (VY CMa)

1. Introduction

VY CMa is one of the most luminous red supergiants in our galaxy. Its luminosity is 4.3×10^5 L$_\odot$ (Humphreys, Helton & Jones 2007) and distance is estimated to be 1.5 kpc (Lada & Reid 1978). VY CMa is ejecting large amount of gas and dust at a high mass loss rate of 4×10^{-4} M$_\odot$ yr^{-1} (Danchi *et al.* 1994). H_2O and SiO masers are commonly found in circumstellar envelopes (CSE) of supergiants. They are important tools to study structures and kinematics of the CSE, the mass-loss process and evolution of massive stars.

2. Parallax measurements of VY CMa

We have observed H_2O masers around the red supergiant VY CMa at 10 epochs with VERA for 13 months. The source J0725-2640, whose angular distanc from VY CMa is 1.059 deg, was observed as a positional reference source. Using a bright H_2O maser spot at the LSR velocity of 0.55 km s^{-1}, we succesfully detected a trigonometric parallax of 0.87 ± 0.08 mas, corresponding to the distance of $1.15\,^{+0.10}_{-0.09}$ kpc. Figure 1 shows the results of our positional measurements for this H_2O maser spot. The astrometric accuracy in right ascension is better than that in declination. A possible reason for this disparity could be uncertainty in the atmospheric zenith delay.

3. H_2O masers and SiO masers around VY CMa

SiO masers are detected in a spherical shell at 2–4 R$_*$ from the central star (Diamond *et al.* 1994) but the H_2O masers are distributed over a wider region than SiO masers. The resulting phase-referencing analysis indicates that the absolute positions for both H_2O masers and SiO masers are measured with an accuracy better than 1 mas. Figure 2

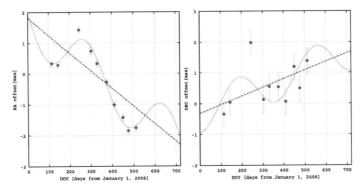

Figure 1. Results of position measurements for H_2O maser spot at the LSR velocity of 0.55 km s^{-1} in VY CMa. *Left panel*: The motion of the maser spot in RA as a function of time over 13 months. *Right panel*: The same as left panel but for Dec. Solid lines represent the best fit model for annual parallax and proper motion of a maser spot. Dotted lines represent proper motion (-2.09 mas yr^{-1} in RA and 1.02 mas yr^{-1} in Dec) and points represent the observed positions of a maser spot with error bars indicating the standard deviation from least-squares analysis (0.10 mas in RA and 0.42 mas in Dec).

shows the distribution of H_2O and SiO maser features. From the position of SiO maser ring, we can estimate the position of stellar component. The H_2O maser features move away from the central star with the velocity of 5–30 km s^{-1}. The diameter of SiO maser feature is about 40 mas, corresponding to 46 AU at the distance of 1.15 kpc. If the SiO masers are formed at 2 stellar radii, then the radius of VY CMa should be about 1200 $R_\odot$.

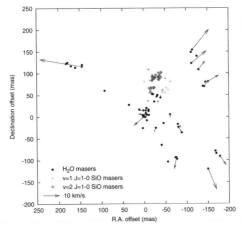

Figure 2. Direct Comparison of absolute positions between H_2O masers and SiO masers are possible with phase-referencing analyses. Closed circle, cross and open diamond represent H_2O masers, v=1 and v=2 J=1-0 SiO masers, respectively. Arrows represent the proper motion of H_2O masers from which the mean motion has been subtracted.

References

Danchi, W. C., Bester, M. Degiacomi, C. G., Greenhill, L. J., & Townes, C. H. 1994, *AJ*, 107, 1469.
Diamond, P. J., Kemball, A. J., Junor, W., *et al.* 1994, *ApJ*, 430, L61.
Humphreys, R. M., Helton, L. A. & Jones, T. J. 2007, *AJ*, 133, 2716.
Lada, C. J. & Reid, M. J. 1978, *ApJ*, 219, 95.

A Giant Step: from Milli- to Micro-arcsecond Astrometry
Proceedings IAU Symposium No. 248, 2007
W. J. Jin, I. Platais & M. A. C. Perryman, eds.

© 2008 International Astronomical Union
doi:10.1017/S1743921308019029

Monitoring the lunar capture of Chang'E-1 satellite by real-time reduction of the instantaneous state vectors

L. Guo[12], J. L. Li[1], S. B. Qiao[123] and F. Tian[12]

[1]Shanghai Astronomical Observatory, Shanghai 200030, China

[2]The Graduate School of the Chinese Academy of Sciences, Beijing 100039, China

[3] Surveying and Mapping Institute, Zhengzhou 450052, China
email: kent-gl@shao.ac.cn

Abstract. Based on our independently developed data processing software, the real-time reduction of the instantaneous state vectors of satellite during the maneuver stage near to perilune is analyzed via experimental observations. Results show that it is a quick and practical method to monitor the orbit evolution and the lunar capture of Chang'E-1 satellite.

Keywords. astrometry, reference systems, Earth, Moon, planets and satellites: general

1. Introduction

As designed in the Chinese lunar exploration project Chang'E-1 (CE-1), the tracking data of satellite consist of range and Doppler measurements from the Chinese United S-Band (USB) network as well as the delay and delay rate from the Chinese VLBI network (CVN). USB tracking stations are Qingdao, Kashi and survey vessels. VLBI antennas are at Shanghai, Urumqi, Beijing and Kunming. We are assigned to process the tracking observations of CE-1 satellite in a real-time way and we have independently developed the data reduction software, which provides the instantaneous state vectors (ISVs) including the three dimensional position and velocity. The software real-timely reads in tracking observations with corrections of clock, instruments and propagation, automatically identifies the central gravitational body within the Earth-Moon system, as well as takes into consideration of the perturbations of non-spherical figure, N-body gravitation, light pressure, atmospheric drag, tidal effects and so on. The satellite ISVs are sequentially reduced with a 5s sampling interval.

From the ISVs it is easy to get the corresponding orbit elements and to predict the satellite ephemeris by orbit integration [1-3]. The ISVs at a specified epoch could be reduced whenever the independent observations related to the wave-front of signal at this epoch are sufficient, that is, enough delay and range observations for the three dimensional position and enough delay rate and Doppler for the velocity. This reduction is geometrically performed rather than applying dynamical constraints on the observations belonging to different wave-fronts at different epochs, and so the length of tracking arc is not a crucial prerequisite. It could be used to monitor the quality of tracking data and to identify the evolution of satellite orbit, which satisfies the needs of efficiency and speediness in the view of the implementation of projects. Comparatively, precise orbit determination requires an enough tracking arc in length and is mainly applied in scientific studies with great precision requirement as in the post analysis stage rather than real-timely.

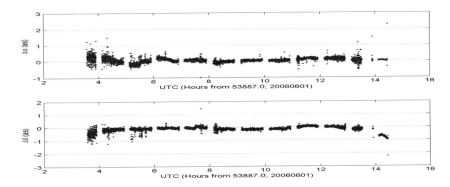

Figure 1. Comparison between the real-time reduced ISVs with the reconstructed ephemeris of a satellite

2. Experimental data verification

Since March of 2005 several tracking experiments of Chinese domestic and abroad satellites have been organized by the orbit monitoring system of CE-1 project and large amount of data have been accumulated. As an example, Figure 1 shows the comparison of the real-time reduced ISVs with the reconstructed ephemeris of Smart-1, a lunar satellite of the European Space Agency. Here reconstructed means that the orbit is post-stage determined and is high in precision. After some outliers are removed the standard deviation of the difference is about 0.1arcsec in both directions of right ascension and declination, which demonstrates that the ISVs are reliable and with sufficient precision for the identification of the lunar capture of CE-1 satellite.

On November 5 of 2007, CE-1 satellite was successfully captured by the Moon. During that period the real-time reduced ISVs by our software very well followed the evolution of the satellite orbit with a delay behind the clock about five minutes, which demonstrated the feasibility of the proposed method and software in this report.

3. Concluding remarks

The real-time reduction of the ISVs of CE-1 satellite is proposed in order to monitor and identify the orbit evolution during the maneuver stage nearby the perilune. In comparison with the precise and short-arc orbit determinations, the sequential ISVs reduction does not set any prerequisites on the length of tracking arc, does not require a precise modelling of various forces exerting on the satellite. This reduction could be real-timely performed and be applied to monitor the quality of tracking data and to identify the orbit evolution of satellite, which is suitable to project needs of great efficiency and speediness.

Acknowledgements

This work is supported by NSFC (*No. 10778635, No. 10173019, No. 10473019*), Chinese lunar exploration project CE-1 and STC of Shanghai Municipality (06DZ22101).

References

Lin, S. Y., Li, Z. J., & He, X. S. 2004, *Missiles and Space Vehicles* 3, 7
Liu, L. & Wang, X. 2002, *Science in China (A)* 32, 1128
Shi, J. & Liu, L. 2004, *Journal of Spacecraft TT&C Technology* 23, 46

A Giant Step: from Milli- to Micro-arcsecond Astrometry
Proceedings IAU Symposium No. 248, 2007
W. J. Jin, I. Platais & M. A. C. Perryman, eds.

The distance to an outer Galaxy star forming region

K. Hachisuka[1,2], A. Brunthaler[2], M. J. Reid[3] and K. M. Menten[2]

[1]Shanghai Astronomical Observatory, 80 Nandan Road, Shanghai, 200030, China
email: khachi@shao.ac.cn

[2]Max-Planck-Institut für Radioastronomie, Auf dem Hügel 69, Bonn, Germany

[3]Harvard-Smithsonian Center for Astrophysics, 60 Garden Street, Cambridge, MA, USA

Abstract. We performed phase-referencing VLBI astrometric observations of the H_2O maser source IRAS 02395+6244, located well beyond the solar circle. We measured its heliocentric distance to be 5.49 ± 0.80 kpc, implying a Galactocentric distance of 12.5 ± 0.5 kpc and a distance of 270 ± 40 pc above the Galactic plane.

Keywords. Galaxy: structure, masers, astrometry

1. Introduction

In the past, the structure and size-scale of our Galaxy have been mainly determined by translating measured radial velocities into distances with the help of a kinematic model – "the rotation curve". (e.g. Nakanishi & Sofue 2003). Such methods originally proved the existence of the Milky Way's spiral structure and are still in use today. However, the resulting "kinematic distances" strongly depend on the Galactic rotation model chosen and peculiar motions may render them questionable (see, e.g., Xu *et al.* 2006). This is particularly true for the outer galaxy, since the Galactic rotation speed there has still a larger uncertainty than in the inner Galaxy (Brand & Blitz 1993). Consequently, many uncertainties remain on the structure and distance scale of the outer Galaxy.

Recently, distances of Galactic maser sources have been determined by annual parallax measurement using the phase-referencing VLBI technique (e.g. Hachisuka *et al.* 2006 and see contribution by M. Reid in these proceedings). This method can measure the annual parallax of Galactic objects out to many kiloparsecs with errors of a few percent at a kpc (scaling with distance), allowing *direct* distance measurements for objects located near the edge of the stellar disk of our Galaxy.

Wouterloot *et al.* (1993) searched for H_2O maser sources in the outer galaxy associated with ^{12}CO emission and found a number of them, for which their kinematic distances indicated large distances from the Galactic center. We selected one of these, associated with IRAS 02395+6244, as the target source for an annual parallax measurement.

2. Observations

We used the VLBA to observe the IRAS 02395+6244 H_2O maser source using phase-referencing VLBI techniques. The observations were performed at 5 epochs spread over a year. For position reference we used an ICRF source separated by 0.5 deg from the target source. See Xu *et al.* (2006) for a description of the data reduction procedures.

3. Results

Several H_2O maser components could be imaged after the calibration. We determined their positions and traced individual maser components between different epochs carefully, monitoring their intrinsic variations. We found an H_2O maser component that was stable over the course of the observations and determined its annual parallax.

Figure 1 shows the change in position versus time of that H_2O maser component. We estimate the annual parallax to be 0.182 ± 0.026 mas, which corresponds to a heliocentric distance of 5.49 ± 0.80 kpc, or 12.5 ± 0.9 kpc from the Galactic center for $R_0 = 8.0$ kpc (Reid 1993). The Galactic latitude of the source is 2.8 deg, hence it is located 270 ± 40 pc above the Galactic plane.

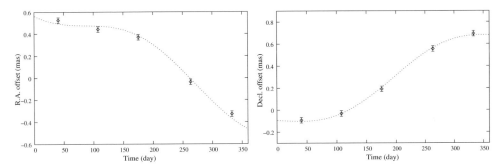

Figure 1. Change of the position with time for an H_2O maser component associated with IRAS 02395+6244 with respect to a background extragalactic source. Displacements in right ascension (left panel) and declination (right panel) are shown. Dotted lines show the best-fit result for a proper motion and an annual parallax.

The kinematic distance of our H_2O maser source with an LSR velocity of -72 km s^{-1} is 8.7 kpc from the Sun, assuming a flat rotation curve and the IAU recommended values of $\Theta_0 = 220$ km s^{-1} and $R_0 = 8.5$ kpc. The kinematic distance is significantly larger than the parallactic distance of 5.49 ± 0.80 kpc. Such discrepancies between kinematic and annual parallaxes have been found for other Galactic maser sources in the outer Galaxy (Hachisuka *et al.* 2006, Xu et al 2006; see contribution by M. Reid). Also, after removing the effects of Galactic rotation, we find a very large peculiar motion at ≈ 30 km s^{-1}. More observations like these should lead to better a understanding of the size, dynamics, and dark matter halo of the Galaxy.

Acknowledgements

This work has been partially supported by the National Natural Science Foundation of China (grants 10573029 and 10625314) and the Knowledge Innovation Program of the Chinese Academy of Sciences (Grant No. KJCX2-YW-T03), and the National Key Basic Research Development Program of China (No. 2007CB815405).

References

Brand, J. & Blitz, L., 1993, *A&A*, 275, 67
Hachisuka, K., Brunthaler, A., Menten, K. M., Reid, M. J. *et al.*, 2006, *ApJ*, 645, 337
Nakanishi, H. & Sofue, Y., 2003, *PASJ*, 55, 191
Reid, M. J., 1993, *ARAA*, 31, 345
Wouterloot, J. G. A., Brand, J., & Fiegle, K., 1993, *A&AS*, 98, 589
Xu, Y., Reid, M. J., Zheng, X. W., & Menten, K. M., 2006, *Science*, 311, 54

A Giant Step: from Milli- to Micro-arcsecond Astrometry
Proceedings IAU Symposium No. 248, 2007
W. J. Jin, I. Platais & M. A. C. Perryman, eds.
© 2008 International Astronomical Union
doi:10.1017/S1743921308019042

Astrometry of Galactic star-forming regions with VERA

M. Honma[1], T. Bushimata[1], Y. K. Choi[1,2], T. Hirota[1], H. Imai[3], K. Iwadate[1], T. Jike[1], S. Kameno[3], O. Kameya[1], R. Kamohara[1], Y. Kan-ya[4], N. Kawaguchi[1], M. Kijima[1], M. K. Kim[1,2], H. Kobayashi[1], S. Kuji[1], T. Kurayama[1], S. Manabe[1], N. Matsumoto[3], T. Miyaji[1], T. Nagayama[3], A. Nakagawa[3], K. Nakamura[3], C.-S. Oh[1], T. Omodaka[3], T. Oyama[1], S. Sakai[1], K. Sato[1], M. Sato[1,2], T. Sasao[5], K. M. Shibata[1], Y. Tamura[1], and K. Yamashita[1]

[1]Mizusawa VERA observatory, NAOJ [2]University of Tokyo, [3]Kagoshima University, [4]Yonsei University, [5]Ajou University,

email: mareki.honma@nao.ac.jp

Abstract. We present an overview of recent astrometric results with VERA. Since 2004, we have been conducting astrometry of tens of Galactic maser sources with VERA, and recently obtained trigonometric parallaxes for several sources, with distances ranging from 180 pc to 5.3 kpc. In this paper, we briefly summarize the results for Galactic star-forming regions, including S269, Orion-KL, NGC 1333, ρ-oph, NGC 281 and others.

Keywords. astrometry, Galaxy: disk, stars: formation

1. VERA project

VERA (VLBI Exploration of Radio Astrometry) is a new VLBI array dedicated to phase referencing VLBI astrometry. This array consists of four 20-m radio telescopes spread over Japan, with a maximum baseline length of 2300 km. VERA's 20m telescopes are operated in the dual beam regime, so that one can simultaneously observe a reference and a target source to cancel out the phase fluctuations caused by tropospheric variations. VERA's main goal is to explore the 3-D structure and dynamics of the Galaxy based on astrometry of ∼1000 of Galactic maser sources (for more details on the VERA project, see Kobayashi *et al.* 2008).

2. Overview on recent results

Since 2004 we have been conducting astrometric observations of tens of maser sources, and recently we have succeeded in measuring trigonometric parallaxes and proper motions of some maser sources. One of the highlights is a parallax measurement of the star-forming region S269 (Honma *et al.* .2007). The parallax of S269 was measured to be $\pi = 189 \pm 8$ μas (D=5.28 ± 0.24 kpc), providing one of the largest distances measured by means of trigonometric parallax. Proper motions of S269 H_2O masers were used to constrain the Galactic rotation velocity at the position of S269 (located at 13 kpc from the Galaxy center), demonstrating that the rotation curve of the Galaxy is basically flat out to 13 kpc (Honma *et al.* .2007).

The other major result recently obtained with VERA is the distance measurement of Orion-KL, which is one of the most important star-forming regions in the Galaxy. Hirota *et al.* (2007) determined the parallax of Orion-KL H_2O maser as π=2.29 ± 0.1 mas

(D=437 ± 19 pc). This value is slightly smaller than previous estimates of ∼ 480 pc, but still is in good agreement with recent measurements using VLBA (Sandstrom *et al.* 2007; Menten *et al.* 2007). The distance of the ρ-Oph cloud was also measured with VERA by Imai *et al.*(2007), providing the distance of 178^{+18}_{-37} pc. This result is consistent with the results obtained by Loinard et al. (2008) based on VLBA parallax measurements of radio-emitting young stars in this region. For other nearby star-forming regions, Hirota *et al.*(2008) obtained the parallax distance of NGC 1333 to be D=235 ± 18 pc.

Another interesting results from VERA is the absolute proper motion measurements for NGC 281 (Sato *et al.* 2007). This star-forming region is associated with an HI "super-bubble", being located ∼300 pc away from the Galactic plan. Sato *et al.*(2007) found that NGC 281 system is moving away from the Galactic plane at the velocity of ∼20 km/s, which strongly supports an idea that a super-bubble was formed by a blow-out from the Galactic plane, most likely by supernovae.

In addition to star-forming regions, VERA also observed masers in late-type stars. Choi *et al.*(2008) recently succeeded in astrometry of a super giant star VY CMa, providing the parallax distance of 1.1 ± 0.1 kpc. They also matched the H_2O maser map with the SiO maser map and determined a position in the H_2O maser map, from which maser spots show rapid expansion away from the central star, explained by the mass loss in this star. Similarly, the study of a semi-regular variable S Crt (Nakagawa *et al.* 2008) provides the parallactic distance of 430 ± 25 pc.

Summarizing the current status, so far we have detected parallaxes for 7 sources, with distances ranging from 180 pc to 5.3 kpc. We are going to observe 70–80 maser sources every year, and in the next 12–15 years VERA will observe ∼1000 Galactic maser sources to precisely locate them in the Galaxy's disk. Similar astrometric studies of methanol maser sources are also being conducted with VLBA (Reid 2008), and finally Gaia (Lindegren et al. 2008) is also going to conduct kpc-scale astrometry by observing billions of stars in the Galaxy. Thus by 2020, when the results of all these projects become available, our knowledge on the Galaxy structure will be revolutionized with the help of high-precision astrometry both in radio and optical wavelengths.

References

Choi, Y. K., *et al.*, 2008, in this volume P.192
Hirota, T., *et al.*, 2007, *PASJ*, 59, 897
Hirota, M., *et al.*, 2008, *PASJ*, 60, 37
Honma, M., *et al.*, 2007, *PASJ*, 59, 889
Imai, H., *et al.*, 2007, *PASJ*, 59, 1107
Kobayashi, H., *et al.*, 2008, in this volume p.148
Lindegren, L., *et al.*, 2008, in this volume p.217
Loinard, L., *et al.*, 2008, in this volume p.186
Menten, K., *et al.*, 2007, *A&A*, 474, 515
Nakagawa, A., *et al.*, 2008, in this volume p.206
Reid, M. 2008, in this volume p.141
Sato, M., *et al.*, 2007, *PASJ*, 59, 743
Sandstrom, M., *et al.*, 2007, *ApJ*, 667, 1161

A Giant Step: from Milli- to Micro-arcsecond Astrometry
Proceedings IAU Symposium No. 248, 2007
W. J. Jin, I. Platais & M. A. C. Perryman, eds.

Distance and kinematics of IRAS 19134+2131 revealed by H$_2$O maser observations

H. Imai[1], R. Sahai[2] and M. Morris[3]

[1]Department of Physics, Kagoshima University, 1-21-35 Korimoto, Kagoshima 890-0065,
Japan, email: hiroimai@sci.kagoshima-u.ac.jp

[2] Jet Propulsion Laboratory, 4800 Oak Grove Drive, Pasadena, CA 91109,USA,
email: raghvendra.sahai@jpl.nasa.gov

[3] Department of Physics and Astronomy, University of California, Los Angeles, CA
90095−1562, USA, email: morris@astro.ucla.edu

Abstract. Using the VLBA, we have observed H$_2$O maser emission in the pre-planetary nebula, IRAS 19134+2131 (I1913), in which the H$_2$O maser spectrum has two groups of emission features separated in radial velocity by $\sim$100 km s^{-1}. The morphology and 3-D kinematics indicate the existence of a fast collimated flow with a dynamical age of only $\sim$40 years. Such a "water fountain" source is a signature of the recent operation of a stellar jet, that may be responsible for the final shape of the planetary nebula into which I1913 is expected to evolve. We have also estimated the distance to I1913 ($\sim$8 kpc) on the basis of an annual parallax and the kinematics of IRAS 19134+2131 in our Galaxy. I1913 may be a component in the "thick disk" or the Galactic "warp", whose kinematics is different from that of the Galactic "thin" disk. These results are reported in Imai, Sahai & Morris (2007).

Keywords. masers, stars: AGB and post-AGB, stars: kinematics, stars: individual (IRAS 19134+2131)

"Water fountain sources" are H$_2$O masers associated with AGB or post-AGB stars, but exhibiting much higher expansion velocities than those observed in classical OH/IR stars. They have been revealed to be highly-collimated (precessing) bipolar jets of molecular gas launched prior to forming the planetary nebulae. So far there are 11 sources identified as water fountain sources. Investigating these sources in more detail using VLBI technique should provide us important clues for elucidating the evolution/devolution of the water fountains and the mechanism of shaping planetary nebulae. Here we report results of VLBA observations of the IRAS 19134+2131 (I1913) H$_2$O masers at 6 epochs during 2003 January–2004 April. All observations have applied the phase-referencing VLBI technique, in which all maser feature positions are determined with respect to the extragalactic reference source J1925+2106 (see figure 1).

In I1913, a clear bipolar flow with high spatial and kinematical collimation was recognized (see figure 1). The dynamical age of the flow is estimated to be $\sim$40 yrs, which is roughly equal to those of other water fountain sources and may give a typical lifetime of water fountains. Note that some water fountains have optical nebulosity while others do not, suggesting that the presence of water fountains may help in the identification of new pre-planetary nebulae, which may be optically invisible due to the heavy circumstellar extinction.

The phase-referencing VLBI technique provides us information on the absolute kinematics of the H$_2$O maser source (see figure 2), giving the annual-parallax distance $D = 8.0^{+0.9}_{-0.7}$ kpc and the mean proper motion. The derived location and the 3D velocity

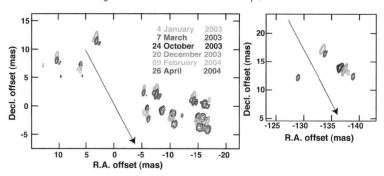

Figure 1. Velocity-integrated images of the H₂O masers in IRAS 19134+2131. An arrow indicates the mean motion in four years.

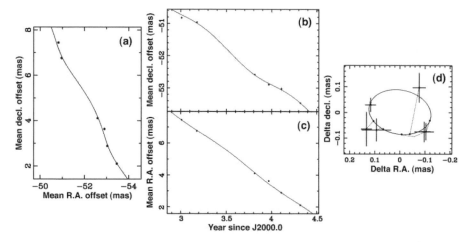

Figure 2. Mean position of the three maser features at all of the observation epochs. (*a*) Mean RA and Dec offsets on the sky with respect to the phase-tracking center. (*b*) and (*c*) Mean RA and Dec offsets against time. (*d*) Relative mean offsets with a position offset and a mean proper motion subtracted. Ellipse indicates the modeled annual-parallax motion. The mean maser position goes clockwise around the ellipse. Each observed point is connected to a relevant point on the ellipse, corresponding to the observation time, with a dotted line.

components of I1913 are $(R, \theta, z) \simeq (7.4 \text{ kpc}, 62°.4, 650 \text{ pc})$ and $(V_R, V_\theta, V_z) \simeq (3, 125, 8)$ [km s⁻¹], respectively. Thus I1913 is located at a large height from the Galactic plane and its kinematics is very different from that expected simply from the Galactic circular rotation ($V_\theta \simeq 220$ km s⁻¹). From the height above the Galactic plane, z, and the velocity component perpendicular to the Galactic plane, V_z, we estimate a rough upper limit of ~9 $M_\odot$ to the stellar mass of I1913's progenitor.

Acknowledgements

The NRAO's VLBA is a facility of the National Science Foundation of the USA, operated under a cooperative agreement by Associated Universities, Inc. H.I. was supported by the Grant-in-Aid for Scientific Research from Japan Society for Promotion Science (18740109).

References

Imai, H., Sahai, R., & Morris, M. 2007, *ApJ* 669, 424

A Giant Step: from Milli- to Micro-arcsecond Astrometry
Proceedings IAU Symposium No. 248, 2007
W. J. Jin, I. Platais & M. A. C. Perryman, eds.

VERA observation of the massive star forming region G34.4+0.23

T. Kurayama

Mizusawa VERA observatory, NAOJ,
Osawa 2-21-1, Mitaka, Tokyo, 181-8588, Japan
email : t.kurayama@nao.ac.jp

Abstract. We observed with VERA the massive star forming region G34.4+0.23, to obtain parallaxes and proper motions. Four infrared dark clouds were observed and water maser were found in two dark clouds, MM1 and MM4. In MM1, the distribution of maser spots shows a "V-shaped" structure and most features co-moving with this structure. Phase-referenced images have peaks and their motion is much larger than the expected parallax. Further analysis is needed to correctly interpret our measurement of parallax.

Keywords. stars: formation, ISM: individual (G34.4+0.23), ISM: kinematics and dynamics, astrometry

1. Introduction

Infrared Dark Clouds (IRDCs) are found in massive star forming regions. They are thought to be an early stage of star formation. One of them is in the region of G34.4+0.23. In this region, four IRDCs (MM1–MM4) are observed, all located in the north-south direction. The IRAS point source IRAS 18507+0121 and a compact HII region are associated with MM2. Water maser was detected in this IRAS source (Scalise Rodíguez & Mendza-Torres (1989)).

Rathmorne *et al.* (2005) derived the masses of IRDCs from absolute luminosities and blackbody temperatures. Absolute luminosities depend on the distances, so the distance information is very important. We can also observe the outflows manifesting as the motions of maser features.

2. Observation

Four stations of VERA were used to observe water masers at 22GHz = 1.3cm. The phase-referencing technique was adopted for VLBI measurements of parallaxes and proper motions. The reference source for phase-referencing was J1855+0251. We have observed at six dates until June 2007 and the monitoring continues. At each date, we observed for nine hours. The observational dates were Oct 21 2006, Nov 16 2006, Jan. 04 2007, Feb 10 2007, Mar 25 2007 and May 4 2007.

3. Results

Figure 1a shows the cross-power spectra for MM1. The time-dependent components are found at 40 and 90 km/s. Since the systemic velocity is about 60 km/s, most components are blueshifted. Figure 1b shows cross-power spectra for MM4 after the phase-referencing.

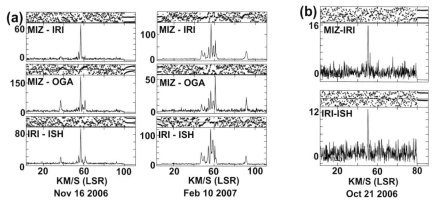

Figure 1. (a) Cross-power spectra for MM1. (b) Cross-power spectra for MM4. (b) is a phase-referenced spectra, while (a) is not.

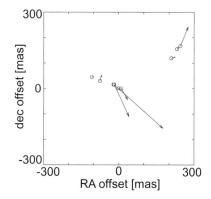

Figure 2. Distribution and linear motions of water masers in G34.4+0.23 between Oct 21 2006 and Feb 10 2007. Arrows represent the motion in 10 years.

Figure 2 shows the distribution and linear motions of maser features. They show a "V-shaped" structure and most of their linear motion is along this "V-shaped" structure.

4. Measurement of parallax

The phase-referenced images, the positions of which are referenced to the extragalactic source, have a few peaks. Their apparent motions are much larger than expected. Therefore, an additional analysis is needed to detect and quantify the true parallactic motion.

References

Rathmorne, J. M. *et al.* 2005, *ApJ*, 630, L181

Scalise, E. Jr., Rodríguez, L. F., & Mendoza-Torres, E. 1989, *A&A*, 221, 105

A Giant Step: from Milli- to Micro-arcsecond Astrometry
Proceedings IAU Symposium No. 248, 2007
W. J. Jin, I. Platais & M. A. C. Perryman, eds.

© 2008 International Astronomical Union
doi:10.1017/S1743921308019078

The 3-mm flux density monitoring of Sagittarius A* with the ATCA

J. Li[1,3], Z. Q. Shen[1,2], A. Miyazaki[4], M. Miyoshi[4], T. Tsutsumi[4], M. Tsuboi[5], L. Huang[1,3], and B. Sault[6]

[1]Shanghai Astronomical Observatory, Chinese Academy of Sciences, Shanghai 200030, China
email: lijuan@shao.ac.cn

[2]Joint Institute for Galaxy and Cosmology of ShAO and USTC, Shanghai 200030, China

[3] Graduate School of the Chinese Academy of Sciences, Beijing 100039, China

[4] National Astronomical Observatory of Japan, 2-21-1 Osawa, Mitaka, Tokyo 181-8588, Japan

[5] Institute of Space and Astronautical Science, Sagamihara, Kanagawa 229-8510, Japan

[6] University of Melbourne, School of Physics, Parkville, Victoria 3052, Australia

Abstract. We have performed monitoring observations of the 3-mm flux density toward the Galactic center compact radio source Sgr A* with the ATCA since 2005 October. It has been found that during several observing epochs Sgr A* was quite active, showing significant intraday variation. Here we report the detection of an IDV in Sgr A* on 2006 August 13, which exhibits a 27% fractional variation in about 2 hrs.

Keywords. Galaxy: center, techniques: interferometric

1. Introduction

There is compelling evidence that Sagittarius A* (Sgr A*), the extremely compact radio source at the dynamical center of the Galaxy, is associated with a massive black hole of 4×10^6 solar masses. Since its discovery in 1974, Sgr A* has been observed with radio telescopes in the northern hemisphere, and temporal flux variations were reported (see, e.g., 7- and 13-mm VLA observations by Yusef-Zadeh *et al.* 2006, 3-mm OVRO by Mauerhan *et al.* 2005, 2-mm NMA by Miyazaki *et al.* 2004, 0.8-mm SMA by Marrone *et al.* 2006). On the other hand, X-ray and infrared flares have also been detected (Genzel *et al.* 2003, Baganoff *et al.* 2001), indicating very short timescales and violent intensity increases. Since Sgr A* is embedded in thick thermal material, it is particularly difficult to observe its structure, thus IDV observation can give indirect constraints on the source emission geometry and emission mechanisms.

Since 2005 October we have performed several epochs of flux density monitoring toward Sgr A* at 3-mm using the ATCA, a five 22-m dish interferometer at Narrabri, Australia where Sgr A* passes almost overhead, allowing a longer observing window (> 8 hr with elevation higher than $40°$). Thus, the ATCA calibrations and flux measurements of Sgr A* are expected to be more accurate. Here we present preliminary result obtained from the observation on 2006 August 13.

2. Observations

On 2006 August 13, we used upper sideband (88.896 GHz) with a bandwidth of 128 MHz for the observation of Sgr A* and other continuum calibrators, and the lower sideband (86.243 GHz) of 32 MHz for SiO maser sources. Quasar 3C 279 was observed as a bandpass calibrator. The instrumental gain and phase were calibrated by alternating

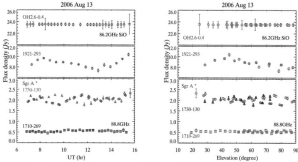

Figure 1. Left: ATCA 3-mm light-curves of Sgr A* ('○' in the lower panel), control source OH2.6−0.4 and other calibrators like PKS 1921−293, PKS 1730−130 ('△' in the lower panel), PKS 1710−269 ('□' in the lower panel) on 2006 August 13. Right: the same plot for flux density as a function of elevation angle. Note that different amplitude scales in different panels.

observations of Sgr A* and calibrators (OH2.6−0.4, PKS 1710−269 and PKS 1730−130). The pointing accuracy was checked by observing VX Sgr every half an hour.

3. Data analysis and preliminary results

All the data processing was conducted using the MIRIAD package. For the amplitude calibration, we first applied the nominal elevation-dependent gains and then used calibrators to determine the necessary corrections. However, calibrators are, in general, variable sources which will cause uncertainties in the gain correction. Therefore, different secondary calibrators (as control sources) were tried to check the consistency of the gain calibrations. Later, we further compare the fractional gain correction for each antenna with the fractional variation in Sgr A* to make sure that the detected variation is real. The calibrated data were then averaged in a 5-minute bins to search for a shorter timescale variability. The flux measurements of Sgr A* were estimated by fitting a point source model to visibility data on the projected baselines longer than 25kλ (or 85 m at 3-mm) to suppress the contamination from the surrounding extended components. Both the fitting error and the rms of the residual were used to get the error estimate.

As a result, we show light-curves of Sgr A* and calibrators on 2006 August 13 at 3-mm in Figure 1. It can be clearly seen that Sgr A* has undergone a flux density variation with a fractional variability of 27% in a timescale of about 2 hrs. This timescale for the rise and fall is consistent with the previous observations at 3-mm (Mauerhan *et al.* 2005).

Acknowledgements

This work was supported in part by the National Natural Science Foundation of China (grants 10573029, 10625314, and 10633010) and the Knowledge Innovation Program of the Chinese Academy of Sciences (Grant No. KJCX2-YW-T03), and sponsored by Program of Shanghai Subject Chief Scientist (06XD14024).

References

Baganoff, F. K., *et al.* 2001, *Nature*, 413, 45
Genzel, R., *et al.* 2003, *Nature*, 425, 934
Marrone, D. P., Moran, J., Zhao, J.-H., & Rao, R. 2006, *ApJ*, 640, 308
Mauerhan, J. C., Morris, M., Walter, F., & Baganoff, F. 2005, *ApJ*, 623, L25
Miyazaki, A., Tsutsumi, T., & Tsuboi, M. 2004, *ApJ*, 611, L97
Yusef-Zadeh, F., *et al.* 2006, *ApJ*, 650, 189

A Giant Step: from Milli- to Micro-arcsecond Astrometry
Proceedings IAU Symposium No. 248, 2007
W. J. Jin, I. Platais & M. A. C. Perryman, eds.
© 2008 International Astronomical Union
doi:10.1017/S174392130801908X

Parallax measurement of the Galactic Mira variables with VERA

A. Nakagawa[1], T. Omodaka[1], K. M. Shibata[2], T. Kurayama[2],
H. Imai[1], S. Kameno[1], M. Tsushima[1], N. Matsumoto[1], M. Matsui[1],
and VERA project team[1,2]

[1]Faculty of Science, Kagoshima University, 1-21-35, Korimoto, Kagoshima, JAPAN
email:nakagawa@astro.sci.kagoshima-u.ac.jp

[2]National Astronomical Observatory of Japan, 2-21-1 Osawa, Mitaka, Tokyo, JAPAN

Abstract. Parallax measurements of the Galactic Mira variables with VERA have started since 2004 to establish their Period-Luminosity (PL) relationship in the Galaxy. Multi-epoch VLBI observations of a semiregular variable S Crt yielded an accurate parallax of 2.27 ± 0.14 mas corresponding to the distance of 441^{+29}_{-24} pc. In addition to the distance, we obtained physical properties of S Crt. Temperature of the photosphere was found to be $\sim$3000 K by fitting the infrared spectrum with a blackbody radiation. The stellar radius was obtained based on the distance, apparent magnitude, and the temperature. Internal proper motions of circularly-arranged maser spots in S Crt were detected for the first time. Observations of the other Mira variables, such as R UMa, SY Scl, AP Lyn, and WX Psc are in progress.

Keywords. astrometry, masers, techniques: high angular resolution, techniques: interferometric, stars: AGB and post-AGB

1. Introduction

Mira variables are pulsating stars with periods of 100 to 1000 days. Although a narrow PL relation for Miras in the Large Magellanic Cloud was confirmed (Feast *et al.* 1989), the same relation for the Galactic Miras (van Leeuwen *et al.* 1997) has not been precisely obtained because of the ambiguity in absolute magnitudes suffering directly from inaccurate distances to each object. Using absolute magnitudes derived from accurate distances measured with VERA, we can derive a much improved PL relation for the Galactic Mira variables.

2. Observations and data reduction

VERA is a Japanese VLBI array dedicated to phase referencing VLBI. Synthesized beam size is typically 1.3 mas at 22 GHz. The dual beam system enables us to observe target and reference sources simultaneously, and make the data acquisition time much longer than that of conventional fast-switching VLBI. From October 2005 through May 2007, each month we observed S Crt and J1147−0724 at 22 GHz. S Crt is a semiregular variable with the pulsation periods of 155 days and have spectral class of M6E-M7E. An extragalactic source J1147−0724 is classified as a candidate source in ICRF catalog, and we used it as a positional and phase reference source. The separation between the target and a reference source is 1.23° at a positional angle of 89°. Most likely the excess of wet atmospheric path residual at zenith causes incoherence to the reduced phase of S Crt. In order to compensate for these residuals and improve the quality of phase-referenced images, we applied trial offsets to the zenith path, ranging from −15 to 15 cm with an interval of 0.5 cm at each station (Fig. 1(b)).

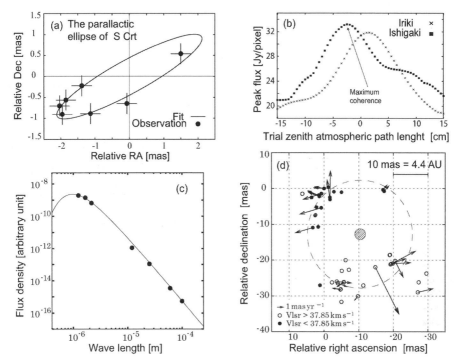

Figure 1. (a)The parallactic ellipse of S Crt. (b)We find the maximum coherence by applying trial offsets. Examples for Iriki and Ishigaki stations are presented. (c)Infrared spectrum of S Crt. The solid line is a best fit blackbody radiation obtained with the unweighted least square fitting. (d)Angular distribution and internal motions of water masers. Open and filled symbols indicate the spots showing red- and blue-shifted components relative to the stellar velocity.

3. Result and discussion

The annual parallax of S Crt, successfully extracted from our observations by tracing the maser spot with the LSR velocity of $34\,\mathrm{mks}^{-1}$, was found to be $2.27\pm0.14\,\mathrm{mas}$ corresponding to the distance of $441^{+29}_{-24}\,\mathrm{pc}$. Figure 1(a) represents the parallactic ellipse. The position error of $213\,\mu\mathrm{as}$ was adopted for all observations. We estimated physical properties of S Crt based on the absolute distance measured with VERA. Infrared spectrum was well fitted with single blackbody radiation providing the temperature (T_{bb}) of $3000\pm163\,\mathrm{K}$ (Fig 1(c)). Using the K-band magnitude ($m_{\mathrm{K}} = 0.79$) and bolometric T_{bb}, we estimated the radius of the photosphere to be $0.67\,\mathrm{AU}$ shown by a shaded-circle in Fig. 1(d), along with the circularly-arranged maser spots. The radius of the maser distribution was found to be $6.6\,\mathrm{AU}$ as indicated by a dashed-circle in the same figure. Arrows indicate proper motions of each spot, corresponding to the outward velocities at $6\text{-}8\,\mathrm{km}\,\mathrm{s}^{-1}$. We found that S Crt differs from typical Mira variables. The maser distribution is more compact, the star's diameter is smaller, and the temperature is higher than those expected for typical Mira variables. Determination of the distances for larger number of samples will help to refine the Galactic PL-relation and study the difference between Miras and semiregular variables.

References

Feast, M. W., Glass, I. S., Whitelock, P. A., & Catchpole, R. M. 1989, *MNRAS* 241, 375
van Leeuwen, F., Feast, M. W., Whitelock, P. A., & Yudin, B. 1997, *MNRAS* 287, 955

A Giant Step: from Milli- to Micro-arcsecond Astrometry
Proceedings IAU Symposium No. 248, 2007
W. J. Jin, I. Platais & M. A. C. Perryman, eds.

© 2008 International Astronomical Union
doi:10.1017/S1743921308019091

Milli-arcsecond binaries

R. M. Torres[1], L. Loinard[1], A. J. Mioduszewski[2] and L. F. Rodríguez[1]

[1]Centro de Radiastronomía y Astrofísica, Universidad Nacional Autónoma de México
Apartado Postal 72-3 (Xangari), 58089 Morelia, Michoacán, México
email: `r.torres,l.loinard,l.rodriguez@astrosmo.unam.mx`

[2]National Radio Astronomy Observatory, Array Operations Center
1003 Lopezville Road, Socorro, NM 87801, USA
email: `amiodusz@aoc.nrao.edu`

Abstract. As part of an astrometric program (see Loinard *et al.* in this volume), we have obtained an unprecedented sample of high-resolution (~ 1 mas) VLBA images of several nearby young stellar systems. As was to be expected, these images revealed interesting new characteristics of these objects. Perhaps the most interesting of these characteristics is the detection of a high rate of very tight binary stars (separations of a few mas) in our sample: of the 9 objects observed to date, 5 are binaries of this type.

Keywords. binaries: general, stars: individual (T Tauri Stars), radiation mechanisms: non-thermal, astrometry

1. Introduction

Multiple studies have led to the well-known conclusion that main-sequence stars in the Solar neighborhood show a high multiplicity rate: $\sim 55\%$ for solar-type stars (Duquennoy & Mayor 1991), and ~ 35 to 42% for M-dwarfs (Reid & Gizis 1997; Fischer & Marcy 1992). Young stars also exhibit a high multiplicity rate: $\sim 50\%$ with separations of between 0.02 and 1 arcseconds (Köhler & Leinert 1997; Duchêne *et al.* 2004; Konopacky *et al.* 2007). Indeed, several star-forming regions appear to show multiplicity rates even higher than those of main sequence stars (Prosser *et al.* 1994; Duchêne et al. 1999; Bouvier *et al.* 1997). This shows that the multiplicity must be already established in the very early phases of star-formation.

2. Observations

We have chosen from the literature a list of 9 young stellar objects associated with non-thermal radio emission: 5 in Taurus, and 4 in the Ophiuchus complex. Those sources are low-mass stars (with the exception of S1 which is a main sequence B star). We make use of a series of six continuum 3.6 cm (8.42 GHz) observations of each source obtained roughly every two or three months. Our calibrators are very compact extragalactic sources whose absolute positions are know to better than about 1 milli-arcsecond. The data were edited and calibrated in a standard fashion using the Astronomical Image Processing System (AIPS).

3. Results and discussion

In this work we find that $\sim 70\%$ of the objects in our sample are multiple stellar systems with separations of a few mas. In the Ophiuchus complex we found a new member in the each of the stellar systems S1, DoAr 21 and VSSG 14, and two new members in WL

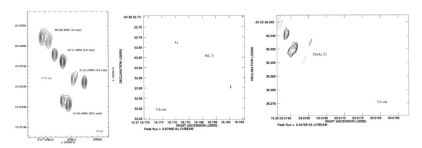

Figure 1. Multiple stellar systems. The first 4 epochs for V773 Tau in Taurus; and the first epoch for WL 5 and DoAr 21 in ρ−Ophiuchus.

5 (see Fig. 1). In addition, we find extended structure around several systems. In the case of Taurus, we knew that T Tau Sb was a member of a triple system, but among these three members we only detect one with a non-thermal radio emission. V773 Tau was previously known to be a quadruple system, and we detect 2 out of the 4 members. Finally, Hubble 4 and HDE 283572, as well as HP Tau/G2 are apparently single.

It is quite unlikely that 70% of all young stellar systems are tight binaries such as those found in our VLBA observations. The high binary rate found here is more likely the result of selection effects. The systems considered here were selected because they were known to be non-thermal emitters, so the high binary rate may indicate that tight binaries are more likely to emit non-thermal radio emission than looser binaries or single stars. This idea is reinforced by the observations of V773 Tau (Fig. 1) where we find that the emitted flux is a clear function of the separation between the two stars (3.8 mJy when they are most separated against 28.3 mJy when they are closest). While the physical mechanism leading to this behavior is not clear, this result certainly suggests a relation between the radio flux and separation. Also, the fraction of tight binaries appears to be larger in ρ−Oph than in Taurus.

4. Perspectives

For pre-main sequence stars, no reliable empirical mass determination is available. For the binary stars detected here, a direct mass determination will be possible using Kepler's third law and the measured relative motions. We are currently obtaining new observations for these multiple sources. New observations will cover complete orbits of the sources and will allow the determination of their orbital parameters and masses.

References

Bouvier, J., Rigaut, F., & Nadeau, D. 1997, *A&A*, 323, 139
Duchêne, G., Bouvier, J., & Simon, T. 1999, *A&A*, 343, 831
Duchêne, G., Bouvier, J., Bontemps, S., Andr, P., & Motte, F. 2004, *A&A*, 427, 651
Duquennoy, A. & Mayor, M. 1991, *A&A*, 248, 485
Fischer, D. A. & Marcy, G. W. 1992 *ApJ*, 396, 178
Köhler, R. & Leinert, C. 1998 *A&A*, 331, 977
Konopacky, Q. M., Ghez, A. M., Rice, E. L., & Duchêne, G. 2007 *ApJ*, 663, 394
Prosser, C. F., Stauffer, J. R., Hartmann, L., Soderblom, D. R., Jones, B. F., Werner, M. W., & McCaughrean, M. J. 1994 *ApJ*, 421, 517
Reid, I. N. & Gizis, J. E. 1997 *AJ*, 113, 2246

A Giant Step: from Milli- to Micro-arcsecond Astrometry
Proceedings IAU Symposium No. 248, 2007
W. J. Jin, I. Platais & M. A. C. Perryman, eds.
© 2008 International Astronomical Union
doi:10.1017/S1743921308019108

A strategy for implementing differential VLBI

J. L. Li[1], J. Wang[1], S. B. Qiao[12] and F. Tian[1]

[1]Shanghai Astronomical Observatory, Shanghai 200030, China

[2]Surveying and Mapping Institute, Zhengzhou 450052, China

email: jll@shao.ac.cn

Abstract. Based on past astrometric and geodetic VLBI observations, a strategy for implementing differential VLBI (DVLBI) is developed by interpolating the non-geometric delay (NGD) at the target's direction using observations of several reference sources spreaded out within a circular ring centered on the target. In contrast to the ordinary approach, in our design the limitations in the angular distance are relaxed and the effects of observational uncertainties in reference sources are reduced. Analysis shows that in our design a precision of the NGD correction in S-band reaches about $1ns$ only. Our design can be adopted for observations of weak sources and in deep space exploration.

Keywords. atmospheric effects, methods: data analysis, reference systems, astrometry

1. Introduction

In order to retain high precision in relative position determination by DVLBI, the angular distance between the reference and target objects should be very small. For single and dual-frequency observations it is usually less than $1°$ and $5°$ respectively (Sovers *et al.* (1998)). There are about several hundreds of sources with high position precision, or about one source per hundred square degrees. In practice a suitable reference source may not be available, thus preventing DVLBI from being widely applied. We developed an implementation strategy for DVLBI by extending the angular distance limitation in order to make this technique more easily applicable.

In the NGD, the clock bias and the instrument delay are relatively stable within a short period of time. They shift the NGD distribution camber rather than change the shape. The ionosphere delay is related to the frequency and direction. At $8.4GHz$ the zenith delay ranges from $0.03m$ to $0.6m$. The troposphere delay is mainly dependent on the elevation. It is about $2m$ at the zenith and $20m$ at $6°$. From the magnitudes and characteristics of all components that shape the NGD distribution camber, we conclude that it is mainly affected by the troposphere delay. Simulations and checks by real observations show that, for a given baseline and observations made within two hours, the NGD distribution camber is smooth, which opens up the possibility for NGD interpolation.

2. A strategic design for the NGD interpolation

For a given baseline and observations made within a short period of time, the NGD can be described as a scalar field on the surface of a half sphere centered at the reference station. Imagine that, for this scalar field with smooth change, the field values at all the points on a great circle passing through target T form a data series. There are various great circles passing through T, which associated with various data series. However, the series value at T for all the series should be equal within the range of uncertainty. Based

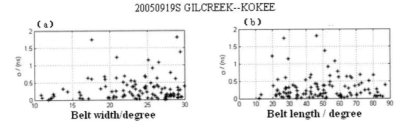

Figure 1. The distribution of σ versus the width and length of the sample ring

on this imagination and in order to sustain sufficient data points in the case of radio sources, a ring with some width and centered along a great circle passing through T could be marked. Neglecting the ring width, and counting from one end, let the accumulated sum of the arc-lengths between neighboring points be the indices of the data series. By smoothing and interpolating on the data series the NGD at T will be resulted. Based on the archive observations of VLBI experiments the following strategy is proposed. (1) Plot the NGD data of a baseline within two hours on the topocentric spherical plane of the reference station. (2) On the plot mark down some great circle rings with some widths. Let each ring contain as many points as possible. Take all the points on a ring as an analysis collection. (3) Assuming a collection contains N sources, perform a poly-fit to the data and the smoothed NGD at each point is obtained as τ_s. (4) Take source 3 through N-2 respectively as target and all the rest as references, thus resulting in N-4 analytical samples. Apply a poly-fit to all the NGD data of references in a sample and then interpolate the NGD at the target according to the arc-length index, taken as τ_i. (5) Take the difference between the corresponding interpolated and smoothed NGD as $\nu_{ims} = \tau_i - \tau_s$. For a collection of N sources, N-4 values of ν_{ims} will result. Define the standard deviation of ν_{ims} as $\sigma = (\Sigma\nu_{ims}^2/(N-5))^{1/2}$, and let σ be the criteria for evaluating the NGD interpolation.

3. Check by real observations and discussion

It is easy to understand that the precision of interpolation of the NGD for the target depends on the number of reference sources, their distribution, and observational uncertainties. Figure 1 shows a check on real NGD observations in the S band for baseline Gilcreek—Kokee on September 9 of 2005. It is the distribution of σ versus the width and length of the ring when the number of sources in each sample is no less than 6. It is shown that when the width and length of the ring are restricted within $30°$ and $90°$ respectively, most of the points are well below $1ns$.

Numerous VLBI experiments have been checked and the situation is similar to Fig. 1. Compared to ordinary DVLBI, in our design the limitation in the angular distance is relaxed and the observation uncertainties of references are reduced. Our design can be adopted in observations of weak sources and in deep space exploration.

Acknowledgements

This work is supported by NSFC (*No. 10778635, No. 10173019, No. 10473019*), Chinese lunar exploration project CE-1 and STC of Shanghai Municipality (06DZ22101).

References

Sovers, O. J., Fanselow, J. L., & Jacobs, C. S. 1998, *Reviews of Modern Physics* 70, 1393

A Giant Step: from Milli- to Micro-arcsecond Astrometry
Proceedings IAU Symposium No. 248, 2007
W. J. Jin, I. Platais & M. A. C. Perryman, eds.

© 2008 International Astronomical Union
doi:10.1017/S174392130801911X

Tropospheric correction in VLBI phase-referencing using GPS data

B. Zhang[12], **X. W. Zheng**[1], **J. L. Li**[2], **Y. Xu**[3] and **J. F. Wu**[4]

[1]Astronomy Department, Nanjing University, Nanjing 210093, P.R. China
[2]Shanghai Astronomical Observatory, Chinese Academy of Sciences, Shanghai 200030, China
[3]Purple Mountain Observatory, Chinese Academy of Sciences, Nanjing 210008, P.R. China
[4]Institute of Surveying and Mapping, PLA Information Engineering University, Zhengzhou 450002, P.R. China
email: bozhang@nju.edu.cn

Abstract. Comparing the tropospheric zenith delays derived from VLBI and GPS data at VLBA stations collocated with GPS antenna, the systematic biases and standard deviations of the difference are both found to be at the level of a sub-centimeter. Based on this agreement, we used GPS data to correct the tropospheric effects in VLBI phase-referencing, resulting in close peak-to-noise ratios of images after tropospheric correction using GPS and VLBI data.

Keywords. atmospheric effects, astrometry

1. Introduction

The dominant error source of astrometry involved in VLBI phase-referencing is the small error of the tropospheric delay model, which is applied by the VLBA correlator at observing frequency greater than 5 GHz (Wrobel *et al.* (1999)). The current method applied to correct this error is using geodetic-like observations to determine the residual (Observation - Model) tropospheric zenith delay (Reid & Brunthaler (2004)). However, this method only recovers the global variation, but neglects the local variation of the residual zenith delays, such as fluctuations which might be significant at some VLBI stations (Keihm *et al.* (2004)). Since the GPS measurements near the VLBI site are subject to the same troposphere, those measurements can be used to estimate the tropospheric contribution to VLBI observables. In the data analysis of project BR100, which determines the parallax and proper motions of 12 GHz methanol masers using VLBI phase referencing by VLBA, we used several geodetic-like observation data sets of phase-referencing observations, and among those observations, about six VLBA sites with VLBI and GPS data are available. To study the feasibility of tropospheric correction in VLBI phase-referencing using GPS data, we compared the tropospheric zenith delays derived from VLBI and GPS data at these VLBA stations, and the PNRs (peak-to-noise ratios) of images with tropospheric effects corrected using only VLBI data or combined GPS and VLBI data.

2. Tropospheric correction using combined VLBI and GPS data

Some VLBI stations are collocated with GPS antennas, and many studies indicate that the difference between tropospheric delays derived from VLBI and GPS data is at the sub-centimeter level (Niell *et al.* (2001); Schuh *et al.* (2004)). This motivates us to calibrate the VLBI data using tropospheric estimates from GPS data. At present only six VLBA sites are collocated with GPS receivers. IGS (the International GNSS Service)

Table 1. Biases (front) and standard deviations (back) in cm between VLBI and GPS (VLBI − GPS) residual zenith delays for VLBA BR100 observations.

Obs. Date	BR	FD	MK	NL	PT	SC
2005.10.20	-1.5 ± 1.0	-1.0 ± 0.5	-0.7 ± 1.0	-1.6 ± 1.0	...	...
2005.10.30	-0.6 ± 0.3	0.3 ± 0.6	-1.2 ± 0.8	-1.1 ± 0.3	-1.1 ± 0.2	-0.6 ± 0.7
2006.01.13	-0.6 ± 0.5	-0.3 ± 0.3	-0.6 ± 0.7	-1.2 ± 0.5	-1.2 ± 0.9	-0.7 ± 0.4
2006.03.16	-1.1 ± 0.3	-1.1 ± 0.6	0.1 ± 0.5	-0.9 ± 0.6	-0.9 ± 0.1	-0.8 ± 0.7
2006.04.07	-0.7 ± 0.6	-0.9 ± 0.4	-1.1 ± 0.5	0.1 ± 1.4	...	0.1 ± 0.5
2006.04.15	-1.3 ± 1.1	-0.9 ± 0.5	0.2 ± 1.5	-1.9 ± 0.7	...	-0.9 ± 0.2
2006.07.23	0.3 ± 0.7	1.4 ± 0.2	...	-0.5 ± 1.0	-0.5 ± 0.6	-0.1 ± 0.8
2006.10.04	-1.0 ± 0.3	-0.2 ± 0.6	-0.6 ± 1.0	-2.2 ± 0.5	-1.2 ± 0.3	-0.6 ± 0.9
2006.10.07	-0.6 ± 0.5	0.6 ± 0.9	-1.2 ± 0.4	-0.8 ± 0.1	-0.9 ± 0.5	-1.8 ± 1.0
2006.10.19	0.5 ± 0.6	-0.3 ± 1.5	-0.8 ± 0.4	-0.8 ± 0.2	...	...

Table 2. Comparison of PNR of image

Source	PNR(VLBI)	PNR(GPS)
G35.20-0.74(Maser)	100.1	103.7
J1855+0215(QSO)	15.1	16.4
J1855+0251(QSO)	15.6	14.1
J1907+0127(QSO)	6.7	6.8

publishes routinely the tropospheric zenith delays derived from GPS measurements of each site, offering us an opportunity to estimate the zenith delay at the corresponding VLBA sites based on GPS data. We compared the residual zenith delays derived from VLBI and GPS data of several geodetic-like observations. Table 1 shows the comparison, with the biases and STDs being at the level of sub-centimeter. Table 2 illustrates the PNRs of images of sources with tropospheric effects corrected, with the second and third columns listing PNRs only using geodetic-like VLBI data and combined GPS and VLBI data respectively. (Note that here sites with available GPS data use GPS data only; otherwise only VLBI data are used.) Differences between the PNRs are found to be small. Based on the agreement noted in the above comparison, we suggest to correct the tropospheric effects in VLBI phase-referencing with combined VLBI and GPS data, in which case the precision of correction will be at the sub-centimeter level.

Acknowledgements

We would like to extend our appreciation to Dr. M. Reid of the Harvard-Smithsonian Center for Astrophysics for his kind help. This work has been partly supported by the National Natural Science Foundation of China (grants 10703010, 10778635, 10673024 and 10473019), and the Science & Technology Commission of Shanghai Municipality (grant 06DZ22101).

References

Keihm S. J., Tanner, & Rosenberg, H., 2004, *IPN progress report*, 42,158

Niell, A. E., Coster, A. J., Solheim, F. S. *et al.*, 2001, *Journal of Atmospheric and Oceanic Technology*, 18, 830

Reid, M. J. & Brunthaler, A., 2004, *ApJ*, 616, 872

Schuh, H., Snajdrova, K., & Bohm, J., 2004, In: Vandenberg, N. R., Baver, K. D., edts., *IVS 2004 General Meeting Proceedings*, 461

Wrobel, J. M., Walker, R. C., Benson, J. M. *et al.*, 1999, *VLBA Scientific Memorandum 24*

A Giant Step: from Milli- to Micro-arcsecond Astrometry
Proceedings IAU Symposium No. 248, 2007
W. J. Jin, I. Platais & M. A. C. Perryman, eds.

The distance to G59.7+0.1 and W3OH

Y. Xu[1,2] †, M. J. Reid[3], K. M. Menten[1], X. W. Zheng[4], A. Brunthaler[1], and L. Moscadelli[5]

[1] Max-Planck-Institute für Radioastronomie, Bonn, Germany
email: xuye@mpifr-bonn.mpg.de

[2] Purple Mountain Observatory, Nanjing, China

[3] Harvard-Smithsonian Center for Astrophysics, USA

[4] Nanjing University, China

[5] Arcetri Obs., Firenze, Italy

Abstract. We have measured the distance to the high-mass star-forming region G59.7+0.1 (IRAS 19410+2336) and W3OH. Their distances, 2.20 ± 0.11 kpc and 1.95 ± 0.04 kpc, respectively, were determined by triangulation using Very Long Baseline Array (VLBA) observations of 12.2 GHz methanol masers phase-referenced to compact extragalactic radio sources. In addition to the distances, we have also obtained their proper motions.

Keywords. ISM: molecules, astrometry, Galaxy: structure

1. Introduction

We are carrying out a large project to study the spiral structure and kinematics of the Milky Way. We will accomplish this by determining distances via trigonometric parallax and proper motions in star forming regions. The target sources are 12 GHz methanol masers. With accurate distance measurements we can locate spiral arms, and with absolute proper motions we can determine the 3-dimensional motions of these regions. Here we report results on G59.7+0.1 and W3OH.

2. Observations

We have conducted phase-referenced observations of G59.7+0.1 and W3OH with the VLBA in order to measure their relative position with respect to extragalactic radio sources. The observing sequence involved rapid switching between the extragalactic radio sources and the maser source. The time period between successive epochs was around three months. The background compact extragalactic sources were from ICRF or/and our VLA surveys (Xu *et al.* 2006a).

3. Results

In Fig. 1 we plot, for each source, the positions of one maser spot relative to the background reference radio sources. The parallax signature is fit by the five required parameters: one parameter for the parallax and two parameters for the proper motion in each of the coordinates. The distances of G59.7+0.1 and W3OH are measured to be 2.20 ± 0.11 kpc and 1.95 ± 0.04 kpc, respectively. The proper motions in the RA and Dec coordinates were -1.63 $\pm$ 0.04 and -5.12 $\pm$ 0.1 mas y^{-1} for G59.7+0.1 and -1.20 $\pm$ 0.02 and -0.15 $\pm$ 0.01 mas y^{-1} for W3OH, respectively.

† Present address: Auf dem Hügel 69 D-53121 Bonn, Germany.

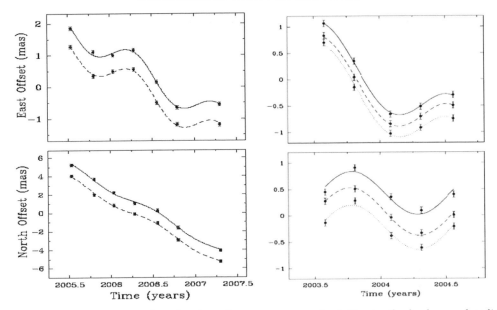

Figure 1. Position versus time for the reference maser spot relative to the background radio sources for G59.7+0.1 (left panel) and W3OH (right panel; Xu *et al.* 2006b, 2008). The top and bottom panels show the eastward and northward offsets, respectively. In both cases, the large difference in position between the maser and each background source has been removed, and the two curves have been offset for clarity.

G59.7+0.1 is not associated with any known arms and is located between the Perseus and the Sagittarius arms (Fig. 2). The distance of 2.2 kpc is close to its near kinematic distance (2.1 kpc). It may be a spur of the Sagittarius arm because it is closer to that arm than to the Perseus arm. Such structures have been observed for many galaxies (La Vigne *et al.* 2006). The spurs may form as a consequence of gravitational instabilities inside spiral arms or/and effects of magnetic fields (Kim & Ostriker 2002). On the other hand, it may be a tail of the Local arm because it looks associated with sources in the Local arm, such as Orion, Cep A and the Sun.

Sources in the Perseus arm (near W3OH) have kinematic distances of about 4.2 kpc, while the luminosity distance of neighboring O-type stars is 2.2 kpc (Humphreys 1978). This discrepancy has fueled debate over whether the Perseus arm is indeed kinematically anomalous or the distances to O-star are inaccurate. We have now resolved this significant discrepancy, showing that the luminosity distance is approximately correct, and W3OH has a large kinematic anomaly.

In order to study the 3-D motion of G59.7+0.1 and W3OH at their positions in the Galaxy, we adopt the IAU standard constants $R_0 = 8.5$ kpc and $\Theta_0 = 220$ km s^{-1}. We also adopt the solar motion values $U = 10.0 \pm 0.40$, $V = 5.25 \pm 0.60$, and $W = 7.17 \pm 0.40$ km s^{-1} from Dehnen & Binney (1998).

For G59.7+0.1, using $V_{LSR} = 22.4$ km s^{-1} from the CS velocity, we have
In the direction of Galactic rotation: 210 ± 10 km s^{-1};
Radial toward the Galactic Center: $+7 \pm 3$ km s^{-1}
Toward North Galactic Pole: -5 ± 1 km s^{-1}.

For W3OH, using $V_{LSR} = 44.2$ km s^{-1} from the maser velocity, we have
In the direction of Galactic rotation: 206 ± 10 km s^{-1};
Radial toward the Galactic Center: $+17 \pm 1$ km s^{-1}
Toward North Galactic Pole: -0.8 ± 0.5 km s^{-1}.

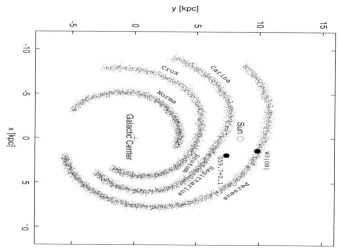

Figure 2. Positions of G59.7+0.1 and W3OH in the Galactic plane. The positions of the spiral arms are taken from Taylor & Cordes (1993). G59.7+0.1 may be located at the tail of Local arm or a spur of the Sagittarius arm. The portion of the Perseus arm traced by W3OH exhibits strong kinematic anomalies.

The errors are mainly caused by large uncertainty in solar motion, galactocentric distance of Sun, and rotation velocity at Sun. The 3-D motion indicates that their peculiar motion, more than 10 km s^{-1} in the galactic plane is quite large, about 3 - 4 times of the velocity dispersion of star-forming clouds. This peculiar motion is in qualitative agreement with spiral density wave theory (Roberts 1972).

This research is supported by NSFC under grants 10673024 and 10621303, and NBRPC (973 Program) under grant 2007CB815403.

References

Dehnen, W. & Binney, J. J., 1998, MNRAS, 298, 387
Humphreys, R. M. 1978, ApJ, 38, 309
Kim, W.-T. & Ostriker, E. C. 2002, ApJ, 570, 132
La Vigne, M. A., Vogel, S. N., & Ostriker, E. C. 2006, ApJ, 650, 818
Roberts, W. W., 1972, ApJ, 259, 283
Taylor, J. H. & Cordes, J. M., 1993, ApJ, 411, 674
Xu, Y., Reid, M. J., Menten, K. M., & Zheng, X. W., 2006a, ApJS, 166, 526
Xu, Y., Reid, M. J., Zheng, X. W., & Menten, K. M., 2006a, *Scince*, 311, 54
Xu, Y., Reid, M. J., Menten, K. M. *et al.*, 2008, in preparation

A Giant Step: from Milli- to Micro-arcsecond Astrometry
Proceedings IAU Symposium No. 248, 2007
W. J. Jin, I. Platais & M. A. C. Perryman, eds.

© 2008 International Astronomical Union
doi:10.1017/S1743921308019133

The Gaia mission:
science, organization and present status

L. Lindegren[1], C. Babusiaux[2], C. Bailer-Jones[3] U. Bastian[4], A. G. A. Brown[5], M. Cropper[6], E. Høg[7], C. Jordi[8], D. Katz[2], F. van Leeuwen[9], X. Luri[8], F. Mignard[10], J. H. J. de Bruijne[11] and T. Prusti[11]

[1]Lund Observatory, Lund University, Sweden, email: lennart@astro.lu.se

[2]Observatoire de Paris, Meudon, France
email: carine.babusiaux@obspm.fr, david.katz@obspm.fr

[3]Max-Planck-Institut für Astronomie, Heidelberg, Germany, email: calj@mpia-hd.mpg.de

[4]Astronomische Rechen-Institut, Heidelberg, Germany, email: bastian@ari.uni-heidelberg.de

[5]Leiden Observatory, Leiden University, The Netherlands, email: brown@strw.leidenuniv.nl

[6]Mullard Space Science Laboratory, Dorking, United Kingdom, email: msc@mssl.ucl.ac.uk

[7]Niels Bohr Institute, Copenhagen, Denmark, email: erik@astro.ku.dk

[8]University of Barcelona – IEEC, Barcelona, Spain, email: carme@am.ub.es, xluri@am.ub.es

[9]Institute of Astronomy, Cambridge, United Kingdom, email: fvl@ast.cam.ac.uk

[10]Observatoire de la Côte d'Azur, Nice, France, email: francois.mignard@obs-nice.fr

[11]ESA, ESTEC, Noordwijk, The Netherlands
email: Jos.de.Bruijne@rssd.esa.int, Timo.Prusti@rssd.esa.int

Abstract. The ESA space astrometry mission Gaia will measure the positions, parallaxes and proper motions of the 1 billion brightest stars on the sky. Expected accuracies are in the 7–25 μas range down to 15 mag and sub-mas accuracies at the faint limit (20 mag). The astrometric data are complemented by low-resolution spectrophotometric data in the 330–1000 nm wavelength range and, for the brighter stars, radial velocity measurements. The scientific case covers an extremely wide range of topics in galactic and stellar astrophysics, solar system and exoplanet science, as well as the establishment of a very accurate, dense and faint optical reference frame. With a planned launch around 2012 and an (extended) operational lifetime of 6 years, final results are expected around 2021. We give a brief overview of the science goals of Gaia, the overall project organisation, expected performance, and some key technical features and challenges.

Keywords. astrometry, Galaxy: kinematics and dynamics, planetary systems, relativity, space vehicles, stars: fundamental parameters, techniques: photometric, techniques: radial velocities

1. Scientific case

The scientific case for Gaia rests on the extremely powerful combination of three distinct qualities in a single mission: (i) the ability to make very accurate (global and absolute) astrometric measurements; (ii) the capability to survey large and complete (flux limited) samples of objects; (iii) the matching collection of synoptic, multi-epoch spectrophotometric and radial-velocity measurements. The range of topics that can be addressed with such a data set is too immense to be described here except in the most superficial and incomplete way. Below we highlight the expectations in four different areas. For more detailed reviews on a much broader variety of topics we refer to the proceedings of the symposium *The Three-Dimensional Universe with Gaia* (Turon *et al.* 2005).

Galactic astrophysics: Large, volume-complete samples allow the determination of spatially resolved statistics such as the luminosity and initial mass functions, star formation rates and stellar multiplicity. Distance information combined with stellar classification and photometry allows unprecedented three-dimensional mapping of interstellar extinction. Gaia is the first survey providing six-dimensional phase space $(\mathbf{r}, \mathbf{v})$ data coupled with photometry for very large, magnitude-limited samples. These are crucial for disentangling the complex relationships between the spatial and kinematic distributions of the stars and their ages and chemical enrichment, which encode the Galactic evolution. The number density and kinematics of tracer stars can be used to map the galactic potential over a large part of the Galaxy, thus determining the distribution of (dark) matter and shedding new light on disk dynamics (bar, spiral structure, warp). In the halo, phase space (or E, L_z) structures may reveal past galactic mergers (Aguilar *et al.* 2006).

Stellar astrophysics: Gaia will provide accurate distances to stellar clusters and individual stars, covering a very wide range of masses and evolutionary stages. Individual distances to better than <1% will be obtained for some 10^7 stars. Together with multi-wavelength spectrophotometry this provides accurate luminosity calibrations and stringent tests of stellar interior models, model atmospheres and stellar evolution. There is clearly an enormous scope for synergy with more specialised ground-based observations (high-resolution spectroscopy, astroseismology) and theoretical investigations. For intrinsically bright objects, a major part of the Galaxy can be reached with moderate distance errors (see Table 2) and fundamental parameters can be obtained for rare objects. For stars closer to the Sun, Gaia will provide a sensitive survey of (visible or invisible) companion objects, including thousands of exoplanets.

Solar system: Gaia will detect and measure asteroids down to 20 mag whenever they enter the fields of view, about 15 times per year, to ~ 1 mas accuracy per epoch. Accurate proper elements can be determined and dynamical families identified. The masses for about 100 asteroids can be determined from close encounters (Mouret *et al.* 2007). Other physical characteristics (size, spin, shape, taxonomic classification) will be obtained for thousands of objects. Observations at high ecliptic latitudes and to within 45° from the Sun will probe populations in exotic orbits and inside the Earth's orbit. The observations of Kuiper Belt objects, where masses may be determined for detected binary objects, will provide a valuable contribution to the study of the outer solar system.

Reference frame and experimental relativity: The many accurate positions and proper motions define a dense (>1500 stars deg^{-2}) net of reference objects. The stellar data are directly linked to the extragalactic reference frame through quasars (about 500,000 observed by Gaia), providing a kinematically non-rotating system to ~ 0.3 μas yr^{-1}. The high accuracy of stellar proper motions means that the reference system is long-lived: e.g. 18 mag positions are good to < 1 mas over the 40-year period 1995–2035. The quasar observations also give of the acceleration of the Solar System in a cosmological frame from the secular change in their stellar aberration (Bastian 1995). General relativity is consistently used for reducing the observations, but possible deviations can be tested through the gravitational light bending by the Sun and planets (sensitive to PPN γ) and the perihelion precession for about 300 asteroids (sensitive to PPN β and solar J_2).

2. Status and organization

The Gaia satellite and mission operation is fully funded by the European Space Agency (ESA). The prime industrial contractor for building the satellite (including its scientific instruments), EADS Astrium, was selected in February 2006. The project passed the Preliminary Design Review in July 2007 and is now in the development and production Phase C/D with a launch target date in December 2011. Transfer to its orbit around the Sun–Earth L_2 Lagrange point (1.5 million km from the Earth) and commissioning will

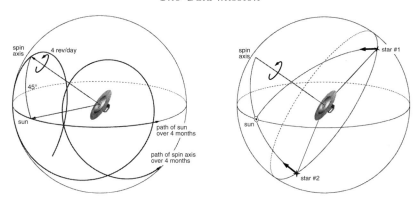

Figure 1. *Left:* Most of the sky is scanned in just a few months by the slowly spinning satellite, thanks to the precession of the spin axis about the satellite–sun direction. *Right:* The parallactic displacement of a star is directed along the great-circle arc from the star to the sun. The observed angle between the indicated stars (#1 and #2) consequently depends on the parallax of #1, but is independent of the parallax of #2. This makes it possible to derive absolute parallaxes.

take up to 6 months, after which the routine science operations phase starts. This phase will last 5 years, with a possible 1 year extension. The data processing will be on-going during the operation phase and for 2–3 years after. The final catalogue is expected around 2021.0, but with intermediate data releases produced during the operational phase. The Data Processing and Analysis Consortium (DPAC) is described elsewhere (Mignard *et al.* 2008).

The overall scientific responsibility for the Gaia mission rests with the ESA Project Scientist, currently T. Prusti; until 2006 M.A.C. Perryman and during 2006 F. Jansen. The Project Scientist interfaces with a number of bodies both within and external to ESA, most importantly the Project Team at ESTEC (responsible for the procurement of the satellite, including its scientific instruments), the flight dynamics and mission operations teams at ESOC, the science operations team at ESAC, and DPAC. The Project Scientist chairs the Gaia Science Team (GST), a senior body in place since 2001 to provide ESA with advice on all aspects of the mission related to its scientific goals, performance and implementation. The present list of authors reflects the current composition of the GST (with J. de Bruijne a member of the Project Scientist's support team at ESTEC), but a very different team will soon become active, following an open invitation by ESA to submit proposals for membership in the GST.

The total human effort expended in the scientific preparations for Gaia is indicated by the list of persons currently involved in the Gaia project; currently (October 2007) this includes some 370 people, not counting the industrial contractors.

3. Observation principle and main technical features

Gaia builds on the proven principles of Hipparcos to determine the astrometric parameters (typically the position, parallax and proper motion of a star) for a large number of objects by combining a much larger number of essentially one-dimensional (along-scan) angular measurements in the focal plane. A continuous scanning motion (Fig. 1, left) similar to that of Hipparcos ensures that every object is observed in at least six distinct epochs per year in patterns that are geometrically favourable for the resolution of the astrometric parameters. The capability to make accurate differential measurements over long ($\sim 90°$) arcs is the key to obtaining absolute parallaxes (Fig. 1, right) as well as a globally consistent reference system for the positions and proper motions. Gaia has all of

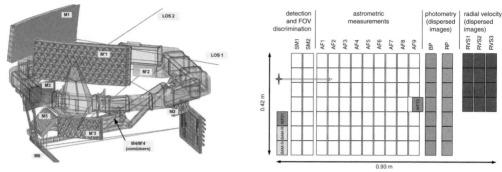

Figure 2. *Left:* The Gaia instrument, comprising two telescopes (mirrors M1–M6 and M'1–M6) with a common focal plane. M4/M'4 is the beam combiner situated at the exit pupil (EADS Astrium). *Right:* Layout of CCDs in the focal plane of Gaia. Images travel from left to right, crossing in turn the skymappers SM, astrometric field AF, blue photometer BP, red photometer RP, and (some of them) the radial-velocity spectrometer (RVS). Also shown are the CCDs for the basic-angle monitor (BAM) and wavefront sensors (WFS, for in-orbit optical adjustment).

these features in common with Hipparcos, although the details differ: the angle between the Sun and spin axis is 45° (compared to 43° for Hipparcos), the spin period is 6 hours (128 min for Hipparcos), and the basic angle is 106.5° (58° for Hipparcos).

The huge gain in accuracy, limiting magnitude and number of objects, compared with Hipparcos, is achieved by a combination of several factors: much larger optics (each of the two fields of view has an entrance pupil of 1.45×0.5 m^2; Fig. 2, left) gives much improved resolution and photon collection; the CCDs have a wider spectral range and much better quantum efficiency than the photomultipliers of Hipparcos, as well as a huge multiplexing advantage, as tens of thousands of objects are observed in parallel. The CCDs have a pixel size of 10×30 μm^2, matched to the optical resolution of the instrument and the effective focal length of 35 m.†

The autonomous on-board processing system detects any object brighter than $\simeq 20$ mag as it enters the focal plane, then tracks the object across a sequence of CCDs dedicated to the astrometric, photometric and radial-velocity measurements (Fig. 2, right). All the CCDs are operated in TDI mode. The astrometric CCDs sample a small window centred on the projected path of the optical image. In contrast to earlier designs of Gaia, the photometry is no longer made using filters, and the photometric and radial-velocity measurements share the same telescope and focal plane as the astrometry. These and other modifications (in early 2006) to the instrument have had a negative impact on the scientific performances, but were considered necessary in order to keep the project within given cost and mass constraints.

For the photometry there are thus two rows of CCDs, referred to as the blue and red photometers (BP and RP). Each of them is a slitless prism spectrophotometer, sampling the dispersed image in some 35 points covering the spectral interval 330–680 nm (BP) and 640–1050 nm (RP); see Fig. 3. The spectral resolution $R = \lambda/\Delta\lambda$ varies from 10 to 30 depending on the wavelength. Accumulated over the mission, any object will be observed ~ 70 times (average value) by each photometer.

Radial-velocity measurements will be made for stars brighter than ~ 16.5 mag using the slitless radial-velocity spectrometer (RVS). Its resolution ($R \simeq 11,500$) and wavelength range (847–874 nm) have been optimised to allow radial velocities to be measured, at

---

† The enormous advantage of using CCDs for direct imaging in a scanning astrometric satellite was shown by Høg (1993) with the Roemer mission design. This efficient use of CCDs, later adopted for Gaia, was far from obvious to anyone at the time (Høg 2008).

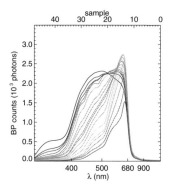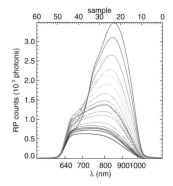

Figure 3. Simulated spectra of 18 main-sequence stars of spectral classes O8–M5, as observed in the blue and red photometers (BP, RP). The photon counts are normalised to the integrated Gaia magnitude $G = 15$. Full PSF smearing effects are taken into account, but no noise has been added. (From Straizys *et al.* 2006.)

the 1–15 km s^{-1} accuracy level, for a wide range of spectral classes. It covers the Ca II triplet, which is strong for most intermediate to late type stars, as well as the hydrogen lines P13–16 dominating the early type spectra, and many weaker lines of various species. For a given object there will be on average ~ 40 transits across the three rows of CCDs used by the RVS.

4. Calibration issues and predicted astrometric performance

Table 1 summarizes the predicted astrometric performance for single, unperturbed solar-type stars. In Table 2 the predicted precision in parallax has been translated into distances for given relative error in parallax (or distance). Note that Gaia will observe virtually *all* stars of a given type and extinction, within their parallax horizon.

The predicted performance was calculated by means of the Gaia Accuracy Analysis Tool (de Bruijne 2005) using current (October 2007) mission and instrument data. The resulting rms errors include a science margin factor of 1.2. The accuracy analysis takes into account all identified relevant error sources, assuming that the engineering specifications of the individual components (CCD quantum efficiency, wavefront errors, etc) are met. While many of the specifications are technologically challenging, they are now all deemed realistic to achieve. However, the performance analysis also includes assumptions concerning the accuracy of critical instrument calibration parameters. These are normally derived as part of the scientific data processing itself, e.g. the precise geometrical layout of the CCDs. In a few cases the associated uncertainties are still considerable, because it has not yet been possible to formulate a complete calibration model.

By far the most difficult calibration issue at present concerns the radiation damage of the CCDs. It is well known that energetic particles in space (mainly protons from solar flares) may cause lattice defects in the silicon material where charges are transported during the CCD readout. As a result, some of the electrons generated by the optical image will be trapped by the defects, and randomly released at a later time. As a result, the electronic image may suffer a distortion, shift and charge loss compared to that of an undamaged CCD. In the RVS and photometers there will be a corresponding degradation in spectral fidelity and resolution. With the expected end-of-mission radiation dose, the systematic centroid shifts of the astrometric images may amount to several milliarcsec, depending in a complex way on many factors including the recent illumination history of the pixel column. The problem is aggravated by Gaia's launch near the next solar

Table 1. Sky-averaged rms errors of the astrometric parameters for G0V stars (no extinction).

V magnitude	6–13	14	15	16	17	18	19	20
parallax [μas]	8	13	21	34	55	90	155	275
proper motion [μas yr^{-1}]	5	7	11	18	30	50	80	145
position at mid-epoch ($\simeq 2015$) [μas]	6	10	16	25	40	70	115	205

Table 2. Parallax horizon of Gaia, i.e. maximum distance (in kpc) for given relative precision in distance (or parallax). Based on current accuracy estimates, with M_V from Cox (2000).

	no extinction ($A_V = 0$)					$A_V = 5$ mag				
Type of star	1%	2%	5%	10%	20%	1%	2%	5%	10%	20%
G0V ($M_V = +4.4$)	0.8	1.1	1.8	2.5	3.5	0.3	0.5	0.7	1.0	1.4
K5III ($M_V = -0.1$)	1.3	2.6	4	7.5	11	1.0	1.5	2.4	3.5	5
Cepheid ($P = 10^{\rm d}$, $M_V = -4.1$)	1.2	2.4	6	12	22	1.2	2.3	3.8	7	10

maximum in 2012, which means that most of the radiation damage is likely to occur early in the mission.

Several methods are used to mitigate the effects of radiation damage, including shielding (which however is difficult to achieve for the expected proton energies over such a large focal plane), optimum choice of the CCD operating temperature ($\simeq 160$ K), and electronic charge injection (and/or a diffuse low-level background illumination of the CCDs) to fill the traps. However, detailed modelling of the effects as part of the data processing and calibration will also be necessary. On the positive side, extensive laboratory tests conducted by EADS Astrium on irradiated CCDs, operated under realistic conditions, have shown that the effects are highly deterministic and therefore possible to calibrate. The accuracy predictions take only part of the radiation damage into account, namely the charge loss. Remaining calibration errors have not been fully quantified yet but are targeted to contribute at modest levels to the end-of-mission performance.

Another critical calibration issue concerns the stability of the basic angle between the two viewing directions – nominally 106.5°. General fluctuations of the basic angle are passively controlled (to < 7 μas rms) by the thermo-mechanical design of the satellite. Moreover, on timescales longer than a few spin periods, the basic angle is accurately determined as part of the data processing. However, on timescales shorter than $\simeq 12$ hours the stability of the basic angle (or knowledge of its variations) is crucial for the global astrometry. In particular, as seen from Fig. 1 (right), a systematic variation in phase with the solar azimuth angle would be very detrimental to the parallax measurements, by producing a global parallax zero-point error that could not be detected from the residuals of the observations. An independent metrology device, the basic-angle monitor (BAM), is therefore in place to measure any short-term variations. It consists of two optical devices, each directing a pair of low-intensity laser beams towards its respective primary mirror. In the focal plane, the beams produce two sets of interference fringes that are detected by a separate CCD (Fig. 2). The devices are optically connected so that the angle between the two pairs of beams is insensitive to the relative orientation of the devices, and therefore to any deformation of the mechanical structure supporting them or the mirrors. The beams consequently form a stable reference against which any opto-mechanical changes can be detected as a relative displacement of the two sets of interference fringes. The BAM is capable of measuring basic-angle variations at the sub-μas level every few minutes. By including these measurements in the data processing chain, the astrometric effects of short-term variations are eliminated. The BAM itself can be calibrated by means of other variations that are not in phase with the solar azimuth angle, and which produce recognizable signatures in the residuals of the observations.

5. Conclusion

Technically and scientifically, the Gaia project is well on track towards its scheduled launch around 2012. It is in many respects a difficult and challenging project, but the rewards in terms of the expected science return are overwhelming.

Acknowledgements

The present advanced state of the Gaia mission and its preparations is the result of a long and fruitful collaboration between many teams and individuals within ESA, industry and the scientific community. This review has freely used material from various sources and documents available within the Gaia community. Financial support by the national governments, research and space organizations to this community is gratefully acknowledged.

References

Aguilar, L. A., Brown, A. G. A., & Velázquez, H. 2006, *Rev. Mexicana AyA* 26, 51
Bastian, U. 1995, in: M. A. C. Perryman & F. van Leeuwen (eds.), *Future Possibilities for Astrometry in Space*, ESA SP-379, p. 99
de Bruijne, J. H. J. 2005, in: C. Turon, K. S. O'Flaherty, & M. A. C. Perryman (eds.), *The Three-Dimensional Universe with Gaia*, ESA SP-576, p. 35
Cox, A. N. (ed.) 2000, *Allen's astrophysical quantities, 4th ed.* (Nex York: AIP Press; Springer)
Høg, E. 1993, in: I. I. Mueller & B. Kołaczek (eds.), *Developments in Astrometry and Their Impact on Astrophysics and Geodynamics*, Proc. IAU Symposium No. 156 (Dordrecht: Kluwer), p. 37
Høg, E. 2008, this volume p.300
Mignard, F., *et al.* 2008, this volume, p.224
Mouret, S., Hestroffer, D., & Mignard, F. 2007, *A&A* 472, 1017
Straižys, V., Lazauskaitė, R., Brown, A. G. A., & Zdanavičius, K. 2006, *Baltic Astronomy* 15, 449
Turon, C., O'Flaherty, K. S., & Perryman, M. A. C. (eds.) 2005, *The Three-Dimensional Universe with Gaia*, ESA SP-576

A Giant Step: from Milli- to Micro-arcsecond Astrometry
Proceedings IAU Symposium No. 248, 2007
W. J. Jin, I. Platais & M. A. C. Perryman, eds.
© 2008 International Astronomical Union
doi:10.1017/S1743921308019145

Gaia: organisation and challenges for the data processing

F. Mignard[1], C. Bailer-Jones[2], U. Bastian[3], R. Drimmel[4], L. Eyer[5],
D. Katz[6], F. van Leeuwen[7], X. Luri[8], W. O'Mullane[9], X. Passot[10], D. Pourbaix[11] and T. Prusti[12]

[1] Observatoire de la Côte d'Azur, Nice, France
email: francois.mignard@oca.eu

[2] Max Planck Institut für Astronomie, Heidelberg, Germany

[3] Astronomisches Recheninstitut (ARI), Heidelberg, Germany

[4] Osservatorio Astronomico di Torino, Torino, Italy

[5] Observatoire de l'Université de Genève, Sauverny, Switzerland

[6] Observatoire de Paris-Meudon, Meudon, France

[7] Institute of Astronomy, Cambridge, England

[8] University of Barcelona, Barcelona, Spain

[9] European Space Astronomy Centre, Madrid, Spain

[10] Centre National d'Etudes Spatiales (CNES), Toulouse, France

[11] Université Libre de Bruxelles, Brussels, Belgium

[12] European Space Agency, ESTEC, Noordwijk, The Netherlands

Abstract. Gaia is an ambitious space astrometry mission of ESA with a main objective to map the sky in astrometry and photometry down to a magnitude 20 by the end of the next decade. While the mission is built and operated by ESA and an industrial consortium, the data processing is entrusted to a consortium formed by the scientific community, which was formed in 2006 and formally selected by ESA one year later. The satellite will downlink around 100 TB of raw telemetry data over a mission duration of 5 years from which a very complex iterative processing will lead to the final science output: astrometry with a final accuracy of a few tens of microarcseconds, epoch photometry in wide and narrow bands, radial velocity and spectra for the stars brighter than 17 mag. We discuss the general principles and main difficulties of this very large data processing and present the organization of the European Consortium responsible for its design and implementation.

Keywords. astrometry, instrumentation: photometers, techniques: spectroscopic, methods: data analysis

1. Introduction

The ESA mission Gaia is a powerful astronomical space project dedicated to high precision astrometry, photometry and spectroscopy. The details of the science objectives together with the principles of the measurements are given in a companion paper (Lindegren *et al.*, 2008) in this proceeding and are not repeated. Gaia will survey the whole sky and detect any sufficiently point-like source brighter than the 20th magnitude. The observations are carried out with a scanning instrument mapping the instantaneous field of view (composed of two disconnected areas on the sky) onto a single mosaic of CCD very precisely mounted on the focal plane. Therefore each detectable source is observed during its motion on the focal plane brought about by the spin of the satellite. On the average a source will transit about 80 times during the five years of the observing mission, leading to about 1000 individual observations per object. They will be more or less regularly distributed over about ~ 40 epochs with about one cluster of ~ 25 individual observations every 6 weeks. These observations consist primarily of a 1D accurate determination of the image location at the transit time on a frame rigidly attached to the payload together with an estimate of the source brightness. The number of sources potentially observable is of the order of one billion, primarily stars. In addition there will be something like $\sim 5 \times 10^5$ QSOs and $\sim 3 \times 10^5$ solar system objects.

2. Challenges with the data processing

Scientifically valuable information gathered by Gaia during its lifetime will be encased in the nearly continuous data stream resulting from the collection of photons in the approximately 100 on-board CCDs in the astrometric, photometric and spectroscopic fields of Gaia. However in its original telemetry format the data are totally unintelligible to scientists, not only because it is squeezed into packets by the numerical coding but, more significantly, because of the way Gaia scans the sky. With a 1D measurement at each source transits, Gaia picks up tiny fragments of the astrometric parameters of the one billion observable sources in the sky. To translate this information into positional and physical parameters, as expected by scientists, a large and complex data analysis must be implemented. How huge and how complex it is, is briefly addressed here.

2.1. *Volume of data*

Data collected by the Gaia detectors will be transmitted by the spacecraft from its position around L2 at a few Mbps for at least five years. Over its lifetime the satellite will deliver to the ground station an uncompressed raw data set of approximately 100 TB, primarily composed of CCD counts collected during the transit of celestial sources on the detector. This will make up the raw data. However the data processing will generate numerous intermediate data to produce the scientifically meaningful data products by 2020. Given these intermediate steps, the need to store the provisional solutions and different backups, it is estimated that the actual size of the Gaia related data will be in the range of a PB (10^{15} Bytes), still a fairly large volume by today's standard, not yet matched by any other space mission, although ground based experiments in particle physics will produce data volume one or two orders of magnitude larger. It is enough to say that to store the text of all the books published each year a storage space of about 1TB is sufficient which could be readily available on anyone's desk. An orderly access to this data is another question. For the astronomers accustomed to working with the CDS on-line data bases, all the catalogues stored in the VizieR service make less than 1 TB in disk space, while the Aladin sky atlas with the images is just above 5 TB. Despite this apparently limited volume, the administration and maintenance of these services require several permanent staff. Therefore, the volume alone is not a major issue, while efficient access to the relevant piece of data is a serious one.

The core of the final data products distributed to the scientific community will be much smaller. Scaling from Hipparcos with 10^5 sources with all the results on a few CDs, this should amount for Gaia to something around 20 TB of data. This depends very much on whether raw or nearly raw data are included in the delivery. If one considers just the final data of astrometry, photometry and spectroscopy (not the spectra but their exploitation in terms of radial velocity and astrophysical diagnostic), all the Gaia data expected for a standard user can be estimated by considering a data base with 10^9 entries, each of them comprising less than 100 fields. The astrometric data will need many significant

digits (or bytes), but most of the other fields will require a limited number of digits
or bits. A volume of 1-3 TB should then be sufficient for the main Gaia product, not
including data produced during the intermediate steps of the data processing. These data
will certainly be made available with calibration data to allow specific reprocessing with
improved models by users.

2.2. *Computational complexity*

There is no all-purpose definition of what a complex computation is. It could be complex
because of the number of elementary operations it requires (time complexity), or because
of the data management (interface complexity), or because of the size of the core memory
needed to perform some parts (space complexity). Gaia can be considered complex at
least on two counts:

• It implies a very large resource of CPU, or more or less equivalently, the number of
basic arithmetical operations is very large by today's standards, and will remain large in
the coming ten years.

• The data are widely interconnected with access sometimes to subsets arranged in
chronological order (observations of a certain type in a certain time interval) or by sources
(all the observations of a subset of sources). This extends to the access of intermediate
data or calibration files.

What is at stake can be easily grasped if we consider that there will be about 10^9
individual sources to be processed. Just spending one second on each of them to get the
calibrations, spacecraft attitude, astrometric parameters, multi-color photometry and
spectra would amount to 30 years of computation. Obviously the overall system must be
very efficient to have a chance to produce the scientific results no longer than in three
years after the end of the operations.

As to the actual computation size, it is not easy to assess with reliability within at least
one order of magnitude as shown already in 2004 by M. Perryman (Perryman, 2004).
However we have now a better view with scalable experiments on simulated data. The
core iterative processing needs (referred as to AGIS) have been evaluated by Lammers
(Lammers, 2006) to 4×10^{19} FLOPs for the last cycle, when all the observation material
is available. Allowing for the intermediate cycles this yields 2×10^{20} FLOPs for the astrometric processing once the image parameters task and source matching are completed.
This latter processing has also been evaluated from realistic simulated telemetry and
should not be larger than 4×10^{18} FLOPs for the full telemetry volume. Regarding the
other processing (complex sources, photometry, spectroscopy and before all the image
updating) the uncertainty remains significant and depends also on how much additional
modelling will be required to account for the radiation effect on the CCDs. A conservative
estimate indicates that the most demanding task will be the image updating (improvement of the image parameters and initial centroiding once a better spacecraft attitude
and calibrations are known). It is estimated to be of the order of 10^{21} FLOPs. But with
numerous small computations repeated on all the sources instead of a global adjustment
requiring a big core memory access, this demanding subsystem could be either distributed
or implemented with a parallel architecture.

The number of 10^{21} FLOPs happens to be very close to the largest computations
undertaken so far, like a search of the largest prime numbers or the location of the
non trivial zeros of the Riemann ζ function. The recent distributed computation for
the protein structure (FAH, 2007) is also in the same area in terms of computing size.
However these computations are very different from Gaia's, since they are distributed

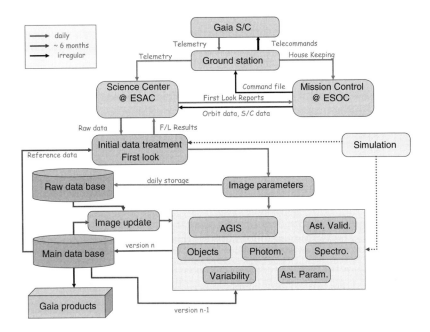

Figure 1. Main structure of the data flow in the processing of the Gaia raw data.

through internet on several thousands processing units of small power (and cheap) and they do not have significant data handling. They are pure computations where each unit is given the task of making a tiny contribution. Most of the Gaia computations cannot be distributed in this way and each computing center will have to access a large amount of data with a significant I/O overhead.

By the time Gaia flies the computing power available to the Gaia Data Processing Centers is expected to be close to 10-20 TFLOP/s on large clusters. In addition one can also count on the part-time availability of the Barcelona Super Computer (Mare Nostrum) which has a today's power of 20 TFLOP/s. A straight evaluation for Gaia tells that about 2 full years of CPU at 20 TFLOP/s will be sufficient to process the data. This is still daunting but tractable, provided the estimate is not wrong by two orders of magnitude.

3. Overview of the data processing

The Gaia data processing is based on an iterative self-calibrating procedure where the astrometric core solution is the single largest treatment and the cornerstone of the whole processing. These iterations reflect the fact that, as a survey satellite reaches the accuracy and completeness levels never obtained before, it has to be self-calibrated in nearly all aspects of its observations. The most obvious in this respect are the astrometric data: the same data that will ultimately form the astrometric catalogue are also used to reconstruct the scan-phase of the satellite (along-the-scan attitude), which is the reference frame for the astrometric measurements. Similar considerations apply to the spectroscopic and photometric data. Additional complications arise from the fact that these three main iterative solutions are in various ways interlinked and rely on common calibration data.

The overall processing starts with an initial data treatment aiming to obtain a first estimate of the star positions and magnitudes together with an initial attitude, primarily based on the on-board crude determination improved on the ground with more accurate (and more computer greedy) algorithms. The next step is the iterative astrometric core solution which provides the calibration data and attitude solution needed for all the other treatments, in addition to the astrometric solution of about 100 million primary sources and the rigid reference system. Once these two steps are performed and their products stored in the central data base, the more specialized tasks are launched with the photometric processing and variability detection, the global analysis of the spectroscopic data and the data treatment for all difficult sources, like planets and multiple stars, that do not fit in the general astrometric or photometric solution of single stars. Two more processing steps close the chain, dealing with the analysis of all types of variable stars and retrieval of their stellar astrophysical parameters like luminosity, temperature or chemical composition. Each step in the processing using earlier data has its own logic regarding the products it will deliver, but that must be seen as part of the validation of the upstream processes. The overall data flow from telemetry to the final products is sketched in Fig. 1. The daily flow follows naturally from the data downlink taking place once per day during the visibility of L2 above the ESA main station in Spain.

The data exchange between the different processing centers (DPCs) takes place via a Main Database (MDB) hosted at a single place. Intermediate processed data from the processing centers flow into the MDB; the processing centers can then extract the data from this Database for further processing. Thus, the Main Database acts as the hub for the Gaia data processing system and must have an efficient and reliable connection with all the DPCs. During operations, the plan is to update this database at regular intervals, typically every six months, corresponding roughly to the period for the satellite to scan the whole celestial sphere. In this way each version of the Main Database is derived from the data in the previous version, supplemented with the processed new observations.

Table 1. Coordination Units of the DPAC and their current leader.

CU	Name	Leader	Affiliation	Location
CU1	System Architecture	W. O'Mullane	ESAC	Madrid
CU2	Simulation	X. Luri	UB	Barcelona
CU3	Core processing	U. Bastian	ARI	Heidelberg
CU4	Object processing	D. Pourbaix	ULB	Brussels
CU5	Photometric processing	F. van Leeuwen	IOAC	Cambridge
CU6	Spectroscopic processing	D. Katz	OBSPM	Paris
CU7	Variability processing	L. Eyer	Obs. Geneva	Geneva
CU8	Astrophysical parameters	C. Bailer-Jones	MPIA	Heidelberg

4. Organization of the scientific community

To cope with the processing challenge, the scientific community, together with ESA, has set up a data processing ground segment comprising a single processing system (no overall duplication) which will deliver the intermediate and final mission science products. Since mission selection, the underlying principles of the data processing have been developed by the Gaia scientific community and individual pieces were successfully tested on small or intermediate size simulations. During this phase one has attempted to identify the critical elements of this processing (size, iterative procedures, instrument calibration, data exchange, human and financial resources, computing power) and assess the risks inherent to an endeavor of that size, unprecedented in astronomy.

Based on these preparatory activities the community has joined forces into a dedicated consortium: the Data Processing and Analysis Consortium (the DPAC). In short, the DPAC is a European collaboration including the ESA Gaia Science Operations Centre (SOC) and a broad, international science community of over 320 individuals, distributed over more than 15 countries, including six large Data Processing Centers (DPCs, Table 2). The Consortium has carefully estimated the effort required and has united in a single organization the material, financial and human resources, plus appropriate expertise, needed to conduct this processing to its completion in around 2020. In 2006 the DPAC has proposed to ESA a complete data processing system capable of handling the full size and complexity of the real data within the tight schedule of the mission. The details of this system, its expected performances, funding, organization and management are described in a document submitted to ESA as a Response to its Announcement of Opportunity for the Gaia data processing (Mignard *et al.*, 2007).

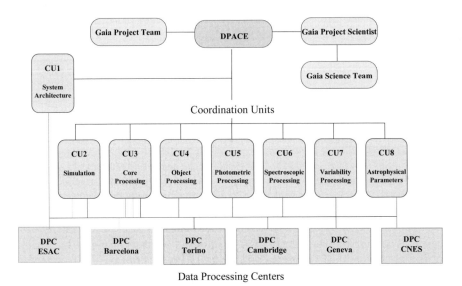

Figure 2. Top level structure of the DPAC with the Coordination Units (CUs) and the data processing centers (DPCs). Colored connectors indicate the link between the CUs and the DPCs.

The Consortium is structured around a set of eight Coordination Units (CUs) each in charge of a specific aspect of the data processing. The CUs are the building blocks of the Gaia Data Processing and Analysis Consortium and they are listed together with their leader's name in Table 1. An additional CU (the CU9) in charge of developing the tools for the Catalogue access by the scientific community is planned in the near future but is not yet activated. The CUs have clearly-defined responsibilities and interfaces, and their boundaries match naturally with the main relationships between tasks and the associated data flow. Responsibilities of the coordination units include: (a) defining data processing tasks and assigning responsibilities; (b) establishing development priorities; (c) optimizing, testing and implementing algorithms; (d) verifying the quality of the science products. Each coordination unit is headed by a scientific manager (the CU leader) assisted by one or two deputies and, where appropriate, a technical manager. The management team of each CU is responsible for acquiring and managing the resources needed for their activities. While the CUs are primarily structured for software development, all of them are closely associated with at least one DPC where the algorithms will be executed for the data processing in the operational phases.

The Consortium is coordinated by the 'Data Processing and Analysis Consortium Executive' (DPACE) committee. This top-level management structure deals with matters that are not specific to the internal management of a CU, defining standards and policies needed to ensure an efficient interaction between all the CUs. Consistent with the Science Management Plan, the DPACE and its chair will serve as an interface between the DPAC and the Project Scientist and the Gaia Science Team. They are ultimately responsible for the data processing carried out by the DPAC. This executive committee is composed at the moment of the DPAC chair and deputy, the leaders of each CU, a representative of the CNES Data Processing Center. The Gaia Project Scientist (an ESA position) has a standing invitation to the DPACE where he has the status of observer.

Table 2. The Data Processing Centres associated to the DPAC, their current manager and involvement.

DPC	Location	Manager	Linked to:
DPC-B	Barcelona	S. Girona	CU2, CU3
DPC-C	CNES	X. Passot	CU2, CU4, CU6, CU8
DPC-E	ESAC	J. Hoar	CU1, CU3
DPC-G	Geneva	M. Beck	CU7
DPC-I	Cambridge	F. van Leeuwen	CU5
DPC-T	Torino	A. Volpicelli	CU3

The DPAC has responded to the Announcement of Opportunity released by ESA on 9 November 2006. The Response document contains, in an hefty volume of more than 650 pages, the overall description of the Gaia data processing to reach the scientific objectives, the organization of the consortium and the funding commitments of the national agencies supporting the DPAC. This response has been reviewed by various ESA advisory committees and after one iteration the DPAC proposal and its selection have been formally endorsed by the ESA Science Program Committee in its meeting of May 2007. In November 2007 the same high level committee has also approved the funding agreement between the national agencies and ESA.

Acknowledgements

This short review of the DPAC organization and activities relies heavily on the contribution of many members of the DPAC who are collectively gratefully acknowledged for their dedication to the project.

References

FAH, 2007, http://folding.stanford.edu/
Lammers, U., 2006, Agis technical report, GAIA-C1-PR-ESAC-UL-018-1.
Lindegren, L., *et al.*, 2008, this volume p.217
Mignard, F., *et al.*, 2007, Proposal for the Gaia Data Processing, GAIA-CD-SP-DPAC-FM-030-2, Mignard, F., Drimmel, eds.
Perryman, M., 2004, Estimation of Gaia Data Processing FLOPs, GAIA-MP-009.

A Giant Step: from Milli- to Micro-arcsecond Astrometry
Proceedings IAU Symposium No. 248, 2007
W. J. Jin, I. Platais & M. A. C. Perryman, eds.

© 2008 International Astronomical Union
doi:10.1017/S1743921308019157

Astrometry with SIM PlanetQuest

M. Shao

Jet Propulsion Laboratory, California Institute of Technology, Pasadena,CA 91007, USA
email: Michael.Shao@jpl.nasa.gov

Abstract. SIM PlanetQuest is a very high accuracy space astrometric instrument based on a long baseline stellar interferometer. For global astrometry SIM was designed to be accurate to $\approx 4\mu as$ (microarcsec) after a 5 year mission. For narrow angle astrometry (≈ 1000 s integration over a $1°$ radius field) SIM is designed for $1\mu as$ precision. The technology program was completed in 2005 and based on laboratory results, the current best estimate of SIM's performance would be 0.6 μas for narrow angle precision and 2.4 μas for global accuracy. This paper describes a variety of science programs that the SIM science team have proposed to conduct from a search for one Earth mass planets in the habitable zone of ($\approx$130) nearby stars to the study of dark matter in the galactic disk, the galactic halo and the local group.

Keywords. techniques: interferometric, astrometry, space vehicles, planetary systems

1. Introduction

SIM PlanetQuest is a long baseline optical interferometer designed for ultra-high accuracy astrometry. This paper describes technology status of the SIM project and a few of the science objectives of the Mission. Other papers at this conference will describe several key science projects in more detail (see Beichman 2007, Majewski 2007, Chaboyer 2007, Unwin 2007).

SIM's goal was to be able to do global astrometry with 4 μas accuracy (5- year mission) and narrow angle astrometry with 1 μas precision over a $\approx 1°$ radius field in ≈ 1000 second measurement. Because this was such a large increase in accuracy over the Hipparcos mission, NASA established a series of eight technical milestones for the project to accomplish prior to the start of building flight hardware. The next section of this paper is a brief summary of these technology milestones, which were completed in 2005–2006. The completion of milestones brought to light a capability that was not recognized at the start. The last part of this paper provides a snap shot of the variety of science that is planned for the SIM mission.

SIM PlanetQuest, as shown in Figure 1, consists of a science interferometer and two guide interferometers. The guide interferometers provide spacecraft attitude information at the microarcsec level. The science interferometer moves from star to star to measure the relative angles between the stars. The science interferometer and guide interferometers are tied together with a laser interferometer metrology truss, with picometer precision. A long baseline interferometer converts an angle between the baseline vector and the star direction to a length through the classic equation X = S*B, where X is the delay position measured by the white light fringe, S is a unit vector to the star and B is the baseline vector, * is a dot product. Many of the technical challenges come from the need to make these length measurements to very high precision. Two stars $15°$ apart would cause the delay (X) to change by ≈ 2.3 meters. A measurement 2.3 m length with an error of 43 picometers is equivalent to a 1 microarcsec angular error.

Figure 1. SIM PlanetQuest Spacecraft

2. Measuring angles to 1 microarcsec

The eight technology milestones were defined around 2001. The early milestones involved component technologies such as laser metrology gauges that had a precision of a few picometers. Some of the intermediate milestones were system level demonstrations (such as Benchmark No.4 for narrow angle measurements) but at a precision that was 3–5 times worse than the final goal. Four of the milestones represent the critical areas needed for SIM to work in space.

The most difficult tasks concerned the ability to make measurements at the microarcsec level. Since no one has flown a long baseline interferometer in space before, we also have requirements to actively control a large flexible space structure with ≈ 10 nanometer stability and to point the interferometer (track the white light fringe) with a precision of ≈ 0.1 milliarcsec. This technology was demonstrated in 2004, with a 3 baseline interferometer and metrology linking the three interferometers. For microarcsec astrometry, the most relevant milestones were the 4 μas precision for wide angle measurements and the 1 μas precision for narrow angle measurements.

3. Global and narrow-angle astrometry

SIM does global astrometry by measuring the relative position of stars over a 15° field, which we call a tile, then covering the sky with overlapping tiles. Typically adjacent tiles overlap by 50 percent. In global astrometry the plan is to cover the sky with overlapping tiles ≈ 20 times over the 5 year mission.

For a technology demonstration, a Micro Arcsecond Metrology testbed(MAM) (see Hines, *et al.* 2003) was constructed, which includes a test interferometer and a pseudo star, which is an interferometer in reverse. The pseudo star could be moved over 15° with laser metrology that would measure this motion within 15°. The test setup is shown in Figure 2.

Global astrometry requires measurements over the full 15° field, while narrow angle measurements were over a 1° field. The major challenge in global astrometry was to

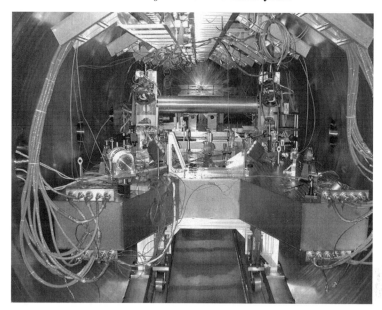

Figure 2. MAM testbed in vacuum chamber

control or model field dependent biases to 4 μas. Concurrent with the laboratory effort was a numerical simulation activity that looked at taking many single tile measurements over the whole sky over 5 years and performing a large least-squares fit of the global parameters in order to derive the position, proper motion and parallax of all objects. In doing the global astrometry simulations several properties were discovered. As suspected a linear scale error in the single tile measurements was not important, as long as they were relatively small. A number of higher order field dependent biases, could be "solved" for in the grid solution if these errors were static over a period of several days. The wide angle astrometry milestone was met in 2005.

The narrow angle milestone was in some way more difficult because the goal was 1 μas in $\approx$20 minutes instead of 4 μas after 5 years. But the small field (1°) means that the major concern in global astrometry, i.e. field dependent biases, was no longer a concern. The major error for narrow angle astrometry turned out to be thermal drift. The laser metrology beam is placed in the center of the white light beam so to first order any motion of optical elements would be monitored. However there were numerous second order effects.

Figure 3 shows the results of the narrow angle tests. For the SIM instrument with 9 m baseline 1 μas is roughly 50pm of optical path. But the allocation to the science interferometer was only 28 pm. One μas (1 sigma) precision means that > 68% of the measurements must have an error within ±28 pm.

Our original plan for exoplanet detection was to visit a target star 25-50 times over 5 years and survey a large number of stars. If we assumed that errors at different epochs were random that was the end of the story. Ideally we should demonstrate in the lab that the systematic error floor was $< 1/\sqrt{25}$. Key to narrow angle measurements is chopping. SIM will move between target and reference star with a $\approx$90 sec period. To demonstrate that the systematic error floor was near 0.2 μas, we conducted a longer test. Figure 4 is a plot of the Allan variance. For integration times of $\approx$ 1 day we had not reached the systematic error floor. The Allan variance had reached 1 picometer in optical path, which

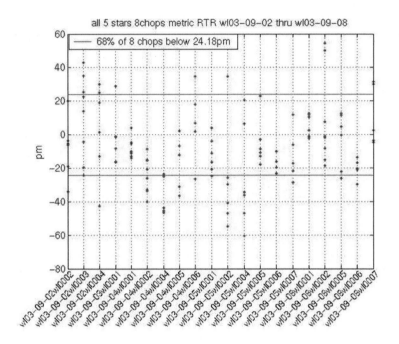

Figure 3. MAM results of 20 narrow angle tests

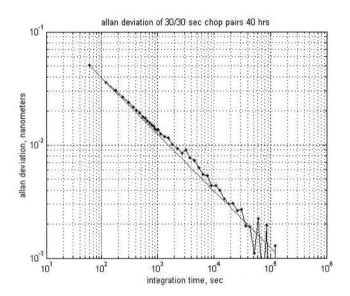

Figure 4. Allan variance of narrow angle measurements

divided by the 9 m baseline is 0.023 μas. The very low systematic error floor enables a different type of exoplanet search.

Table 1. Number of detectable Earth-like planets versus the mission systematic noise floor

Noise floor	No. of targets
0.2 μas	2
0.1 μas	4
0.075 μas	14
0.05 μas	43
0.04 μas	80
0.03 μas	198

4. Science for searching exoplanets

Over 250 exoplanets have been discovered. The SIM tier-1 exoplanet project is to look for terrestrial planets in the habitable zone. If we are limited to looking for astrometric signatures $> 1.0\mu$as, there are only two stars around which an Earth clone ($1M_{Earth}$, 1 $AU * \sqrt{L}$, where L is luminosity) can be found, i.e. two stars of the wide binary α Cen. But if instead of just ≈ 50 visits per star and a total of 14 hours of observing time in 5 years, we increase the total time to 200-250 hours, the minimum detectable astrometric signature would drop by a factor of 5–6, and the number of stars around which an Earth clone could be detected would grow substantially. The Table 1 lists the number stars around which an Earth clone can be found versus the mission systematic noise floor.

SIM PlanetQuest with its very low systematic noise floor is limited by the total observing time. The current best estimate (with no margin) is that a target star of 7 mag with 9.6 mag reference stars could be measured with 0.7 μas in a 1000 s measurement. Getting to a mission accuracy of 0.03 μas means over 150 hours spent per target. One hundred such targets would occupy two full years of integration. If we were to conduct a search down to 1 M_{Earth}, 1AU, the nearest stars would not need 150 hours each. But the majority of the stars would. Going through the list of nearest stars, SIM PlanetQuest can conduct a search of ≈ 130 stars for an Earth clone, using 40% of its 5 year mission.

While we have demonstrated that instrumental errors can be controlled at levels of well below 1μas, we need to take a careful look at astrophysical noise or error sources. Two major astrophysical errors are: 1) star spots or more generally, stellar variability that results in a displacement of the center of light with respect to the center of mass, and 2) the presence of planets around the reference stars used for narrow angle astrometry. These topics will be addressed in much more details in a separate paper underway. A typical star spot on the Sun at 10pc that causes a $5 * 10^{-4}$ photometric variation which produces a maximum astrometric shift of 0.12 μas. Just as important, at periods longer than the 30 day rotation period of the Sun the astrometric noise from star spots averages out (spot on the left drifts to the right side of the star) as $\sqrt{N}$, where N is the number of epochs where the separation between epochs of observation is greater than the rotation period of the star. In our simulations, putting in an 11 year sunspot cycle shows no astrometric noise with an 11 year period.

The second astrophysical concern is planets around the reference stars. The reference stars are typically 9–10 mag K giants at a distance of 700–1000 pc. A Jupiter in a 1 AU orbit around a one solar mass K giant would have an astrometric wobble of 1 μas. If the instrumental noise floor was 0.03 μas, these planets with a 1 μas signature are easily detected and their effect can be removed from the "local" reference frame. Much smaller planets ($< 0.05\mu$as) would be simply absorbed as noise. Intermediate sized planet (signature between 0.3 μas and 0.05 μas) could not be reliably detected but would produce wobbles that are not ignorable. On average only $\approx$5% of stars would have such

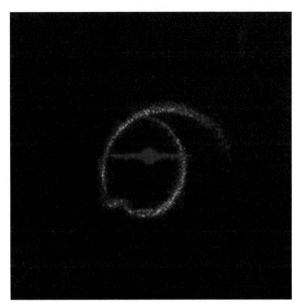

Figure 5. A demonstration of the sensitivity of SIM to the Galactic potential using stellar steams

planets. And if only 1 of 4 reference stars are like this, the end result is only a minor adjustment in the mission noise floor.

5. Astrophysics with SIM

When the SIM science program was selected in 2001-2002 through a call for proposals, Gaia was already a well established mission concept under study at ESA. Consequently, most of the SIM astrophysics is in areas outside of the main Gaia science program. The principle area for SIM astrophysics (see Unwin, S. *et al.* 2007) is moderately high accuracy ($\approx 10\mu$as) astrometry of faint stars ($17-20$ mag).

The major themes for SIM astrophysics are dark matter, Hubble constant, and the masses of post main sequence objects like neutron stars and stellar mass black holes in binaries. Dark matter will be studied by several key projects. One key project with Andrew Gould as the PI would followup microlensing events both astrometrically and photometrically with SIM ($\approx$0.5AU from the Earth) to tightly constrain the mass of the lensing object. Another with Steve Majewski would measure the parallax and proper motion of faints stars in tidal tails of dwarf galaxies ripped apart by the gravity gradient of the Milky Way. The motion of the stars in the tidal tales will let us map the dark matter distribution in the halo of our galaxy as shown in Figure 5. The third project in this area is the measurement of the proper motion of galaxies in the local group, to get a handle on the dark matter in the local group shown in Figure 6.

Proper motion of stars in a few nearby spiral galaxies will enable us to measure the distances to these galaxies via a dynamical parallax technique. This distance could be used to calibrate the zero point of the Cepheid period-luminosity-metalicity relation.

Last of all SIM will be able to measure the orbits and hence masses of neutron star and black hole binaries throughout the galaxy. Many of these object are both faint and have short periods which require a short but concerted observing campaign.

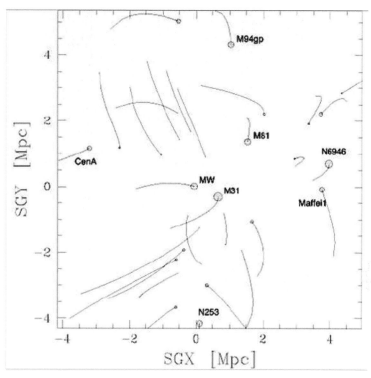

Figure 6. The Trajectories of nearby galaxies and groups going out to the distance of the Virgo Cluster from a NAM calculation

6. Summary

SIM Planet Quest is a mission whose very ambitious technology has been completed after a decade of effort. This technology will enable a wide range of science from the detection of Earths around nearby stars to the study of dark matter in the halo of our galaxy.

The research described in this paper was carried out at the Jet Propulsion Laboratory, California Institute of Technology, under a contract with the National Aeronautics and Space Administration.

References

Beichman, C. 2008, *Proceedings of IAU Symposium 248, in this volume p.238*
Chaboyer, B. 2008, *Proceedings of IAU Symposium 248, in this volume p.440*
Hines, B., *et al.* 2003, *SPIE, vol. 4852, p45*
Majewski, S. 2008, *Proceedings of IAU Symposium 248, in this volume p.450*
Unwin, S. 2008, *Proceedings of IAU Symposium 248, in this volume p.288*
Unwin, S. *et al.* 2007, *PASP, accepted*

A Giant Step: from Milli- to Micro-arcsecond Astrometry
Proceedings IAU Symposium No. 248, 2007
W. J. Jin, I. Platais & M. A. C. Perryman, eds.

© 2008 International Astronomical Union
doi:10.1017/S1743921308019169

Astrometric planet searches with SIM PlanetQuest

C. A. Beichman[1], S. C. Unwin[2], M. Shao[2], A. M. Tanner[1,2], J. H. Catanzarite[2] and G. W. Marcy[3]

[1] Michelson Science Center, California Institute of Technology, 770 S. Wilson Ave., Pasadena, CA 91125
email: chas@ipac.caltech.edu

[2] Jet Propulsion Laboratory, California Institute of Technology, 4800 Oak Grove Drive, Pasadena, CA 91109

[3] University of California, Berkeley, 417 Campbell Hall, Berkeley, CA 94720

Abstract. SIM will search for planets with masses as small as the Earth's orbiting in the 'habitable zones' around more than 100 of the nearest stars and could discover many dozen if Earth-like planets are common. With a planned "Deep Survey" of 100–450 stars (depending on desired mass sensitivity) SIM will search for terrestrial planets around all of the candidate target stars for future direct detection missions such as Terrestrial Planet Finder and Darwin. SIM's "Broad Survey" of 2100 stars will characterize single and multiple-planet systems around a wide variety of stellar types, including many now inaccessible with the radial velocity technique. In particular, SIM will search for planets around young stars providing insights into how planetary systems are born and evolve with time.

Keywords. astrometry, planets and satellites: formation, solar system: formation

1. Introduction

The existence of planets orbiting around other stars has been transformed from a philosophical question (Epicurus, ca. 300 BCE) and a scientific aspiration (Struve 1952) to an astronomical industry with more than 250 planets discovered via radial velocity measurements (Mayor *et al.* 1995; Marcy *et al.* 2005; Butler *et al.* 2006), transits (Henry *et al.* 2000) and microlensing (Beaulieu et al 2006). Each technique contributes to the long term goal of detecting and characterizing earth-like planets and ultimately finding signatures of life beyond our solar system (Beichman et al. 2007). This paper highlights astrometric searches for planets and the possibilities for developing a fuller understanding of the existence, formation and evolution of planetary systems using the Space Interferometer Mission (SIM PlanetQuest). An overview and a complete description of SIM's astrometric goals in all areas of astrophysics can be found in Shao *et al.* (this volume) and in Unwin *et al.* (2007), respectively. As will be described below, SIM will carry out three distinct programs: a deep search for rocky planets, a broad search for more massive planets orbiting a wide variety of host stars, and a search for planets around young stars. A modestly descoped version of SIM, nominally called SIM-Lite, can achieve the same measurement goals as SIM/PlanetQuest but for only half the number of objects.

2. The search for potentially habitable planets

SIM measures three crucial characteristics of a planet that suggest the circumstances of its birth, define its long term fate, and hint at its suitability for the existence of life: mass,

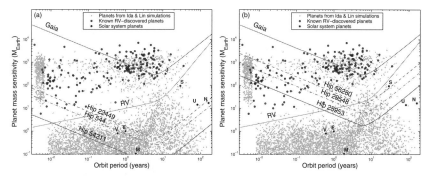

Figure 1. a, left) The discovery space for rocky Earth-like ($\sim1-10M_\oplus$) planets in the habitable zone (~ 0.7–1.5 AU for a G star), for an "Earth Analog Survey" of 129 stars; b, right) a comparable plot for the "Broad Survey" obtained with 4 μas accuracy. In both cases, the small dots represent a theoretical planet distribution Ida & Lin (2005) for planets of $0.1-3000M_\oplus$. Exoplanets discovered before early 2007 and with semimajor axes > 0.03 AU are shown as filled circles. Labeled curves represent the estimated sensitivity limits of indirect detection methods: for radial velocity method (RV at $1\,\mathrm{m\,s^{-1}}$), and astrometry with SIM and Gaia. The SIM sensitivity in this space is a broad band, defined by the three lowest curves (labeled with specific Hipparcos star numbers). The lowest curve shows the 'best' star (as computed from star mass and distance); the middle curve represents the median star; and the upper curve shows the least favorable star in the sample. Also shown is the effective sensitivity of Gaia for stars at 50 pc, a typical distance for Gaia targets.

the size and shape of its orbit, and its orbital inclination (where the latter quantity allows us to assess the critical question of co-planarity relative to other planets in the system). Perhaps the most important characteristic is mass. During the formation phase, an object with a mass greater 10 $M_\oplus$ will probably rapidly accrete enough volatiles to become an icy or gas giant. An object with a mass less than that of Mars is unlikely to retain volatiles to form an atmosphere or harbor oceans of liquid water. A planet's semi-major axis is also critical: too close to its parent star, stellar insolation and a runaway greenhouse will produce a Cytherean Hell, too far from the star will result in a world of perpetual ice. Secondary effects such as high eccentricity or interactions with other planets will adversely affect short and long term climatic stability and even continued membership in the planetary system. By finding planets of terrestrial mass in the habitable zones of their parent stars (Kasting *et al.* 1993) SIM will make a dramatic step toward realizing the goal of identifying other habitable worlds (Beichman *et al.* 2007).

The radial velocity (RV) technique will have great difficulty in detecting planets of $1\,M_\oplus$ orbiting solar type stars near 1 AU, as the RV signature of an earth analog is ~0.1 m s^{-1}compared with stellar surface jitter of ~1 m s^{-1}. While ESA's Gaia mission will detect many exoplanets (Lindegren and Sozzetti, this volume), Gaia's discovery space is that of gas-giant planets (Fig. 1). Furthermore, Gaia will achieve its highest precision for stars only fainter than ~7 mag due to detector effects, thereby eliminating the brightest, closest stars from the search for the lowest mass planets. *SIM offers a unique opportunity to detect Earth analogs, planets of one Earth mass in the habitable zones of nearby Solar-type stars.*

2.1. *Astrometric detection of terrestrial planets*

SIM's narrow-angle astrometry of each target star will be made relative to at least 3 reference stars selected to surround the target star within $1°\sim5°$. The reference stars are K giants brighter than $V = 10$ mag, within roughly 600 pc, so that the astrometric 'noise' due from planets orbiting their reference stars is minimized (Frink *et al.* 2001). A single

Table 1. Planet Mass-Limited Surveys with SIM

	Mass sensitivity	Number of stars surveyed
Survey 1	1.0 $M_\oplus$	129
Survey 2	2.0 $M_\oplus$	297
Survey 3	3.0 $M_\oplus$	465

measurement will achieve $0.85\,\mu$as differential measurement precision for $V = 7$ stars, including instrumental and photon-limited errors compared with the $0.3\ \mu$as signature of an Earth analog orbiting a solar type stars at 10 pc. Recent laboratory measurements indicate that SIM's systematic noise floor is below $0.1\ \mu$as after many repeated measurements, opening up the possibility of detecting Earth mass planets around over 100 of the closest stars. Starspots and other manifestations of stellar variability can affect the astrometric centroid resulting in a jitter similar to the spectroscopic noise that limits radial velocity precision. Models of astrometric jitter based on many years of precision photometry of the Sun indicate that these effects will not be the primary limitation for the detection of terrestrial planets around quiescent solar type stars (Unwin et al. 2007; Catanzarite, Shao & Law, in preparation). More realistic modeling for a variety of spectral types and stellar ages will be possible using data from the CoRoT and Kepler satellites.

2.2. *The SIM Deep Survey*

In an "Earth Analog Survey" SIM will study up to several hundred stars located within 30 pc in a manner designed to detect the lowest possible mass planets in the habitable zone of each star, $r_{HZ} \sim 1\ (L/L_\odot)^{0.5}$ AU. We allocate to each target enough integration time to allow a planet of $1\,M_\oplus$ to be detected at the radius of its mid-habitable zone (as determined from its spectral type). With an assignment of 40% of the observing time available in a 5-year mission, SIM can probe the mid-habitable zone of 129 stars for $1.0\,M_\oplus$ planets (Table 1). Over 450 stars can be observed in this same way to the $3.0\,M_\oplus$ level. For an alternative strategy based on allocating equal time to all stars see Catanzarite et al. (2006).

The discovery space (planet mass vs. orbit radius) for the "Earth Analog Survey" is shown in Figure 1a, with the 129 stars filling a band in the lower portion of the plot. Distributing the observing time over a larger target list allows one to detect more terrestrial planets, albeit at higher masses. Table 1 shows the expected SIM yield for three different values of the search depth. A survey to a sensitivity of $3M_\oplus$ would encompass more nearby stars than would likely be observed by the Terrestrial Planet Finder (TPF) Mission (Turnbull & Tarter 2003). In each survey, the mass sensitivity improves with orbit radius, out as far as orbits with periods $\lesssim 5$ yr (Fig. 1a).

Based on assumptions discussed in Unwin et al. (2007) and including estimates of the incidence of planets of various masses and orbital radii from Cumming et al. (2007), Figure 2a shows histograms for the input vs detected numbers of terrestrial planet masses in the "Earth Analog" survey. The histogram shows counts based on statistics of 1000 simulated surveys. In the habitable zone, SIM would detect 61% of all the terrestrial planets, including almost half all planets with masses in the range $1-1.5M_\oplus$, and nearly every planet of higher mass.

3. The SIM Broad Survey

In a "Broad Survey" using about 4% of a 5-year mission, some 2000 stars will be measured 100 times each with a single measurement accuracy of $4\,\mu$as. This census will

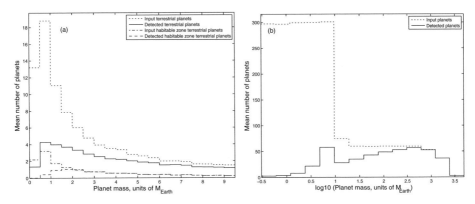

Figure 2. Histograms of the expected yield of terrestrial (a, left) and icy/gas giant (b, right) planets for the SIM Earth Analog and Broad Surveys. The histograms are the mean of 1000 simulated surveys in which geometric parameters of the model orbits were randomized. The mean input distributions of terrestrial ($M < 10M_\oplus$) and giant planets are shown as dotted curves, and the mean number of planets detected by SIM as solid curves. For terrestrial planets in the habitable zone, the corresponding curves are shown as dash-dotted and dashed respectively.

include over 2000 stars of all spectral types (including O,B,A and early F, which are not accessible to RV measurements), binary stars, stars with a broad range of age and metallicity, stars with dust disks, evolved stars, white dwarfs, and stars with planets discovered with RV surveys. Each class addresses specific features of the planet-formation process: *Are metals necessary for giant planet formation? Does the number of planets decline slowly with time due to dynamical evolution? What is the relation between dust disks and planets?*. The Broad Survey is expected to yield a large sample of hot, cold, rocky, ice giant and gas giant planets, as well as multiple-planet systems for tests of planet-formation theories. Figure 2b shows that relative to the assumed input population of planets (Unwin *et al.* 2007) SIM's Broad Survey will find 7% of the terrestrial planets, 2% of all terrestrial planets in the habitable zone, 47% of the Neptune-class planets and 87% of the Jupiter-class planets.

4. The SIM Young Planets survey

The "Young Planets" survey, targeted toward 150–200 stars with ages from 1 Myr to 100 Myr, will help us understand the formation and dynamical evolution of gas giant planets. The vast majority of the known planetary host stars are mature main sequence stars chosen for quiescent photospheres suitable for sensitive RV measurements. Similarly, stellar photospheres must be quiescent at the milli-magnitude level for transit detections. Since young stars have RV fluctuations or rotationally broadened line widths of hundreds of m s^{-1} and brightness fluctuations of many percent, optical RV or transit observations cannot be used to detect planets around young stars. While near-IR RV measurements may detect a few "hot Jupiters" and coronagraphic imaging may find a few distant Jupiters (> 25 pc), it is a fact that as a result of the (irreducible) sensitivity limits of these other techniques, *we know almost nothing about the incidence of planets within the region where these objects are actually thought to be forming.* SIM will take advantage of the fact that a Jupiter orbiting at 1 AU around a 0.8 $M_\odot$ star at the distance of the youngest stellar associations (1–10 Myr) such as Taurus (140 pc) and Chamaeleon would produce an astrometric amplitude of 8 μas and up to 20 μas at the 25–50 pc distance

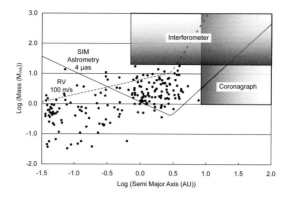

Figure 3. Planet mass detection sensitivity for the SIM Young Planet Survey (solid curve) in M_J versus orbital semi-major axis. Estimated capabilities for a large ground-based coronagraph (taken to be a diameter $d = 30$ m telescope at $\lambda = 1.6\,\mu$m operating an angular distance of $4\lambda/d$) and a near-IR interferometer (85-m baseline at $\lambda = 1.6\,\mu$m) are shown as shaded boxes. Also plotted are the properties of the known radial velocity (RV) detected planets (dots). The RV accuracy for young stars (dashed curve) is limited to about $100\,\mathrm{m\ s^{-1}}$. The astrometric and imaging sensitivity limits assume a distance of 140 pc.

of the nearest young stars (10–50 Myr) such as members of the β Pic and TW Hya groups. While some of the closest planetary systems will be detectable by Gaia, many of the youngest stars (<10 Myr) will require SIM's measurement accuracy. Figure 3 shows SIM's expected astrometric accuracy for the SIM "Young Planet Survey" as a function of planet mass and semi-major axis.

A SIM survey of 200 young stars could find anywhere from 10-20 (assuming that only the presently known fraction of stars, 5–10%, have planets) to 200 (all young stars have planets) planetary systems. These observations, when combined with the results of planetary searches of mature stars, will allow us to test theories of planetary formation and early Solar System evolution. By searching for planets around pre-main sequence stars carefully selected to span an age range from 1 to 100 Myr, we will learn at what epoch and with what frequency giant planets are found at the water-ice 'snowline' where they are expected to form (Pollack et al. 1996). This will provide insight into the physical mechanisms by which planets form and migrate from their place of birth, and about their survival rate.

Young stars are photospherically more active than main sequence stars, magnifying the concern expressed about starspots cited above for searches for terrestrial planets. A Monte Carlo analysis based on a starspot model (Beichman *et al.* 2001; Tanner *et al.* 2007) shows that for a typical T Tauri star radius of 3 $R_\odot$, the astrometric jitter is less than 3 μas for an R-band variability less than 0.05 mag. A selection of target stars and a precursor observing program to validate them (Tanner *et al.* 2007) has identified more than 75 stars well suited for observation by SIM out of an initial target list of 200 objects; variability greater than 0.1 mag removed many of the potential targets. More stars will be added to the precursor program to bring the total up to the desired number of 150−200 stars. SIM will make 100 visits to each star (up to 50 2-D visits) which, spread over 5 years, will be adequate to identify and characterize up to 3 planets per star having periods ranging from less than a year up to 2.5 years.

5. Conclusions

We have described the compelling use of the Space Interferometer Mission (SIM Planetquest) for the detection of planets beyond our solar system. SIM can survey over 100 nearby stars for 1 $M_\oplus$ planets in the habitable zone of their parent stars and over 300 stars at a sensitivity adequate to find 3 $M_\oplus$ planets. A descoped version of SIM, SIM-Lite, will be able to achieve these same goals around a smaller number of stars. No other mission is technologically ready to pursue such dramatic goals along the path to identifying other habitable planets around nearby solar analogs. SIM can lead the way for direct detection missions such as the Terrestrial Planet Finder and Darwin, identifying specific host stars and target planets, as well as providing essential information such as mass needed for detailed characterization. In addition to its unique role in finding and characterizing terrestrial planets, SIM will make unique contributions to the study of gas and icy planets around a wide variety of stars as well as enabling a study of newly formed planets that may help to unravel the processes of planetary system formation.

Acknowledgments

We gratefully acknowledge the contributions of the many dedicated scientists, technologists, engineers and managers. who have worked on SIM over the years. The research described in this paper was carried out at the Jet Propulsion Laboratory, California Institute of Technology, under contract with the National Aeronautics and Space Administration. This research has made use of the NASA/IPAC Infrared Science Archive, which is operated by the Jet Propulsion Laboratory, California Institute of Technology, under contract with the National Aeronautics and Space Administration.

References

Beaulieu, J.-P. *et al.* 2006, Nature, 439, 437.

Beichman, C. A., Fridlund, M., Traub, W. A., Stapelfeldt, K. R., Quirrenbach, A., Seager, S. 2007 in *Protostars and Planets V*, B. Reipurth, D. Jewitt, and K. Keil (eds.), University of Arizona Press, Tucson, p. 915.

Beichman, C. A. 2001, in *Young Stars Near Earth: Progress and Prospects,* ASP Conference Series Vol. 244. Edited by Ray Jayawardhana and Thomas Greene. San Francisco: Astronomical Society of the Pacific, p. 376.

Butler, R. P., Wright, J. T., Marcy, G. W., Fischer, D. A., Vogt, S. S., Tinney, C. G., Jones, H. R. A., Carter, B. D., Johnson, J. A., McCarthy, C., & Penny, A. J. 2006, *ApJ*, 646, 505

Catanzarite, J., Shao, M., Tanner, A., Unwin, S., & Yu, J. 2006, *PASP*, 118, 1322

Cumming, A., Marcy, G. W., Butler, R. P., *et al.* 2007, *ApJ*, in preparation

Frink, S., Quirrenbach, A., Fischer, D., Röser, S., & Schilbach, E. 2001, *PASP*, 113, 173

Henry, G. W., Marcy, G. W., Butler, R. P., Vogt, S. S. 2000, *ApJ*, 529, L41.

Ida, S., & Lin, D. N. C. 2005, *ApJ*, 626, 1045

Kasting, J. F., Whitmire, D. P., Reynolds, R. T. 1993, *Icarus*, 101, 108.

Marcy, G., Butler, R. P., Fischer, D., Vogt, S., Wright, J. T., Tinney, C. G., & Jones, H. R. A. 2005, Progress of Theoretical Physics Supplement, 158, 24

Mayor, M. & Queloz, D. 1995, Nature, 378, 355.

Pollack, J., Hubickyj, O., Bodenheimer, P., Lissauer, J. J., Podolak, M. Greenzweig, Y. 1996, Icarus, 124, 62

Struve, O. 1952, The Observatory, 72, 199.

Tanner, A., *et al.* 2007, *PASP*, 119, 747

Turnbull, M. C., & Tarter, J. C. 2003, *ApJ(S)*, 145, 181

Unwin, S. C., and the SIM Science Team 2007, *PASP*, in press.

A Giant Step: from Milli- to Micro-arcsecond Astrometry
Proceedings IAU Symposium No. 248, 2007
W. J. Jin, I. Platais & M. A. C. Perryman, eds.

© 2008 International Astronomical Union
doi:10.1017/S1743921308019170

Measuring proper motions of galactic dwarf galaxies with *Hubble Space Telescope*

S. Piatek[1] and C. Pryor[2]

[1] Department of Physics, New Jersey Institute of Technology,
Newark, NJ 07102-1982, USA
email: piatek@physics.rutgers.edu

[2] Dept. of Physics and Astronomy, Rutgers, the State University of New Jersey
136 Frelinghuysen Rd.
Piscataway, NJ 08854–8019, USA
email: pryor@physics.rutgers.edu

Abstract. Over the past several years, our research group has been measuring proper motions for nearby dwarf satellite galaxies using data taken with the *Hubble Space Telescope*. In order to measure proper motions with an expected size of several tens of milliarcseconds per century using a time baseline of 2-4 years, our work required that positions of stars and QSOs be measured to an accuracy of ∼0.25 mas (∼0.005 pixel). This contribution reviews the scientific justification of this work and our methodology. It concludes with a few general results and future directions.

Keywords. astrometry, galaxies: dwarf, Local Group, methods: data analysis

1. Introduction

The Milky Way and Andromeda are the two most luminous and massive galaxies in the Local Group (LG). Nearly all of the remaining galaxies are dwarfs whose luminosities and masses are orders of magnitude smaller. They tend to be concentrated around the two most massive members, implying that many of them are bound satellites. This hierarchical clustering is predicted by the ΛCDM models for galaxy formation. However, there have been proposals that at least some of these satellites are the remnants of more massive progenitors tidally disrupted by the Milky Way. Lynden-Bell & Lynden-Bell (1995) observed that some dwarf galaxies and globular clusters lie on great circles in the sky and, thus, proposed that they have similar orbits — forming streams of tidal debris. Kroupa et al. (2005) noted that 11 of the closest dwarf galaxies lie in a common plane and argue that this is inconsistent with the prediction of the ΛCDM model at the 99.5% level (though see Kang, et al. (2005) and Zentner *et al.* (2005) for an opposing view).

There are nearly two dozen known dwarf galaxies within 250 kpc of the Milky Way, with about half of them identified in the Sloan Digital Sky Survey (SDSS) data only in the past couple of years. At least a factor of two more dwarf galaxies are expected to be discovered when similar surveys are conducted in the southern hemisphere. Even if the number doubles or triples, their count would still be lower than several hundred dwarf galaxies predicted by ΛCDM simulations of the formation of the LG (Moore *et al.* 1999, Klypin *et al.* 1999).

Dwarf galaxies in the LG exhibit a wide range of properties. Their luminosities range from that of the least-luminous globular clusters to a factor of ∼10^6 greater. Stellar populations range from being single-age and metal-poor, such as that in Draco, to having multiple ages and a spread in metallicity, such as that in Carina. Deep imaging of the outer regions of dwarf galaxies suggests that the Galactic tidal field has disturbed their structure. These tidal effects range from minor, as in Carina and Ursa Minor, to severe, as in Sagittarius. If the destruction and eventual incorporation of Sagittarius into the Galactic halo is not a unique event, the disruption of dwarf satellites has contributed to the formation of the Galactic stellar halo. Tidal effects may also help to explain the wide range of properties of the dwarfs.

Knowing the orbits of dwarf galaxies around the Milky Way would help to clarify some of the issues discussed above and, in particular, test models for the formation of the LG. One of the most difficult steps in obtaining an orbit is to measure the proper motion: the change in the angular position of the galaxy over time.

2. Observational Challenge

Measuring a proper motion, μ, of a galaxy is very difficult for a number of reasons. Mathematically, $\mu = (46 \text{ milliarcsec century}^{-1})(v_t/220 \text{ km s}^{-1})(100 \text{ kpc}/d)$, where v_t is the tangential velocity of the galaxy and d is its heliocentric distance. Owing to the blurring caused by the Earth's atmosphere, time baselines of several tens of years are required to achieve a proper motion with an accuracy of about 10 mas century^{-1} with ground-based telescopes. Such long time baselines create their own problems: an astrometrist must contend with images taken with different telescopes, detectors, pointings, and atmospheric conditions. Despite these challenges, measurements based on ground-based imaging exist in the peer-reviewed literature for the Fornax (Dinescu *et al.* 2004), LMC (e.g., Pedreros *et al.* 2006), Sagittarius (Dinescu *et al.* 2005), Sculptor (Schweitzer *et al.* 1995), and SMC (e.g., Kroupa & Bastian 1997) dwarf galaxies. Measurements for a few additional dwarf galaxies are likely in the near future. Using ground-based data, reducing the uncertainties of these measured proper motions by extending time baselines is daunting and independent checks are impossible. Adaptive optics can reduce the atmospheric blurring, but it does not eliminate atmospheric refraction and the small corrected-field-of-view makes it difficult to tie such images to the earlier data.

The launch of the *Hubble Space Telescope (HST)* opened up a new avenue for measuring proper motions. Freed from atmospheric effects, the resolution of *HST* is diffraction limited and so the required time baselines were reduced from several tens of years to only a few years. See, for example, Kallivayalil *et al.* (2006a) and Piatek *et al.* (2007). Nevertheless, there are serious scientific, engineering, and economic trade offs. The small field of view of *HST* means that: only a few tens to a few hundred stars enter the analysis; a single QSO per field provides the standard of rest; and corrections for the internal motions of the galaxy may be necessary. Being in space, any technical problems with the telescope or the detectors are difficult to solve and, in fact, the best onboard imager, the Advanced Camera for Surveys (ACS), is currently out of service. Operating and servicing *HST* is costly. Although its life will likely be extended for another decade by the next servicing mission, it is finite and there are currently no plans for a comparable replacement. In the future, space-based astrometry will face the same problems of combining data from different detectors and telescopes that complicate ground-based astrometry.

3. Observing Method and Data Analysis

Observational setups have varied and the reader should refer to the corresponding published article for a given galaxy. However, a typical data set has the following features. The imaged field is centered on a known QSO which serves as an extragalactic standard of rest. The accuracy of the proper motion is almost always determined by the positional uncertainty of the QSO. In our programs, a $S/N \simeq 100$ has achieved an accuracy of about 0.25 mas (0.005 pixel). The exposure time is typically set so that the brightest pixel in the QSO's image is about a third full, providing some protection against the variability of QSOs. Since the FWHM of a stellar PSF is comparable to the size of a pixel in the CCD, the images are dithered to reduce the effects of undersampling. Dithering also prevents the complete failure of the measurement caused by the QSO always falling on a bad pixel. Typically there are at least 8 exposures per epoch, though the actual number ranges between 4 and 40.

We have used the WFPC2, STIS, ACS/HRC, and ACS/WFC imagers. All of these cameras introduce significant geometrical distortions which pose a problem for measuring the average motion of stars distributed throughout a field with respect to the QSO. The effect of errors in the distortion correction on the location of the QSO is minimized by using the same camera, filter, pointing, and orientation at all epochs for a given field. Unfortunately, the execution was never ideal: the pointing of the telescope could differ by several tens of pixels and the LMC and SMC programs allowed large rotations between epochs. With the demise of ACS and STIS, some of the fields are now being imaged with different detectors at different epochs. Although potentially helpful, multi-color imaging is not necessary. However, imaging in multiple fields — each centered on its own distinct QSO — is vital as this provides an independent check on the measured proper motions. Averaging the different measurements also reduces the final uncertainty.

The data analysis employed in our work has evolved over time and it has been adapted to the idiosyncrasies of particular data sets. Nevertheless, there are several canonical steps. 1) Derive the initial (x, y) coordinates for the stars and the QSO in each exposure of each epoch using a standard software package. 2) At each epoch, iteratively derive a single, constant effective point-spread function (ePSF) and apply it to produce progressively more accurate (x, y) coordinates of stars and the QSO for each exposure. It would be desirable to construct an ePSF that varies with position in the image for each exposure. However, our fields do not contain enough stars with sufficient S/N to do so. 3) Transform the (x, y) coordinates from the different exposures at a given epoch to a fiducial coordinate system and calculate the mean coordinate of every object. 4) Correct the coordinates of objects for the known geometrical distortions using available prescriptions. 5) Transform the mean coordinates of objects at the different epochs to a common coordinate system which is co-moving with the stars of the galaxy. Objects that are not members of the galaxy have a linear motion fitted in the transformation. The proper motion of the galaxy derives from the fitted motion of the QSO.

The last step above may also require correcting for the effects of the degrading charge transfer efficiency (CTE). A close inspection of images taken with *HST* shows "tails" behind stars, all pointing away from the serial register. They are one manifestation of the imperfect charge transfer between pixels during the readout, caused when some of the transferring charge is trapped for a period of time and then released. Since the amount of trapped charge depends on the number and distribution of "charge traps," the fractional loss decreases with increasing brightness of the star or QSO. Since the number of traps increases with time because of the damage from cosmic radiation, the effect also depends on the length of time the detector was in space. It is this time dependence that makes correcting for CTE necessary for precise astrometry. One method of correction is to "restore" images to their pre-readout state. Bristow & Alexov (2002) developed such software for images taken with STIS; unfortunately, no comparable software exists for other detectors. When using other detectors, the correction is made by including a term depending on time, S/N, and distance from the serial register in the transformation between epochs. This semi-empirical correction does not capture all of the physics of

CTE and so the authors think that a concerted effort is needed to understand CTE effects and to derive better ways of correcting them.

4. Results and Future Work

We have published proper motions measured with *HST* for the Carina, Fornax, Sculptor, and Ursa Minor galaxies. We have confirmed the Kallivayalil *et al.* (2006a, 2006b) measurements for the LMC and SMC, respectively. Our analysis detected the rotation of the LMC for the first time using proper motions. A paper giving a proper motion for Sagittarius derived from archival *HST* data is in preparation. Finally, second-epoch *HST* data for Draco and Sextans will have been taken by the end of 2007.

The most notable result so far from the above measurements is that orbits of the dwarf galaxies around the Milky Way tend to be polar, coplanar, and to have the same sense. However, there are exceptions. For example, the orbit of Sculptor is in the same plane but of the opposite sense. These results need to be strengthened with observations for more galaxies. Measurements for the galaxies recently discovered in the SDSS data will be particularly interesting, as they may explain the cause of their low surface brightnesses and luminosities. More measurements should also provide tighter constraints on both the shape of the Galactic potential and on our models for the formation of the LG.

Measurements for the newly-discovered galaxies will by challenging because they contain few stars and a search in the SDSS archive for bright QSOs behind them produced only a handful of possible candidates. We are searching for fainter AGN using longer exposures. An alternative approach is to use faint compact galaxies as reference points, which will necessitate much longer exposure times using *HST*.

The effects caused by degrading CTE will only get worse as the cameras on the *HST* age. We plan to develop better methods for correcting these effects. Ideally, such a software package would restore any image taken with *HST* given only the date of the exposure. We will also develop methods for combining data taken with different cameras, which will be increasingly more important as the *HST* archive grows and new cameras are installed.

References

Bristow, P. & Alexov, A. 2002, ST-ECF Instrument Science Report CE-STIS – 2002-001

Dinescu, D. I., Girard, T. M., van Altena, W. F., & Lopez, C. E. 2005, *ApJ*, 618, L25

Dinescu, D. I., Kinney, B. A., Majewski, S. R., & Girard, T. M. 2004, *AJ*, 128, 687

Kallivayalil, N., van der Marel, R. P., Alcock, C., Axelrod, T., Cook, K. H., Drake, A. J., & Geha, M. 2006a, *ApJ*, 638, 772

Kallivayalil, N., van der Marel, R. P., & Alcock, C. 2006b, *ApJ*, 652, 1213

Kang, X., Mao, S., Gao, L., & Jing, Y. P. 2005, *A&A*, 437, 383

Kroupa, P. & Bastian, U. 1997, *NewA*, 2, 77

Kroupa, P., Theis, C., & Boily, C. M. 2005, *A&A*, 431, 517

Klypin, A., Kravtsov, A. V., Valenzuela, O., & Prada, F. 1999, *ApJ*, 522, 82

Lynden-Bell, D. & Lynden-Bell, R. M. 1995, *MNRAS*, 275, 429

Moore, B., Diemand, J., Madau, P., Zemp, M., & Stadel, J. 2006, *MNRAS*, 368, 563

Pedreros, M., Costa, E., & Mendez, R. A. 2006, *AJ*, 131, 146

Piatek, S. *et al.* 2007, *AJ*, 133, 818

Schweitzer, A. E., Cudworth, K. M., Majewski, S. R., & Suntzeff, N. B. 1995, *AJ*, 110, 2747

Zentner, A. R., Kravtsov, A. V., Gnedin, O. Y., & Klypin, A. A. 2005, *ApJ*, 629, 219

A Giant Step: from Milli- to Micro-arcsecond Astrometry
Proceedings IAU Symposium No. 248, 2007
W. J. Jin, I. Platais & M. A. C. Perryman, eds.

Infrared space astrometry project JASMINE

N. Gouda[1], Y. Kobayashi[1], Y. Yamada[2], T. Yano[1] and JASMINE Working Group

[1]National Astronomical Observatory of JAPAN, Mitaka, Tokyo, 181-8588, JAPAN
email:naoteru.gouda@nao.ac.jp, yuki@merope.mtk.nao.ac.jp, yano.t@nao.ac.jp

[2]Department of Physics, Kyoto University, Kyoto, 606-8502, JAPAN
email:yamada@scphys.kyoto-u.ac.jp

Abstract. A Japanese plan of an infrared (z-band:0.9 μas or k-band:2.2 μas) space astrometry (JASMINE - project) is introduced. JASMINE (Japan Astrometry Satellite Mission for INfrared Exploration) will measure distances and tangential motions of stars in the bulge of the Milky Way. It will measure parallaxes, positions with an accuracy of 10 μas and proper motions with an accuracy of 10 μas/year for stars brighter than z=14 mag or k=11 mag. JASMINE will observe about ten million stars belonging to the bulge component of our Galaxy. With a completely new "map" of the Galactic bulge, it is expected that many new exciting scientific results will be obtained in various fields of astronomy. Presently, JASMINE is in a development phase, with a targeted launch date around 2016. Science targets, preliminary design of instruments, observing strategy, critical technical issues in JASMINE and also Nano-JASMINE project are described in this paper.

Keywords. infrared: general, space vehicles, astrometry, Galaxy: bulge

1. Outline and scientific targets of JASMINE mission

1.1. JASMINE project

JASMINE is an astrometric mission that observes in an infrared (z-band: central wavelength is 0.9 μm, or k-band: central wavelength is 2.2 μm). The k-band is a preferable candidate at the present time. It is designed to perform a survey towards the Galactic bulge, determining positions and parallaxes accurate to $\sim$10 μas for stars brighter than z=14 mag or k=11 mag, and proper motion errors of $\sim$ 10 μas/yr. JASMINE was originally proposed to survey the whole Galactic plane ($360° \times 7°$) with both a primary mirror of $\sim$ 1.5 m and a beam combiner(Gouda *et al.*(2004)). Now, a new version of JASMINE is proposed that only performs a survey towards the Galactic bulge, with a single-beam telescope. Investigations of the Galactic bulge are very important to clarify the structure and formation history of the Galaxy, the evolution of types of galaxies (*e.g.* spiral galaxy versus elliptical galaxy), the activity of the Galactic center and so on.

JASMINE can detect about one million bulge stars with $\sigma_\pi/\pi \leqslant 0.1$, about 1000 times more than the number of stars measured by Hipparcos (about 400 stars) in the survey area of JASMINE (see Fig. 1). Here π is a parallax of a star and σ_π is the observational error of the parallax. This was derived using our Galactic model whose original version was made by Cohen and his collaborators(Wainscoat *et al.*(1992)).

JASMINE will adopt a "frame-link" method which is explained in section 3. This method can be used when the number of stars in each field-of-view(small-field) is so large that a frame of the whole survey region can be made by the combined small-fields with the required accuracy. The star density in the Galactic bulge is found to be sufficient to apply this method.

There exist a number of major technical problems, such as telescope pointing stability, and thermal stability of the instruments. Investigations proceed in collaboration with JAXA(Japanese Aerospace Exploration Agency). A Lissajous orbit around the Sun-Earth Lagrange point L2 is a preferred option because this region provides a very stable thermal environment, minimization of eclipses, and so on. Also, other candidate orbits such as HCPO (High altitude Circular Polar Orbit) are studied. We hope that JASMINE will be launched around 2016. The mission lifetime will be 6 years.

1.2. *Science targets*

JASMINE will provide distances and transverse motions of stars in the Galactic bulge. Other space astrometry missions will accurately provide the structure of halo and disk of the Galaxy. Thus JASMINE will be complementary to Gaia, SIM and OBSS(refer to each URL in the reference list).

Bulges in spiral galaxies, similar in many ways to elliptical galaxies, are the key to study galaxy formation and the evolution of galaxy types. However, the size, shape and kinematical properties of the bulge in our Galaxy are pending problems. It is therefore important to investigate the 3-dimensional positions and motions of stars in the Galactic bulge.

Some observations(*e.g.* Eckart & Genzel(1996)) suggest that a super massive black hole exists in the center of the Galaxy; its growth seems strongly related to the growth of the bulge (*e.g.* Gebhardt *et al.*(2000)). Thus it is very interesting and important to clarify the growth of the bulge in the Galactic bulge. We believe these problems can be addressed, using the dynamical and kinematical structures obtained by JASMINE.

Furthermore the astrometric parameters of the many stars in the bulge, obtained by JASMINE, will have a large impact on the formations and evolutions of stars, and dust information in the Galactic bulge.

2. Preliminary instrument design

A candidate for the optics of the JASMINE telescope is a modified Korsch system with three mirrors and four folding flats to fit the focal length into the available volume . The telescope has a circular primary mirror with $\sim$ 75 cm diameter and 22.5 m(for z-band) or 12.3 m(for k-band) focal length.

The telescope provides a flat image plane that contains an array of large format detectors with a field of view of 0.7° × 0.7°. A total of 81 2k×2k CCDs(for z-band) or 16 2k×2k HgCdTe(for k-band) are read out.

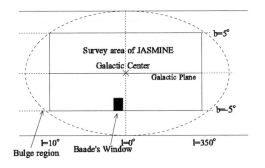

Figure 1. Survey area of JASMINE.

3. Observing strategy

3.1. *Procedures of observation*

JASMINE has a single telescope with a $0.7° \times 0.7°$ field of view. One field of view will be observed during a few seconds. One set of stellar images observed in a few seconds is called a "small-frame". In such a small-frame the relative positions (centroids) of stars will be calculated using an algorithm that determines the weighted mean of the positions of the photons registered by 5×5 pixels.

The telescope moves toward an adjacent field of view (small-frame) overlapping the previous small-frame (the overlap area is about a half of the frame size). In about 10 hours the telescope takes stellar images over the whole survey region ($20° \times 10°$), covering it by about 2000 small-frames. The small-frames are linked together by the many stars in the overlap regions. The whole survey region, linked together by small-frames is termed "a large-frame". The random error caused by linking small-frames is proportional to the number of small-frames in one-direction and inversely proportional to the square root of the number of common stars in the overlap region of each two adjacent small-frames. It is numerically found that the number density of measurable stars by JASMINE is large enough to attain the targeted accuracy of the link.

The above procedure is repeated during the whole mission life of about 6 years and finally about 10000 large-frames will be observed. Each large-frame covers the common survey area of $20° \times 10°$. Combining these large-frames allows to determine the 5 astrometric parameters with the targeted accuracy.

It should be noted that about a quarter of a year the direction towards the Sun from the spacecraft overlaps with that of the Galactic bulge. Observing the Galactic bulge is thus restricted to about 75 % of the total mission life.

3.2. *Systematic errors and technical issues*

In this sub-section, systematic errors that may affect JASMINE are briefly reviewed. Large systematic errors can occur in making a large-frame because small-frames may have different sizes and distortions due to variations of the telescope geometry according to temperature variations around the telescope. Hence very small temperature variations around the telescope and also minimal changes in telescope's geometry are mandated to reduce the systematic error. The use of QSOs measurable by JASMINE in the survey region will relax the requirement to the satellite system and the telescope. However variations of the telescope geometry smaller than about 30 pm within around 10 hours are required. This requirement will be reached by using ultra-low expansion optics and thermal structure of the satellite that ensure high stability of temperature around the telescope. These technical issues are now under investigation. Furthermore it is necessary to monitor small variations of the telescope geometry with high accuracies in order to verify the required small variations. Such monitoring systems are developed using laser interferometer technology.

Long term-contributors to JASMINE's error budget include changes in the scale of large-frames. This effect can be calibrated by QSOs.

There exist other systematic errors caused by radiation damage to the detectors, aberration, multiple stars, gravitational lensing effects, but these effects can be calibrated.

One QSO measurable by JASMINE has been already found. We have 14 QSO candidates measurable by JASMINE. Their true nature will be confirmed by planned spectroscopic observations.

4. Nano-JASMINE project

A Nano-JASMINE project is planned to demonstrate space astrometry in Japan and to examine some technical issues for JASMINE (Kobayashi *et al.*(2006);Suganuma *et al.*(2006)).Nano-JASMINE uses a nano-satellite whose size and weight are about 40 cm^3 and 14 kg, respectively. The size of the telescope is reduced to a 5cm aperture and a focal length of about 1.7m. One CCD with 1k × 1k pixels for z-band will be put in the focal plane. A candidate orbit for Nano-JASMINE is a sun-synchronous orbit. We will examine damage due to radiation on the CCD, on-board processing, thermal variations of the optical system and so on. The cost of Nano-JASMINE is low and it is expected to be launched around 2009. The development of the spacecraft is ongoing, in collaboration with Prof. Nakasuka and his group at the University of Tokyo.

5. Management and schedule

The establishment of a JASMINE working group at the JAXA was approved in 2003 by a science committee of ISAS (Institute of Space and Astronautical Science) of JAXA. JASMINE project office has been established at the NAOJ (National Astronomical Observatory of Japan) in 2004. We aim at a proposal for the JASMINE mission to JAXA, to get the launch approval and the required budget from the Japanese government about 4 years later. We hope that JASMINE will be launched around 2016. It may take long time to get the accurate astrometric parameters in the Galactic bulge. We would like to ask you for your continuous encouragement and cooperation.

References

Eckart, A. & Genzel, R. 1996, *Nature* **383**, 415.

Gebhardt, K., Karl, B., Ralf, B., Gary, D., Alan, F., Faber, S. M., Filippenko, A. V., Green, R., Grillmair, C., Carl, H., Luis, C., Kormendy, J., Lauer, T. R., Magorrian, J., Pinkney, J., Richstone, D., & Tremaine, S. 2000, ' *ApJ*, **539**, L13, 2000.

Gouda, N., Yano, T. , Kobayashi, Y., Yamada, Y., Tsujimoto, T., Nakajima, T., Suganuma, M., Matsuhara, H., Ueda, S. & JASMINE Working Group. 2004, in:D.W.Kurtz(eds.), *Transits of Venus: New Views of the Solar System and Galaxy*, Proc.IAU Symposium No.196(Cambridge University Press), p. 455.

Kobayashi, Y., Gouda, N., Tsujimoto, T., Yano, T., Suganuma, M., Yamauchi, M., Takato, N., Miyazaki, S., Yamada, Y., Sako, N., & Nakasuka, S. 2006, in *Proceedings of the SPIE*, **6265** J. Mather etc., ed., 626544-1–626544-10.

Suganuma, M., Kobayashi, Y., Gouda, N., Yano, T., Yamada, Y., Takato, N., & Yamauchi, M. 2006, in *Proceedings of the SPIE*, **6265** J. Mather etc., ed., 626545-1–626545-12.

Wainscoat, R. J., Cohen, M., Volk, K., Walker, H. J. & Schwartz, D. E. 1992, *ApJS*, **83**, 111–146.

GAIA:http://www.rssd.esa.int/index.php?project=GAIA

SIM:http://planetquest.jpl.nasa.gov/SIM/

OBSS:http://ad.usno.navy.mil/OBSS/

JASMINE:http://www.jasmine-galaxy.org/index.html

A Giant Step: from Milli- to Micro-arcsecond Astrometry
Proceedings IAU Symposium No. 248, 2007
W. J. Jin, I. Platais & M. A. C. Perryman, eds.

© 2008 International Astronomical Union
doi:10.1017/S1743921308019194

Determination of the barycentric velocity of an astrometric satellite using its own observational data

A. G. Butkevich[1]† and S. A. Klioner[2]

Lohrmann Observatory, Dresden Technical University,
01062 Dresden, Germany
[1]email: alexey.butkevich@tu-dresden.de
[2]email: sergei.klioner@tu-dresden.de

Abstract. The problem of determination of the orbital velocity of an astrometric satellite from its own observational data is studied. It is well known that data processing of microarcsecond-level astrometric observations imposes very stringent requirements on the accuracy of the orbital velocity of satellite (a velocity correction of 1.45 mm/s implies an aberrational correction of 1 μas). Because of a number of degeneracies the orbital velocity cannot be fully restored from observations provided by the satellite. Seven constraints that must be applied on a velocity parameterization are discussed and formulated mathematically. It is shown what part of velocity can be recovered from astrometric data using a combined fit of both velocity parameters and astrometric parameters of the sources. Numerical simulations show that, with the seven constraints applied, the velocity and astrometric parameters can be reliably estimated from observational data. It is also argued that the idea to improve the velocity of an astrometric satellite from its own observational data is only useful if a priori information on orbital velocity justifies the applicability of the velocity constraints. The proposed model takes into account only translational motion of the satellite and ignores any satellite-specific parameters. Therefore, the results of this study are equally applicable to both scanning missions similar to Gaia, and pointing ones like SIM, provided that enough sources were observed sufficiently uniformly.

Keywords. astrometry, methods: data analysis, ephemerides, reference systems

1. Introduction

The astrometric accuracy of several microarcseconds announced for three space missions (Gaia, SIM and Jasmine) implies that the velocity of the satellite should be known within several mm/s. Such a high precision is a challenge to standard orbit determination techniques.

Another way to proceed was discussed in the Gaia community since 2001 and formulated in the written form by Klioner (2005). The idea is to use Gaia's own astrometric data to fit a correction to the Gaia velocity. The velocity correction $\delta\mathbf{v}$ is the difference between the real velocity of the satellite and its ephemeris velocity available a priori. A straightforward theoretical analysis of this idea (Butkevich 2006) demonstrated that velocity can be determined from observations with the required precision, provided that all other parameters are exactly known. Some criticism against this approach has been formulated by Bastian (2004a) who argued that the velocity obtained from observations will strongly correlate with the source parameters. A detailed theoretical study of this problem has been done by Klioner & Butkevich (2007), who explicitly demonstrated that the velocity correction to be fitted from the data must satisfy some constraints.

---

† On leave from Pulkovo Observatory, 196140 Saint-Petersburg, Russia.

2. What part of observer's velocity can be restored from astrometric data?

Let us consider the case when the velocity correction has a constant component:

$$\delta \mathbf{v} = \mathbf{v}_0 = \text{const}, \tag{2.1}$$

then this additional velocity leads to the following aberrational correction

$$\delta \mathbf{u} \approx -\mathbf{u} \times [\mathbf{u} \times \mathbf{v}_0/c]. \tag{2.2}$$

This correction depends on the stellar position $\mathbf{u}$, but for each star it is constant. Such a signal in $\delta \mathbf{v}$ is equivalent to a constant change of positions, which cannot be detected from observations since it cannot be distinguished from their *real* change.

Similarly, for a velocity correction changing linearly with time,

$$\delta \mathbf{v} = \mathbf{a}_0 t \quad \mathbf{a}_0 = \text{const}, \tag{2.3}$$

we have

$$\delta \mathbf{u}(t) \approx \boldsymbol{\mu}_0 t, \tag{2.4}$$

where

$$\boldsymbol{\mu}_0 = -\mathbf{u} \times [\mathbf{u} \times \mathbf{a}_0/c]. \tag{2.5}$$

This correction is equivalent to an additional constant proper motion for each star. Such a correction again cannot be detected from observations since it cannot be distinguished from a *real* change of the proper motion parameters for each source.

If $\delta \mathbf{v}$ is exactly proportional to the barycentric position of the satellite $\mathbf{r}(t)$:

$$\delta \mathbf{v}(t) = \alpha_0 \mathbf{r}(t) \quad \alpha_0 = \text{const}, \tag{2.6}$$

the corresponding first-order aberrational correction reads

$$\delta \mathbf{u} \approx -\mathbf{u} \times [\mathbf{u} \times \alpha_0 \mathbf{r}/c]. \tag{2.7}$$

On the other hand, the aberrational effect caused by a global offset of parallaxes $\delta \pi$ is

$$\delta \mathbf{u} = \mathbf{u} \times [\mathbf{u} \times \delta \pi \mathbf{r}/\text{AU}]. \tag{2.8}$$

These effects are indistinguishable provided that

$$\delta \pi = -\alpha_0 \, \text{AU}/c. \tag{2.9}$$

Thus the problem has seven free parameters (α_0 and six components of $\mathbf{v}_0$ and $\mathbf{a}_0$), which correlate with some astrometric information, i. e. it has seven degrees of freedom. This rank deficiency makes the direct determination of velocity impossible. It can be, however, demonstrated that the degeneracy can be eliminated if the following constraints would be imposed onto the solution:

$$\int_0^T \delta \mathbf{v}(t)\, \mathrm{d}t = 0 \quad \text{to remove} \quad \delta \mathbf{v} = \mathbf{v}_0, \tag{2.10}$$

$$\int_0^T (t - T/2)\, \delta \mathbf{v}(t)\, \mathrm{d}t = 0 \quad \text{to remove} \quad \delta \mathbf{v} = \mathbf{a}_0 t, \tag{2.11}$$

$$\int_0^T \frac{\delta \mathbf{v}(t)\, \mathbf{r}(t)}{|\mathbf{r}(t)|^2}\, \mathrm{d}t = 0 \quad \text{to remove} \quad \delta \mathbf{v} = \alpha_0 \mathbf{r}. \tag{2.12}$$

Table 1. Singular values of the normal matrix.

n	1	:	7	8	:	max
σ_n	$9.6 \cdot 10^{-8}$	:	$9.0 \cdot 10^{-6}$	2.1	:	1045

Table 2. Error in parameters.

Parameter	α	δ	μ_α	μ_δ	π	δv_x	δv_y	δv_z
Error	1.6 μas	1.5 μas	1.3 μas/yr	1.2 μas/yr	1.5 μas	1.8 mm/s	1.5 mm/s	1.6 mm/s

These constraints guarantee that the solution does not contain the relevant signals in the sense of least-squares.

Although it can hardly be proven analytically that the problem has no other degrees of freedom, it can be checked numerically. We calculated singular value decomposition (SVD) of a normal matrix and found only seven small singular values shown in Table 1. This indicates that no other degree of freedom exists

3. Legitimacy of the constrained velocity

The constraints may only be applied when a priori accuracy of the ephemeris guarantees that no signal of a given kind can exist in real velocity, or, strictly speaking, it is so small that any effect due to it can be neglected.

The *real* $\delta\mathbf{v}_{\mathrm{real}}$ velocity correction can be represented as a sum of the two different components

$$\delta\mathbf{v}_{\mathrm{real}} = \delta\mathbf{v} + \mathbf{R}(t) , \qquad (3.1)$$

where $\delta\mathbf{v}$ is the component that can be fitted and $\mathbf{R}(t)$ is the component violating the seven constraints. The fitted component $\delta\mathbf{v}$ is useful if and only if the uncertainty of the ephemeris velocity is such that it guarantees that

$$|\mathbf{R}(t)| < \epsilon \qquad (3.2)$$

at any instant of time. Here ϵ is a required velocity accuracy (for Gaia, $\epsilon = 1$ mm/s).

Fortunately, this can be demonstrated for Gaia (Klioner & Butkevich, 2007) but cannot be guaranteed for other missions.

4. Results of numerical simulations

To study the problem of the velocity determination numerically, we have implemented a simple simulator of Gaia observations - Dresden Gaia Simulator (DGS) - that includes many basic features of a real mission. Table 2 shows the results of a simulation run with 2048 stars covering 5 years of observations (Butkevich & Klioner 2007a). The accuracy of an individual observation was chosen to be 30 μas. The small errors found suggest that the constrained solution allows one to achieve a precise and reliable determination of velocity and source parameters.

5. Assessment of the accuracy of velocity determination

Our simulation has one serious drawback – it can only handle limited datasets and never approaches the data volume near to the expected Gaia output. A straightforward

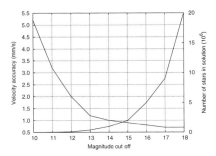

Figure 1. Dependence of the velocity accuracy (the descending curve) and the number of stars in the astrometric solution (the ascending curve) on the limiting magnitude.

estimate can be obtained using simple statistical considerations (Butkevich & Klioner 2007b). One of the critical parameters in the velocity determination is the number of stars used in the solution, which depends on the apparent magnitude cut-off. The estimated velocity accuracy together with relevant star counts are shown in Fig. 1 for the limiting V magnitudes. Besides the number of stars, the accuracy also depends on the temporal resolution of the fitted velocity correction. The accuracy obviously degrades when a finer resolution is used. The time scale of velocity variations was chosen to be 6 hours, close to the rotation period of Gaia. From our estimates, we conclude that at least 10^6 stars are needed to obtain velocity with an accuracy of 1 mm/s.

6. Satellite specific parameters

The discussed model takes into account only translational motion of the satellite and ignores any satellite-specific calibration parameters (e. g. attitude parameters). The situation may become more complicated when those other parameters are also considered. Our analysis shows that for Gaia the scientifically important parameters can be successfully restored even in this case.

Acknowledgements

We acknowledge useful discussions with Ulrich Bastian and Lennart Lindegren. This work was partially supported by the BMWi grant 50 QG 0601 awarded by the Deutsche Zentrum für Luft- und Raumfahrt e.V. (DLR).

References

Bastian, U. 2004a, Improving Gaia's orbit with Gaia's astrometry?, GAIA-ARI-BAS-007

Butkevich, A. G. 2006, On the velocity determination from observational data, GAIA-CA-TN-LO-AGB-001-1

Butkevich, A. G. & Klioner, S. A. 2007a, On the simultaneous determination of velocity correction and source parameters, GAIA-CA-TN-LO-AGB-002-1

Butkevich, A. G. & Klioner, S. A. 2007b, Assessing the accuracy of the velocity determination, GAIA-CA-TN-LO-AGB-005-1

Klioner, S. 2005, On the possibility to improve the velocity of Gaia from the Gaia's own astrometric data, available from Gaia Livelink

Klioner, S. A. & Butkevich, A. G. 2007, What part of observer's velocity can be restored from astrometric data?, GAIA-CA-TN-LO-SK-001-2

A Giant Step: from Milli- to Micro-arcsecond Astrometry
Proceedings IAU Symposium No. 248, 2007
W. J. Jin, I. Platais & M. A. C. Perryman, eds.

Testing planet formation models
with Gaia μas astrometry

A. Sozzetti[1,2], S. Casertano[3], M. G. Lattanzi[1], A. Spagna[1], R. Morbidelli[1], R. Pannunzio[1], D. Pourbaix[4] and D. Queloz[5]

[1]INAF- Astronomical Observatory of Torino, via Osservatorio 20, I-10025, Pino Torinese
(TO), Italy
email: sozzetti@oato.inaf.it

[2]Harvard-Smithsonian Center for Astrophysics, 60 Garden Street, Cambridge, MA 02138, USA

[3]Space Telescope Science Institute, San Martin Drive, Baltimore, MD 21218, USA

[4]Institut d'Astronomie et d'Astrophysique, Université Libre de Bruxelles, CP. 226, Boulevard
du Triomphe, 1050 Bruxelles, Belgium

[5]Observatoire de Genève, 51 Ch. de Maillettes, 1290 Sauveny, Switzerland

Abstract. In this paper, we first summarize the results of a large-scale double-blind tests campaign carried out for the realistic estimation of the Gaia potential in detecting and measuring planetary systems. Then, we put the identified capabilities in context by highlighting the unique contribution that the Gaia exoplanet discoveries will be able to bring to the science of extrasolar planets during the next decade.

Keywords. planetary systems, astrometry, methods: data analysis, methods: numerical, stars: statistics

1. Introduction

Despite a few important successes (e.g., Bean *et al.* 2007, and references therein), astrometric measurements with a mas precision have so far proved to be of a limited value when employed as either a follow-up tool or for independent searches of planetary-mass companions orbiting nearby stars (e.g., Sozzetti 2005, and references therein).

In several past exploratory works (Casertano *et al.* 1996; Lattanzi *et al.* 1997, 2000; Sozzetti *et al.* 2001, 2003), we have shown in some detail what space-borne astrometric observatories with μas-level precision, such as Gaia (Perryman *et al.* 2001), can achieve in terms of search, detection and measurement of extrasolar planets in the mass range from Jupiter-like to Earth-like. In those studies we adopted a qualitatively correct description of the measurements that each mission will carry out, and we estimated detection probabilities and orbital parameters using realistic, non-linear least squares fits to those measurements.

Those exploratory studies, however, need updating and improvements. In the specific case of planet detection and measurement with Gaia, we have thus far largely neglected the difficult problem of selecting adequate starting values for the non-linear fits, using perturbed starting values instead. The study of multiple-planet systems, and in particular the determination of whether the planets are coplanar—within suitable tolerances—is incomplete. The characteristics of Gaia have changed, in some ways substantially, since our last work on the subject (Sozzetti *et al.* 2003). Last but not least, in order to render the analysis truly independent from the simulations, these studies should be carried out in double-blind mode.

We present here a substantial program of double-blind tests for planet detection with Gaia (preliminary findings were presented by Lattanzi *et al.* (2005)), with the three-fold goal of obtaining: a) an improved, more realistic assessment of the detectability and measurability of single and multiple planets under a variety of conditions, parametrized by the sensitivity of Gaia; b) an assessment of the impact of Gaia in critical areas of planet research, and dependence on its expected capabilities; and c) the establishment of several Centers with a high level of readiness for the analysis of Gaia observations relevant to the study of exoplanets.

2. Double-blind tests campaign results

We carry out detailed simulations of Gaia observations of synthetic planetary systems and develop and utilize in double-blind mode independent software codes for the analysis of the data, including statistical tools for planet detection and different algorithms for single and multiple Keplerian orbit fitting that use no a priori knowledge of the true orbital parameters of the systems.

Overall, the results of our earlier works (e.g., Lattanzi *et al.* 2000; Sozzetti *et al.* 2001, 2003) are essentially confirmed, with the fundamental improvement due to the successful development of independent orbital fitting algorithms applicable to real-life data that do not utilize a priori knowledge of the orbital parameters of the planets. In particular, the results of the T1 test (planet detection) indicate that planets down to astrometric signatures $\alpha \simeq 25$ μas, corresponding to ~ 3 times the assumed single-measurement error, can be detected reliably and consistently, with a very small number of false positives (depending on a specific threshold for detection). The results of the T2 test (single-planet orbital solutions) indicate that: 1) orbital periods can be retrieved with very good accuracy (better than 10%) and small bias in the range of $0.3 \lesssim P \lesssim 6$ yr. In this period range the other orbital parameters and the planet's mass are similarly well estimated. The quality of the solutions degrades quickly for periods longer than the mission duration, and the fitted value of P is systematically underestimated; 2) uncertainties in orbit parameters are well understood; 3) nominal uncertainties obtained from the fitting procedure are a good measure of the actual errors in the orbit reconstruction. Modest discrepancies between estimated and actual errors arise only for planets with extremely good signal (errors are overestimated) and for planets with very long period (errors are underestimated); such discrepancies are of interest mainly for a detailed numerical analysis, but they do not address adequately the assessment of Gaia's ability to find planets and our preparedness for the analysis of perturbation data. The results of the T3 test (multiple-planet orbital solutions) indicate that 1) over 70% of the simulated orbits under the conditions of the T3 test (for every two-planet system, periods shorter than 9 years and differing by at least a factor of two, $2 \leqslant \alpha/\sigma_\psi \leqslant 50$, $e \leqslant 0.6$) are correctly identified; 2) favorable orbital configurations (both planets with periods $\leqslant 4$ yr and astrometric signal-to-noise ratio $\alpha/\sigma_\psi \geqslant 10$, redundancy of over a factor of 2 in the number of observations) have periods measured to better than 10% accuracy over 90% of the time, and comparable results hold for other orbital elements; 3) for these favorable cases, only a modest degradation of up to 10% in the fraction of well-measured orbits is observed with respect to single-planet solutions with comparable properties; 4) the overall results are mostly insensitive to the relative inclination of pairs of planetary orbits; 5) over 80% of the favorable configurations have $i_{\rm rel}$ measured to better than 10 degrees accuracy, with only mild dependencies on its actual value, or on the inclination angle with respect to the line of sight; 6) error estimates are generally accurate, particularly for fitted parameters, while modest discrepancies (errors are systematically underestimated) arise between formal and actual errors on $i_{\rm rel}$.

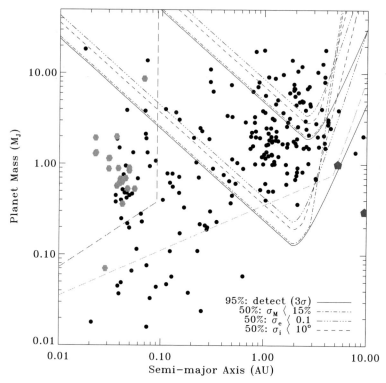

Figure 1. Gaia discovery space for planets of given mass and orbital radius compared to the present-day sensitivity of other indirect detection methods, namely Doppler spectroscopy and transit photometry. Red curves of different styles (for completeness in planet detection and orbit measurement to given accuracy) assume a 1-$M_\odot$ G dwarf primary at 200 pc, while the blue curves are for a 0.5-$M_\odot$ M dwarf at 25 pc. The radial velocity curve (pink line) is for detection at the $3 \times \sigma_{RV}$ level, assuming $\sigma_{RV} = 3$ m s^{-1}, $M_\star = 1M_\odot$, and 10-yr survey duration. For transit photometry (green curve), $\sigma_V = 5$ milli-mag, $S/N = 9$, $M_\star = 1$ $M_\odot$, $R_\star = 1$ $R_\odot$, uniform and dense (> 1000 data points) sampling. Black dots indicate the inventory of exoplanets as of October 2007. Transiting systems are shown as light-blue filled pentagons. Jupiter and Saturn themselves are shown as red pentagons.

3. The Gaia expectations

In Fig. 1 we show Gaia's discovery space in terms of detectable and measurable planets of a given mass and orbital separation around stars of a given mass at a given distance from Earth (see caption for details). From this Figure, one would then conclude that Gaia could discover and measure massive giant planets ($M_p \gtrsim 2 - 3$ M_J) with $1 < a < 4$ AU orbiting solar-type stars as far as the nearest star-forming regions, as well as explore the domain of Saturn-mass planets with similar orbital semi-major axes around late-type stars within 30-40 pc. These results can be turned into a number of planets detected and measured by Gaia, using Galaxy models and the current knowledge of exoplanet frequencies. By inspection of Tables 1 and 2, we then find that Gaia could measure accurately thousands of giant planets, and accurately determine coplanarity (or not) for a few hundred multiple systems with a favorable configuration.

In conclusion, Gaia's main strength continues to be the ability to measure actual masses and orbital parameters for possibly thousands of planetary systems. The Gaia data have the potential to a) significantly refine our understanding of the statistical properties of extrasolar planets: the predicted database of several thousand extrasolar planets with

Table 1. Number of giant planets that could be detected and measured by Gaia, as a function of increasing distance. Star counts are obtained using the Besancon model of stellar population synthesis (Bienaymé *et al.* 1987), while the Tabachnik & Tremaine (2002) model for estimating planet frequency as a function of mass and orbital period is utilized.

Table 2. Number of multiple-planet systems that Gaia could potentially detect, measure, and for which coplanarity tests could be carried out successfully.

Δd (pc)	$N_\star$	Δa (AU)	ΔM_p (M_J)	N_d	N_m
0-50	1×10^4	1.0 - 4.0	1.0 - 13.0	1400	700
50-100	5×10^4	1.0 - 4.0	1.5 - 13.0	2500	1750
100-150	1×10^5	1.5 - 3.8	2.0 - 13.0	2600	1300
150-200	3×10^5	1.4 - 3.4	3.0 - 13.0	2150	1050

Case	Systems
Detection	~ 1000
Orbits and masses to better than 15-20% accuracy	$\sim 400 - 500$
Successful coplanarity tests	~ 150

well-measured properties will allow for example to test the fine structure of giant planet parameters distributions and frequencies, and to investigate their possible changes as a function of stellar mass with unprecedented resolution; b) help crucially test theoretical models of gas giant planet formation and migration: for example, specific predictions on formation time-scales and the role of varying metal content in the protoplanetary disk will be probed with unprecedented statistics thanks to the thousands of metal-poor stars and hundreds of young stars screened for giant planets out to a few AUs; c) improve our understanding of the role of dynamical interactions in early as well as long-term evolution of planetary systems. For example, the measurement of orbital parameters for hundreds of multiple-planet systems, including meaningful coplanarity tests will allow to discriminate between various proposed mechanisms for eccentricity excitation; d) aid in the understanding of direct detections of giant extrasolar planets. For example, actual mass estimates and full orbital geometry determination for suitable systems will inform direct imaging surveys about where and when to point, in order to estimate optimal visibility, and will help in the modeling and interpretation of giant planets' phase functions and light curves; e) provide important supplementary data for the optimization of the target selection for Darwin/TPF: for example, all F-G-K-M stars within the useful volume (~ 25 pc) will be screened for Jupiter- and Saturn-sized planets out to several AUs, and these data will help probing the long-term dynamical stability of their Habitable Zones, where terrestrial planets may have formed, and maybe found.

References

Bean, J. L., *et al.* 2007, *AJ*, 134, 749
Bienaymé, O., Robin, A. C., & Crézé, M. 1987, *A&A*, 180, 94
Casertano, S., Lattanzi, M. G., Perryman, M. A. C., & Spagna, A. 1996, *AP&SS*, 241, 89
Lattanzi, M. G., Spagna, A., Sozzetti, A., & Casertano, S. 2000, *MNRAS*, 317, 211
Lattanzi, M. G., Casertano, S., Jancart, S., Morbidelli, R., Pannunzio, R., Pourbaix, D., Sozzetti, A., & Spagna, A. 2005, *ESA SP*-576, 251
Perryman, M. A. C., *et al.* 2001, *A&A*, 369, 339
Sozzetti, A., Casertano, S., Lattanzi, M. G., & Spagna A. 2001, *A&A* (Letters), 373, L21
Sozzetti, A., Casertano, S., Lattanzi, M. G., & Spagna A. 2003, *ESA SP*-539, 605
Sozzetti, A. 2005, *PASP*, 117, 1021
Tabachnik, S., & Tremaine, S. 2002, *MNRAS*, 335, 151

A Giant Step: from Milli- to Micro-arcsecond Astrometry
Proceedings IAU Symposium No. 248, 2007
W. J. Jin, I. Platais & M. A. C. Perryman, eds.

A Gaia oriented analysis of a large sample of quasars

A. H. Andrei[1,2,3], M. Assafin[2], C. Barache[3], S. Bouquillon[3], G. Bourda[4], J. I. B. Camargo[2], J.-F. le Campion[4], P. Charlot[4], A.-M. Gontier[3], S. Lambert[5], J.J. Pereira Osório[6], D.N. da Silva Neto[2,7], J. Souchay[3] and R. Vieira Martins[1]

[1] Observatório Nacional/MCT
R. Gal. JosCristino 77, RJ, Br asil
email: oat1@on.br

[2] Observatório do Valongo/UFRJ, Brasil

[3] SYRTE/Observatoire de Paris, France

[4] Observatoire de Bordeaux, France

[5] Observatoire Royal de Belgique

[6] Observatório Astronico da Universidade do Porto, Portugal

[7] Universidade Estadual da Zona Oeste/RJ, Brasil

Abstract. Gaia photometric capabilities should distinguish quasars to a high degree of certainty. With this, they should also be able to deliver a clean sample of quasars with a negligible trace of stellar contaminants. However, a purely photometric sample could miss a non negligible percentage of ICRF sources counterparts - and this interface is required to align with the ICRS and de-rotate the GCRF (Gaia Celestial Reference Frame), on grounds of continuity. To prepare a minimum clean sample forming the initial quasar catalogue for the Gaia mission, an all sky ensemble was formed containing 128,257 candidates. Among them there is at least one redshift determination for 98.75%, and at least one magnitude determination for 99.20% of the targets. The sources were collected from different optical and radio lists. We analyze the redshift, magnitude, and color distributions, their relationships, as well as their degree of completeness.

Complementary, the candidate sources enable to form an optical representation of the ICRS from first principles, namely, kinematically non-rotating with respect to the ensemble of distant extragalactic objects, aligned to the mean equator and dynamical equinox of J2000, and realized by a list of adopted coordinates of extragalactic sources.

Keywords. astronomical data bases: miscellaneous, astrometry, reference systems, quasars: general

1. Gaia's Initial QSO catalogue and clean sample

The establishment of an initial quasar catalogue for the Gaia mission must fulfill criteria of quantity and sky distribution. The expected number of quasars to be detected by Gaia is about 400,000 objects. That is about four times the number of QSOs presently known. A much smaller number, between 6,000 to 10,000, is required for definition of the fundamental frame. This sets the minimum requirements for the initial quasar catalogue clean sample, that is to zero level of contaminants. Since the all sky distribution (exempting the galactic disk) is also desirable, direct observations are out of question and the pre-existing quasar catalogues ought to be scrutinized.

As the first step the common entries are sorted and a consolidated list is produced, including a reliability estimator and the radio astrometric accuracy. This consolidated list, plus information on redshift, multi-band magnitude and radio flux is the aim of the VLQAC (Souchay *et al.*, 2007, and references therein)

Since the consolidated list in none of its constituents derives from apparent magnitude limits in the G band, a check must then be performed to test each source G magnitude. The B1.0 and GSC2.3 catalogs present the whole sky complete up to magnitude V=20 (though reaching beyond in some zones). For most of the sky they contain the B, R, and I magnitudes, and so the G magnitude can be acceptably derived. With this the list is trimmed of its weaker sources (which will nevertheless be flagged as such and then kept), which would give rise to poor Gaia centroid determination.

Next, from available optical images, the objects PSF is compared to the stellar neighbors, in order to reject no pointlike objects. Finally, the objects observational history is followed, to the effect of retaining a core of (at least) double-checked quasars. The objects passing through all checks form the clean sample. They are further flagged according to their history of stable radio emission.

However, the clean sample does not meet the even sky distribution desirable. In order to densify the initial quasar list the sources failing to enter the clean sample are included, and accordingly flagged to, based on two additional tests. The assured quality of the single observation reported. The placement in the QSO color loci. These additional criteria involve a one-by-one examination of sources resulting to a slower pace of each decision. Currently about one thousand of objects are being examined.

2. Optical celestial reference frame

The construction of the OCRF (Optical Celestial Reference Frame, Andrei *et al.*, 2007, and references therein) starts from the updated and presumably complete VLQAC (Very Large Quasar Catalog) list of QSOs, initial optical positions for those quasars are found in the USNO B1.0 and GSC 2.3 catalogues, and from the SDSS Data Release 5. The initial positions are next placed onto UCAC2 based reference frames, following by an aligment to the ICRF, as represented by the optical counterpart of ICRF sources as well as of the most precise sources from the VLBA calibrator list and from the VLA calibrator list. Finally the OCRF axis are surveyed through spherical harmonics, contemplating right ascension, declination and magnitude terms.

The OCRF contains J2000 refered equatorial coordinates of 105,000 quasars, well represented on all-sky basis, from −88.5 to +89.5 degree of declination, and with 0.5deg as the average distance between adjacent elements. The global alignment to the ICRF is of 1.5mas, and the individual position accuracies are represented by 80mas + 0.1R (where R is the red magnitude). As a by product, significant equatorial corrections appear for all the used catalogues (but the SDSS DR5); an empirical magnitude correcition can be considered for the GSC23 intermediate and faint magnitude range; both the 2MASS and the preliminary northernmost UCAC2 positions show consistent astrometric precision; and small harmonic terms.

References

Andrei, A. H., Souchay, J., Zacharias, N., Smart, R. L., de Cameargo, J. I. B., da Silva Neto, D. N., Vieira Martins, R., Assafin, M., Barache, C., Bouquillon, S., & Penna, J. L. 2007, *AA*, in preparation.

Souchay, J., Andrei, A. H., Barache, C., Bouquillon, S., Suchet, D., Baudin, M., Gontier, A.-M., Lambert, S., Le Poncin-Lafitte, C., Taris, F., & Arias, F. E. 2007, *AA*, in preparation.

A Giant Step: from Milli to Micro-arcsecond Astrometry
Proceedings IAU Symposium No. 248, 2007
W. J. Jin, I. Platais & M. A. C. Perryman, eds.
© 2008 International Astronomical Union
doi:10.1017/S1743921308019224

Gaia: how to map a billion stars with a billion pixels

J. H. J. de Bruijne[1]

[1]ESA/ESTEC/RSSD/SCI-SA (Astrophysics & Fundamental Physics Missions Division),
Postbus 299, 2200 AG Noordwijk, The Netherlands
Email: jdbruijn@rssd.esa.int

Abstract. Gaia, ESA's ambitious star-mapper mission due for launch late-2011, will provide multi-epoch micro-arcsecond astrometric and milli-magnitude photometric data for the brightest one billion objects in the sky, down to at least magnitude 20. Spectroscopic data will simultaneously be collected for the subset of the brightest 100 million stars, down to about magnitude 17. This massive data volume will allow astronomers to reconstruct the structure, evolution and formation history of the Milky Way. It will also revolutionize studies of the solar system and stellar physics and will contribute to diverse research areas, ranging from extra-solar planets to general relativity.

Underlying Gaia's scientific harvest will lie in a Catalogue, built on the fundamental space-based measurements. During the 5-year nominal operational lifetime, Gaia's payload, with its CCD mosaic containing 1 billion pixels, will autonomously detect all objects of interest and observe them throughout their passage of the focal plane. This paper discusses the workings of the Gaia instrument, details its payload, and discusses in depth how the scientific measurements will be collected. It addresses issues like maintenance of the scanning law, on-board data processing, the detection and confirmation of objects (single and multiple stars), the detection and rejection of cosmic rays and solar protons, the fundamental science measurements themselves composed of windows of CCD samples (pixels), and special strategies employed to maximize the science return for moving (i.e., solar-system) objects. The paper also explains how an on-board priority scheme will ensure catalogue completeness down to the faintest magnitudes possible, despite the limited ground-station availability and the enormous data volume that will be sent to the ground.

Keywords. space vehicles: instruments, surveys, astrometry, instrumentation: detectors, instrumentation: photometers, instrumentation: spectrographs

1. Introduction and conclusion

Gaia has received ample attention at this conference. Invited talks on Gaia's science and the Gaia Data Processing and Analysis Consortium (DPAC) have been delivered by L. Lindegren and F. Mignard (see Lindegren 2008 and Mignard 2008). Given the limited space allocated to this paper in these proceedings, we have decided to present our poster in Figure 1 and to refer the interested reader to the above-mentioned contributions in these proceedings and, of course, to the Gaia website at http://www.rssd.esa.int/Gaia for more details about the Gaia mission.

Acknowledgements

We would like to thank the symposium organizers for a very pleasant and instructive meeting. We would like to acknowledge the useful comments we received on the poster from K. O'Flaherty, T. Prusti, and A. Short. The poster layout was kindly provided by

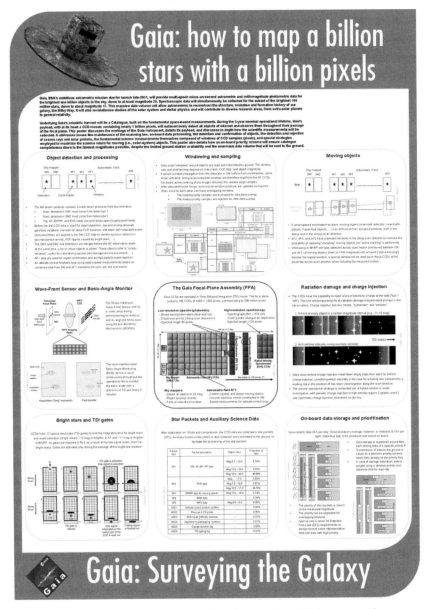

Figure 1. Poster presented at IAU Symposium 248 by the author.

J. Douglas. The images used in the poster are courtesy of EADS Astrium, M. Biermann, A. Short, and F. Mignard.

References

Lindegren, A. 2008, in: Wenjing Jin, *et al.* (eds), *A Giant Step: from Milli- to Micro-arcsecond Astrometry*, Proc. IAU Symposium No. 248 (Cambridge University Press), this volume p. 217

Mignard, F. 2008, in: Wenjing Jin, *et al.* (eds), *A Giant Step: from Milli- to Micro-arcsecond Astrometry*, Proc. IAU Symposium No. 248 (Cambridge University Press), this volume p. 224

A Giant Step: from Milli- to Micro-arcsecond Astrometry
Proceedings IAU Symposium No. 248, 2007
W. J. Jin, I. Platais & M. A. C. Perryman, eds.

© 2008 International Astronomical Union
doi:10.1017/S1743921308019236

To determining the mass of a gravitational body with micro-arcsecond astrometric data

H. Cheng, Z. H. Tang and C. Huang

Shanghai Astronomical Observatory, Chinese Academy of Science,
Nandan Road 200030, Shanghai, China
email: chz@shao.ac.cn zhtang@shao.ac.cn

Abstract. A beam of light from a background source to an observer is deflected by a gravitational body when travelling through the space in its vicinity. The deflection changes with configuration of the background source, the gravitational body and the observer, and the amount of deflection is dependent on the gravitational body's mass. It is anticipated that its mass could be determined if the shift can be measured by future precise astrometric projects such as Gaia or SIM-PlanetQuest.

Keywords. stars: fundamental parameters (masses), stars: kinematics, gravitation

1. Introduction

Determining a gravitational body's mass with its parallax and proper motion data is a method that could be used in the near future, since Gaia or SIM-PlanetQuest would provide micro-arcsecond data from which we can detect tiny shifts in the optical positions of stable background sources, the radiation of which is deflected by nearby gravitational bodies.

New methods for determining the masses of stars by micro-lensing events were developed nearly twenty years ago. The time scale of event and the change in luminosity of the farther star could be used to determine the mass of the nearer star (Gould *et al.* 1993). In addition, the radius and shape of Einstein rings could be used to determine the deflector's mass (Jiang *et al.* 2004).

These methods are based on the same principle that relies on the gravitational deflection of light. Although such a deflection always exists, it couldn't be tested until the configuration of the source, deflector and observer changes. In Hosokawa (1993), the optical position of the background source changes when the observer moves (annual parallax), as the light path from the source to the observer changes. Then an expression for the deflector's mass can be obtained. However, this method is only applicable to some nearby stars with large masses. A method suggested by Paczynski (1995) determines the mass of a star by the deflection of light caused by the proper motion of a star. This method is based on the idea similar to Hosokawa's. When compared to parallax, proper motion often causes bigger shifts of deflection per year, and the effects could be added up if the observations last several years, so that the masses of more stars could be obtained.

2. Calculating the mass with all factors considered

In theoretical considerations, the shifts caused by both parallax and proper motion should be jointly considered, especially for some nearby stars.

A star with a mass of $1M_\odot$, which is 1kpc away from the observer and 8 arcsec away from the background source in celestial background, would cause a $1\mu as$ shift.

Here we suppose that the background source is affected by a single star, since even for Gaia normally there will be less than 1 star in $10 \times 10 as^2$. Projecting the observer(O),the gravitational body (GB) and the background source (BS) to the celestial background, the following plot is obtained.

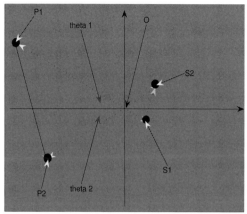

In the above plot, the observer was projected to O, the same point coinciding with the actual position of the background source (BS). P_1, P_2 represent two positions of the gravitational body at time t_1, t_2, while S_1, S_2 represent the optical positions of BS at t_1, t_2, and D is the distance of the observer from the gravitational body. The final expression of the mass (M) of the gravitational body is given by:

$$ M = \frac{D}{4G} \frac{P_1 O * P_2 O * S_1 S_2}{P_1 P_2} $$

Here G is the gravitational constant. To get Fig.1, we should use two overlapping photographs in which shifts caused by other factors had been effaced.

3. Simulation and Discussion

In our calculation, the shifts caused by annual parallax can be neglected when the star is more than 100pc away, or is observed in the same season of different years.

If every star in the Hipparcos catalogue would have a mass of $1 M_\odot$, there was an ideal background source on the inverse line of its proper motion, the source was 5 arcsec away from any star at the original time t_0, and the observation lasted for three years. Then the shifts due to gravitational deflection can be calculated. About 8000 Hipparcos stars were found to have the shifts larger than a $1 \mu as$.

Here sources which have no stars affecting them can be confirmed as motionless. Then the triangular arc method should be applied to confirm positions of sources affected by stars. However, it may be problematic to confirm positions so precisely.

References

M. Hosokawa, K. Ohnishi, T. Fukushima, & M. Takeuti, 1993 *A&A*. 278, L27-L30.
Paczynski, 1995 *astro-ph.* /9504099 vl 27 apr.
Andrew Gould, David. P. Bennett, & David R. Alives, 2004 *ApJ*. 614:404-411.
Jiang Guangfei, DePoy D. L., Gal-Yam A. *et al.*, 2004 *ApJ*. 617, 1307J.

A Giant Step: from Milli- to Micro-arcsecond Astrometry
Proceedings IAU Symposium No. 248, 2007
W. J. Jin, I. Platais & M. A. C. Perryman, eds.

© 2008 International Astronomical Union
doi:10.1017/S1743921308019248

Reference frame linking and tests of GR with Gaia astrometry of asteroids

D. Hestroffer[1], S. Mouret[1], J. Berthier[1], F. Mignard[2] and P. Tanga[2]

[1]IMCCE, Observatoire de Paris, CNRS,
77 av. Denfert-Rochereau, F-75014, Paris, France
email: hestro@imcce.fr

[2]Laboratoire Cassiopée, observatoire de la Cote d'Azur, CNRS
Mont Gros, F-06000, Nice, france
email: mignard@oca-nice.fr

Abstract. The Gaia satellite, an ESA cornerstone mission to be launched at the end of the year 2011, will observe a large number of celestial bodies including also small bodies of the solar system. Albeit spread from the inner to the outer regions of the solar system, these are mainly near-Earth objects and main-belt asteroids. All objects brighter than magnitude $V \leqslant 20$ that cross the field of view (i.e. with solar elongation $45° \leqslant L \leqslant 135°$) of the survey-mode scanning telescope will be observed. The mission will provide, over its 5 years duration, high precision photometry and astrometry with an unprecedented accuracy ranging roughly from 0.3 to 3 milli-arcsecond on the CCD level, and depending on the target's magnitude. In addition, several hundreds of QSOs directly observed by Gaia will provide the kinematically non-rotating reference frame in the visible light, resulting in the construction of a 'Gaia-ICRF'.

The positions of the asteroids hence enable to relate the dynamical reference frame—as defined by the equations of motion—to the kinematic one, and to further check the non-rotating consistency between both frames' definition. Here we show the results of a variance analysis obtained from a realistic simulation of observations for such a link. The simulation takes into account the time sequences and geometry of the observations that are particular to Gaia observations of solar system objects, as well as the instrument sensitivity and photon noise. Additionally, we show the achievable precision for the determination of a possible time variation of the gravitational constant $\dot{G}/G$. Taking into account the non-completeness of the actually known population of NEOs, we also give updated values for the nominal precision of the joint determination of the solar quadrupole J_2 and PPN parameter β.

Keywords. asteroids, space vehicles, astrometry, reference systems

1. Introduction

Ten years have passed now since the publication of the Hipparcos catalogue. The next generation of space astrometry mission Gaia will overpass the already important results achieved or enabled by the Hipparcos/Tycho astrometric catalogues. In addition to the observations of the stars in our Milky-Way, Gaia will also detect and observe a large number of 'secondary' sources: the distant quasars, and the nearby solar system objects. Indeed, Gaia will observe almost all celestial bodies brighter than $V \leqslant 20$, hence $\approx 300{,}000$ asteroids; mainly near-Earth objects (NEOs) and main-belt asteroids (MBAs). Among the many scientific outputs from Gaia observations of asteroids (see e.g. Mignard *et al.* 2007) the highly accurate astrometry of asteroids will enable dramatic orbit improvements, and also detection of such a subtle effect as the orbit perturbations. These concern the positioning of the equinox and ecliptic in the optical-ICRS, and the test of a possible net rotation between the two supposedly non-rotating frames. It also concerns a test of GR from relativistic effects, and variation of the constant of gravity.

2. Results

All observed directions are related to the Gaia reference frame which at the end is itself materialized by primary sources and the QSOs. The computed positions on the other hand are given by the equations of motion of the target and the Gaia platform, all connected to the dynamical reference frame materialized by the ecliptic and equinox. By definition both frames are non-rotating. Differences between the observed and computed directions can arise from errors in the dynamical model (initial, conditions, masses, orbit of the Earth, etc.) and/or from the fact that they are not expressed in the same frame. One possible parameterization is to introduce a rotation Ω and rotation rate $\dot{\Omega}$ between the two frames. Similarly a temporal change of the constant of gravity G will perturb all orbits. Simulation over 180,000 retained objects yields a formal precision of $\sigma_\Omega \sim 15\,\mu\text{as}$ and $\sigma_{\dot{\Omega}} \sim 6\,\mu\text{as/year}$, corresponding to testing a possible rotation rate at a $3 \times 10^{-11}\,\text{rad/year}$ level.

The major effect of GR on the orbit is a precession of the perihelion. In the parameterized post-Newtonian (PPN) formalism, this effect depends on the parameter β when γ is supposed to be known with enough accuracy by other means. Taking into account their eccentricities, it appears that highly eccentric NEOs are as much sensitive to this effect than Mercury is. However not all NEOs are currently known and more are to be discovered. Samples of hypothetical populations can then be created following the un-biased orbital and size distribution given by Bottke *et al.* (2002). It turns out that more highly eccentric orbits are likely to be found prior to Gaia launch. Another effect coupled with the relativistic one, arises from the unknown solar quadrupole J_2, since both will provide secular drift of the perihelion. Nevertheless this effect can, in theory, be decoupled from the relativistic one by observing a large number of bodies well distributed in the (a, e) plane. We have generated three different samples of possible NEOs and computed either separately or simultaneously the PPN parameter β and the solar J_2. As a result β and J_2 can be derived *simultaneously* with a precision of $\sigma_\beta \sim 5 \times 10^{-4}$ and $\sigma_{J_2} \sim 1.5 \times 10^{-8}$. Since the correlation is not too high (corresponding to the case where both parameters cannot be obtained simultaneously), these results will be slightly improved when one considers only one of the parameters in the fitting procedure.

3. Discussion

These results are encouraging and will provide complementary information to those obtained by other techniques (Lunar Laser Ranging, planetary ephemerides, helioseismology) and involving other models. While the catalogue of known MBAs is almost complete down to magnitude 20, more NEOs and inner-Earth orbits are still to be discovered with current surveys and Gaia. One should note also that systematic effects due to non-modeled non-gravitational effect could degrade these results. Further, similar studies could be applied to test the SEP through a measure of Nordtvedt parameter η, and post-Einsteinian metric theories of relativity (Jaekel & Reynaud 2006), by involving the motions of Trojans, dormant comets and Centaurs in particular.

References

Mignard, F., Cellino, A., Muinonen, K., Tanga, P., Delbò, M., Dell'Oro, A., Granvik, M., Hestroffer, D., Mouret, S., Thuillot, W., & Virtanen, J. 2007, *Earth, Moon, and Planets*, 101, 97–125.

Bottke, W. F., Morbidelli, A., Jedicke, R., Petit, J.-M., Levison, H. F., Michel, P., & Metcalfe, T. S. *Icarus*, 156, 399.

Jaekel, L. & Reynaud, S. 2006, *Class. Quantum Grav.*, 23, 777–798.

A Giant Step: from Milli- to Micro-arcsecond Astrometry
Proceedings IAU Symposium No. 248, 2007
W. J. Jin, I. Platais & M. A. C. Perryman, eds.

© 2008 International Astronomical Union
doi:10.1017/S174392130801925X

Gaia and the Astrometric Global Iterative Solution

D. Hobbs[1], L. Lindegren[1], B. Holl[1], U. Lammers[2] and W. O'Mullane[2]

[1]Lund Observatory, Lund, Sweden
email: david@astro.lu.se, lennart@astro.lu.se, berry@astro.lu.se

[2]European Space Astronomy Center, ESA, Spain
email: Uwe.Lammers@esa.int, william.omullane@esa.int

Abstract. Gaia is an ESA space astrometry mission due for launch in 2011–12. We describe part of the work carried out in the Gaia Data Processing and Analysis Consortium, namely the Astrometric Global Iterative Solution (AGIS) currently being implemented at the European Space Astronomy Center (ESAC) in Spain and largely based on algorithms developed at Lund Observatory. Some provisional results based on simulated observations of one million stars are presented, demonstrating convergence at microarcsec level independent of initial conditions.

Keywords. astrometry, methods: data analysis, methods: numerical, reference systems

1. The challenge

The immense volume of data created by Gaia and especially their complex relationships make the data processing requirements amongst the most challenging even by the standards of computational power in the next decade. The raw science data amount to some 100 Terabytes and the total data archive may surpass one Petabyte. The required numerical processing is of the order of 10^{21} floating-point operations.

To meet this challenge, the Gaia Data Processing and Analysis Consortium (DPAC) was formed in 2006 (Mignard *et al.* 2008). A central part of the processing is to determine the accurate spacecraft attitude, geometric instrument calibration and astrometric reference data for a well-behaved subset of all the objects. This core astrometric processing falls within the responsibility of coordination unit CU3 of the DPAC, and is currently being implemented and will be run at the European Space Astronomy Centre (ESAC) near Madrid in Spain. Lund Observatory and ESAC are jointly responsible for providing the framework and algorithms for the implementation of AGIS. Working versions of all the main algorithms are already in place but require further refinement.

ESAC is one of the main data processing centres for DPAC. The current system, a cluster of 18 dual-3.6 GHz Xeon processors, gives an overall performance of 140 Gflops. Many AGIS test runs have already been made on this system, using simulated data for 10^6 objects. Up until Gaia's launch (end of 2011) the system will be scaled and exercised with progressively more complex and realistic simulations of up to 10^8 objects.

2. The Astrometric Global Iterative Solution (AGIS)

The core astrometric processing will determine the positions, parallaxes and proper motions for a subset of 'primary sources' (bona-fide single stars and quasars), together with the spacecraft attitude as function of time, and a large number of instrument calibration parameters. These data must be estimated to microarcsecond accuracy, essentially by a least-squares adjustment to the measured image positions on the CCDs. The total number of unknowns is of the order of 10^9. Although the resulting system of equations is very sparse, a direct solution taking fully into account the dependencies between all the unknowns seems to be impractical by many orders of magnitude. Consequently, iterative solution methods are used. The currently adopted 'Astrometric Global Iterative Solution' (AGIS) consists of four blocks executed cyclically until convergence is achieved:

- Source Update − determination of the five astrometric parameters (α, δ, π, $\mu_{\alpha*}$, μ_δ) of each primary source
- Attitude Update − determination of the celestial orientation of the instrument axis as a function of time (using spline representations of the attitude quaternion)
- Calibration Update − determination of the geometric calibration parameters (basic angle, CCD geometry, etc.)
- Global Update − determination of a relatively small number of model parameters that are constant throughout the entire mission (e.g., the PPN parameter γ).

In each block, updates are computed based on the results from the other blocks in this and the previous iteration. The system is data driven: each observation is read from disk precisely once per iteration and updated values passed to the other processes sequentially.

The positions and apparent proper motions of quasars are used to connect the solution to the extragalactic reference frame and to determine the (galactocentric) acceleration of the solar system in the cosmological frame from the secularly changing aberration effect.

AGIS will be run in six-month cycles over the Gaia time-line, using progressively larger sets of observational data. Other processes (e.g., photometry) are run in parallel, on synchronised cycles, and their results fed to the astrometric processing of the next cycle (and vice versa) via a common database. The final iterations therefore incorporate data from the whole mission as well as updated information from all the parallel processes.

3. Some preliminary results

A number of AGIS tests have been executed on simulated data for 10^6 primary sources and a mission length of 5 years. This typically involved 5×10^6 astrometric parameters (unknowns), 2×10^7 attitude parameters, and some 2×10^5 geometric calibration parameters. All the parameters were given initial (random and/or systematic) errors up to 0.1 arcsec, which were brought down to levels compatible with the expected overall mission accuracy in some 40–50 AGIS iterations. The iterations were interrupted when the mean update in parallax was below 1 μas. Remaining errors in the astrometric parameters do however show patterns at the level of 10–20 μas, which continue to shrink with further iterations. This behaviour is characteristic for the simple iterative scheme used by AGIS, which leads to the errors (and updates) eventually decreasing exponentially by a factor equal to the largest eigenvalue of the iteration matrix. The relatively significant systematic errors still observed after ~50 iterations reflect the eigenvector associated with the largest eigenvalue, and disappear with further iterations. This property will be used to accelerate the convergence rate of AGIS and to derive alternative iteration methods.

Acknowledgements

This work is strongly dependent on parallel developments in other coordination units within DPAC, in particular CU1 (System architecture) and CU2 (Data simulations).

References

Mignard, F., *et al.* 2008, this volume p.224

A Giant Step: from Milli- to Micro-arcsecond Astrometry
Proceedings IAU Symposium No. 248, 2007
W. J. Jin, I. Platais & M. A. C. Perryman, eds.

© 2008 International Astronomical Union
doi:10.1017/S1743921308019261

The current status of the Nano-JASMINE project

Y. Kobayashi[145], N. Gouda[15], T. Yano[1], M. Suganuma[1], M. Yamauchi[14], Y. Yamada[3], N. Sako[2] and S. Nakasuka[2]

[1] National Astronomical Observatory of Japan, 2-21-1 Osawa, Mitaka, Tokyo, 181-8588 Japan

[2] School of Engineering, The University of Tokyo, 7-3-1 Bunkyoku, Hongo, Tokyo, Japan

[3] Faculty of Science, Kyoto University Kitashirakawa, Oiwakecho, Sakyoku, Kyoto, Japan

[4] Department of Astronomy, The University of Tokyo, 7-3-1 Bunkyoku, Hongo, Tokyo, Japan

[5] The Guraduate University for Advanced Studies 2-21-1, Osawa, Mitaka, Tokyo, Japan

Abstract. Nano-JASMINE is a nano-size astrometry satellite that will carry out astrometry measurements of nearby bright stars for more than one year. This will enable us to detect annual parallaxes of stars within 300 pc from the Sun. We expect the satellite to be launched as a piggy-back system as early as in 2009 into a Sun synchronized orbit at the altitude between 500 and 800 km. Being equipped with a beam combiner, the satellite has a capability to observe two different fields simultaneously and will be able to carry out HIPPARCOS-type observations along great circles. A 5 cm all aluminum made reflecting telescope with a aluminum beam combiner is developed. Using the on-board CCD controller, experiments with a real star have been executed. A communication band width is insufficient to transfer all imaging data, hence, we developed an onboard data processing system that extracts stellar image data from vast amount of imaging data. A newly developed 2K × 1K fully-depleted CCD will be used for the mission. It will work in the time delayed integration(TDI) mode. The bus system has been designed with special consideration of the following two points. Those are the thermal stabilization of the telescope and the accuracy of the altitude control. The former is essential to achieve high astrometric accuracies, on the order of 1 mas. Therefore relative angle of the beam combiner must be stable within 1 mas. A 3-axes control of the satellite will be realized by using fiber gyro and triaxial reaction wheel system and careful treatment of various disturbing forces.

Keywords. space vehicles, instrumentation: detectors

1. Introduction

It is possible to realize a low cots satellite mission within a short development period, if the satellite is physically small and if a piggy-back launching is employed. A two-dimensional array detector enables us to perform efficient observations, and a larger number of stars can be observed than that observed by Hipparcos. By combining the data in the Hipparcos catalog, we can determine proper motions with accuracies as high as 0.1 mas yr^{-1} because the temporal span between Hipparcos and Nano-JASMINE will exceed 15 years.

2. Current status

We have performed evaluation experiments for a prototype system. By using the Nano-JASMINE simulator, we have evaluated altitude control strategies in both the initial and normal operation phases. The results of thermal analysis for the Nano-JASMINE structure will be used to improve the final design of the satellite.

The optics for the 5-cm telescope are manufactured by using super-precision diamond-turning machines. These optics form a Ritchey-Chrétien-type optical system that has a composite F-ratio of 33, which matches the pixel size of a CCD. We have achieved sufficient optical performance for this telescope(Suganuma *et al.* 2008). A fully-depleted CCD that has been developed by Hamamatsu Photonics K.K. will be used for Nano-JASMINE. We have been developing a sufficiently small controller that can control the CCD in the TDI operation mode. Since the data transfer rate is limited, we have been developing a hardware stellar image extractor that can work sufficiently fast to be operated in real time. The system is constructed on a FPGA system, and a new algorithm has been developed for this system(Yamauchi *et al.* 2008). The mission control computer will be constructed on a FPGA system that has a CPU core. The prototype system has been subjected to integral experiments.

The entire satellite system is controlled in the concentrated mode by a FPGA controller that has a CPU core. Li-ion batteries are used for the power unit and a mutual surveillance system maintains its stability. A large area is reserved for solar batteries so that the required power can be maintained; this increases the size of the satellite. The satellite has magnetic sensors, fiber optical gyros, a star tracker for altitude sensing, and a magnetic torquer and reaction wheel system for the altitude actuator. A data transmitter will support more than 3 kbps in the worst case.

Since the structural deformation of optics due to the changes in temperature will degrade the accuracy of astrometric measurements, we have performed thermal analysis. Reducing the temperature gradient in the telescope structure is most important. We have taken care of the telescope's thermal insulation and have increased its capacity to endure a large amount of heat conduction. Our analysis shows that the deformation of the beam combiner, which is the most sensitive component with respect to measurement accuracy, is less than 0.7 mas. The thermal radiation efficiency is also important to maintain the CCD at low temperatures.

The altitude control of Nano-JASMINE is one of the major challenges. We have studied the methods to form an operational sequence by using the Nano-JASMINE simulator. In the first stage of operations, when the satellite altitude is significantly displaced from the expected position, we will operate the Magnetic Torquer with reference to the outputs of magnetic sensor. In the next stage, we will use the Fiber Optical Gyro and Star Tracker as altitude sensors and a 3-axes reaction wheel to control the altitude. In the final stage, we will use the CCD outputs. Elongated stellar images will indicate on the altitude error.

We have not yet finalized the launcher. Among others HIIa and Cyclone4 are being considered for the launcher. The satellite will be launched into a sun-synchronized orbit at the altitude range between 500 km and 800 km. The other orbital parameters depend on the major satellite requirements. We will optimize the survey parameters for different orbital parameters. The date of the launch is not yet fixed. It can be as early as 2009.

References

Suganuma, M. *et al.* 2008, *Proc. IAU symposium No. 248*, this volume p. 284
Yamauchi, M. *et al.* 2008, *Proc. IAU symposium No. 248*, this volume p. 294

A Giant Step: from Milli- to Micro-arcsecond Astrometry
Proceedings IAU Symposium No. 248, 2007
W. J. Jin, I. Platais & M. A. C. Perryman, eds.

© 2008 International Astronomical Union
doi:10.1017/S1743921308019273

CTE in Space Astrometry

V. Kozhurina-Platais[1], M. Sirianni[1,2] and M. Chiaberge[1,2]

[1] Space Telescope Science Institute,
3700 San Martin Dr. Baltimore, 21904, USA
email: verap@stsci.edu

[2] European Space Agency
email: sirianni@stsci.edu,marcoc@stsci.edu

Abstract. An imperfect CTE in CCD detectors is one of the most important instrumental issues affecting both photometry and astrometry, especially in space-based observations. We discuss the CTE effect in the images taken with the Hubble Space Telescope's, Advanced Camera for Surveys in Wide Field Channel (ACS/WFC). ACS is the only imaging instrument capable of delivering sub-mas astrometry from a single observation, and it is important to take into account any instrumental systematic in positions such as the CTE effect.

Keywords. astrometry, techniques: image processing, instrumentation: detectors

There is a significant Charge Transfer Efficiency (CTE) effect for all CCD detectors used in HST instruments; WFPC2: Dolphin (2000), STIS: Goudfrooij *et al.*(2006), ACS: Riess & Mack (2004), where the CTE effect was discussed in the terms of CTE-induced photometric losses in aperture photometry. However, the images taken with HST in particular with ACS, nominally yield sub-mas precision in differential astrometry. As shown by Kozhurina-Platais *et al.* (2007), the CTE-induced centroid shift depends on the magnitude of a star and its position on the CCD. In the worst case the CTE effect may induce a centroid shift exceeding 0.1 pixel (or 5 mas), that is not acceptable in high-precision astrometry. The *effective* PSF (ePSF) method in combination with distortion solution derived by Anderson & King (2006) for ACS/WFC can reach precisions at 1-2 mas level, which is significantly lower than CTE-induced systematics.

The ACS/WFC detector employs a mosaic of two CCD chips, the each with an array of $2K \times 2K$ pixels. Because of a large number of transfers, the WFC detector is expected to be more affected by CTE. To study the CTE-induced centroid shifts (Δ Y), observations of 47 Tuc taken with ACS/WFC with the F606W filter were used. The residuals in X and Y positions between long and short exposures (Δ X and Δ Y) derived with ePSF and corrected for geometry distortion, are plotted as a function of position for different ranges of instrumental magnitudes (Fig.1, plot on the left). Δ Y on this plot shows a sharply defined discontinuity at Y = 2048. This is a clear indication of CTE-induced centroid shifts in the direction away from the read-out registers, which are located at the top of WFC1 and at the bottom of WFC2. It also shows that the amplitude of the centroid shift depends on the brightness of a star. The centroid shift is negligible for bright stars (mag $\lesssim -6.0$), while it reaches up to 0.15 pixels for faint stars (mag $\gtrsim -4.0$), if they are located far away from the read-out amplifier. As seen in the middle plot of Fig.1, the centroid shift (Δ Y) is linear with instrumental magnitude for both chips.

Thus, CTE-induced centroid shift (ΔY) can be presented as a linear function of magnitude and Y, namely: Δ Y $= a_0 + a_1 \times Mag + a_2 \times Y$. The numerical implementation of the 2-D functional fit was realized by employing a non-linear least-square fit. As a result,

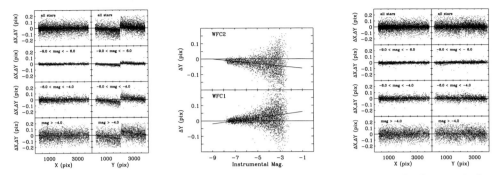

Figure 1. *The left plot* is the difference in X and Y positions between the long and short exposures. From top to bottom, the residuals ΔX (left panel) and ΔY (right panel) are shown for different ranges of magnitudes. The over-plotted lines in the right panels indicate the trend of CTE-induced centroid shift (ΔY) in the direction away from the read-out registers for WFC2 ($Y \lesssim 2048$ pixels) and for WFC1 ($Y \gtrsim 2048$ pixels). *The middle plot* is the centroid shift (ΔY) as function of instrumental magnitude for ACS CCD chips (WFC1 & WFC2 from top to bottom, respectively). The over-plotted solid line represents a linear fit of ΔY vs. instrumental magnitude. *The right plot* is residuals of X (left pannel) and Y (right pannel) after the correction was applied, to be compared with plot on the left.

the post-correction residuals show only a small 0.8 mas scatter as opposed to the 10 mas scatter from uncorrected positions (Fig.1, the plot on the right).

The coefficients a_0, a_1, a_2 derived from observations of 47 Tuc with ACS/WFC show the effect of CTE degradation with time (Fig.2). For two ACS WFC CCD chips, a_0, a_1 have a symmetric appearance, whereas a_2 are identical. This is due to the fact that CTE-induced centroid shift depends on the number of transfers (i.e. Y-2048) which are symmetrical for each chip.

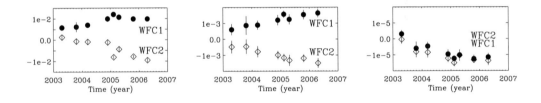

Figure 2. Coefficients a_0, a_1, a_2 (from left to the right) show the CTE degradation with time.

While empirical CTE corrections for the photometric losses have been developed in the past few years for all HST CCD instruments, this is the first attempt to characterize an empirical CTE corrections for astrometry.

Acknowledgments. V.K.-P. thanks J. Anderson for the centering and distortion codes. V.K.-P. is greatly appreciated for useful discussion and suggestions to I. Platais (JHU), T. Girard and W. van Altena (Yale U.).

References

Anderson, J. & King, I. 2006, *ACS Instrument Science Report*, 2006-01, (Baltimore:STScI)
Dolphin, A. E. 2000, *PASP*, 112, 1397
Goudfrooij, P., Bohlin, R. C., & Maíz-Apellániz, J., 2006, *PASP*, 118, 1455
Kozhurina-Platais, V., *et al.*, 2007, *ACS Instrument Science Report*, 2007-04, (Baltimore:STScI)
Riess, A. & Mack, J. 2004, *ACS Instrument Science Report*, 2004-06, (Baltimore:STScI)

A Giant Step: from Milli- to Micro-arcsecond Astrometry
Proceedings IAU Symposium No. 248, 2007
W. J. Jin, I. Platais & M. A. C. Perryman, eds.

Design of a compact astrometric instrument for the GAME mission

D. Loreggia, M. Gai, A. Vecchiato, D. Gardiol, S. Ligori and M. G. Lattanzi

All Authors are from INAF - Astronomical Observatory of Turin, St. Osservatorio, 20, 10025
Pino Torinese, Torino, Italy.
email: loreggia@oato.inaf.it

Abstract. We present the design of a Fizeau interferometer to be implemented for the GAME mission. The aim is to measure the PPN γ parameter with the same technique used for the first time by Dyson, Eddington *et al.*, but at a 10^{-6} accuracy level. GAME will observe about 10^6 sufficiently bright stars at about $2°$ from the Sun. A dedicated space mission has the advantage of observing the light bending without waiting for an eclipse.

Keywords. gravitational lensing, space vehicles, instrumentation: interferometers, techniques: high angular resolution

1. Introduction

The *Gamma Astrometric Mission Experiment* - GAME (Vecchiato, *et al.*2008) will observe two symmetric fields, each about $2°$ from the Sun. The relative position of the targets is measured again 6 months later, without the Sun in between, giving a direct estimation of the light bending and the γ PPN parameter with an expected accuracy of 10^{-6}. Hereafter we briefly discuss a possible telescope design, preserving some requirements for a small mission.

2. Optical design

A 5-aperture Fizeau interferometer design is well suited for our propose of merging good performance with some geometrical constraints such as a distance between the main mirrors not larger that 1.5m, the size of the pupil compatible with diffraction limited (DL) images and with the size and the geometry of the interferometric apertures. These requirements must match the best sampling condition of the fringes (Loreggia, *et al.*), assuming a pixel size of $15\mu m$ in the high resolution direction (HRD) and asking for a baseline between each couple of apertures B=2L, where L the aperture side in HRD. The aperture area dimensioning sets directly all the instrument parameters and is driven by the photons collecting needs and the final encumber limitations. At λ_{ref}=650nm, the best solution is obtained with apertures of 40x220mm over a pupil of 660mm, giving an EFL = 20600mm and a total baseline (between the outer apertures) B_t=320mm, as shown in Fig.1.

GAME will look at two lines of sight very close to the Sun, which combination and folding is implemented by mean of a beam combiner in front of the primary mirror. A baffling array is needed and is under study; we tested that looking at about $2°$ from the Sun, a typical baffling length may be $\sim$4.2m, which can be achieved with the folded configuration presented.

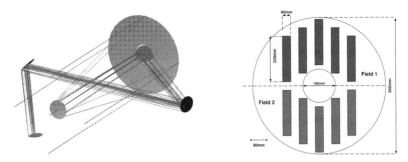

Figure 1. Telescope 3D layout (left); 5 apertures interferometric mask (right)

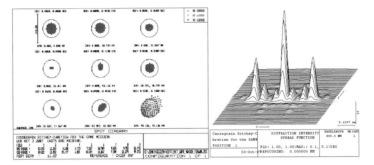

Figure 2. Spot diagram for the full aperture: circle dimension gives the diffraction limit equal to $50\mu m$ (left). Fringe pattern for the representative field (0.12,0.12)deg (right)

3. Results

In Fig (2) the spot diagrams of the full aperture are given for some reference fields over a FOV of 7'x7'. The outer field (0.12, 0.12)deg is not DL but when the interferometric mask is applied the expected reducing of aberration makes the fringes quality good over the whole FOV, with a quite constant V=**0.95**. The accuracy of an astrometric measurement is linked to instrument gain parameter, the SNR and the X_{RMS} equivalent aperture of the interferometer. It can be shown that for the 5 apertures interferometer geometry we are considering, $X_{RMS} = 114mm$. Assuming to work in the V band with a sufficiently narrow FWHM and an exposure of 100sec, the SNR is equal to 385 (no extinction, instrument transmission T and QE = 90%) and we get an accuracy on the single measurement $\sigma = \mathbf{0.58mas}$. This precision scales with the magnitude approximatively as $10^{0.2(m-m_0)}$. At 2° from the Sun the light deflection is $\Delta\alpha \cong 0.2$" and the error $\sigma_{\Delta\alpha}$ is related to the accuracy of the γ parameter estimation by (Vecchiato, *et al.*2008):

$$\frac{\sigma_{\Delta\alpha}}{\Delta\alpha} \cong \frac{\sigma_\gamma}{2\gamma} \quad \Rightarrow \quad \sigma_\gamma = 2\gamma\sigma_{\Delta\alpha}/\Delta\alpha = \sigma_{\Delta\alpha} \cdot \mathbf{10^{-2}} \qquad (3.1)$$

This means the need to observe at least 10^6 objects to have the expected final accuracy of 10^{-6}, an accessible result from this mission (Vecchiato, *et al.*2008).

References

Loreggia, D., Gardiol, D., Gai, M., Lattanzi, M. G., & Busonero, D. 2004, *Apl.Opt* Vol.53 No. 4., 721-28;

Vecchiato, A., Lattanzi, M. G., Gai, M., & Morbidelli, R., *Gamma Astrometric Mission Experiment*, Proc. IAU Symposium No.248, in this volume, p.290

A Giant Step: from Milli- to Micro-arcsecond Astrometry
Proceedings IAU Symposium No. 248, 2007
W. J. Jin, I. Platais & M. A. C. Perryman, eds.

© 2008 International Astronomical Union
doi:10.1017/S1743921308019297

JStuff - a preliminary extragalactic model for the ESA-Gaia satellite simulation framework

A. G. O. Krone-Martins[1,2], C. Ducourant[1], R. Teixeira[2] and X. Luri[3]

[1]Observatoire de Sciences d'Aquitaine, Université de Bordeaux I
2, rue de l'Observatoire, Floirac, France
email: krone@obs.u-bordeaux1.fr

[2]Inst. de Astron. Geof. e Ciên. Atm., Universidade de São Paulo,
Rua do Matão, 1226, 05508-900, Cidade Universitária, São Paulo, Brazil

[3]Dept. d'Astronomia i Meteorologia, Universitat de Barcelona, Martí Franquès S/N, Barcelona,

Abstract. In this work we describe the JStuff, a preliminary generator of mock catalogues of extragalactic objects based on the Stuff code. This version is being implemented in the Java language for the ESA-Gaia satellite simulation framework. We also compare some results obtained with both versions of the simulator.

Keywords. catalogs, galaxies: fundamental parameters, galaxies: luminosity function

1. Introduction

The ESA-Gaia mission, Perryman (2005), is one of the most ambitious projects of the modern astronomy. With the satellite launch date fixed at the end of 2011, the final all-sky catalog shall be released sometime before 2020. The satellite observations and the reduction procedures will result into an impressive terabyte class all-sky catalog, complete up to G=20, with more than one billion objects of all types with amazing astrometric precision (7μas at V<13, 25μas at V<15 and a fraction of mas at V<20, 11 million paralaxes with errors smaller than 1%, and 150 million in the 10% range), spectrophotometric data spread over 66 pixels (330-1000nm) and high-resolution spectrometric data in the CaII-triplet region (847-874nm, 0.026nm/pix).

In order to design the reduction and analysis algorithms, it is necessary to implement a realistic mission simulation environment. The GaiaSimulator, Altamirano *et al.* (2005), is the official framework where all the simulation packages are built-in: it includes a pixel-level simulation package, a telemetry stream simulation package, a statistically simulated catalog generator, and the universe model simulators: Galaxy, Extragalactic, Solar System. JStuff is the first implementation in Java of the code to generate catalogs of unresolved galaxies.

2. The simulation algorithm

This version of JStuff is heavily based on the Stuff code first described in Bertin & Arnouts (1996). It simulates a mock catalog of galaxies at a given region of the sky, at this time using a positional sampling that is uniformly random. Nonetheless it is planned as a possible future improvement, the implementation of a two-point correlation function.

In order to simulate the galaxies parameters, the code first sample randomly, from a poissonian distribution, the number of galaxies of Hubble type (E, S0, Sab, Sbc, Scd, and Irr are implemented), assuming Schechter (1976) luminosity functions. Then, each galaxy

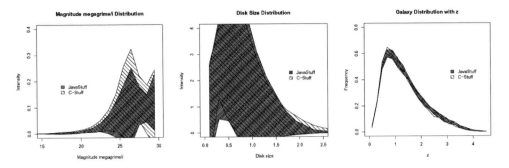

Figure 1. Comparison of the results obtained with the JStuff and the Stuff codes for three parameters. The confidence intervals are at the three sigma level.

is assembled as a sum of two components, a spheroidal (following a de Vaucouleurs law) and a disk-like one.

For elliptical galaxies, the adopted relation between the absolute B magnitude and effective radius is the one described in Binggelli *et al.* 1984. The spheroid flattening is randomly selected from a normal distribution, from Sandage *et al.* (1970). The disk component is simulated with an exponential profile. The central surface brightness is randomly selected from a gaussian distribution, and the disk inclination and position angle are also randomly selected.

The code simulates the internal disk extinction in the observed galaxy, using a curve obtained for the LMC. We apply Metcalfe *et al.* (1991) K+e corrections with polynomial fits. Finally, each galaxy is placed at a redshift (and its bolometric luminosity and angular diameter distance are calculated). Spectra is generated as well, using a small library for selected disk and spheroidal components by combining these spectra and re-scaling to give the magnitude sampled from the luminosity function.

3. Firsts tests

In order to verify how well JStuff is performing in relation to the original E. Bertin's C version, we simulated a small dataset, containing about 150,000 galaxies with the C version's default parameters. This first simulation indicates that our code is reproducing well most aspects of its predecessor (Fig. 1). The code is being modified to use a spectral library of galaxies, simulated with the PEGASE code at the University of Athens. As soon as this integration is finished, a strictier validation, using a larger simulated dataset will be performed. At this validation a comparison with real data is planned.

References

Altamirano, P., Babusiaux, C., Luri, X., & Masana, E. 2005, Technical Report GAIA-SWG-004
Bertin, E. & Arnouts, S. 1996, *A&AS*, 117, 393
Binggeli, B., Sandage, A., & Tarenghi, M. 1984, *AJ*, 89, 64
Erben, T. Van Waerbeke, L., Bertin, E., Mellier, Y., & Schneider, P. 2001, *A&A*, 366, 717
Metcalfe, N., Shanks, T., Fong, R., & Jones, L. R. 1991, *MNRAS*, 249, 498
Perryman, M. 2005, *The Three-Dimensional Universe with Gaia*, ESA/SP-576
Sandage, A., Freeman, K. C., & Stokes, N. R. 1970, *ApJ*, 160, 831
Schechter, P. 1976, *ApJ*, 203, 297
Simien, F. & de Vaucouleurs, G. 1986, *ApJ*, 302, 564

A Giant Step: from Milli- to Micro-arcsecond Astrometry
Proceedings IAU Symposium No. 248, 2007
W. J. Jin, I. Platais & M. A. C. Perryman, eds.
© 2008 International Astronomical Union
doi:10.1017/S1743921308019303

Simulating Gaia observations and on-ground reconstruction

E. Masana, C. Fabricius, J. Torra, J. Portell and J. Catañeda

Departament d'Astronomia i Meteorologia. Universitat de Barcelona
Av. Diagonal 647, E 08028 Barcelona, Spain
email: emasana@am.ub.es

Abstract. On a daily basis the Gaia telemetry data (some 30 GB) must be stored and treated in order to reconstruct the actual observations. This initial data treatment processes all newly arrived telemetry and various pieces of auxiliary data. The first part of the process is merely a reformatting to create raw objects for permanent storage in the raw data base (some 40 TB at the end of the mission). The next part is to analyze the data to derive initial values for the observables, e.g. transit times and fluxes, producing intermediate objects. Finally, the intermediate objects are matched with sources in the data base, linking all the observations of a given source.

To check the initial data treatment algorithm we use simulations of the telemetry stream provided by GASS, the Gaia System Simulator. GASS simulates astrometric, photometric and radial velocity data, using models of the satellite and on-board instruments, as well as the models of different of objects observed by Gaia (stars, galaxies, solar system objects, ...). On the other hand, the initial data treatment allows us to validate the data generated by GASS, which are used too to check other algorithms like the First Look or the Astrometric Global Iterative Solution (AGIS).

Keywords. astrometry, methods: data analysis

1. Introduction

The daily Gaia telemetry stream is reformatted and stored in the raw data base as raw objects (about 40 TB at the end of mission). The observations are cross-matched and stored in the Main Data Base (MDB). Different algorithms, such as Astrometric Global Iterative Solution (AGIS), Photometric Global Iterative Solution (PGIS) or specific algorithms from the Coordination Units of the Data Processing and Analysis Consortium (DPAC), work with the objects stored in the MDB to achieve the final results.

2. The Gaia System Simulator

The Gaia System Simulator (GASS) is part of the Gaia Simulator, a set of three data generators designed to cover the simulation needs of the Gaia community. GASS simulates astrometric, photometric and radial velocity telemetry data, using models of the satellite and on-board instruments, as well as the models of objects observed by Gaia such as stars, galaxies, solar system objects, quasars.

3. The Initial Data Treatment

Initial Data Treatment (IDT) is a data processing system that receives raw telemetry data from the spacecraft (or from the Gaia System Simulator during the design phase) and outputs basic image parameters of the observations. IDT computes:

- Refined attitude data generated from raw attitude.
- Image parameter: transit times, centroids and fluxes.
- Raw astrometric, photometric and spectrometric data.
- Approximate celestial coordinates for source identification.

The data processing chain that forms the core of IDT includes also calculations such as the background of the measured objects and their colors.

4. Validation of on-ground reconstruction

An automatic validation tool has been developed to compare the results of IDT with the ideal values provided by GASS. It has demonstrated the feasibility of the on-ground sky reconstruction using real mission observations.

The first trial with a dataset with 10 millions of sources observed over 400 days shows a correct reconstruction of the celestial coordinates and photometry of the sources, as well as validates the high-precision observations, required for the Astronomical Iterative Solution.

5. Acknowledgements

This project is supported by the Spanish MEC under contract PNE2006-13855-C02-01.

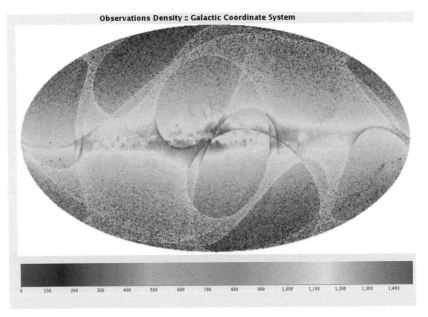

Figure 1. On-ground reconstruction of the sky: density of observations per square degree for a six months simulation. The observed pattern is a combination of the Galaxy Model and the Nominal Scanning Law of Gaia.

A Giant Step: from Milli- to Micro-arcsecond Astrometry
Proceedings IAU Symposium No. 248, 2007
W. J. Jin, I. Platais & M. A. C. Perryman, eds.
© 2008 International Astronomical Union
doi:10.1017/S1743921308019315

Laser interferometric high-precision geometry (angle and length) monitor for JASMINE

Y. Niwa[1], K. Arai[1], A. Ueda[1], M. Sakagami[2], N. Gouda[1], Y. Kobayashi[1], Y. Yamada[3], and T. Yano[1]

[1]National Astronomical Observatory Japan,
2-21-1 Osawa, Mitaka, Tokyo, 181-8588, JAPAN email: `kazin.niwa@nao.ac.jp`

[2]Cosmology and Gravity group Dept. of Fundamental Science Fac. of Integrated Human Studies, Kyoto University Kyoto 606-8501, JAPAN

[3]Theoretical Astrophysics Group Department of Physics, Kyoto University Kyoto 606-8502, JAPAN

Abstract. The telescope geometry of JASMINE should be stabilized and monitored with the accuracy of about 10 to 100 pm or 10 to 100 prad of rms over about 10 hours. For this purpose, a high-precision interferometric laser metrology system is employed. Useful techniques for measuring displacements on extremely small scales are the wave-front sensing method and the heterodyne interferometrical method. Experiments for verification of measurement principles are well advanced.

Keywords. instrumentation: interferometers, techniques: interferometric, telescopes

1. Introduction

Next-generation astrometry satellite missions will measure parallaxes, positions with the accuracy of 10 microarcsec. So the optical component of their telescope should be stabilized and its fluctuations should be monitored with high accuracy. In JASMINE(Japan Astrometry Satellite Mission for Infrared Exploration) which is one of next generation astronomical satellite missions astronomical parameter is derived by the frame linking method, and It is necessary to suppress fluctuations of frame expansion or distortion according to the temperature changing during the observation as much as possible; the telescope geometry, the distance of primary mirror to secondary mirror and the angle between two mirrors, should be stabilized with the accuracy of about 10 to 100 pm or 10 to 100 prad in root-mean-square over about 10 hours; moreover, the fluctuations should be monitored with such accuracy. For this purpose, a high-precision interferometric laser metrology system is employed.

2. Concept of high-precision geometry monitor

One of the available techniques for measuring the fluctuations of the angle is a method known as the gwave front sensingh using a Fabry-Perot type laser interferometer. One of the advantages of the technique is that the sensor is made to be sensitive only to the relative fluctuations of the angle which the JASMINE wants to know and to be insensitive to the common one; in order to make the optical axis displacement caused by relative motion enhanced the Fabry-Perot cavity is formed by two mirrors which have long radius of curvature. The heterodyne interferometrical method is useful for the measurement of longitudinal fluctuations. Moreover, this technique easily can measure displacements of two or more degree of freedom because the measurement signals can be detected without controlling optical path length with actuators. Therefore, displacements in some parts of telescope can be measured at the same time using this method.

3. Performance tests

To verify the principles of these ideas, the experiments were performed.

(*a*) The wave front sensing output signal   The experiment was performed using a 0.1m-length Fabry-Perot cavity with the mirror curvature of 20m. The mirrors of the cavity were artificially actuated in either relative way or common way, and a grate difference between the response to relative motion and the response to common one could be seen. The ratio of response to relative motion to response to common one was about 85 times; the sensor had good response only to relative motion.

(*b*) The heterodyne interferometer output signal   The calibration of the PZT actuator movements which was attached to the target mirror was done by using four 2.5m heterodyne interferometers to test the operation with them. The calibration values for longitudinal, horizontal and vertical motions could be measured with the same accuracy to about nm.

(*c*) The noise behavior of the interferometer   Figure.1 shows the wave front sensor and heterodyne interferometer noise spectrum measured in the atmosphere without temperature control system. The smooth lines show the measurement value for root-mean-square from 1kHz to each frequency. The geometry monitor stability was 13.5 nrad in the direction of the angle and 10.4 nm in the direction of the length in 1000s. The monitor instability depends on a thermal drift in low frequency band, so we plan to make interferometers compact and to introduce a temperature control system.

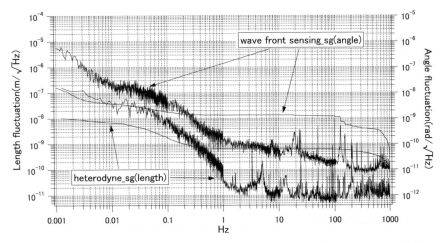

Figure 1. Two types Interferometers noise spectrums: the wave front sensor and heterodyne interferometer. The smooth lines represent root-mean-square values from 1kHz to each frequency.

References

Euan Morrison, Brian, J. Meers, David, l. Robertson, & Henry Ward 1994, *Appl. Opt.*, 33, 5041
Euan Morrison, Brian, J. Meers, David, l. Robertson, & Henry Ward 1994, *Appl. Opt.*, 33, 5037
G Heinzel, C. Braxmaier, R. Schilling, A Rudiger, D. Robertson, M. te Plate, V. Wand, K Arai, U Johann, & K. Danzmann 2003, *Classical and Quantum Grav.*, **20** (2003) S153-S161

A Giant Step: from Milli- to Micro-arcsecond Astrometry
Proceedings IAU Symposium No. 248, 2007
W. J. Jin, I. Platais & M. A. C. Perryman, eds.

Gaia Science Operations Centre

W. O'Mullane, U. Lammers, J. Hoar, J. Hernandez

European Space Astronomy Centre, Avda. de los Castillos s/n E-28692, Villanueva de la
Canada (Madrid)

Abstract. The Gaia Science Operations Centre (SOC), based at ESAC near Madrid, is building
up to play an important role in Gaia operations and data processing.

Keywords. space vehicles, methods: data analysis, surveys, astrometry, astronomical data
bases: miscellaneous

1. Introduction

The SOC hosts the data hub for the Gaia data processing (Sect. 4) and must carefully
balance ESA style standards (and bureaucracy) with the scientific communities creativity
and dedication. SOC is closely involved with the DPAC (Sect 2) and is the communities
interface to the satellite through MOC. SOC takes a direct development role not just
in the operations software such as POS (Sect. 3) but in some of the science processing
software such as the Astrometric Global Iterative Solution (AGIS, see Hobbs *et al.* in this
volume) SOC is a major contributor to the GaiaTools library (Sect. 5). During mission
operations SOC will run both operations type software as well as science processing
systems such as IDT (Sect 6) and AGIS. Some further details are given bellow.

2. Data Processing and Analysis Consortium (DPAC)

In May 2007 with approval from ESA's Science Programme Committee (SPC) DPAC
was formalized as a pan-European consortium in charge of designing, implementing, and
operating the data processing system needed to create the Gaia catalogue. DPAC adopted
the ECSS (European Cooperation on Space Standardization) software engineering standards and practices tailored for the Gaia project while following a cyclic development
model in which software modules are incrementally built. The time to launch is subdi-
vided into phases of six months durations. A large emphasis is placed on unit/integration
and end-to-end system testing.

3. Payload Operations System (POS)

The function of the POS is to support all payload operations activities performed at the
Gaia Science Operations Centre (SOC). The POS shall support the following activities:
- Monitoring the execution of the Nominal Scanning Law
- Calculation of transition times between the Nominal and Modified Scanning Law
- Generation of the Science Schedule containing the predicted on-board data rate
according to the Scanning Law, galaxy and instrument models
- Tracking the current status and history of configuration parameters of payload
- Creating and dispatching of Scientific Operations Requests, including the generation
of updates to configuration parameters of payload systems

• Tracking the execution of Scientific Operations Requests through the Mission Timeline and Telecommand History

• Processing of optical observations of Gaia received from DPAC into the required format for the MOC (for improving Gaia position and velocity data from standard ranging measurements)

4. Gaia Main Database (MDB)

The MDB is the central repository for all the data produced by Gaia and the DPAC. The system shall be operated at ESAC but its development and definition involves all of the the CUs. The MDB is the hub for all Gaia processing in that it provides the input data for, and receives the generated output from, all processing systems. The details of these relations are specified in the MDB Interface Control Document (ICD). The MDB will hold a range of different data types and the total data volume that the MDB will ultimately contain is still somewhat uncertain but it shall be of the order of hundreds of Terabytes and could even reach the Petabyte range.

The MDB will contain a number of version-controlled results data bases that will form a sequence $V_0, V_1 \ldots V_n$. Each V_m comprises three main parts, viz. *raw data*, coming from the satellite, *intermediate data* generated by the pre-processing pipeline and *reduced data*, coming from the various processing systems. The reduced data in any V_m is only produced from data contained in the previous V_{m-1}. The versions will be produced cyclically at regular intervals, e.g. every 6 months.

5. GaiaTools library and Parameter DB

GaiaTools is a common software library collaboratively written in Java to support and facilitate the software development in DPAC. The principle reasons for GaiaTools are to avoid duplication of effort and reduce errors in common routines. Policy decisions are taken by a GaiaTools committee composed of one representative from each CU All code development is done by DPAC members in a distributed manner following DPAC's software engineering guidelines. Functionality in GaiaTools range from the Data Access Layer (DAL) and plotting through general numerical routines (Vector+Matrix algebra, interpolation etc) to Solar System ephemeris access.

The Gaia Parameter Database (GPDB) is a central, searchable repository of all mathematical, physical, mission, satellite, and payload design parameters with relevance for Gaia. A Java export of the GPDB is distributed together with the GaiaTools library.

6. Initial Data Treatment

IDT can be regarded as a pre-processing pipeline converting raw telemetry from the satellite into higher level data products for downstream software (like AGIS). It will be operated in a quasi-continuous manner at SOC to process telemetry in near-real time. Telemetry data will go into a *Raw Database* - processed data into a IDT/FL database and from there into the Main Data Base (Sect. 4. The primary IDT outputs will be: Extracted images parameters (centroids, fluxes), Match-table linking observations to sources, and refined on-board attitude.

The system is developed by DPAC within CU3 with major contributions from University of Barcelona, Spain.

A Giant Step: from Milli- to Micro arcsecond Astrometry
Proceedings IAU Symposium No. 248, 2007
W. J. Jin, I. Platais & M. A. C. Perryman, eds.

© 2008 International Astronomical Union
doi:10.1017/S1743921308019339

Development of a very small telescope for a milli-arcsec space astrometry

M. Suganuma[1]†, Y. Kobayashi[1], N. Gouda[1], T. Yano[1], Y. Yamada[2], N. Takato[3] and M. Yamauchi[4]

[1] National Astronomical Observatory of Japan, 2-21-1 Osawa, Mitaka, Tokyo 181-8588, Japan
email: suganuma@merope.mtk.nao.ac.jp

[2] Department of Physics, Kyoto University, Oiwake-cho Kita-shirakawa, Sakyo-ku, Kyoto 606-8502, Japan

[3] Subaru Telescope, National Astronomical Observatory of Japan, 650 North A'ohoku Place, Hilo, HI 96720, USA

[4] Department of Astronomy, Graduate School of Science, The University of Tokyo, 7-3-1 Hongo, Bunkyo-ku, Tokyo 113-0033, Japan

Abstract. We are preparing a reflecting telescope for Nano-JASMINE, a very small satellite for global space astrometry of milli-arcsecond accuracy. The telescope has a 5-cm diameter primary mirror and a beam-combiner in front of it. It occupies only about 12x12x17cm and is entirely made out of aluminum alloy. The telescope and its surrounding structures are carefully designed for thermal stability of the optics, especially to control changes in the relative angle of the beam-combiner.

Keywords. astrometry, space vehicles: instruments, telescopes

1. The Nano-JASMINE Telescope

Nano-JASMINE satellite, which weighs about 14kg, will survey all-sky in wavelength around z-band using a CCD in time-delayed-integration (TDI) mode. In this mission, we are going to demonstrate a global astrometry observation with a small satellite. In the case of success, a new astrometric catalog with a few mas accuracies could produce proper motions accurate to 0.1 mas yr^{-1}, when combining with Hipparcos catalog (Kobayashi *et al.* 2006).

For Nano-JASMINE, we are developing a very small telescope that is specified in Table 1. Similar to Hipparcos, we place a dual-angled flat mirror, called beam-combiner, in front of the primary mirror so that we could simultaneously expose two different field-of-views separated by 99.5 degrees in order to execute wide-field astrometry.

2. Developement and current status

We made all telescope parts, including the mirrors, out of aluminum alloy using an ultra-precise milling machine, however, we could not fabricate aspherical mirrors satisfactory (for details, see Suganuma *et al.* 2006). Figure 1 shows a picture of a proto-model of the telescope, together with its optical layout. A diffraction-limited performance of the telescope optics was confirmed both by wavefront measurements and imaging experiments.

† Present address: National Astronomical Observatory of Japan, 2-21-1 Osawa, Mitaka, Tokyo 181-8588, Japan.

Table 1. Specifications of Nano-JASMINE Telescope

Effective Aperture	$\phi=5$cm, divided into two by a beam-combiner.
Focal Length	167cm (F/33)
Optics Type	Ritchey-Chretien type, followed by three folding mirrors.
Field of View	0.5 × 0.5 deg
Basic Angle	99.5 deg
Wavelength	z-band $(\lambda \sim 0.9\mu)$
CCD	1024×1024 pix (1.76 arcsec/pix)
Operating Temperature	-50 - -100°C

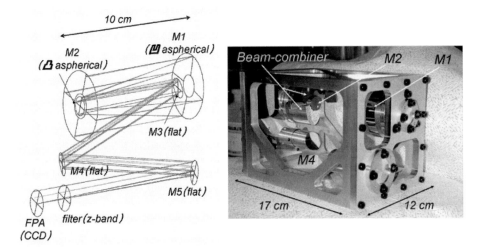

Figure 1. *Left:* Optical layout of Nano-JASMINE telescope. A beam-combiner that should appear around the M2 is omitted here. *Right:* Assembled proto-model of the telescope. All parts are figured out of aluminum alloy except the gold coat on the optical reflecting surfaces.

The surrounding structures are designed to cool down the telescope and the CCD radiatively below -50°C . Also, the telescope is well-insulated from exterior structures to minimize the thermal gradients in the optics and their time variation during the orbital period of the satellite. We made a realistic thermal design that passively controls the changes in the relative angle of the beam-combiner within a sub-milliarcsecond.

Acknowledgements

The Nano-JASMINE project is collaborated with Intelligent Space Systems Laboratory, Univ. of Tokyo. Our development is supported by the Advanced Technology Center at the National Astronomical Observatory of Japan. The practical machining process was carried out by Corporate Manufacturing Engineering Center of TOSHIBA Co., Ltd.. We thank IHI Aerospace Engineering Co., LTD. for their thermal analysis and constructive suggestions on the thermal design. We also acknowledge colleagues of JASMINE working group for helpful discussions.

References

Kobayashi, Y. *et al.* 2006, *Proc. SPIE* 6265, 626544

Suganuma, M., Kobayashi, Y., Gouda, N., Yano, T., Yamada, Y., Takato, N., & Yamauchi, M. 2006, *Proc. SPIE* 6265, 626545

A Giant Step: from Milli- to Micro-arcsecond Astrometry
Proceedings IAU Symposium No. 248, 2007
W. J. Jin, I. Platais & M. A. C. Perryman, eds.

© 2008 International Astronomical Union
doi:10.1017/S1743921308019340

Astrometry by small ground-based telescopes

W. Thuillot[1], M. Stavinschi[2], M. Assafin[3] and the IAU Working Group ASGBT †

[1]Institut de mécanique céleste et de calcul des éphémérides, IMCCE-Paris Observatory
77, av. Denfert Rochereau, 75014, Paris, France
email: thuillot@imcce.fr

[2]Astronomical Institute of the Romanian Academy, AIRA,
Str. Cutitul de Argint, 5 , Bucharest, Romania
email: magda@aira.astro.ro

[3]Observatório Nacional, MCT,
R. Gal. José Cristino, 77 20921-400, So Cristóvo, Rio de Janeiro, Brazil
email: massaf@ov.ufrj.br

Abstract. Many small ground-based telescopes (with diameter less than 2m) allow us to perform programs of observations well adapted to astrometric measurements. The improvement of limiting magnitudes thanks to the use of CCD detector and their availability make them very useful for follow-up programs or observations on alert. This communication gives several examples of research carried out by members of the IAU working group "Astrometry by small ground-based telescopes". We also propose setting up of a network of observers for the Gaia follow-up observations.

Keywords. astrometry, surveys, telescopes, asteroids, occultations

1. Introduction

What is the interest to use small telescopes for astrometry nowadays when we can access new technologies on large telescopes or in space? What will be the use of small telescopes for astrometry once the Gaia mission provides a variety parameters for astrophysical objects and Solar System bodies with a never reached accuracy? Our Working Group intends to give the answers to these questions, by collecting information on research programs carried out and on scientific results obtained thanks to these instruments. The WG would also like to promote new astrometric programs.

2. Overview

Small telescopes (up to 2 m diameter) are numerous and generally easier to access than the larger ones. These characteristics are precious and we can get benefit from them by performing well adapted programs, in particular, observations on alert or long term programs. A census organized in 2006 by our working group showed that many programs are active. Some of them will be no more relevant after the Gaia mission but may

† Astrometry by Small Ground-Based Telescopes Working Group members: Andrei A., Arlot, J.-E., Pinigin G. , Bazey N., Gontcharov G., Gumerov R., Jin Wenjing, Muinos Haro J., Niarchos P., Pereira Osorio J., Pascu D., Pauwels Th., Prostyuk Y., Pugliano A., Rafferty Th., Russell J. L., Rylkov V., Sanchez M., Shulga A., Souchay J., Tang Z. H., Teixeira R., Upgren A., van Altena W., Vieira Martins R., Zacharias N

nevertheless contribute to its preparation. Our website at http://www/imcce.fr/astrom provides this list.

3. Several international cooperations

Several projects are based on the use of the small telescopes leading to international cooperation. We can for example mention the following examples. Astrometry for the prediction of stellar occultation can be a key program in order to organize the campaign for these events. The small telescopes can carry out the photometric observations and guarantee the success thanks to the longitude and latitude coverage. Such events have been successfully predicted and observed recently by Sicardy *et al.* (2006),in particular thanks to 60 cm to 1.6 m size telescopes of the Laboratorio Nacional de Astrofysica (LNA, Itajuba, Brazil) by the Rio Group (Assafin M., Andrei A., Vieira-Martins R., Veiga C. and colleagues from Observatorio Nacional/MCT and Observatorio do Valongo). Small telescopes can perform the photometry of mutual events of natural satellites which are currently predicted for the Galilean system, but also the Saturnian and the Uranian satellites. A campaign is organized by Arlot *et al.* (2006) and lightcurves will give highly accurate astrometric measurements. Mutual events of satellites of asteroids can also be performed to get the measurement of shape, size, and the determination of orbital elements (Descamps *et al.* 2007). Another international collaboration is involved in the CCD astrometry of ICRF radio sources. The Rio Group and astronomers from the Astronomical Institute of the Romanian Academy and USNO currently use small size telescopes in this work (Assafin *et al.* 2003).

4. A ground-based network for a Gaia follow-up

Small telescopes are involved in the organization of a ground-based network for the Gaia follow-up. This network will have the goal to supplement some Gaia observations for specific Solar System objects (Thuillot 2005). Supplementary astrometric observations could be necessary for some fast moving Near Earth Asteroids, Inner-Earth orbit asteroids, for improvement of orbits and photometric measurements could be necessary for objects suspected in cometary activity. Furthermore, astrometric observations will be useful to supplement the observations of asteroids by Gaia gravitationally deflected during asteroid/asteroid encounters on dates close to the edge of the mission (Mouret *et al.* 2007). Setting up such a ground-based network of observers dedicated to the Gaia follow-up is in progress. Several candidate sites are already identified but new locations are welcome in order to cover a large span of longitudes and latitudes.

References

Arlot, J.-E., Lainey, V., & Thuillot, W. 2006, *AA*, 456, 1173
Assafin, M., Monken Gomes, P. T., da Silva Neto, D. N., Andrei, A. H., Vieira Martins, R., Camargo, J. I. B., Teixeira, R. & Benevides-Soares, P. 2005, *AJ*, 129, 2907
Descamps, P., Vieira Martins, R. (2007), *Icarus*, 187, 482
Mouret, S., Hestroffer, D. & Mignard, F. 2007, *AA*, 472, 1017.
Sicardy, B., Bellucci, A., Gendron, E., Lacombe, F., Lacour, S., Lecacheux, J., Lellouch, E., Renner, S., Pau, S., Roques, F. & 35 coauthors 2006, *Nature*, 439, 52
Thuillot, W. 2005, in C. Turon, K. S. O'Flaherty & M. A. C. Perryman (Eds.), *Proc. Gaia Symposium The Three-Dimensional Universe with Gaia* (ESA SP-576), p. 317

A Giant Step: from Milli- to Micro arcsecond Astrometry
Proceedings IAU Symposium No. 248, 2007
W. J. Jin, I. Platais & M. A. C. Perryman, eds.

Quasar astrophysics with the Space Interferometry Mission

S. C. Unwin[1], A. E. Wehrle[2], D. L. Meier[1], D. L. Jones[1], and B. G. Piner[3]

[1] Jet Propulsion Laboratory, California Institute of Technology, 4800 Oak Grove Drive, Pasadena, CA 91109; email: stephen.unwin@jpl.nasa.gov

[2] Space Science Institute, 4750 Walnut Street, Suite 205, Boulder, CO 80301

[3] Whittier College, Dept. Physics & Astronomy 13406 E. Philadelphia St., Whittier, CA 90608

Abstract. Optical astrometry of quasars and active galaxies can provide key information on the spatial distribution and variability of emission in compact nuclei. The Space Interferometry Mission (SIM PlanetQuest) will have the sensitivity to measure a significant number of quasar positions at the microarcsecond level. SIM will be very sensitive to astrometric shifts for objects as faint as $V = 19$. A variety of AGN phenomena are expected to be visible to SIM on these scales, including time and spectral dependence in position offsets between accretion disk and jet emission. These represent unique data on the spatial distribution and time dependence of quasar emission. It will also probe the use of quasar nuclei as fundamental astrometric references. Comparisons between the time-dependent optical photocenter position and VLBI radio images will provide further insight into the jet emission mechanism. Observations will be tailored to each specific target and science question. SIM will be able to distinguish spatially between jet and accretion disk emission; and it can observe the cores of galaxies potentially harboring binary supermassive black holes resulting from mergers.

Keywords. astrometry, instrumentation: interferometers, quasars: general, radiation mechanisms: general

---

Astrometry at microarcsecond accuracy is a new probe of the central engine of quasars. With flexible scheduling and single-measurement accuracy of $10-15\,\mu$as, SIM (see Unwin *et al.* 2007) is an ideal instrument to coordinate with ground instruments (optical monitoring and VLBI imaging) and space instruments (IR and X-ray) to monitor positions of blazars during outbursts. Position shifts of variable quasar nuclei relative to a local reference frame are expected, even though they will not be spatially resolved by the SIM interferometer. Specific questions that can be probed, using observations of a small number of highly variable objects, include the following: What are the sizes and geometric relations between the components of the core region (jets, accretion disk, hot corona)? How are galaxy mergers related to the AGN phenomenon - do binary black holes result from mergers and how common are they? Does the most compact non-thermal optical emission from an AGN come from an accretion disk or from a relativistic jet? Does the separation of the radio core and optical photocenter of the quasars used for the reference frame tie change on the timescales of their photometric variability, or is the separation stable? Do the cores of galaxies harbor binary supermassive black holes remaining from galaxy mergers?

In radio-loud quasars, powerful (and often relativistic) jets are ejected from the central engine. From jet models, e.g., Königl (1981) we can estimate that there will be astrometric offsets between radio and optical (up to about 70 μas, using the quasar 3C 345 at $z = 0.6$ as an example), and across the optical band (up to 30 μas). Figure 1 shows a sketch of the region where jet acceleration begins, showing the various physical components. SIM can probe the sub-parsec structure using time-dependent and color-dependent astrometry. The blue accretion disk and red corona are very compact and spatially coincident, so SIM would not expect to detect time- or color-dependent astrometric shifts in these components. However, if the red jet dominates, then a color shift should be seen, with the shift aligned with the jet, and any astrometric variability should also be co-aligned.

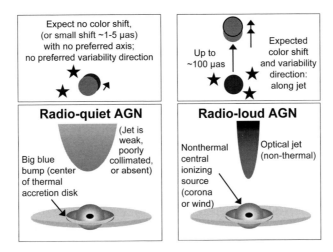

Figure 1. Schematic diagram of the structure of a typical quasar on scales $\lesssim 1$ pc. The lower left panel illustrates a radio-quiet AGN where the emission is dominated by the corona and/or accretion disk. As shown by the upper left panel, variability is not expected to be manifested as an astrometric shift, nor would the shift be color-dependent. For a radio-loud AGN or a blazar (right panels) variability (which may be color dependent) in the jet will cause a shift in photocenter which is aligned with the jet axis.

SIM can also search for evidence of binary black holes which may be end-products of galaxy mergers. If massive binary black holes are found, we have a new means of directly measuring the masses and estimating the coalescence lifetimes of the binaries. One promising candidate is OJ 287 ($z = 0.3$), based on brightness variations with 12-year periodicity Kidger (2000), Valtonen *et al.* (2006). About 14 μas of orbital motion may be expected during a five-year span.

Acknowledgements

The authors would like to thank their many colleagues, too numerous to mention, whose vision and determination over the course of more than a decade have brought SIM PlanetQuest to its current mature design. Through their efforts, we are now poised, technology in hand, to build the next generation of astrophysics instruments based on interferometry. The research described in this paper was carried out at the Jet Propulsion Laboratory, California Institute of Technology, under contract with the National Aeronautics and Space Administration.

References

Kidger, M. 2000, *AJ*, 119, 2053
Königl, A. 1981, *ApJ*, 243, 700
Unwin, S. C., *et al.* 2007, *PASP*, in press
Valtonen, M., *et al.* 2006, *ApJ* (Letters), 6423, L9

A Giant Step: from Milli- to Micro-arcsecond Astrometry
Proceedings IAU Symposium No. 248, 2007
W. J. Jin, I. Platais & M. A. C. Perryman, eds.

Gamma Astrometric Measurement Experiment: testing General Relativity with a small mission

A. Vecchiato, M. G. Lattanzi, M. Gai and R. Morbidelli

INAF- Astronomical Observatory of Torino,
via Osservatorio 20, I-10025, Pino Torinese (TO), Italy
email: vecchiato@oato.inaf.it

Abstract. GAME (Gamma Astrometric Measurement Experiment) is a concept for an experiment whose goal is to measure from space the γ parameter of the Parameterized Post-Newtonian formalism, by means of a satellite orbiting at 1 AU from the Sun and looking as close as possible to its limb. This technique resembles the one used during the solar eclipse of 1919, when Dyson, Eddington and collaborators measured for the first time the gravitational bending of light. Simple estimations suggest that, possibly within the budget of a small mission, one could reach the 10^{-6} level of accuracy with $\sim 10^6$ observations of relatively bright stars at about $2°$ apart from the Sun. Further simulations show that this result could be reached with only 20 days of measurements on stars of $V \leqslant 17$ uniformly distributed. A quick look at real star densities suggests that this result could be greatly improved by observing particularly crowded regions near the galactic center.

1. Introduction

The very first experiment devoted to the testing of General Relativity was based on measures of the bending of a light path due to the gravitational pull of massive bodies. This experiment was conducted during the solar eclipse of 1919, by Dyson, Eddington and collaborators and confirmed the predictions of GR within a 10% accuracy. In the PPN formalism, this is equivalent to say that the parameter γ is equal to 1 ± 0.1.

The same technique was used for several decades after 1919 but, despite the many further attempts, its accuracy could not be improved (Soffel, 1989), so modern experiments use different observing conditions or different kind of observables (Viking, VLBI, Cassini). The best measure of the PPN-γ parameter achieved so far was done using the Cassini data (10^{-5} level of accuracy) while, limiting ourselves to astrometric measurements, the most promising effort presently under development is represented by the Gaia mission (10^{-7} level by the end of the next decade (Vecchiato *et al.*, 2003)).

Modern technology can solve the technical difficulties which made impossible to pursue the solar eclipse tests, i.e. the short time available for the observations (limited by the eclipse duration) and the background noise due to the solar corona.

A dedicated satellite, possibly within the budget of a small mission, looking close to the Sun could compete with the present and future astrometric measurements of γ.

2. Preliminary mission constraints

After a preliminary assessment driven by the constraints of a small mission, we focused on the following measurement concept (more technological details in Loreggia *et al.* (2008)): a) observing at $\sim 2°$ away from the Sun with a FoV of $\sim 7' \times 7'$; b) reconstructing

Table 1. Estimated mission performances from simulation results.

M_{max}	mean n. of obs.	$\sigma_{\gamma*-\bar{\gamma}}$
14.5	203214	$3.95 \cdot 10^{-6}$
15.0	419309	$3.16 \cdot 10^{-6}$
16.0	964938	$2.85 \cdot 10^{-6}$
17.0	1835608	$2.61 \cdot 10^{-6}$

the attitude w.r.t. the Sun with a final accuracy $\lesssim 1''$; c) accuracy for the stellar positions at the level of the milli-arcsecond. This would suggest that $\sim 10^6$ single measurements of relatively bright stars could give a final result of $\Delta\gamma \sim 10^{-6}$. Estimates of the possible number of observations based on the star counts of the GSC-II catalog suggest that one could reach one million of observations in a reasonable amount of time.

3. First simulation results

We used the astrometric model developed in Vecchiato *et al.* (2003), which is based on a PPN Schwarzschild metric for the Sun and considers as observable the arc between two stars, to give a more reliable assessment of the mission capabilities for the measurement of γ.

The "Universe model" considers a sky with uniform spatial distribution and magnitude distribution compatible with the total GSC-II star counts, and the observation concept can be summarized as follows: a) the satellite observes simultaneously the stars in two Fields Of View (FoVs) with the Sun in between and with an exposure time of 100 s; b) the two FoVs are about $2°$ away from the Sun and have an amplitude of $7' \times 7'$; c) the observations are subsequently repeated, at intervals of 120 s, keeping the Sun between the FoVs, for about 20 days; d) the same FoVs are observed six months later (i.e. without the Sun in between).

As regards the data reduction scheme, the observables are arcs between two stars, and for each couple of FoVs, they are formed coupling each star in the upper FoV with the brightest star of the lower FoV, and vice versa.

The value of the PPN-γ parameter is estimated by the differences between the arcs measured with the Sun in between and six months later, and the results of a series of 50 Monte-Carlo runs with different magnitude limits indicate that after 20 days of measurements it seems possible to reach the 10^{-6} level of accuracy for $\delta\gamma$ (Tab. 1).

Simulations considered a uniform stellar density on the sky, but obviously that is not realistic. If we consider real star densities instead, it can be estimated that, choosing appropriate sky regions, the measurement of the γ parameter could be improved by an order of magnitude.

References

Loreggia, D., Gai, M., Lattanzi, M. G., & Vecchiato, A. 2008, *in this volume*, p.274

Soffel, M. 1989, *Relativity in Astrometry, Celestial Mechanics and Geodesy* (Springer-Verlag, Astronomy and Astrophysics Library series)

Vecchiato, A., Lattanzi, M. G., Bucciarelli, B., Crosta, M., de Felice, F. & Gai, M. 2003, *A&A*, 399, 337

A Giant Step: from Milli- to Micro-arcsecond Astrometry
Proceedings IAU Symposium No. 248, 2007
W. J. Jin, I. Platais & M. A. C. Perryman, eds.

© 2008 International Astronomical Union
doi:10.1017/S1743921308019376

Gaia Data Flow System (GDFS) Project: the UK's contribution to Gaia data processing

N. A. Walton[1], M. Cropper[2], G. Gilmore[3], M. Irwin[3], and F. van Leeuwen[3]

[1] Institute of Astronomy, University of Cambridge, Madingley Road, Cambridge, CB24 5JE, UK
email: naw@ast.cam.ac.uk

[2] MSSL, University College London, Dorking, Surrey, UK

[3] Institute of Astronomy, University of Cambridge, Madingley Road, Cambridge, CB24 5JE, UK

Abstract. Gaia is an ESA cornerstone mission which will observe some billion stars in the galaxy enabling micro-arcsec astrometric catalogues to be constructed. In addition Gaia will produce high quality photometric and spectroscopic catalogues.

The data processing tasks are large and complex. A European consortium has been formed - the Gaia Data Processing and Analysis Consortium (DPAC). This paper describes the form of the UK Gaia Data Flow System Project contribution to the DPAC.

Keywords. space vehicles, methods: data analysis

1. Introduction

Gaia (see http://www.rssd.esa.int/gaia) is an approved ESA cornerstone project, which entered its B2 phase in 2006 for launch in late 2011. Gaia will provide photometric, positional, spectroscopic and radial velocity measurements with the accuracies needed to produce a stereoscopic and kinematic census of about one billion stars in both our Galaxy and the Local Group, addressing its core science goals to quantify the formation- and assembly history of a large spiral galaxy, the Milky Way. Gaia will achieve this by obtaining a six-dimensional (spatial & kinematic) phase-space map of the Galaxy. This will be complemented by an optimised high-spatial resolution multi-colour photometric survey, coupled with the largest stellar spectroscopic and radial velocity surveys ever made.

The Gaia data set will be constructed from the many repeat observations of some billion objects. The analysis task is a complex one, involving both real-time and end-of-mission data products.

This paper notes the UK GDFS activities as part of the European wide Gaia Data Analysis and Processing Consortium. We briefly describe the data processing challenges that need to be overcome to meet the heavy demands placed by Gaia.

2. Gaia Data Processing & Analysis Consortium (DPAC)

Gaia is a mission of extremes (see Lindegren, these proceedings). This is reflected in the range of data reduction techniques required: from extreme-accuracy global astrometry to large scale photometry and spectroscopy, all based on data gathered in a constantly

Table 1. DPAC Coordination Units

Unit	Name	Leading Institute	Notes
CU1	System design	ESAC	
CU2	Data simulations	Barcelona	
CU3	Core processing	ESAC	
CU4	Object processing	CNES	
CU5	Photometry	Cambridge	UK Lead
CU6	Spectroscopy	Paris	UK major partner
CU7	Variability analysis	Geneva	
CU8	Astrophysical parameters	Heidelberg	

moving focal plane. The volume of data is large, covering nearly 2×10^{12} individual observations, but more importantly, nearly the entire volume of data needs to be considered in some of the data processing.

The Gaia data processing forms the link between the measurements produced by the satellite payload and the various catalogues with scientific results that will be the products of the mission. Unlike the design and construction of the actual satellite and payload, which is entirely funded by the European Space Agency (ESA), the Gaia data processing is the responsibility, and is being funded by, the wider scientific community. The DPAC (see Mignard, these proceedings) is the pan-European consortium that will construct the data analysis system for Gaia and deliver the science data products after launch.

To manage the various tasks and their interconnectivity, the DPAC data processing is being organised around a small number of coordination units (see Table 1) which cover all data processing and essential analysis software developments.

3. GDFS

The UK Contribution to these efforts is through the Gaia Data Flow System (GDFS) project which is funded by the UK's Science and Technology Facilities Council (http://www.stfc.ac.uk). The groups involved in the GDFS are led by the Institute of Astronomy, University of Cambridge, and include, School of Engineering and Design, Brunel University; Institute for Astronomy, University of Edinburgh; Department of Physics and Astronomy, University of Leicester; MSSL, UCL; Rutheford Appleton Laboratory, STFC.

UK participation in the construction of the Gaia Data Flow system is concentrated in the areas of CU5 - Photometric processing (where the IoA, Cambridge is the lead institute). MSSL, UCL lead the significant UK involvement in the CU6 Spectroscopy unit.

The UK GDFS project investment is £10M for all development activities to launch. All staff will be in place by March 2008. The UK team is led by Gilmore at the IoA, and managed by van Leeuwen.

The GDFS project is ensuring that its processing system is compatible with relevant Virtual Observatory standards (see e.g. http://www.ivoa.net). Further, members of the DPAC and UK GDFS projects are playing an active role in the further development of the standards process through involvement in appropriate International Virtual Observatory Alliance technical working groups, and interaction with both the UK's AstroGrid (http://www.astrogrid.org) and Euro-VO (http://www.euro-vo.org) projects.

A Giant Step: from Milli- to Micro-arcsecond Astrometry
Proceedings IAU Symposium No. 248, 2007
W. J. Jin, I. Platais & M. A. C. Perryman, eds.

A Star Image Extractor for the Nano-JASMINE satellite

M. Yamauchi[1][2], N. Gouda[1], Y. Kobayashi[1], T. Tsujimoto[1], T. Yano[1], M. Suganuma[1], Y. Yamada[3], S. Nakasuka[2], N. Sako[2] *et al.*

[1]National Astronomical Observatory of Japan,
2-21-1 Osawa, Mitaka, Tokyo, Japan

[2]University of Tokyo,
7-3-1 Hongo, Bunkyo-ku, Tokyo, Japan

[3]Kyoto University,
Yoshida-Honmachi, Sakyo-ku, Kyoto, Japan
email: yamauchi@merope.mtk.nao.ac.jp

Abstract. We have developped a software of Star-Image-Extractor (SIE) which works as the on-board real-time image processor. It detects and extracts only the object data from raw image data. SIE has two functions: reducing image data and providing data for the satellite's high accuracy attitude control system.

Keywords. methods: data analysis, techniques: image processing

We have developed a software of Star Image Extractor (SIE) which works as the on-board real-time image processor. It detects and extracts only an object data from raw image data. SIE will be equipped to the Nano-JASMINE satellite. Nano-JASMINE is a small astrometry satellite that that observe objects in our galaxy. It will be launched in 2009 for a two year mission. Nano-JASMINE observes an object in the Time Delayed Integration (TDI) mode. TDI is one of the operation modes of CCD. The data are obtained, by reading out the CCD at a rate synchronized with a vertical charge transfer of CCD. The image data are sent thorough SIE to the Mission-controller. We show the data flow of the imaging system in Fig. 1

SIE has two purposes. One of the purposes is reducing image data. The original data rate of image is 2 Mb per second. The amount of raw images is too large to be transmitted to the ground, so we use only extracted images for the transmission. SIE reduces the data rate from 2 Mbps to N×50 bps, where N is the averaged number of detected objects in one field of view. Nano-JASMINE should be able to detect a few stars each second. Another purpose is to make the data available for the satellite's high accuracy attitude control system. The extracted image is used as a rotation sensor, because the point-spread function of an object will be elongated more in the TDI direction.

A software simulator of SIE has been developed to test the star extracting algorithm. FITS formatted files are used for input and output data of the simulation.Two kinds of input data are used for the simulation. One is the image that obtained by the imaging experiments using the hardware of Nano-JASMINE imaging system. Another is the image, produced by the software simulator, which has no background noise. The simulation experiments have shown that we can extract stars from the input images successfully. Results of our simulations are shown in Fig. 2. The software simulator has been finished. Now the logic circuit is under development. The logic circuit will be written on a Field Programmable Gate Array (FPGA) device.

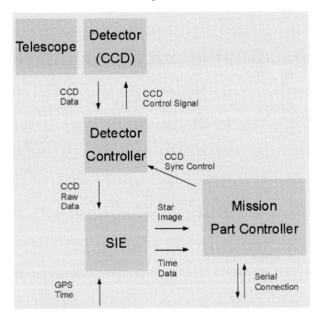

Figure 1. Data flow of imaging system

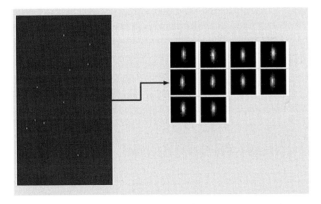

Figure 2. Simulation result (Extracting stars)

References

Hirte, S. & Scholz, R. 2002, *Die Verarbeitung des Rohdatenflusses eines Weltrauminterferome-ters*, Schlussbericht zur Studie, Astrophysikalisches Institut Potsdam

Kovalevsky, J. 2001, *Modern Astrometry(Second Edition)*, Springer

Takagi, M. & Shimoda, H. 2004, *Handbook of Image Analysis*, University of Tokyo Press

A Giant Step: from Milli- to Micro-arcsecond Astrometry
Proceedings IAU Symposium No. 248, 2007
W. J. Jin, I. Platais & M. A. C. Perryman, eds.

Space Astrometry JASMINE

T. Yano[1], N. Gouda[1], Y. Kobayashi[1], Y. Yamada[2], T. Tsujimoto[1], M. Suganuma[1], Y. Niwa[1], M. Yamauchi[3]

[1] National Astronomical Observatory, Mitaka, Tokyo 181-8588, Japan
email: `yano.t@nao.ac.jp`, `naoteru.gouda@nao.ac.jp`, `yuki@merope.mtk.nao.ac.jp`,
`taku.tsujimoto@nao.ac.jp`, `suganuma@merope.mtk.nao.ac.jp`, `kazin.niwa@nao.ac.jp`

[2] Graduate School of Science, Kyoto University, Sakyo-ku, Kyoto 606-8502, Japan
email: `yamada@scphys.kyoto-u.ac.jp`

[3] Department of Astronomy, School of Science, University of Tokyo, Tokyo 113-0033, Japan
email: `yamauchi@merope.mtk.nao.ac.jp`

Abstract. JASMINE is the acronym of the Japan Astrometry Satellite Mission for INfrared (z-band :0.9 micron) Exploration, and is planned to be launched around 2017. The main objective of JASMINE is to study the fundamental structure and evolution of the Milky Way bulge components. In order to accomplish these objectives, JASMINE will measure trigonometric parallaxes, positions and proper motions of about ten million stars in the Galactic bulge with a precision of 10 microarcsec at $z = 14$mag.

The primary mirror for the telescope has a diameter of 75cm with a focal length of 22.5m. The back-illuminated CCD is fabricated on a 300 micron thick substrate which is fully depleted. These thick devices have extended near infrared response. The size of the detector for z-band is 3cm×3cm with 2048×2048 pixels. The size of the field of view is about 0.6deg×0.6deg by using 64 detectors on the focal plane. The telescope is designed to have only one field of view, which is different from the designs of other astrometric satellites. JASMINE will observe overlapping fields without gaps to survey a total area of about 20deg×10 deg around the Galactic bulge. Accordingly we make a "large frame" of 20deg×10 deg by linking the small frames using stars in overlapping regions. JASMINE will observe the Galactic bulge repeatedly during the mission life of about 5 years.

Keywords. space vehicles, astrometry, Galaxy: bulge

---

1. Summary of the observational procedures

The procedures of our method for measuring astrometric parameters of Galactic bulge stars can be summarize in the following three steps.

1. Centroiding of stars

The centroids of stars in a field of view are determined using a photon weighted means of stars Yano *et al.* (2004) Yano *et al.* (2006). Consequently positions of stars in the field of view (hereafter we call small frame) are obtained.

2. Construction of a large frame

JASMINE will take overlapping fields of view without gaps to survey the area of about 20°×10°. During this time (about 10h) variations or distortions of the optical equipments must be small (order of 10~100pm). In order to confirm the stability of the optical equipments, they are monitored using a laser interferometer. Then, we connect all small frames using positions of stars in the overlapping areas to form a large region (hereafter large frame).

3. Estimating the parallaxes of stars

We continue the above procedure about 10^4 times during the mission life. We derive the astrometric parameters such as parallaxes of stars using the large frames. In connecting all large frames we make use of QSOs as fixed points. As a consequence we obtain the positions of stars directly into the International Celestial Reference System (ICRS).

2. QSO

We plan to make use of QSOs in order to obtain astrometric parameters and fix to ICRS (International Celestial Reference System). There are many bright QSO candidates in the survey area of JASMINE. There are over 50 candidates in the survey area, and there are 14 candidates brighter than K⩽13. These candidates are selected using the photometric variability during a few years in OGLE (Optical Gravitational Lensing Experiment). Although we do not know how many candidates are QSOs, we expect that about 40-70 % of candidates are QSOs. Up to now one bright QSO is confirmed in the survey area.

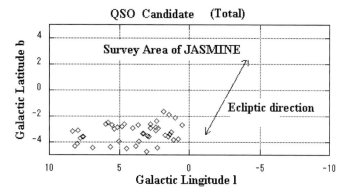

Figure 1. QSO candidates in the survey area.

3. Thermal environment

In order to satisfy the objective of high accuracy measurement, the optical equipment (optics, detectors, and so on) must be stabilized. SiC is one of the candidates for material of optics (mirrors and structures). In this case the thermal capacity decreases as the temperature decreases. The expansion coefficient also decreases as the temperature decreases. The effect of the thermal expansion exceeds that of the thermal capacity. As a consequence, optical equipment is stabilized as the temperature decreases. However the values of the thermal expansion at low temperatures are uncertain. Then it is important to investigate thermal expansion of materials at low temperature.

References

Yano, Taihei, Gouda, Naoteru, Kobayashi, Yukiyasu, Tsujimoto, Takuji, Nakajima, Tadashi, Hanada, Hideo, Kan-ya, Yukitoshi, Yamada, Yoshiyuki, Araki, Hiroshi, Tazawa, Seiichi, Asari, Kazuyoshi, Tsuruta, Seiitsu, & Kawano, Nobuyuki 2004, *PASP*, 116, 667
Yano, Taihei, Araki, Hiroshi, Gouda, Naoteru, Kobayashi, Yukiyasu, Tsujimoto, Takuji, Nakajima, Tadashi, Kawano, Nobuyuki, Tazawa, Seiichi, Yamada, Yoshiyuki, Hanada, Hideo, Asari, Kazuyoshi, & Tsuruta, Seiitsu 2006, *PASP*, 118, 1448

A Giant Step: from Milli- to Micro-arcsecond Astrometry
Proceedings IAU Symposium No. 248, 2007
W. J. Jin, I. Platais & M. A. C. Perryman, eds.

© 2008 International Astronomical Union
doi:10.1017/S1743921308019406

Time transfer by laser link between China and France

C. Zhao[1,2] W. T. Ni[1] and E. Samain[3]

[1]Center for Gravitation and Cosmology, Purple Mountain Observatory, Chinese Academy of Sciences, Nanjing 210008, China.
email: zhaocheng@pmo.ac.cn, wtni@pmo.ac.cn

[2]Graduate University of Chinese Academy of Sciences, Beijing 100049, China.
[3]Observatoire de la Côte d'Azur, UMR Gemini, Caussols, 06460 France.
email: etienne.samain@obs-azur.fr

Abstract. To advance from milli-arcsecond to micro-arcsecond astrometry, time keeping capability and its comparison among different stations need to be improved and enhanced. The T2L2 (Time transfer by laser link) experiment under development at OCA and CNES to be launched in 2008 on Jason-2, allows the synchronization of remote clocks on Earth. It is based on the propagation of light pulses in space which is better controlled than propagation of radio waves. In this paper, characteristics are presented for both a common view and non-common view T2L2 comparisons of clocks between China and France.

Keywords. time, standards, astrometry

1. Introduction

T2L2 on Jason-2 will permit to synchronize remote ground clocks and compare their frequency stabilities using laser telemetry with a performance of 1-2 orders of magnitude better than present (Samain & Weick (2006)). T2L2 will allow to measure the stability of remote ground clocks over continental distances with 1ps over 1000s. China possesses five permanent laser ranging stations and a mobile station – TROS (Yang. (2001)), their geography allows common view comparisons among each other, and non-common view comparisons between Chinese and French stations.

2. T2L2 principle and performance budget

A given light pulse is emitted from station A at time t_s, and received in Jason-2 at time t_b. This pulse is also bounced from the retroreflector on Jason-2 and received at station A at time t_r. The synchronization X_{AS} between the ground clock A and the satellite clock S is:

$$X_{AS} = \frac{t_s + t_r}{2} - t_b + \tau_{Geometry} + \tau_{Atmosphere} + \tau_{Relativity} \qquad (2.1)$$

Also we can get X_{BS}, the synchronization between another ground clock B and the satellite clock. So, the time transfer between A and B can simply defined by:

$$X_{AB} = X_{AS} - X_{BS} \qquad (2.2)$$

The performance budget is shown in Fig. 1 (Guillemot & Samain(2006)). In the common view configuration, T2L2 should reach the majority of current atomic clocks for integration times exceeding 1000s. In non-common view, with limitations imposed by the onboard clock, T2L2 will offer an interesting alternative in calibration campaigns of radio frequency and time transfer systems based on transportable stations.

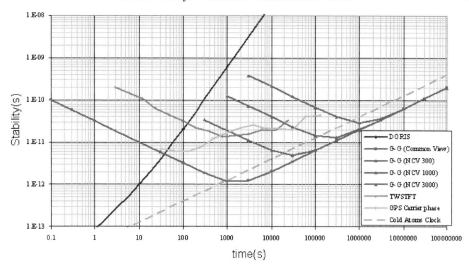

Figure 1. T2L2 ground-to-ground time stability in common view (CV) and non-common view (NCV) configuration with various interval ("NCV 300" means non-common view comparison with 300s interval between two stations).

Table 1. Grasee(France) - Chinese Stations Time Transfer in Non Common View

Station	Total Access	Good Access	Average Duration [sec] Grasse	China	Interval [sec] Average	Maximum	Minimum
Grasse-Beijing	35	25	972.92	923.11	392.94	554.95	345.52
Grasse-Changchun	32	24	986.73	973.47	401.20	620.41	345.81
Grasse-Kunming	26	18	958.15	897.78	536.88	763.02	456.57
Grasse-Shanghai	29	20	1030.85	893.98	605.99	823.41	531.02
Grasse-Wuhan	29	21	1021.22	883.47	560.09	867.04	489.76

3. Time transfer between Grasse and Chinese stations

T2L2, a passenger instrument of Jason-2, will be launched by a Delta II 7320 into an earth orbit with period 6745.72s, inclination 66° and apogee 1,336km high. The orbit drifts to the east 39.5° every revolution and returns to the original orbit every cycle (9.9156 days). This orbit will allow common view comparisons with about 3000km baseline and six repeating ground tracks per day over the ground stations.

For common view time transfer, all five fixed laser ranging stations in China are able to track the Jason-2 synchronously with a long common duration. We expect abundant time transfer results in common view between these stations.

For non-common view time transfer, as Table 1 shows, there are around 20 good passes that each Chinese station can access Jason-2 after several minutes when Grasse station loses its sight. This is quite good for non-common view time transfer.

References

Guillemot, P. & Samain, E. *et al.* 2006, *in proc of the IFCS 2006.*, p. 771
Samain, E. & Weick, J. *et al.* 2006, *International Journal of Modern Physics D*, in press.
Yang, F.M 2001, *Surveys in Geophysics*, D22, p. 465

A Giant Step: from Milli- to Micro-arcsecond Astrometry
Proceedings IAU Symposium No. 248, 2007
W. J. Jin, I. Platais & M. A. C. Perryman, eds.
© 2008 International Astronomical Union
doi:10.1017/S1743921308019418

From the Roemer mission to Gaia

E. Høg

Niels Bohr Institute, Copenhagen University, Juliane Maries Vej 30, 2100 Copenhagen Ø,
Denmark
email: erik@astro.ku.dk

Abstract. At the IAU symposium in Shanghai September 1992 the present author made the first proposal for a specific mission concept post-Hipparcos, the first scanning astrometry mission with CCDs in time-delayed integration mode (TDI). Direct imaging on CCDs in long-focus telescopes was described as later adopted for the Gaia mission. The mission called Roemer was designed to provide accurate astrometry and multi-colour photometry of 400 million stars brighter than 18 mag in a five-year mission. The early years of this mission concept are reviewed.

Keywords. space vehicles, astrometry, instrumentation: photometers

---

When Hipparcos was launched in August 1989 the Hipparcos Science Team (HST) was present in Kourou and we were greatly relieved seeing the take-off after the many years of preparation. But that changed to grim disappointment the next day when we learned that the apogee boost motor had not started so that the satellite was stuck in an elliptical transfer orbit instead of the intended geostationary. This endangered the whole mission and we would possibly only get a much shorter set of poor observations, perhaps only months and not the planned three years. Passing through the radiation belts every few hours could soon destroy the electronics and solar cells.

In this mood, but optimistic as always, I presented the Hipparcos mission on behalf of Michael Perryman who could not be present, and the Tycho project at the IAU Symposium No. 141 October 17–21, 1989 in Leningrad (now St. Petersburg). The audience was full of high hopes for Hipparcos - hopes which were in fact justified as we should later see. We noticed that Soviet (later Russian) colleagues presented ideas at the Symposium for a successor to Hipparcos. They themselves had three projects on the drawing boards: AIST/STRUVE, LOMONOSOV, and REGATTA-ASTRO. The basic idea was to reobserve the 120 000 Hipparcos stars and utilize the positions from Hipparcos and those from a new epoch to get much better proper motions than Hipparcos alone would achieve, even if its severe problems would be cured.

Such ideas were far beyond the horizon of anyone in the Hipparcos team, busy as we were to get our mission to work and to perform the very complex data analysis. I was myself leader of one of the two Hipparcos data analysis teams and of the Tycho team and thus had more than enough to look after.

Shortly later I was invited to lecture about Hipparcos at the Pulkovo Observatory in Leningrad, the Mission Control Center in Moscow, and the Kislovodsk Observatory in Caucasus. I was accompanied on the journey in August 1990 by M.S. Chubey, V.V. Makarov, and V.N. Yershov so we had plenty of time for discussions. I wanted to understand how their AIST project functioned, but unexpectedly, after a day I was more occupied by designing a second Hipparcos myself, realizing that it could easily be made ten times more efficient in utilizing the star light, mainly by employing more detectors, while keeping the same telescope aperture of 0.29 m.

In June 1991 an International Symposium "Etalon" Satellites was held in Moscow where I presented a paper with Mark Chubey "Proposal for a second Hipparcos", but the proceedings were not published. If launched ten years after Hipparcos the mission could obtain proper motions for the 120 000 Hipparcos stars with an accuracy 10 times better than expected from Hipparcos as well as 1 mas accuracy for all astrometric parameters of some 400 000 stars and four-colour photometry for two million stars. This proposal was considered by the Mission Control Centre in Moscow.

During 1990-91 we met many times for discussion of our ideas as they developed, and Lennart Lindegren joined us. At the HST meetings I got a few minutes to present the progress, somewhat reluctantly allowed by the chairman who probably thought we had enough to do with one mission for the time being.

In 1991 I had left the study of photon counting techniques as in Hipparcos and tried to use CCDs, a completely new technique for me. I learnt it from our engineer, R. Florentin Nielsen, and designed a detector system using a modulating grid as in Hipparcos. The result was 1000 times better light efficiency than Hipparcos (see Høg & Lindegren in IAU Symposium 156, 1993).

Having done that I dropped the modulating grid and tried direct imaging on the CCDs imployed in drift-scan mode or time-delayed integration (TDI). That design was called Roemer and gave 100 000 times better light efficiency with the same telescope aperture (0.29 m), but a very long focal length was needed, 5 m instead of the 1.4 m in Hipparcos (see Høg 1993). Both systems were presented at the IAU Symposium 156 in Shanghai September 15–19, 1992.

The Roemer design was proposed in June 1993 for the Third Medium Size ESA Mission (M3) by a team mainly from the HST. The proposal got a high rating in the ESA selection committee, but was not finally selected because it was considered to come too early after Hipparcos. This view was not shared by the proposers, but in hindsight it was a wise decision because it gave us time for much development in the subsequent years.

Interferometry was proposed at the IAU Symposium No. 166 in August 1994 by Lindegren & Perryman "A small interferometer in space for global astrometry: The Gaia concept", stating the "very strong scientific case for global optical astrometry at the 20 microarcsec accuracy level." The satellite should contain three Fizeau-type interferometers with 2.5 m baselines.

At the same IAU Symposium a 10 microarcsec mission (Roemer+) with 9-colour intermediate- and wide-band filter photometry was proposed by the present author. The better performance was obtained with two telescopes of larger apertures of 70 cm instead of 29 cm. Picometer gauges were adopted to monitor the alignment of the telescopes.

The development of instrument ideas had mainly three scientific goals: higher astrometric accuracy of 10 microarcsec instead of the 100 microarcsec envisaged in Roemer, measurement of radial velocities for the brighter stars with the satellite, and better multicolour photometry. These improvements were considered crucial for an ambitious ESA mission aiming for understanding the details of our Galaxy. Thorough assessment of the scientific goals and the data analysis was also made. – Finally, Gaia is now scheduled for launch in 2011 on a 5 year mission to measure 1000 million stars brighter than 20 mag.

2. Why interferometry - and why not?

There was a widespread belief at the time of the Roemer proposal, Lindegren *et al.* (1993), that interferometry could give better astrometry from space, and a section was included: "Towards 10 microarcsec astrometry: The FIZEAU option". It was not part of the baseline Roemer proposal, but was meant "to point out a possible development towards a scanning satellite with ten times the angular accuracy of Roemer", and the enourmous scientific benefits of such an accuracy for millions of stars were outlined.

A Fizeau principle was subsequently used in several proposals for scanning astrometric satellites, e.g., the Gaia concept of 1994 mentioned above, FAME and DIVA.

I agreed that interferometric options should be deeply studied as they in fact were during the following years. Perhaps the complications of interferometry could be alleviated, or at least the fallout from studies could bear fruit in other (unforeseen) contexts. These studies always focussed on a scanning astrometry satellite similar to Hipparcos because a systematic scanning of the sky was considered the only way to measure the millions of stars required for our scientific goal. A pointing satellite could never do that, but would of course have the advantage of allowing longer integration time on any selected area.

My own preference in instrument design has always been to identify and focus on difficulties and try to solve or circumvent them. So I believed more in direct imaging on CCDs from full-aperture telescopes than in the diluted apertures required for interferometry. The Roemer+ design of 1994 used full apertures and obtained 10 microarsec, but it required picometer gauges to monitor the alignment of telescopes, a technique nearly always required in interferometric options.

In 1995 we designed an interferometric option later published by Høg et al. (1997). It used a beam-combiner of 150 cm aperture and a simple telescope, basically an aplanatic Gregorian system. A prism provided a low dispersion perpendicular to the scanning direction so that spectrophotometry could be obtained. This new option of Gaia was adopted in ten times smaller size for the proposal by Röser et al. (1997). This was a small German astronomy satellite, DIVA, planned for launch in 2003 to measure about 40 million stars as a fore-runner for Gaia. But funding did not follow suit.

The ESA studies of the interferometric option are described at length in a section (pp.331-338) of ESA (2000) and complete references are given. The history of the development of Gaia is briefly summarized in Perryman et al. (2001). One of the problems was that the split pupil of an interferometer did not allow accurate measurement of the stars about 20 mag required for the ambitious scientific goal, but only about 17 mag. Another problem came from the required data rate to be transmitted from the satellite. An interferometric image requires a lot more data points to cover the fringes of a star than a direct stellar image from a full aperture. The higher data rate could well be accepted from a geostationary orbit, but the thermal control during eclipses would jeopardize the instrument stability, so the orbit around L2 was required for thermal stability. Here the data rate of one Megabit per second for the full aperture option was acceptable, but not the higher rate for interferometry. Other problems of interferometry were indentified and in the end the full aperture could be selected and we were sure that all had been done to investigate both options, based on industrial studies by Matra Marconi Space for the baseline design and Alenia Aerospazio for the interferometric.

References

ESA 2000 *Gaia: Composition, Formation and Evolution of the Galaxy* Technical Report ESA-SCI(2000)4

Høg E. 2007, This poster on 6 pages, including the complete list of references at www.astro.ku.dk/~erik/ShanghaiPoster.pdf, and some slides at www.astro.ku.dk/~erik/ShanghaiHoeg.pdf

IAU Symposium No. 141, 1990, J. H. Lieske & V. K. Abalakin (eds.) *Inertial Coordinate System of the Sky.*

IAU Symposium No. 156, 1993, I. I. Mueller & B. Kolaczek (eds.) *Developments in Astrometry and their Impact on Astrophysics and Geodynamics.*

IAU Symposium No. 166, 1995, E. Høg & P. K. Seidelmann (eds.) *Astronomical and Astrophysical Objectives of sub-milliarcsecond Optical Astrometry.*

A Giant Step: from Milli- to Micro-arcsecond Astrometry
Proceedings IAU Symposium No. 248, 2007
W. J. Jin, I. Platais & M. A. C. Perryman, eds.

Maintenance and densification: current proper-motion catalogs

T. M. Girard

Department of Astronomy, Yale University,
P.O. Box 208101, New Haven, Connecticut, 06520-8101, USA
email: girard@astro.yale.edu

Abstract. An overview of currently available, large-area, proper-motion catalogs is presented. These include the well-known catalogs based on historical Schmidt-telescope surveys as well as other projects that make use of observational material the primary purpose of which, from inception, was the determination of proper motions. The various catalogs are characterized and compared, with an emphasis on their limitations and their appropriateness for various astrophysical uses.

In addition to allowing for the maintenance of a practical celestial reference system, absolute proper-motion surveys provide the raw material from which a better understanding of our Galaxy's structure and kinematics can be built. Several examples will be cited in which large proper-motion surveys are used to probe and describe the distinct stellar components that comprise our Milky Way Galaxy.

Keywords. catalogs, astrometry

---

1. Introduction

Proper-motion measures perform the glamourless, utilitarian function of maintaining positional reference catalogs into the future. Beyond this obvious and important role they also provide more direct astrophysical functions. Among these are their use as a distance/luminosity estimator, thus providing a means of discrimination between classes of stellar objects, (e.g. via reduced proper motions). An even more direct astrophysical use of proper-motion measures is in the construction of transverse velocity distributions, a key tool in the study of the kinematics and equilibrium structure of the various Galactic components, (i.e., thin disk, thick disk, and halo).

Not all proper-motion catalogs are designed to accomplish each of the functions mentioned above. It is important to select the appropriate proper-motion catalog for the proper-motion task at hand. This may appear obvious, but the literature is littered with studies that have misused available proper-motion catalogs, overstretching their limits, by mistaking precision for accuracy or assuming $\sqrt{N}$ can lead to perfection.

It is thus the purpose of this paper to give an overview of *some* of the currently available proper-motion catalogs and their approximate characteristics such that non-astrometrists can choose and use them wisely. The following section lists and describes the most oft-used as well as some lesser-known proper-motion catalogs. In each of the subsequent sections examples are given of particular catalogs appropriate to the three tasks mentioned earlier; *i)* reference frame utility, *ii)* stellar distance estimation and luminosity class discrimination, and *iii)* kinematic study of Galactic components.

Table 1. *Some* Currently available proper-motion catalogs

"The Big Three"

#. Catalog	N	Coverage	Dens	V_{comp}	σ_μ (r/a)	Source Material/Notes
01. Hipparcos	118K	all-sky	3	7.5	0.9 / 0.25	fully satellite-based
02. Tycho-2	2.5M	all-sky	62	11.5	2.5/ ?	Hipp. starmapper + grnd-based cats
03. UCAC2	48M	$\delta < +45°$	1.6K	16.5	~3/ ~3	recent CCD + grnd-based cats

Schmidt survey-based

#. Catalog	N	Coverage	Dens	V_{comp}	σ_μ (r/a)	Source Material/Notes
04. USNO-B1.0	1046M	all-sky	25K	~21	~5/ -	POSS1, POSS2, ESO, AAO, SERC
05. GSC 2.3	945M	all-sky	23K	~21	??/ ?	POSS1, POSS2, ESO, AAo, SERC
06. SuperCOSMOS	>1000M	$\delta < +3°$	46K	~21	10-50/ <1 (?)	POSS1, POSS2, UKST, ESO

Schmidt-derivative

#. Catalog	N	Coverage	Dens	V_{comp}	σ_μ (r/a)	Source Material/Notes
07. SDSS-USNOB1	~8M	2K sq deg	3.8K	19.7	3-3.5/ <0.5(?)	SDSS minus USNOB1

Proper-motion threshold

#. Catalog	N	Coverage	Dens	V_{comp}	σ_μ (r/a)	Source Material/Notes
08. rNLTT	59K	44% sky	3	~18	3-5/	NLTT+POSS1+2MASS
09. LSPM	122K	[all-sky]	3	21	~5/	DSS's and SUPERBLINK software $\mu > 150$ mas/yr

Rank-merged

#. Catalog	N	Coverage	Dens	V_{comp}	σ_μ (r/a)	Source Material/Notes
10. NOMAD	1100M	all-sky	27K	~21	1-5/ 0.2 - ?	Hipp.+Tycho2+UCAC2+USNOB1

Astrograph-based proper-motion surveys

#. Catalog	N	Coverage	Dens	V_{comp}	σ_μ (r/a)	Source Material/Notes
11. NPM	0.4M	$\delta > -23°$	13	—	~5/ ~1	astrograph w/ obj. grating
12. SPM	11M	$-30°$ band (3700 sq°)	2.9K	17.5	3-5/ 0.4	astrograph w/ obj. grating

CdC-based

#. Catalog	N	Coverage	Dens	V_{comp}	σ_μ (r/a)	Source Material/Notes
13. PM2000	2.7M	$+15°$ band	0.5K	15.4	TBD	Bordeaux Meridian Circle (M2000) +CdC+AC2000+USNOA2+YS3
14. CdC-SF1	0.5M	$-6°$ band (1080 sq°)	0.5K	15.1	1.1-3/ ?	San Fernando CdC + UCAC2 (commercial flatbed scanner)

01. ESA 1997, (ESA SP-1200) (Noordwijk: ESA)
02. Høg *et al.* 2000, AA 355, 27
03. Zacharias, *et al.* 2004, AJ 127, 3043
04. Monet *et al.* 2003, AJ 125, 984
05. Spagna *et al.* 2006, Mem. S. A. It. 77, 1166 (see also Bucciarelli in these proceedings)
06. Hambly *et al.* 2001, MNRAS 326, 1315
07. Munn *et al.* 2004, AJ 127, 3034
08. Salim & Gould 2003, ApJ 582, 1011
09. Lépine & Shara 2005, AJ 129, 1483 (see also Lépine in these proceedings)
10. Zacharias *et al.* 2004, AAS 205.4815
11. Hanson *et al.* 2004, AJ 128, 1430
12. Girard *et al.* 2004, AJ 127, 3060
13. Ducourant *et al.* 2006, AA 448, 1235
14. Vicente *et al.* 2007, AA 471, 1077

2. Proper-motion catalogs

Table 1 lists some of the more useful proper-motion catalogs currently available to the astronomy community. The catalogs have been divided into groups in a somewhat arbitrary fashion, primarily by source material, although three catalogs have been singled out as being the most often used and (generally) rightly so. The columns of the table list the catalog name; the number of objects it contains; the sky coverage; the density in objects per square degree; a rough estimate of the V-band completeness limit if it has one; the estimated proper-motion uncertainty per star (in mas/yr) both in a relative sense (r) and in an absolute sense (a), that is the limit of the proper-motion system of the catalog and thus the "floor" one should expect when considering ensembles of stars; and finally a description of the source material from which the catalog was constructed and/or other notes. References for the catalogs are given in the tablenotes.

The Hipparcos and Tycho-2 catalogs need no introduction. These are superb resources in general as well as specifically with respect to their proper motions. Although ostensibly on the system of the ICRS via Hipparcos, the uncertainty of the absolute proper-motion system of Tycho-2, particularly at the faint end, is not well-established due to the use of early-epoch ground-based plate catalogs and the everpresent question of residual magnitude equation, as discussed in Section 5. The UCAC2, with its density of 1600 stars per square degree, is an extremely useful astrometric reference. In its current version its proper motions suffer from systematics that are on the order of 3 mas/yr. Addressing this, as well as issues of completeness and the inclusion of the north celestial cap, will be major improvements in the next UCAC release expected in 2008.

Among the catalog products of the major Schmidt-survey plate digitization programs, the USNO-B1 is probably the most often used. Its proper motions are not absolute. More precise and accurate are the proper motions derived from the USNO-B1 in combination with other databases, such as the Sloan Digital Sky Survey.

Some other catalogs worthy of mention are those specifically geared to find stars with high proper motions, above some cutoff threshold. Naturally, these are most useful in studies seeking to detect nearby stars. Examples of such studies are mentioned briefly in Section 4.

NOMAD is a simple rank merging of existing astrometric catalogs. The proper-motion systems of the various sources should not be expected to be homogeneous. On the other hand, absolute proper motions on an absolute frame were the design goal of the NPM and SPM programs. The NPM, however, is not complete to any magnitude, measuring just a fraction of the stars on the plates due to time constraints. The most recent catalog of the SPM program, the SPM3, is nominally complete to $V=17.5$.

Finally, there are recent proper-motion programs that attempt to exploit the largely neglected turn-of-the-century Carte du Ciel plates. The San Fernando project being noteworthy for its successful use of a commercial flatbed scanner in digitizing the original plate material.

3. Utilitarian use

Perhaps the most common use of astrometric catalogs is as a reference system upon which to express the celestial coordinates of a target or targets of interest. This may, for example, be to facilitate cross-identification by position, to provide input coordinates for follow-up fiber-optic spectroscopic observations, or to calibrate the optical field angle distortion of a new telescope and/or instrument, (the last being the most demanding of these examples). In these cases the most important aspects of the reference catalog to consider are its precision at epoch and whether or not the catalog provides a sufficient number of stars that are in a measurable magnitude range. The proper motions serve the purpose of lessening the degradation of positional precision at the epoch of interest.

Figure 1 illustrates the tradeoff between precision at a recent epoch and magnitude of available reference stars, which of course is related to the density of reference stars (see Table 1) for the most commonly used astrometric reference catalogs. These include "the big three" along with the USNO-B1, (which one might consider the "the big fourth"), and the 2MASS point source catalog (Skrutskie *et al.* 2006). While lacking proper motions and not originally intended as an astrometric catalog, 2MASS can provide a dense set of reference stars at a modern epoch. The utility of 2MASS and the USNO-B1 cannot be overestimated for cases when deep and dense astrometric reference fields are needed.

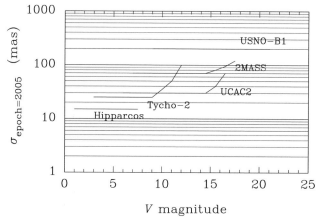

Figure 1. A comparison of useful astrometric catalogs. Approximate positional uncertainties at epoch 2005 are shown schematically as a function of magnitude.

4. Distance/luminosity estimation

Proper motions can provide a practical estimate of distance, in a statistical sense. A clever use of proper motions in this regard is to separate a sample of stars into known luminosity classes by means of the reduced proper motion. The reduced proper motion is a proxy of sorts for absolute magnitude given by the following expression,

$$H_V = V + 5log\mu + 5,$$

where V is the apparent magnitude in some passband (e.g. Johnson V) and μ is the total proper motion in arcsec/yr. The technique relies on stars in a sample sharing a common and relatively narrow tangential velocity distribution and the simple relation between tangential velocity and proper motion, i.e. the ratio of these being proportional to distance. In practice the technique works because of the large range in stellar luminosities and distances relative to their variation in velocities. In particular, the reduced proper motion diagram, in which H is plotted versus a color index, can provide an excellent discriminant of luminosity class, especially for nearby star samples.

A reduced proper motion diagram, taken from Finch *et al.* (2007), is reproduced in the left panel of Figure 2. White dwarf candidates are separated nicely from main sequence stars in this manner using proper motions derived from multiple epochs of Schmidt survey plate scans and software developed as part of the SuperCOSMOS-RECONS project, (Henry *et al.* 2004). An example of the astrophysical knowledge that can be gleaned from

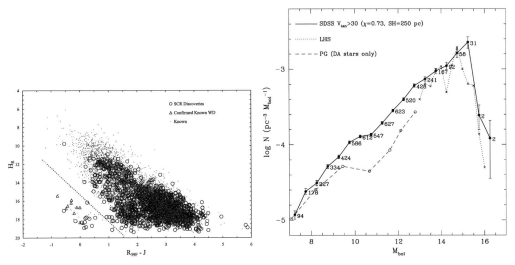

Figure 2. Examples of the use of reduced proper motions. *Left:* This reduced proper-motion diagram of Finch *et al.* (2007, their Fg. 5) was constructed using their SuperCOSMOS-RE-CONS proper-motion determinations. The white dwarfs are well separated from main-sequence dwarfs. *Right:* In a similar study, the white-dwarf luminosity function determined by Harris *et al.* (2006, their Fig. 4) was based on reduced-proper-motion selection using SDSS-USNOB1 proper motions.

this type of investigation is given in the right panel of Figure 2. It shows the white-dwarf luminosity function based on a similar study that made use of proper motions derived from SDSS-USNOB1, (see Harris *et al.* 2006). In this and the Finch *et al.* study a deep sample of precise proper motions is the key to success. Accuracy of the proper-motion system is not critical.

5. Kinematic description of Galactic components

Both precision and accuracy of a proper-motion catalog are important in studies of Galactic kinematics. As an example, the Yale/San Juan Southern Proper Motion (SPM) program, the southern-sky counterpart to the Lick NPM program, was designed from inception to yield precise and accurate proper motions over a wide magnitude range. These photographic surveys relied on an objective grating and two separate exposures per field to not only increase the effective dynamic range of the plates but also to provide a means of correcting for the bane of the photographic astrometrist - magnitude equation. Magnitude equation is the term given to the magnitude-dependent bias in image centers that is caused by the combination of non-linear detector and asymmetric image profiles. It is present to some degree in virtually all photographic measures and, to a generally lesser extent, in CCD positions as well. As demonstrated by Girard *et al.* (1998), the grating images on the SPM plates allow a self-calibration for magnitude equation to be bootstrapped from each plate and an individual correction applied.

Using SPM3 proper motions and 2MASS J, K photometry, Girard *et al.* (2006) isolated a sample of thick-disk giants at the South Galactic Pole (SGP) in order to ascertain the kinematical nature of the Galactic thick disk. Distances were estimated photometrically from the 2MASS photometry in a Monte Carlo fashion that automatically incorporated and modelled the Malmquist and Lutz-Kelker type biases. (See Girard *et al.* 2006 for details.)

The resulting *observed* thick-disk velocity distributions are shown in Figure 3. The Monte-Carlo modeling allows us to infer the *intrinsic* distributions, assuming the

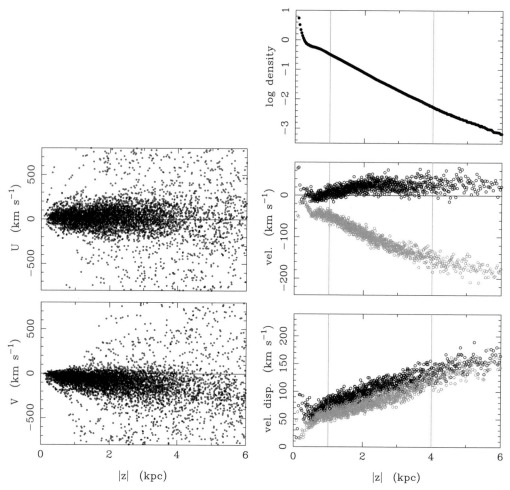

Figure 3. Transverse velocity distributions, as derived from SPM3 proper motions, for a predominantly thick-disk sample of ~ 1100 giants at the SGP. *Left:* U,V velocities and distances are
estimated in a Monte-Carlo fashion, integrating over an assumed absolute magnitude distribution. Dwarf stars, incorrectly classified as giants, are easily identified by their unrealistically high
velocities. *Right:* Observed number density, transverse velocity, and velocity dispersion profiles
as a function of distance for this sample. In the lower two panels black symbols illustrate the
U profiles while gray symbols signify the V profiles. The sample is considered well-measured in
the range 1 to 4 kpc.

proper-motion errors are well-understood. Notably, we derive a scale height for the thick
disk of 783 ± 48 pc; a vertical gradient to the V velocity (shear) of -30 ± 3 km/s/kpc;
and a vertical gradient for the velocity dispersions in both the U and V components of
$\sim 9 \pm 3$ km/s/kpc. It is possible to show that for reasonable assumptions as to the form
of the Galactic potential, the slight gradient in σ_U is consistent with the large gradient
in V within a condition of dynamical equilibrium.

6. Conclusion

"The right tool for the job," is an expression familiar to weekend handymen and
professionals alike. The present paper has attempted to stress the importance of utilizing
the appropriate proper-motion catalog for the task at hand. A listing of some of the more

useful such catalogs has been provided, as have several examples of fruitful science that results when proper-motion measures are properly considered and used. For diplomatic purposes, it was thought best not to include here examples of the *improper* use of proper-motion data. The reader is warned, however, that the literature is not without such occurences.

One minor instance, drawn from personal experience and involving an anonymous reviewer of a research proposal for which this author was a co-investigator, will perhaps make the point. It was stated by the reviewer that it was wasteful to fund a project seeking to measure and use, for Galactic kinematic studies, absolute proper motions down to $V=17.5$ based on a specially designed program to determine such (i.e. the SPM) when the USNO-B1 catalog already gives proper motions to a similar precision and to a deeper magnitude limit. Figure 4 illustrates the folly of this statement. The USNO-B1 is a tremendous resource with a multitude of important uses but it is not the appropriate tool to use to discern the detailed nature of Milky Way kinematics. The following year the research proposal was resubmitted with the new inclusion of Figure 4. It can be happily reported that the proposal was accepted and funded, leading to the completion of the SPM observing program and the soon-to-be-released SPM4 southern-sky proper-motion catalog.

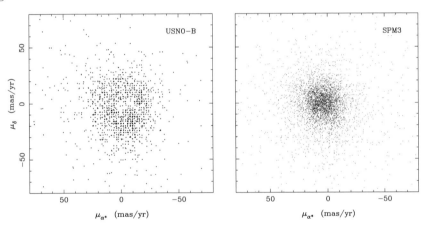

Figure 4. The central $2° \times 2°$ region of an arbitrary southern-sky SPM field. The vector point diagram constructed using proper motions from the USNO-B1 *(left panel)* is compared to that of the SPM3 *(right panel)* for stars in common.

References

Finch, C. T., Henry, T. J., Subasavage, J. P., Jao, W.-C., & Hambly, N. C. 2007, *AJ*, 133, 2898

Girard, T. M., Korchagin, V. I., Casetti-Dinescu, D. I., van Altena, W. F., López, C. E., & Monet, D. G. 2006, *AJ*, 132, 1768

Girard, T. M., Platais, I., Kozhurina-Platais, V., van Altena, W. F., & López 1998, *AJ*, 115, 855

Harris, H. C., Munn, J. A., Kilic, M., Liebert, J., Williams, K. A., von Hippel, T., Levine, S. E., Monet, D. G., & 11 coauthors 2006, *AJ*, 131, 571

Henry, T. J., Subasavage, J. P., Brown, M. A., Beaulieu, T. D., Jao, W.-C., & Hambly, N. C. 2004, *AJ*, 128, 2460

Skrutskie, M. F., Cutri, R. M., Stiening, R., Weinberg, M. D., Schneider, S., & 26 coauthors 2006, *AJ*, 131, 1163

A Giant Step: from Milli- to Micro-arcsecond Astrometry
Proceedings IAU Symposium No. 248, 2007
W. J. Jin, I. Platais & M. A. C. Perryman, eds.

© 2008 International Astronomical Union
doi:10.1017/S1743921308019431

Dense optical reference frames: UCAC and URAT

N. Zacharias

U. S. Naval Observatory, 3450 Mass. Ave. NW, Washington, DC 20392, USA
email: nz@usno.navy.mil

Abstract. A series of ground-based, dedicated astrometric, observational programs have been performed or are in preparation which provide a dense and accurate optical reference frame. Integral to all these programs are new observations to link the Hipparcos Celestial Reference Frame (HCRF) to the International Celestial Reference Frame (ICRF), based on compact, extragalactic radio sources.

The U.S. Naval Observatory CCD Astrograph Catalog (UCAC) 3rd release is in preparation. A pixel re-reduction is in progress to improve astrometric and photometric accuracy as well as completeness of this all-sky reference catalog to 16th magnitude. Optical counterparts of ICRF radio sources have been observed with 0.9-meter telescopes contemporaneously. Scanning of over 5000 early-epoch astrograph plates on StarScan has been completed. These data will improve the proper motions of stars in the 10 to 14 mag range for the UCAC3 release.

A 111 million-pixel CCD was successfully fabricated in 2006 and test observations at the USNO astrograph are underway. Four of such detectors will be used for the USNO Robotic Astrometric Telescope (URAT) focal plane assembly. Phase I of URAT will use the astrograph to reach 18th magnitude, while the new 0.85-meter telescope with a 4.5 deg diameter field of view will reach 21st magnitude. The URAT primary mirror has been fabricated.

Keywords. astrometry, reference systems, catalogs, instrumentation: detectors

1. Introduction

The primary, optical reference frame currently is the Hipparcos Celestial Reference Frame, HCRF (IAU 2000), which is a subset of the Hipparcos Catalogue (ESA 1997). The HCRF contains about 100,000 selected stars from the brightest to about V = 12 magnitude, with most stars around V = 8 to 9. The Tycho-2 Catalogue (Høg, Fabricius, Makarov *et al.* 2000) contains an almost complete set of the 2.5 million brightest stars to about V = 11.5 and is based on both the Hipparcos space mission Tycho instrument data and early-epoch, ground-based catalogs, with the AC2000 (Urban, Corbin, Wycoff *et al.* 2003) playing a major role. The large epoch difference between these ground-based catalogs and the Hipparcos mission allowed to derive reliable proper motions for Tycho-2 stars at the 1 to 2 mas yr^{-1} accuracy level.

However, even the Tycho-2 Catalogue can not be utilized as an astrometric standard in most practical applications due to the smallness of the field of view and saturation limiting magnitude of typical sky images taken with today's telescopes. The United States Naval Observatory (USNO) has engaged in programs to further densify the optical reference frame. The USNO CCD Astrograph Catalog (UCAC) project is a completed observing program and is discussed in the next section. The future USNO Robotic Astrometric Telescope (URAT) and its first phase (nicknamed "U-mouse") is presented in Sections 3 and 4, respectively, followed by a brief discussion of merged data products.

There are other dense optical reference star catalogs, which are not discussed here. The USNO-B catalog (Monet, Levine, Canzian *et al.* 2003) is widely used when deeper data

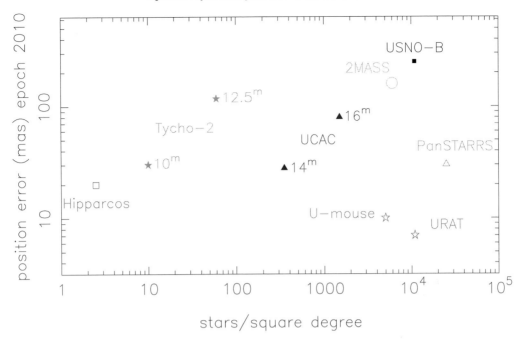

Figure 1. The global positional accuracies of some optical reference star catalogs are plotted as a function of the density (number of stars per square degree). For Tycho-2 and UCAC a range is given for positional errors depending on the magnitude as indicated. See text for further explanation.

than 16th magnitude are needed. The Deep Astrometric Standard (DAS) project (Platais, Wyse, Zacharias 2006) is well underway to reach 25th magnitude in 4 selected areas in the sky (see also these proceedings). The Bordeaux Observatory M2000 (Ducourant, Le Campion, Rapaport *et al.* 2006) and Carlsberg Meridian Telescope catalogs (Evans, Irwin, Helmer 2002) are examples of other dense optical reference catalogs covering specific declination zones with positional accuracies comparable to those of UCAC. The IAU Working Group on densification of the optical, celestial reference frames is working toward bringing all those catalogs onto a common system.

For comparison, the positional accuracies of current and near future major star catalogs are presented in figure 1. Given are mean standard errors of absolute positions per coordinate, propagated to the epoch of 2010 whenever proper motions are available. The value for PanSTARRS is actually achieved with commissioning data, while the goal is about 10 mas (E. Magnier, private com., see also these proceedings). Values for URAT and U-mouse are conservative estimates based on the UCAC experience.

2. UCAC

The UCAC project began in 1996 with the automation of the USNO 8-inch Twin Astrograph. Its visual corrected lens was used for guiding, while a new 5-element "redlens" imaged 1.0 square degree sections of the sky onto a 4k×4k CCD. Over 277,000 such images were taken between 1997 and 2004 at the Cerro Tololo Interamerican Observatory (CTIO) and the Naval Observatory Flagstaff Station (NOFS). The first data release in 2000 covers a significant fraction of the southern hemisphere, while the second release, the UCAC2 catalog (Zacharias, Urban, Zacharias, *et al.* 2004), covers the sky area from the south celestial pole to about +50° declination.

The UCAC2 is widely used because of its density (about 50 million stars) and high astrometric accuracy of about 20 mas for stars in the 10 to 14 mag range. A single bandpass (579 to 642 nm) was utilized. All stars in UCAC2 have proper motions which were derived by combining the new epoch CCD data with early-epoch transit circle and photographic data. The UCAC2 was reduced to the International Celestial Reference Frame (ICRF) using Tycho-2 reference stars.

A complete re-processing of the UCAC data is in progress, with a planned final catalog release (UCAC3) in 2008. Major improvements upon the UCAC2 release are: a) complete sky coverage, b) improved completeness, by including for example many double stars, previously skipped "problem" cases and high proper motion stars, c) smaller positional errors at the epoch of observation due to better modeling, d) improved proper motions due to inclusion of new or newly reduced early-epoch data, and e) better photometry due to better modeling and use of a real-aperture photometry.

The test re-processing of UCAC pixel data resulted in the detection of many more faint stars not included in the UCAC2 release. This improved limiting magnitude is obtained by utilizing a more sophisticated dark frame subtraction algorithm than was used in the previous catalog construction. Re-processing of all 4.5 TB compressed raw pixel data is in progress.

Since the release of UCAC2 the StarScan plate measuring program (Zacharias et al. 2008) has been completed. All available (over 1900) plates of the AGK2 project (Schorr, Kohlschütter 1951) as well as the 2300 Hamburg Zone Astrograph and 900 USNO Black Birch program plates were measured. A re-processing of the Northern (Klemola, Jones, Hanson 1987) and Southern (Platais, Girard, Kozhurina-Platais et al. 1998) Proper motion (NPM, SPM) program plates are in progress as well, in a joint effort with Yale University and Lick Observatory.

An integral part of the UCAC program is a direct link to the ICRF radio sources which define the primary celestial reference frame. Images of the ICRF optical counterparts (mostly quasars) were observed mainly with the Kitt Peak National Observatory (KPNO) and CTIO 0.9-meter telescopes with contemporaneous coverage of the same fields by specific astrograph observations, not included in the UCAC2 public release. All relevant extragalactic data are being reduced (Zacharias & Zacharias 2005) for a check on the optical to radio reference frame link (see also these proceedings).

3. URAT

The URAT project will conduct a very accurate (5 mas level for 13 to 18 mag stars) astrometric survey of the entire sky extending to a magnitude limit of 21, obtaining positions, proper motions and parallaxes based on new observations with a dedicated 0.85 meter aperture and 3.6 meter focal length telescope. The optical design of that telescope with a 4.5° diameter field of view (Laux, Zacharias 2005) was completed by the EOST company in 2005. The excellent quality primary mirror was delivered to USNO in November 2008. The completion of the telescope optics, tube assembly, and mount depends on future funding.

The development of the URAT focal plane detectors progressed in 2006 by successful fabrication of a single large-format CCD containing 111 million pixels (Bredthauer, Boggs, Bredthauer 2007, Zacharias, Dorland, Bredthauer, et al. 2007). A "10k" camera based on a thinned, back-illuminated version of this STA1600 chip was completed in 2007 by Semiconductor Technology Associates (STA) and saw first light at the USNO astrograph on October 9 (Fig. 2). The URAT camera will consist of 4 of these CCDs plus 4 smaller, rectangular CCDs used for guiding and focus control (Fig. 3). USNO

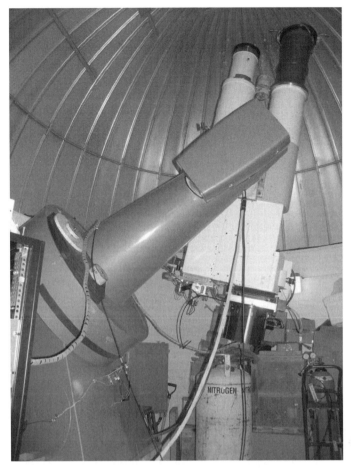

Figure 2. 10k camera with a single STA1600 chip mounted at the USNO astrograph, currently located at the Flagstaff station.

funded this URAT focal plane assembly in 2008. More details about the URAT project and telescope can be found elsewhere (Zacharias 2005).

4. U-mouse

With the URAT telescope not complete yet, the URAT focal plane assembly can be put to good use for the "redlens" of the USNO astrograph. The pixel size of the STA1600 chip (9 μm) is the same as that of the 4k camera used for the UCAC project, and the field of view of the "redlens" is 9° (324 mm) in diameter, a perfect match for the "4-shooter" URAT camera. The resulting project is the first phase of URAT ("U-mouse") and aims at an all-sky survey to 18th magnitude. The gain of 2 magnitudes over the UCAC project using the same telescope comes from a higher quantum efficiency (back vs. front illumination), and the use of longer exposure times (300 sec vs. 125 sec) in the new survey. Exposure times in the UCAC survey were limited by a high dark current, set to mitigate the CTE effect. Going from 16th magnitude (UCAC) to 18th (U-mouse) gives about a factor of 10 more stars and makes all the difference for projects like PanSTARRS (Magnier 2007) with the anticipated saturation limit near 16th magnitude.

A single exposure of the "4-shooter" camera at the astrograph will cover 27 square degrees of sky and thus will enable several complete overlaps of the accessible sky per

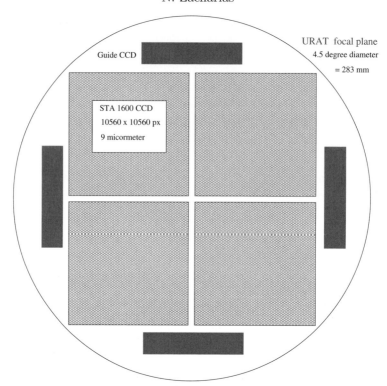

Figure 3. Layout of the 4-shooter focal plane with 4 main CCDs (each 10,560 by 10,560 pixels) and additional, smaller, rectangular CCD chips for guiding and focus control. The same focal plane assembly can be used at both the refurbished USNO astrograph and the future URAT 0.85-meter telescope.

year. Operations are planned to start in 2009 at CTIO after a major re-build of the astrograph (single lens tube assembly) has been completed, which started in 2007.

5. NOMAD

The Naval Observatory Merged Astrometric Dataset (www.nofs.navy.mil/nomad) is a first step in combining various star catalogs for dense optical reference frame applications. NOMAD covers all-sky and includes stars from the brightest to the limit of the Schmidt plate surveys. Astrometric information is picked from the Hipparcos, Tycho-2, UCAC, "Yellow-Sky" (NPM, SPM data) and the USNO-B catalogs by a priority algorithm. NOMAD is not a compiled catalog and only a single entry per unique star is picked from the parent catalogs, inheriting also the systematic errors of a particular catalog. The NOMAD astrometric data are supplemented by available optical and 2MASS photometry. Data extracts for small areas in the sky can be obtained online through the above link. The complete NOMAD dataset consists of 90 GB in binary format.

USNO has recently started a long-term project to put all such data and reduction procedures together into a dynamical universal star catalog (S. Levine, private comm). However, it will be a major effort to characterize even the existing star catalogs and remove systematic errors before combining astrometric data. A major step in this direction will be the final UCAC3 release, which is anticipated to be used as a dense optical reference frame for analyzing other star catalogs like the Sloan Digital Sky Survey (SDSS)

(Pier, Munn, Hindsley *et al.* 2003) and the Guide Star Catalog (GSC) data (Bucciarelli, Lattanzi, McLean *et al.* 2008).

Acknowledgements

Thanks to all members of the UCAC team, the staff at the Cerro Tololo Interamerican Observatory, and the Washington DC and Flagstaff based Naval Observatory personnel involved in these astrometric catalog efforts.

References

Bredthauer, R., Boggs, K., & Bredthauer, G. 2007, *2007 Image Sensor Workshop*, p.170

Bucciarelli, B., Lattanzi, M. G., McLean, B. *et al.* 2008, GSC-II catalog release GSC2.3, *these proceedings*

Ducouraunt, C., Le Campion, J. F., Rapaport, M. *et al.* 2006, *A&A* 448, 1235

European Space Agency 1997, *The Hipparcos Catalogue*, SP-1200

Evans, D. W., Irwin, M. J., & Helmer, L. 2002, *A&A* 395, 347

Høg, E., Fabricius, C., Makarov, V. V. *et al.* 2000, *A&A* 355, L27

IAU Resolution B1.2 2000, *BAAS* 127, 3043, and in proceedings IAU 24th GA, Manchester, p.36

Klemola, A. R., Jones, B. F., & Hanson, R. B. 1987, *AJ* 94, 501

Laux, U. & Zacharias, N. 2005, in Proceedings of "Astrometry in the Age of the Next Generation of Large Telescopes" Eds: Kenneth Seidelmann and Alice K. B. Monet, *ASP-CS* 338, 106

Magnier, E. 2007, *ASP-CS* 364, 153

Monet, D. G., Levine, S. E., Canzian, B. *et al.* 2003, *AJ* 125, 984

Pier, J. R., Munn, J. A., Hindsley, R. B. *et al.* 2003, *AJ* 125, 1559

Platais, I., Girard, T. M., Kozhurina-Platais, V. *et al.* 1998, *AJ* 116, 2556

Platais, I., Wyse, R. F. G., & Zacharias, N. 2006, *PASP* 118, 107

Schorr, R. & Kohlschütter, A. 1951, "Zweiter Katalog der Astronomischen Gesellschaft", AGK2, Volume I, Hamburg

Urban, S. E., Corbin, T. E.; Wycoff, G. L. *et al.* 2003, *BAAS* 33, 1494 (199th AAS meeting, abstr. 129.04)

Zacharias, N., Urban, S. E., Zacharias M. I., Wycoff G. L., Hall D. M., Monet D. G., & Rafferty T. J. 2004, *AJ* 127, 3043

Zacharias, N. 2005, in Proceedings of "Astrometry in the Age of the Next Generation of Large Telescopes" Eds: Kenneth Seidelmann and Alice K. B. Monet, *ASP-CS* 338, 98

Zacharias M. I. & Zacharias N. 2005, in P. Kenneth Seidelmann & Alice K. B. Monet (eds.), *Astrometry in the Age of the Next Generation of Large Telescopes*, Proc. meeting at Lowell Observatory, Flagstaff, USA (San Francisco: *ASP-CS* 338) p. 184

Zacharias, N., Dorland, B., Bredthauer, R., Boggs, K., Bredthauer, G., & Lesser, M. 2007, *SPIE* 6690, paper 8

Zacharias, N., Winter, L., Holdenried, E. R., De Cuyper, J.-P., Rafferty, T. J., Wycoff, G. L., 2008, in press, *PASP*

A Giant Step: from Milli- to Micro-arcsecond Astrometry
Proceedings IAU Symposium No. 248, 2007
W. J. Jin, I. Platais & M. A. C. Perryman, eds.

© 2008 International Astronomical Union
doi:10.1017/S1743921308019443

The GSC-II catalog release GSC 2.3: description and properties

B. Bucciarelli[1], M. G. Lattanzi[1], B. McLean[2], R. Drimmel[1], G. Greene[2], C. Loomis[2], R. Morbidelli[1], R. Pannunzio[1], R. L. Smart[1] and A. Spagna[1]

[1]INAF, Astronomical Observatory of Torino
10025 Pino Torinese, TO, Italy

[2]Space Telescope Science Institute
3700 San Martin Drive, Baltimore MD, USA

Abstract. The GSC 2.3 is a current catalog release extracted from the Guide Star Catalog II database, which is maintained at the Space Telescope Science Institute in Baltimore, USA. The catalog contains astrometry, multi-band photometry (B_J, R_J, I_N) and star/non-star classification for 945,592,683 objects down to the magnitude limit of the survey plates. We review the performance of stellar parameters, anticipating the improvements in astrometric accuracy foreseen by its recalibration with the newly available catalog in the UCAC series.

Keywords. astrometry, catalogs, reference systems, surveys, techniques: miscellaneous

1. Overview and applications

The Guide Star Catalog II (GSC-II) - a joint effort of the Space Telescope Science Institute (ST ScI) and the Torino Astronomical Observatory (INAF-OATo), with additional resources from international institutions - is an all-sky astro-photometric catalog derived from the digitization of the Palomar and UK Schmidt surveys. Its latest release, GSC 2.3, (Lasker *et al.* 2007) is the operational catalog for HST starting from GO Cycle 15 and is one of the primary data sources for preparation of the Initial Source List of the Gaia reduction pipeline; it is also planned to be an input catalog for the Large Sky Area Multi-Object Fiber Spectroscopic Telescope (LAMOST) of China.

Given its completeness, depth, and multi-color characteristics, GSC-II has also been used for investigations on Galactic structure and stellar populations through proper motions (see, e.g., Sciortino *et al.* 2000, Vallenari *et al.* 2006). The latter, which are not included in the current public version of the catalog, have been obtained from the GSC-II database by ad-hoc algorithms, developed at OATo, linking measured positions (x,y) of the same objects on overlapping plates from different epochs. While the intrinsically precise x,y measurements should allow to obtain the proper motion accuracies of the order of 3-5 mas yr^{-1}, the procedure cannot be automatized easily, and it resorts to the presence of field galaxies - not available in high-extinction regions - for the link to the absolute reference frame (Spagna *et al.* 1996). Alternatively, proper motions can be derived in a straightforward manner from multi-epoch plate-based sky positions relying on external reference catalogs. In this case, their accuracy reflects possible deficiencies of a single-plate astrometric model and therefore can suffer from the well known problems that have plagued the Schmidt plate astrometry.

Although the overall quality of GSC 2.3 astrometry is quite good, small systematics are still present. In the following sections we summarize the properties of the current catalog and outline the foreseen ways of improvements which will lead to the publication of a new release with the inclusion of proper motions.

2. Catalog description

The GSC-II is derived from the uncompressed Digitized Sky Survey (DSS) material scanned at the STScI with two heavily modified PDS machines, named the GAMMA. Image cutouts of all detected objects, and their associated features, are extracted from each digitized plate by pipelined tasks (*sky background determination, object identification, blend resolution, centroiding, classification*), and outputted for the subsequent plate calibration algorithms performing: a) star/non-star classification via an oblique-decision-tree algorithm (Murthy *et al.* 1994) trained by 5334 hand-classified objects; b) photometric calibrations; and c) astrometric calibrations.

Photometric reductions are based on Johnson-Kron-Cousins B,V,R standards from the GSPC-I (Lasker, Sturch *et al.* 1988) and GSPC-II (Bucciarelli *et al.* 2006), plus additional B_T and V_T phtometry from TYCHO stars needed to constrain the bright range of the density-to-intensity relation of the photographic emulsion. Magnitudes are computed in the natural passbands of the plates, defined by the emulsion/filter combination. The instrumental parameter of choice is the integrated density and its fit to standard magnitudes is performed via Chebyshev polynomials.

Stellar positions are directly tied to the International Celestial Reference System (ICRS) by the use of the ACT and TYCHO-2 as reference catalogs. A polynomial model is applied to map the measured x, y coordinates of reference stars (pre-corrected for refraction) onto their standard coordinates (ξ, η). Then, for each survey, the $\Delta\xi$, $\Delta\eta$ residuals are accumulated in a plate-based grid of 4.4×4.4 mm^2 bins to build the so-called astrometric masks. The resulting patterns in these masks are used to remove the systematic errors left out by the model.

All the calibrated plates are loaded into COMPASS, the GSC-II database built on *Objectivity* - an Object-Oriented Database Management System - (Greene *et al.* 1998), which sets a unique correspondence between the plate-based data and partitions of the celestial sphere. For every new plate which is loaded into COMPASS, a matching task identifies and labels unique objects on different plates.

After the GSC 2.2 release in July 2001, the production of a GSC 2.3 export catalog extracted from the GSC-II database was mainly motivated by the desire to release a large and valuable data set to the international astronomical community, and also by the needs to improve the planning of observations and operations of HST (McLean 2007). GSC 2.3 contains nearly all objects down to the magnitude limit of the plates (a total of 945,592,683 entries), so that it is expected to be complete to at least $R_F = 20$. This limit is confirmed at high galactic latitudes; however, in very crowded fields near the Galactic plane it may suffer incompleteness at magnitudes brighter than $R_F \approx 18$ (Drimmel *et al.* 2007) .

Extraction rules have been set to minimize the deficiencies of the calibration algorithms at the plate borders where highly non-linear effects are dominant. In particular: a) positions are measured on the red IIIaF plates, and multiple measurements are solved by selecting the closest solution to the plate center. If the object is not detected on any IIIaF plate, then its position from the blue IIIaJ, visual IIaD, and infrared IV-N plates is used, following this order of priority: b) photographic magnitudes (B_J, R_F, I_N,) are provided, if available, for all exported objects; additional V_{12} (Palomar Quick-V) and O (POSS-II) magnitudes are also given. When different plates of the same band overlap, the observation closest to the plate center is selected without applying any averaging procedure.

Also, to improve the completeness and accuracy at the brightest magnitudes, GSC 2.3 has been supplemented with data from the TYCHO-2 and SKY2000 (Myers *et al.* 2002) catalogs, that substitute, when feasible, the GSC-II data for any given object.

3. Catalog properties

The main properties of GSC 2.3 are summarized in Table 1. Performance of the different stellar parameters has been assessed via comparisons to external reference catalogs; in particular, the astrometric and photometric errors reported in the table are global averages against the fifth data release of the Sloan Digital Sky Survey (SDSS DR5, Adelman-McCarthy 2007).

Photometric errors are only reported for stellar sources, as the GSC-II pipeline is not optimized for galaxy photometry. In GSC 2.3 extended objects appear systematically brighter than from the SDSS, although the effect decreases towards fainter objects until becoming negligible at around 18th magnitude.

In terms of astrometry, the point-like sources have lower errors. Larger errors are present for the extended objects ("non-stars") – which may include galaxies and nebulae, as well as blends and unresolved binaries – mainly due to saturation and sampling problems for very bright and very faint objects, respectively.

Table 1. Overview of GSC 2.3 Global Properties

Total number of objects	945,592,683 (206,688,844 point-like; 738903839 extended)
Magnitude limit	$B_J = 22.5$, $R_F = 20.5$, $I_N = 19.5$
Mean epoch of positions	1992.5
Reference frame	ICRF
Astrometric reference catalogs	ACT + TYCHO2
Average positional accuracy:	
–stellar objects	$0\rlap{.}''30$
–extended objects	$0\rlap{.}''40$
Average positional precision:	
–stellar objects	$0\rlap{.}''22$
–extended objects	$0\rlap{.}''35$
Photometric reference catalogs	GSPC1 + GSPC2 + TYCHO
Average photometric accuracy(stellar sources)[1]:	
at $B_J \approx 20$	0.18 mag
at $R_F \approx 19$	0.15 mag
at $I_N \approx 18$	0.20 mag
Stellar Classification reliability:	at least 90%[2]
Completeness:	> 98%[3]

Notes:
[1] Photometry of non-stellar sources suffers from systematic errors which, for very bright objects, can be as high as ≈ 2 mag.
[2] Percentage of stars correctly classified as such; the success ratio decreases at magnitudes fainter than $R_F = 19.5$
[3] up to $R_F = 20$ and $l > 30°$; incompleteness at brighter magnitudes is strictly related to the object field density

4. Magnitude-dependent errors and future re-calibration

Further astrometric tests of GSC 2.3 have detected a magnitude-dependent systematic error reaching a level of the order of $\approx 0\rlap{.}''10$, also dependent on object's position on the plate. As shown in Fig. 1, which plots the right ascension differences against the UCAC 2 catalog for the bright and faint magnitude samples as a function of declination, the effect is more pronounced for faint stars, farthest from the magnitude range of the TYCHO 2 to which GSC 2.3 has been tied. Whatever the cause of this effect, we are confident that it can be cured by using fainter astrometric calibrators, such as those available from the UCAC catalog, whose CCD stellar images are are optimally exposed as they fall in the linear range of the photographic emulsion sensitivity. Preliminary tests have already showed the potential astrometric improvements (Tang *et al.* 2008), and we are in the process of setting up a fully operational procedure to perform a new astrometric calibration of the entire catalog, which will allow us to obtain scientifically valuable proper motions over the whole sky.

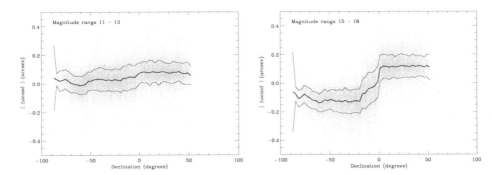

Figure 1. GSC 2.3 - UCAC2 RA differences as a function of declination for the magnitude ranges 11-13 (left panel) and 15-16 (right panel). The central solid line marks the median for the data binned in declination, while the external ones trace the RMS deviations about each median.

References

Adelman-McCarthy, J. K. *et al.* 2007 *ApJS, preprint doi:10.1086/518864*

Bucciarelli *et al.* 2006, *GSPC V2.4, VizieR Online Data Catalog 2272:0+*

Drimmel, R., Bucciarelli, B., Lattanzi, M. G., Spagna, A., Jordi, C., Robin, A. C., Reyle, C., & Luri, X. 2007, *The Three-Dimensional Universe with GAIA*, ESA Special Publication vol. 576, 163

Greene, G. R., McLean, B. J., Lasker, B. M., Wolfe, D., Morbidelli, R., & Volpicelli, C. A. 1998, McLean B. J. *et al.* eds., *IAU Symp. 179*, 474-477

Lasker, B. M., Lattanzi, M. G., McLean B. *et al.* 2008, accepted for publication in *ApJ*

McLean, B. 2005, *The 2005 HST Calibration Workshop*, A.M. Koekemoer, P. Goudfrooij, and L. L. Dressel, eds., p413

Myers, J. R., Sande, C. B., Miller, A. C., Warren, J. W. H., & Tracewell, D. A., 2002 *Sky2000 master catalog, version 4*.ftp://cdsarc-ustrasb.fr/pub/cats/V/109

Sciortino, S., Micela, G., Favata, F., Spagna, A., & Lattanzi, M. G. 2000, *A&A*, 357, 460

Spagna, A., Lattanzi, M. G., Lasker, B. M., McLean, B. J., Massone, G., & Lanteri, L. 1996, *A&A*, 311,758

Tang, Z., Qi, Z., Smart, R. L., Bucciarelli, B., Lattanzi, M. G., & Spagna, A. 2008, *in this volume*, p.334

Vallenari, A., Pasetto, S., Bertelli, G., Chiosi, C., Spagna, A., & Lattanzi, M. G. 2006, *A&A*, 451, 125

A Giant Step: from Milli- to Micro-arcsecond Astrometry
Proceedings IAU Symposium No. 248, 2007
W. J. Jin, I. Platais & M. A. C. Perryman, eds.

Deep Astrometric Standards

I. Platais[1], A. L. Fey[2], S. Frey[3], S. G. Djorgovski[4], C. Ducourant[5], Ž. Ivezić[6], A. Rest[7], C. Veillet[8], R. F. G. Wyse[1] and N. Zacharias[2]

[1] Johns Hopkins University, Department of Physics & Astronomy, Baltimore, MD 21218, USA
email: `imants,wyse@pha.jhu.edu`

[2] US Naval Observatory, Washington DC, USA

[3] FÖMI Satellite Geodetic Observatory, Hungary; [4] Caltech, Pasadena, CA 91125, USA

[5] Observatoire de Bordeaux, Floirac, France

[6] University of Washington, Seattle, WA 98155, USA

[7] Harvard University, Cambridge, MA 02138, USA; [8] CFHT, Kamuela, HI 96743, USA

Abstract. The advent of next-generation telescopes with very wide fields-of-view creates a need for deep and precise reference frames for astrometric calibrations. The Deep Astrometric Standards (DAS) program is designed to establish such a frame, by providing absolute astrometry at the 5–10 mas level in four 10 $\deg^2$ Galactic fields, to a depth of $V = 25$. The source of our basic reference frame is the UCAC2 catalog, significantly improved by additional observations, and new VLBI positions of radio-loud and optically visible QSOs. We describe all the major steps in the construction of the DAS fields and provide the current status of this project.

Keywords. astrometry, catalogs

1. Introduction

When astronomers refer to the term "astrometric standard", there could be differences in the perception of what this entails. For a quick astrometric calibration of images (i.e., deriving celestial coordinates in the system of ICRS), there are a few all-sky optical catalogs readily available. For many applications in astronomy these are sufficient, except for astrometry itself. Why? Because none of these catalogs can match the accuracy of formal image centering, now reaching 5–10 mas with the best ground-based imaging telescopes, and around 1 mas with the Hubble Space Telescope. This argument gains another dimension when confronted with large aperture telescopes and huge CCD mosaic cameras, such as the planned LSST facility (Tyson 2002). For these new facilities, some of the key science requirements aimed at unlocking the secrets of dark matter and dark energy mandate that this accuracy of image centering is *not lost* in manipulation of images, such as co-adding images obtained with significant dithers. From the standpoint of astrometry, there are two main issues with modern imaging facilities: 1) the fragmented detector plane (a mosaic of solid-state devices) and 2) the considerable geometric distortions in the focal plane. The combination of these two issues requires a very complex plate model involving hundreds of unknowns. Such a complex model is not readily described by *ad hoc* solutions, particularly at high Galactic latitudes where few reference stars are available. To address these issues, we proposed in 2006 to set up Deep Astrometric Standards (DAS; Platais *et al.* 2006) which should allow us a significant simplification of the plate model, reducing the number of unknowns from hundreds potentially down to only three basic constants – zeropoint, scale (stretch), and rotation. This enables robust and reliable equatorial solutions anywhere on the sky. In this contribution we report on the status of this project.

2. Properties of the DAS fields

Considering the specifications of Pan-STARRS (Kaiser 2004) and LSST facilities and their operational needs, four DAS fields have been established (Platais *et al.* 2006). Each DAS field covers a ~10 deg^2 semi-circular area with a projected depth down to $V = 25$. Our goal is to reach an accuracy of 5–10 mas in positions and 2 mas yr^{-1} for absolute proper motions. At optical wavelengths, such high positional accuracies over a relatively large area at $V > 12$ are yet to be achieved. Our strategy in reaching of this goal is to tie the optical positions directly into the International Celestial Reference Frame (ICRF) by densifying locally the ICRF itself. This provides an exceptional opportunity to ensure that metrics of a reference frame defined by the DAS is invariant to one part in ~100,000 over one degree of arc.

The specified depth of DAS fields requires at least 3–4m class telescopes equipped with a CCD mosaic imager, now a commonly available detector for astrometry. Platais *et al.* (2002) have demonstrated that the National Optical Astronomy Observatory's (NOAO) CCD mosaic imagers are suitable for astrometry (despite some initial scepticism in the community), being stable at the 10 mas level over a single lunation period. This level of stability is one of the key factors to make the DAS project feasible. Another critical component in the realization of DAS is access to a good reference frame. Currently the best relatively dense optical reference frame is the UCAC2 (Zacharias *et al.* 2004) but even this should be locally improved by acquiring new observations with automated meridian circles and other means. One of the biggest concerns is that the reference frame represented by the UCAC2 could be locally biased in some way that would then propagate into the DAS positions. New ICRF sources in the DAS fields should play a key role in identifying, quantifying and eliminating this kind of problem.

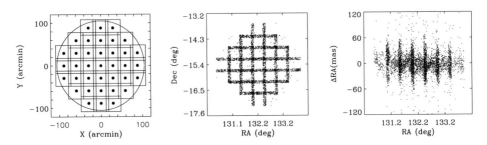

Figure 1. *Left panel:* Pointing layout of a DAS field with the NOAO 4m telescopes and their CCD mosaic imagers. In total, 37 partially overlapping (overlap $\Delta = 6'$) pointings are necessary to fill in a 3°.5-diameter field; *Middle panel:* Spatial distribution of stars in the frame overlap area (a few points elsewhere are artifacts); *Right panel:* Coordinate differences of the overlap region stars in the Hydra field from one complete pass (37 pointings) in a recent run. The rms of the scatter is about 25 mas, still dominated by residual systematic errors that have yet to be accounted for.

3. Optical imaging

The locations of the DAS fields require observations from both hemispheres. The first DAS field in Hydra was observed in February 2006 with the Cerro Tololo Inter-American

Observatory's (CTIO) Blanco 4m telescope (FOV = $36' \times 36'$). In the same year another DAS field in Sagittarius was completed at CTIO. As indicated in Fig. 1, a total of 37 partially overlapping pointings are necessary to fill a DAS field. The wide DAS magnitude range of $10 < V < 25$ means that a set of 10 s, 100 s, and 600 s exposures are required at each pointing in the Sloan g and i filters. In total, the average number of CCD frames per field is about 350. In a similar manner we have completed the northern DAS field in Ophiuchus, using the Kitt Peak Mayall 4m telescope. The remaining field in Gemini-Orion-Taurus is scheduled for observations in early 2008 with the 3.6m Canada-France-Hawaii Telescope (CFHT). We have obtained the pixel positions for objects in all Hya and Sgr mosaic frames. We use the software developed for NOAO surveys such as MSCPIPE and PHOTPIPE; these include standard manipulations for CCD data and DoPHOT object detection, photometry, and astrometry (Schechter *et al.* 1993) – all streamlined and with minimal supervision. Preliminary results are shown in Fig. 1.

4. Local improvements to the UCAC

Our plan is to use a dense optical reference frame – the UCAC – to calibrate our CCD mosaic frames and refine the constants of the geometric distortions across the field. This catalog is known to have small systematic errors, on the order of 5–10 mas, at the epoch of the CCD observations around yr 2000. More important, however, are the proper motions needed to extrapolate to current epochs. As indicated by Zacharias *et al.* (2004), there are problems with the positions from photographic plates, which are used to derive proper motions at $V > 12$. To avoid such complications, we are obtaining new epoch observations with a monolithic 111M pixel CCD camera attached to the USNO Twin Astrograph's "red lens" tube. In parallel, we have already observed two DAS fields with the Bordeaux automatic CCD meridian circle, in a time delay integration mode (e.g., Rapaport *et al.* 2001). Except for the final coordinate zeropoint, such observations are independent of the reference catalog and should help to minimize local systematics in UCAC. The coordinate zeropoint will be set by the optical counterparts of new ICRF sources, which will also provide the means to check the system of equatorial coordinates.

5. Ancillary observations

Conventional reductions of observations at two or more epochs produce *relative* proper motions. To make them *absolute*, the positions should be linked to an extragalactic reference frame represented by QSOs and/or galaxies with well-defined cores. Unfortunately, background QSOs are rare even with wide fields-of-view. We have initiated a new search for bright QSOs in the DAS fields, using the Palomar-Quest 161M-pixel camera, operated in a drift scan mode, through SDSS *griz* filters (see http://palquest.org). In color-color diagrams the QSO candidates show up as outliers from the stellar sequence, although this domain is also occupied by hot white dwarfs and some other Galactic objects with peculiar spectral energy distributions. Spectroscopic confirmation is therefore required to establish a list of *bona fide* QSOs and for this we will utilize the Palomar 200 inch telescope, with the Double Spectrograph.

The DAS fields, with one exception, lack previous observations in the SDSS *griz* filters. Secondary standards of calibrated *griz* photometry are now being set up in the DAS fields using the 0.5m Photometric Telescope at Apache Point Observatory, New Mexico, and the US Naval Observatory's 1m telescope at the Flagstaff station. These standards will be used to calibrate the DAS instrumental g, i magnitudes.

6. Local densification of the ICRF

One of the most innovative aspects of DAS is its direct link to the ICRF. This requires considerable effort to find optically visible compact radio sources and to measure their positions using the technique of VLBI phase referencing. As a first step, in each DAS field we select ∼25 unresolved strong sources (S_{20cm}>50 mJy) from the 20-cm NRAO VLA Sky Survey (NVSS; Condon *et al.* 1998). Each source is then observed with the Very Large Array (VLA) at 3.5 and 6 cm, to measure its radio spectrum and verify that it is indeed compact. Extragalactic radio sources show a flat-spectrum core, with extended emission in the form of steep-spectrum jet components. Typically about 50% of our initial sample survives after the VLA observations. Secondly, the surviving compact sources are observed with a VLBI facility, such as the European VLBI Network (EVN). Our first phase-referencing EVN observations at 6 cm were conducted in November 2006. In the Gemini-Orion-Taurus field they produced seven *bona fide* detections with peak brightness ranging from ∼5 mJy to 46 mJy per beam (Frey *et al.* 2007). The few existing CFHT deep frames indicate that four EVN detections have a reasonably bright optical counterpart but for two other EVN detections any optical counterpart is fainter than the limiting magnitude. We plan to extend the search for more sources in the GOT field at lower flux densities (50 >S_{20cm}>20 mJy) since the south-east portion of this field so far has produced no VLBI detections.

7. Project status

With lots of enthusiasm but with essentially zero dedicated funding we were able to obtain the first epoch optical imaging in all four DAS fields and test all vital steps of the DAS concept. The VLA observations of candidate compact sources are completed in three DAS fields. One lesson we have learned from the VLA and VLBI observations is that the fainter radiosources (S_{20cm}>20 mJy) should also be considered, as indicated by successful VLBI detections of additional sources from the GB6 catalog (Gregory *et al.* 1996). The first DAS catalog is expected to appear in the second part of 2008.

8. Acknowledgements

IP gratefully acknowledges support from the US National Science Foundation through grant AST 04-06689 to Johns Hopkins University. SF acknowledges support from the Hungarian Scientific Research Fund (OTKA T046097). We thank Allyn Smith and Douglas Tucker for obtaining initial data for the secondary SDSS standard in a DAS field.

References

Condon, J. J., Cotton, W. D., Greissen, E. W., *et al.* 1998, *AJ*, 115, 1693

Frey, S., Platais, I., & Fey, A. L. 2007, in *Proc. 18th European VLBI for Geodesy and Astrometry Working Meeting*, eds. J. Boehm, A. Pany, H. Schuh, p. 111

Gregory, P. C., Scott, W. K., Douglas, K., & Condon, J. J. 1996, *ApJS*, 103, 427

Kaiser, N. 2004, *Proc. SPIE*, 5489, 11

Platais, I., Wyse, R. F. G., & N. Zacharias 2003, *PASP*, 118, 107

Platais, I., Kozhurina-Platais, V., Girard, *et al.* 2002, *AJ*, 124, 601

Rapaport, M., Le Campion, J.-F., Soubiran, C., *et al.* 2001, *A&A*, 376, 325

Schechter, P., Mateo, M., & Saha, A. 1993, *PASP*, 105, 1342

Tyson, J. A. 2002, *Proc. SPIE*, 4836, 10

Zacharias, N., Urban, S. E., Zacharias, M. I., Wycoff, G. L., Hall, D. M., Monet, D. G., & Rafferty, T. J. 2004, *AJ*, 127, 3043

A Giant Step: from Milli- to Micro-arcsecond Astrometry
Proceedings IAU Symposium No. 248, 2007
W. J. Jin, I. Platais & M. A. C. Perryman, eds.
© 2008 International Astronomical Union
doi:10.1017/S1743921308019467

VLBI observations of weak extragalactic radio sources for the alignment of the future Gaia frame with the ICRF

G. Bourda[1], P. Charlot[1], R. Porcas[2] and S. Garrington[3]

[1]Laboratoire d'Astrophysique de Bordeaux, Université Bordeaux 1, CNRS, Floirac, France
email: bourda@obs.u-bordeaux1.fr

[2]Max Planck Institute for Radio Astronomy, Bonn, Germany

[3]Jodrell Bank Observatory, The University of Manchester, Macclesfield, UK

Abstract. The space astrometry mission Gaia will construct a dense optical QSO-based celestial reference frame. For consistency between the optical and radio positions, it will be important to align the Gaia frame and the International Celestial Reference Frame (ICRF) with the highest accuracy. Currently, it is found that only 10% of the ICRF sources are suitable to establish this link, either because they are not bright enough at optical wavelengths or because they have significant extended radio emission which precludes reaching the highest astrometric accuracy. In order to improve the situation, we have initiated a VLBI survey dedicated to finding additional high-quality radio sources for aligning the two frames. The sample consists of about 450 sources, typically 20 times weaker than the current ICRF sources, which have been selected by cross-correlating optical and radio catalogues. This paper presents the observing strategy and includes preliminary results of observation of 224 of these sources with the European VLBI Network in June 2007.

Keywords. astrometry, reference systems, quasars: general

The ICRF (International Celestial Reference Frame) is the fundamental celestial reference frame adopted by the International Astronomical Union (IAU) in August 1997 (Ma *et al.*, 1998; Fey *et al.*, 2004), currently based on the VLBI (Very Long Baseline Interferometry) positions of 717 extragalactic radio sources. The European space astrometry mission Gaia, to be launched by 2011, will survey about one billion stars in our Galaxy and 500,000 Quasi Stellar Objects (QSOs), brighter than magnitude 20 (Perryman *et al.*, 2001). Unlike Hipparcos, Gaia will construct a dense optical celestial reference frame directly in the visible, based on the QSOs with the most accurate positions (i.e. with magnitude $V \leqslant 18$; Mignard 2003). In the future, the alignment of the ICRF and the Gaia frame will be crucial for ensuring consistency between the measured radio and optical positions. This alignment, to be determined with the highest accuracy, requires several hundreds of common sources, with a uniform sky coverage and very accurate radio and optical positions. Obtaining such accurate positions implies that the link sources must have $V \leqslant 18$ and no extended VLBI structures. In a previous study, we investigated the current status of this link based on the present list of ICRF sources (Bourda *et al.*, 2008). We found that although about 30% of the ICRF sources have an optical counterpart with $V \leqslant 18$, only one third of these are compact enough on VLBI scales for the highest astrometric accuracy. Overall, only 10% of the current ICRF sources ($\simeq$70 sources) are thus available today for the alignment of the Gaia frame. This highlights the need to identify additional suitable radio sources, which is the purpose of the project described here.

Searching for additional sources suitable for aligning accurately the ICRF and the Gaia frame implies going to weaker radio sources having flux densities typically below

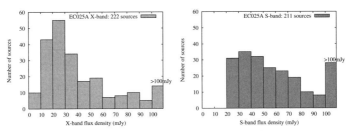

Figure 1. Flux density distribution (units in mJy) for the sources detected in our initial experiment EC025A (June 2007). On average, the flux density of these sources is 20 times and 7 times weaker than that of the ICRF and VCS sources (Kovalev *et al.*, 2007 and references therein).

100 mJy. This can be realized owing to recent increases in the VLBI network sensitivity (e.g., recording now possible at 1 Gb/s) and by using a network with large antennas like the EVN (European VLBI Network). A sample of about 450 radio sources, for which there are no published VLBI observations, was selected for this purpose by cross-identifying the NRAO VLA Sky Survey (NVSS; Condon *et al.*, 1998), a deep radio survey (complete to the 2.5 mJy level) that covers the entire sky north of $-40°$, with the Véron-Cetty & Véron (2006) optical QSOs catalogue. This sample is based on the following criteria: (i) $V \leqslant 18$ (to ensure very accurate positions with Gaia), (ii) $\delta \geqslant -10°$ (for possible observing with northern VLBI arrays), and (iii) NVSS flux density $\geqslant 20$ mJy (for possible VLBI detection). The observing strategy to identify appropriate link sources in the sample includes three successive steps: (1) to determine the VLBI detectability of these weak radio sources; (2) to image the sources detected in the previous step; and (3) to determine an accurate astrometric position for the most compact sources of the sample.

The first observations for this project (experiment EC025A) were made in June 2007, with a network of four EVN telescopes. Half of our sample (i.e. 224 target sources, most of which belonging to the CLASS catalogue; Myers *et al.*, 2003) was observed during this experiment to determine their VLBI detectability (step 1 described above). The rest of the sources have been observed in October 2007. Our results for EC025A indicate excellent detection rates of 99% at X band and 95% at S band, with 222 sources and 211 sources detected at X and S bands, respectively. The mean correlated flux densities have a median value of 32 mJy at X band and 55 mJy at S band (Fig. 1). The spectral index α of the 211 radio sources detected at both frequencies was also investigated. Its median value is -0.3 and most of the sources have $\alpha > -0.5$, hence indicating that they must have a dominating core component, which is very promising for the future stages of this project.

References

Bourda, G., Charlot, P., & Le Campion, J.-F. 2008, *A&A* (submitted)

Condon, J. J., Cotton, W. D., Greisen, E. W., Yin, Q. F., Perley, R., *et al.* 1998, *AJ* 115, 1693

Fey, A. L., Ma, C., Arias, E. F., Charlot, P., Feissel-Vernier, M., *et al.* 2004, *AJ* 127, 3587

Kovalev, Y., Petrov, L., Fomalont, E., & Gordon, D. 2007, *AJ* 133, 1236

Ma, C., Arias, E. F., Eubanks, T. M., Fey, A. L., Gontier, A.-M., *et al.* 1998, *AJ* 116, 516

Mignard, F. 2003, in: R. Gaume, D. McCarthy & J. Souchay (eds.), *IAU 25 Joint Discussion 16: The International Celestial Reference System, Maintenance and Future Realizations*, 133

Myers, S. T., Jackson, N. J., Browne, I. W. A., de Bruyn, A. G., Pearson, T. J., *et al.* 2003, *MNRAS* 341, 1

Perryman, M. A. C., de Boer, K. S., Gilmore, G., Hog, E., Lattanzi, *et al.* 2001, *A&A* 369, 339

Véron-Cetty, M.-P. & Véron, P. 2006, *A&A* 455, 773

A Giant Step: from Milli- to Micro-arcsecond Astrometry
Proceedings IAU Symposium No. 248, 2007
W. J. Jin, I. Platais & M. A. C. Perryman, eds.

© 2008 International Astronomical Union
doi:10.1017/S1743921308019479

The Infrared Astrometry today

A. S. Kharin, I. P. Vedenicheva and A. V. Zolotukhina

Main Astronomical Observatory, National Academy of Sciences of Ukraine
27 Akademika Zabolotnoho Str., 03680 Kiev, Ukraine
email: `kharin@mao.kiev.ua`

Abstract. Infrared Astrometry is one of the seven subdivisions of astrometry that detects emission in the range from 0.7 to 350 μm. Specific features and some problems of this one are discussed here and more details are provided on our web site "INFRARED ASTROMETRY" with the following URL `http://www.mao.kiev.ua/IR`. This web page includes 7 sections that will be expanded and regularly updated.

Keywords. infrared: general, catalogs, astrometry

What is Infrared Astrometry? Entire definition of astrometry was presented by Kovalevsky in his book "Modern Astrometry" (Kovalevsky 2002). One can only add that astrometry as well as astronomy can be divided, depending on the range of electromagnetic spectrum where the objects are detected, or photons energy, into the following seven astronomical or astrometrical subdivisions. They are as follows:

(1) Gamma-ray Astrometry,
(2) X-ray Astrometry,
(3) Ultraviolet Astrometry,
(4) Visual Astrometry,
(5) Infrared Astrometry,
(6) Submillimetr Astrometry,
(7) Radio Astrometry.

Thus, the Infrared Astrometry is one of the above seven subdivisions that detects emission in range from 0.7 to 350 μm. Evidently, it is a new direction in Astrometry.

The terrestrial atmosphere is more opaque for IR radiation than in visual optics. Beginning from 0.7 mkm it is transparent only within some bands because the absorption by water vapour and carbon dioxide. Due to this and some other circumstances, the ground-based IR observations are more complicated in comparison with the visual ones.

The Space observations are free from these deficiencies. The advantages of the space IR observations, in comparison with the ground-based ones, were demonstrated by the satellite IRAS survey. This advances in space IR observations were connected tightly with the progress in technology of IR sensors.

Successful ground-based near infrared observations (DENIS and 2MASS surveys) became possible due to a good progress in CCD and IR array detectors, as well as the progress in the full automation of telescope control and the computer data processing. But the next advance in infrared astrometry, the same as in optical one, will be connected probably with an implementation of two space-based interferometer projects Gaia, and SIM, which promise manifold increasing of the accuracy of observed stars and other objects. If these projects are successful they would extend infrared coordinate system to ICRS.

The IR Reference Catalogue which contains precise positions and proper motions is necessary both for presenting coordinate system and organizing various observations in this spectral range. Now such a catalogue can be created by identification of IR stars from the catalogues IRAS PSC, DENIS, and 2MASS with their optical counterparts from precise astrometrical space catalogues HIPPARCOS and TYCHO or ACT and TYCHO2.

Specific features of this new direction, its 25 years history and some problems in more detail are discussed on the web page "INFRARED ASTROMETRY" created at the web site of the Main Astronomical Observatory of the National Academy of Sciences of Ukraine. Internet address of this one is `http://www.mao.kiev.ua/IR`. Now this web-page includes 7 sections:

(1) Inevitability of Infrared Astrometry,
(2) Short History,
(3) Peculiarity of IR Observations,
(4) Infrared Astrometry Program,
(5) Designing and Manufacturing of IR Astrometric Instruments,
(6) Creating and Extending of an IR Reference Catalogue,
(7) Connection of the Optical, IR and Radio Coordinate System.

It is proposed that this one will be expanded and regularly updated. Main goal of this page is to develop a new direction in Astrometry – "IR Astrometry".

Conclusion

The projects Gaia and SIM, when they are realized after 10 or 20 years, will help to solve many problems of Astrometry and Astronomy at the level of some microarcsecond accuracy. Practically it is Astrometry of the far future.

But the today Infrared Astrometry based on identification method, which help to improve the bad positions in the infrared catalogues, is appeared only 25 years ago. The first problem was solved by this one is the problem of the IRAS PSC position melioration. In the result the catalogue CPIRSS (Hindsley & Harrington 1994) and some others (Kharin 1992, Kharin 1997) were compiled on the basis of identifications of the IRAS PSC infrared sources with their optical counterparts from precise astrometrical catalogues in the FK5 system.

The today Infrared Astrometry is presented now by two ground-based catalogues. These are the DENIS and the 2MASS. Their positional accuracies in α and δ are 0.3 and 0.2, respectively. It is the level of the visual optical astrometrical accuracy at the end of the last century.

Improvement of bad position above two catalogues may be also achieved by identifying them with contemporary precise catalogues which are in ICRS coordinate system. On the basis of such identification the IR Reference Catalogue containing precise positions and proper motions of sources may be compiled. It is necessary both for presenting ICRS coordinate system in infrared and organizing various observations in this spectral range.

Acknowledgements

A.S.K. thanks to the Shanghai Astronomical Observatory for its invitation and financial support. Personally thanks to Profs. Shuhua Ye and Wenjing Jin, Drs. Zhenghong Tang and Dr. Shuhe Wang for their sincere friendship.

References

Kovalevsky, J. 2002, Modern Astromenry (Heidelberg: Springer Verlag Berlin), 2002
Hindsley, R. & Harrington, R. 1994, *AJ*, 201, 280
Kharin, A. S. 1992, *Kinematics and Physics of Celestial Bodies*, 8, 4, 67
Kharin, A. S. 1997, *Baltic Astronomy*, 6, 344

A Giant Step: from Milli- to Micro-arcsecond Astrometry
Proceedings IAU Symposium No. 248, 2007
W. J. Jin, I. Platais & M. A. C. Perryman, eds.

Asteroid astrometry as a link between ICRF and the Dynamical Reference Frames

D. A. Nedelcu[1], J. Souchay[2], M. Birlan[3], P. P. Popescu[1], P. V. Paraschiv[1] and O. Badescu[1]

[1] Astronomical Institute of the Romanian Academy,
Cutitul de Argint 5, RO-040557, Bucharest, Romania
email: [nedelcu, petre, paras, octavian]@aira.astro.ro
[2] SYRTE Observatoire de Paris,
61, avenue de l'Observatoire 75014, Paris, France
email: jean.souchay@obspm.fr
[3] IMCCE Observatoire de Paris,
61, avenue de l'Observatoire 75014, Paris, France
email: mirel.birlan@obspm.fr

Abstract. Highly accurate astrometry of asteroids in the frame of QSOs will provide the direct link between the Dynamical Reference Frame and the International Celestial Reference Frame. We propose a procedure that implies a selection of events for asteroids with accurately determined orbits crossing the CCD field containing the selected quasars. For asteroid ephemerides, a numerical integration method coupled with precise modelling of asteroid brightness will be used for analyzing our observations. A list of predictions for this type of "close approaches" will be presented.

Keywords. astrometry, minor planets, asteroids, reference systems

1. Introduction

In this paper we propose a project on the link between Dynamical Reference Frame (DRF) and the International Celestial Reference Frame (ICRF) by means of close angular approaches of asteroids and quasars. ICRF is the radio realization of International Celestial Reference System and is formed by VLBI radio positions of 212 extragalactic radiosources (quasars, galaxies, BL Lac objects) distributed over the entire sky and selected by astrometrical stability criteria. Realization of ICRS at optical wavelengths is given by Hipparcos Celestial Reference Frame (HCRF). The DRF is a realization of a dynamical reference system and it is defined by the ephemerides of one or more solar system bodies. The curent ICRF with 717 radiosources has an average sky density of ≈ one source per $9° \times 9°$. This makes difficult direct observations of a quasar and an asteroid in the same field of view (FOV). Thus, in order to investigate the link between the DRF and ICRF we need a much larger sample of extragalactic objects directly accesible to observations. The obvious choice is the Vèron-Cetty & Vèron list of 85221 quasars (Fig. 1).

2. Methods

Accurate positions for quasars could be obtained : 1) from USNO B1.0 plates using UCAC2 and 2MASS as astrometric reference catalogs (Andrei *et al.* 2004) or 2) by direct astrometry of quasars in the optical domain (Assafin *et al.* 2007). While the first method is fully automated the second one allow us to directly obtain astrometric positions for

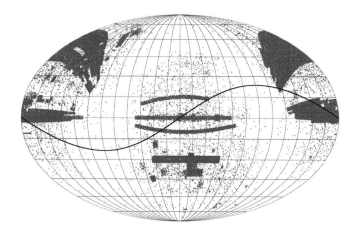

Figure 1. All sky distribution of Veron-Cetty & Veron 85221 quasars. Image center has $\alpha = 0h, \delta = 0°$ coordinates. Black line represents the ecliptic.

quasars and the nearby asteroids using the same set of reference stars. In 2004 at the Belogradchick Observatory ($\lambda = -22°40'30''; \phi = +43°37'22''$; h = 650m), Bulgaria we started an observational program aiming at densifying the northern hemisphere coverage with precise astrometric ICRF source positions. Observations were carried out using a 0.60m Zeiss telescope (f/12.5) equipped with an Apogee 47P CCD detector of 1024 × 1024 pixels of 13 μm size. The detector was operated in a 2 × 2 binned mode resulting on a FoV of 6.16' × 6.16' with an image scale of 0".72/pixel. From a set of 59 northern ICRF sources (their optical counterparts) we obtained the average and standard deviation for the optical minus radio position offsets of +6mas (51mas) and +7mas (57mas). The optical CCD source positions were referred to the UCAC2 catalog - the best HCRF representative available to date in terms of stellar density, position and proper motion accuracy. Asteroid positions are obtained in the framework of an accurate, fully relativistic solar system integration for the 8 planets, Pluto and the largest 3 asteroids (Ceres, Pallas, Vesta). The integrator uses a 12-order Bulirsch-Stoer method of Bader and Deuflhard with automatic step size control provided by Gnu Scientific Library (GSL). The initial conditions for the asteroid are obtained from the latest osculating elements (Bowell *et al.* 1994). Beside selecting asteroids with a good orbit-quality parameter (OQP), we also consider the effect of photocenter displacement (for the minor bodies of known sizes) by numerically computing the surface luminosity distribution, corresponding to asteroids obeying two different scattering laws (Lommel-Seeliger and Minnaert). Finally we run a search for all the forthcoming close approaches between selected asteroids and quasars in 2008. For these events to be observable by our instrumentation we impose two conditions: a maximum magnitude of 18.0 for the quasar and a maximum angular separation for the two objects of 10'. The complete list of events is available at http://observer.astro.ro/QA.

References

Andrei, A. H., Fienga, A., Assafin, M., Penna, J. L., da Silva Neto, D. N., & Vieira Martins, R. 2004, *Journées 2004 - systèmes de reféréncе spatio-temporels.*, p. 228

Assafin, M., Nedelcu, D. A., Badescu, O., Popescu, P., Andrei, A. H., Camargo, J. I. B., da Silva Neto, D. N., & Vieira Martins R. 2007, *A&A*, in press

Bowell, E., Muinonen, K., & Wasserman, L. H. 1994, *Asteroids, comets, meteors 1993*, p. 477

Vèron-Cetty M.-P. & Vèron P. 2006, *A&A*, 455, 733

A Giant Step: from Milli- to Micro-arcsecond Astrometry
Proceedings IAU Symposium No. 248, 2007
W.J. Jin, I. Platais & M.A.C. Perryman, eds.
© 2008 International Astronomical Union
doi:10.1017/S1743921308019492

Selection of stable sources from VLBI observations from 1984 to 2006†

S. B. Qiao[12]**, J. L. Li**[1]**, B. Zhang**[1] **and F. Tian**[1]

[1]Shanghai Astronomical Observatory, Shanghai 200030, China
[2]Surveying and Mapping Institute, Zhengzhou 450052, China
email: qsb@shao.ac.cn

Abstract. Based on the Calc/Solve system and aiming at the selection of stable sources, we have obtained a series of global solutions to astrometric/geodetic VLBI observations from 1984 to 2006 by changing the settings of the control parameters. After comparing the solutions and performing statistical analysis we proposed a list of 173 candidate stable sources. We also compared the list with those recommended by others authors.

Keywords. methods: data analysis, methods: statistical, catalogs, astrometry, reference systems

1. Introduction

The International Celestial Reference Frame (ICRF) (Ma *et al.* 1998) contains three categories of sources: defining, candidate and others. The latest version of ICRF is ICRF-ext2 (Fey *et al.* 2004), including 717 sources among which 212 are defining. The defining sources are used for the definition of ICRF and are the most important. They should be selected to be relatively stable and compact with comprehensive consideration of various kind of information. There are three representative schemes, namely, Ma *et al.*(1998), Fey & Charlot(1997) and Feissel (2003). The three resulting lists of stable sources are quite different from each other. In this report we will further discuss the selection of stable sources.

2. Our selection scheme

We prepared a working list of sources using the Calc/Solve system based on a global solution of VLBI data acquired from 1984 to 2006, with all ICRF-ext2 sources serving as the global parameters and the rest – as arc parameters. In the solution, all sources with coordinate uncertainties larger than 1mas and/or the coordinate correlation coefficient larger than 0.3 are removed. The 826 remaining objects form a working list of stable sources.

The working list is sorted by declination and then is divided into six groups. Mathematically, the ith source belongs to group j where $j = mod(i/6)$, and if $j = 0$ then set $j = 6$. By doing so the six groups have a similar sky coverage. Take each of the six groups in turn as arc parameters and all the others as global parameters to obtain the global solution of the VLBI data. In each of the six solutions, the frame orientation is constrained to ICRF while different solutions may still have their own orientations that are slightly different from each other. We unify the orientations of all six solutions by

† This work is supported by the Chinese Natural Science Foundation Committee (*No. 10778635, No. 10173019, No. 10473019*) and Chang'E-1 project.

Table 1. Criteria of source index.

Index	1	2	3	4
LV	$\leqslant 50\mu as/yr$	$(50\mu as/yr, 60\mu as/yr]$	$(60\mu as/yr, 80\mu as/yr]$	$> 80\mu as/yr$
WRMS	$\leqslant 150\mu as/yr$	$(150\mu as/yr, 300\mu as/yr]$	$(300\mu as/yr, 450\mu as/yr]$	$> 450\mu as/yr$
ASV	$\leqslant 150\mu as/yr$	$(150\mu as/yr, 300\mu as/yr]$	$(300\mu as/yr, 450\mu as/yr]$	$> 450\mu as/yr$

performing small angle rotations in order to let all the arc parameters of the six solutions be referred to a unique system.

The working list is then further cleaned up with the following criteria: (1) The arc positions of sources are rejected if the uncertainties of their coordinates are larger than 1mas; (2) A source is deleted if the standard deviation of the mean of its arc positions is larger than 0.5mas; (3) A source is removed if the observations were made in less than 20 sessions and/or the epoch coverage is less than 2 years.

The above exercise produced 344 sources out of the 826 available ones, among which 116, 108, 76 and 16 are respectively the ICRF defining, candidate, other, and new sources. With the selected 344 well observed sources, we calculate their yearly mean of right ascension and declination and then compose a time series of coordinates for each source.

The source index is classified as shown in Table 1 by referring to the following quantities:

(1) **Linear velocity (LV)**: the yearly average velocity of source coordinates derived by least-squares (LS) estimate.

(2) **Weighted root mean square (WRMS)**: $WRMS = \sqrt{\sum_{i=1}^{N}(\frac{x_i - \bar{x}}{\sigma_i})^2 / \sum_{i=1}^{N} \frac{1}{\sigma_i^2}}$

(3) **Allan standard variance (ASV)**: $\sigma_A^2(\tau) = \frac{1}{2N}\sum_{i=1}^{N}(x_{i+1} - x_i)^2$

Sources with index 1 or 2 are considered to be stable but 3 or 4 as unstable. We rejected the sources with index 4. Due to the page limit we omit detailed information about these sources. Among the above 316 sources, 235, 259 and 225 have indices of 1 or 2 for the LV, WRMS and ASV criteria respectively. Furthermore, 173 sources have indices of 1 or 2 under all the three criteria and we consider them as stable. Comparison shows that among our list of stable sources there are respectively 64, 60, 41 and 8 ICRF defining, candidate, other and new sources. While in Feissel's list the stable, unstable and new sources are 119, 25 and 29 respectively, but in Fey & Charlot's list, there are 111, 32 and 30 such sources.

Some sources could be apparently stable within relative short time spans, but with accumulation of new data they may exhibit instability. Noting that the data we analyzed cover more than 20 years, refinement of the list of stable sources is a long term pursuit.

References

Feissel-Vernier, M. 2003, *A&A*, 403, 105

Fey, A. L. & Charlot, P. 1997, *ApJS*, 111, 95

Fey, A. L., Ma, C., Arias, E. F., Charlot, P., Feissel-Vernier, M., Gontier, A. M., Jacobs, C. S., Li, J., & MacMillan, D. S. 2004, *AJ*, 127, 3587

Ma, C., Arias, E. F., Eubanks, T. M., Fey, A. L., Gontier, A.-M., Jacobs, C. S., Sovers, O. J., Archinal, B. A., & Charlot, P. 1998, *AJ*, 116, 516

A Giant Step: from Milli- to Micro-arcsecond Astrometry
Proceedings IAU Symposium No. 248, 2007
W. J. Jin, I. Platais & M. A. C. Perryman, eds.

© 2008 International Astronomical Union
doi:10.1017/S1743921308019509

CTIO 0.9m observations of ICRF optical counterparts

M. I. Zacharias and N. Zacharias

U. S. Naval Observatory, 3450 Mass. Ave. NW, Washington, DC 20392, USA
email: miz@usno.navy.mil, nz@usno.navy.mil

Abstract. We present astrometric results from 7 observing runs at the Cerro Tololo Interamerican Observatory (CTIO) 0.9m telescope of 197 extragalactic reference frame sources, selected from the original International Celestial Reference Frame (ICRF) catalog. This is part of the U.S. Naval Observatory (USNO) reference frame link program. Contemporaneous to the CTIO deep imaging, wide-field CCD data were taken with the USNO Twin Astrograph to provide accurate secondary reference stars in the 13 to 16 mag range. The optical positions are on the Hipparcos system (via Tycho-2 stars). The unweighted, mean RMS of positional differences 'optical−radio' for a single source is 28 and 25 mas for RA and Dec, respectively.

Keywords. astrometry, reference systems, catalogs, quasars: general

1. Astrograph reference stars

This reference frame program links Hipparcos and Tycho stars to the defining, optically faint sources in a two-step approach, deep observing (0.9m) and wide-field observing with the USNO Twin Astrograph. For each observing run an individual reference star catalog was constructed with these dedicated, unpublished observations from the USNO Twin Astrograph, with the following advantages with respect to the general UCAC2 (Zacharias, Urban, Zacharias, *et al.* (2004)): there are many more frames per source, the frames are centered on the source, and the telescope was on the east and west side of the pier to compensate for residual systematic errors.

2. Deep frame observations and reductions

Deep frames were observed with the CTIO 0.9m telescope. A summary of deep optical imaging observing runs can be found in Zacharias, Zacharias, Rafferty (2003). A customized filter was used to match the spectral bandpass of the USNO Twin Astrograph. At least 4 frames were taken per source. In addition, calibration fields were taken with offsets. So far 128 sources are classified as "good," 18 are optically faint, 27 are potential problem sources (identification confirmed but position offset larger than expected), 23 are empty fields (no optical counterpart visible at the corresponding radio position), and 1 yielded no result (only observing attempt). A faint optical source has a signal/noise ratio of 5 or less. For a potential problem source, the (optical−radio) position difference is greater than 3-sigma of the total estimated errors.

Residuals from the reductions of deep CCD frames were collected, binned and smoothed to establish the field distortion pattern. These data were then used to correct the x,y data of each deep frame. A linear plate model was adopted for the final, weighted least-squares adjustment of each frame similar to the procedures used for the Kitt Peak National Observatory (KPNO) 2.1m data (Zacharias & Zacharias (2005)). Roughly 20 to 900 reference stars are available on a single frame.

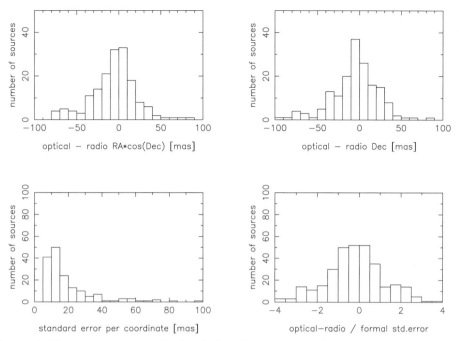

Figure 1. Histograms showing the (optical−radio) position difference distribution and the distribution of total optical position errors.

3. Optical−radio results

The unweighted mean RMS position difference optical−radio for a single source is 28 and 25 mas for RA and Dec, respectively. The mean total error of 146 sources (good and faint) is about 20 mas per coordinate (Fig. 1).

Acknowledgements

The deep frame observing is based on observations obtained at the Cerro Tololo Inter-American Observatory, a division of the National Optical Astronomy Observatories, which is operated by the Association of Universities for Research in Astronomy, Inc. under cooperative agreement with the National Science Foundation.

References

Zacharias M. I., Zacharias N., & Rafferty T. J. 2003, in Ralph Gaume, Dennis McCarthy, Jean Souchay (eds.), *The International Celestial Reference System, Maintenance and Future Realizations*, Proc. of IAU General Assembly XXV, Joint Discussion 16, Sydney, AUS (Washington) p. 188

Zacharias M. I. & Zacharias N. 2005, in P. Kenneth Seidelmann & Alice K. B. Monet (eds.), *Astrometry in the Age of the Next Generation of Large Telescopes*, Proc. meeting at Lowell Observatory, Flagstaff, USA (San Francisco: ASP-CS 338) p. 184

Zacharias, N., Urban, S. E., Zacharias M. I., Wycoff G. L., Hall D. M., Monet D. G., & Rafferty T. J. 2004, *AJ* 127, 3043

A Giant Step: from Milli- to Micro-arcsecond Astrometry
Proceedings IAU Symposium No. 248, 2007
W. J. Jin, I. Platais & M. A. C. Perryman, eds.

© 2008 International Astronomical Union
doi:10.1017/S1743921308019510

Re-calibration of GSC2.3 with UCAC2

Z. H. Tang[1], Z. X. Qi[1,2], Y. Yu[1], R. L. Smart[3], B. Bucciarelli[3], M. G. Lattanzi[3] and A. Spagna[3]

[1]Shanghai Astronomical Observatory, Chinese Academy of Sciences, Shanghai 200030, China
email: zhtang@shao.ac.cn

[2]Graduate School of the Chinese Academy of Sciences, Beijing 100039, China

[3]INAF-Astronomical Observatory of Turin, Pino-Torinese 2010025, Italy

Abstract. To make GSC-II more accurate and useful, it is necessary to re-calibrate GSC-II when a better reference catalogue (UCAC2) is available. With UCAC2 as the reference, preliminary re-calibration of some sample plates from GSC2.3 were carried out with different methods, such as Global, Mask and Filter. The results indicate that a 7th-order polynomial is sufficient to account for the influence of Schmidt plate deformation on the measured coordinates of stars. The magnitude equation can be eliminated after correcting for a common magnitude equation. The RMS of the re-calibrated data is around ±0.2~0.3 arcsec.

Keywords. astrometry, catalogs, surveys

1. Introduction

Over the last five decades, Schmidt telescopes have been frequently used as astrometric and astro-photometric survey instruments. And the surveys in various bandpass have already covered the entire sky several times in different epoch. The GSC-II is one of the largest all-sky astro-photometric catalog which is derived from the digitization of the Palomar and UK Schmidt survey plates today. The 2.3.2 version (about 1 billion objects) was released in October 2005, GSC-II is widely used in many fields. However, our research shows that there are some systematic residuals of position and magnitude equation still existing in this catalog. The main cause is that the plate reduction method did not account for the specific properties of plate which is bent during exposure. But sometimes the method we choose depends on the reference catalog. When GSC2.3 was calibrated, the unique and best choice of reference catalog was TYCHO-2, whose density is very low, magnitude range is narrow ($70\mathrm{deg}^{-2}$, V<11.5) and star images are over-exposure with large centroiding error in the plate. In 2003, the second version of USNO CCD Astrograph Catalog (UCAC2) was released, which is a high density ($1300\mathrm{deg}^{-2}$, R<16), high accurate astrometric catalog, and helping us to improve the GSC2.3 to a new level. In the following sections, we will talk about the property of former GSC2.3 and the results of re-calibration.

2. Astrometric property of GSC2.3

We use UCAC2 as the 'standard catalog' to inspect the systemic deformations related to location of the target on the plate, since its density, precision and accuracy of position and proper motion are enough for this purpose. To inspect magnitude equation, SDSS is an unique catalog to be used because of its accuracy and wide magnitude range which is around 10 to 21 mag. We can see from figure 1.1 that the plate deformation of the GSC2.3 still exists, and its bigger than 0.5 arc-sec near the plate edges. There is also a boundary whose radius is around 2.5° from the plate corner. The main reason for those

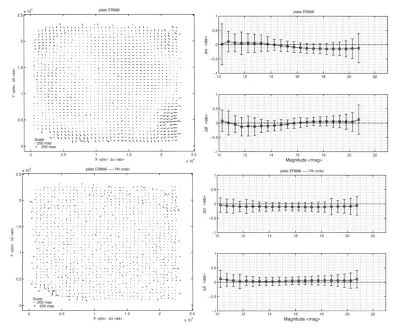

Figure 1. `Upper left`: Fig.1.1 Vector distribution of positional differences between GSC2.3 and UCAC2 on one plate. `Lower left`: Fig.1.3 Vector distribution of position differences between re-calibrated GSC2.3 and UCAC2 on the same plate. `Upper right`: Fig.1.2 Systematic and random errors of position differences between GSC2.3 and SDSS. `Lower right`: Fig.1.4 Systematic and random errors of position differences between the re-calibrated GSC2.3 and SDSS.

is the Schmidt plates are bent during exposure and expanded after exposure. But the former method of plate reduction, which is a second order polynomial, can not account for the specific properties. We can see from figure 1.2 that the well-known magnitude equation still exists, moreover it is nonlinear. The peak-to-peak amplitude is around 0.2 arc-sec. The possible physical factor of magnitude equation is the systematical influences of the mechanical and optical systems of telescopes on stars with different brightness.

3. Astrometric re-calibration of GSC2.3

3.1 The main processing pipelines of re-calibration with Global method are as follows:
- Equidistant projection
- 7th order plate model (`complete 7th order polynomial`)
- Magnitude equation correction

To account for the influence of Schmidt plate deformation on the measured coordinates of stars, after our study we find that a 7th order polynomial can be introduced when it is reduced against UCAC2. Figure 2.1 displays different order polynomial fitting to the standard coordinates as a function of measured coordinates for more than 100 plates, these plates are from Palomar and UK Schmidt telescopes with different sky cover, epoch, emulsion and filter. There are two inflection points, one is at the 5th order and the other is 7th. It is stable for both cases after the 7th order fitting. Figure 1.3 displays the vector distribution of position differences between re-calibrated GSC2.3 and UCAC2, we can find that the plate deformation is removed. We find that the tendency and the grade of the magnitude equation are similar for the plates in a same survey(e.g. `XP, XJ, XI, XO, ER, IS`) from the same telescope. Figure 2.2 displays the common magnitude

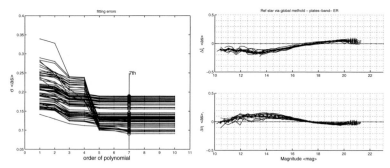

Figure 2. `Upper left`: Fig.2.1 Fitting error with different polynomial order for various plates. `Upper right`: Fig.2.2 Magnitude equation of ER plates, created from 13 plates of the UK Schmidt telescope ER survey.

equation of ER source(`ER surveys`), created from 13 ER plates, which is used to correct magnitude equation of ER plates. From Figure 1.4 we can find that the well-known magnitude equation has disappeared after correcting the common magnitude equation. And the off-set between the new GSC2.3 and SDSS is caused by the systemic error between the UCAC2 (`the reference catalog of the re-calibrated GSC2.3`) and SDSS.

3.2 The main processing pipelines of re-calibration with 'filter+mask' method (`double filter`) are as follows:
- Equidistant projection
- Obtain the mask from reference stars which has been filtered on the plate
- Filter the filed stars
- Apply the mask to the filed stars
- Magnitude equation correction

Just like using 7th order Global method, we can also remove the plate deformation via the 'double filter' method, but considering the speed and accuracy of these two methods we propose to use the Global method.

4. Conclusion

Preliminary tests show that the Schmidt plate deformation and magnitude equation can be nearly eliminated after re-calibration. The RMS of the re-calibrated data is around $\pm 0.2 \sim 0.3$ arc-sec for stars of different brightness. Some further tests are ongoing, which will help us to learn more about the factors in astrometric calibrations. After the tests, an astrometric re-calibration of the entire catalog will be performed and then the GSC-II should become more accurate.

References

Zacharias, *et al.* 2004, *AJ*, 127, 3043
Jeffrey R. P., *et al.* 2003, *AJ*, 125, 1559
Dick, W. R., *et al.* 1991, *AN*, 312, 113, 120
Bucciarelli, B., *et al.* 2008, *this volume*, p. 316

A Giant Step: from Milli- to Micro-arcsecond Astrometry
Proceedings IAU Symposium No. 248, 2007
W. J. Jin, I. Platais & M. A. C. Perryman, eds.
© 2008 International Astronomical Union
doi:10.1017/S1743921308019522

The second realization of
the ICRF with VLBI

C. Ma

NASA Goddard Space Flight Center
Code 698, Greenbelt, Maryland, 20771, USA
email: Chopo.Ma@nasa.gov

Abstract. The ICRF derived from VLBI observations of extragalactic radio sources up to 1995.6 and effective since 1998.0 was a radical change from the FK5 stellar/equinox celestial system. Since then the number of geodetic/astrometric VLBI observations has tripled and the number of radio sources with astrometrically useful data has quadrupled. These data along with advances in modeling and estimation will be used to generate the next ICRF realization in the microwave band. Analysis of source position time series and source structure evolution will be used to select better "defining" sources. Working groups have been established by the IAU, IERS and IVS with the goal of presenting the second realization of the ICRF at the IAU General Assembly in 2009.

Keywords. : astrometry, reference systems, catalogs, surveys

1. Introduction

Since 1998 the basis for celestial positioning has been the ICRS (International Celestial Reference System) and the ICRF (International Celestial Reference Frame). The following descriptions are taken from the NFA IAU 2000 Glossary developed by the IAU Working Group "Nomenclature for Fundamental Astronomy" (Capitaine *et al.* (2007)):

The ICRS is the idealized barycentric coordinate system to which celestial positions are referred. It is kinematically nonrotating with respect to the ensemble of distant extragalactic objects. It has no intrinsic orientation but was aligned close to the mean equator and dynamical equinox of J2000.0 for continuity with previous fundamental reference systems. Its orientation is independent of epoch, ecliptic or equator and is realized by a list of adopted coordinates of extragalactic sources.

The ICRF is a set of extragalactic objects whose adopted positions and uncertainties realize the ICRS axes and give the uncertainties of the axes. It is also the name of the radio catalog whose 212 defining sources are currently the most accurate realization of the ICRS. Note that the orientation of the ICRF catalog was carried over from earlier IERS radio catalogs and was within the errors of the standard stellar and dynamical frames at the time of adoption. Successive revisions of the ICRF are intended to minimize rotation from its original orientation. Other realizations of the ICRS have specific names (e.g., the Hipparcos Celestial Reference Frame).

The ICRF catalogue was generated from 2/8 GHz VLBI data and the analysis available in mid 1995. The error for individual source positions has an estimated floor of 250

337

microarcseconds and the accuracy of the realized ICRS coordinate axes is ∼30 microarc-seconds based on the 212 defining sources. The ICRF was updated to ICRF-Ext.2 (Fey *et al.* (2004)) using additional data through 2002, but the axes are unchanged since the same defining sources and their ICRF positions are retained.

2. Organization of the second VLBI realization

The ICRS/ICRF was adopted by a resolution of the IAU General Assembly in Kyoto in 1997 acting for the whole astronomical community. The groundwork had been laid through a series of IAU working groups and colloquia held over nearly a decade that addressed both the relativistic foundation and the suitability of compact extragalactic radio sources as fiducial objects. The maintenance of the ICRS/ICRF was delegated by the IAU to the IERS (International Earth Rotation and Reference Systems Service). As the ICRF was derived from VLBI observations, the IVS (International VLBI Service for Geodesy and Astrometry), one of the independent Technique Centers of the IERS, has the primary responsibility for continuing observations and analysis to support advance-ment of the ICRF at radio frequencies. There are two working groups involved in the second ICRF realization, one established by the IAU to provide oversight for the wide astronomical community and the other established jointly by the IERS and IVS to carry out the actual analysis. The charter of the IERS/IVS Working Group states:

> The purpose of the working group is to generate the second realization of the ICRF from VLBI observations of extragalactic radio sources, consistent with the current realization of the ITRF and EOP data products. The working group will apply state-of-the-art astronomical and geophysical models in the analysis of the entire relevant S/X astrometric and geodetic VLBI data set. The working group will carefully consider the selection of defining sources and the mitigation of source position variations to improve the stability of the ICRF. The goal is to present the second ICRF to relevant authoritative bodies, e.g. IERS and IVS, and submit the revised ICRF to the IAU Division I working group on the second realization of the ICRF for adoption at the 2009 IAU General Assembly.

3. VLBI observations to improve the ICRF

One difficulty faced by the original ICRF analysis was the fact that the overall VLBI data set was ∼95% from geodetic observing sessions, which used a limited set of "geode-tic" sources. The geodetic sources are generally the strongest "compact" sources, in-cluding in the early years of geodetic/astrometric VLBI some sources with considerable structure like 3C273B. Figure 1 shows that the imbalance between the number of sessions for geodetic and non-geodetic sources is still present. However, the peak at 11-30 sessions reflects a specific IVS observing program to enhance the astrometric data set based on Feissel-Vernier's analysis of annual source position time series (Feissel-Vernier (2003)), which identified sources with stable and unstable positions. Figure 2 shows stable and potentially stable sources separated between ICRF defining and non-defining sources. Potentially stable sources did not have sufficient data in 1989-2002 to fully categorize stability but had no indication of instability. Figure 3 shows that the current set of ICRF defining sources is probably not ideal, some being unstable and some having minimal data during 1989-2002. The stable, potentially stable and non-geodetic ICRF defining sources have been observed regularly since 2004 in the IVS CRF monitoring program, which uses a small fraction of the time during selected geodetic observing sessions (Fig. 4). The goal is to observe each source at least semiannually. The data will be used in the selection of defining sources for the second realization. Figures 5 and 6 show that the CRF monitoring sources and the ICRF defining sources have been observed, but much more data would be desirable. Southern sources are observed in specific astrometric sessions, but the networks are small.

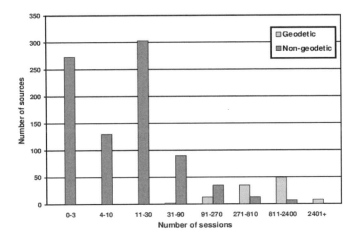

Figure 1. Sources' session participation, 1979–2007.

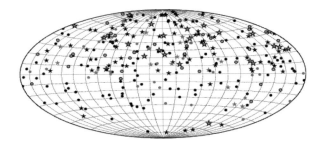

Defining: ★81 Stable: F.-V.A&A 2003 ☆20 Stable: F.-V. additional ✪25 Potentially stable
Other: •117 Stable: F.-V.A&A 2003 ◦33 Stable: F.-V. additional ○36 Potentially stable

Figure 2. Stable and potentially stable sources.

4. Position time series and source structure

Position time series will be one of the means for selecting the defining sources for the next VLBI ICRF. New time series have been generated by members of the IERS/IVS Working Group for this purpose. Figure 7 shows the sum of the wrms scatter for the right ascension and declination time series as a vector at the position of each source. Only sources with summed wrms < 0.5 mas are included. Although there are probably enough sources to select good defining sources, it can be seen that there is a data deficiency below −20° declination in both quantity and quality. This deficiency may require flexible criteria for defining sources to achieve more uniform spatial distribution.

Source structure index information will also be used to select the defining source (See Charlot, this volume). Structure index 1 and 2 indicate minimal structure while index 3

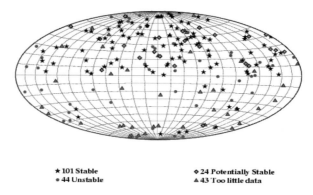

★ 101 Stable   ◇ 24 Potentially Stable
● 44 Unstable   ▲ 43 Too little data

Figure 3. ICRF defining sources.

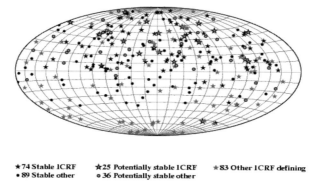

★ 74 Stable ICRF   ★ 25 Potentially stable ICRF   ★ 83 Other ICRF defining
● 89 Stable other   ⊙ 36 Potentially stable other

Figure 4. Total CRF monitoring sources.

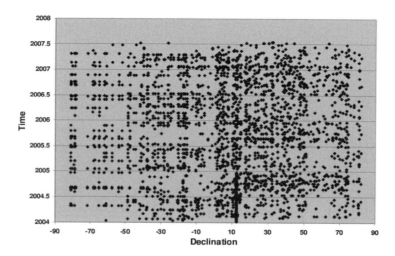

Figure 5. Observations of CRF Monitoring Sources, 2004–2007.

and 4 reflect undesirable complexity. Figures 8 and 9 show there is a relationship between structure index and scatter of position time series. Index 1 and 2 sources generally have less scatter than index 3 and 4 sources. There are, however, outliers that need to be examined.

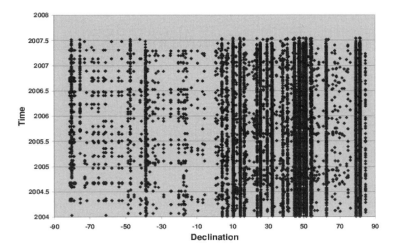

Figure 6. Observations of ICRF Defining Sources, 2004–2007.

Sources (285) with Arc WRMS < 0.5 mas

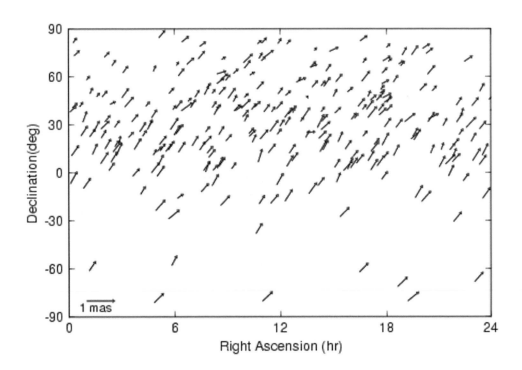

Vector components are [wrms(RAcosDec), wrms(DEC)]

Figure 7. Summed wrms scatter (for sums < 0.5 mas only).

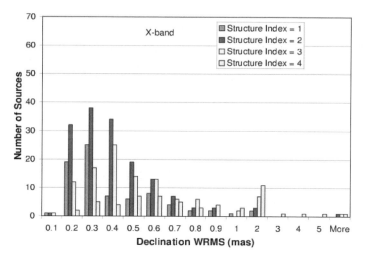

Figure 8. Distribution of sources for declination scatter (categorized by structure index).

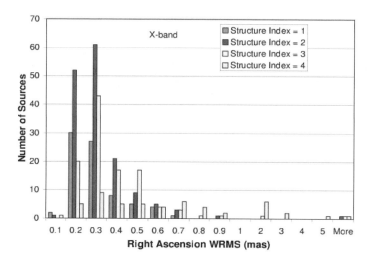

Figure 9. Distribution of sources for right ascension scatter (categorized by structure index).

5. Timetable for IERS/IVS Working Group

Spring 2007	Working Group meeting, Vienna
Fall 2007	Generation, comparison of time series Working Group meeting, Paris
Spring 2008	Analysis of time series Working Group meeting, St. Petersburg
Mid 2008	Defining source criteria
Fall 2008	Selection of defining sources, analysis configuration
Spring 2009	Generation of ICRF-2 catalogue, presentation to IVS, IERS, and IAU working group

6. VLBA Calibrator Survey (VCS)

In addition to the geodetic and astrometric sessions scheduled by the IVS and earlier international space geodesy collaborations, a series of observations which began in 1994 using the VLBA has provided the much larger VCS catalogue (Beasley *et al.* (2002), Fomalont *et al.* (2003), Petrov *et al.* (2005), Petrov *et al.* (2006), Kovalev *et al.* (2007)) with good astrometric positions (Figure 10). Each target source, however, is observed in only one session, so the quality of positions is somewhat worse than from the IVS data set and there is no temporal variation information. These sources will be included in the analysis for the second ICRF realization.

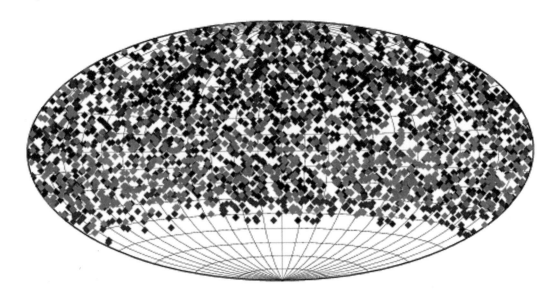

◆1576 VCS1 ◆241 VCS2 ◆308 VCS3
◆261 VCS4 ◆590 VCS5 ◆215 VCS6

Figure 10. VCS sources.

References

Beasley, A. J., Gordon, D., Peck, A. B., Petrov, L., MacMillan, D. S., Fomalont, E. B., & Ma, C. 2002, *ApJ*, Supp. 141, 13

Capitaine *et al.*, IAU Working Group "Nomenclature for Fundamental Astronomy" 2007, *http://syrte.obspm.fr/iauWGnfa/*

Feissel-Vernier, M. 2003, *A&A*, 403, 105

Fey, A. L., Ma, C., Arias, E. F., Charlot, P., Feissel-Vernier, M., Gontier, A.-M., Jacobs, C. S., Li, J., & MacMillan, D. S. 2004, *AJ*, 127, 3587

Fomalont, E., Petrov, L., MacMillan, D. S., Gordon, D., & Ma, C. 2003, *AJ*, 126 (N5), 2562

Kovalev, Y., Petrov, L., Fomalont, E., & Gordon, D. 2007, *AJ*, 133, 1236

Petrov, L., Kovalev, Y., Fomalont, E., & Gordon, D. 2005, *AJ*, 129, 1163

Petrov, L., Kovalev, Y., Fomalont, E., & Gordon, D. 2006, *AJ*, 131, 1872

A Giant Step: from Milli- to Micro-arcsecond Astrometry
Proceedings IAU Symposium No. 248, 2007
W.J. Jin, I. Platais & M.A.C. Perryman, eds.

Source structure: an essential piece of information for generating the next ICRF

P. Charlot[1], **A. L. Fey**[2], **A. Collioud**[1], **R. Ojha**[2], **D. A. Boboltz**[2] **and J. I. B. Camargo**[3]

[1]Laboratoire d'Astrophysique de Bordeaux, Université Bordeaux 1, CNRS
BP 89, 33270 Floirac, France
email: charlot@obs.u-bordeaux1.fr

[2]U.S. Naval Observatory
3450 Massachusetts Ave., NW, Washington, DC 20392-5420, USA

[3]Observatorio do Valongo, Universidade Federal do Rio de Janeiro
Ladeira do Pedro Antônio, 43, Rio de Janeiro, RJ, Brazil

Abstract. The intrinsic radio structure of the extragalactic sources is one of the limiting factors in defining the International Celestial Reference Frame (ICRF). This paper reports about the ongoing work to monitor the structural evolution of the ICRF sources by using the Very Long Baseline Array and other VLBI telescopes around the world. Based on more than 5000 VLBI images produced from such observations, we have assessed the astrometric suitability of 80% of the ICRF sources. The number of VLBI images for a given source varies from 1 for the least-observed sources to more than 20 for the intensively-observed sources. Overall, we identify a subset of 194 sources that are highly compact at any of the available epochs and which are prime candidates for the realization of the next ICRF with the highest accuracy.

Keywords. astrometry, reference systems, quasars: general

1. Introduction

The International Celestial Reference Frame (ICRF), which has been the official IAU reference frame in use since 1 January 1998, is currently based on the VLBI positions of 717 extragalactic radio sources. Of these, 608 sources are from the original ICRF built in 1995, with a categorization that comprises 212 well-observed *defining* sources (which served to set the axes of the frame), 294 less-observed *candidate* sources, and 102 *other* sources showing coordinate instabilities (Ma *et al.* 1998). The accuracy of the individual ICRF source positions has a floor of 250 μas, while the axes of the frame are stable to about 20 μas. Since then the positions have been improved for the non-defining sources and the frame has been extended by 109 *new* sources in ICRF-Ext.1 and ICRF-Ext.2 using additional data acquired in the period 1995–2002 (Fey *et al.* 2004).

At the IAU XXVIth General Assembly in Prague (August 2006), the community decided to engage in realization of the successor to ICRF, to be presented at the next IAU General Assembly in 2009. The motivation for generating this new celestial frame is to benefit from recent improvements in VLBI modeling (e.g. for the troposphere) and to take advantage of the wealth of VLBI data that have been acquired since the time the ICRF was established. A major issue to be addressed in this new realization is the revision of source categorization, in particular the choice of the defining sources. Such a revision is necessary because some of the original ICRF defining sources are found to have extended structures (Fey & Charlot 2000) or position instabilities (e.g. MacMillan 2006), and are therefore inadequate for defining the celestial frame with the highest accuracy.

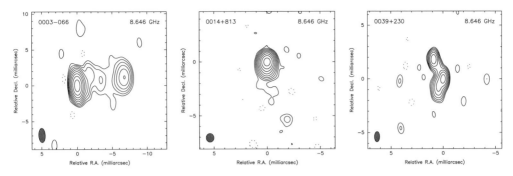

Figure 1. VLBI images at X band (8 GHz) for three ICRF sources (0003−066, 0014+813, 0039+230) as derived from the data of a RDV session conducted on 2003 December 17. The three sources were selected randomly according to increasing right ascension starting at RA=00h.

The selection criteria that are considered by the working group in charge of the next ICRF realization are based either on source structure information (VLBI images), to evaluate the compactness and astrometric suitability of the sources, or on time series of source coordinates, to assess the source position stability. In this paper, we discuss only the former. There are now more than 5000 VLBI images available to evaluate source quality, whereas these were less than a hundred at the time the ICRF was built, thereby permitting considerable progress. In Sect. 2, we present the observational data and analysis method used in this work. Our results are discussed in Sect. 3 with emphasis on statistics of source quality for each ICRF source category and for the overall ICRF. We also draw prospects for the further improvements in assessing the source quality before the next ICRF is generated.

2. Observational data and analysis method

The VLBI maps used in our analysis were produced from a total of 38 VLBI sessions conducted between 1994 and 2007, which were imaged either at USNO or at Bordeaux Observatory. These comprise:

• 8 dedicated dual-frequency (8 GHz/2 GHz) imaging sessions conducted with the Very Long Baseline Array (VLBA) between July 1994 and January 1997 (Fey *et al.* 1996, Fey & Charlot 1997, Fey & Charlot 2000);

• 23 dual-frequency (8 GHz/2 GHz) Research & Development VLBA (RDV) sessions conducted between January 1997 and June 2007; these sessions include the 10 VLBA stations and up to 10 additional geodetic telescopes;

• 5 dedicated southern-hemisphere 8 GHz imaging sessions conducted between July 2002 and April 2004 with the Australian Long baseline Array, augmented by radio telescopes in South-Africa, Hawaii, and Japan (Ojha *et al.* 2004, Ojha *et al.* 2005);

• 2 VLBA sessions at 2 GHz/8 GHz/24 GHz and 8 GHz/24 GHz conducted in February 2004 and August 2005 as part of a project to extend the ICRF to higher frequencies (Lanyi *et al.* 2007).

Altogether, this represents a total of 2697 maps at X band (8 GHz) from 577 ICRF sources and 2388 maps at S band (2 GHz) from 492 ICRF sources. Less sources have been imaged at S band because the southern-hemisphere sessions did not observe at this band. There are up to 28 VLBI epochs available for the most intensively-observed source, whereas only one epoch is available for the least-observed sources. A sample of X-band VLBI images for three ICRF sources, as derived from the data of a RDV session conducted in December 2003, is shown in Fig. 1. From these images, only one source

(0014+813) is relatively compact, while the other two (0003−066 and 0039+230) show extended structures. This fraction of compact sources is consistent with that found for the whole ICRF, as discussed in Sect. 3.

Based on the VLBI images, we derived the expected effects of intrinsic source structure on the VLBI delay astrometric quantities, following the algorithm of Charlot (1990). We then used the "structure index" indicator to define the astrometric source quality, as devised by Fey & Charlot (1997). Structure index values of 1 and 2 point to excellent and good astrometric suitability, respectively, while values of 3 and 4 indicate a poor suitability. A given source may have differing structure indices at X band and S band, depending on properties of the brightness distribution at each band. The structure index may also vary with time because of possible temporal evolution of the brightness distribution.

For each source, we obtained a series of structure index at each band according to the number of available VLBI images (e.g., Charlot *et al.* (2006)). Adopting a conservative approach, we chose the maximum value of structure index as the source quality indicator when multi-epoch structure indices are available. In other words, if a source shows a structure index value of 3 or 4 at one or more epochs, it should be regarded as unsuitable for highly-accurate astrometry even though it has the structure index values of 1 or 2 at some other epochs. This multi-epoch structure index value is the criterion used for drawing statistics on the source quality.

3. Results

As noted above, we have obtained structure indices for 577 sources in X band and 492 sources in S band (representing 80% and 69% of the current 717 sources of the ICRF, respectively). The X-band structure index distribution (Fig. 2) shows that 197 sources (or 34% of all sources) are astrometrically-suitable at this frequency according to our criterion (structure index is either 1 or 2). In S band, 86% of the sources have a structure index either 1 or 2 (Fig. 2). This indicates that the contribution of the S-band structure to the dual-frequency (S/X) calibrated delay is usually smaller compared to the contribution by X-band structure, as already noted in Fey & Charlot (1997). Comparing the X- and S-band structure indices for each source shows that, with three exceptions, all sources that have a S-band structure index of either 3 or 4 have also a X-band structure index of 3 or 4. Based on the S-band structure index, we thus exclude only 3 additional sources, which leaves a total of 194 ICRF sources astrometrically-suitable at both frequencies.

In Fig. 3, the X-band structure index distribution is compared for each ICRF source category. As expected, the distribution is somewhat better for the defining sources than for the candidate and "other" sources. However, only about 40% of the ICRF defining sources have a structure index value of either 1 or 2. The fraction of suitable sources drops down to 32% for the candidate sources and 22% for the "other" sources, while it is 48% for the "new" sources. Overall, these results confirm that revision of source categories is mandatory for the next ICRF.

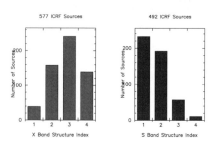

Figure 2. The structure index distribution at X band (8 GHz) and S band (2 GHz) for all ICRF sources that have a structure index available at these frequencies.

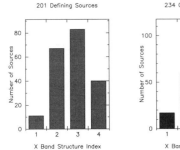
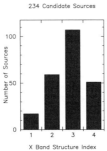
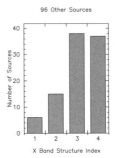
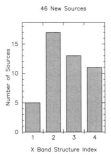

Figure 3. Distribution of the X-band (8 GHz) structure indices in each ICRF source category (defining, candidate, "other", "new"). The 577 ICRF sources with currently available structure indices are included.

4. Conclusions

We have evaluated the astrometric suitability of 80% of the sources in the ICRF based on multi-epoch VLBI maps of their structures. From this analysis, a sample of 194 astrometrically-suitable ICRF sources that have compact or very compact structures according to our "structure index" indicator has been identified. It is anticipated that the remaining 20% of ICRF sources for which the astrometric suitability has not been assessed (mostly in the southern sky) will be imaged in the near future through further VLBI observing programs in the southern hemisphere. The astrometric suitability of the sources already imaged, and discussed in this paper, will also be refined as new VLBI sessions are processed and maps become available. This information will be essential for selecting the proper defining sources and generating the next ICRF by 2009.

References

Charlot P. 1990, *AJ* 99, 1309

Charlot, P., Fey, A. L., Ojha, R., & Boboltz, D. A. 2006, in: International VLBI Service for Geodesy and Astrometry 2006 General Meeting Proceedings, Eds. D. Behrend and K. D. Baver, NASA/CP-2006-214140, p. 321.

Fey, A. L., Clegg, A. W., & Fomalont, E. B. 1996, *ApJS* 105, 299

Fey, A. L. & Charlot, P. 1997, *ApJS* 111, 95

Fey, A. L. & Charlot, P. 2000, *ApJS* 128, 17

Fey, A. L., Ma, C., Arias, E. F., Charlot, P., Feissel-Vernier, M., Gontier, A.-M., Jacobs, C. S., Li, J., & MacMillan, D. S. 2004, *AJ* 127, 3587

Lanyi, G. E., Jacobs, C. S., Naudet, C. J., Zhang, L. D., Boboltz, D. A., Fey, A. L., Charlot, P., Fomalont, E. B., Geldzahler, B., Gordon, D., Ma, C., Romney, J. E., & Sovers, O. J. 2007, *AJ* (in preparation)

Ma, C., Arias, E. F., Eubanks, T. M., Fey, A. L., Gontier, A.-M., Jacobs, C. S., Sovers, O. J., Archinal, B. A., & Charlot, P. 1998, *AJ* 116, 516

MacMillan, D. S. 2006, in: International VLBI Service for Geodesy and Astrometry 2006 General Meeting Proceedings, Eds. D. Behrend and K. D. Baver, NASA/CP-2006-214140, p. 274.

Ojha, R., Fey, A. L., Johnston, K. J., Jauncey, D. L., Reynolds, J. E., Tzioumis, A. K., Quick, J. F. H., Nicolson, G. D., Ellingsen, S. P., Doodson, R. G., & McCulloch, P. M. 2004, *AJ* 127, 3609

Ojha, R., Fey, A. L., Charlot, P., Jauncey, D. L., Johnston, K. J., Reynolds, J. E., Tzioumis, A. K., Quick, J. F. H., Nicolson, G. D., Ellingsen, S. P., McCulloch, P. M., & Koyama, Y. 2005, *AJ* 130, 2529

A Giant Step: from Milli- to Micro-arcsecond Astrometry
Proceedings IAU Symposium No. 248, 2007
W. J. Jin, I. Platais & M. A. C. Perryman, eds.
© 2008 International Astronomical Union
doi:10.1017/S1743921308019546

Opacity in compact extragalactic radio sources and its effect on radio-optical reference frame alignment

Y. Y. Kovalev[1,2]**, A. P. Lobanov**[1]**, A. B. Pushkarev**[1,3,4]**, J. A. Zensus**[1]

[1] Max-Planck-Institut für Radioastronomie, Auf dem Hügel 69, 53121 Bonn, Germany
e-mail: ykovalev, alobanov, apushkar, azensus@mpifr-bonn.mpg.de
[2] Astro Space Center of Lebedev Physical Institute,
Profsoyuznaya 84/32, 117997 Moscow, Russia
[3] Pulkovo Astronomical Observatory, Russia; [4] Crimean Astrophysical Observatory, Ukraine

Abstract. Accurate alignment of the radio and optical celestial reference frames requires detailed understanding of physical factors that may cause offsets between the positions of the same object measured in different spectral bands. Opacity in compact extragalactic jets (due to synchrotron self-absorption and external free-free absorption) is one of the key physical phenomena producing such an offset, and this effect is well-known in radio astronomy ("core shift"). We have measured the core shifts in a sample of 29 bright compact extragalactic radio sources observed by Very Long Baseline Interferometry (VLBI) at 2.3 and 8.6 GHz. We report the results of these measurements and estimate that the average shift between radio and optical positions of distant quasars could be of the order of 0.1–0.2 mas. This shift exceeds the expected positional accuracy of Gaia and SIM. We suggest two possible approaches to carefully investigate and correct for this effect in order to align accurately the radio and optical positions. Both approaches involve determining a Primary Reference Sample of objects to be used for tying the radio and optical reference frames together.

Keywords. galaxies: active, galaxies: jets, radio continuum: galaxies, astrometry, reference systems

1. Introduction

Extragalactic relativistic jets are formed in the immediate vicinity of the central black holes in galaxies, at distances of the order of 100 gravitational radii, and they become visible in the radio at distances of about 1000 gravitational radii (Lobanov & Zensus 2007). This apparent origin of the radio jets is commonly called the "core". In radio images of extragalactic jets, the core is located in the region with an optical depth $\tau_s \approx 1$. This causes the absolute position of the core, $r_{\rm core}$, to vary with the observing frequency, ν, since the optical depth profile along the jet depends on ν: $r_{\rm core} \propto \nu^{-1/k_{\rm r}}$ (Blandford & Königl 1979). Variations of the optical depth along the jet can result from synchrotron self-absorption (Königl 1981), pressure and density gradients in the jet and free-free absorption in the ambient medium most likely associated with the broad-line region (BLR) (Lobanov 1998).

The core shift is expected to introduce systematic offsets between the radio and optical positions of reference sources, affecting strongly the accuracy of the radio-optical matching of the astrometric catalogues. The magnitude of the core shift can exceed the inflated errors of the radio and optical positional measurements by a large factor. This makes it necessary to perform systematic studies of the core shift in the astrometric samples in order to understand and remove the contribution of the core shift to the errors of the radio-optical position alignment.

Measurements of the core shift have been done so far only in a small number of objects (e.g., Marcaide *et al.* 1994; Lara *et al.* 1994; Porcas & Rioja 1997; Lobanov 1996, 1998; Ros & Lobanov 2001; Kadler *et al.* 2004; Sokolov & Cawthorne 2007). In this paper, we present results for 29 compact extragalactic radio sources used in VLBI astrometric studies and discuss the core shift effect on the alignment of radio and optical reference frames.

2. Core shift measurements between 2.3 and 8.6 GHz

We have imaged and analyzed 277 sources from ten 24 hr-long geodetic RDV (Research and Development VLBI experiments, see, e.g., Gordon 2005) observations obtained in 2002 and 2003. Geodetic RDV sessions feature simultaneous observations at 2.3 GHz and 8.6 GHz (S and X bands) with a global VLBI network which for each session includes the VLBA and up to nine other radio telescopes around the world.

This long-term RDV program is one of the best opportunities for a large project to measure two-frequency core shifts for several reasons: (i) it is optimized to have a good (u,v)-coverage, (ii) it has the maximum possible resolution for ground-based VLBI at these frequencies, (iii) the frequency ratio between the simultaneously observed bands is high (3.7), and (iv) the core shift per unit of frequency between 2.3 and 8.6 GHz is larger than that at higher frequencies (see, e.g., Lobanov 1998). A dedicated multi-band VLBI project covering a wider frequency range would certainly provide results of a better quality, but it is extremely time consuming.

We have measured the frequency-dependent core shift between 2.3 and 8.6 GHz by model-fitting the source structure with two-dimensional Gaussian components (Pearson 1999) and referencing the position of the core component to one or more jet features, assuming the latter to be optically thin and having frequency-independent peak positions.

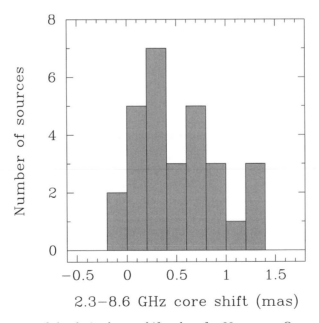

Figure 1. Histogram of the derived core shift values for 29 sources. One average core shift value per source is used. The median value for the distribution is equal to 0.44 mas.

The shifts were successfully measured in 29 Active Galactic Nuclei (AGN), with the values of a shift ranging between -0.1 and 1.4 mas and the median value for the sample of 0.44 mas (Fig. 1). In this study, typical errors of the core shift measurements are about or less than 0.1 mas. For 90% of the 277 objects imaged, no reliable estimates of core shifts have been obtained by this method.

3. Radio-optical alignment of astrometric positions

The core shift issue gains specific importance when the radio reference frame needs to be linked to an optical reference frame. So far the link is based on the study of some radio stars which are seen both by *Hipparcos* and VLBI. These measurements are based on stars with large radiospheres and with big spots at their optical surfaces; both effects may introduce large uncertainties (see, e.g., Lestrade *et al.* 1995; Boboltz 2005; Ros 2005). Future accurate alignment of the frames has to rely on compact radio sources such as distant quasars.

Below, we discuss the alignment problem for compact extragalactic radio sources. We assume that the dominating component in both the radio and optical bands is the synchrotron self-absorbed compact jet origin (core). Broad-band modeling of blazar spectral energy distribution supports this hypothesis (see, e.g., recent review by Böttcher 2007). High-resolution VLBI observations of nearby AGN imply that the jet is formed and emitting in the radio already at distances of $\leqslant 1000$ gravitational radii from the central engine (e.g. Junor *et al.* 1999; Kadler *et al.* 2004). Thus the physical offset between the jet base and the central nucleus can be much smaller than the positional shift due to opacity in the jet (the latter can be larger than 1 pc). This implies that the offset between radio and optical positions of reference quasars will be dominated by the core shift even if the optical emission comes from the accretion disk around the central nucleus.

The magnitude of the core shift, Δr, between two arbitrary frequencies ν_1 and ν_2 ($\nu_1 > \nu_2$) caused by synchrotron self-absorption can be predicted for an object with known synchrotron luminosity, L_{syn}, of a compact jet (Lobanov 1998). If not corrected for, the core shift will introduce an additional error factor in the alignment of radio and optical reference frames. For typical parameters of relativistic jets we estimate the average shift between 8.6 GHz and 6000 Å for a complete sample of compact extragalactic sources to be on the order of 0.1–0.2 mas.

4. Summary

Measurements of frequency-dependent shifts in the parsec-scale jet cores in AGN are reported for 29 bright extragalactic radio sources. It is shown that the shift can be as high as 1.4 mas between 2.3 and 8.6 GHz. We have shown that the core shifts are likely to pose problems for linking radio and optical reference frames. We have estimated theoretically the average shift between the radio (4 cm) and optical (at 6000 Å) bands to be of an order of 0.1 mas for a complete sample of radio selected AGN.

The estimated radio-optical core shift exceeds the positional accuracy of Gaia and SIM. It implies that the core shift effect should be carefully investigated, and corrected for, in order to align accurately the radio and optical positions. Based on our study, we suggest two possible approaches, both involving the Primary Reference Sample of objects to be used for tying the radio and optical reference frames: 1) Firstly, multi-frequency VLBI measurements can be used for calculating the projected optical positions, assuming that the radio and optical emission regions are both dominated by a spatially compact component marginally resolved with VLBI and SIM and are point-like for Gaia. The discrepancies between the measured optical and radio positions can then be corrected for the predicted shifts, and the subsequent alignment of the radio and optical reference frames can be done using standard procedures. 2) Secondly, more conservative approach may also be applied, by employing the VLBI observations to identify and include in the Primary Reference Sample only those quasars in which no significant core shift has been detected in multi-epoch experiments. Either of the two approaches should lead to a substantial improvement of the accuracy of radio-optical positional alignment.

Acknowledgements

These proceedings are based on a paper by Y. Y. Kovalev *et al.* (A&A, 2007, submitted). The raw VLBI data were provided to us from the open NRAO archive. The National Radio Astronomy Observatory is a facility of the National Science Foundation operated under cooperative agreement by Associated Universities, Inc. Y. Y. Kovalev is a Research Fellow of the Alexander von Humboldt Foundation. Y. Y. Kovalev was supported in part by the Russian Foundation for Basic Research (project 05-02-17377) while working in Moscow in the first half of 2006. We would like to thank Patrick Charlot, Ed Fomalont, Leonid Petrov, Richard Porcas, Eduardo Ros as well as the NASA GSFC VLBI group and the MOJAVE team for fruitful discussions.

References

Blandford, R. D. & Königl, A. 1979, *ApJ*, 232, 34

Boboltz, D. A. 2005, in: *ASP Conf. Ser. 340, Future Directions in High Resolution Astronomy*, J. Romney & M. Reid (eds.), p. 342

Böttcher, M. 2007, *Ap&SS*, 309, 95

Gordon, D. 2005, in: *ASP Conf. Ser. 340, Future Directions in High Resolution Astronomy* J. Romney & M. Reid (eds.), p. 496

Junor, W., Biretta, J. A., & Livio, M. 1999, *Nature*, 401, 891

Kadler, M., Ros, E., Lobanov, A. P., Falcke, H., & Zensus, J. A. 2004, *A&A*, 426, 481

Königl, A. 1981, *ApJ*, 243, 700

Lara, L., Alberdi, A., Marcaide, J. M., & Muxlow, T. W. B. 1994, *A&A*, 285, 393

Lestrade, J.-F., *et al.* 1995, *A&A*, 304, 182

Lobanov, A. & Zensus, J. A. 2007, in: *Exploring the Cosmic Frontier, ESO Astrophysics Symposia*, A. P. Lobanov, J. A. Zensus, C. Cesarsky, & P. J. Diamond (eds.), p. 147

Lobanov, A. P. 1996, PhD thesis, New Mexico Institute of Mining and Technology, Socorro, NM, USA (1996)

Lobanov, A. P. 1998, *A&A*, 330, 79

Marcaide, J. M., Elosegui, P., & Shapiro, I. I. 1994, *AJ*, 108, 368

Pearson, T. J. 1999, in: *ASP Conf. Ser. 180, Synthesis Imaging in Radio Astronomy II*, G. B. Taylor, C. L. Carilli, & R. A. Perley (eds.), p. 335

Porcas, R. W. & Rioja, M. J. 1997, in: *Proceedings of the 12th working meeting on European VLBI for Geodesy and Astrometry*, B. R. Pettersen (ed.), p. 133

Ros, E. 2005, in: *ASP Conf. Ser. 340, Future Directions in High Resolution Astronomy*, J. Romney & M. Reid (eds.), p. 482

Ros, E. & Lobanov, A. P. 2001, in: *Proceedings of the 15th Workshop Meeting on European VLBI for Geodesy and Astrometry*, D. Behrend & A. Rius (eds.), p. 208

Sokolov, A. & Cawthorne, T. V. 2007, in: *ASP Conf. Ser., Extragalactic Jets: Theory and Observations from Radio to Gamma Rays*, T. A. Rector & D. S. De Young (eds.), in press

A Giant Step: from Milli- to Micro-arcsecond Astrometry
Proceedings IAU Symposium No. 248, 2007
W. J. Jin, I. Platais & M. A. C. Perryman, eds.

Multi-wavelength VLBI phase-delay astrometry of a complete sample of radio sources

I. Martí-Vidal, J. M. Marcaide and J. C. Guirado

Dpt. Astronomia i Astrofísica, Universitat de València, C/ Dr. Moliner 50, 46100 Burjassot,
Valencia (SPAIN)
email: i.marti-vidal@uv.es

Abstract. We report on the first global high-precision (differential phase-delay) astrometric analyses performed on a complete set of radio sources. We have observed the S5 polar cap sample, consisting of 13 quasars and BL Lac objects, with the VLBA at 8.4, 15, and 43 GHz. We have developed new algorithms to enable the use of the differential phase-delay observable in global astrometric observations. From our global analyses, we determine the relative positions between all pairs of sources with typical precisions ranging from 10 to 200 μas, depending on observing frequency and source separation. In this paper, we discuss the impact of this observable in the enhancement of the astrometric precision. Since a large fraction of the S5 polar cap sources are ICRF defining sources, this may result in a test of the ICRF stability. Our multi-epoch / multi-frequency approach will also provide both absolute kinematics and spectral information of all sources in the sample. In turn, this will provide an important check on key predictions of the standard jet interaction model.

Keywords. astrometry, techniques: interferometric, quasars: general, BL Lacertae objects: general, radio continuum: general

1. Introduction

Over the last years, we have carried out a series of VLBI observations, using the Very Long Baseline Array (VLBA) at 8.4, 15.4, and 43 GHz, aimed at studying the absolute kinematics of a complete sample of extragalactic radio sources using astrometric techniques. The target of our programme is the "S5 polar cap sample" (see Fig. 1 (a)), consisting of 13 radio sources from the S5 survey (Kühr, Witzel, & Paulini-Toth 1981, Eckart, Witzel & Biermann 1986). All sources in this sample have flux densities larger than 0.2 Jy at 8.4 and 15 GHz (0.13 Jy at 43 GHz) at the epochs of our observations and have well-defined ICRF positions. Most of the S5 polar cap sample sources have a structure that changes with time. For a reliable study of absolute kinematics of their components we need to refer the source positions and their changes with time and frequency to stable (or *fixed*) positions in the sky. The use of phase-delays allows us to refer the positions of the sources to the *phase centers* of the maps, providing a suitable reference for the source structure. In principle, the effect of source structure can be removed from the group delays, but large and rapid changes in the phase structure for this observable makes this approach complicated.

2. Observations

We observed all 13 members of the sample with the VLBA at 12 epochs (4 for each observing frequency) between years 1997 and 2004. We observed the sources in duty cycles, about 4 minutes long, using a multiple triangulation approach (see Martí-Vidal *et al.* 2008 for details). We indicate one such duty cycle in Fig. 1 (a), where the arrows mark the directions of the (antenna) slewings. The structures of the S5 polar cap sources for two epochs (years 1999.57 and 2000.46) at 15.4 GHz and two epochs (years 1997.93 and 1999.41) at 8.4 GHz, were discussed by Pérez-Torres *et al.* (2004) and Ros *et al.* (2001), respectively. At the time of writing this contribution, we have finished the astrometric analysis of the two 15 GHz epochs discussed in Pérez-Torres *et al.* (2004), and we have obtained preliminary results for the 8.4 GHz epochs on years 2001.09 and 2004.53, and for the 43 GHz epoch on year 2004.62.

3. The phase connection

Phase connection is the process by which phase-delays are converted into a non-ambiguous observable. The procedure of phase connection in our astrometric analysis is described in Martí-Vidal *et al.* (2008). It is based on previous approaches (e.g. Shapiro *et al.* 1979, Guirado *et al.* 1995), but with some substantial differences such as the use of an in-house developed automatic connection algorithm, described in Martí-Vidal *et al.* (2008), that corrects unmodelled phase-delay cycles imposing the nullity of all the closure phases. The use of this algorithm, or a similar one, is mandatory for a correct phase connection, given the large amount of data from many sources and antennas involved in our observations.

4. The University of Valencia Precision Astrometry Package (UVPAP)

For the astrometric fits and the phase-connection process we used the *University of Valencia Precision Astrometry Package*, an extensively improved version of the VLBI3 program (Robertson 1975). See Martí-Vidal *et al.* (2008), and references therein, for details of the UVPAP model. The main improvements of UVPAP include the use of the JPL ephemeris binary tables and the upgrade of relativistic effects of the Solar System bodies, computed using the Consensus Model (McCarthy & Petit 2003). The main advantage of UVPAP with respect to other software packages is the ability to perform multi-source differential phase-delay astrometric fits. The differential phase-delays are largely free from unmodelled effects of the troposphere, ionosphere, and antenna electronics (e.g. Marcaide *et al.* 1994). Thus, their use brings more precision to astrometric measurements. In Fig. 1 (b), we show the residuals of undifferential (upper plot) and differential (lower plot) phase-delays corresponding to the baseline Hancock − Kitt Peak and to ALL sources observed on the epoch 2000.46 at 15 GHz. As seen, all unmodelled contributions present in the residuals of undifferential delays cancel out when we compute the differential delays between all the source pairs (notice that the dashed lines in the lower plot mark the delays corresponding to ±1 phase cycle at 15 GHz).

5. Differential phase-delay astrometry vs. phase-reference mapping

In essence, our global differential phase-delay astrometry is the same as the phase-reference astrometry (Beasley & Conway 1995), since in both approaches the main observable used is the differential phase (delay) between the signals coming from different sources. Nevertheless, there is an important distinction between our astrometric approach and the phase-reference technique: in the phase-reference astrometry, the coordinates of one source (the target source) are determined with respect to the coordinates of another

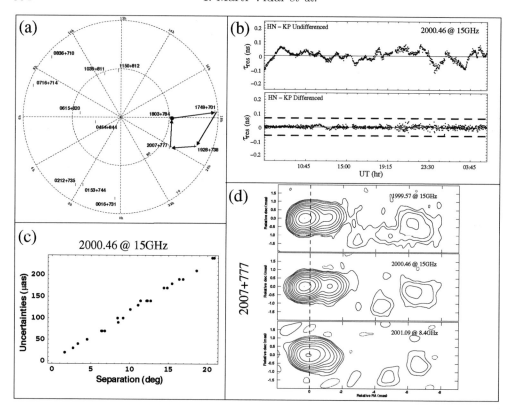

Figure 1. Several figures related to the astrometric analyses here reported (see text).

source (the reference source). *In our global analysis, we use data from all the 13 sources of the S5 polar cap simultaneously, in a single fit, thus putting more constrains on the solution and increasing the astrometric precision.* Of course, the corrections corresponding to the source pairs with large separations have relatively large uncertainties, but such sources will have more precise position estimates with respect to closer reference sources, as we show in Fig. 1 (c). In this figure, we plot the uncertainties of the source angular separations as a function of separation. The dependence is roughly linear, as predicted by Shapiro *et al.* (1979), and later found in numerical simulations by Pradel, Charlot, & Lestrade (2006). Comparing our results to those reported by Pradel, Charlot, & Lestrade (2006) using the phase-referenced technique, with our global analysis we increase the precision ~10 times.

6. Absolute kinematics

The complete analysis of our multi-epoch and multi-frequency observations of the S5 polar cap sample will allow us to study absolute kinematics of the components of all observed quasars and BL Lac objects. We will be able to check some key predictions of the standard jet interaction model, the synchrotron opacity effects and stationarity of the source cores. As an example of the results that we have from our astrometric analysis, Fig. 1 (d) shows preliminary results of the position changes of source 2007+777 for the epochs 1999.57 and 2000.46 (both at 15 GHz) and 2001.09 at 8.4 GHz. The contours are at −0.1, 0.1, 0.2, 0.4, 1, 2, 4, 8, 16, 32, 64, and 99 percent of the brightness peak at each epoch. The dashed line marks the right ascension of 2007+777 according to the ICRF-Ext.2 positions (see Fey *et al.* 2004). As we can see in Fig. 1 (d), there is a clear indication of a shift in the peak of brightness at 8.4 GHz towards the West, with respect to the brightness peak at 15 GHz.

7. Conclusions

We have performed the first high-precision wide-field multi-source astrometric analyses using the phase-delay observable. We have developed new algorithms and updated existing software to create UVPAP, a package that allows us to solve automatically for the ambiguities of the phase-delays and perform a multi-source differential phase-delay astrometric analysis. Such analysis brings nearly an order of magnitude higher precision than the commonly used phase reference technique.

Other wide-field high-precision astrometric analyses, similar to those reported here, are currently under way and will eventually provide spectral information and absolute kinematics of all sources in the S5 polar cap sample. Ultimately, we expect to provide a definitive test on the stationarity of the radio source cores and some stability checks for this subset of ICRF sources.

Acknowledgements

This work has been partially funded by Grants AYA2004-22045-E, AYA2005-08561-C03, and AYA2006-14986-C02 of the Spanish DGCYT. The National Radio Astronomy Observatory is a facility of the National Science Foundation operated under cooperative agreement by Associated Universities, Inc.

References

Beasley, A. J. & Conway, J. E. 1995, *ASP Conf. Ser. 82: Very Long Baseline Interferometry and the VLBA* 82, 328

McCarthy, D. D. & Petit, G. 2003, *The International Celestial Reference System: Maintenance and Future Realization, 25th meeting of the IAU* Joint Discussion 16, 22 July 2003, Sydney, Australia

Eckart, A., Witzel, A., Biermann, P., *et al.* 1986, *A&A* 168, 17

Fey, A. L., Ma, C., Arias, E. F., *et al.* 1986 *AJ*, 127, 3587

Guirado, J. C., Marcaide, J. M., Elosegui, P., *et al.* 1995, *A&A* 293, 613

Kühr, H., Witzel, A., Pauliny-Toth, I. I. K., *et al.* 1981, *A&AS* 45, 367

Marcaide, J. M., Elósegui, P., & Shapiro, I. I. 1994, *AJ* 108, 368

Martí-Vidal, I., Marcaide, J.M., Guirado, J.C. *et al.* 2008, *A&A*, 478, 267

Pérez-Torres, M. A., Marcaide, J. M., Guirado, J. C., *et al.* 2004, *A&A* 428, 847

Pradel, N., Charlot, P., & Lestrade, J.-F. 2006, *A&A* 452, 1099

Robertson, D. S. 1975, *Geodetic and astrometric measurements with Very Long Baseline Interferometry* Ph. D. Thesis, MIT

Ros, E., Marcaide, J. M., Guirado, J. C., *et al.* 2001, *A&A* 376, 1090

Shapiro, I.I., Wittels, J.J., Counselman, C.C. *et al.* 1979, *AJ* 84, 1459

A Giant Step: from Milli- to Micro-arcsecond Astrometry
Proceedings IAU Symposium No. 248, 2007
W. J. Jin, I. Platais & M. A. C. Perryman, eds.

© 2008 International Astronomical Union
doi:10.1017/S174392130801956X

Relativistic astrometry
and astrometric relativity

S. A. Klioner

Lohrmann Observatory, Dresden Technical University, 01062 Dresden, Germany
email: Sergei.Klioner@tu-dresden.de

Abstract. The interplay between modern astrometry and gravitational physics is very important for the progress in both these fields. Below some threshold of accuracy, Newtonian physics fails to describe observational data and the Einstein's relativity theory must be used to model the data adequately. Many high-accuracy astronomical techniques have already passed this threshold. Moreover, modern astronomical observations cannot be adequately modeled if relativistic effects are considered as small corrections to Newtonian models. The whole way of thinking must be made compatible with relativity: this starts with the concepts of time, space and reference systems.

An overview of the standard general-relativistic framework for modeling of high-accuracy astronomical observations is given. Using this framework one can construct a standard set of building blocks for relativistic models. A suitable combination of these building blocks can be used to formulate a model for any given type of astronomical observations. As an example the problem of four dimensional solar system ephemerides is exposed in more detail. The limits of the present relativistic formulation are also briefly summarized.

On the other hand, high-accuracy astronomical observations play important role for gravitational physics itself, providing the latter with crucial observational tests. Perspectives for these astronomical tests for the next 15 years are summarized.

Keywords. gravitation, relativity, astrometry, celestial mechanics, reference systems, time

1. Introduction

The tremendous progress in technology, which we have been witnessing during the last 30 years, has led to enormous improvements of accuracy in the disciplines of astrometry and time. A good example here is the growth of accuracy of positional observations in the course of time: during the 25 years between 1988 and 2013 we expect the same gain in accuracy (4.5 orders of magnitude) as that realized during the whole previous history of astrometry, from Hipparchus till 1988 (over 2000 years). Observational techniques like Lunar and Satellite Laser Ranging, Radar and Doppler Ranging, Very Long Baseline Interferometry, high-precision atomic clocks, etc., have already made it possible to probe the kinematical and dynamical properties of celestial bodies to unprecedented accuracy. Microarcsecond astrometry projects like Gaia (Lindegren 2008) and SIM (Shao 2008) will open fascinating possibilities for obtaining important physical information on celestial objects using their astrometry.

The goal of this review is to stress that the fascinating potential possibilities of high accuracy astronomical observations can only be realized if data modeling and analysis are made fully compatible with general relativity. It is not sufficient to consider the relativistic effects as small corrections to some Newtonian picture. The basic concepts of a reference system, moment of time, simultaneity, etc. are fundamentally different from their Newtonian counterparts. Although the relativistic (at least post-Newtonian) data modeling is rather simple conceptually, it requires a different way of thinking compared to that of a typical Newtonian physicist or astronomer.

2. Relativistic astrometry

In 2000 the International Astronomical Union has adopted the standard general-relativistic framework for modeling the high-accuracy astronomical observations. This framework allows one to construct a standard set of building blocks for relativistic models, a suitable combination of which can be used to formulate a model for any given type of astronomical observations. The IAU has adopted two reference systems defined in the mathematical language of general relativity. These reference systems are the Barycentric Celestial Reference System (BCRS) and the Geocentric Celestial Reference System (GCRS). The BCRS is the fundamental reference system covering the solar system and observed sources. The center of the BCRS lies in the barycenter of the Solar system. The word "celestial" in the name of BCRS is used to underline that the BCRS does not rotate with the Earth and that remote sources (e.g., quasars) can be assumed to be at rest with respect to the BCRS in some averaged sense. The BCRS is used to model the dynamics of the solar system as a whole and to describe light propagation between the source and the observer. The coordinate time of the BCRS is called Barycentric Coordinate Time (TCB). The Parameterized Post-Newtonian (PPN) version of the BCRS valid for certain class of metric theories of gravity has been also considered by a number of authors.

The GCRS is constructed in such a way that the gravitational fields generated by other bodies are reduced to tidal potentials and are thus effaced as much as it is possible according to general relativity. The coordinate time of the GCRS is called Geocentric Coordinate Time (TCG). Its scaled version, called Terrestrial Time (TT), is a physical model of TAI. The GCRS is a reference system physically suitable for modeling of physical processes in the vicinity of the Earth (e.g. Earth rotation or motion of an Earth's satellite).

The theory of the local reference systems like GCRS can be applied to any massive or massless bodies of the solar system. In particular, a GCRS-like reference system can be constructed for an observing satellite (Klioner 2004). That reference system is physically adequate to model any physical processes occurring within or in the immediate vicinity of a satellite (for example, the process of registration of incoming photons and the rotational motion of satellite).

Using these standard reference systems any kind of astronomical observations can be modeled in the way depicted on Figure 1. A suitably chosen reference system allows one to derive four building blocks. The equations of motion of the observed object, the observer and the electromagnetic signal relative to the chosen reference system should be derived and a method to solve these equations should be found. The equations of motion of an object and the observer and the equations of light propagation enable one to compute positions and velocities of the object, observer and the photon (light ray) with respect to that particular reference system at a given moment of the coordinate time, provided that the positions and velocities at some initial epoch are known. As the last step the observable quantities should become relativistic definitions. This part of the model allows one to compute a coordinate-independent theoretical prediction of observables starting from the coordinate-dependent quantities mentioned above. These four components can now be combined into relativistic models of observables. The models give an expression for relevant observables as a function of a set of parameters. These parameters can then be fitted to observational data using some kind of parameter estimation scheme. The

358 S. A. Klioner

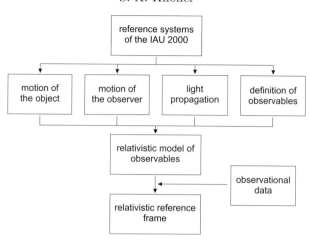

Figure 1. General scheme of relativistic modeling of astronomical observations.

sets of certain estimated parameters appearing in the relativistic models of observables represent astronomical reference frames.

Most of the relativistic models used in practice are fully compatible with the IAU framework sketched above. These are the models for VLBI, SLR, LLR, pulsar timing, time keeping and time transfer, GPS, and the models used for the development of solar system ephemerides. Some of these models will soon need some refinement because of the expected increase of accuracy. The only operational model which is not compatible with the IAU framework is that of the rotational motion of the Earth (and other planets). The rotational motion of celestial bodies is currently modeled in a purely Newtonian framework with the main relativistic effect - the geodetic precession - added in a non-consistent way. This situation must be improved in the nearest future. Some recent progress in relativistic modeling of Earth rotation can be found in (Klioner *et al.* 2007).

3. Towards four-dimensional solar system ephemerides

One of the questions which triggers a lot of difficulties and controversies among astronomers are relativistic time scales and their interrelation. Since the relativistic framework involves several relativistic reference systems (at least, BCRS and GCRS), several coordinate time scales must be considered (at least *TCB* and *TCG*). One important implication of relativity is that the transformation of time moments from one reference system to another is only possible if the position of the event is specified. So to say, without knowing "where" one cannot know "when". This is a natural consequence of the four-dimensionality of the coordinate transformations in relativity. In general the transformations of coordinate times T and t of two relativistic reference systems $(t, \boldsymbol{x})$ and $(T, \boldsymbol{X})$, $\boldsymbol{x}$ and $\boldsymbol{X}$ being the spatial coordinates, is a part of four-dimensional transformations between these two reference system:

$$T = T(t, \boldsymbol{x}),$$
$$\boldsymbol{X} = \boldsymbol{X}(t, \boldsymbol{x}). \tag{3.1}$$

The details of this transformation for BCRS and GCRS are given, e.g., by Soffel *et al.* (2003). An important *special case* of this transformation is to specify a particular spatial location (maybe as a function of time in order to calculate the transformation along some worldline): for any t one has $\boldsymbol{x} \equiv \boldsymbol{x}_\mathrm{o}(t)$. In this case T becomes a function of t only:

$T = T(t, \boldsymbol{x}) = T(t, \boldsymbol{x}_\mathrm{o}(t)) = T(t)$. Trajectory $\boldsymbol{x}_\mathrm{o}(t)$ may be chosen to coincide with the coordinates of a observing astrometric satellite like Gaia or SIM, or with the geocenter. The latter case is especially important for practice. Let us consider it in more detail. The well-known differential relation between $T = TCG$ and $t = TCB$ at the geocenter reads

$$\frac{dT}{dt} = 1 + \frac{1}{c^2}\,\alpha(t) + \frac{1}{c^4}\,\beta(t) + \mathcal{O}(c^{-5}). \tag{3.2}$$

Here $\alpha(t)$ and $\beta(t)$ are known functions of time t. There functions involve masses, positions and velocities of the massive bodies of the solar system and can be computed as

$$\alpha = -\frac{1}{2}\,v_E^2 - \sum_{A \neq E} \frac{GM_A}{r_{EA}}, \tag{3.3}$$

$$\begin{aligned}
\beta = &-\frac{1}{8}\,v_E^4 + \left(\beta - \frac{1}{2}\right)\left(\sum_{A \neq E}\frac{GM_A}{r_{EA}}\right)^2 + (2\beta - 1)\sum_{A \neq E}\left(\frac{GM_A}{r_{EA}}\sum_{B \neq A}\frac{GM_B}{r_{AB}}\right) \\
&+ \sum_{A \neq E}\frac{GM_A}{r_{EA}}\left(2(1+\gamma)\boldsymbol{v}_A \cdot \boldsymbol{v}_E - \left(\gamma + \frac{1}{2}\right)v_E^2 - (1+\gamma)v_A^2 \right. \\
&\left. + \frac{1}{2}\boldsymbol{a}_A \cdot \boldsymbol{r}_{EA} + \frac{1}{2}(\boldsymbol{v}_A \cdot \boldsymbol{r}_{EA}/r_{EA})^2\right),
\end{aligned} \tag{3.4}$$

where capital latin subscripts A, B and C enumerate massive bodies, E corresponds to the Earth, M_A is the mass of body A, $\boldsymbol{r}_{EA} = \boldsymbol{x}_E - \boldsymbol{x}_A$, $r_{EA} = |\boldsymbol{r}_{EA}|$, $\boldsymbol{v}_A = \dot{\boldsymbol{x}}_A$, $\boldsymbol{a}_A = \dot{\boldsymbol{v}}_A$, a dot signifies time derivative with respect to TCB, and $\boldsymbol{x}_A$ is the BCRS position of body A. The PPN parameters β and γ (both equal to 1 in general relativity) are given here for completeness, and normally can be put to 1 for practical calculations. Introducing two functions $\Delta t(t)$ and $\Delta T(T)$

$$T = t + \Delta t(t), \tag{3.5}$$
$$t = T - \Delta T(T). \tag{3.6}$$

one gets two ordinary differential equations for $\Delta t(t)$ and $\Delta T(T)$

$$\frac{d\Delta t}{dt} = \frac{1}{c^2}\,\alpha(t) + \frac{1}{c^4}\,\beta(t) + \mathcal{O}(c^{-5}), \tag{3.7}$$

$$\frac{d\Delta T}{dT} = \frac{1}{c^2}\,\alpha(T - \Delta T) + \frac{1}{c^4}\left(\beta(T - \Delta T) - \alpha^2(T - \Delta T)\right) + \mathcal{O}(c^{-5}). \tag{3.8}$$

Initial conditions for these two differential equations are given by the IAU definitions of TCB and TCG: $TCB = TCG = 32.184$ s on 1977, January 1, $0^h\ 0^m\ 0^s$ TAI at the geocenter: for $JD_{TCB} = 2443144.5003725$ one has also $JD_{TCG} = 2443144.5003725$. Any reasonable numerical integrator can be used to integrate these two differential equations with these initial conditions and using any given solar system ephemerides. Moreover, the numerical integration of $\Delta t(t)$ and $\Delta T(T)$ can be performed simultaneously with the equations of motion of the solar system which are routinely integrated during the process of ephemeris construction. Eq. (3.2) and, correspondingly, Eqs. (3.7)–(3.8) can be modified in an obvious way to relate any reasonable pair of the time scales TCG, TCB, TT and TDB, the last two being fixed linear functions of the first two (Soffel *et al.* 2003; IAU 2006). For TDB the initial conditions have to be selected according to IAU (2006): for $JD_{TT} = 2443144.5003725$ one has $JD_{TDB} = 2443144.5003725 - 6.55 \times 10^{-5}/86400$.

It is natural to include the results of the numerical integration of (3.7)–(3.8) into the standard distribution of the ephemerides. Functions $\Delta t(t)$ and $\Delta T(T)$ can be easily presented in the same Chebyshev polynomial representation as the rest of the ephemeris data. Each ephemeris has slightly different numerical values of $\Delta t(t)$ and $\Delta T(T)$ exactly in the same way as each ephemeris has its own numerical values of positions and velocities of the solar system bodies. This naturally leads to four-dimensional space-time ephemerides of the solar system as it has been implied by general relativity from the very beginning.

4. Limitations of the current relativistic framework

The standard relativistic framework of the IAU will have to be extended when more accurate data models will be required. Currently the relativistic framework is formulated in the post-Newtonian approximation of general relativity. This means that the relativistic effects are taken into account to the lowest order in c^{-2}. In principle, the extension to the next order, the post-post-Newtonian approximation, is straightforward, although not all aspects of the formalism are understood with the same level of details as in the post-Newtonian approximation. However, the post-Newtonian approximation scheme just operates with analytical orders of magnitude of various terms and not with their numerical magnitude. Taking into account all terms of order c^{-4} we would do a lot of unnecessary work since only a few of those terms are numerically important. A lot of work should be done to identify which post-post-Newtonian terms should be accounted for and which can be safely neglected in various situations.

Another restriction of the current formalism is the assumption that the solar system is isolated. This means that all gravitational fields generated outside of the solar system are ignored. Those external gravitational fields are only interesting if they produce time-dependent effects (the time-independent part is absorbed by the source positions just like secular aberration). The main time-dependent effects of this kind is the gravitational light deflection caused by (a) weak microlensing on the stars of the Galaxy (Belokurov & Evans 2002), and (b) lensing on gravitational waves (both primordial ones and those from compact sources). All these effects can be taken into account by a simple additive extension of the standard model since at the required accuracy the external gravitational fields can be linearly superimposed on the solar system gravitational field. The only exception could be the effects of cosmological background, but a preliminary study by Klioner & Soffel (2005) shows that even here the coupling of the local solar system fields and the external ones can be neglected.

5. Astrometric relativity

Since the formulation of general relativity in 1915, astronomical observations have played a very important role for testing this theory. Three of the four classical tests of General Relativity are based on astronomical observations. Although General and Special Relativity have been tested with a good precision, some ideas in the field of gravitational physics suggest that a deviation from general relativity maybe expected at the level of 10^{-5} to 10^{-8} (see, Damour, Piazza & Veneziano(2002) and references therein). This level is still beyond the possibilities of the available tests. On the other hand, independent of these arguments it is clearly the basic principle of natural sciences to test the suggested theories as accurately as possible.

It is expected that by 2020 the main PPN parameters β and γ will be measured with 6-7 meaningful digits. Considering that at such a high level of accuracy interpretation

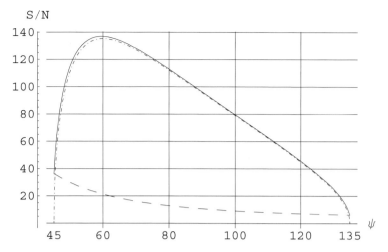

Figure 2. Signal-to-noise ratio in the Gaia observations of the gravitational light deflection from the Sun as function of the angular distance ψ (in degrees) between the Sun and the observed source for bright stars ($V = 13$ and brighter). Lower (dashed) line is the signal-to-noise ratio in the across scan data, the dot-dashed line is that in the along scan data and the solid line is the total signal-to-noise ratio. The across scan data contribute significantly only close to the maximal and minimal values of ψ where the along scan observations are insensitive to the gravitational light deflection due to the Sun.

of astronomical data becomes increasingly complicated, special care should be taken to avoid and/or understand all possible sources of systematic errors. It is important to realize that a detection of any (even very small) violation of General and Special Theories of Relativity in the weak field of solar system would have profound consequences for physics and astrophysics as a whole. For example, our understanding of physical properties of black holes and gravitational radiation may need a revision.

The most stringent relativistic test which can be expected from optical astrometry in the next 15 years is the measurement of the PPN parameter γ by astrometry mission Gaia. The expected accuracy here is $10^{-6}-5{\cdot}10^{-7}$. Due to high sensitivity of observations in a wide range of angular distances from the Sun (see Fig. 2), it can be expected that besides γ (that gives just the amplitude of a particular light deflection pattern) it will be possible to verify that light deflection indeed follows the general-relativistic deflection law for different angular distances. That latter test is beyond the scope of conjunction experiments like Cassini, BepiColombo or LATOR that can provide estimates of γ in a very narrow range of angular distances.

The PPN parameter β will be measured by Gaia from observations of asteroids with an accuracy of $< 10^{-4}$ (Hestroffer *et al.* 2007), but the most stringent test of β will be done by BepiColombo with expected accuracy of $2 \cdot 10^{-6}$ (Milani *et al.* 2002).

Besides these "major" tests, a series of other tests are possible (Klioner 2007). Let us give a few examples.

• Three components of the gravitational light deflection due to giant planets will be measured (Anglada *et al.* 2007): (1) the monopole gravitational deflection (with a precision of 10^{-3} for Jupiter), (2) the deflection due to translational motion of the planets (with a precision of $2 \cdot 10^{-3}$ for Jupiter), and (3) the deflection due to the quadrupole gravitational field of Jupiter (with a precision of > 0.08).

• Gaia will be able to measure the acceleration of the solar system's barycenter with respect to remote sources (quasars) with an accuracy of about $1.8 \cdot 10^{-11}$ ms^2 (Mignard

& Klioner 2007). An acceleration causes a specific change of secular aberration and thus can be measured as a specific pattern of proper motions. The amplitude of that pattern is related to the absolute value of acceleration. The acceleration of the solar system relative to the galactic center, as expected from the standard orbit of the solar system, should produce a proper motion of up to 4.5 μasyr^{-1}. Gaia will be able to measure this with the accuracy of 10%. This will allow one to verify, for the first time, the accuracy claims of the binary pulsar tests of relativity that assume the value of the acceleration from the standard model of the Galaxy.

• Another pattern in the proper motions allows one to constrain (Gwinn *et al.* 1997) the energy flux of gravitational waves with frequencies $\omega < 3 \cdot 10^{-9}$ Hz (see Mignard & Klioner (2007) for further details).

• Finally, one more group of tests deals with specially selected, relativistically interesting objects. An outstanding example here is given by compact binaries with one component being a black hole candidate. Combining usual Doppler measurements of these objects with Gaia astrometry one can derive the mass of the invisible companion without any further, model-dependent assumptions (Fuch & Bastian 2005). For example, for the well-known system Cyg X-1 the astrometric wobble of the visible companion is expected to attain 25 μas and can be measured by Gaia. Not only Gaia, but also ground-based interferometers have a good chance since narrow-field differential astrometry with those interferometers seems to be sufficient to detect the wobble with a well-known period.

References

Anglada-Escudé, G., Klioner, S. A., & Torra, J. 2007, Proc. of the 11th Marcel Grossmann Meeting on General Relativity, H. Kleinert, R. T. Jantzen & R. Ruffini (eds.), World Scientific, Singapore, in press

Belokurov, V. A. & Evans, N. W. 2002, *Mon.Not.R.astr.Soc*, **331**, 649

Damour, T., Piazza, F., & Veneziano, G. 2002, *Phys.Rev.D*, **66**, 046007

ESA, 2000, GAIA: Composition, Formation and Evolution of the Galaxy, Concept and Technology Study Report, ESA-SCI(2000)4 (Noordwijk: European Space Agency)

Fuch, B. & Bastian, U., 2005, In: The Three-Dimensional Universe with Gaia, ESA SP-576, 573

Gwinn, C., Eubanks, T., Pyne, T., Birkinshaw, M., & Matsakis, D. 1997, *Astrophys. J.*, **485**, 87

Hestroffer, D., Mouret, S., Mignard, F., Berthier, J., & Tanga, P. 2008, *this* volume, p.266

IAU, 2006, "Re-definition of Barycentric Dynamical Time, *TDB*", Resolution adopted by the XXVIth General Assembly

Klioner, S. 2007, In: Lasers, Clocks and Drag-Free: Exploration of Relativistic Gravity in Space, H. Dittus, C. Lämmerzahl, S. G. Turyshev (eds.), Astrophysics and Space Science Library **349**, Springer, Berlin, 399

Klioner, S. A. 2004, *Phys.Rev.D*, **69**, 124001

Klioner, S. A., & Soffel, M. H. 2005, In: The Three-Dimensional Universe with Gaia, ESA SP-576, 305

Klioner, S. A., Soffel, M., & Le Poncin-Lafitte, C. 2007, In: The Celestial Reference Frame for the Future, Proceedings of Journées'2007, N. Capitaine (ed.), Paris Observatory, Paris, in press

Lindegren, L. 2008, *this* volume, p.217

Mignard, F. & Klioner, S. A. 2007, Proc. of the 11th Marcel Grossmann Meeting on General Relativity, H. Kleinert, R. T. Jantzen, & R. Ruffini (eds.), World Scientific, Singapore, in press

Milani, A., Vokrouhlicky, D. Villani, D., Bonanno, C., & Rossi, A. 2002, *Phys.Rev.D*, **66**, 082001

Shao, M. 2008, this volume, p.231

Soffel, M., Klioner, S. A., Petit, G. *et al.* , 2003, *Astron. J.* **126**, 2687

A Giant Step: from Milli- to Micro-arcsecond Astrometry
Proceedings IAU Symposium No. 248, 2007
W. J. Jin, I. Platais & M. A. C. Perryman, eds.

© 2008 International Astronomical Union
doi:10.1017/S1743921308019571

Asteroid mass determination with the Gaia mission

S. Mouret[1], D. Hestroffer[1] and F. Mignard[2]

[1]IMCCE/CNRS, UMR 8028, Paris observatory, Denfert-Rochereau 77, 75014 Paris, France
emails: `mouret@imcce.fr`, `hestroffer@imcce.fr`

[2]OCA/Cassiopée/CNRS, UMR 6202, Nice observatory, Le Mont Gros, BP 4229, 06304 Nice
Cedex 4, France
email: `francois.mignard@obs-nice.fr`

Abstract. The ESA astrometric mission Gaia, due for launch in late 2011, will observe a very large number of asteroids ($\sim$ 350,000 down to the magnitude 20), most from the main belt, with an unprecedented positional precision (at the sub-milliarcsecond level). Such high-precision astrometry will enable to considerably improve the orbits of a large number of objects, and also to determine the masses of the largest asteroids by analyzing their gravitational pull during close encounters with smaller ones. A global solution involving simultaneously all the perturbers and the smaller targets should yield about hundred masses with a precision better than 30 percent. The knowledge of these masses will be a rich source of information on the physics of main-belt asteroids and will increase the accuracy of modern solar system ephemerides. We outline the principle of the mathematical method based on variational equations developed to solve for the orbital parameters and the masses of the largest bodies. Calculations have been performed by taking into account realistic simulation of the Gaia observations such as geometry, time sequence, magnitude and by considering possible close approaches among 350,000 asteroids. We give a list of asteroids whose mass can be estimated along with its formal precision.

Keywords. astrometry, asteroids, solar system: general

1. Introduction

Gaia is an astrometric cornerstone mission of the European Space Agency. With a launch due in late 2011, Gaia will have much more ambitious goals than its precursor Hipparcos: to obtain a "3D census" of our galaxy with astrometric, photometric, and spectroscopic observations. It will pinpoint the objects with an unprecedented positional precision (at a sub-milliarcsecond level for a single observation) allowing to observe about 350,000 asteroids (mainly main-belt asteroids) down to magnitude 20. The high precision astrometry will enable to considerably improve the orbits of almost all observed asteroids, yielding the masses of the largest ones from mutual perturbations during close approaches with smaller asteroids. The objective of this work is determination of the masses of a small subset of minor planets from the perturbations on smaller objects during close encounters. Not only does the method consider multiple close encounters with different targets, but it also simultaneously includes all perturbing asteroids (*perturbers*) together with their target asteroids. Here, we briefly present the different steps of this method, and give a list of asteroids for which the mass can be derived. The expected precision of masses by the time of mission's completion is estimated using realistic simulations of the Gaia observations (geometry, time sequence, magnitude).

The analysis of the gravitational perturbations during a close encounter between two asteroids, has been the most productive method to derive the masses. However, its application to ground-based observations has been limited in most cases to rare, particularly favorable encounters between a perturber and a small and faint target. This can be explained in part by the accuracy of observations allowing to analyse solely the very significant gravitational signature. As a consequence, only 21 asteroids have their mass estimates with a precision as good as 10% and the other 41 – at a 50% mark. Moreover, these estimates are often corrupted by systematic errors, which do not appear in the post-fit standard deviations. Thus, our knowledge of asteroid masses is still very poor, while their masses are of great interest for planetary ephemerides and the physics of asteroids.

Currently, the factor limiting the precision of the ephemerides for Mars is the uncertainty on the masses of the largest asteroids (Standish *et al.* 2002). The accuracy of their derivation will play a major role in the precision of ephemerides for the other inner planets when space missions will provide more precise observations of these asteroids. Regarding the physics of asteroids, the knowledge of mass and size (the latter being given by IRAS and many other techniques such as occultations, high angular resolution, but also from Gaia) of an asteroid leads to their bulk density, or equivalently to the amount of matter making up the body and the space occupied by its pores and fractures. The comparison with the grain density of meteorites with analogous composition, allows to estimate the porosity, a parameter related to the collisional history of asteroids and to their internal structure (Britt *et al.* 2002). In addition, a measurement of the bulk density will shed light on the relationship, if any, between the spectroscopic taxonomic class and the density, or possibly the porosity. Ultimately, this is related to the origin of asteroids and the formation process of the Solar System (Zappalà *et al.* 2002). If a well-defined relationship can be found, this could be used to derive density estimate for objects with a known taxonomic class. Eventually a reassessment of this relationship, used in planetary theories to estimate the masses of asteroids from their volume, will benefit to the accuracy of the ephemeris for the inner planets.

2. The mass determination method

The observed minus calculated positions (**O-C**) for each observation of minor planets expressed in Gaia longitude λ projected over a given Great Circle, can be linearized and solved by the least-squares method †,

$$\mathbf{O} - \mathbf{C} = \mathbf{A} \begin{pmatrix} \delta\mathbf{u_0} \\ \delta\mathbf{m_p} \end{pmatrix} \Rightarrow \begin{pmatrix} \delta\mathbf{u_0} \\ \delta\mathbf{m_p} \end{pmatrix} = (\mathbf{A}^t\mathbf{A})^{-1}\mathbf{A}^t(\mathbf{O} - \mathbf{C}) \qquad (2.1)$$

where, $\delta\mathbf{u^0} = (\delta\mathbf{u_1^0}, .., \delta\mathbf{u_n^0})^t$ are the corrections to the initial conditions of the n asteroids with $\delta\mathbf{u_k^0} = (\delta x^0, \delta y^0, \delta z^0, \delta\dot{x}^0, \delta\dot{y}^0, \delta\dot{z}^0)$ standing for the corrections to the position and velocity of the asteroid k at the reference time, and $\delta\mathbf{m_p} = (\delta m_1, .., \delta m_p)^t$ are the corrections to the masses of the p perturbers.

The matrix $\mathbf{A}$ in Eq.(2.1) represents the partial derivatives of the longitudes λ of the minor planets with respect to their initial parameters,

$$[\mathbf{A}]_{i,j} = \sum_{q=1}^{3} \frac{\partial\lambda_i}{\partial x_q}\frac{\partial x_q}{\partial C_j^0} \quad \text{where} \quad \begin{cases} \mathbf{x}_{q=1,...,3} = (x, y, z) \\ \mathbf{C} = (\mathbf{u_1^0}, .., \mathbf{u_n^0}, m_1, .., m_p) \\ \mathbf{u_k^0} = (x^0, y^0, z^0, \dot{x}^0, \dot{y}^0, \dot{z}^0) \text{ for the } k^{th} \text{ asteroid} \\ \lambda \text{ the observed longitudes for all asteroids.} \end{cases}$$

$$(2.2)$$

† Weighting of the equations has been omitted for the sake of brevity.

The expression $\partial\lambda_i/\partial x_q$ is determined analytically, while the $\partial x_q/\partial C_j$ are evaluated numerically. This is done by expressing the variations of the rectangular heliocentric coordinates $x_{q=1,...,3}$ (in an inertial frame) of each asteroid with respect to the unknowns by integrating the variational equations simultaneously with the equations of motion (Herget 1968).

From Eq.(2.1), it is possible to obtain the precision $\sigma(\delta\mathbf{m_p})$ attainable for the masses as a function of the observational accuracy. Each line of the matrix $\mathbf{A}$ corresponding to an asteroid k for an observation i is weighed according to the error on the position $\sigma_{k,i}$ (depending on the magnitude) by a constant σ_0, and so, the measurements of the weighted $(\mathbf{O\text{-}C})$ have the same variance σ_0^2. In addition, we make the assumption that the measurements of positions are independent, and, consequently, $\mathrm{cov}(\mathbf{O} - \mathbf{C}) = \sigma_0^2 Id$ and we obtain,

$$\mathrm{cov}\left(\delta\mathbf{u^0}, \delta\mathbf{m_p}\right)^t = \sigma_0^2(\mathbf{A}^t\mathbf{A})^{-1} \tag{2.3}$$

More details are given in Mouret *et al.* (2007).

3. The selection of the target asteroids

An important step prior to this solution is the indentification of suitable targets significantly perturbed during close approaches by larger asteroids. In the simulation for Gaia, the target asteroids are selected among the first 350,000 numbered asteroids from a systematic search of all close approaches with a set of bodies over a prescribed duration. The perturbers are chosen among the first 20,000 numbered asteroids as the bodies having a mass larger than a given threshold (here set at $10^{-13} M_\odot$). A close approach is considered meaningful if the impact parameter b (the minimum distance between the two asteroid trajectories in the case where we do not take into account their mutual perturbations) is smaller than 0.5 AU and the deflection angle θ greater than 1 *mas* which is estimated from the impulse approach by:

$$\tan\frac{\theta}{2} = \frac{G(m + M)}{v^2 b}, \tag{3.1}$$

where, G is the gravitational constant, M is the mass of the perturber, m is the mass of the target asteroid, and v is the relative velocity of the encounter.

For the period of time covering the Gaia mission—from year 2010.5 to 2016—the overall statistics are given in Table 1, where the second column indicates the number of bodies which are in the perturber list, but appears occasionally as target of a larger body.

Table 1. Results of the close approach simulations.

	Number of		
perturbers	602	perturbers and targets	
target asteroids	43513	simultaneously	434

4. Results

We have produced realistic simulations of the Gaia observations (geometry, time sequence, magnitude) and the set of close approaches (Table 1) to evaluate the final precision achievable on the masses of the perturbers (Eq. 2.3). The main result is shown in

Table 2 (*left*) with the statistical distribution of the relative precision on the masses. The right part of the Table 2 lists individual objects for which we obtained the best relative precision. The number of target asteroids for each solved mass is also listed together with the formal precision $\sigma(m)$, the reference mass m of the perturber, and the relative precision $\sigma(m)/m$.

Table 2. Number of masses determined in each class of the relative precision (*left*). The formal precision on the perturbers masses (*right*). The given masses m are the ones used in our simulation. The new masses for which no direct measurement is known are marked by ⋆⋆.

Number of perturbers		n°	Asteroid name	number of target asteroids	$\sigma(m)$ [M$_\odot$]	mass m [M$_\odot$]	$\sigma(m)/m$ [%]
Total	602	1	Ceres	6007	3.19×10^{-13}	4.75×10^{-10}	0.067
$\sigma(m)/m < 0.1\%$	2	4	Vesta	6685	1.23×10^{-13}	1.36×10^{-10}	0.091
$\sigma(m)/m < 1\%$	3	10	Hygiea	3180	3.42×10^{-13}	4.54×10^{-11}	0.753
$\sigma(m)/m < 10\%$	36	14	Irene⋆⋆	1973	1.53×10^{-13}	1.41×10^{-11}	1.090
$\sigma(m)/m < 15\%$	59	16	Psyche	4699	3.91×10^{-13}	3.38×10^{-11}	1.160
$\sigma(m)/m < 20\%$	75	27	Euterpe⋆⋆	1053	6.30×10^{-14}	5.38×10^{-12}	1.170
$\sigma(m)/m < 30\%$	106	7	Iris	1821	1.80×10^{-13}	1.41×10^{-11}	1.270
$\sigma(m)/m < 40\%$	135	2	Pallas	1194	1.35×10^{-12}	1.00×10^{-10}	1.350
$\sigma(m)/m < 50\%$	149	9	Metis	3046	2.10×10^{-13}	1.45×10^{-11}	1.450
		15	Eunomia	1859	2.82×10^{-13}	1.64×10^{-11}	1.720

5. Conclusions

The results of our simulations are very encouraging. Analyzing the close approaches between 602 perturbers and 43,513 target asteroids between 2010.5 and 2016, we find 36 asteroids for which the mass can be estimated to better than 10% and this number rises to 149 for a 50% precision. Gaia will improve most of current the measurements of masses thanks to a very complete dynamical modelling which limit the effects of systematic errors. Gaia will also yield several new masses, among which we expect a very precise mass estimate for (14) Irene and (27) Euterpe. As shown in Mouret (2007), the well-planned ground-based observations before and after the Gaia mission should improve these results. In the end, the number of well-determined masses after the mission's completion could be larger than the above estimate.

References
Britt, D. T., Yeomans, D., Housen, K., & Consolmagno, G. 2002, in: W. F. Bottke, A. Cellino, P. Paolicchi & R. P. Binzel (eds.), *Asteroids III*, (University of Arizona Press, Tucson), p. 485
Herget, P. 1968, *AJ*, 99, 225
Mouret, S., Hestroffer, D., & Mignard, F. 2007, *A&A*, 472, 1017
Standish, E. M.Jr & Fienga, A. 2002, *A&A*, 384, 322
Zappalà, V. & Cellino, A. 2002, in: O. Bienaymé & C. Turon (eds.), GAIA: a European space project, (EAS Publication Series), 2, 343
Mouret, S., "Investigations on the dynamics of minor planets with Gaia: orbits, masses and fundamental physics", Doctoral thesis, IMCCE, Paris observatory, 2007.

A Giant Step: from Milli- to Micro-arcsecond Astrometry
Proceedings IAU Symposium No. 248, 2007
W. J. Jin, I. Platais & M. A. C. Perryman, eds.

Definition and realization of the celestial intermediate reference system

N. Capitaine

Observatoire de Paris, SYRTE, 61 avenue de l'Observatoire, 75014, Paris, France
email: n.capitaine@obspm.fr

Abstract. The transformation between the International Terrestrial Reference System (ITRS) and the Geocentric Celestial Reference system (GCRS) is an essential part of the models to be used when dealing with Earth's rotation or when computing directions of celestial objects in various systems. The 2000 and 2006 IAU resolutions on reference systems have modified the way the Earth orientation is expressed and adopted high accuracy models for expressing the relevant quantities for the transformation from terrestrial to celestial systems. First, the IAU 2000 Resolutions have refined the definition of the astronomical reference systems and transformations between them and adopted the IAU 2000 precession-nutation. Then, the IAU 2006 Resolutions have adopted a new precession model that is consistent with dynamical theories and have addressed definition, terminology or orientation issues relative to reference systems and time scales that needed to be specified after the adoption of the IAU 2000 resolutions. These in particular provide a refined definition of the pole (the Celestial intermediate pole, CIP) and the origin (the Celestial intermediate origin, CIO) on the CIP equator as well as a rigorous definition of sidereal rotation of the Earth. These also allow an accurate realization of the celestial intermediate system linked to the CIP and the CIO that replaces the classical celestial system based on the true equator and equinox of date. This talk explains the changes resulting from the joint IAU 2000/2006 resolutions and reviews the consequences on the concepts, nomenclature, models and conventions in fundamental astronomy that are suitable for modern and future realizations of reference systems. Realization of the celestial intermediate reference system ensuring a micro-arc-second accuracy is detailed.

Keywords. standards, astrometry, ephemerides, reference systems, time, Earth

1. Introduction

The accurate realization of the celestial and terrestrial reference systems as well as the celestial orientation of the Earth is essential for the reduction of astronomical observations and scientific exploitation. Determining and providing that orientation is coordinated at the international level by the International Service for Earth Rotation and Reference systems (IERS). The IERS products, i.e. the International Terrestrial System (ITRS), the Celestial Reference Systems (ICRS), and the Earth Orientation Parameters (EOP), are based on data provided by the international services (IVS, ILRS, IGS, IDS). Those data are derived from observations by various modern techniques, namely, Very Long Baseline Interferometry (VLBI) of extragalactic radio sources for the IVS, laser ranging of artificial satellites and the Moon for the ILRS, observations with the GNSS systems for the IGS, and observations with the DORIS system for the IDS. Each of these techniques have specific potential for Earth orientation determination.

The transformation between the celestial and terrestrial systems is based on IAU and IUGG standards and models, plus IERS Earth Orientation parameters (EOP). These include (i) polar motion (represented by the x and y coordinates of the direction of the pole in the terrestrial system), which is quasi-periodic and essentially unpredictable,

(ii) Universal Time, UT1 that provides the variations in the Earth's diurnal angle of rotation and (iii) small adjustments (denoted dX and dY) to the celestial direction of the pole as predicted by the a priori precession-nutation model.

Recently, IAU and IUGG resolutions have been passed that modify the way the Earth orientation is expressed and adopted the high accuracy models for expressing the EOPs. This has important consequences on the concepts, the nomenclature and the models for fundamental astronomy that are used for modern realization of reference systems.

2. The IAU 2000/2006 Resolutions on reference systems

A major change in astronomy was the adoption by the IAU in 1997, of the International Celestial Reference Frame (ICRF) (Ma *et al.* 1998) that realizes the International Celestial Reference System (ICRS) and, at the same time, of the Hipparcos catalogue, as the realization of the ICRS at optical wavelengths. The transition from the FK5, based on stellar positions and proper motions, to the ICRS, based on observed extragalactic radio-sources, has made it possible to access celestial reference frames at a submilliarc-second accuracy. Then, several IAU resolutions on reference systems have been passed in 2000 and 2006 that were endorsed by the IUGG in 2003 and 2007, respectively.

2.1. *The IAU 2000 Resolutions*

- IAU 2000 Resolution B1.3 specifies that the systems of space-time coordinates for the solar system and the Earth within the framework of General Relativity are named the Barycentric Celestial Reference System (BCRS) and the Geocentric Celestial Reference System (GCRS), respectively. The corresponding time-coordinates are the Barycentric Coordinate Time (TCB) and the Geocentric Coordinate Time (TCG), respectively. This resolution also provides a general framework for expressing the metric tensor and defining coordinate transformations at the first post-Newtonian level (see Soffel *et al.* 2003) between the BCRS and the GCRS.

- IAU 2000 Resolution B1.5 provides an extended relativistic framework for time transformation in order to give a set of formulas for practical transformations between relativistic time scales (see Soffel *et al.* 2003).

- IAU 2000 Resolution B1.6 recommends the adoption of the new precession-nutation model that is designated IAU 2000 (version A corresponding to the model of Mathews *et al.* (2002), of 0.2 mas accuracy and version B corresponding to its shorter version (McCarthy and Luzum 2002) with an accuracy of 1 mas).

- IAU 2000 Resolution B1.7, specifies that the pole of the nominal rotation axis is the Celestial Intermediate Pole (CIP), which is defined as being the intermediate pole, in the ITRS to GCRS transformation, separating its GCRS motion (nutation) from polar motion by a specific convention in the frequency domain.

- IAU 2000 Resolution B1.8 recommends using the "non-rotating origin" (Guinot, 1979), designated Celestial and Terrestrial Ephemeris Origins, as origins on the CIP equator in the celestial and terrestrial reference systems, respectively, and defines UT1 as linearly proportional to the Earth Rotation Angle (ERA) between those origins on the CIP equator (Capitaine *et al.* 2000). This resolution recommends that the ITRS to GCRS transformation be specified by the positions of the CIP in the GCRS and the ITRS, and the ERA. It also recommends that the IERS continue to provide data and algorithms for the transformations referred to the equinox.

- IAU 2000 Resolution B1.9 provides a re-definition of Terrestrial time (TT) through a conventional linear relation between TT and TCG.

IAU 2000 Resolutions B1.6, B1.7 and B1.8 came into force on 1 January 2003. At that time, the models, procedures, data and software to implement these resolutions operationally had been made available by the IERS Conventions 2003 and the Standards Of Fundamental Astronomy (SOFA) activities (Wallace 1998). These include the procedure based on non-rotating origins, but also the equinox-based procedure, both being delivered with equal precisions.

2.2. *The IAU 2006 Resolutions*

The precession part in the IAU 2000A model consists only of corrections to the precession rates of the IAU 1976 precession and hence does not correspond to a dynamical theory. This is why IAU 2000 Resolution B1.7 that recommended the IAU 2000A precession-nutation model, recommended at the same time the development of new expressions for precession consistent with dynamical theories and with IAU 2000A nutation. The 2003-2006 IAU Working Group on "Precession and the Ecliptic" (P&E) looked at several solutions and recommended (Hilton *et al.* 2006) the adoption of the P03 precession theory (Capitaine *et al.* 2003); this was endorsed by IAU 2006 Resolution B1. In parallel, the new terminology associated with the IAU 2000 resolutions, along with some additional definitions related to them, were recommended by the 2003-2006 IAU Working Group on "Nomenclature for Fundamental Astronomy" (NFA) (Capitaine *et al.* 2006) and endorsed by IAU 2006 Resolutions B2 and B3. In summary:

- IAU 2006 Resolution B1 recommends the adoption (from 2009) of the P03 Precession as a replacement to the precession part of the IAU 2000A precession-nutation in order to be consistent with both dynamical theories and the IAU 2000 nutation. It also clarifies the definition of the precession and of the ecliptic.

- IAU 2006 Resolution B2, which is a supplement to the IAU 2000 Resolutions on reference systems, consists of two recommendations:
  (1) harmonizing "intermediate" to the pole and the origin (i.e. celestial and terrestrial intermediate origins, CIO and TIO instead of CEO and TEO, respectively) and defining the celestial and terrestrial "intermediate" systems;
  (2) fixing the default orientation of the BCRS and GCRS, which is, is unless otherwise stated, assumed to be oriented according to the ICRS axes.

- IAU 2006 Resolution B3 recommends a re-definition of the Barycentric Dynamical Time (TDB) through a conventional linear relation between TDB and TCB.

3. Consequences of the IAU resolutions on the concepts and definitions

3.1. *The Barycentric and Geocentric celestial reference systems*

The IAU 2000/2006 Resolutions have provided clear procedures for theoretical and computational developments of the space-time coordinates to be used in the framework of General Relativity. The BCRS, which can be considered to be a global coordinate system for the Solar System, has to be used (with TCB) for planetary ephemerides. In contrast, the GCRS, which can only be considered as a local coordinate system for the Earth, has to be used (with TCG) for Earth rotation, precession-nutation of the equator, motion of Earth's satellite, etc. The spatial orientation of the GCRS is derived from that of the BCRS. Therefore, the GCRS being "dynamically non-rotating", Coriolis terms (that come mainly from geodesic precession) have to be considered when dealing with equations of motion in that system. For all practical applications, unless otherwise stated, the BCRS (and hence GCRS) is assumed to be oriented according to the ICRS axes. The expression of the transformation between the barycentric and geocentric coordinates (i.e. an extension of the Lorentz transformation) is defined at the first post-Newtonian level for space coordinates and at the extended level for the time coordinates.

3.2. *The Terrestrial Time and Barycentric Dynamical Time*

The IAU 2000/2006 Resolutions have clarified the definition of the two time scales TT and TDB. TT has been re-defined as a time scale differing from TCG by a constant rate, which is the defining constant. In a very similar way, TDB has been re-defined as a linear transformation of TCB, the coefficients of which are the defining constants. The practical consequences is that TT (or TDB) is for some practical applications more convenient to use than TCG (or TCB) and can be used instead of TCG (or TCB) with the same accuracy. This applies in particular to satellite orbit computations for TT and solar system ephemerides, or analysis of pulsars timings, for TDB.

3.3. *The Celestial Intermediate Pole*

The definition of the CIP has refined the 1980 definition of the Celestial Ephemeris pole in order to realize the pole in the high frequency domain. The CIP has been defined as the intermediate pole, in the transformation from the GCRS to the ITRS, the motion of which is described as follows in the frequency domain. The celestial motion of the CIP includes by convention all the terms with periods greater than 2 days in the GCRS (i.e. frequencies between -0.5 cycles per sidereal day (cpsd) and $+0.5$ cpsd); the terrestrial motion of the CIP, includes by convention all the terms outside the retrograde diurnal band in the ITRS (i.e. frequencies less than -1.5 cpsd or greater than -0.5 cpsd).

```
frequency in ITRS _ _ _ _|_____ |_____ |_____ |_____ |_____ |_____ |_ _ _ _
                        -3.5    -2.5    -1.5    -0.5    +0.5    +1.5   (cpsd)
                        _____ polar motion|        |polar motion _____
frequency in GCRS  _ _ _|_____ |_____ |_____ |_____ |_____ |_____ |_ _ _ _
                        -2.5    -1.5    -0.5    +0.5    +1.5    +2.5   (cpsd)
                                         |nutation|
```

The GCRS position of the CIP replaces the classical precession and nutation quantities (see Fig. 1). The coordinates X, Y of the CIP can be provided by expressions as function of time of the rectangular coordinates, $X = \sin d \cos E$; $Y = \sin d \sin E$ of the GCRS direction of the CIP unit vector, which include precession and nutation and the frame bias between the pole of the GCRS and the CIP at J2000.0.

3.4. *The Earth rotation angle*

The diurnal rotation is expressed through the following conventional linear transformation of UT1 (Capitaine *et al.* 2000), called Earth Rotation Angle (ERA):

$$\text{ERA(UT1)} = 2\pi[0.7790572732640$$
$$+1.00273781191135448 \, (\text{JulianUT1date} - 2451545.0)] \tag{3.1}$$

The linear relationship between ERA and UT1 is a consequence of the kinematically non-rotating nature of the origins to which the ERA refers.

4. New nomenclature for fundamental astronomy

The IAU 2000/2006 Resolutions have provided a new paradigm for the GCRS to ITRS transformation, with associated terminology for the pole, the Earth's angle of rotation, the longitude origins and the related reference systems.

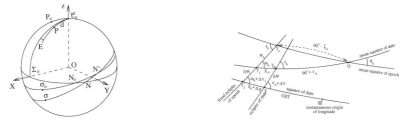

Figure 1. The precession-nutation of the equator with respect to the GCRS: the CIP coordinates (left figure where P is the CIP) versus the classical precession-nutation angles (right figure)

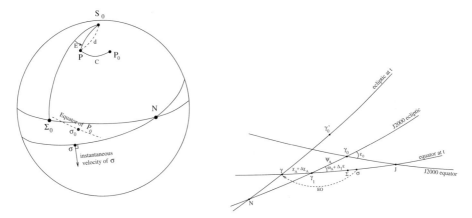

Figure 2. Origin for the Earth angle of rotation with respect to the GCRS: the kinematical definition of the CIO versus the geometrical definition of the classical equinox

4.1. *Terminology choices*

The IAU WG on "Nomenclature for Fundamental Astronomy" (NFA) proposed two IAU 2006 Resolutions, made 14 recommendations, and produced the "IAU 2006 Glossary" including definitions corresponding to the IAU 2000 resolutions as well as new definitions proposed by the WG (including those formally endorsed by the IAU in 2006 or by the IUGG in 2007). In particular, it has been recommended: 1) harmonizing the name of the pole and the origin to "intermediate" and therefore using "Celestial Intermediate Origin" (CIO) and "Terrestrial Intermediate Origin" (TIO), respectively instead of CEO/TEO; 2) using "equinox based" and "CIO based" for referring to the classical and new paradigms, respectively; 3) using "intermediate" to describe the moving geocentric celestial reference system containing the CIP and the CIO (with its terrestrial counterpart containing the TIO); 4) choosing "equinox right ascension" (or "RA with respect to the equinox") and "intermediate right ascension" (or "CIO right ascension", or "RA with respect to the CIO"), for the azimuthal coordinate along the equator in the classical and new paradigms, respectively; 5) giving the name "equation of the origins" (EO) to the distance between the CIO and the equinox along the intermediate equator, the sign of this quantity being such that it represents the CIO right ascension of the equinox, or equivalently, the difference between the Earth Rotation Angle and Greenwich apparent sidereal time. This has provided a new nomenclature associated with the use of the new origins and also associated with the equatorial coordinates used in an extended way, such that it can be referred to any equator (i.e. GCRS equator or CIRS equator, etc.) and to any origin on those equators (i.e. GCRS origin, CIO, equinox, etc.).

4.2. The celestial intermediate reference system

The celestial intermediate reference system, can be derived from the GCRS by using the GCRS CIP coordinates X, Y and the CIO "locator" (positioning the CIO). Note that the CIO at present is very close to GCRS longitude zero and almost stationary in longitude. The CIO based procedure allows a clear separation between (i) the GCRS motion of the CIP, which is dependent on the precession-nutation model and (ii) the Earth's angle of rotation, ERA, which is not model-dependent. In contrast, precession and nutation are mixed up with Earth's rotation into the expression of Greenwich sidereal time (GST) as function of UT1, which can easily be derived from the following relationship: GST(UT1, TT) = ERA − EO, where EO represents the accumulated precession and nutation in right ascension, directly dependent on the precession-nutation models.

5. Consequences of the IAU resolutions on high accuracy models

The IAU 2000/2006 Resolutions have adopted a high precision model for precession and nutation in two successive steps.

The first step was the adoption of the IAU 2000 precession-nutation of Mathews *et al.* (2002), which has been implemented in the IERS Conventions 2003. This model is composed of a nutation part and a precession part. In addition to the IAU 2000A model are frame bias values between the J2000 mean pole and equinox and the GCRS. The IAU 2000A nutation includes 1365 luni-solar and planetary terms, which are based on rigid Earth nutation transformed with the transfer function based on basic Earth parameters (BEP) that have been fitted to VLBI data. That model is expected to have an accuracy of about 10 μas for most of its terms. In contrast, the so-called free core nutation (FCN), which is due to geophysical effects and is largely unpredictable, is not part of the model. The precession part, which consists only in corrections to the precession rates of the IAU 1976 precession, was known to not following the predictions of a dynamical theory. The second step in the improvement of the IAU precession-nutation was the adoption (IAU 2006 Resolution B1) of the P03 Precession as a replacement to the precession part of the IAU 2000A precession-nutation in order to be consistent with both dynamical theory and the IAU 2000 nutation. The P03 model has provided improved expressions for both the precession of the ecliptic and the precession of the equator. The latter has taken into account the Earth's J_2 rate effect, mostly due to the post-glacial rebound. The P03 precession polynomial developments provide separately the developments for the basic quantities for the ecliptic and the equator that are directly solutions of the dynamical equations, and derived quantities, such as those for the GCRS coordinates of the CIP. These also include expressions for the P03 sidereal time that can been derived form the expression of the Earth Rotation Angle, which is independent of the precession-nutation model and the expression for the equation of the origin, which is directly model-dependent. The various ways of forming the precession-nutation matrix in the new IAU framework have been discussed in Capitaine & Wallace (2006) and the precession-nutation procedures consistent with IAU 2006 resolutions have been provided in Wallace & Capitaine (2006).

6. Conclusions

The 2000 and 2006 IAU resolutions on reference systems adopted greatly improved numerical models, but also a new paradigm, with associated terminology for the pole, the Earth's angle of rotation, the longitude origins and the related reference systems.

They have improved the definition and realization of different reference systems and time scales. They have in particular modified the way the Earth orientation is expressed and adopted high accuracy models for expressing the relevant quantities for the transformation from terrestrial to celestial systems. This provides a new framework for dealing with Earth's rotation or computing directions of celestial objects or Earth's satellites in various systems.

A very important effort has been made within the international astronomical community in order to prepare and then to best implement these resolutions, especially through specific IAU Working Groups, the IERS Conventions and the set of software developed by SOFA.

The concepts, nomenclature, models and conventions in fundamental astronomy based on the IAU 2000/2006 Resolutions are suitable for modern and future realizations of the reference systems. This allows the highest accurate realization of the celestial intermediate system linked to the Celestial intermediate pole and origin that replaces the classical celestial system based on the true equator and equinox of date.

The definition and the high accuracy realization of the celestial intermediate reference system based on the IAU 2000/2006 IAU Resolutions is consistent with a micro-arcsecond accuracy. This will allow us to best benefit of the accuracy that is expected to be achieved with future projects related to the celestial reference system, such as Gaia or the VLBI 2010 Observing Programme.

References

Capitaine, N., Guinot, B., & McCarthy, D.D., 2000, *A&A* 355, 398

Capitaine, N., Wallace, P. T., & Chapront, J., 2003, *A&A* 412, 567

Capitaine, N. & Wallace, P. T., 2006, *A&A*, 450, 855

Capitaine & the IAU NFA WG, 2007, in *Transactions of the IAU* XXVIB, van der Hucht, K. A. (ed)

Guinot, B., 1979, in *Time and the Earth's Rotation*, D. D. McCarthy and J. D. Pilkington (eds), D. Reidel Publishing Company, 7

Hilton, J., Capitaine, N., Chapront, J., *et al.*, *Celest. Mech. Dyn. Astr.* 94, 3, 351

IAU, 1998, *Transactions of the IAU* XXIIIB, Anderson, J. (ed), 40

IAU 2000, *Transactions of the IAU* XXIVB; Manchester, Rickman. H. (ed), Astronomical Society of the Pacific, Provo, USA, 2001, pp. 34-58

IAU 2006, *Transactions of the IAU* XXVIB; van der Hucht, K. A. (ed)

IUGG 2007, IUGG Resolutions, http://www.iugg.org/resolutions/perugia07.pdf .

IERS Conventions (2003), *IERS Technical Note 32*, D. D. McCarthy and G. Petit (eds), Frankfurt am Main: Verlag des desamts für Kartographie und Geodäsie, 2004

Ma, C., Arias, E. F., Eubanks *et al.*, 1998, *A&A* 116, 516

Mathews, P. M., Herring, T. A. & Buffett B. A., 2002, *J. Geophys. Res.* 107, B4, 10.1029/2001JB000390

McCarthy, D. D., Luzum, B. J., 2003, *Celest. Mech. Dyn. Astr.* 85, 37

Soffel, M., Klioner, S. A., Petit *et al.*, 2003, *AJ* 126, 6, 2687

Wallace P. T., 1998, in *Highlights of Astronomy* Vol. 11A, J. Andersen (ed.), Kluwer Academic Publishers, 11, 191

Wallace P. T., Capitaine, N., 2006, *A&A* 459, 3, 981

A Giant Step: from Milli- to Micro-arcsecond Astrometry
Proceedings IAU Symposium No. 248, 2007
W. J. Jin, I. Platais & M. A. C. Perryman, eds.

Solved and unsolved questions in the non-rigid Earth nutation study

C. L. Huang

Shanghai Astronomical Observatory, Chinese Academy of Sciences,
80 Nandan Rd., Shanghai 200030, China. Email: clhuang@shao.ac.cn

Abstract. At the IAU 26th GA held in Prague in 2006, a new precession model (P03) was recommended and adopted to replace the old one, IAU1976 precession model. This new P03 model is to match the IAU2000 nutation model that is for anelastic Earth model and was adopted in 2003 to replace the previous IAU1980 model. However, this IAU2000 nutation model is also not a perfect one for our complex Earth, as stated in the resolution of IAU nutation working group. The Earth models in the current nutation theories are idealized and too simple, far from the real one. They suffer from several geophysical factors: the an-elasticity of the mantle, the atmospheric loading and wind, the oceanic loading and current, the atmospheric and oceanic tides, the (lateral) heterogeneity of the mantle, the differential rotation between the inner core and the mantle, and various couplings between the fluid outer core and its neighboring solids (mantle and inner core). In this paper, first we give a very brief review of the current theoretical studies of non-rigid Earth nutation, and then focus on the couplings near the core-mantle boundary and the inner core-outer core boundary, including the electro-magnetic, viscous, topographic, and gravitational couplings. Finally, we outline some interesting future studies.

Keywords. reference systems, standards, magnetic fields, Earth, methods: analytical

1. Introduction

Nutation is the periodic motion of a conceptual axis, called the Celestial Intermediate Pole (CIP), or previously, the Celestial Ephemerids Pole (CEP), in space with periods between 2 days and 18.6 years as seen from space. To replace the previous IAU1980 model, the new nutation model adopted by the IAU in 2000 is the so-called MHB2000 model (Mathews *et al.* 2002) that is for anelastic Earth model. At the IAU 26th GA held in Prague in 2006, a new precession model, P03 (Capitaine *et al.* 2003), as well as corresponding definition and terminology, was recommended and adopted to replace the old one, IAU1976 precession model, to match the IAU2000 nutation model (see also contribution by Capitaine (this volume) for a review of definition and realization of the celestial intermediate reference system).

Observation on nutation Space geodetic techniques especially VLBI have developed very fast in the past 3 decades and have played the most important role in observation of nutation, while LLR have also contributed a lot to the long period terms (like 18.6 yr), and GPS may have also helped to determine short period nutations (<9d). Generally speaking, nutations can be determined by observation with uncertainties at the level of 40 and 10 μas for 18.6-year and other terms respectively. For determination of rotational normal modes, their uncertainties can reach approximately 0.1, 0.1, and 5 day for Chandler wobble (CW), free core nutation (FCN), and free inner core nutation (FICN) respectively. Besides these high precisions, another achievement is that they can provide the nutation solution in very short intervals, say 1 hour. However, there are still some discussions on the problems of VLBI observation on nutation: for example, the VLBI

374

network geometry and observation strategy, radio source structure, and the software for solving EOP, etc.

Steps to study non-rigid earth nutation (NREN) Generally, in order to calculate a non-rigid earth nutation, we first need a rigid earth nutation (REN) model that is usually derived from celestial mechanics, by the torque, Hamilton or tidal potential method with ephemerides of the Sun, the Moon and planets. Second, from the equations of infinitesimal elastic gravitational motion for a rotating, slightly elliptical earth, Poisson equation, stress-strain relation equation, a set of boundary conditions, and a given earth model that gives the internal profiles of density, elastic coefficients and so on, we can get two products: one is the normal modes, the other is the so-called earth transfer function (ETF) that represents the response of a non-rigid earth (NRE) to outer force or excitation. Third, convolving the REN series with this ETF, we get a NREN series. Finally, incorporating the effects of other geophysical factors including surface fluids (atmosphere and ocean) and dynamic processes inside the Earth (eg. various coupling at core-mantle-boundary (CMB) and inner-core-boundary (ICB)), with correction afterward by torque or angular momentum methods, or via boundary conditions in the second step, we get a new nutation model of a more realistic earth.

Theoretical approaches and models Generally, we can classify available methods for calculating non-rigid earth nutation into 3 categories: (1)numerical integration or displacement field approach used by Smith, Wahr, Dehant, Schastok, Huang, and else; (2)angular momentum approach used in SOS theory by Sasao *et al.* and MHB model by Mathews *et al.*; (3)Hamiltonian approach that was first used in REN theory, and then developed by Getino & Ferrándiz *et al.* and applied to NREN study.

On the one hand, all the above theoretical nutation models are very comparable to each other although they use different approaches; on the other hand, one apparent 'advantage' of MHB2000 over the other models is that the observed differences in the principal nutation terms (mainly the retro-annual (-1yr) term) with respect to VLBI observations are reduced by accounting for the effects of electro-magnetic coupling (EMC) of the FOC to the mantle and the SIC (Buffett *et al.*, 1992, 2002), in which the coupling constants at CMB and ICB and related compliance parameters are fitted to the VLBI nutation observations.

2. Phenomena

Comparing between theoretical nutation models and observation, or between different nutation models, the main problematic terms become obvious: 18.6yr & -1yr nutations, and the out-of-phase (op) components of some terms as related to dissipation. The second phenomenon is about the global dynamic ellipticity (H) that will be discussed later. The third concerns FCN that depends very intensively on the physical and dynamical properties near CMB. From observation, its period is about 430 day, while from PREM, it is about 460 day. Moreover, with more and more observation data accumulated, it tends to agree that the amplitude and period of FCN vary with time. Is it true, and why do they change? The biggest effect of these phenomena is on -1yr nutation due to strong resonance. The fourth phenomenon concerns FICN, which is also important for geophysicist as it reflects the physics and dynamics near the ICB. But since it is very faint, the question "is the 'observed' FICN a true one?" is still uncertain.

There are many celestial mechanical and geophysical (both inside and outside the Earth) factors in computing theoretical nutation. We can classify them into 4 groups: (1)the direct and indirect gravitational forces from the sun, the moon and the planets; (2)excitations of the surface geophysical fluids; (3)properties of the Earth itself, i.e. the earth model; (4)dynamical processes inside the earth, including various couplings between its 3 layers, and large scale convection in the mantle, and so on.

REN model: All the REN models can be regarded as perfect compared with the NREN models. They can be consistent with each other in $10\mu as$ or better, although there are still some unresolved issues, such as general relativistical effects, and all 2nd-order (in)direct effects.

Surface geophysical fluids: In regard to contributions from surface geophysical fluids, for both oceans and atmosphere, and including both load (or pressure) and motion terms, one can use effective angular momentum method, torque method, or direct integration method by introducing an outer surface boundary condition (Huang *et al.*, 2001). They influence mostly 18.6yr, 1yr and 0.5yr nutations, as well as prograde annual nutation caused by thermal S1 atmosphere tide with ip of 18 μas and op of 114 μas (Yseboodt *et al.*, 2002). Generally, the diurnal atmospheric excitation is less than half that required to explain the observed FCN amplitude (Lambert, 2006). In addition to previous studies focusing on resonance at FCN, Dehant *et al.* (2005) show that resonance at FICN induced by surface geophysical fluids only induces a very small signal, with a maximum of $20\mu as$ at 880-day prograde term.

Earth model: Next, although seismology has presented many earth models, the most often used one in nutation study is still the PREM model. There are many shortcomings in PREM, such as (1)it is a 1-dimension model depending on the spherical radii only; (2)the Earth is assumed to be in hydro-static equilibrium (HSE); (3)the medium is isotropic; and (4)how to treat the ocean layer is also a major problem. However, currently we do not have a choice. For example, using the PREM model, we can get the ellipticity of the CMB and the period of FCN at about -458 days. However, if we assume there is a deviation from HSE near the CMB, and increase the equator's radius at the CMB by 400 meter, then we get the FCN at -432 days, which is closer to observation (Huang et al., 2001).

Global dynamic flattening (H): By definition, $H = \frac{C - \frac{1}{2}(A+B)}{C}$, where A, B and C are the three principal moments of inertia of the Earth. Therefore H depends on the density distribution inside the Earth, and is related to the lunar-solar main precession, main nutation, tilt-over-mode (TOM) and so on. Precession observations give $H_{obs} \approx 1/305.5$, while it is approximately $1/308.8$ at the first order accuracy from the PREM model. This 1.1% difference has stimulated many interesting discussions. Using a more precise potential theory to third-order accuracy in HSE, Liu & Huang (2008, this Volume) re-calculate the geometrical flattening (f) profile of the Earth interior from PREM and obtain $H_{PREM} = 1/308.5$. However, After replacing the homogenous outermost crust and oceanic layers in PREM with some real surface layers data, such as topography+ocean, topography+ocean+bathmetry+upper crust data, or CRUST2.0 model, down to depth of 5.615km, 10.376km, 70.137km respectively, they obtain $1/H$ =318.14, 320.22, 310.70 respectively. These results deviate from the observed value more than H_{PREM} and verify the isostasy theory indirectly. They may imply that 'positive' effects arising from, for example, mantle circulation associated with the density anomalies may be larger than what was discussed before.

3. Dynamical processes inside the earth

There may exist 4 kinds of angular momentum or torque coupling near the CMB and ICB: gravitational, electro-magnetic (EM), viscous, and topographic. And there may also be deviation from HSE.

Gravitational coupling: This mechanism can be used to fill all main gaps between the theoretical model and observation (eg., Jault & LeMouel, 1989), but the key parameters used are too arbitrary.

EM coupling: As mentioned above, it is used by Buffett *et al.* in their MHB model to explain the op component of -1yr nutation, but it is open to question. For example, taking a direct numerical integration approach, Huang *et al.*(2006) show that, even using the same values of EM properties as the MHB model did, possible contribution of EM coupling to -1yr nutation is only about one-tenth of what is required. The authors of the MHB model themselves also realized that the contribution of EM coupling maybe not so significant as they declared before, so they and other colleagues divert their interest to viscous coupling at CMB (Mathews & Guo, 2005; Deleplace & Cardin, 2006; Buffett & Christensen 2007).

Viscous coupling: In order to fill the gap for -1yr nutation. the effective viscosity of the core fluid is required to be at the level of $0.03 m^2/s$. However, if based on laboratory and physical consideration of the fluid viscosity, the eddy viscosity is smaller than $10^{-4} m^2/s$ and therefore may be also too small for the -1yr nutation.

Topographic torque at CMB: It is also a potential mechanism for explaining the dissipation in the -1yr nutation. It is related to core angular momentum exchange and pressure at the CMB. It also depends very intensively on topography at the CMB. For example, Wu & Wahr (1997) concluded that the non-Y_2^0 parts of the topography of the CMB within 3.5, 4-5, or 6-7 km (rms) may contribute to the -1yr nutation by 0.55, 0.77 or 2 *mas* respectively. However, the seismology tomography data shows that the topography at the CMB is smaller than 2 km, and the above results seem to lack support from seismology. Therefore it is still a difficult, challenging and controversial topic but cannot be completely ruled out (see also Mound & Buffett, 2003).

4. Some other discussions

Lateral heterogeneity: There was some research on the effects of lateral heterogeneity on earth tides based on a 3D earth model. For example, it may change the gravity tidal admittance δ_{M2} by up to 0.5%(Wang, 1991), and the Love number k_{M2} by about 0.19% (Li *et al.*,1996). But so far there is not any study in nutation, maybe due to the difficulty in theoretical work.

$2^n d$ *order theory and terms:* Although there are some theoretical studies of the effects of the truncated $2^n d$ order terms on nutation (Huang, 2001; Van Hoolst & Dehant, 2002; Rogister & Rochester, 2004), there is not any numerical result of its direct effect so far.

Lambert (2006) (see also Lambert & Mathews, 2006) calculated zonal tides indirect effects that are 37 μas and 1 μas for the $\Delta\psi$ and $\Delta\varepsilon$ of 18.6yr nutation and -518 $\mu as/c$ for precession in ψ. Folgueira *et al.*(2007) and Dehant *et al.*(2007) also discussed tidal Poisson terms, which are periodic in amplitude linearly dependent on time, and showed that it may change the nutation by approximately several tens μas or several μas for a very long period (10467.6 years) term.

Differential rotation: The possible differential rotation of the inner core with respect to the mantle ($\Delta\Omega_{IC-M}$) may contribute a little to 18.6-yr nutation. For example, if $\Delta\Omega_{IC-M} = 1^o/yr$, the biggest change will be $\delta\varepsilon_{ip}^{18.6y} \approx 0.01 mas$ (Huang & Dehant, 2002). Although it is very small, it may be detectable by nutation observation in future.

5. Short remarks

From the above brief review, we see that nutation study suffers mostly from the rough information of the earth model (mostly from the CMB to the outer surface) and the dynamical processes near the CMB. Therefore, precise nutation observation provides another way to study the earth interior besides seismology.

Acknowledgements

This work is supported by NSFC (10773025/10633030), CAS(KJCX2-YW-T13), and Science & Technology Commission of Shanghai Municipality (06DZ22101/06ZR14165).

References

Buffett, B. A. 1992, *J. Geophys. Res.*, 97(B13), 19581

Buffett, B. A., Mathews, P. M., & Herring, T. A. 2002, *J. Geophys. Res.*, 107(B4), DOI: 10.1029/2000JB000056

Buffett, B. A., & Christersen, U. R. 2007, *Geophys. J. Int.*, 171, 145

Capitaine, N., Chapront, J, Lambert, S., & Wallance, P. T. 2003, *Astron. & Astrophys.*, 400, 1145, doi: 10.1051/0004-6361:20030077

Dehant, V., de Viron, O., & Greff-Lefftz, M. 2005, *Astron. & Astrophys.*, 438, 1149, doi: 10.1051/0004-6361:20042210

Dehant, V., Folgueira, M., Rambaux, N., & Lambert, S. B. 2007, *Proc. IUGG XXIVth GA* (in press)

Deleplace, B. & Cardin, P. 2006, *Geophys. J. Int.*, 167, 557

Folgueira, M., Dehant, V., Lambert, S. B., & Rambaux, N. 2007, *Astron. & Astrophys.*, 469, 1197, doi: 10.1051/0004-6361:20066822

Huang, C. L. 2001, *Earth, Moon, and Planets*, 84, 125

Huang, C. L., Jin, W. J., & Liao, X. H. 2001, *Geophys. J. Int.*, 146, 126

Huang, C. L. & Dehant, V. 2002, in: N. Capitaine (ed.), *Proceedings of Journées 2001*, 20

Huang, C. L., Dehant, V., Liao, X. H., de Viron, O., & Van Hoolst, T. 2006, in: *IAU XXVIth GA Abstract book*, 404

Jault, D., Le Mouël, J. L. 1989, *Geophys. & Astrophys. Fluid Dyn.*, 48(4), 273, DOI:10.1080/03091928908218533

Lambert, S. 2006, *Astron. & Astrophys.*, 457, 717, doi: 10.1051/0004-6361:20065813

Lambert, S., & Mathews, P. M. 2006, *Astron. & Astrophys.*, 453, 363, doi: 10.1051/0004-6361:20054516

Li, G. Y., Peng, L. H., & Xu, H. Z. 1996, *Acta Geophys. Sinica*, 39, 672

Mathews, P. M., Herring, T. A., & Buffett, B. A. 2002, *J. Geophys. Res.*, 107(B4), DOI: 10.1029/2001JB000390

Mathews, P. M., & Guo, J. Y. 2005, *J. Geophys. Res.*, 110(B), DOI: 10.1029/2003JB002915

Mound, J. E. & Buffett, B. A. 2003, *J. Geophys. Res.*, 108(B7), DOI: 10.1029/2002JB002054

Rogister, Y. & Rochester, M. 2004, *Geophys. J. Int.*, 159, 874

Van Hoolst, T., & Dehant, V. 2002, *Phys. Earth Planet.Inter.*, 134, 17

Wang, R. J. 1991, *Tidal deformation on a rotating, spherically asymmetric, visco-elastic and laterally heterogeneous Earth* (Frankfurt am Main: Peter Lang)

Wu, X. P. & Wahr, J. 1997, *Geophys. J. Int.*, 128, 18

Yseboodt, M., de Viron, O., Chin, T. M., & Dehant, V. 2002, *J. Geophys. Res.*, 107(B2), DOI: 10.1029/2000JB000042

A Giant Step: from Milli- to Micro-arcsecond Astrometry
Proceedings IAU Symposium No. 248, 2007
W. J. Jin, I. Platais & M. A. C. Perryman, eds.

© 2008 International Astronomical Union
doi:10.1017/S1743921308019601

Deep-space laser-ranging missions ASTROD and ASTROD I for astrodynamics and astrometry

W. T. Ni and the ASTROD I ESA COSMIC VISION 2015-2025 TEAM

Center for Gravitation and Cosmology, Purple Mountain Observatory,
Chinese Academy of Sciences, Nanjing, 210008, China,
email: wtni@pmo.ac.cn

Abstract. Deep-space laser ranging will be ideal for testing relativistic gravity, and mapping the solar-system to an unprecedented accuracy. ASTROD (Astrodynamical Space Test of Relativity using Optical Devices) and ASTROD I are such missions. ASTROD I is a mission with a single spacecraft; it is the first step of ASTROD with 3 spacecraft. In this talk, after a brief review of ASTROD and ASTROD I, we concentrate on the precision of solar astrodynamics that can be achieved together with implications on astrometry and reference frame ties. The precise planetary ephemeris derived from these missions together with second post-Newtonian test of relativistic gravity will serve as a foundation for future precise astrometry observations. Relativistic frameworks are discussed from these considerations.

Keywords. astrometry, gravitation, relativity, reference systems, celestial mechanics

1. Introduction

Improvement of the accuracy of Satellite Laser Ranging (SLR) and Lunar Laser Ranging (LLR) has a great impact on the geodesy, reference frames and test of relativistic gravity. At present, the accuracy for 2-color (2-wavelength) satellite laser ranging reaches 1 mm and that for 2-color lunar laser ranging are progressing towards a few millimeters. The present SLR and LLR are passive ranging; the intensity received at the photodetector is inversely proportional to the fourth power of the distance; for LLR, the received number of photons is low. However, for active ranging in which the laser light is received at the other end of ranging, the intensity received is inversely proportional to the second power of distance. Even at interplanetary distances, the received photons would be abundant and deep-space laser ranging missions are feasible. The primary objective for the long-term ASTROD (Astrodynamical Space Test of Relativity using Optical Devices) concept is to maintain a minimum of 3 spacecrafts in orbit within our solar system using laser interferometric ranging to ultimately test relativity and to search for gravitational waves. This is divided into 3 distinct stages, each with increasing order of scientific benefits and engineering milestones. The first stage, ASTROD I, will comprise a single spacecraft communicating with ground stations using laser pulse ranging. Fig. 1 (left) gives a schematic of the proposed orbit design. The second phase, ASTROD II, also called ASTROD since it is a testbed of the general ASTROD mission concept, will consist of three spacecrafts (2 in solar orbits, one at the Sun-Earth Lagrangian point L1 [or L2]) communicating with 1-2 W CW lasers. Fig. 1 (right) depicts the orbit design and configuration of the spacecraft 700 days after launch. The third and final stage, ASTROD III or Super ASTROD, will then explore the possibility of larger orbits in an effort to detect lower frequency primordial gravitational waves.

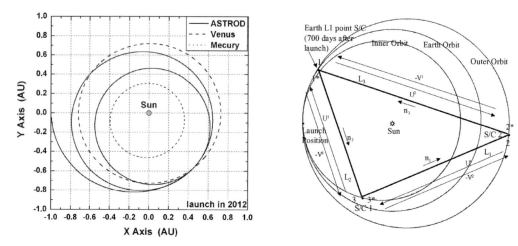

Figure 1. (left) A 2012 orbit in heliocentric ecliptic coordinate system for ASTROD I spacecraft; (right) A schematic ASTROD configuration (baseline ASTROD after 700 days from launch).

Table 1. Summary of the scientific objectives of the ASTROD I mission.

Effect/Quantity	Present accuracy	Projected accuracy
PPN parameter β	2×10^{-4}	3×10^{-8}
PPN parameter γ	4.4×10^{-5}	3×10^{-8}
Lense Thirring Effect	0.1	0.1
(dG/dt)/G	$10^{-12} \ \mathrm{yr}^{-1}$	$3 \times 10^{-14} \ \mathrm{yr}^{-1}$
Anomalous Pioneer acceleration A_a	$(8.74 \pm 1.33) \times 10^{-10} \, \mathrm{m/s^2}$	$0.7 \times 10^{-16} \ \mathrm{m/s^2}$
Determination of solar quadrupole moment	$(1-3) \times 10^{-7}$	1×10^{-9}
Determination of planetary masses and orbit parameters	(depends on object)	1-3 orders better
Determination of asteroid masses and densities	(depends on object)	2-3 orders better

2. ASTROD I

The basic scheme of the ASTROD I space mission concept as proposed to ESA Cosmic Vision 2015-2025 is to use two-way laser pulse ranging between the ASTROD I spacecraft in solar orbit and deep space laser stations on Earth to improve the precision of solar-system dynamics, solar-system constants and ephemeris, to measure the relativistic gravity effects and test the fundamental laws of spacetime more precisely, and to improve the measurement of the time rate of change of the gravitational constant. A summary of ASTROD I goals is compiled in Table 1 (Appourchaux *et al.* (2007)).

To achieve these goals, the timing accuracy for ranging is required to be less than 3 ps and the drag-free acceleration noise at 100 μHz is required to be below 3×10^{-14} m s^{-2}Hz$^{-1/2}$. Figure 2 (left) shows the drag-free acceleration noise requirements for AS-TROD I, the LTP, LISA and ASTROD for comparisons. To separate astrodynamic effect from relativistic-gravity test, the ASTROD I orbit is designed to reach the opposite side of the Sun in about 365 day and 680 day from launch. The apparent angles of the spacecraft during the two solar oppositions are shown in Fig. 2 (right) for the 2012 trajectory in Fig. 1. The maximum one-way Shapiro time delays near the two solar oppositions are

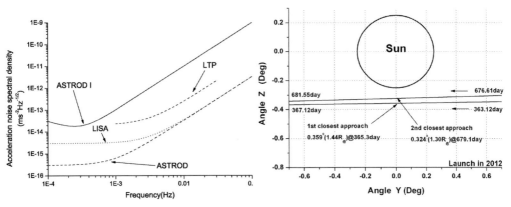

Figure 2. (left) A comparison of the target acceleration noise curves of ASTROD I, the LTP, LISA and ASTROD; (right) Apparent angles during the two solar opposition for the ASTROD I orbit in figure 1.

Table 2. Summary of the astrodynamical and astrometric objectives of the ASTROD I and ASTROD missions.

Effect/Quantity	ASTROD I	ASTROD
Timing accuracy	3 ps	1 ps or better
Timing precision	3 ps	0.1 ps
Ranging accuracy	0.9 mm	0.3 mm or better
Ranging precision	0.9 mm	$3 - 10~\mu$m
Acceleration noise at 100 μHz	3×10^{-14} m s^{-2} Hz$^{-1/2}$	3×10^{-16} m s^{-2} Hz$^{-1/2}$
Derived angle accuracy	$< 10~\mu$as	1 μas
Dynamical frame accuracy	$< 100~\mu$as	$< 10~\mu$as

0.1172 ms (in 365.3 day) and 0.1196 ms (in 679.1 days), respectively. The Shapiro time delay can be measured precisely to give a good test of relativistic gravity.

Table 2 summarizes the astrodynamical and astrometric objectives of the ASTROD I and ASTROD missions.

The angle accuracy depends on (i) derived angle accuracy from ranging of ASTROD I spacecraft and (ii) laser pointing measurement noise. With a ranging accuracy of 0.9 mm and acceleration noise as quoted, the derived azimuthal range accuracy would be much better than 1 m. This corresponds to 1.3 μas in angle at a distance of 1 AU. For laser pointing measurement accuracy, we assume that it is below 10 μas. This assumption gives the derived angle accuracy below 10 μas (Table 2). For the dynamical frame accuracy, we conservatively down grade it by a factor of 10 to below 100 μas.

Just like LLR, ASTROD I (and ASTROD) will contribute to the determination of celestial reference frame, terrestrial frame and earth rotation. A dynamical realization of the International Celestial Reference System (ICRS) by the lunar orbit has been obtained from LLR to an uncertainty about 1 mas. A similar determination from ASTROD I orbit would lead to an uncertainty below 100 μas. The simultaneous data-fitting solution for the coordinates and velocities for the laser ranging stations will contribute to the realization of international terrestrial reference frame. Ranging to ASTROD I would contribute to the determination of nutation parameters due to the dynamical stability of ASTROD I orbit.

3. ASTROD

ASTROD (ASTROD II) mission concept is to have two spacecraft in separate solar orbits carrying a payload of a proof mass, two telescopes, two 1 - 2 W lasers, a clock and a drag-free system, together with a similar spacecraft near Earth at one of the Lagrange points L1/L2 (Bec-Borsenberger et al. 2000; Ni et al. 2004). The three spacecrafts range coherently with each other using lasers to map the solar-system gravity, to test relativistic gravity, to observe solar g-mode oscillations, and to detect gravitational waves. Distances between spacecraft depend critically on the solar-system gravity (including gravity induced by solar oscillations), underlying gravitational theory and incoming gravitational waves. A precise measurement of these distances as a function of time will determine these causes. After 2.5 years, the inner spacecraft completes 3 rounds, the outer spacecraft 2 rounds, and the L1/L2 spacecraft (Earth) 2.5 rounds. At this stage two spacecraft will be on the other side of the Sun, as viewed from the Earth, for conducting the Shapiro time delay experiment efficiently. The spacecraft configuration after 700 days from launch is shown in Figure 1(right).

For a mission like ASTROD II within the next 10 - 20 years, the timing accuracy better than 1 ps (300 μm in terms of range) can be anticipated. In coherent interferometric ranging, timing events need to be generated by modulation/encoding technique or by superposing timing pulses on the CW laser light. The interference fringes serve as consecutive time marks. With timing events aggregated to a normal point using an orbit model, the precision can reach 30 μm in range. The effective range precision for parameter determination could be better, reaching $3-10$ μm by using orbit models. Since ASTROD range is typically of the order of $1-2$ AU ($(1.5-3)\times10^{11}$ m), a range precision of 3 μm will give a fractional precision of distance determination of 10^{-17}. Therefore, the desired clock accuracy/stability is 10^{-17} over 1000 s of travel time. Optical clocks with this accuracy/stability are under development. The space optical clock is under development for the Galileo project. This development would pave the road for ASTROD to use optical clocks. With these anticipations, we list the astrodynamical and astrometric objectives in Table 2.

4. Relativistic frameworks

For the relativistic framework to test relativity and to do astrodynamics and astrometry, we use the scalar-tensor theory of gravity including intermediate range gravity which we used in obtaining the second post-Newtonian (2 PN) approximation (Xie et al. 2007). For the light deflection in the 2 PN approximation in the solar field, please see Dong & Ni (2007). To extend to multi-bodies with multipoles in the solar system, we will extend our calculation following the work of Kopeikin & Vlasov (2004).

We thank National Natural Science Foundation of China (Grant Nos 10475114 and 10778710) and Foundation of Minor Planets of Purple Mountain Observatory for support.

References

Appourchaux, T., et al. 2007, ASTROD I: proposal for ESA Cosmic Vision 2015-2025
Bec-Borsenberger, A., et al. 2000, ASTROD—A Proposal for Two ESA Flexi-Mission F2/F3
Dong, P. & Ni, W.-T. 2008, Light deflection in the second post-Newtonian approximation of scalar-tensor theory of gravity, this volume, p.401
Kopeikin, S. & Vlasov, I. 2004, Phys. Rep., 400, 209
Ni, W.-T., Shiomi, S., & Liao, A.-C. 2004, Class. Quantum Grav., 21, S641
Xie, Y., Ni, W-T., Dong, P., & Huang, T-Y. 2007, Adv. Spac. Res., doi: 10.1016/j.asr.2007.09.022; arXiv:0704.2991

A Giant Step: from Milli- to Micro-arcsecond Astrometry
Proceedings IAU Symposium No. 248, 2007
W. J. Jin, I. Platais & M. A. C. Perryman, eds.

© 2008 International Astronomical Union
doi:10.1017/S1743921308019613

Radio interferometric tests of general relativity

E. B. Fomalont[1] and S. Kopeikin[2]

[1]National Radio Astronomy Observatory,
520 Edgemont Road, Charlottesville, VA 22903, USA
email: efomalon@nrao.edu

[2]Dept. of Physics and Astronomy,
University of Missouri, Columbia, MO 65211, USA
email: kopeikins@missouri.edu

Abstract. Since VLBI techniques produce a microarcsecond positional accuracy of celestial objects, tests of GR using radio sources as probes of a gravitational field have been made. We present the results from two recent tests using the VLBA: in 2005, the measurement of the *classical* solar deflection; and in 2002, the measurement of the retarded gravitational deflection associated with Jupiter. The deflection experiment measured γ to an accuracy of 3×10^{-4}; the Jupiter experiment measured the retarded term to 20% accuracy. The controversy over the interpretation of the retarded term is summarized.

Keywords. gravitation, techniques: interferometric, astrometry

1. Introduction

The theory of general relativity (GR) describes the interaction of matter and light with a gravitational field; hence, any accurate measurement of this interaction is a test of GR. The simplest test was the GR prediction of the angular deflection of starlight passing near the limb of the sun, first performed in 1916 during a solar eclipse. The results of other experiments, most using radio light rather than star light, agreed with the GR prediction to < 0.1% accuracy. Departures of γ from unity are expected at the 10^{-6} level, and more accurate deflection observations will continue.

Other properties of gravity can be measured with different experiments. For example, the perihelion shift of Mercury is a measure of the non-linearity of the GR. The second measurement described in this paper, that of the retarded deflection of light caused by the motion of the gravitating body, obtained results in agreement with GR, but its interpretation with the property of gravity that is constrained by this experiment is controversial.

2. The 2005 VLBA Deflection Experiment

We believed that a new, well-designed, VLBA experiment could significantly improve upon the accuracy of previous solar radio deflection experiments for two reasons. With a stable electronic system, sensitive receivers and accurate astrometric/tropospheric modeling, the relative position of sources separated by a few degrees in the sky were now being routinely measured by the VLBA with about 20 μas accuracy. Secondly, the VLBA was now operating routinely at 43 GHz where solar coronal effects are relatively small. With a typical troposphere coherence time of 1 min (maximum integration time per observation) and the compactness of most quasars, several groups of quasars within a five degree

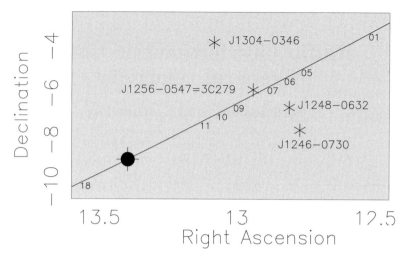

Figure 1. Source configuration and observation dates for the 2005 October deflection experiment:

region near the ecliptic were easily detectable. The source configuration that we selected for the October experiment is shown in Fig. 1.

We observed on 8 days, each for a period of 5 hours. On October 1 and 18, the relative gravitational and coronal bending was less than 2 μas so the undeflected relative positions for the sources could be determined. Also, any significant structural changes of the quasars over the experimental period could be ascertained (none were seen). Significant relative gravitational bending was obtained on the other observing days. The observations on each day were identical. Each source was observed for 45 sec with a switching time of about 25 sec, in the sequence: 3C 279, J1304, 3C 279, J1256, 3C 279, J1258, 3C 279, J1304, etc. As long as the phase between subsequent 3C 279 observations could be connected, this observing scheme accurately determined the source relative positions.

In order to determine the coronal refraction at 43 GHz, we time-shared the observations on each day among 43, 23, and 15 GHz. This was done by using the source observing sequence noted above, but switching frequencies every 25 minutes. In this manner we determined the relative position of the sources at any one frequency in each 25-min block to an accuracy of ~ 100 μsec. By comparing the source positions as a function of frequency, we determined any significant frequency dependent position change. However, when a source was closer than about 2.2° of the sun, the coronal refraction *vibrated* the source position at 43 GHz more than 50 μsec over a time scale of a few seconds, making the phase tracking impossible to follow. Hence, all data were phase unstable for October 7 and several sources could not be used on October 5, 6 and 9. An additional 10% of the data were lost for some VLBA telescopes during periods of poor weather.

As a rough guide to the experiment sensitivity, at 43 GHz the typical accuracy for the relative position between two sources over one five hour period was about 0.07 mas. With a relative gravitational deflection of 100 mas for most observing days, we obtained a deflection accuracy of 1 part in 1400 (0.0007) for each source-pair and day. When all source-pairs and are averaged at 43 GHz, we obtain $(\gamma - 1) = -0.00070 \pm 0.00040$ (rms error). If we include the 23 GHz and 15 GHz data to remove the small coronal bending, we obtain $(\gamma - 1) = -0.00006 \pm 0.00027$, our best estimate.

We believe that we can increase the experiment accuracy by a factor two to three. First, a group of sources in May when the sun is at a more northern declination will improve the astrometric quality of the data. Second, the most accurate results are obtained when the sources are between 3° to 5° from the sun. Third, the 2005 experiment scheduled too much time with sources too near the sun and not enough time at these intermediate solar separations. Finally, most of the observation time should be made at 43 GHz, with perhaps 25% of the time at 23 GHz to remove the significant coronal bending.

3. The 2002 Jupiter Deflection Experiment

On September 8, 2002, Jupiter passed within 3.7′ of the quasar J0842+1835. Such a close passage of Jupiter with a bright quasar (0.4 Jy, among the brightest 1000 sources) occurs on average once every 20 years. A similar encounter in 1988 with a different bright quasar was observed, and the gravitational deflection of about 1 mas was detected by Treuhaft & Lobe (1991). Our goal was to measure not only the radial deflection, but the retarded component as well. The relevant parameters at closest approach are shown in Fig. 2. The maximum radial deflection was 1190 μas and occurred at 16:30 UT on September 8. Because of the motion of Jupiter, there is a retarded deflection component of 51 μas opposite to the direction of motion of Jupiter, and it is within reach with the VLBA.

The details of the experiment and the results have been published by Fomalont & Kopeikin (2003). The VLBA observed at 8.4 GHz and observations switched between J0842+1835 with another quasar about 0.8° east J0839+1802, and with J0854+2006, about 3.4° to the west. A complete cycle took about 5 minutes. Five observing days, each 7 hours long, were made on September 4, 7, 8, 9, 12. Since the retarded term was significant only on September 8, the other four days were used to measure the undeflected position of the sources and to determine realistic errors.

These three quasars were positioned nearly linearly in the sky; hence, the appropriate combination of the measured phases for J0839 and J0854 not only removed the temporal troposphere and ionosphere refraction changes, but also the effect quasi-stationary phase gradients in the sky among the three sources. These systematic phase gradients are caused by many small astrometric effects (antenna location offsets, earth-orientation modeling errors), as well as troposphere and ionospheric structure from two to twenty degrees in the sky. The use of two calibrators rather than one reduced the residual position uncertainty per day from about 25 μas to about 10 μas (Fomalont (2003)).

Both the radial and retarded deflections were easily detected. Analysis of the radial deflection gave $\gamma = 1.01 \pm 0.03$. The GR prediction of the retarded deflection varied between 41 and 51 μas on September 8, and the ratio of the measured retarded deflection to the GR prediction (assuming that the velocity associated with the retardation is the speed of light) was 0.98 ± 0.19.

The experiment confirms the GR prediction for the retarded deflection at the 20% level. A useful question is: what property of gravity is constrained by this experiment? Our interpretation is that the retardation is a measure of the propagation speed of gravity, and is related to a gravito-magnetic field associated with currents (motion) of matter (Kopeikin & Fomalont (2007)).

A summary of other interpretations has been compiled by Will (2008). The basic disagreement is whether the speed of propagation of light or gravity (c) can be manifested in (v_j/c) terms or only in $(v_j/c)^2$ terms, where v_j is the velocity relevant to the object. More specifically, Asada (2002) claims that the speed of light was determined. Will (2003) believes that the PPN parameter α_1 was measured, albeit poorly. Carlip (2004) finds the distinction between the speed of gravity and the speed of light somewhat ill-posed. Stuart (2004) believes that the experiment measured nothing useful.

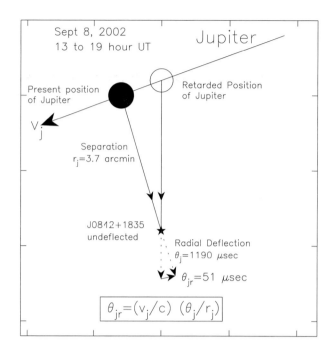

Figure 2. The Retarded Deflection of the Jovian Experiment in 2002: The line shows the path of Jupiter during the Sept 8 observations, with closest approach at 16:30 hr UT. The relevant deflection angles are shown.

4. Summary

The two experiments demonstrate that the micro-arcsecond positional accuracy of the VLBA produces significant tests of GR. The goal of the tests was to measure important parameters and their limits, and to foster discussion concerning the generalization of GR and its interaction with light and mass.

The National Radio Astronomy Observatory is a facility of the National Science Foundation operated under cooperative agreement by Associated Universities, Inc.

References

Asada, H. 2002, *ApJ*, 574, L69

Carlip, S. 2004, *Class. Quant. Grav.*, 21, 3803 (arXiv:gr-qc/0403060v3)

Fomalont, E. B., 2003, *Future Directions in High Resolution Astronomy*, Ed. J. D.Romney and M. J. Reid, Socorro, NM, p55

Fomalont, E. B. & Kopeikin, S. M. 2003, *ApJ*, 598, 704

Kopeikin, S. M. & Fomalont, E. B. 2007, *Gen. Rel. and Grav.*, 39, 1583 (arXiv:gr-qcd/0510077v4)

Stuart, S. 2004, *Int. J. Mod. Phys.*, D13, 1753.

Treuhaft, R. N. & Lowe, S. T. 1991, *AJ*, 102, 1879

Will, C. M., 2003, *ApJ*, 590, 683

Will, C. M., 2008, http://wugrav.wustl.edu/people/CMW/SpeedofGravity.html

A Giant Step: from Milli- to Micro-arcsecond Astrometry
Proceedings IAU Symposium No. 248, 2007
W. J. Jin, I. Platais & M. A. C. Perryman, eds.

© 2008 International Astronomical Union
doi:10.1017/S1743921308019625

Astrometric microlensing with the RadioAstron space mission

A. F. Zakharov [1,2,3,4]

[1] National Astronomical Observatories of Chinese Academy of Sciences, 20A Datun Road, Chaoyang District, Beijing, 100012, China;

[2] Institute of Theoretical and Experimental Physics, B. Cheremushkinskaya, 25, 117259, Moscow, Russia;

[3] Center of Advanced Mathematics and Physics, National University of Sciences and Technology, Rawalpindi, Pakistan;

[4] Bogoliubov Laboratory for Theoretical Physics, Joint Institute for Nuclear Research, 141980 Dubna, Russia

Abstract. According to a revised schedule of the Russian Space Agency, in October 2008 the 10 m space telescope RadioAstron will be launched in a high eccentric orbit around the Earth. Acting together with ground based radio telescopes, the VLBI interferometer with a ground-space arm will operate. The interferometer will have extraordinary angular resolution of a few microarcsecond (μas) at the shortest wavelength (1.35 cm). Since typical angular scales for gravitational microlensing are at the μas level for cosmological locations of sources and microlenses, in principle there is a chance to resolve microimages and (or) at least, detect astrometrical shift of bright point like images. In particular, gravitationally lensed systems, such as B1600+434, where in radio band a signature of microlensing is found, look suitable for direct observations of microlensing, since microlensing with the RadioAstron interferometer may be detected in the future (considering its high angular resolution and a relatively high sensitivity and assuming a ground support by the advanced radio telescopes).

Keywords. gravitational lensing, techniques: high angular resolution, astrometry, instrumentation: high angular resolution, instrumentation: interferometers

---

The RadioAstron space telescope is the high priority space project among Russian scientific space projects. According to the formal schedule of the Russian Space Agency (ROSCOSMOS), the space radio telescope RadioAstron should be launched in 2008.† This space based 10-meter radio telescope will be used for space – ground VLBI measurements. For observations four wavelength bands will be used. The bands correspond to $\lambda = 1.35$ cm, $\lambda = 6.2$ cm, $\lambda = 18$ cm, $\lambda = 92$ cm. If the interferometer operates with the 130-m Radio Telescope (EVLA)‡ as a ground telescope, the minimal correlated flux can be at the levels of about 10, 1.3, 1.4, 3.2 mJy (RMS) at 0.327 GHz (P-band), 1.665 GHz (L-band), 4.830 GHz (C-band) 18.392-25.112 GHz (K-band), respectively, at the 8σ level with an integration time about 300 s.¶ Although we understand that, perhaps, the estimated sensitivity levels give an optimistic forecast (moreover, clearly the RadioAstron–ground interferometer will not operate at the maximum level of its sensitivity, since, obviously, the EVLA is not a dedicated facility for the RadioAstron space mission), we will use the estimates for our analysis as an initial point to demonstrate

---

† See the official web-site of the agency http://www.federalspace.ru/science0615E.asp, but now the mission was postponed (re-scheduled) to launch the spacecraft in October 2008, as it was indicated at the web-site of the RadioAstron project http://www.asc.rssi.ru/radioastron/.
‡ http://www.aoc.nrao.edu/evla/pbook.shtml
¶ http://www.asc.rssi.ru/radioastron/description/observations_eng.htm

a potential possibility of discovering phenomenon such as splitting the images due to microlensing.

Assuming the concordant cosmological model parameters ($\Omega_{\text{tot}} = 1, \Omega_{\text{matter}} = 0.3, \Omega_\Lambda = 0.7$), a typical length scale for microlensing (the Einstein – Chwolson radius) is

$$R_E = \sqrt{2r_g \cdot \frac{D_s D_{ls}}{D_l}} \approx 3.4 \cdot 10^{16} \sqrt{\frac{M}{M_\odot}} h_{65}^{-0.5} \, \text{cm},$$

where "typical" microlens and source redshifts are assumed to be $z_l = 0.5$, $z_s = 2$ (Treyer & Wambsganss 2004), $r_g = \dfrac{2GM}{c^2}$ is the Schwarzschild radius corresponding to microlens mass M, $h_{65} = H_0/((65 \text{ km/sec})/\text{Mpc})$ is the dimensionless Hubble constant, and the distance between two images is about $2R_E$. We have a significant magnification A (gravitational focusing), so that if the impact parameter b is less than R_E ($b < R_E$), then $A > 3\sqrt{5}/5 \approx 1.34$.

The corresponding angular scale is

$$\theta_0 = \frac{R_E}{D_s} \approx 2.36 \cdot 10^{-6} \sqrt{\frac{M}{M_\odot}} h_{65}^{-0.5} \, \text{arcsec}.$$

If $M = 10^{12} M_\odot$ (typical galactic masses), typical angular distances between images are about a few arcseconds, and it is called macrolensing. On the other hand, if $M = M_\odot$, typical angular distances between images are about microarcseconds, corresponding to the case of microlensing.

Using this length scale and a velocity scale such as an apparent velocity β_{app}, one could calculate the standard time scale corresponding to the time to cross the Einstein – Chwolson radius

$$t_E = (1 + z_l) \frac{R_E}{v_\perp} = \begin{cases} \approx 2\sqrt{\dfrac{M}{M_\odot}} \beta_{\text{app}}^{-1} h_{65}^{-0.5} \text{ weeks}, & \text{if } v_\perp = c\beta_{\text{app}}, \\[2ex] \approx 27\sqrt{\dfrac{M}{M_\odot}} v_{600}^{-1} h_{65}^{-0.5} \text{ years}, & \text{if } v_\perp \sim 600 \text{ km/c}. \end{cases}$$

Here we assume time scales are determined by an apparent velocity or a typical transverse velocity ($v_{600} = v_\perp/(600 \text{ km/c})$), respectively. The time scale t_E corresponds to the approximation of a point mass lens and a small size source in comparison with the Einstein – Chwolson radius. The approximation and the time scale estimate could be used if microlenses are distributed freely at cosmological distances and a fixed Einstein – Chwolson cone is actually located far enough from the other ones.

For gravitationally lensed systems, if microlenses form a significant part of a gravitational lens mass, a caustic network generated by microlenses and photometric microlensing (fluctuations in light curves corresponding to different images) arise if a source crosses these caustics. If we use the simple caustic microlens model (like the straight fold caustic model), there are two time scales, depending on the "caustic size" and source radius R_{source}, respectively. Under the condition that the source radius is larger or about the same as the "caustic size" r_{caustic} (if we use the following approximation for the magnification near the caustic $\mu = \sqrt{\dfrac{r_{\text{caustic}}}{y - y_c}}$ where $y > y_c$ and y is the perpendicular direction to the fold caustic), if $R_{\text{source}} \gtrsim r_{\text{caustic}}$, then the relevant time scale is the "crossing caustic time" (Treyer & Wambsganss 2004)

$$t_{\text{cross}} = (1 + z_l) \frac{R_{\text{source}}}{v_\perp (D_s/D_l)} \approx 0.62 \, R_{15} v_{600}^{-1} h_{65}^{-0.5} \text{ years} \approx 226 \, R_{15} v_{600}^{-1} h_{65}^{-0.5} \text{ days},$$

where D_l and D_s correspond to $z_l = 0.5$ and $z_s = 2$ respectively and $R_{15} = R_{\text{source}}/10^{15}$ cm.

However, if the source radius R_{source} is much smaller than the "caustic size" r_{caustic} ($R_{\text{source}} \ll r_{\text{caustic}}$), one could use the "caustic time", namely the time when the source is located in the area near the caustic, and the time scale then corresponds to

$$t_{\text{caustic}} = (1 + z_l)\frac{r_{\text{caustic}}}{v_\perp(D_s/D_l)} \approx 0.62 \; r_{15} v_{600}^{-1} h_{65}^{-0.5} \text{ years} \approx 226 \; r_{15} v_{600}^{-1} h_{65}^{-0.5} \text{ days},$$

where $r_{15} = r_{\text{caustic}}/10^{15}$ cm.

These time scales t_{cross} and t_{caustic} could be about days (or even hours) if $v_\perp$ is determined by an apparent motion of superluminal motion of knots in jet.

Thus, t_{cross} could be used as a lower limit for typical time scales for the simple caustic microlens model, but there are two length parameters in the problem. In general we do not know their values, and so we could not evaluate R_{source} only from the time scales of microlensing because time scales could correspond to two different length scales. However, if we take into account variation amplitudes of luminosity, one could say that, in general, t_{cross} corresponds to smaller variation amplitudes than t_{caustic}. This is because if the source area is large there is a "smoothness" effect — since only a small fraction of source area is located in the high amplification region near the caustic.

In conclusion, first, one could point out that gravitational lensed systems are the most perspective objects to search for microlensing. Astrometric microlensing could be detected in the gravitational lens system such as B1600+434 in case proper motions of source, lens and an observer are generated mostly by a superluminal motion of knots in jet. But in this case, with the density flux estimates obtained by Koopmans & de Bruyn (2000), one could say that based on the optimistic estimates of the RadioAstron interferometer acting with ground support of the EVLA (or similar ground based radio telescope), such motions in knots may be detectable.

Second, in case there is microlensing in the B1600+434 system, for example, then astrometric microlensing in the system could be about $\theta_{\text{shift}} \sim 20 - 40 \; \mu as$ (Treyer & Wambsganss 2004). In the best case the RadioAstron interferometer will have enough sensitivity to detect such an astrometric displacement as it was noted earlier, and moreover, as estimates showed, even splitting of the microimages could be detectable.

Third, in principle microlensing for distant sources could be the only tool used to evaluate Ω_L (including dark matter component) from event rate for microlensing. To solve this problem with the RadioAstron interferometer one should analyze variabilities of compact sources with a core size $\lesssim 40 \; \mu as$ and with high enough flux densities. We should apply the most reliable model for variabilities of the sources such as scintillations, microlensing etc. A fraction of the sources in the list of extragalactic targets for VSOP and RadioAstron is suitable for this analysis. If the analysis would indicate that other explanations (such as scintillations) are preferable and future observations with RadioAstron interferometer would show that there are no features for astrometric microlensing, one could conclude that the Hawkins claim (supporting microlensing as a source of variabilities for a significant part of distant quasars) should be ruled out. But if an essential fraction of variability could be explained by microlensing, the sources could serve as first order candidates to search for astrometric microlensing. Therefore, one could say that astrometric microlensing (or direct image resolution with RadioAstron interferometer) is the crucial test to confirm (or rule out) the microlens hypothesis for gravitational lensed systems and for point like distant objects. Astrometric microlensing due to MACHO action in our Galaxy is not very important for observations with the space interferometer RadioAstron. This is because first, such probabilities are not high and secondly, typical time scales are longer than the estimated life time of the RadioAstron space mission.

Thus, right after the RadioAstron launch, we should take the first opportunity to detect microlensing (splitting of the micro-images) in a direct way. So the main objective of this paper to pay attention to such a challenging possibility because, preflight time is very short now and perspective targets should be analyzed carefully in advance by observational and theoretical means. Since there are only few point like bright sources at cosmological distances and gravitational lensed systems with point like components demonstrating microlens signatures, these sources should be analyzed with great care way to search for candidates for which the microlens model is preferable in comparison with alternative explanations of variabilities.

We would like to point out also that there are some non-negligible chances of observing mirages (shadows) around the black hole at the Galactic Center and in nearby AGNs in the radio-band (or in mm–band) using RadioAstron (or Millimetron) facilities (see the detailed discussion by Zakharov et al. (2005a,b)). Since a shadow size should be about 50 μas for the black hole in the Galactic Center, analyzing the shadow size and shape, one could enable us to evaluate a spin and a charge of the black hole.

Finally, we note that the next space Russian dish for VLBI (ground–space) facilities will be in action after about one decade, since the Russian Space Agency and the Russian Academy of Sciences have planned to launch the cryogenic telescope Millimetron for observations at mm (and sub-mm) wavelength bands in 2016.† This instrument will be placed near the Lagrange point L_2 of the Sun – Earth system. The Millimetron itself will have bolometrical sensitivity of about 5×10^{-9} Jy at 0.3 mm wavelength with an integration time of about 1 hour. The ALMA – Millimetron interferometer with the space-ground arm will have an angular resolution of 10^{-9} arcseconds with a sensitivity of about 0.1 mJy level at 0.5 mm wavelength with an integration time of about 300 s. This will give us an opportunity to analyze not only microlensing but also nanolensing due to actions of gravitating bodies with planet masses.

Acknowledgments

The author is grateful to the National Natural Science Foundation of China (NNSFC) (Grant # 10233050) and the National Key Basic Research Foundation (Grant # TG 2000078404) for a partial financial support of the work and profs. J.-X. Wang and J. Zhang for fruitful discussions.

References

Koopmans, L. V. E., & de Bruyn, A. G., 2000, A&A, 358, 793
Popović L. C. Jovanović, P., Mediavilla, E., et al. ApJ, 2006, 637, 620
Treyer, M. & Wambsganss, J., 2004, A&A, 416, 19
Zakharov, A. F., 1995, A & A, 293, 1
Zakharov, A. F., 1997a, Ap & SS, 252, 369
Zakharov, A. F., 1997b, Gravitational Lensing and Microlensing, Janus-K, Moscow
Zakharov, A. F., 2003, Publ. Astron. Obs. Belgrade 75, 27; astro-ph/0212009
Zakharov, A. F., 2005, in A.I. Studenikin (ed.), Proc. of the Eleven Lomonosov Conference on Elementary Particle Physics, (World Scientific, Singapore) p. 106; astro-ph/0403619
Zakharov, A. F., 2006a, Astron. Reports, 50, 79
Zakharov, A. F., 2006b, Physics of Particles and Nuclei Lett., (in press); astro-ph/0610857
Zakharov, A. F., Nucita, A. A., De Paolis, F., & Ingrosso, G., 2005a, New Astron., 10, 479
Zakharov, A. F., Nucita, A. A., De Paolis, F., & Ingrosso, G., 2005b, A & A, 442, 795
Zakharov, A. F., Popović, L. Č., & Jovanović, P., 2004, A & A, 420, 881
Zakharov, A. F. & Sazhin, M. V., 1998, Physics-Uspekhi, 41, 945

† http://www.federalspace.ru/science0615E.asp.

A Giant Step: from Milli- to Micro-arcsecond Astrometry
Proceedings IAU Symposium No. 248, 2007
W. J. Jin, I. Platais & M. A. C. Perryman, eds.

© 2008 International Astronomical Union
doi:10.1017/S1743921308019637

Gravitational bending of light by planetary multipoles and its measurement with microarcsecond astronomical interferometers

S. Kopeikin[1] and V. Makarov[2]

[1]Department of Physics & Astronomy, University of Missouri-Columbia, Columbia, MO 65211
email: kopeikins@missouri.edu

[2]Michelson Science Center, California Technology Institute, Pasadena, CA 91125
email: vvm@caltech.edu

Abstract. General relativistic deflection of light by mass, dipole, and quadrupole moments of gravitational field of a moving massive planet in the Solar system is derived in the approximation of the linearized Einstein equations. All terms of the order of 1 μas and larger are taken into account, parameterized, and classified in accordance with their physical origin. We discuss the observational capabilities of the near-future optical and radio interferometers for detecting the Doppler modulation of the radial deflection, and the dipolar and quadrupolar light-ray bending by Jupiter and Saturn.

Keywords. gravitation, gravitational lensing, astrometry, reference systems, planets and satellites: individual (Jupiter), Saturn, techniques: interferometric

Attaining the level of a microarcsecond (μas) positional accuracy and better will completely revolutionize fundamental astrometry by merging it with relativistic gravitational physics. Beyond the microarcsecond threshold, one will be able to observe a new range of celestial physical phenomena caused by gravitational waves from the early universe and various localized astronomical sources, space-time topological defects, moving gravitational lenses, time variability of gravitational fields of super-massive binary black holes located in quasars, and many others (Kopeikin *et al.* 1999, Wex & Kopeikin 1999, Kopeikin & Gwinn 2000, Kopeikin & Makarov 2006). Furthermore, it will allow us to test general theory of relativity in the Solar system in a dynamic regime, that is when the velocity- and acceleration-dependent components of gravitational field (the metric tensor) of the Sun and planets bring about observable relativistic effects in the light deflection, time delay and frequency to an unparalleled degree of precision (Kopeikin 2001, Fomalont & Kopeikin 2003, Kopeikin *et al.* 2007).

Preliminary calculations (Brumberg, Klioner & Kopeikin 1990) reveal that the major planets of the Solar system are sufficiently massive to pull photons by their gravitational fields, which have significant multipolar structures, in contrast with the Sun whose quadrupole moment is only $J_{2\odot} \leqslant 2.3 \times 10^{-7}$ (Pitjeva 2005). Moreover, in the case of a photon propagating near the planet the interaction between the gravitational field and the photon can no longer be considered static, because the planet moves around the Sun as the photon traverses through the Solar system(Kopeikin & Schäfer 1999, Kopeikin & Mashhoon 2002). The optical interferometer designed for the space astrometric mission SIM (Laskin 2006) is capable of observing optical sources fairly close in the sky projection to planetary limbs with a microarcsecond accuracy. Similar resolution can be achieved for radio sources with the Square Kilometer Array (SKA) (Carilli & Rawlings 2004) if it is included in the inter-continental baseline network of VLBI stations (Fomalont & Reid 2004). The Gaia (Perryman 2005) and JASMINE (Gouda *et al.* 2006) astrometric projects represent another alternative path to microarcsecond astrometry. It is a challenge for the SIM and SKA interferometers as well as for Gaia and JASMINE to measure the gravitational bending of light caused by various planetary multipoles and the orbital motion of the planets. This measurement, if successful, will be a cornerstone step in further deployment of theoretical principles of general relativity to fundamental astrometry and navigation at a new and exciting technological level.

The first detection of gravitational bending of light by Jupiter was conducted in 1988 (Schuh *et al.* 1988, Treuhaft & Lowe 1991), and the deflection term associated with the monopole field of Jupiter was determined to an accuracy of $\simeq 15\%$ to be in agreement with Einstein's general relativity theory. Later on, the Hubble Space Telescope was used to measure the gravitational deflection of light of the bright star HD 148898 as it passed within a few seconds of arc near Jupiter's limb on 24 September 1995 (Whipple *et al.* 1996). Kopeikin (2001) proposed to use Jupiter's orbital motion in order to measure the retardation effect in the time of propagation of the dynamic part of gravitational force of Jupiter to photon, that appears as a small excess to the Shapiro time delay and should be interpreted as a gravimagnetic dragging of light ray caused by the orbital motion of Jupiter (Kopeikin & Fomalont 2006, Kopeikin & Fomalont 2007). This proposal was executed experimentally in 2002 September 8, and the gravimagnetic dragging of light was measured to $\simeq 20\%$ accuracy (Fomalont & Kopeikin 2003) thus, complementing the LAGEOS measurement of the gravimagnetic field induced by intrinsic rotation of the Earth (Ciufolini 2007).

Crosta & Mignard (2006) proposed to measure the deflection-of-light term associated with the axisymmetric (quadrupolar) part of Jupiter's gravitational field. Detection and precise measurement of the quadrupolar deflection of light in the Solar system is important for providing an independent experimental support for detection of dark matter via gravitational lensing by clusters of galaxies (Schneider, Ehlers & Falco 1992). The work by Crosta & Mignard (2006) can be extended in several directions (Kopeikin & Makarov 2007). First, it assumes that light propagates in the field of a static planet while Jupiter moves on its orbit as light traverses the Solar system toward the observer. Second, Crosta & Mignard (2006) implicitly assumed that the center of mass of the planet deflecting light rays coincides precisely with the origin of the inertial coordinate system on the sky used for interpretation of the apparent displacements from the gravity-unperturbed (catalogue) positions of stars. This makes the dipole moment, I^i, of the gravitational field of Jupiter vanish, which significantly simplifies the theoretical calculation of light bending. However, the assumption of $I^i = 0$ is not practical because the instantaneous position of the planet's center of mass on its orbit is known with some error due to the finite precision of the Jovian ephemeris limited to a few hundred kilometers (Pitjeva 2005). The ephemeris error will unavoidably bring about a non-zero dipole moment that must be included in the multipolar expansion of the gravitational field of the plan*et al.*ong with its mass and the quadrupole moment.

The dipolar anisotropy in the light-ray deflection pattern is a coordinate-dependent effect and, hence, should be properly evaluated and suppressed as much as possible by fitting the origin of the coordinate system used for data analysis to the center of mass of the planet. Until the effect of the gravitational dipole is properly removed from observations it will forge a model-dependent quadrupolar deflection of light because of the translational change in the planetary moments of inertia – the effect known as the parallel-axis theorem (Arnold 1978). This translation-induced quadrupolar distortion of the light-ray deflection pattern should be clearly discerned from that caused by the physical quadrupole moment of the planet J_2.

Perhaps, the SIM, which is a Michelson-type interferometer with articulating siderostat mirrors, holds the best prospects for precision tests of general relativity in the Solar system through gravitational bending effects. In these experimental-gravity applications the advantages of the SIM facility are as follows:

(*a*) SIM is a pointing mission.

(*b*) In the differential regime of operation the SIM interferometer is expected to achieve an unprecedented accuracy of 1 μas in a single observation on stars separated on the sky by $\sim$2 deg.

(*c*) The baseline of SIM can be rotated by 90 degrees for a dedicated observation. Two-dimensional observations on a given set of stars are crucial for unambiguous disentangling the dipole and quadrupole deflection patterns.

(*d*) SIM will self-calibrate its 15°-wide field of view. This dramatically reduces the correlated and/or systematic errors.

(*e*) SIM can observe stars and quasars as close as several arcseconds from the planetary limb.

Extensive discussion of various fascinating science drivers and of the evolving technical possibilities has led to a concept for the Square Kilometer Array (SKA) and a set of design goals (Carilli & Rawlings 2004). The SKA will be an interferometric array of individual antenna stations, synthesizing an aperture with a diameter of up to several 1000 kilometers. A number of configurations will distribute the 1 million square meters of collecting area. These include 150 stations each with the collecting area of a 90 m telescope and 30 stations each with the collecting area equivalent to a 200 meters diameter telescope. The sensitivity and versatility of SKA can provide $\sim$ 1 μas astrometric precision and high quality milliarcsec-resolution images by simultaneously detecting calibrator sources near the target source if an appreciable component of SKA is contained in elements which are more than 1000 km from the core SKA (Fomalont & Reid 2004).

Measurement of the light bending by a moving planet with microarcsecond accuracy requires a continuous phase-referencing observation of the target and the calibrating radio sources (Fomalont & Kopeikin 2003). The main limitation of the accuracy is the tropospheric refraction which affects radio observations. The large-scale tropospheric refraction can be estimated by observing many radio sources over the sky in a short period of time. At present the determination of the global troposphere properties can only be estimated in about one hour, and smaller angular-scale variations can not be determined in most cases. However, the SKA, by using observations in ten sub-arrays, on strong radio sources across the sky, will determine the tropospheric properties on time-scales which may be as short at five minutes.

Quasars as astrometric calibrators have one peculiar property: they are variable. The massive outflows and shocks in the jet change the intensity and the structure of the radio emission. Hence, the position of the quasar reference point is variable by about 0.05 mas in most quasars. Thus, the calibrators used to determine the SKA astrometric precision to better than 10 μsec have a jitter which is somewhat larger. In order to reach the intended angular precision, the change in the position of calibrators must be determined.

In addition to various special and general relativistic effects in the time of propagation of electromagnetic waves from a quasar to the SKA-VLBI antenna network, we must account for the effects produced by the planetary magnetosphere. The magnetospheric deflection estimate reveals that a single frequency observation of the light deflection will be affected by the magnetosphere at the level exceeding 1 μas. This assumes that we should observe at two widely spaced frequencies to determine and eliminate the magnetospheric effects. The noise due to turbulence in the magnetosphere (and the Earth ionosphere) may also be a limiting factor.. However, this rapidly fluctuation model is fairly pessimistic and unlikely, and would probably average out to a steady state model.

Further particular details of the theoretical study of the deflection of light by quadrupole and higher gravitational multipoles can be found in papers (Kopeikin, Korobkov & Polnarev 2006, Kopeikin & Makarov 2007) and references therein.

Acknowledgements

We thank the National Science Foundation for supporting our research with travel grant AST0726470.

References

Arnold, V. I. 1978, *Mathematical Methods of Classical Mechanics* (Springer-Verlag: Berlin)

Brumberg, V. A., Klioner, S. A., & Kopeikin, S. M. 1990, In: *Inertial Coordinate System on the Sky*, Proc. of the IAU Symp. 141, Eds. J. H. Lieske & V. K. Abalakin (Kluwer: Dordrecht), pp. 229–240

Carilli, C. & Rawlings, S. (eds.) 2004, "Science with the Square Kilometre Array", *New Astron. Rev.*, Vol. 48, pp. 979–1606

Ciufolini, I., 2007, *Nature*, 449, 41

Crosta, M. T. & Mignard, F. 2006, *Class. Quant. Grav.*, 23, 4853

Fomalont, E. B. & Kopeikin, S. M. 2003, *Astrophys. J.*, 598, 704

Fomalont, E. & Reid, M. 2004, *New Astron. Rev.*, 48, 1473

Gouda, N., Kobayashi, Y., Yamada, Y., Yano, T., Tsujimoto, T., Suganuma, M., Niwa, Y., Yamauchi, M., Kawakatsu, Y., Matsuhara, H., Noda, A., Tsuiki, A., Utashima, M.; Ogawa, A.; Sako, N.; JASMINE working group 2006, Memorie della Societa Astronomica Italiana, 77, 1185

Kopeikin, S. M. 2001, *Astrophys. J. Lett.*, 556, 1

Kopeikin, S. M. & Schäfer, G. 1999, *Phys. Rev. D*, 60, 124002

Kopeikin, S. M., Schäfer, G., Gwinn, C. R., & Eubanks, T. M. 1999, *Phys. Rev. D*, 59, 084023

Kopeikin, S. M. & Gwinn, C. R. 2000, In: *Towards Models and Constants for Sub-Microarcsecond Astrometry*, Proc. of IAU Coll. 180, Washington DC: U.S. Naval Obs., 2000. Eds K. J. Johnston, D. D. McCarthy, B. J. Luzum and G. H. Kaplan., pp. 303–307

Kopeikin, S. & Mashhoon, B. 2002, *Phys. Rev. D*, 65, 064025

Kopeikin, S. M. & Makarov, V. 2006, *AJ*, 131, 1471

Kopeikin, S., Korobkov, P., & Polnarev, A. 2006, *Class. Quant. Grav.*, 23, 4299

Kopeikin, S. M. & Fomalont, E. B. 2006, *Found. of Phys.*, 36, 1244

Kopeikin, S. M. & Fomalont, E. B. 2007, *Gen. Rel. Grav.*, 39, 1583

Kopeikin, S. M. & Makarov, V. V. 2007, *Phys. Rev. D*, 75, 062002

Kopeikin, S., Schäfer, G., Polnarev, A., & Vlasov, I. 2006, Phys. Lett. A, 367, 276

Laskin, R. A. 2006, In: *Advances in Stellar Interferometry* Eds. J. D. Monnier, M. Schöller & W. C. Danchi. Proc. of the SPIE, vol. 6268, p. 65

Perryman, M. A. C. 2005, In: *Astrometry in the Age of the Next Generation of Large Telescopes*, Eds. P. K. Seidelmann & A. K. B. Monet (San Francisco: Astron. Soc. of the Pacific) ASP Conference Series, vol. 338, p. 3

Pitjeva, E. V. 2005, *Astron. Lett.*, 31, 340

Schneider, P., Ehlers, J., & Falco, E. E. 1992, *Gravitational Lenses*, (Springer: Berlin)

Schuh, H., Fellbaum, M., Campbell, J., Soffel, M., Ruder, H., & Schneider, M. 1988, *Phys. Lett. A*, 129, 299

Treuhaft, R. N. & Lowe, S. T. 1991, *Astron. J. (USA)*, 102, 1879

Wex, N. & Kopeikin, S. M., 1999, *Astrophys. J.*, 514, 388

Whipple, A. L., Jefferys, W. H., Benedict, *et al.* 1996, *BAAS*, 28, 1187

A Giant Step: from Milli- to Micro-arcsecond Astrometry
Proceedings IAU Symposium No. 248, 2007
W. J. Jin, I. Platais & M. A. C. Perryman, eds.
© 2008 International Astronomical Union
doi:10.1017/S1743921308019649

The restoration of the quadrupole light bending

M. T. Crosta, D. Gardiol, M. G. Lattanzi and R. Morbidelli

INAF-OATo, Strada Osservatorio 20, Pino Torinese, Torino, Italy, email:crosta@oato.inaf.it

Abstract. The ESA astrometric mission Gaia will be able to put to test General Relativity thanks to differential astrometric measurements. The differential experiment, GAREX, implemented in the form of repeated Eddington-like measurement, aims at measuring the quadrupole light bending due to an oblate planet by comparing the evolution of relative distances in stellar fields in the vicinity of it. Simulations which utilize (i) selected fields extracted from the GSCII data base, (ii) a realistic error model as function of the star's magnitude and distance from Jupiter's edge, show the real best scenarios and how to improve the Gaia ability to detect this relativistic effect.

Keywords. astrometry, relativity, solar system: general, catalogs

1. The GAia Relativistic EXperiment

The light deflection produced by an oblate planet on grazing photons (hereafter named the *q-effect*) was simulated for a Gaia-like mission for the first time in Crosta and Mignard (2006). This investigation was the first step of a wider project called GAia Relativistic EXperiment (GAREX), which aims at testing General Relativity (GR) with highly accurate astrometric differential measurements. The present study is more complicated than the case discussed in paper above (GAREX-I), based on a very crude Galaxy model, and taken as an "ideal experiment". This tells us to what extent the q-effect is measurable once we deal with a more realistic observational scenario. The q-effect has been parameterized by introducing a new parameter ϵ, equal to one if GR predictions are true. This secondary deflection has a very specific pattern as a function of (i) the position of the star with respect to the oblate deflector and (ii) the orientation of its spin axis. After a set of Monte-Carlo runs, the results of GAREX-I showed that the q-effect is detectable within a 3-σ confidence level, but a single experiment on specific bright stars around Jupiter can do almost as well as a 5-year mission. Therefore, we designed a new GAREX experiment, GAREX-II, focused on single epochs of observations generated by using the GSCII data base, all along the mission life time and including realistic satellite observations of Jupiter. From the generated epochs we have selected the favorable ones, taking into account the astrometric accuracies deduced from the current error budget. The F-band magnitudes were chosen (Johnson R), since they are close to the Gaia's G-band magnitudes. Among the 8552 stellar fields investigated, we rejected those where the axis of Jupiter aligns approximately with the line of sight and we kept those with σ_ϵ below 0.5 in order to investigate more cases close to the reference value of GAREX-I.

GAREX-II has two major improvements. (i) The new sets of simulation utilize *real* star counts (ten thousand contiguous fields, 3-4 observations per day) from 2011 to 2020 using Jupiter ephemeris as observed from the Lissajous orbit of Gaia, in areas of about 0.5×0.5 square deg (approximately the size of the field of view of Gaia) and centered around the epoch equatorial coordinates of Jupiter. This choice assures a sufficiently fine sampling for searching the best candidate scenarios and, consequently, to place requirements on the initial phase of the scanning law in order to observe the selected fields as close as possible to the optimal epoch.

(ii) The model of astrometric errors as a function of magnitude is obtained from a simulation, taking into account the most relevant noise sources. However, the most important effect for bright magnitudes is CCD saturation. At magnitude 12 (13 in the selected Gaia configuration) the PSF starts to become saturated. For this reason part of the signal is lost, and the astrometric error increases with respect to the un-saturated case. Monte-Carlo simulations show that using an appropriate centroiding algorithm it is still possible to achieve good performances on partially saturated images. In this case the astrometric error can be described, as function of magnitude, by the following approximated formula $\sigma = 10^{g(F)}$. The function $g(F)$ will be published in a forthcoming paper (Crosta *et al.*). For a complete transit we have 9 independent measurements and the final error is divided by 3 times the square root of the mean number of observations per star. The background noise appears to be properly accounted by the expression *log f (F,r)*, where f is a complex function of the F magnitude and the angular distance from Jupiter's limb (r). The exact values provided by *log f* depend on several assumptions on the actual Gaia (stray-light, astrometric algorithm, and measurement process).

2. Results and open questions

GAREX-II proves the importance of having a real sky, as realized by the GSCII, to define the best epochs for the experiment. In fact, it appears that the best way to detect the quadrupole light bending effect is to choose optimal configuration during the mission operational life, confirming the statistical results already obtained in Crosta and Mignard: it is sufficient to select background fields which include a few bright stars close to Jupiter to produce the best results. If we include the background noise in the error model, the experiment is still possible, but not closer than about one Jupiter's radius.

Compared to the case without background, several good fields will be lost. For example, in 2011 the number of good cases decreases to four from an initial list of eleven. The experiment can be performed very early in the mission's lifetime and then repeated in 2012, 2013, and 2014. It means that *detecting the quadrupole light deflection of Jupiter as predicted by GR could be the first remarkable scientific result well before the end of the mission!*

Future work will take into account a further improved description of the observing scenario by including the details of the instrumental/technical effects (e.g., how to compare the two observations with/without Jupiter), and those associated with the stellar fields because of proper motions. The final task will be to apply the complete relativistic model, which includes all relevant relativistic effects at the level of Gaia's accuracy.

Last, but not least, the results of the experiment will contribute to checking the performance of the mission during its operational life time. The observation simulator assumes that the instrument is performing according to specifications. Degradation will be a big factor, therefore the good fields should be taken early into the mission.

References

Crosta, M. T. & Mignard, F., 2006,*Class. Quantum Grav.* 23,4853-4871
Crosta, M. T., Gardiol, D., Lattanzi, M. G., & Morbidelli, R. *The restoration of the quadrupole light bending by Gaia observations*, to be submitted

A Giant Step: from Milli- to Micro-arcsecond Astrometry
Proceedings IAU Symposium No. 248, 2007
W. J. Jin, I. Platais & M. A. C. Perryman, eds.

The RAMOD astrometric observable and the relativistic astrometric catalogs

M. T. Crosta[1], B. Bucciarelli[1], F. de Felice[2], M. G. Lattanzi[1] and A. Vecchiato[1]

[1]INAF-OATo, Strada Osservatorio 20, Pino Torinese, Torino, Italy
[2]University "G.Galileo", Dept. of Physics, via Marzolo 8, Padova, Italy

Abstract. We describe a way to compare current relativistic astrometric models accurate to the micro-arcsecond level. The observed stellar direction can be written as a function of several parts, linking the astrometric observables to the relativistic effects associated to the stellar kinematical properties and distances as seen inside the gravitational field of our Solar System, i.e. the so called *relativistic astrometric parameters*, providing a tool for comparing the RAMOD framework to the pM/pN approaches.

Keywords. astrometry, general relativity, solar system: general, catalogs

1. Relativistic astrometric models

It has been shown (Kopeikin and Mashhoon 2002; Klioner 2003; de Felice *et al.* 2006 and references therein) that, at the micro-arcsecond level of accuracy, astrometry needs General Relativity (GR) to trace back the stellar position and take into account all the effects due to the gravitating bodies and their tidal stresses on the background geometry at the same level of accuracy. There are three main approaches to deal with the problem of the relativistic sphere reconstruction. But ways to implement the current astrometric models in the context of Gaia mission and optimal strategies for their detailed comparisons should still be devised. This contribution goes toward this effort. (i) As a Parameterized Post Newtonian (PPN) extension of a seminal study conducted in the framework of post-Newtonian (pN) approximation of GR (Klioner & Kopeikin, 1992), Klioner (2003) produced a PPN formulation of a model for relativistic astrometry, accurate to 1 μas, in which the finite distance and the angular momentum of the gravitational deflector are included, linked to the motion of the observer and the source in order to consider the effects of parallax, aberration, and proper motion. This model is considered the baseline for the Gaia data reduction (GREM). (ii) Kopeikin & Schäfer (1999), using the pM approximation, solved the metric tensor in retarded Lienard-Weichert potentials and later Kopeikin & Mashhoon (2002) included all relativistic effects related to the gravitomagnetic field, produced by the translational velocity/spin-dependent metric terms. Both studies rewrite the null geodesic as a function of two independent parameters and solve the light trajectory as a straight line (Euclidean geometry) plus corrections in the form of integrals, containing the perturbations encountered, from a source at an arbitrary distance to an observer located within the Solar System. (iii) The same parameterization can be obtained in RAMOD3 (de Felice *et al.*, 2004), where it is always possible to map the null geodesic onto hypersurfaces of simultaneity with the epoch of observation (Crosta, 2003). RAMOD is a well-established framework of general relativistic astrometric models which can be extended to any accuracy and physical requirements. It tries to be as close as possible to the concept of curved geometry of GR in order to reconstruct all *simultaneous observations* in a curved spacetime (de Felice & Clarke 1990). Recent developments have included the retarded distance effects, due to the moving bodies of solar system, in the process of evaluation of astrometric observable. The major difficulty is to analytically integrate a set of non linear coupled differential equations which allow to trace back the star positions and which include, a priori by definition, the background property of a curved space-time. At present, we can integrate the equations of the model numerically. A semi-analytical solution has been discussed in de Felice & Preti (2006). Indeed, whether analytical or numerical, the solution of those equations contains "globally" all relativistic perturbations to a photon moving along its trajectory. A boundary conditions fixed by the astrometric observables as a function of analytical relativistic description of the satellite solves the Cauchy problem and allows a unique prediction for the stellar location (Bini *et al.* 2003). From the theoretical point of view, RAMOD is complete and ready to be implemented in the end-to-end simulation of a Gaia mission, aimed to estimate the astrometric parameters of celestial objects from a well-defined set of measured quantities.

2. The pM/RAMOD comparison model

A first comparison between different approaches can be done by using the GREM definition of the observed stellar direction together with the light deflection terms computed in the pM approach. In the latter the hyperbolic character of field equations is preserved and the positions of the bodies are a priori functions of retarded time as in RAMOD. More generally, the proper stellar direction can be expressed "globally" as a sum of a set of *relativistic astrometric parameters* (RAPs) for the Gaia-like catalogue and can be directly compared with the general expression that will be soon derived by the RAMOD-like models. The astrometric parameters depend on which part of the Galaxy Gaia will be observing, as they simultaneously link all possible "astrometric" relativistic effects related to the light propagation. A complete set of RAPs have already been computed for stars in the solar vicinity (Crosta, 2003), i.e. the expressions for the *aberration, barycentric direction, deflection, parallaxes and, proper motion* parameters accurate up to a 0.1 μas ($\sim (v_{planet}/c)^3$). Part of the relativistic effects could be induced by the adopted approximation scheme and relativistic coordinate transformations which utilize several expansions with respect to a small parameter $\epsilon \sim (v_{planet}/c)$. Future work will include the analysis of each term's significance in order to investigate whether all these terms are physically related to the stellar kinematical properties. Term by term comparisons using the final solution of the covariant approach of RAMOD, as well as the GREM formalism developed in the Gaia context, should provide a definite answer.

References

Bini, D., Crosta, M. T., De Felice, F., 2003,*Class.Quantum Grav.*, 20, 4695-4706
Crosta, M. T., 2003, *Methods of Relativistic Astrometry for the analysis of astrometric data in the gravitational field of the Solar System*, CISAS-University of Padova
de Felice & Clarke, 1990, *Relativity on curved Manifolds*, Cambridge University press
de Felice, F., Crosta, M. T., Vecchiato, A., Bucciarelli, B., Lattanzi, M. G., 2004, *ApJ*, 607(1), 580-595
de Felice, F., Vecchiato, A., Crosta, M. T., Bucciarelli, B., Lattanzi, M. G., 2006, *ApJ*, 653, 1552-1565
de Felice, F., & Preti, G., 2006, *Class. Quantum Grav.*,23, 5467-5476
Klioner,, S.,& Kopejkin, S. M.,1992, *AJ*, 104 (2), 897
Klioner, S., 2003, *AJ*, 125, 1580
Kopeikin, S., & Schäfer,G., 1999, *Phy.Rev.D*, 60, 124002
Kopeikin, S. M., & Mashhoon, B., 2002, *Phy.Rev.D*, 65, 064025

A Giant Step: from Milli- to Micro-arcsecond Astrometry
Proceedings IAU Symposium No. 248, 2007
W. J. Jin, I. Platais & M. A. C. Perryman, eds.
© 2008 International Astronomical Union
doi:10.1017/S1743921308019662

Preliminary investigation of the gravitomagnetic effects on the lunar orbit

X. M. Deng

Purple Mountain Observatory, Nanjing 210008, China
email: dengxuemei403@163.com

Abstract. Gravitomagnetic effects are studied in this paper. Starting from the metric in the BCRS and then using the matching method, the metric in the GCRS are derived. Furthermore, we give some estimation of the order of the gravitomagnetic effects on the lunar orbit.

Keywords. reference systems, Moon

1. Introduction

The magnetic field is produced by the motion of electric-charge. The close analogy between the Einstein's field equation and the Maxwell's equation led many authors to investigate whether or not the mass current would produce what is so-called the gravitomagnetic field. But, first of all, we must clarify the definition of the gravitomagnetic field. Generally, one of the Maxwell's equation has the following form

$$\nabla \times \vec{B} = \vec{J} + \frac{\partial \vec{E}}{\partial t} \tag{1.1}$$

where not only the electric-charge but also the variation of $\vec{E}$ with respect to time can produce the magnetic field. So, not only the mass current (the spins of celestial bodies) but also the potentials's change with time such as the third body's effect can produce the gravitomagnetic field. Recently, Murphy *et al.* (2007a, 2007b) and Kopeikin (2007) have a go-round about whether the Lunar Lase Ranging (LLR) is currently capable to detect the gravitomagnetic effect. We will give some results of our work about this problem.

2. Our works

In our model, the Sun and the Earth are massive point-particles. The Moon is considered as a massless test particle. Based on the International Astronomical Union (IAU) resolution for the relativistic reference systems in 2000, we employ the barycentric celestial reference systems (BCRS) and the geocentric celestial reference systems (GCRS) with two PPN parameters β and γ and the harmonic gauge (Klioner and Soffel, 2000). The metric respectively in the BCRS and GCRS is

$$
\begin{aligned}
g_{00}(t, \vec{x}) &= -1 + 2\epsilon^2 w(t, \vec{x}) - 2\beta\epsilon^4 w^2(t, \vec{x}) + \mathcal{O}(\epsilon^5), \\
g_{0i}(t, \vec{x}) &= -2(1+\gamma)\epsilon^3 w^i(t, \vec{x}) + \mathcal{O}(\epsilon^5), \\
g_{ij}(t, \vec{x}) &= \delta_{ij}(1 + 2\epsilon^2\gamma w(t, \vec{x})) + \mathcal{O}(\epsilon^4),
\end{aligned}
\tag{2.1}
$$

and

$$
\begin{aligned}
G_{00} &= -1 + 2\epsilon^2 W(T, \vec{X}) - 2\epsilon^4\beta W^2(T, \vec{X}) + \mathcal{O}(\epsilon^5), \\
G_{0a} &= -2\epsilon^3(1+\gamma)W^a(T, \vec{X}) + \mathcal{O}(\epsilon^5),
\end{aligned}
$$

$$G_{ab} = \delta_{ab}(1 + 2\epsilon^2 \gamma W(T, \vec{X})) + \mathcal{O}(\epsilon^4), \tag{2.2}$$

where $\epsilon = 1/c$. Procedures we adopt are as the following:

(a) Derive the potentials w and w^i in $g_{\mu\nu}$ under the BCRS;

(b) Match the potentials between the BCRS and the GCRS and obtain W and W^a in the GCRS;

(c) Derive the equation of motion for the moon in the GCRS by using the geodesic principle;

(d) Derive the ranging time from the light equation;

(e) Compute the observable quantity – proper laser ranging time.

Among them, (d) and (e) are in workings.

3. Discussion

Based on our current works, the gravitomagnetic acceleration produced by the Earth's spin is

$$\vec{a}_{GMES} = (1 + \gamma) G\left[\frac{\vec{V} \times \vec{J}_\oplus}{r^3} + 3\frac{\vec{r} \times \vec{V}}{r^5}(\vec{r} \cdot \vec{J}_\oplus)\right] \sim 10^{-16}\vec{a}_{New}, \tag{3.1}$$

the gravitomagnetic acceleration caused by the Sun's angular momentum is

$$\vec{a}_{GMSS} \sim -2(1 + \gamma)\frac{G}{R^3}(\vec{V} \times \vec{J}_\odot) \sim 10^{-14}\vec{a}_{New}, \tag{3.2}$$

and the gravitomagnetic acceleration due to the third body's effect from the Sun is

$$\vec{a}_{GMSP} \sim (1 + \gamma)\left[2\frac{Gm_\odot}{R^3}v^2\vec{r} - 3G\frac{\vec{r} \times \vec{J}_\odot}{R^5}(\vec{R} \cdot \vec{v})\right] \sim 10^{-11}\vec{a}_{New}, \tag{3.3}$$

where the Newtonian term is

$$\vec{a}_{New} = -\frac{Gm_\oplus}{r^3}\vec{r}. \tag{3.4}$$

and v is the Earth's velocity around the Sun; the Moon's velocity around the Earth is V; the mass of the Sun is $m_\odot$; the mass of the Earth is $m_\oplus$; the distance between the Sun and the Earth is R; the distance between the Earth and the Moon is r; $J_\oplus$ and $J_\odot$ are respectively the spins of the Earth and the Sun.

Although $\vec{a}_{GMSP}$ has some effect on the Earth-Moon distance on the order of a millimeter, which could be measured by the next generation LLR, it is a coordinate-dependent quantity instead of an observable one. To remove the residual gauge freedom, we will calculate the proper time for the LLR in the next move.

Acknowledgements

This work is supported by the National Natural Science Foundation of China under Grant No. 10475114. I appreciate very much Prof. Kopeikin's clarification about the definition of gravitomagnetic effect.

References

Klioner, S. A. & Soffel, M. H., 2000, *Phys. Rev. D* 62, 024019
Kopeikin, S. M., 2007, *Phys. Rev. Lett.* 98, 229001
Murphy, T. W., Nordtvedt, K., & Turyshev, S. G. 2007, *Phys. Rev. Lett.* 98, 071102
Murphy, T. W., Nordtvedt, K., & Turyshev, S. G. 2007, *Phys. Rev. Lett.* 98, 229002
Soffel, M., *et al.* 2002 *A & A* 126, 2687

A Giant Step: from Milli- to Micro-arcsecond Astrometry
Proceedings IAU Symposium No. 248, 2007
W. J. Jin, I. Platais & M. A. C. Perryman, eds.
© 2008 International Astronomical Union
doi:10.1017/S1743921308019674

Light deflection in the second post-Newtonian approximation of scalar-tensor theory of gravity

P. Dong[1,2] and W. T. Ni[1]

[1] Center for Gravitation and Cosmology, Purple Mountain Observatory,
Chinese Academy of Sciences, Nanjing, 210008, China
emails: `dongpeng@pmo.ac.cn`, `wtni@pmo.ac.cn`

[2] Graduate University of the Chinese Academy of Sciences, Beijing, 100049, China

Abstract. In this paper, we use the metric coefficients and the equation of motion in the 2nd post-Newtonian approximation in scalar-tensor theory including intermediate range gravity to derive the deflection of light and compare it with previous works. These results will be useful for precision astrometry missions like Gaia, SIM, and LATOR (Laser Astrometric Test Of Relativity) which aim at astrometry with micro-arcsecond and nano-arcsecond accuracies and a need for the 2nd post-Newtonian framework and ephemeris to determine the stellar and spacecraft positions.

Keywords. astrometry, reference systems, time

1. Introduction

The relativistic light deflection passing near the solar rim is 1.75 as (arcsec). The first post-Newtonian approximation is valid to 10^{-6} and the second post-Newtonian is valid to 10^{-12} of relativistic effects such as light deflection in the solar system. For astrometry mission to measure angles with an accuracy in the range of nano-as to μas, the 2nd post-Newtonian approximation of relevant theories of gravity is required for both the angular measurement and relativistic gravity tests. The scalar-tensor theory is widely discussed and used in tests of relativistic gravity. In order to confront the predictions of scalar-tensor theory with experiment in the solar system, it is necessary to compute its second post-Newtonian approximation and certain gravitational effects such as deflection of light, time delay of light and perihelion shift. The 2nd post-Newtonian contribution for light ray has been discussed by and Epstein & Shapiro (1980), Richter & Matzner (1982) and by others. In this paper, we use the metric coefficients we obtained earlier (Xie *et al.* 2007) to compute the deflection in the second post-Newtonian approximation considering the velocity of the observer (spacecraft).

2. Metric coefficients

The calculation of light deflection to 2PN approximation requires knowledge of terms in the metric to order $(v/c)^4$. For the scalar-tensor theory, the metric coefficients are

$$g_{00} = 1 - 2U + 2(1 + \bar{\beta})U^2, \quad g_{ij} = -\delta_{ij}[1 + 2(1 + \bar{\gamma})U + \frac{3}{2}(1 + \Lambda)U^2] \quad (2.1)$$

in the global coordinates for the static case. Note that U is given by

$$U = \int \frac{\rho(\vec{x}', t)}{|\vec{x} - \vec{x}'|} d^3 x' - \frac{\bar{\gamma}\xi_1}{4\pi} \int \frac{d^3 x'}{|\vec{x} - \vec{x}'|} \int \frac{\rho(\vec{x}'', t)}{|\vec{x}' - \vec{x}''|} \exp[-\xi_1(\vec{x}' - \vec{x}'')] d^3 x'' \quad (2.2)$$

and the parameters $\bar{\gamma}$, $\bar{\beta}$, ξ_1 and Λ are given in Xie *et al.* (2007) by

$$\bar{\gamma} = -\frac{1}{\omega_0 + 2}, \bar{\beta} = \frac{\omega_1}{(2\omega_0 + 3)(2\omega_0 + 4)^2}, \Lambda = \frac{15}{6}\bar{\gamma} + \frac{4}{3}(\bar{\gamma}^2 + \bar{\beta}), \xi_1 = -4\frac{\bar{\gamma}\lambda_2\phi_0^2}{2 + \bar{\gamma}} \quad (2.3)$$

3. Deflection angle

The basic equations of a light ray trajectory are

$$g_{\mu\nu}k^\mu k^\nu = 0, \ dk^\mu/d\lambda + \Gamma^\mu_{\rho\sigma}k^\rho k^\sigma = 0 \quad (3.1)$$

where $k^\mu \equiv dx^\mu/d\lambda$ and λ is an affine parameter. Consider a light signal emitted at $(\vec{x}_0, t_0)$ in the initial direction described by a unit vector $\hat{n}$ satisfying $\hat{n} \cdot \hat{n} = 1$ and let it have the form

$$\vec{x}(t) = \vec{x}_0 + \hat{n}(t - t_0) + \vec{x}_p(t) + \vec{x}_{pp}(t) \quad (3.2)$$

where $\vec{x}_p(t)$ and $\vec{x}_{pp}(t)$ are the first and second post-Newtonian correction respectively. We obtain the solution needed for the second-order approximation by iterative methods.

Consider an observer (satellite, spacecraft) with the four-velocity u^μ who receives the signals from two different sources. The angle α between the directions of two incoming photons is given by the following expression:

$$\cos\alpha = k_I^\mu P_\mu^\rho k_{II}^\sigma P_{\rho\sigma} |k_I^\mu P_\mu^\rho|^{-1} |k_{II}^\sigma P_{\rho\sigma}|^{-1} = f(g_{\mu\nu}, u^\sigma, k_I^\rho, k_{II}^\tau), \quad (3.3)$$

where $P_{\mu\nu} = g_{\mu\nu} + u_\mu u_\nu$ is a projection operator. Defining the angle $\delta\alpha$ to be the deflected angle from the original angle α_0, and expanding $\cos\alpha$ around α_0 to the second order, we have

$$\delta\alpha_p = \cot\alpha_0 - f\csc\alpha_0, \delta\alpha_{pp} = \cot\alpha_0 - f\csc\alpha_0 - \frac{1}{2}(\delta\alpha_p)^2\cot\alpha_0 - \delta\alpha_p \quad (3.4)$$

where $\delta\alpha_p$ and $\delta\alpha_{pp}$ are the deflection angles for the first and second post-Newtonian approximations. After a straightforward but lengthy calculation, we have

$$\delta\alpha = 2(2 + \bar{\gamma})(M/R) + (2 + \bar{\gamma})\bar{\gamma}\xi_1(M/2\pi R) - (2 + \bar{\gamma})(MR/2r_{os}^2)$$
$$+ [(30 + 31\bar{\gamma} + 8\bar{\gamma}^2)\pi - 16(2 + \bar{\gamma})^2](M^2/8R^2) + 2(2 + \bar{\gamma})J_2(M/R), \quad (3.5)$$

for light passing the solar limb in the equatorial plane from outside the solar system. Here M is the solar mass, R is the solar radius and J_2 is the solar quadrupole moment parameter. The second term comes from the intermediate-range force and other terms agree with the former works.

The 2PN light trajectory obtained here is useful for obtaining 2PN range of deep space laser ranging missions ASTROD I and ASTROD (Ni 2004). A detailed paper on this topic will be presented in the future.

Acknowledgement

We thank the National Natural Science Foundation (Grant Nos 10475114 and 10778710) and the Foundation of Minor Planets of purple Mountain Observatory for support.

References

Epstein, R. & Shapiro, I. I. 1980, *Phys. Rev.*, D22, 2947; and references therein
Richter, G. W. & Matzner, R. A. 1982, *Phys. Rev.*, D26, 1219; and references therein
Xie, Y., *et al.* 2007, *Adv. Sp. Res.*, doi: 10.1016/j.asr.2007.09.022, arXiv:0704.2991v2
Ni, W. T., *et al.* 2004, *Class. Quantum Grav.*, 21, S641

A Giant Step: from Milli- to Micro-arcsecond Astrometry
Proceedings IAU Symposium No. 248, 2007
W. J. Jin, I. Platais & M. A. C. Perryman, eds.
© 2008 International Astronomical Union
doi:10.1017/S1743921308019686

Direct contribution of the surface layers to the Earth's dynamical flattening

Y. Liu[1,2] and C. L. Huang[1]

[1]Shanghai Astronomical Observatory. China
80 Nandan Road, Shanghai 200030, China
email: yuliu@shao.ac.cn

[2]Graduate University of Chinese Academy of Science

Abstract. The global dynamic flattening (H) is an important quantity in research of rotating Earth. Precession observations give $H_{obs} = 0.0032737 \approx 1/305.5$. We recalculate the geometrical flattening profile of the Earth interior from potential theory in hydrostatic equilibrium. Results coincide with that of Denis (1989). We derive expression for H to the third-order accuracy and obtain $H_{PREM} = 1/308.5$. This matches similar studies, in which there is a difference about 1% between this and the observed value. In order to understand where this difference comes from, we replace the homogenous outermost crust and oceanic layers in PREM with some real surface layers data, such as oceanic layer (ECCO) , topography data (GTOPO30), crust data (CRUST2.0) and mixed data (ETOPO5). Our results deviate from the observed value more than H_{PREM}. These results verify the isostasy theory indirectly and may imply that the "positive" effects from such as mantle circulation associated with the density anomalies maybe larger than thought before.

Keywords. reference systems, standards, magnetic fields, Earth, methods: analytical

1. Introduction

The global dynamic flattening (H) is an important quantity in research of rotating Earth. It can be related with the luni-solar precession, nutation(18.6 yr terms), tilt-over mode and so on. From observations, the value of H is approximately 1/305.5 (Dehant, 1997). But the H of PREM model (Dziewonski, 1981) is about 1/308.8 at the first order accuracy. There is about 1% difference between them, which is an interesting question we discuss in this paper.

We recalculate the geometrical flattening profile of the Earth interior from potential theory at 3rd accuracy. The flattening profile matches that of Denis (1989). We obtain the dynamic flattening at the 3rd accuracy by using the geometrical flattening profile. Comparing our results with the others, we find that adopting a high order accuracy can reduce the difference between H_{PREM} and H_{obs} , but the effect is too small to remove the difference.

At the same time, we note that the oceans and lands are distributed uniformly in the PREM model. The real oceans and lands are distributed nonuniformly obviously. What effect the real oceans and land make?

2. Dynamic flattening of modified Earth model with real surface layers (ocean and topography)

According to the depth range of each real surface layers model, we construct three Earth models with real surface data form PREM, ECCO, GTOPO30 as follows: (a)

PREM without 5.615 km depth surface layer(PREM_NS_5) + real ocean data(ECCO) + real topography data (GTOPO30); (b) PREM without 10.376 km depth surface layer(PREM_NS_10) + real ocean and topography data (ETOPO5); (c) PREM without 70.137 km depth surface layer(PREM_NS_70) + real crustal data (CRUST2). We assume simply that the shapes and so on of PREM_NS_5 and PREM_NS_10 in our model are as the same as in PREM. The following table shows all the results.

Model	A $10^{37}\ kg \cdot m^2$	B $10^{37}\ kg \cdot m^2$	C $10^{37}\ kg \cdot m^2$	1/H
PREM(Full)	8.0115651	8.0115651	8.0376170	308.52
PREM_NS_5	7.9966287		8.0226249	
ECCO(KF049f)	0.0132681	0.0128369	0.0125290	
GTOPO30	0.0008458	0.0006980	0.0005578	
PREM+E+G	8.0107426	8.0101637	8.0357116	318.14
PREM_NS_10	7.9788034		8.0047331	
ETOPO5	0.0312185	0.0305047	0.0300231	
PREM+ETOPO5	8.0100220	8.0093081	8.0347563	320.22
PREM_NS_70	7.7087284		7.7336553	
CRUST2	0.3008642	0.3008114	0.3017731	
PREM+CRUST2	8.0095926	8.0095399	8.0354284	310.70

3. Discussion

The mass of surface layer is less than 0.1% of the whole Earth. But the real surface layer can reduce the global dynamic flattening from 1/308.53 to 1/318.14 (about 3%). It is a large effect, because the surface layer is the outermost layer of the Earth.

H deviates more with the depth of surface layer displaced by real data being deeper, untill somewhere deeper than 10.376 km under the mean sea level. The value of H of the deepest model C is enlarged and deviates less than the above two models. The isostasy theory can explain why this is happening and our results provide an indirect evidence for the theory of isostasy.

The effects from nonuniform distribution of oceans and lands are "negative". They make H deviate from the observed value even further. Although the theory of isostasy can explain the difference among the three models, there is still a difference between H and the observed value. This may imply that, in order to balance the "negative" effect from real oceans and lands, the "positive" effectssuch as the mantle circulation associated with the density anomalies maybe larger than what was thought before.

Acknowledgements

This work is supported by NSFC (10773025/10633030) and Science & Technology Commission of Shanghai Municipality (06DZ22101/06ZR14165).

References

Dehant, V. & Capitaine, N. 1997, *CeMDA*, 65, 439
Denis, C. 1989, *Physics and Evolution of the Earth's Interior*, 4, 111
Dziewonski, A. M. & Anderson, D. L. 1981, *PEPI*, 25, 297

A Giant Step: from Milli- to Micro-arcsecond Astrometry
Proceedings IAU Symposium No. 248, 2007
W. J. Jin, I. Platais & M. A. C. Perryman, eds.

© 2008 International Astronomical Union
doi:10.1017/S1743921308019698

Second post-Newtonian approximation of light propagation in Einstein-Aether theory†

Y. Xie[1]and T. Y. Huang[2]

Astronomy Department, Nanjing University,
Nanjing 210093, 22 Hankou Road, Nanjing, China
[1]email: yixie@nju.edu.cn [2]email: tyhuang@nju.edu.cn

Abstract. Future deep space laser ranging missions together with astrometry missions will be able to test relativistic gravity to an unprecedented level of accuracy and will require second post-Newtonian approximation of relevant theories of gravity. Einstein-aether theory is adopted as the theory of gravity and the second post-Newtonian approximation of light propagation is studied.

Keywords. astrometry, reference systems

1. Introduction

Future deep space laser ranging missions and astrometry missions will be able to test relativistic gravity to an unprecedented level of accuracy. More precisely, these missions will enable us to test relativistic gravity to $10^{-6} - 10^{-9}$ and will require second post-Newtonian (2PN) approximation of relevant theories of gravity.

Recently, Einstein-Aether theory (ae-theory), which is a kind of vector-tensor theory with a unit time-like vector, is widely discussed in its violation of Lorentz symmetry and broadly applied to astrophysics (Jacobson & Mattingly 2001, 2004, Eling & Jacobson 2006, Garfinkle *et al.* 2007, etc). Furthermore, a vector-tensor theory can show a more comprehensive picture of 2PN approximation than the scalar-tensor theories (Xie *et al.* 2007) and provides new information for the future parametrized 2PN metric.

In this paper, we adopt ae-theory as the gravitational theory and derive its 2PN approximation and equations of light propagation in a N point masses model as a simplified treatment for a real solar system in barycentric reference system.

2. 2PN Approximation of Ae-theory

In a general tensor-vector theory of gravity, the Lagrangian scalar density involves the metric $g_{\mu\nu}$ and a 4-vector field K_{μ}. The action defining the theory reads

$$S = \frac{c^3}{16\pi G} \int \left[R - c_1 K^{\mu;\nu} K_{\mu;\nu} - c_2 K^{\mu}_{;\mu} K^{\nu}_{;\nu} - c_3 (K^2) K^{\mu;\nu} K_{\nu;\mu} \right.$$

$$\left. + c_4 (K^2) K^{\lambda} K^{\kappa}_{;\lambda} K^{\rho} K_{\kappa;\rho} + \lambda (K^2 + 1) \right] \sqrt{-g} \mathrm{d}^4 x + S_m (\psi, g_{\mu\nu}), \quad (2.1)$$

where $g = \det(g_{\mu\nu}) < 0$ is the determinant of the metric tensor $g_{\mu\nu}$, R is the Ricci scalar, ψ denotes all the matter fields, $K^2 \equiv K^{\lambda} K_{\lambda}$, $-\delta^0_{\mu}$, which is Kronecker δ, is the asymptotic value of K_{μ} and c_i ($i = 1, 2, 3, 4$) are constants. The Lagrange multiplier λ constrains the vector field K^2 to be -1. Here, we respect the Einstein equivalence principle so that the

† This work was funded by the Natural Science Foundation of China.

matter fields ψ do not interact with the vector field, i.e. the action of matter, S_m, is the function of ψ and $g_{\mu\nu}$ only.

In PN approximation, we consider an asymptotically flat spacetime, whose metric $g_{\mu\nu}$ has the form to second order as

$$g_{00} = -1 + \epsilon^2 N + \epsilon^4 L + \epsilon^6 Q + \mathcal{O}(\epsilon^8) \tag{2.2}$$

$$g_{0i} = \epsilon^3 L_i + \epsilon^5 Q_i + \mathcal{O}(\epsilon^7), \tag{2.3}$$

$$g_{ij} = \delta_{ij} + \epsilon^2 H_{ij} + \epsilon^4 Q_{ij} + \mathcal{O}(\epsilon^6), \tag{2.4}$$

and an asymptotically unit time-like vector field, which is

$$K_0 = -1 + \epsilon^2 \overset{(2)}{K_0} + \epsilon^4 \overset{(4)}{K_0} + \epsilon^6 \overset{(6)}{K_0} + \mathcal{O}(\epsilon^8), \tag{2.5}$$

$$K_i = \epsilon^3 \overset{(3)}{K_i} + \epsilon^5 \overset{(5)}{K_i} + \mathcal{O}(\epsilon^7), \tag{2.6}$$

where $\epsilon \equiv 1/c$. With the constraints of $K^2 = -1$ and the harmonic gauge

$$(\sqrt{-g}g^{\mu\nu})_{,\nu} = 0, \tag{2.7}$$

the 2PN metric coefficients for N point masses have been worked out following the procedures proposed by Chandrasekhar (1965). The details can be found in Xie & Huang (2007).

3. 2PN light propagation

Light equations are (Will, 1993)

$$g_{\mu\nu} \frac{dx^\mu}{dt} \frac{dx^\nu}{dt} = 0, \tag{3.1}$$

and

$$\frac{d^2 x^i}{dt^2} = \left(\frac{1}{c} \Gamma^0_{\mu\nu} \frac{dx^i}{dt} - \Gamma^i_{\mu\nu} \right) \frac{dx^\mu}{dt} \frac{dx^\nu}{dt}. \tag{3.2}$$

Assuming

$$\vec{x} = \vec{x}_0 + c\vec{n}(t - t_0) + \vec{x}_1 + \vec{x}_2, \tag{3.3}$$

where $\vec{x}_1$ and $\vec{x}_2$ are 1PN and 2PN corrections, We have obtained the relevant light equation up to the 2PN approximation. If all the planets are ignored, the deflection of light to second order is

$$\Delta\phi = \frac{4\mathcal{G}M_\odot}{c^2 d} + \left[\frac{(15 + c_{14})\pi}{4} - 8 \right] \frac{\mathcal{G}^2 M_\odot^2}{c^4 d^2}, \tag{3.4}$$

where d represents the coordinate radius at the point of closest approach of the ray, $\mathcal{G} = 2G/(2 - c_{14})$ and $c_{14} = c_1 + c_4$.

References

Chandrasekhar, S. 1965, *ApJ*, 142, 1488
Eling, C. & Jacoboson, T. 2006, *Class. Quant. Grav.*, 23, 5625; *Class. Quant. Grav.*, 23, 5643
Garfinkle, D., Eling, C., & Jacobson, T. 2007, *Phys. Rev. D*, 76, 024003
Jacoboson, T. & Mattingly, D. 2001, *Phys. Rev. D*, 64, 024028; 2004, *Phys. Rev. D*, 70, 024003
Will, C. 1993, *Theory and Experiment in Gravitational Physics*, Cambridge Univ. Press.
Xie, Y., Ni, W.-T., Dong, P., & Huang T.-Y. 2007, *Adv. Sp. Res.*, accepted, doi:10.1016/j.asr.2007.09.022
Xie, Y. & Huang T.-Y. 2007, in preparation.

A Giant Step: from Milli- to Micro-arcsecond Astrometry
Proceedings IAU Symposium No. 248, 2007
W. J. Jin, I. Platais & M. A. C. Perryman, eds.

JASMINE data analysis

Y. Yamada[1], N. Gouda[2], T. Yano[2], Y. Kobayashi[2], Y. Niwa[2,3], and
JASMINE Working group

[1]Department of Physics, Kyoto University,
Oiwake-cho Kita-Shirakawa Kyoto 606-8502 JAPAN
email:yamada@amesh.org

[2]National Astronomical Observatory of JAPAN,
National Institutes of Natural Sciences, Mitaka, Tokyo, 181-8588, JAPAN
email:naoteru.gouda@nao.ac.jp,yuki@merope.mtk.nao.ac.jp,t.yano@nao.ac.jp

[3]Graduate School of Human and Environment Study,
Kyoto University Kyoto 606-8501, JAPAN
email:kazin.niwa@nao.ac.jp

Abstract. Japan Astrometry Satellite Mission for Infrared Exploration (JASMINE) aims to construct a map of the Galactic bulge with a 10 μas accuracy. We use z-band CCD or K-band array detector to avoid dust absorption, and observe about 10×20 degrees area around the Galactic bulge region.

In this poster, we show the observation strategy, reduction scheme, and error budget. We also show the basic design of the software for the end-to-end simulation of JASMINE, named JASMINE Simulator.

Keywords. methods: data analysis

1. Introduction

JASMINE will observe bulge stars with a single telescope and create accurate map using the block-adjustment algorithm (see Eichhorn & Clary(1974) and Zacharias *et al.* (1992)). Each observation corresponds to a "Small Frame". By adjusting several thousand Small Frames from about 10-hrs of observations, we obtain a "Large Frame". By adjusting about 2000 Large Frames, we can get astrometric parameters of bulge stars with high accuracy. In the plate mapping method, exposure time can be adjusted accordingly for each FoV in order to minimize the errors within the limits of hardware and mission's life. JASMINE will observe about 1 million bulge stars with an accuracy of $\sigma/\pi < 0.1$.

2. Instrument and observation strategy overview

The JASMINE telescope is a three mirror system. The diameter of the primary mirror is 75 cm, and the focal length is 22.5m in z-band option or 13.5m in K-band option. Observations are performed within $20° \times 10°$ region using $0.7° \times 0.7°$ FoVs. Because the stellar density is very high, we can apply block adjustment method and the beam combiner which is used in HIPPARCOS and Gaia is not needed. Our satellite has single telescope without beam combiner and like the ordinary space telescope. The telescope has very long focal length because the PSF size should be larger than several pixels for centroiding. There are 4 plane mirrors to fold the long optical path.

The constructing CCD detector which is sensitive to z-band(0.9μm) is almost completed in collaboration of Subaru telescope and Japanese company (see Kamata *et al.* (2006)). Other candidate of detector is HgCdTe K-Band array detector.

In the galactic bulge, we can adjust each frames with very high accuracy from the simple statistical considerations because density of stars are very high as the block adjustment algorithms for astrometry. About 5000 stars are observed within each "Small frame", we may expect that the accuracy of determining plate parameters is about $30 \sim 50$ times better than that of each stellar observation in the statistical sense.

Diffraction limit of our telescope will be about 100 milli arc-sec. From the picture of FoV, relative distance between stars in the FoV are calculated about 1/200 diffraction limit accuracy. By adjusting 10^7 FoVs, we can solve the astrometric parameters of 10 μas accuracy.

In the process of spacial connection of each "Small Frame" to construct "Large Frame", we estimate both the location and size of each "Small Frames". Higher order mapping parameters such as distortion cannot be estimated within the limitation of aperture size. So, optics deformation should be less than that the distortion and other higher order mapping parameters changes are not essential for position determination. By using Laser Interferometer technique which is developed in the gravitational wave detection, deformations of instruments are monitored.

JASMINE mission areas are very close to the ecliptic, errors along the ecliptic longitude direction is important and may affect the precision of astrometric parameter estimation. By optimizing reduction processes and appropriate choice of coordinates, accumulating errors of ecliptic longitude direction by adjustment processes can be $0.1\sigma \sim 0.2\sigma$ in "Large frame" , where σ is the accuracy of centroiding in each star. So, it is very important to prevent or monitor the deformation of mirror and telescope frames.

It is considered that very long time (order of an year) calibration of scale is needed in the estimation of astrometric parameters. Because the calibration of the hardware in such a long time is impossible, HIPPARCOS satellite uses beam combiner. But it is needed that only linear and periodic with one year period variation may cause the degeneracy to astrometric parameters. So we only need to estimate above two modes in enough accuracy. Any other variation modes can be considered to be random errors and which may not be a bias in the estimation of astrometric parameters. For final calibration, we use QSOs. There are about 40 QSO candidates in our mission area, and 3 to 5 QSOs will be able to be observed by JASMINE.

3. Constructing end-to-end simulation tool

We are now constructing end-to-end simulation software. The basic design and implementation is completed. The software is coded by using event-driven architecture. The simulator is applied for developing and checking algorithms for attitude control in Nano-JASMINE satellite.

References

Eichhorn, H., & Clary, W. G. 1974, *MNRAS*, 166, 425

Kamata, Y., Miyazaki, S., Nakaya, H., Tsuru, T. G., Takagi, S., Tsunemi, H., Miyata, E., Muramatsu, M., Suzuki, H., & Miyaguchi, K. 2006, *High Energy, Optical, and Infrared Detectors for Astronomy II. Edited by Dorn, David A.; Holland, Andrew D.. Proceedings of the SPIE,* , 6276, 62761U.

Zacharias, N., de Vegt, C., Nicholson, W., & Penston, M. J. CPC2 - the Second Cape Photographic Catalog II. Conventional plate adjustment and catalog construction. 1992 *A&A* 254, 397

A Giant Step: from Milli- to Micro-arcsecond Astrometry
Proceedings IAU Symposium No. 248, 2007
W. J. Jin, I. Platais & M. A. C. Perryman, eds.

© 2008 International Astronomical Union
doi:10.1017/S1743921308019716

Preliminary result of the Earth's free oscillations by Galerkin method

M. Zhang[1], B. Seyed-Mahmoud[2], C. L. Huang[1]

[1]Shanghai Astronomical Observatory, Chinese Academy of Sciences,
80 Nandan Rd., Shanghai 200030, China. Email: jitai@shao.ac.cn
[2]Department of Physics, the University of Lethbridge,
University Drive, Lethbridge, Albert, Canada, T1k 3m4

Abstract. We use a Galerkin method to compute the eigenfunctions and eigenperiods of some of the Earths spheroidal and toroidal modes. The boundary conditions are treated using a Tau method. We show that for a realistic Earth model the difference between the computed and observed periods is less than 1.4%. We conclude that a Galerkin method may be an effective tool for the studies of the Earth's normal modes.

Keywords. methods: analytical, Earth

1. Method

Galerkin method is an efficient method to convert an operator problem to a discrete problem (Li 2006; Seyed-Mahmoud 1994). Consider

$$L[\chi(x)] + \phi(x) = 0 \text{ over the interval } a \leqslant x \leqslant b . \quad (1.1)$$

where L is a linear differential operator, and χ and ϕ are linear functions. Let $S = \{y_i(x)\}_{i=1}^{\infty}$ define the set of all linear independent functions. Any function $\chi(x)$ can then be written uniquely as a linear combination of $\chi(x) = \sum_{i=1}^{N} a_i y_i(x)$. Application of a Galerkin then makes the RHS of equation (1.1) as null as possible by requiring that

$$\int_{a}^{b} y_j(x) \left\{ L\left[\sum_{i=1}^{N} a_i y_i(x) \right] + \phi(x) \right\} dx = 0 \text{ for j=1,...,N} \quad (1.2)$$

This leads to a system of N equations in N unknowns (a_i) which we can solve uniquely.

As a test in this work, we consider a spherical non-rotating elastic isotropic (SNREI) earth model in hydrostatic equilibrium. In solid layers, linear controlling equations are:

$$\rho_0 \omega^2 \vec{u} + \rho_0 \nabla V_1 + \rho_0 \nabla(\vec{u} \cdot \vec{g_0}) - \rho_0 \vec{g_0}(\nabla \cdot \vec{u}) + \nabla \cdot \tilde{\Gamma} = 0 \quad (1.3)$$

and for the isotropic small oscillations of an inviscous liquid core:

$$\rho_0 w^2 \vec{u} - \nabla p_1 + \rho_0 \nabla V_1 + \rho_1 \vec{g_0} = 0 \quad (1.4)$$

$$\nabla^2 V_1 + 4\pi G \rho_1 = 0 \quad (1.5)$$

$$\rho_1 = -\nabla \cdot (\rho_0 \vec{u}) \quad (1.6)$$

$$p_1 = -\vec{u} \cdot \nabla P_0 + \alpha^2 \rho_1 + \alpha^2 \vec{u} \cdot \nabla \rho_0 \quad (1.7)$$

where, the displacement vector field $\vec{u}$ is expanded to spheroidal and toroidal fields.

$$\vec{u}(r,\theta,\phi) = \sum_{n,m} \{u_n^m(r)Y_n^m(\theta,\phi)\hat{r} + v_n^m(r)\nabla_1 Y_n^m(\theta,\phi) - w_n^m(r)\hat{r} \times \nabla_1 Y_n^m(\theta,\phi)\} \quad (1.8)$$

And the boundary conditions for $\overrightarrow{u}$, the stress field $\tilde{\Gamma}$, and the incremental potential V_1 and its gradiant are: $\hat{n} \cdot \tilde{\Gamma}$, $\hat{n} \cdot (\nabla V_1 - 4\pi G \rho_0 \overrightarrow{u})$, V_1 be continuous on all boundaries; $\overrightarrow{u}$ be continuous cross welded boundaries, and $\hat{n} \cdot \overrightarrow{u}$ be continuous across solid-fluid boundary. We use a Tau method to solve boundary conditions.

2. Results & conclusion

Figure 1 are some results of the eigenfunctions (displacements) of the simple Earth as a solid sphere:

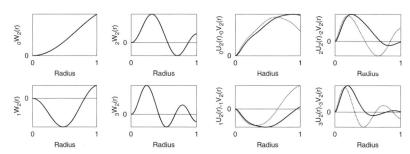

Figure 1. Eigenfunctions (u,v,w) of fundamental (n=0) and three overtones (n=1,2,3) of a solid sphere. The radius are normalized to be 1 at surface. In the right half 4 subplots, bold and dotted lines are for u(r) and v(r), respectively.

The eigen-periods are calculated for two earth models and listed in Table 1. The preliminary reference Earth model(PREM) is used in which the Earth is treated as inner solid core, fluid outer core and solid mantle. There is a maximum 4.1% or 1.4% difference if we treat the mantle(+crust) as 1 or 10 layers, respectively.

Table 1. Periods of toroidal and spheroidal modes in minutes, the observed values are from Lapwood & Usami (1981).

Modes		Toroidal			Spheroidal		
Legendre degree	Overtone	Observed	Calculated		Observed	Calculated	
			1-layer	10-layer		1-layer	10-Layer
2	0	44.01	45.53	43.63	53.89	51.84	53.43
	1	12.61	13.06	12.53	24.51	25.35	24.18
3	0	28.43	29.58	28.15	35.56	36.36	35.26
	1	11.59	11.90	11.48	17.68	18.06	17.54

Conclusion: As more realistic Earth model is considered, the numerical results for the periods of the Earth's acoustic modes converge to those of the observed ones. The remained 1.4% discrepancies between the numerical and observed periods may be reduced quickly by a more realistic Earth model (such as rotation or/and ellipticity). We conclude that the Galerkin method is a powerful tool for the studies of the Earth's normal modes.

Acknowledgements: This work is supported by NSFC (10773025/10633030) and Science & Technology Commission of Shanghai Municipality (06DZ22101/06ZR14165).

References

Seyed-Mahmoud, B. 1994, *Wobble/nutation of a rotating ellipsoidal Earth with liquid core*, M. Sc. thesis, Memorial University of Newfoundland.

Lapwood, E. R. & Usami, T. 1981, *Free oscillations of the Earth*, Cambridge University Press.

Li, R.-H. 2006, *Galerkin method of boundary problems*, Science Press.

A Giant Step: from Milli- to Micro-arcsecond Astrometry
Proceedings IAU Symposium No. 248, 2007
W. J. Jin, I. Platais & M. A. C. Perryman, eds.
© 2008 International Astronomical Union
doi:10.1017/S1743921308019728

Stars in the age of micro-arc-second astrometry

Y. Lebreton

GEPI, UMR 8111, Observatoire de Paris-Meudon, 92195 Meudon, France
email: Yveline.Lebreton obspm.fr

Abstract. The understanding and modelling of the structure and evolution of stars is based on statistical physics as well as on hydrodynamics. Today, a precise identification and proper description of the physical processes at work in stellar interiors are still lacking (one key point being that of transport processes) while comparison of real stars to model predictions, which implies conversions from the theoretical space to the observational one, suffers from uncertainties in model atmospheres. This results in uncertainties on the prediction of stellar properties needed for galactic studies or cosmology (as stellar ages and masses). In the next decade, progress is expected from the theoretical, experimental and observational sides. I illustrate some of the problems we are facing when modelling stars and possible ways toward their solutions. I discuss how future observational ground-based or spatial programs (in particular those dedicated to micro-arc-second astrometry, asteroseismology and interferometry) will provide precise determinations of the stellar parameters and contribute to a better knowledge of stellar interiors and atmospheres in a wide range of stellar masses, chemical composition and evolution stages.

Keywords. stars: interiors, stars: evolution, stars: fundamental parameters, stars: oscillations

1. Stellar internal structure and evolution studies: goals and tools

Major goals of stellar structure and evolution studies are (i) to characterize and describe the physics of matter in the extreme conditions encountered in stars, and (ii) to determine stellar properties (like age and mass) that trace the history and evolution of galaxies and constrain cosmological models. To achieve these goals, we rely on numerical stellar models based on input physics that integrate the results of recent theoretical studies, numerical simulations and laboratory experiments. The model input and output are chosen and/or validated by comparison with accurate astronomical observations.

Numerical 2D and 3D hydrodynamical simulations of limited regions of stellar interiors and atmospheres are now under reach of computers. They provide valuable constraints and data for the current standard (1D) stellar models: abundances, convection, rotationally induced instabilities and mixing, magnetic fields (see Asplund 2005; Talon 2007; Zahn 2007, for reviews). In parallel, the physics of stellar plasma is studied in the laboratory with (i) fluid experiments (study of turbulence in rotating, magnetic fluids, see e.g. Richard & Zahn 1999), (ii) particle accelerators (nuclear reaction cross sections) and, (iii) the so-called high energy-density facilities (based on high power lasers or z-pinches) which aim at exploring the high temperature and high density regimes found in stars, brown dwarfs and giant planets to get information on the equation of state (EOS), opacities or thermonuclear reactions (see Remington *et al.* 2006).

Modern ground-based and spatial telescopes equipped with high quality instrumentation are in use or under development (VLT-VLTI, JWST, etc.). They provide very accurate data which, after treatment, give access to stellar global parameters: luminosity, radius, mass, effective temperature $T_{\rm eff}$, gravity $\log g$, abundances. On the other hand, seismic data (such as oscillation frequencies or amplitudes) are being obtained from velocities (e.g. Bedding & Kjeldsen 2007) and from photometry by the space missions MOST (Walker *et al.* 2003) and CoRoT (Michel *et al.* 2006). In the next decade, valuable observational data are expected. For instance Gaia (ESA 2000; Perryman *et al.* 2001), to be launched in 2011, will make astrometric measurements, at the micro-arc second level together with photometric and spectroscopic observations of a huge number of stars covering the whole range of stellar masses, compositions and evolution stages while the Kepler mission, to be launched in 2009, will provide the opportunity to make asteroseismic observations on a wide range of stars (Christensen-Dalsgaard *et al.* 2007).

   In the following, I discuss the different aspects of stellar modeling, the problems encountered, and the perspectives.

2. Stellar models: input parameters and observational constraints

2.1. *Input physics for stellar models*

Stellar model calculations require a good description of the physical processes at work. Microscopic physics (opacities, EOS, nuclear reaction rates, atomic diffusion) are now rather well described which improves agreement between models and observations. However, difficulties still arise in the modelling of e.g. (i) cold, dense stars (molecular opacities, non-ideal effects in the EOS), (ii) advanced stages of evolution (nuclear reaction rates), and (iii) stars to be modelled very accurately like the Sun (see Sect. 3.1). On the other hand, despite important recent progress, macroscopic processes (convection, transport of chemicals and angular momentum related to differential rotation and the role of magnetic fields and internal gravity waves) are not yet fully understood (see e.g. Talon 2007).

2.2. *Input from observations and from model atmospheres*

Stellar model calculations involve input and constraints derived from observation. Observational data (e.g. magnitudes, colors, spectra, light and velocity curves, astrometric data) must be treated to get stellar parameters (luminosity, mass, radius, abundances, etc.). Model atmospheres are a crucial step in the analysis, for instance, to predict fluxes in different bands, synthetic spectra, or limb-darkening coefficients. They also provide the boundary conditions for the interior model as well as bolometric corrections and color-temperature conversions. Recently, model atmospheres have been improved by bringing in better atomic and molecular data and by applying more sophisticated algorithms. The output of 2-3D MHD simulations, including NLTE effects are being used to derive abundances, T_{eff} or $\log g$ from high quality spectra (Asplund 2005; Zahn 2007).

   Although internal errors in the determination of stellar parameters are becoming quite small for dwarfs, subgiants, and giants of spectral types A to K (see Table 1 in Lebreton 2005), large systematic errors still remain for cool and hot stars. For instance, the systematic errors on metallicity of metal poor dwarfs and cool giants amount to 0.2-0.3 dex, that is ten times more than the typical internal errors, while differences in T_{eff}-scales can reach up to 200–400 K (Gustafsson 2004; Allende Prieto 2006).

   Oscillations have been detected in many stars: solar-like, δ Scuti, β Cephei, γ Dor, Cepheids, RR Lyrae, SPB, WD, etc. They result from propagation of acoustic pressure or gravity waves, depending on the mass, evolution stage, chemical composition and excitation mechanism. Valuable constraints can be drawn from the frequencies or their combination (see, e.g. Christensen-Dalsgaard 1988; Roxburgh & Vorontsov 2003, and references therein).

3. Modeling calibrators

Only a few stars for which we have strong or numerous observational constraints are used as calibrators, i.e. they serve to validate the models and learn the physics. What is learned from calibrators can then be applied to stars with incomplete or less accurate observations. This implies extrapolation to compositions, masses or evolution stages not covered by calibrators. Quantities of astrophysical interest such as age, helium content or distance scale can then be derived for large numbers of stars.

3.1. *Solar modeling: AGS05 mixture and seismology*

The solar photospheric abundances have been re-determined recently on the basis of 3D radiative-hydrodynamical model atmospheres including better atomic data and NLTE effects (Asplund 2005). The new mixture, referred to as AGS05, gives a global metallicity and abundances of C, N, O smaller by 30-40% than those given by the GN93 mixture (Grevesse & Noels 1993) derived from 1D hydrostatic models. As a result, the interior opacity is reduced and the solar model no more satisfies the helioseismic constraints (e.g., convection zone depth, sound speed profile). This problem is especially acute in the region between $\sim$0.4 and $\sim$0.7$R_\odot$ (upper part of the radiative zone and shear region below the convective zone, i.e. tachocline). Several authors have shown that the present uncertainties in the input physics of the solar model, in particular the opacities or the atomic diffusion velocities, can hardly explain the differences (see e.g. Montalbán *et al.* 2006). On the other hand, it has been suggested that the neon abundances could be in error by a factor of at least 2, but the problem is still open (Grevesse *et al.* 2007).

3.2. *Binary system modeling: the RS Cha and α Centauri binary systems*

The modelling of a binary system should reproduce the observed constraints under the assumption that the two stars have the same age and initial composition. This may allow to infer the values of the model unknowns: age, initial helium, and physical parameters such as the mixing-length parameter for convection or overshooting parameter.

The binary system RS Cha is an interesting SB2 eclipsing system whose components are A-type oscillating stars in a PMS evolution stage corresponding to the onset of the earliest reactions of the CNO cycle. The modeling of the system by Alecian *et al.* (2007a,b) in the light of new accurate observations (masses, radii, metallicity) has shown that, to get agreement between the model and observations, carbon and nitrogen must be depleted with respect to their values in the GN93 mixture. The AGS05 mixture fulfills this condition but to assess this result and the values of the system's age and helium abundance derived from the calibration, it is now necessary to further improve the observational data. In particular, the [Fe/H] value should be redetermined using 3D model atmospheres and it would be valuable to get individual abundances of major elements and better seismic data, the present ones being too crude to provide useful constraints.

The binary system α Centauri is the closest and best-known one. The observed global parameters are accurate (in particular the accuracy on the interferometric radii is at 1%) and seismic observations of both stars have allowed to measure the frequencies of several low degree p-modes with the accuracy of $\sigma_\nu \simeq$ 0.3-2.0 μHz. Several authors have performed calibrations of the system (e.g., Miglio & Montalban 2005). They conclude that it is difficult to find a set of parameters that satisfies simultaneously the global and seismic constraints. As a result, the constraints on the physics of models remain loose, age and initial helium abundance are still poorly determined. It would be interesting to better assess the radii and masses by confirming the parallax of the system (different values were obtained from the analysis of Hipparcos data) and to further improve and enlarge the seismic data that are still scarce and coarse.

3.3. *Modeling stars in open clusters*

Members of stellar clusters can also serve as calibrators. They can be studied under the assumption that they all have the same age and initial composition but different mass. The initial helium abundance and age of a cluster can be derived from comparison of a model isochrone with observations of cluster stars in the color-magnitude diagram (see Fig. 3 in Lebreton 2005). The surface He abundance Y can be derived from the position of the lower main sequence (MS) and the age from the MS turn-off. Such studies require accurate observations (parallax, magnitude, T_{eff}, [Fe/H]). The metallicity uncertainty affects the Y estimate because of the helium-metallicity degeneracy in the H–R diagram. Also, uncertainties on Y and on age come from a poor knowledge of input physics, like envelope or core convection, rotational mixing or atomic diffusion (Lebreton *et al.* 2001).

The observations of binary stars in a cluster can provide additional constraints. For instance, the position of the low-mass, non-evolved stars in the mass-luminosity plane is related to their helium abundance while physics can be constrained if several binaries spanning a large mass range can be observed. Again the accuracy of the parallax and metallicity are crucial. A study of the Hyades (the only cluster where individual distances have been obtained by Hipparcos and masses measured for a few stars) has shown the limitation of the method due to the uncertainties on metallicity and input physics, and the small number of stars with accurate mass determination (Lebreton *et al.* 2001).

Investigations by e.g. Mazumdar & Antia (2001); Basu *et al.* (2004); Piau *et al.* (2005) have shown that the seismic analysis of different modes of oscillations should help to probe their inner properties such as the outer convection zone depth and helium content or the convective core boundary, and to estimate their mass and age. For instance, in solar-type stars, the higher the helium abundance in the convective envelope, the deeper the depression in the adiabatic index Γ_1 in the region of second helium ionisation. As shown by Basu *et al.* (2004), the helium abundance in the envelope of low-mass stars could be derived using the signature of this depression in the p-mode frequencies. This would require that low degree p-modes are observed with a frequency accuracy of 0.01% and that the mass or radius of a star is known independently.

4. Deriving astrophysical parameters: the example of stellar ages

In our Galaxy, the ages of A and F stars are crucial inputs for studying the disc while those of old metal poor stars and globular clusters provide valuable constraints for cosmology. The uncertainty on age depends on many factors (precision of the position in the H–R diagram, abundances, knowledge of the model input physics such as convection, rotational mixing, atomic diffusion). The case of globular clusters is discussed by Chaboyer (these proceedings). I focus here on the ages of A-F stars.

4.1. *Ages of A and F stars and the size of their mixed cores*

A-F stars have convective cores on the MS and may be fast rotators in the δ Scuti instability strip. Therefore, to model these stars, we are faced with difficulties in describing the central extra-mixing by core convection overshoot and rotationally induced mixing in the radiative zone. We have to estimate these effects of rotation from photometric data. These processes modify either the stellar models or the position of the observed star in the H–R diagram that, in turn, affects the age determination.

The efficiency of mixing in the stellar core determines the quantity of fuel available to a star with a crucial impact on its lifetime. Overshooting of the convective cores produces an extra-mixing. In model calculations, this extra-mixing is usually crudely parameterized with a coefficient α_{ov}, the value of which probably depends on mass, composition and evolution stage as shown by empirical calibrations based on binaries and the MS width observations (Ribas *et al.* 2000; Young *et al.* 2001; Cordier *et al.* 2002). Rotationally induced mixing can also bring extra fuel to the stellar engine. Recently, progress has been made in modeling the transport of angular momentum and chemicals resulting from differential rotation (see the review by Mathis *et al.* 2007).

Goupil & Talon (2002) have calculated models of a typical A-star including either overshooting or rotational mixing and have found that these distinct processes cannot be discriminated in the H–R diagram where they have similar signatures. Goupil & Talon also showed that if the A-star pulsates as a δ Scuti star then the signature of the mixing process could be seen in the oscillations frequencies, provided enough modes are observed and identified. We estimate that an uncertainty on age ranging from 13 to 24% results from a poor knowledge of the inner mixing processes (Lebreton *et al.* 1995). We expect that in the near future, the improvement of the observed H–R diagram (in particular, the luminosities from micro-arc-second astrometry) and the availability of precise seismic data will allow us to reduce this uncertainty down to 3-5% (Lebreton 2005).

5. Perspectives in the context of micro-arc-second astrometry

Today very few calibrators are available to probe the physics of stellar interiors. After Hipparcos, $\sim$200 stars have distances accurate to better than 1%. Also, the sample of stars with masses and radii measured with the accuracy better than 1% remains small. Concerning open clusters, Hipparcos provided precise individual distances only for the Hyades, while the binaries have been analysed in the Hyades and Pleiades only. There is no cluster star known with the solar-like oscillations being detected. Concerning A-F stars, Hipparcos determined the distances of 10^3 A-F stars with an accuracy better than $\sim$10%, while the seismic data have been obtained for quite some stars, but they are often coarse.

In the next decade, we expect that observations of stars will increase both in numbers and quality. Missions dedicated to global astrometry like Gaia (ESA 2000; Perryman *et al.* 2001) and SIM (Unwin *et al.* 2007) will measure the parallax with accuracy better than 10 micro-arcseconds. In parallel, the measurements of stellar parameters (magnitudes, temperatures, abundances, masses, radii etc.) are expected to be much improved due to the high resolution spectroscopy and interferometry. High-quality seismic data are expected from several missions: CoRoT will reach an accuracy of 0.1μHz on frequencies for about 50 targets (mainly solar-like oscillators, β Cephei and δ Scuti) while Kepler will enlarge the seismic sample to thousands of stars with a frequency accuracy of 0.1-0.3μHz.

5.1. *Expected returns from Gaia*

Gaia's astrometry will be complemented by photometry and spectroscopy allowing most masses and evolution stages to be precisely documented. The number of calibrators for stellar physics will be drastically increased and homogeneous global parameters will be provided, e.g., magnitudes, masses and abundances. The parallax of 7×10^5 (21×10^6) stars will be measured with accuracies of at least 0.1% (1%) and the mass of stars in 17 000 binary systems will be obtained with accuracies better than 1%. About 120 open clusters (up to 1 kpc) will be brought to a level of precision better than now existing for the Hyades. Parallax measurements, accurate to 0.5%, will be provided for 5×10^5 A stars and 3×10^6 F stars. Furthermore, while Hipparcos yielded direct distances with accuracies better than 12% for only 11 subdwarfs and 2 subgiants, Gaia will provide (i) precise direct distances for very large samples of subdwarfs and for all subgiants up to 3kpc and (ii) individual distances with an accuracy better than 10% for stars in $\sim$20

416 Y. Lebreton

globular clusters. We will have access to very precise H–R diagrams of very large stellar samples with complementary data (as mass, radius, detailed abundances and seismic data) for various subsamples of stars. The interpretation of these data in the light of future improvements on theoretical, numerical and experimental physics will certainly bring further insights in the understanding of stellar interiors and evolution.

References

Alecian, E., Goupil, M.-J., Lebreton, Y., Dupret, M.-A., & Catala, C. 2007a, *A&A*, 465, 241

Alecian, E., Lebreton, Y., Goupil, M.-J., Dupret, M.-A., & Catala, C. 2007b, *A&A*, 473, 181

Allende Prieto, C. 2006, in: S. J. Kannappan, S. Redfield, J. E. Kessler-Silacci, M. Landriau, & N. Drory (eds.), *New Horizons in Astronomy: F. N. Bash Symp.*, ASP Conf. Ser.,352, 105

Asplund, M. 2005, *ARAA*, 43, 481

Basu, S., Mazumdar, A., Antia, H. M., & Demarque, P. 2004, *MNRAS*, 350, 277

Bedding, T. R. & Kjeldsen, H. 2007, *CoAst*, 150, 106

Christensen-Dalsgaard, J. 1988, in: J. Christensen-Dalsgaard & S. Frandsen (eds.), *Advances in Helio- and Asteroseismology*, IAU Symp. 123, p. 295

Christensen-Dalsgaard, J., Arentoft, T., Brown, T. M., *et al.* 2007, *CoAst*, 150, 350

Cordier, D., Lebreton, Y., Goupil, M.-J., *et al.* 2002, *A&A*, 392, 169

ESA. 2000, White-Book, Gaia - Composition, Formation and Evolution of the Galaxy, Concept and Technology Study Report (ESA-SCI(2000)4), 1–381

Goupil, M. J. & Talon, S. 2002, in: C. Aerts, T. R. Bedding, & J. Christensen-Dalsgaard (eds.), *Radial and Nonradial Pulsations as Probes of Stellar Physics*, ASP Conf. Ser., 259, 306

Grevesse, N., Asplund, M., & Sauval, A. J. 2007, *Space Sci. Revs*, 105

Grevesse, N. & Noels, A. 1993, in: N. Prantzos, E. Vangioni-Flam and M. Casse (eds.), *Origin and Evolution of the Elements*, CUP, p. 14

Gustafsson, B. 2004, in: A. McWilliam & M. Rauch (eds.), *Origin and Evolution of the Elements*, CUP, p. 104

Lebreton, Y. 2005, in: C. Turon, K. S. O'Flaherty, & M. A. C. Perryman (eds.), *The Three-Dimensional Universe with Gaia*, ESA SP 576, 493

Lebreton, Y., Fernandes, J., & Lejeune, T. 2001, *A&A*, 374, 540

Lebreton, Y., Michel, E., Goupil, M. J., Baglin, A., & Fernandes, J. 1995, in: E. Hog & P. K. Seidelmann (eds.), *Astronomical and Astrophysical Objectives of Sub-Milliarcsecond Optical Astrometry*, IAU Symp. 166, 135

Mathis, S., Eggenberger, P., Decressin, T., *et al.* 2007, in: C. Straka, Y. Lebreton, M. Monteiro (eds.), *Stellar Evolution and Seismic Tools for Asteroseismology*, EAS Publi. Ser. 26, 65

Mazumdar, A. & Antia, H. M. 2001, *A&A*, 377, 192

Michel, E., Baglin, A., Auvergne, M., *et al.* 2006, in: M. Fridlund, A. Baglin, J. Lochard, & L. Conroy, (eds.), *The CoRoT Mission*, ESA SP 1306, 39

Miglio, A. & Montalbán, J. 2005, *A&A*, 441, 615

Montalbán, J., Miglio, A., Theado, S., Noels, A., & Grevesse, N. 2006, *CoAst*, 147, 80

Perryman, M. A. C., de Boer, K. S., Gilmore, G., *et al.* 2001, *A&A*, 369, 339

Piau, L., Ballot, J., & Turck-Chièze, S. 2005, *A&A*, 430, 571

Remington, B. A., Drake, R. P., & Ryutov, D. D. 2006, *Rev. Mod. Phys.*, 78, 755

Ribas, I., Jordi, C., & Giménez, Á. 2000, *MNRAS*, 318, L55

Richard, D. & Zahn, J.-P. 1999, *A&A*, 347, 734

Roxburgh, I. W. & Vorontsov, S. V. 2003, *A&A*, 411, 215

Talon, S. 2007, ArXiv e-prints, 708

Unwin, S. C., Shao, M., Tanner, A. M., *et al.* 2007, ArXiv e-prints, 708

Walker, G., Matthews, J., Kuschnig, R., *et al.* 2003, *PASP*, 115, 1023

Young, P. A., Mamajek, E. E., Arnett, D., & Liebert, J. 2001, *ApJ*, 556, 230

Zahn, J.-P. 2007, in: F. Kupka, I. Roxburgh, & K. Chan (eds.), *Convection in Astrophysics*, Proc. IAU Symp. 239, p. 517

A Giant Step: from Milli- to Micro-arcsecond Astrometry
Proceedings IAU Symposium No. 248, 2007
W. J. Jin, I. Platais & M. A. C. Perryman, eds.

The ESPRI project:
narrow-angle astrometry with VLTI-PRIMA

R. Launhardt[1,*], T. Henning[1], D. Queloz[2], A. Quirrenbach[3], F. Delplancke[4], N.M. Elias II[1,3], F. Pepe[2], S. Reffert[3], D. Ségransan[2], J. Setiawan[1], R. Tubbs[1] and the ESPRI consortium[1,2,3,4]

[1]Max-Planck-Institut für Astronomie, Königstuhl 17, 69117 Heidelberg, Germany

[2]Observatoire Astronomique de l'Universitè de Genève, 1290 Sauverny, Switzerland

[3]Landessternwarte Königstuhl, 69117 Heidelberg, Germany

[4]European Southern Observatory (ESO), Karl-Schwazschild-Str. 2, 85748 Garching

*email: rlau@mpia.de

Abstract. We describe the ongoing hardware and software developments that shall enable the ESO VLTI to perform narrow-angle differential delay astrometry in K-band with an accuracy of up to $10\,\mu$arcsec. The ultimate goal of these efforts is to perform an astrometric search for extrasolar planets around nearby stars.

Keywords. instrumentation: interferometers, techniques: high angular resolution, techniques: interferometric, astrometry, planetary systems, infrared: stars

1. Overview: VLTI, PRIMA, and the ESPRI project

The ESO Very Large Telescope Interferometer (VLTI) consists of four stationary 8.2-m VLT "Unit Telescopes" (UTs), four movable 1.8-m "Auxiliary Telescopes" (ATs), and six long-stroke dual-beam delay lines (DLs). It provides baselines of up to 200m length and covers a wavelength range that extends from the near infrared ($1\,\mu$m) up to $13\,\mu$m.

PRIMA, the instrument for Phase Referenced Imaging and Micro-arcsecond Astrometry at the VLTI, is currently being developed at ESO. It will implement the dual-feed capability for two UTs or ATs to enable simultaneous interferometric observations of two objects that are separated by up to $1'$. PRIMA is designed to perform narrow-angle astrometry in K-band with two ATs as well as phase-referenced aperture synthesis imaging with instruments like Amber (Petrov *et al.* 2000) and Midi (Leinert *et al.* 2003). PRIMA is composed of four major sub-systems: Star Separators (STS), a laser metrology system (PRIMET), Fringe Sensor Units (FSUs; see Sahlmann *et al.* 2008), and Differential Delay Lines (DDLs) (Quirrenbach *et al.* 1998; Delplancke *et al.* 2000; Derie *et al.* 2003). The first three subsystems are currently being tested at ESO, but the fate of the DDLs, which are crucial for reaching high astrometric precision, was insecure until 2004.

In order to speed up the full implementation of the $10\,\mu$arcsec astrometric capability and to carry out a large astrometric planet search program, a consortium lead by the Observatoire de Genéve (Switzerland), Max Planck Institute for Astronomy, and Landessternwarte Heidelberg (both Germany), is currently building the DDLs for PRIMA and develops the astrometric observation preparation and data reduction software. The facility is planned to become fully operational in early 2009. In return for its effort, the consortium has been awarded guaranteed observing time with PRIMA and two ATs to carry out a systematic astrometric Exoplanet Search with PRIMA (ESPRI).

2. The method: narrow-angle astrometry with PRIMA

A two-telescope interferometer measures the delay between the wavefront sections from a star as they arrive at the telescopes. However, atmospheric piston perturbations usually prohibit accurate measurements of this delay in absolute terms. To circumvent this problem, a dual-star interferometer like PRIMA measures the differential delay between two stars. When their angular separation is smaller than the isoplanatic angle of the atmosphere ($\approx 10''$ in K-band), the piston perturbations of the two wavefronts are correlated and the differential perturbations (ΔOPD_{turb}) average to zero rapidly (Shao & Colavita 1992). If one of the stars is bright enough to measure its fringe phase within the atmospheric coherence time, it can be used to stabilize the fringes on the other star (fringe-tracking), thus allowing for much longer integrations and hence increasing the number of observable objects. In order to obtain fringes on the detector, the external delay difference, which is directly related to the angular separation ($\Delta\alpha$) via the interferometer baseline (B), must be compensated with optical Delay Lines in the interferometer. At zero fringe position external and internal delays are equal. For astrometry, the two stars are supposed to have intrinsic phase (Φ) zero. The laser-monitored internal delay (ΔOPD_{int}) together with the residual differential fringe phase (ΔOPD_{FSU}) is then the primary observable of the interferometer (see Fig. 1).

In PRIMA, the dual feed is realized with Star Separators (STS) at the Coudé foci of the telescopes, which pick up two sub-fields with the target and astrometric reference star and feed them as separate beams to the DLs. In order to minimize the effects of air turbulence in the tunnels, the two star beams are sent parallel through one main DL. The large optical path difference (OPD) between the two telescopes, that is to first order common to both stars, is thus compensated without introducing further differential perturbations. Due to the non-zero angular separation between the two stars and the diurnal motion, there is however also a variable differential OPD between the two stars. After the main DLs, the beams of two stars are therefore sent to separate DDLs that operate in vacuum and provide a much smaller stroke ($\leqslant 60$ mm). On a 100 m baseline, 10 μarcsec correspond to 5 nm OPD, which defines the total error budget for DDLs, fringe detection, and metrology. The beams from the two telescopes are then interferometrically combined in the PRIMA Fringe Sensor Units (FSU). The measured fringe phase of the brighter star is used to control the DLs and DDLs and to stabilize the fringes of the fainter star. A second FSU, that can now integrate much longer, measures the fringe phase of the fainter star. An approximation to the internal OPD in the interferometer is measured by the on-axis end-to-end laser metrology system PRIMET.

Based on the error budget and preliminary exposure time calculator, the minimum K-band brightness of reference stars required to reach 10 μarcsec is K $\leqslant 14$ mag (Tubbs *et al.* 2008). The maximum separation between target and reference star is $\approx 15''$. Figure 2 shows the dependence of astrometric accuracy on stellar brightness for $10''$ separation.

3. Differential delay lines and astrometric data reduction software

Hardware developments: The design of the DDLs has been developed by the consortium in close collaboration with ESO. The DDLs consist of Cassegerain-type, all-aluminum retro-reflector telescopes (cat's eyes) with ≈ 20 cm diameter that are mounted on stiff linear translation stages. A stepper actuator at the translation stage provides the long stroke of up to 60 mm, while a piezo actuator at M3 in the cat's eye provides the 1 nm resolution fine stroke over ≈ 10 μm. Together with an internal metrology system, the DDLs are mounted on a custom-made optical bench in non-cryogenic vacuum vessels. The cat's eye optics has been successfully tested at MPIA and is currently being integrated with the other DDL components and prepared for acceptance tests in Geneva (see Fig. 3).

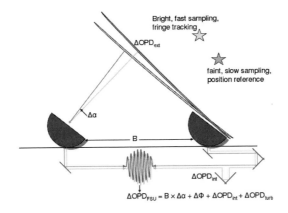

Figure 1. Measurement principle: astrometry with differential delay interferometry.

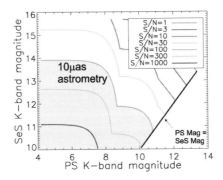

Figure 2. Dependence of astrometric accuracy of PRIMA with two ATs on stellar brightness for $10''$ separation between primary (PS) and secondary (SeS) star and 30 min integration.

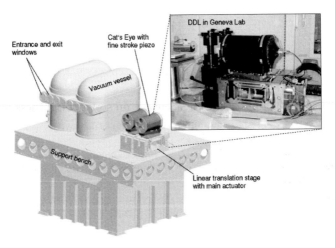

Figure 3. Final design of the Differential Delay Lines and photograph of the first DDL cat's eye on the translation stage in the Lab at Geneva observatory (setup for acceptance tests).

Software developments by the ESPRI consortium include Observation Preparation Software for PRIMA astrometry and the complete astrometric data reduction system (see Fig. 4). Data reduction from raw instrumental data to calibrated delays will proceed fully automatically with two pipelines and a set of calibration parameters that is re-derived every few months from all available PRIMA astrometry data (Elias *et al.* 2008). The software packages will be delivered to ESO prior to commissioning of the instrument and will be available to all users. The conversion of calibrated delays into astrophysical quantities like, e.g., planet orbits, is the responsibility of the user.

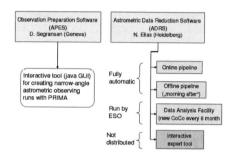

Figure 4. Overview of astrometric software developments for PRIMA.

4. Science goals: the Exoplanet Search program

Starting in spring 2009, we want to carry out a systematic astrometric Exoplanet Search with PRIMA (ESPRI), which will address the following key questions: *(i)* Precise determination of the planetary mass distribution, *(ii)* Detection of new Saturn- down to Uranus-mass planets around nearby stars, *(iii)* Formation and evolution of multiple planetary systems, and *(iv)* Exploring planet formation as a function of stellar age and mass. With these objectives in mind, we have defined three lists of potential targets, containing in total nearly 900 stars: *1.* Stars with known radial velocity planets, *2.* Nearby main-sequence stars within 15 pc around the Sun, and *3.* Young stars with ages 5-300 Myr within 100 pc around the Sun. We are currently carrying out an extensive preparatory observing program to identify suitable astrometric reference stars and to characterize the planet search target stars (Geisler *et al.* 2008). With a final detection rate for reference stars of 10-15%, we will then monitor $\approx 100-150$ stars for astrometric signatures of extrasolar planets. The scientific program and detection spaces are described in more detail in Launhardt *et al.* (2008 - IAU 249).

References

Delplancke, F., Lévêque, S. A., Kervella, P., *et al.* 2000, *Proc. SPIE*, 4006, 365
Derie, F., Delplancke, F., Glindemann, A., *et al.* 2003, *ESA SP-522*
Elias, N. M., Tubbs. R. N., Köhler, R., *et al.* 2008, *IAUS*, 249, *in press*
Geisler, R., Setiawan, J., Launhardt, R., *et al.* 2008, *IAUS*, 249, *in press*
Launhardt, R., Henning, Th., Queloz, D., *et al.* 2008, *Proc. IAU Symp.*, 249, *in press*
Leinert, Ch.; Graser, U.; Przygodda, F., *et al.* 2003, *Ap&SS*, 286, 73
Petrov, R. G., Malbet, F., Richichi, A., *et al.* 2000, *Proc. SPIE*, 4006, 68
Quirrenbach, A., Coude Du Foresto, V., Daigne, G., *et al.* 1998, *Proc. SPIE*, 3350, 807
Sahlmann, J., *et al.* 2008, *IAUS*, 248, in this volume p. 124
Shao, M. & Colavita, M. M. 1992, *A&A*, 262, 353
Tubbs, R., Elias II, N. M., Launhardt, R., *et al.* 2008, *IAUS*, 248, in this volume p. 132

A Giant Step: from Milli- to Micro-arcsecond Astrometry
Proceedings IAU Symposium No. 248, 2007
W. J. Jin, I. Platais & M. A. C. Perryman, eds.

© 2008 International Astronomical Union
doi:10.1017/S1743921308019741

Spying on your neighbors with ultra-high precision

W. C. Jao[1], T. J. Henry[1], J. P. Subasavage[1], P. A. Ianna[2], E. Costa[3], R.A. Méndez[3] and the RECONS Team

[1]Department of Physics and Astronomy, Georgia State University,
Atlanta, GA 30302, USA
email: `jao, thenry, subasavage@chara.gsu.edu`

[2]University of Virginia,
Charlottesville, VA 22903, USA
email:pai@virginia.edu

[3]Universidad de Chile, Santiago, Chile
email:costa, rmendez@das.uchile.cl

Abstract. We are entering the era of microarcsecond astrometric accuracy. Breaking the milliarcsecond barrier will lead to consequent leaps in astronomical understanding of diverse topics. Here we review some current ground-based trigonometric parallax efforts and their recent scientific results. We highlight the current status of nearby star research, including the RECONS census of stars within a 10 pc horizon, white dwarfs and cool subdwarfs, and the push to detect substellar objects via astrometry. We also provide details about recent improvements in the methodology that have permitted the determination of parallaxes with ∼1 milliarcsecond accuracy, and what might be done to push routinely into the sub-millarcsecond regime.

Keywords. astrometry, stars: kinematics, white dwarfs, subdwarfs

1. Introduction

The nearest stars provide astronomers with much of our understanding of stellar astronomy. For most types of stars, the fundamental efforts of stellar astronomy are built upon direct measurements of luminosities, colors, temperatures, radii, and masses for nearby stars. With a robust sample of nearby stars, we can also determine the total contribution of stellar mass to the Galaxy.

Only ∼480 total trigonometric parallaxes have been published since the Yale Parallax Catalog (van Altena *et al.* 1995, hereafter YPC) and the *Hipparcos* astrometry mission (ESA 1997). Since 1995, parallax contributions have been made by Dahn *et al.* (2002), Ianna *et al.* (1996), Smart *et al.* (2007), Tinney *et al.* (1995), Vrba *et al.* (2004) and Weis *et al.* (1999), among others. RECONS (Research Consortium on Nearby Stars) initiated its Cerro Tololo Inter-american Observatory Parallax Investigation (CTIOPI) in August 1999 under the NOAO Surveys program to search for missing nearby stars in the southern sky. As of February 2003, CTIOPI has continued as part of the Small and Moderate Aperture Research Telescope System (SMARTS) Consortium. Both the CTIO 0.9m and 1.5m telescopes were used during NOAO time, and the program continued on the 0.9m during SMARTS time. To date, we have published 136 parallaxes (28% of all parallaxes, and the most of any group since 1995) in four different papers (Jao *et al.* 2005, Costa *et al.* 2005, Costa *et al.* 2006 and Henry *et al.* 2006).

A total of 440 targets are being or have been observed for parallaxes during CTIOPI. Here we present the current census status for red dwarfs, white dwarfs (WDs) and cool subdwarfs, and some of the astrometric perturbations we have detected. In the final

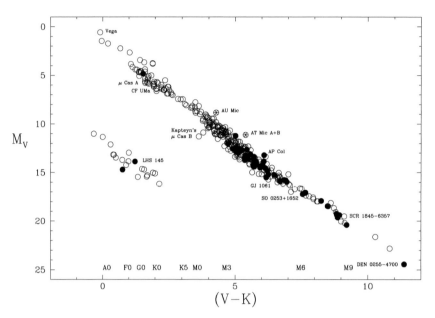

Figure 1. Members of the current RECONS 10 pc sample are shown on the HR diagram, $M_V vs.(V-K)$. Filled circles indicate stars and brown dwarfs for which RECONS has determined the first accurate parallaxes placing them within 10 pc. The two asterisk points are for the three components of the AU/AT Mic system that is only $\sim 8--20$ Myr old. Other labeled stars are discussed in the text.

section, we discuss a few techniques that have recently been employed to reduce trigonometric parallax errors, and offer a few additional techniques that may improve the errors to sub-milliarcsecond accuracy.

2. Census status

2.1. *The RECONS 10 parsec sample*

One of the main science goals of the CTIOPI project is to develop a complete census of stars within the RECONS horizon of 10 pc. In order for a stellar system to be included in the RECONS sample, it must have trigonometric parallax larger than 100 mas with an error less than 10 mas. As of January 1, 2000, there were 215 stellar systems containing 295 objects known within 10 pc.

Efforts by RECONS and other groups continue to fill in the census. The 2000 census list is based almost entirely on the combination of ground-based parallaxes included in the YPC and the results from the *Hipparcos* mission. New candidate members are selected from proper motion and photometric studies, from which the best candidates are usually revealed using a suite of photometric distance estimate relations (Henry *et al.* 2004). Including our own parallax results as well as the updates by other groups, there are now 260 systems containing 365 objects, constituting a 21% increase in systems (24% in objects) since 2000. Of the 45 new systems, 40 have entered the 10 pc sample via RECONS efforts.

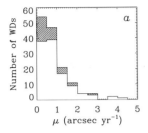 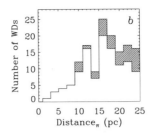 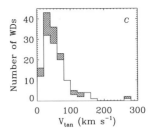

Figure 2. Histograms of astrometric properties of the new and known 25 pc WD samples. Plot (*a*) shows the proper motion distribution, binned by 0.5″ yr^{-1}. Plot (*b*) shows the distance distribution, binned by 2 pc. Plot (*c*) shows the tangential velocity distribution, binned by 20 km s^{-1}. Shaded regions indicate the new WDs already with reliable parallaxes from CTIOPI placing them within 25 pc. White regions indicate the 110 known WD systems within 25 pc prior to this effort.

Figure 1 shows the current HR diagram for stars in the RECONS 10 pc sample. The main sequence stretches from Vega at the blue end to stars with $M_V \sim 20$, $(V - K) \sim 9$ at the red end. The three points in the lower right are L dwarfs that are likely brown dwarfs. Precise distances of the three subdwarf systems, μ Cas AB, CF UMa, Kapteyn's Star, and the $\sim$8–20 Myr old system AU/AT Mic allow us to map accurately the location of these relatively rare objects in the HR diagram. Also highlighted are several of the more compelling red and brown dwarfs belonging to the 10 pc sample because of the RECONS efforts that include: AP Col (an X-ray active, likely young star), GJ 1061 (the 20th nearest system), SCR 1845-6397 (the 24th nearest system, which has a T dwarf companion that promises an accurate mass determination), and DEN 0255-4700 (to our knowledge, the faintest object for which M_V has been determined outside our Solar System). SO 0235+1652 (Teegarden *et al.* 2003) at 3.8 pc with μ=5″1 yr^{-1} is the 23rd nearest stellar system and the highest proper motion star to be found in the past several decades, ranking as the eighth fastest system overall. Also noteworthy is LHS 145, a DA8 white dwarf for which we have determined its first trigonometric parallax, which places it at 9.6 pc.

2.2. *White dwarfs*

The end product of stellar evolution for stars with masses less than $\sim$8 $M_\odot$ are WDs. The study of nearby WD population provides insight into their space density, formation rate, and evolution. A volume-limited sample provides unbiased statistics that can be applied in broader context to the Galaxy as a whole. In addition, the WD contribution to halo dark matter can be probed by identifying candidate halo WDs that happen to be passing through the solar neighborhood.

We can make an estimate of the number of WDs missing from current compendia by extrapolating the space density from the 18 WDs known within 10 pc, to 25 pc. Within this sphere we anticipate that there should be 281 WDs, yet only 110 are known. Thus, a staggering 61% of the nearby WD population may be undetected. Some will be unseen companions, while others will be free floating in the field, and yet unidentified.

The reduced proper motion (RPM) diagram is a powerful way to find WD candidates (see Figure 5 in Subasavage *et al.* 2005 as an example). The RPM diagram separates high proper motion stars into three different populations – dwarfs, subdwarfs and WDs. Follow-up spectroscopic observations are needed to confirm the luminosity classes of objects in each category. Once the candidates are spectroscopically confirmed, accurate distance estimates are made so that the WDs most likely to be within 25 pc can be targeted by CTIOPI. Currently, there are a total of 139 WDs within 25 pc horizon, 110 with parallaxes from YPC and the *Hipparcos* mission, and 29 with parallaxes from CTIOPI, constituting an increase of 26% in the WD population. There are additional 14 WDs on CTIOPI possibly within 25 pc based on their photometric distances, but we do not yet have enough data for these WDs to calculate a definitive trigonometric parallax. Of these, 11 were selected from our SuperCOSMOS-RECONS (SCR) proper motion surveys and three were discovered by other authors (Kawka *et al.* 2004, Kawka *et al.* 2007 and Wegner 1973).

Figure 2 illustrates several important aspects of the WD 25 pc sample as it stands now. Perhaps to be expected, the majority of our new 25 pc members have $\mu < 1\!''\!0$ yr^{-1} (Figure 2*a*) because WD candidates with $\mu \geqslant 1\!''\!0$ yr^{-1} were historically high priority targets for characterization. Figure 2*b* shows that most of the nearby WDs yet to be found are anticipated to be beyond 10 pc (although a few are being revealed within 10 pc). Finally, with precise proper motions and distances, tangential velocities can be determined, as shown in Figure 2*c*. A few possible halo WDs have been identified: WD 1339−340 at ∼22 pc that has $V_{tan} \sim 260$ km s^{-1} (Lépine *et al.* 2005 found that this object's orbit is nearly perpendicular to the Galactic plane so it is likely a halo WD), and WD 1756+827 at 15.6 pc that has $V_{tan} = 266$ km s^{-1}.

2.3. *Cool subdwarfs*

The HR diagram is the most important map of stellar astronomy. It provides a relatively straightforward method for separating different stellar luminosity classes, e.g. super-giants, bright giants, giants, subgiants, main sequence dwarfs, and white dwarfs, using their colors and luminosities. However, the combination of spectroscopic and trigonomet-ric parallax results has revealed a seventh distinct stellar luminosity class — subdwarfs — that lie below the main sequence dwarfs in the HR diagram.

Cool subdwarfs comprise a relatively new species in the HR diagram. Historically, they have been missed during nearby star searches because of their intrinsic faintness and rarity. Consequently, only 97 K and M type subdwarfs are known within 60 pc, compared to tens of thousands of their red dwarf cousins. Revealing nearby cool subdwarfs is one of the goals of CTIOPI. In total, we have made spectroscopic observations for 28 cool subdwarfs within 60 pc, and parallax observations for 16. Reasons for building a census of nearby cool subdwarfs within 60 pc include 1) developing a complete volume-limited sample for population studies, and 2) understanding how their metallicities affect their location in the HR diagram.

Jao *et al.* (2007) discuss different metallicity and gravity effects that lead to differences in the observed spectra of cool subdwarfs (see Figure 3). According to Gaia synthetic spectra, metallicity affects cool subdwarfs' spectra between 6000Å and 8300Å, while gravity affects only the CaHn (n=1–3) bands (Brott & Hauschildt 2005). The traditional spectral classification methods for cool subdwarfs (Gizis 1997, Lépine *et al.* 2007) do not take into account gravity effects, so different spectral sub-types may be assigned to two objects, even though their overall spectra are the same (see the gravity effects in Figure 3). We suggest using the continuum region between 8300Å and 9000Å to yield more consistent spectral sub-types, and to mesh these types with main sequence stars.

Because some of the targets discussed in Jao *et al.* (2007) have trigonometric paral-laxes, we can plot these stars in the HR diagram (M1.0VI is shown in Figure 3 as an example). We found that decreasing metallicities generally make subdwarfs bluer (smaller $V - K_s$) and brighter (smaller M_K), while increasing gravities generally make them redder and fainter. These trends are shown at each sub-type we have, but more trigonometric

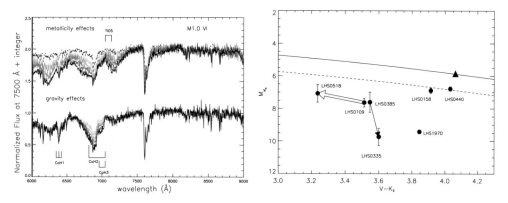

Figure 3. Left: M1.0VI subdwarf spectra are plotted at different metallicities (top spectra) and gravities (bottom spectra). The top spectra show metallicities decreasing from bottom to top (black to gray, where the lowest metallicity is plotted as a dashed line). There are seven different metallicity scales shown (0–6), including a black spectrum (scale 0) that represents our M1.0V dwarf spectral standard. The bottom spectra show the effects of gravity, decreasing from bottom to top (black to gray). All of these spectra have been selected from metallicity scale 4. Right: The HR diagram is shown for the M1.0VI subdwarfs in our sample. Filled circles indicate M1.0VI subdwarfs and a filled triangle represents the M1.0V spectroscopic standard star. The hollow arrow illustrates the shift on the HR diagram caused by decreasing metallicity (bluer stars). The solid arrow illustrates the shift caused by higher gravity (fainter stars). The solid line represents a fit to main sequence stars, while the dashed line is one magnitude fainter than the solid line.

parallaxes of later type subdwarfs (later than M5) are necessary to make a more complete map of their locations in the HR diagram.

3. Perturbation results from CTIOPI

We initiated ASPENS (Astrometric Search for Planets Encircling Nearby Stars) in 1999 with two dozen targets, and have been building the program since 2003, when D. Koerner (Northern Arizona University) joined the effort. The targets are primarily red dwarfs and WDs within 10 pc that are south of $\delta=0$ deg and fainter than $VRI=9$. We have just reached 100 targets in the ASPENS program, which includes 86 red dwarfs and 14 WDs. The main goal is to have the most comprehensive astrometric search for companions to red dwarfs and WDs in the southern sky.

In Figure 4, we present two of the new definite perturbation detections from ASPENS. These initial results show stars with ∼8 years of astrometric coverage. The left plot is for a binary M dwarf system with orbital period 6.8 years and photocentric semimajor axis 28 mas. The right plot is for an M/L dwarf pair. The orbital motion has not yet wrapped in our dataset, but the perturbation is clear.

Since the days of van de Kamp (1982) the discovery and confirmation of a planet orbiting a nearby star via astrometry has been elusive, although using $HST - FGS$, Benedict *et al.* (2002) have detected one of the planets orbiting GJ 876, which was first revealed using radial velocities. As shown in Figure 5, we are on the brink of detecting our first planetary companion. The astrometry indicates a possible companion, although the lack of a clear signature in declination is worrisome. However, we are heartened by the supportive, but not yet conclusive, radial velocity results from four epochs that hint at a confirmation.

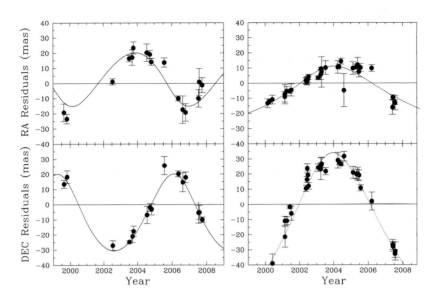

Figure 4. Astrometric residuals in the RA and DEC directions are shown for two definite perturbations, after solving for parallax and proper motion. Each point represents typically 5–10 frames taken on a single night.

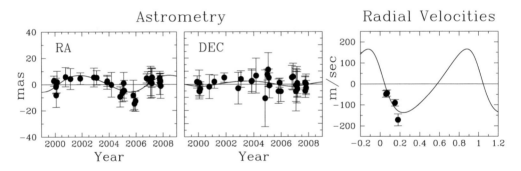

Figure 5. Left and center panels: Astrometric residuals in the RA and DEC directions are shown for a red dwarf in ASPENS, after solving for parallax and proper motion. Each point represents typically 5–10 frames taken on a single night. The orbital fit shown has a period of 6.6 years and photocentric semimajor axis of 6.5 mas. Right panel: Using the orbit derived from the astrometry, we plot the four nights of radial velocity data acquired so far using the HET. Each point represents 2–3 spectra taken on a single night. The shape of the orbit is set entirely by the astrometry, while the curve slides up/down depending on the systemic velocity. Although the confirmation is tentative, there does appear to be a change in velocity consistent with the astrometric predictions.

4. How to reduce parallax errors

Parallax errors have dropped dramatically moving from photographic plates to modern CCD observations. Since the era of CCD observations began, astrometrists have used several modifications to parallax observations and data analysis to reduce errors, including 1) using a finer pixel scale CCD imagers or improved stellar centroiding methods to yield better PSF centers (Pravdo *et al.* 2005), or 2) using improved global iterative methods to reduce parallax errors (Ducourant *et al.* 2008). These different attempts have decreased the parallax errors from several milliarcseconds to roughly 1 mas.

Other possible solutions that can be used to decrease errors in ground based trigono-metric parallaxes include 1) adding tip/tilt or adaptive optic systems to improve image quality and stability, 2) building telescopes with better optical designs to produce finer images and reduce field distortions, and 3) setting "no-fly" zones over the observing sites to reduce the effects of airplane contrails. Additional error-reducing strategies have focused on reducing the effects of the Earth's atmosphere, arguably one of the greatest re-maining causes of parallax errors. Parallax observations made in the near infrared (Vrba *et al.* 2004) have effectively eliminated the need for differential color refraction (DCR) corrections, while strictly observing targets within ± 15 minutes of meridian (Dahn *et al.* 2002) reduces the DCR effect even at optical wavelengths. In CTIOPI, we typically ac-quire data within 30 minutes of the meridian in the VRI filters, and use the method of Monet *et al.* (1992) to correct for DCR effects. As demonstrated in Jao *et al.* (2005), we can also successfully reduce the DCR effect to manageable levels for images taken at high airmass. However, as Monet *et al.* (1992) pointed out, the DCR corrections can be improved further with better models that mimic "real" atmosphere refraction and by using a better choice of colors, e.g. $(V - K)$, to compute the refraction coefficients.

A few large scale observational efforts promise to provide large number of parallaxes in the future. On the ground, Pan-STARRS is likely to be the first sky survey effort to yield parallaxes, while LSST is the future for the southern sky. In space, Gaia and SIM will not likely provide definite parallaxes until 2015–2020. All these projects will provide us with astrometric data of unprecedented precision. Thus, the promise for compelling results on nearby stars, white dwarfs, and cool subdwarfs is high, and we are poised to see a whole new view of our Galaxy. In addition, by breaking through the barrier of milliarcsecond astrometry into the realm of microarcsecond astrometry, new research fields will undoubtedly open up, thereby providing a wealth of scientific opportunities in the decades to come.

References

Benedict, G. F., *et al.* 2002, *ApJ*, 581, L115

Brott, I. & Hauschildt, P. H. 2005, *The Three-Dimensional Universe with Gaia*, 576, 565

Costa, E., Méndez, R. A., Jao, W.-C., Henry, T. J., Subasavage, J. P., Brown, M. A., Ianna, P. A., & Bartlett, J. 2005, *AJ*, 130, 337

Costa, E., Méndez, R. A., Jao, W.-C., Henry, T. J., Subasavage, J. P., & Ianna, P. A. 2006, *AJ*, 132, 1234

Dahn, C. C., *et al.* 2002, *AJ*, 124, 1170

Ducourant, C., Ramachrisna, T. & Gael, C. 2008, *in this volume*, p.98

Gizis, J. E. 1997, *AJ*, 113, 806

Henry, T. J., Subasavage, J. P., Brown, M. A., Beaulieu, T. D., Jao, W.-C., & Hambly, N. C. 2004, *AJ*, 128, 2460

Henry, T. J., Jao, W.-C., Subasavage, J. P., Beaulieu, T. D., Ianna, P. A., Costa, E., & Méndez, R. A. 2006, *AJ*, 132, 2360

Ianna, P. A., Patterson, R. J., & Swain, M. A. 1996, *AJ*, 111, 492

Jao, W.-C., Henry, T. J., Subasavage, J. P., Brown, M. A., Ianna, P. A., Bartlett, J. L., Costa, E., & Méndez, R. A. 2005, *AJ*, 129, 1954

Jao, W. C., *et al.* 2007, *AJ*, submitted

Kawka, A., Vennes, S., Schmidt, G. D., Wickramasinghe, D. T., & Koch, R. 2007, *ApJ*, 654, 499

Kawka, A., Vennes, S., & Thorstensen, J. R. 2004, *AJ*, 127, 1702

Lépine, S., Rich, R. M., & Shara, M. M. 2005, *ApJ*, 633, L121

Lépine, S., Rich, R. M., & Shara, M. M. 2007, *ApJ*, 669, 1235

Monet, D. G., Dahn, C. C., Vrba, F. J., Harris, H. C., Pier, J. R., Luginbuhl, C. B., & Ables, H. D. 1992, *AJ*, 103, 638

Perryman, M. A. C., & ESA 1997, *ESA Special Publication*, 1200

Pravdo, S. H. *et al.* 2005, *Astrometry in the Age of the Next Generation of Large Telescopes*, 338, 288

Smart, R. L., Lattanzi, M. G., Jahreiß, H., Bucciarelli, B., & Massone, G. 2007, *A&A*, 464, 787

Subasavage, J. P., Henry, T. J., Hambly, N. C., Brown, M. A., & Jao, W.-C. 2005, *AJ*, 129, 413

Teegarden, B. J., et al. 2003, *ApJ*, 589, L51

Tinney, C. G., Reid, I. N., Gizis, J., & Mould, J. R. 1995, *AJ*, 110, 3014

Wegner, G. 1973, *MNRAS*, 163, 381

Weis, E. W., Lee, J. T., Lee, A. H., Griese, J. W., III, Vincent, J. M., & Upgren, A. R. 1999, *AJ*, 117, 1037

van Altena, W. F., Lee, J. T., & Hoffleit, D. 1995, *VizieR Online Data Catalog*, 1174

van de Kamp, P. 1982, *Vistas in Astronomy*, 26, 141

Vrba, F. J., *et al.* 2004, *AJ*, 127, 2948

A Giant Step: from Milli- to Micro-arcsecond Astrometry
Proceedings IAU Symposium No. 248, 2007
W. J. Jin, I. Platais & M. A. C. Perryman, eds.

L and T dwarfs in Gaia/SIM

R. L. Smart[1], B. Bucciarelli[1], M. G. Lattanzi[1] and H. R. A. Jones[2]

[1]Istituto Nazionale di Astrofisica (INAF), Osservatorio Astronomico di Torino, Strada
Osservatorio 20, I-10025 Pino Torinese, Italy
emails: `smart/bucc/lattanzi@to.astro.it`

[2]Centre for Astrophysics Research,University of Hertfordshire College Lane, Hatfield AL10
9AB, UK
email: `H.R.A.Jones@herts.ac.uk`

Abstract. We discuss the role of distances for understanding brown dwarfs and estimate the contribution expected by Gaia. We show that Gaia will only observe 25% of L and T dwarfs within 50pc which, at a conservative estimate, amounts to less than 400 objects. We discuss how Gaia results will nevertheless aid the ground-based programs providing reliable, bias free constraints for the calculation of parallaxes in an absolute system. We list the current ground-based programs underway and the possibilities for future all sky survey programs.

Keywords. stars: low-mass, brown dwarfs, stars: distances

1. Introduction

The first brown dwarf, GD 165B, was found almost 20 years ago as a red companion to the white dwarf GD 165 (Becklin & Zuckerman 1988) . At first this object was considered an anomaly but after the discovery of similar objects it was realized that spectral types beyond M were required to classify them, hence the introduction of L and T spectral types (see Kirkpatrick 2005 for review). These objects from around mid-L are not massive enough to burn hydrogen so for simplicity we shall call them brown dwarfs. Since GD 165B over 600 brown dwarfs have been discovered primarily in the large infrared surveys and the Sloan survey. These objects form the link between stars and planets, they will probably prove to be more common than stars and are extremely long lived hence provide insights into the history of our Galaxy.

For those objects in the solar neighborhood the determination of useful parallaxes, e.g. with relative errors less than 10%, is within the range of ground-based programs; indeed over 40 are already determined to this precision. Here we discuss the importance of their distance determination, the contributions of the Gaia and SIM missions and the future for ground-based programs.

2. Distances for brown dwarfs

Since brown dwarfs are fundamentally different to stars, it is useful to examine why their distances are important. The atmospheres of these objects are very complex, being dominated by methane, clouds and dust. The spectral types, especially for T dwarfs, are more indications of the cloud and dust processes in their atmospheres rather than indications of changing in their effective temperature. From their birth these objects are continually cooling hence their luminosity and spectral type are a strong function of age. For example, a .05 $M_\odot$ objects starts its life as a late M dwarf, from .1 to 1 Gyr it transverses the L dwarf spectral types and then continues into the late T types.

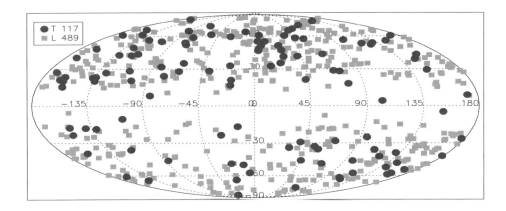

Figure 1. The distribution of known L and T dwarfs in galactic coordinates.

Finally, as with stars, the physical properties of brown dwarfs also depend on chemical composition.

The distance, in combination with a given apparent magnitude, directly provides a luminosity. Since the colors are not good indicators of the luminosity, large numbers of these distances are needed to calculate a statistically significant luminosity function. With a distance estimate under-luminous objects can be identified as possible sub-dwarfs e.g. not part of the local neighborhood but of another galactic component. When we understand better the evolution of these objects, the fact that they visibly change with time will be an important tool in understanding the history of our Galaxy. Finally distances, proper motions and luminosity also allow us to isolate binary systems that are particularly interesting for mass and age determinations.

3. Gaia/SIM and currently known L/T dwarfs

In the online compendium of L and T dwarfs (www.dwarfarchive.org) as of 5 October 2007 there are 606 L/T objects whose galactic distribution is shown in Fig. 1. Recent work by Metchev *et al.* (2007) in the SDSS footprint implies that this sample is 50% complete outside of the galactic plane. Extrapolating the area of the sky examined to the whole sky we can conservatively estimate that this sample is at least 25% of the L and bright T dwarfs within 50pc.

Using i,z' apparent magnitudes when available, or the J/K to i/z' relations for these stars from Hawley *et al.* (2002) when not, we can estimate Gaia G magnitudes using internal transformations (Jordi private communication). Using the spectral type to absolute magnitude relation in Hawley *et al.* (2002) we have calculated the spectro-photometric distances of each object. In Fig. 2 we plot the distribution of the 493 objects within 50pcs in bins of 5 parsecs.

If we assume the limiting magnitude of Gaia is $G = 20$ only the objects shaded as black on the histrogram in Fig. 2 will be observed by Gaia. Hence of the 606 currently known L and T dwarfs Gaia will observe less than 100. In conjunction with the estimated completeness, this implies that in total Gaia will observe less than 400 objects directly and most of those at its limiting magnitude. Gaia will also indirectly observe many other L and T dwarfs that are members of binary systems with a brighter companion, but the number of expected objects observed this way is currently not possible to estimate.

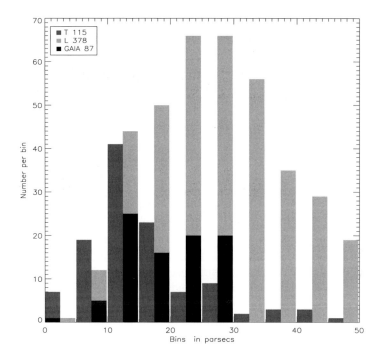

Figure 2. The distribution of L and T dwarf photometric distances within 50pcs. In the legend is shown the color code of the different objects as well as the total number of objects. Gaia will observe the objects distributed in the black histrogram, e.g. only 87 of the total 493 known objects.

While SIM and JWST will go fainter, the investment of observing time will probably only be made for those extremely interesting candidates such as binary systems.

4. The future from the ground

While Gaia will not observe many of these objects directly the observations of anonymous reference stars will help ground-based observations immensely. The two biggest sources of error in ground-based observations are the correction from relative to absolute parallaxes and the calibration of instrumental effects. Gaia will provide data for the anonymous reference stars with a higher precision that will be possible from the ground, hence an application of a full adjustment with constraints will allow us to approach the limit given by random errors. Simulations have shown that given 20 reference stars with Gaia parallaxes, proper motions and a nominal CCD measuring error of 4mas, a precision of better than 1mas is possible after just 30 observations spread over 2.5 years. Sub-mas precisions have also been obtained from the ground Harris *et al.* (2005) and the expectation is that the constraints from Gaia will allow a comparable increase in the accuracy. Given this it is not unrealistic that ground-based parallaxes will be able to provide distances with precisions of 1% for a statistically significant sample of these objects. Having the parameters of the reference stars with such high precision will allow a shortening of the observational campaign and also the ability to calibrate instrumental variations should they occur during the campaign. Gaia parallaxes will allow us to find absolute parallaxes without having to rely on photometric distances, that are prone to

Table 1. Current programs to determine distances of L and T dwarfs.

Program PI	Telescope + Detector	Objects under study
Ducourant, OB France	ESO NTT + SUSI2	10 ultracool TW Hydrae members
Faherty, AMNH USA	1.3m Cerro Tololo+ANDICAM	30 bright L and T dwarfs
Penna, ON Brazil	ESO 2.2m + WFI	80 L dwarfs
Röser, ARI Germany	Calar Alto 3.5m + Omega2000	10 ultracool (>sdM7) subdwarfs
Smart, OATo Itlay	TNG+NICS, UKIRT+WFCAM	20 peculiar or very cool T dwarfs
Tinney, UNSW Australia	AAO 3.9m AAT+WFI	30 bright L dwarfs
Tinney, UNSW Australia	AAO 3.9m AAT+IRIS2	15 T-dwarfs

errors in adopted absorptions, or spectroscopic distances, which are both time consuming and also prone to systematic errors.

Currently there are 7 active programs known to the authors which are attempting to find distances of L and T dwarfs, these are listed in table 1. The formerly largest program, that of the USNO, was stopped because of problems with the IR camera, in any case to observed objects fainter than Gaia will require a larger telescope than used in that program. This is where access to 4m class telescopes with IR detectors will play an important role. Working in the IR has three advantages over optical: the seeing is better, the differential reddening correction is smaller and the observations are quicker so we can make more observations per visit thus reducing random errors. Finally, future sky survey programs, PanSTARRS - LSST - SKYMAPPER, offer the possibility to determine parallaxes of these objects to a reasonable precision as a matter of routine.

5. Conclusions

Gaia will observe less than 400 L and T dwarfs and all of those within 50pcs. Distances of fainter objects within the 50pc and all objects outside this limit will not be observed. Parallaxes for the fainter objects can be determined from the ground and dedicated programs are already underway. The results of Gaia will be fundamental for these programs, as they will allow a high precision control of the instrument and provide constraints to directly find absolute parallaxes without the necessity to resort to galactic models or photometric parallaxes.

References
Becklin, E. E. & Zuckerman, B. 1988, *Nature*, 336, 656
Harris, H. C., Canzian, B., Dahn, C. C., *et al.* 2005, in: P. K. Seidelmann & A. K. B. Monet, (eds.) *Astrometry in the Age of the Next Generation of Large Telescopes, Astronomical Society of the Pacific Conference Series*, Vol. 338, p. 122
Hawley, S. L., Covey, K. R., Knapp, G. R., *et al.* 2002, *AJ*, 123, 3409
Kirkpatrick, J. D. 2005, *ARAA*, 43, 195
Metchev, S., Kirkpatrick, J. D., Berriman, G. B., & Looper, D. 2007, ArXiv e-prints

A Giant Step: from Milli- to Micro-arcsecond Astrometry
Proceedings IAU Symposium No. 248, 2007
W. J. Jin, I. Platais & M. A. C. Perryman, eds.

© 2008 International Astronomical Union
doi:10.1017/S1743921308019765

Open clusters: their kinematics and metallicities

L. Chen[1], J. L. Hou[1], J. L. Zhao[1] and R. de Grijs[2]

[1]Shanghai Astronomical Observatory, Chinese Academy of Sciences, 80 Nandan Road,
Shanghai 200030, China
email: chenli@shao.ac.cn, houjl@shao.ac.cn, jlzhao@shao.ac.cn

[2]Department of Physics and Astronomy, University of Sheffield, Hicks Building, Hounsfield
Road, Sheffield S3 7RH, UK; and National Astronomical Observatories, Chinese Academy of
Sciences, 20A Datun Road, Chaoyang District, Beijing 100012, China
email: r.degrijs@sheffield.ac.uk

Abstract. We review our work on Galactic open clusters in recent years, and introduce our proposed large program for the LOCS (LAMOST Open Cluster Survey). First, based on the most complete open clusters sample with metallicity, age and distance data as well as kinematic information, some preliminary statistical analysis regarding the spatial and metallicity distributions is presented. In particular, a radial abundance gradient of -0.058 ± 0.006 dex kpc^{-1} is derived. By dividing clusters into the age groups we show that the disk abundance gradient was steeper in the past. Secondly, proper motions, membership probabilities, and velocity dispersions of stars in the regions of two very young open clusters are derived. Both clusters show clear evidence of mass segregation, which provides support for the "primordial" mass segregation scenario. Based on the advantages of the forthcoming LAMOST facility, we have proposed a detailed open cluster survey with LAMOST (the LOCS). The aim, feasibility, and the present development of the LOCS are briefly summarized.

Keywords. Galaxy: disk, open clusters and associations: general, open clusters and associations: individual (NGC2244), open clusters and associations: individual (NGC6530)

1. The open cluster system and the observational database

Open clusters (OCs) are considered excellent laboratories for studies of stellar evolution. Studies in the research area dealing with OCs show a rapid growth in 1990's and this area continues to develop vigorously. There may be several reasons for this recent growth of OC studies. New techniques are greatly beneficial to the OC observations, including, for example, the application of wide-field, high quality CCD cameras and, more recently, multi-object spectrographs. In addition, OC studies play a very important and unique role in determining the structure and evolution of the Galactic disk.

OCs have long been used to trace the structure and evolution of the Galactic disk. From the observational point of view, there are some important advantages of using OCs as opposed to field stars. In OCs, we deal with groups of stars of nearly the same age, a similar composition and at a similar distance. We can observe OCs to large distances, and the photometric distances to most of the OC sample have already been derived. In particular, OCs have a relatively stable orbital motion, which can be used as a better tracer of the Galactic disk structure. OCs also have a wide range of ages, so that – combined with their spatial distribution and kinematic information – we can study the effects of dynamical evolution. Very young OCs are ideal objects for studies of the stellar initial mass function. Furthermore, when combined with abundance data, we can

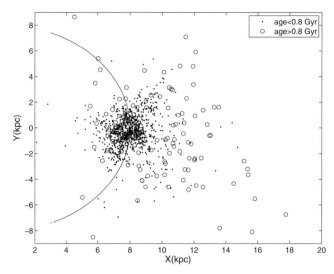

Figure 1. Spatial distribution of OCs in the Galactic (X, Y) plane. The open circles and dots are the older and younger clusters, respectively. The large semi-circle has a radius of 8.0 kpc, centered on the Galactic center.

investigate the chemical evolution history of the Galactic disk (Carraro *et al.* 1998, Chen *et al.* 2003).

At present, the total number of detected clusters and associations is around 1700 (Dias *et al.* 2002), about 60% of which have distance and age information and for about half of them have proper motions available. Less than one-fourth of the sample have both proper motion and radial velocity parameters, and only a small portion (about 8%) of the OCs have well-determined mean metallicities. We have compiled an updated OC catalog (Chen et al., in prep.), for which data have been extracted from various sources – mainly from (Dias *et al.* 2002), (Kharchenko *et al.* (2005a), Kharchenko *et al.* (2005b)), and the WEBDA database (Mermilliod & Paunzen 2003). Based on this large sample, we can embark on statistical studies of the Galactic OC system.

2. Spatial and kinematic properties of Galactic Open clusters

Using our updated database for 993 OCs with distance and age data, we plotted the cluster positions on an (X, Y) coordinate system, with a zero point at the Galactic Center (where $R_\odot = 8.0$ kpc), as shown in Fig. 1. The arc in Fig.1 represents the solar circle about the Galactic Center. One can see from this figure that there are very few OCs at the galactocentric distances less than 5 kpc. Furthermore, while young clusters (with ages less than 0.8 Gyr; see Phelps *et al.* 1994) are distributed quite uniformly around the Sun, most of the older OCs are distributed in the outer part of the disk. Such apparent distributions are partially due to the much higher extinction in the direction of Galactic Center and the deficiency of older clusters in the inner part of the disk, that has been attributed to the preferential destruction of these clusters by giant molecular clouds, which are primarily found in the inner Galaxy.

Regarding the spatial distribution perpendicular to the Galactic plane, most OCs represent the typical thin-disk population, with a scale height of about 66 pc. However, the subsample of old OCs, most of which are found in the outer disk, has a much larger scale height of 221 pc. These scale heights are in excellent agreement with the earlier results of Janes *et al.* (1988) and Janes & Phelps (1994).

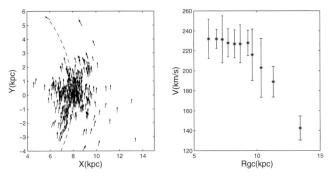

Figure 2. *Left:* Velocity projection on the Galactic plane of 369 clusters. The Sun is located at $(X = 8, Y = 0)$ kpc. *Right:* Rotation velocity of the OCs as a function of galactocentric distance

In our kinematic sample, we have compiled a total of 369 OCs for which a distance and both proper motion and radial velocity data are available. In left-hand panel of Fig. 2, we plot their projected velocity onto the disk, whilst the right-hand panel shows the averaged rotation velocity of OCs as a function of the galactocentric distance.

Based on our kinematic dataset we can also constrain the Galactic differential rotation, as well as the Galactic radial motion parameters in the solar neighborhood. We have compiled a subsample of 117 clusters with heliocentric distances of $0.5 - 2.0$ kpc and ages younger than 0.8 Gyr, which can be considered as typical thin-disk objects in the Galaxy, for which the Oort's theory is applicable. The kinematical parameters determined from these clusters can be used to represent the kinematic properties of thin-disk objects in the vicinity of the Sun. Thus, we deduced the following Galactic parameters: (i) the mean heliocentric velocity of the OC system, $(u1, u2, u3) = (-16.1 \pm 1.0, -7.9 \pm 0.4, -10.4 \pm 1.5)$ km s^{-1}, (ii) the characteristic velocity dispersions, $(\sigma_1, \sigma_2, \sigma_3) = (17.0 \pm 0.7, 12.2 \pm 0.9, 8.0 \pm 1.3)$ km s^{-1}, (iii) the Oort constants, $(A, B) = (14.8 \pm 1.0, -13.0 \pm 2.7)$ km s^{-1} kpc^{-1}, and (iv) the radial motion parameters, $(C, D) = (1.5 \pm 0.7, -1.2 \pm 1.5)$ km s^{-1} kpc^{-1} (Zhao *et al.* 2006). The parameters determined from these clusters have accuracies significantly better than those obtained from the other groups of clusters.

3. The disk abundance gradient based on open clusters

OCs can also be used as a powerful tool to understand whether and how the spatial abundance distribution changes with time, because OCs have formed at all epochs and since their ages, distances, and metallicities can be derived more reliably than the equivalent parameters of field stars.

We have compiled a full OC sample, containing 144 objects, with metallicity, distance and age parameters. From this sample, we obtain a radial metallicity gradient of -0.058 ± 0.006 dex kpc^{-1} (Chen *et al.*, in prep.), for galactocentric distances ranging from about 7 kpc to 17 kpc. By dividing the clusters into young and old subsamples (e.g., Chen *et al.* 2003), we find that the corresponding gradients are significantly different, as shown in the upper panel of Fig. 3. That is, the gradient is steeper in the past, and shallower for younger clusters. In the bottom panel of Fig. 3, the gradients of the inner and outer subsamples have similar values.

These abundance gradient results are consistent with those from the HII regions (Deharveng *et al.* 2000) or planetary nebula data (Maciel *et al.* 2006).

In a recent review, Maciel *et al.* (2006) combined abundance data from planetary nebula and OC samples to provide observational constraints on disk chemical evolution models.

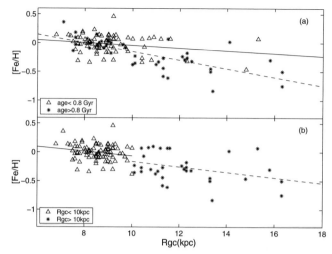

Figure 3. *(a)* Time evolution of the [Fe/H] gradient. Triangles and stars represent younger and older OCs, respectively. *(b)* Gradients for inner-disk and outer disk clusters. (Chen *et al.* 2003)

As shown in Fig. 10 of their paper, these authors investigated the abundance gradient as a function of time by combining the results derived from OC data by Friel *et al.* (2002), and by Chen *et al.* (2003). The theoretical tracks of Chappini *et al.* (2001) and Hou *et al.* (2000), are also shown, respectively. The differences between these two model predictions are rather large. In the model by Hou *et al.* (2000), based on an exponentially decreasing infall rate and an "inside-out" scenario for the formation of the Galactic disk, a rapid increase of metal abundances at early times in the inner disk caused a steep gradient. As the star formation activity migrated to the outer disk, the abundances got enhanced in that region as well, so that the gradients flatten out. In the model used by Chiappini *et al.* (2001), two infall episodes are assumed to form halo and the disk. The disk is also formed in an "inside-out" formation scenario, in which the time-scale is a linear function of the galactocentric distance. In general, some steepening of the gradient is predicted by that model. It seems that the Hou *et al.* (2000) model is better supported by the observational data.

Thus, based on OC data we may constrain the disk chemical evolution such that in the early stages of disk formation it showed a steeper abundance gradient, while later on the gradient became flatter. These inferences are not very conclusive, since we still do not have a sufficiently large sample of outer disk clusters. The compilation of such a sample is critical for these statements to be made firmer, and we propose to focus on obtaining such a sample as a priority in this field.

As regards the overall gradient along the disk, Twarog *et al.* (1997) proposed a step function for the disk abundance. That is, inside 10 kpc, there is a shallow gradient, while beyond 10 kpc the sample is too small to determine, if a gradient exists at all. They used a sample of 14 OCs in the outer disk. However, considering that we now have 34 OCs beyond 10 kpc, our conclusion of a continuous linear gradient is more likely.

4. Very young open clusters and the effects of mass segregation

In some previous work on young open clusters, including the Orion Nebula Cluster (Hillenbrand 1997) and R136 (e.g. Cambell *et al.*1992, Brandl *et al.* 1996) the authors found evidence of mass segregation. However, the question is, for such young star clusters,

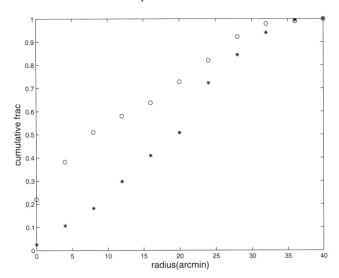

Figure 4. Normalized cumulative radial number density profile for NGC 6530 members with $m_B \leqslant 12$ (o) and $m_B > 12$ (∗). Result from Chen *et al.* (2007)

whether these effects are mainly the result of dynamical evolution or possibly primordial in nature.

Recently, we investigated two very young open clusters, NGC 2244 and NGC 6530, both with ages younger than 5 Myr. As an example, here we show some results for NGC 6530 (Chen *et al.* 2007).

The observational data were taken from our historical photographic plates at Shanghai Observatory, with a time baseline of 87 years. The old photographic plates were taken by the 40-cm refractor at Sheshan observational station.

From these photographic data, we determined proper motions for 364 stars in a one square degree region centered on the cluster, and estimated their membership probabilities. Since these cluster stars are located at the same distance, and because most of them are main sequence stars, we can use luminosity instead of mass to study the (mass/luminosity) segregation effects. Figure 4 shows the cumulative radial number density profile for bright and faint cluster stars, which shows that brighter (or massive) stars are more concentrated towards the inner part of the cluster. this indicates an evident luminosity or mass segregation effect. A similar result was obtained for our other sample cluster, NGC 2244 (Chen *et al.* 2007).

We conclude that both OCs, NGC 6530 and NGC 2244, show clear evidence of mass (luminosity) segregation. These two clusters are too young for them to have been shaped by dynamical relaxation processes. The observed segregation might be due to the combination of both initial conditions and dynamical evolution. Therefore, our results support the "primordial mass segregation" scenario, such as the "competitive accretion" model. (see Larson 1991, Bonnell *et al.* 2001a, Bonnell *et al.* 2001b)

5. The LOCS – LAMOST Open Cluster Survey project

Regarding the present status of our OC database, we need a much larger sample with both abundance and three-dimensional spatial motion data in order to study the global disk dynamical and chemical evolution. In particular, to study the disk chemical evolution, we need more outer-disk cluster data so that we can determine the abundance gradient evolution decisively. Equally important, uniform observational data sets for different clusters are very much needed. As for a single cluster, in many cases only a few stars are observed spectroscopically to derive the average metallicity and mean radial velocity. On the whole, a large and homogenous spectroscopic data set for OCs is non-existent yet.

To improve the situation, there are some important OC projects ongoing or in the planning phases. These includes, for example, WOCS (The WIYN Open Cluster Study; Mathieu 2000), which aims to obtain comprehensive and definitive astrometric, photometric and spectroscopic databases for several fundamental clusters; SOCS (SEGUE Open Cluster Survey), which, as part of the AS-2 project, is designed to obtain radial velocities and metallicities for more than 100 OCs. The Italian BOCCE project (Bragaglia 2006), by obtaining photometric and high-resolution spectroscopic data of 30 clusters, is performing a detailed investigation of their ages, distances, and abundances.

Recently we proposed the LOCS project – the LAMOST Open Cluster Survey. The Large Sky Area Multi-Object Fiber Spectroscopic Telescope (LAMOST) project (see http://www.lamost.org/en/) is one of the National Major Scientific Projects undertaken by the Chinese Academy of Science. LAMOST is a 4-meter aperture quasi-meridian reflecting Schmidt telescope, with a large field of view of about 20 square degree, and 4000 fibers.

The telescope is located at the Xinglong Observing Station of the National Astronomical Observatories. At present, a small system with an effective 2-meter aperture instrument has already obtained first light. The full system is scheduled to be finished in early 2009.

For the LOCS, the main advantage of using LAMOST is that it can be most efficient for an OC survey. With its multiple fibers and large field of view, we can complete spectroscopic survey observation for at least one cluster per observing night. This type of survey observations will have deep enough magnitude limits to reach most of the old clusters within $5 - 8$ kpc.

We expect the LOCS, in about 4-years of observations, to survey a total of amount six hundred OCs, each covering one square degree of the sky. We plan to obtain homogeneous metallicities and radial velocities of stars in each cluster. This will lead to reliable membership determinations, which will, in turn, form a fundamental basis for OC studies. Obtaining many more abundance and kinematic properties of old and distant OCs provides crucial constraints for modelling of the structure and evolution of the Galaxy.

From the selected OC targets, we choose 150 OCs with suitable ages and distances as the first-priority sample for LOCS observations. Another 13 well-studied OCs have been chosen as standard objects for comparison. We have already prepared input catalogs for stars in the selected clusters, and a preliminary membership estimation for candidate clusters is being derived.

Acknowledgements

LC, JLH and JLZ are supported by the NSFC (Grant Nos. 10773021, 10573028,10333050, 10333020,10573022) and NKBRSFG 2007CB815402 and 403. RdG was partially supported by an "International Joint Project" grant, jointly funded by the Royal Society in the UK and the Chinese National Science Foundation (NSFC). This research has made use of the WEBDA database, operated at the Institute for Astronomy of the University of Vienna.

References

Bonnell, I. A., Bate, M. R.,Clarke, C. J., & Pringle, J. E. 2001a, *MNRAS*, 323, 785

Bonnell, I. A., Clarke, C. J., Bate, M. R., & Pringle,J. E. 2001b, *MNRAS*, 324, 573

Brandl, B., Sams, B. J., Bertoldi, F., Eckart, A.,Genzel, R., Drapatz, S., Hofmann, R., Löwe, M., & Quirrenbach, A. 1996, *ApJ*, 466, 254

Bragaglia, A. & Tosi, Monica. 2006, *AJ*, 131, 1544

Campbell, B., *et al.* 1992, *AJ*, 104, 1721

Carraro, G., Ng, Y. K., & Portinari, L. 1998, *MNRAS*,296, 1045

Chen, L., de Grijs, R., & Zhao, J. L. 2007, *AJ*, 134, 1368

Chen, L., Hou, J. L., & Wang, J. J. 2003, *AJ*, 125, 1397

Chiappini, C., Matteucci, F., & Romano, D. 2001, *ApJ*, 550, 1044

Deharveng, L., Peña, M., Caplan, J., Costero, R. 2000, *MNRAS*, 311, 329

Dias, W. S., Alessi, B. S., Moitinho, A., & Lépine, J. R. 2002, *A&A*, 389, 871

Friel, E. D., Janes, K. A., Tavarez, M., *et al.* 2002, *AJ*, 124, 2693

Hillenbrand, L. A. 1997, *AJ*, 113, 1733

Hou, J. L., Prantzos, N., & Boissier, S. 2000, *A&A*, 362, 921

Janes, K. A., Tilley, C., & Lyngå, G.1988, *AJ*, 95, 771

Janes, K. A. & Phelps, R. L. 1994, *AJ*, 108, 1773

Kharchenko, N. V., Piskunov, A. E., Roeser, S., Schilbach, E., & Scholz, R.-D. 2005, *A&A*, 438, 1163

Kharchenko, N. V., Piskunov, A. E., Roeser, S., Schilbach, E., & Scholz, R.-D. 2005, *A&A*, 440, 403

Larson, R. B. 1991, in: Falgarone, E., Boulanger, F., & Duvert, G., (eds.), *Fragmentation of Molecular Clouds and Star Formation, IAU Symp. 147* (Dordrecht: Kluwer), p. 261

Maciel, W. J., Lago, L. G., & Costa, R. D. D. 2006, *A&A*, 453, 587

Mathieu, R. D. 2000, in: R. Pallavicini, G. Micela, & S. Sciortino, (eds.), *Stellar Clusters and Associations: Convection, Rotations, and Dynamos, ASP Conference Series*, p. 517

Mermilliod, J.-C. & Paunzen, E. 2003, *A&A*, 410, 511

Phelps, R. L., Janes, K. A., & Montgomery, K. A. 1994, *AJ*, 107, 1079

Twarog, B. A., Ashman, K. A., & Anthony-Twarog, B. J. 1997, *AJ*, 114, 2556

Zhao, J. L., Chen, L., & Zu, Z. L. 2006, *Chin. J.Astron.Astrophysics*, 6, 287

A Giant Step: from Milli- to Micro-arcsecond Astrometry
Proceedings IAU Symposium No. 248, 2007
W. J. Jin, I. Platais & M. A. C. Perryman, eds.

© 2008 International Astronomical Union
doi:10.1017/S1743921308019777

Distances and ages of globular clusters

B. Chaboyer

Department of Physics and Astronomy, Dartmouth College, Hanover, NH 03755, USA
email: Brian.Chaboyer@Dartmouth.edu

Abstract. As the oldest objects whose ages can be accurately determined, Galactic globular clusters can be used to establish the minimum age of the universe (and hence, to constrain cosmological models) and to study the early formation history of the Milky Way. The largest uncertainty in the determination of globular cluster ages is the distance scale. The current uncertainty in the distances to globular clusters is $\sim 6\%$, which leads to a 13% uncertainty in the absolute ages of globular clusters. I am the PI on a SIM-Planetquest key project to determine the distances of 21 globular clusters with an accuracy of ranging from 1 to 4%. This will lead to age determinations accurate to $5 - 9\%$. The mean age of the oldest, most metal-poor globular clusters will be determined with an accuracy of $\pm 3\%$.

Keywords. globular clusters: general

1. Introduction

A typical globular cluster contains several hundred thousand stars, all with the same age and composition. As a result, globular clusters provide an excellent test of stellar evolution models and represent the oldest objects in our galaxy for which accurate ages can be obtained. There is still considerable uncertainty about the distance scale to globular clusters, and Population II stars in general. For example, the review by Krauss & Chaboyer (2003) reviews 8 different methods to determine the distances to globular clusters, and concludes that these various methods imply an uncertainty in the distance modulus to globular clusters of ± 0.12 mag, or $\pm 6\%$ in the distance. The nearest globular cluster is approximately 2 kpc distant, and there are only a few globular clusters within 5 kpc of the Sun. Significant progress in refining the distance scale to globular clusters will require parallaxes accurate at the micro-arcsecond level.

2. Globular cluster ages

The ages of globular clusters are determined by comparing the results of stellar evolution calculations to the observed properties of stars within a globular cluster. There has been some recent progress in using white dwarfs to determine the age of the nearest globular cluster (Hansen *et al.* 2007), but white dwarfs are faint and it is extremly difficult to observe the entire white dwarf cooling sequence in a globular cluster. As a result, the main sequence turn-off region is the diagnostic which is usually used to determine globular cluster ages.

The calculations of stellar evolution models require a knowledge of the physical processes which occur in high temperature plasma. This includes such things as nuclear reaction rates, opacities, the equation of state and turbulent convection. These input ingredients to stellar evolution calculations are not known exactly, but the uncertainty in this calculations can be quantified, which in turn allows for a reliable estimate in the uncertainty in the determination of globular cluster ages. The stellar evolution Monte Carlo simulation of Krauss & Chaboyer (2003) demonstrated that, for absolute ages, the absolute magnitude of the main sequence turn-off region was the best diagnostic of the absolute age of globular clusters. Of course, determining the absolute magnitude of the turn-off requires a knowledge of the distance to a globular cluster. Krauss & Chaboyer (2003) show that the current uncertainties in the distance scale to globular clusters and in the stellar evolution calculations leads to an uncertainty of $\pm 13\%$ in the determination of absolute age of the oldest globular clusters.

3. Globular cluster distances with SIM-Planetquest

NASA had an open call for proposals for SIM key science projects in 2000, and as a result 10 key projects were selected. The importance of improving our knowledge of the globular cluster distance scale lead to the selection of our proposal to NASA for a SIM key science project to determine parallaxes to a number of globular clusters. The design goal of SIM-Planetquest is the determination of stellar parallaxes with an accuracy of ± 4 micro-arcsecond. As the current distance scale to globular clusters is accurate to $\pm 6\%$, one would like to have individual parallaxes accurate to $\pm 4\%$ in order to make a significant improvement in our knowledge of the globular cluster distance scale. This implies observing globular clusters which are within 10 kpc of the Sun.

There are of order 90 globular clusters within 10 kpc of the Sun (Harris 1996). However, many of these globular clusters are located in the bulge of the Milky Way, and are highly reddened. Determination of the absolute magnitude of the turn-off requires a knowledge of the absorption along the line of sight in addition to the parallax. Going to the infrared can significantly reduce the uncertainty in the absorption along the line sight. However, age determinations which use the absolute magnitude of the main sequence turn-off in the K band are a factor of two more sensitive to distance uncertainties than those ages determined in the V band. For example, an uncertainty of ± 0.06 mag in the K band leads to a 12% uncertainty in the age determination, while an uncertainty of ± 0.06 mag in the V band leads to a $\pm 6\%$ uncertainty in the age determination. As a result, for accurate age determination, one needs to use globular clusters which have low interstellar absorptions.

Our SIM key project 'Anchoring the Population II Distance Scale: Distances and Ages of Globular Clusters' selected 21 globular clusters within 10 kpc of the Sun for observation by SIM. These clusters are listed in Table 1. In this table, the distance from the Sun ($R_\odot$), reddening (E(B – V)), metallicity ([Fe/H]), and magnitude of the horizontal branch (V_{HB}) are taken from the 2003 online version of the Harris (1996) catalog of Milky Way globular clusters. The estimated percent uncertainty in the parallax to each cluster is listed under σ_π, while σ_{age} gives the percent uncertainty in the age. The uncertainty in the age includes uncertainties in the distance, reddening, composition and stellar evolution models. It is based upon a Monte Carlo simulation which takes into account all known errors in globular cluster age determination, as discussed by Krauss & Chaboyer (2003).

Of key interest to cosmology is the absolute age of the oldest globular clusters. There are 5 metal-poor globular clusters which are believed to be among the first objects formed in the halo of the Milky Way. Given that these clusters are scattered over the sky, the error in the distance uncertainty will not be correlated, allowing one to average together the age determinations and determine the mean age with a lower uncertainty. Of course, the possible errors in stellar evolution models are the same for all clusters, so one must take into account this correlated error when determining the mean age of the oldest, most metal-poor globular clusters. Taking this into account, parallaxes from SIM-Planetquest will allow our key project to determine the absolute age of the oldest globular clusters with an uncertainty of $\pm 3\%$.

B. Chaboyer

Table 1: Target Globular Clusters

NGC	Name	$R_\odot$ (kpc)	E(B − V)	V_{HB}	[Fe/H]	σ_π (%)	σ_{age} (%)
			[Fe/H]< −1.80				
6397	–	2.2	0.18	12.87	−1.95	1	6
6809	M55	5.3	0.07	14.40	−1.81	2	5
6541	–	7.4	0.12	15.30	−1.83	3	6
7099	M30	7.9	0.03	15.10	−2.12	3	6
6341	M92	8.1	0.02	15.10	−2.29	3	6
4590	M68	10.1	0.04	15.68	−2.06	4	7
			−1.80 < [Fe/H] < −1.45				
6752	–	3.9	0.04	13.70	−1.55	2	5
6218	M12	4.7	0.19	14.60	−1.48	2	7
3201	–	5.1	0.21	14.80	−1.48	2	8
5139	ω Cen	5.1	0.12	14.53	−1.62	2	9
6205	M13	7.0	0.02	14.90	−1.54	3	6
5272	M3	10.0	0.01	15.65	−1.57	4	8
			−1.45 < [Fe/H] < −1.00				
5904	M5	7.3	0.03	15.07	−1.29	3	6
6362	–	7.5	0.09	15.34	−1.06	3	7
288	–	8.1	0.03	15.30	−1.24	3	7
362	–	8.3	0.05	15.43	−1.16	3	7
6723	–	8.6	0.05	15.50	−1.12	3	7
			[Fe/H]> −1.00				
6838	M71	3.8	0.25	14.44	−0.73	2	6
104	47 Tuc	4.3	0.05	14.06	−0.76	2	6
6352	–	5.6	0.21	15.13	−0.70	2	8
6652	–	9.4	0.09	15.85	−0.85	4	8

This significant reduction in the uncertainty in the absolute age of the oldest globular clusters will provide a strong consistency check on cosmological models. For example, an age of 15.5 ± 0.4 Gyr for the oldest clusters would be inconsistent at the 4σ with the currently favored cosmological model, which implies an age of 13.7 ± 0.2 Gyr (Spergel *et al.* 2003). Alternatively, if the absolute globular cluster ages are consistent with cosmology, then one can use the absolute ages to infer the redshift of formation for the oldest stars in the Milky Way. For example, an age determination of 11.3 ± 0.3 Gyr for the oldest globular clusters would imply that these clusters formed at a redshift of $z \simeq 3$.

The launch of SIM-Planetquest has been delayed a number of times, and currently there is no firm launch date. The earliest possible launch date is 2012. Parallaxes accurate to 4 micro-arcseconds will require 5 years of observing by SIM-Planetquest. Thus, it will be some time before we will obtain accurate parallaxes to globular clusters. When the data do arrive, they will lead to an important advance in our understanding of globular cluster ages, and provide new constraints on galaxy formation and cosmological models.

References

Hansen, B. M. *et al.* 2007, ApJ, 671, 380
Harris, W. E. 1996, AJ, 112, 1487
Krauss, L. M. & Chaboyer, B. 2003, Science, 299, 65
Spergel, D. N. *et al.* 2003, ApJS, 148, 175

A Giant Step: from Milli- to Micro-arcsecond Astrometry
Proceedings IAU Symposium No. 248, 2007
W. J. Jin, I. Platais & M. A. C. Perryman, eds.

© 2008 International Astronomical Union
doi:10.1017/S1743921308019789

Modelling the Galaxy from survey data

A. C. Robin[1], C. Reylé[1] and D. Marshall[1,2]

[1]CNRS-UMR6213, Institut Utinam, Observatoire de Besançon, BP1615, F-25010 Besançon
cedex, France
email: annie.robin@obs-besancon.fr

[2]Département de Physique, Génie Physique et Optique, Université Laval, Québec, QC, G1K
7P4, Canada

Abstract. Recent optical and near-infrared surveys have considerably improved our knowledge of galactic structure and galactic evolution. Two ways can be used to infer this knowledge from datasets: either inversing the data to get parameters describing the Galaxy, or using a synthetic approaches to test scenarios of formation and theoretical models for star and galaxy formation and evolution, both approaches being complementary. Using the synthetic approach the Besancon Galaxy model allows to test scenarios for the structure and evolution of the Galaxy by comparing simulations with the survey data. Examples are given using the 2MASS survey. Future uses of astrometric survey data are shown to be able to efficiently constrain the kinematics and dynamics of the Galaxy.

Keywords. Galaxy: stellar content, Galaxy: evolution, Galaxy: kinematics and dynamics

1. Introduction

Constraints on Galactic structure and evolution come from a wide variety of data sets, such as multi-wavelength photometry, kinematics, microlensing events, among others. Scenarios of Galaxy formation and evolution are inferred from these constraints. An ultimate test of these scenarios can be done using these constraints piece by piece in order to build a population synthesis model which predictions can be directly compared with observations. This synthetic approach ensures that biases have been correctly taken into account and that the scenario is compatible with a variety of constraints. In recent years wide surveys have been obtained from optical and near-infrared photometry thanks to the wide-field CCD mosaic cameras, multi-object spectrometers, and the dedicated ground- and space-based telescopes. Astrometric accuracy has been largely improved and catalogues with accurate proper motions are available, well-calibrated by Hipparcos and Tycho data. The homogeneity of these data sets is a great help for avoiding systematic bias percolating from one set to another, which have created difficulties in data interpretation in the past. All these data sets benefit to Galactic evolution studies and provide constraints on the population synthesis approach.

Here we report on the development of the Besançon Galaxy model based on this synthesis approach. We overview the basic scheme and input (Sect. 2) and we describe new results obtained by analysis of large data sets like the 2MASS surveys, in particular of the Galactic central region and on the disk external structure (Sect. 3). We then describe (Sect. 4) which constraints on kinematics have to be taken into account by any dynamical model and how the future astrometric surveys will help to build a fully self-consistent model of the Galaxy.

2. The population synthesis approach

The population synthesis approach aims at assembling current scenarios of galaxy formation and evolution, theories of stellar formation and evolution, models of stellar atmospheres and dynamical constraints, in order to make a consistent picture explaining currently available observations of different types (photometry, astrometry, spectroscopy) at different wavelengths. The validity of a Galactic model is always questionable, as it describes a smooth Galaxy, while inhomogeneities exist, either in the disk or the halo. The issue is not to make a perfect model that reproduces the known Galaxy at all scales. Rather one aims to produce a useful tool to compute the probable stellar content of large data sets and therefore to test the usefulness of such data to answer a given question in relation to Galactic structure and evolution. Modelling is also an effective way to test alternative scenarios of galaxy formation and evolution.

The originality of the Besançon model, as compared to a few other population synthesis models presently available for the Galaxy, is in its dynamical self-consistency. The Boltzmann equation allows the scale height of an isothermal and relaxed population to be constrained by its velocity dispersion and the Galactic potential (Bienaymé et al. 1987). The use of this dynamical constraint eliminates a set of free parameters which are difficult to determine: the scale height of the thin disc at different ages. It gives the model an improved physical credibility. However this constraint is only applied at the solar position and perpendicular to the plane. A fully consistent dynamical model would apply self-consistency constraints at any position in the Galaxy.

The main scheme of the model is to reproduce the stellar content of the Galaxy, using some physical assumptions and a scenario of formation and evolution. We assume that stars belong to four main populations : the thin disk, the thick disk, the stellar halo (or spheroid), and the outer bulge. The modelling of each population is based on a set of evolutionary tracks, assumptions on density distributions, constrained either by dynamical considerations or by empirical data, and guided by a scenario of formation and evolution, that is to assume certain initial mass function (IMF) and star formation rate (SFR) for each population. More detailed descriptions on these constraints can be found in Haywood et al. (1997) for the thin disk, Reylé & Robin (2001) for the thick disk, Robin et al. (2000) for the spheroid, and Picaud & Robin (2004) for the outer bulge.

The Galactic model has been developed to return the results in the near-infrared and visible filters, but it has been extended to predict the stellar content in the X-ray domain (Guillout et al. 1996). More recently, the Hipparcos mission and large scale surveys in the optical and the near-infrared have led to new physical constraints improving our knowledge of the overall structure and evolution of the Galaxy. These new constraints are now included in the version of the model described in Robin et al. (2003).

2.1. Limitations

The model essentially produces a smooth Galaxy, which is certainly an over-simplification of reality. Inhomogeneities exist which cannot be easily modelled (clusters, associations, star forming regions). Major streams could be modelled, however, it might not be an objective of a model to reach such a degree of realism. To search for streams in observations, it may be more useful to produce simulations without streams and subtract them from the observations in order to amplify the contrast of the stream. A homogeneous model is also a good tool to estimate the degree of homogeneity present in real data with respect to the Poisson noise included in the smooth model.

3. 2MASS survey

The first effect visible in the star density distribution near the Galactic plane is due to the dust, even at near-infrared wavelength. Without a good estimate of the extinction and of its distance it is nearly impossible to understand the structure in the disk. Hence we have first attempted to build a 3D model of extinction in the plane. The 2MASS survey allows to study large scale structures in the Galaxy, particularly in the Galactic plane because the NIR data are well suited to study stellar populations in regions of medium to high extinction. In this survey the star counts appear to be dominated by red clump giants. Here we show how it can be used to determine the 3D distribution of the extinction and then to constrain structure of the disk.

3.1. *Extinction model*

Extinction is so clumpy that it drives the number density of stars more than any other large scale stellar structure. However, photometry and star counts contain information about the dust extinction. Marshall *et al.* (2006) have shown that the 3D extinction distribution can be inferred from the stellar colour distributions from the 2MASS survey. Using stellar colours in J-K as extinction indicators and assuming that most of the model prediction deviations in the observed colours arise from the variation of extinction along the line of sight, they built a 3D extinction model of the galactic plane ($-10 < b < 10$ deg and $-90 < l < 90$). The resolution in longitude and latitude is 15 arcmin and the resolution in distance varies between 100 pc to 1 kpc, depending on stellar density and the density of dust along the line of sight. The resulting 3D extinction model provides an accurate description of the large scale structure in the disk of dust.

3.2. *Comparing 2MASS with model star counts*

A comparison between the basic model described in Robin *et al.* (2003) – modified by the 3D extinction model of Marshall *et al.* (2006) – and 2MASS star counts (Fig. 1) shows that the model reproduces the data with a high degree of realism in the plane, even without modeling the spiral arms. It means that either the spiral structure does not have a high contrast or that the 2MASS data are not a good tracer for these arms. This is expected as the model estimates that the 2MASS counts are dominated by red clump giants which are old stars (90% have ages >1 Gyr, having made at least 4 revolutions around the Galaxy) and have lost the memory of their place of birth. However, two non-axisymmetric features are not well modelled : the external bulge and the warp. This is easily seen in Fig. 2. The triaxial bulge model has been adjusted by Picaud & Robin (2004) in the central region limited in longitude from -10 to +10 and -4 and +4 in latitude using, the DENIS survey data (Epchtein *et al.* 1997). Comparisons between predictions and 2MASS data in Fig. 2 show that this region is well reproduced but the more external bulge is not : differences appear at latitudes $4 < |b| < 10$ and at longitudes $5 < |l| < 10$. We are presently investigating this problem. Our preliminary conclusion is that the central part of the Galaxy ($-5 < l < 5$) is well fitted by the triaxial old bulge model with a small angle between the major axis and the sun-center axis, as found by Picaud & Robin (2004), while the external part in the plane $5 < |l| < 10$ is better fitted by a bar with a higher angle, the bar being also a more elongated structure than the bulge with a smaller scale height of its minor axis.

3.3. *The Galactic warp*

Warps may have originated from interactions between the disk and (i) the dark halo (if the angular momenta are not aligned), (ii) nearby satellite galaxies, such as the Sgr dwarf

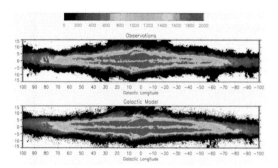

Figure 1. 2MASS star counts per square degree and model simulations in the Galactic plane as a function of longitude and latitude. The scale is number of stars per square degree. Top: Data, Bottom: Model.

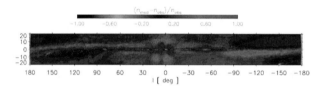

Figure 2. Relative difference between 2MASS star counts per square degree and standard model simulations in the Galactic plane (simulated minus observed, divided by observed star counts).

or the Magellanic Clouds, (iii) infalling intergalactic gas. Details of these scenarios can be found in Bailin (2003), Garcia-Ruiz (2002) or López-Corredoira *et al.* (2002).

For simulations we use a tilted ring model such as by Porcel *et al.* (1997). The free parameters are the galactocentric radius where the warp starts, the slope and the azimuthal angle of the line of nodes. Following Burton (1988), we assume that the Sun lies approximately on the line of nodes of the warp ($\theta_{\mathrm{max}} = 90°$). We adopt a slope similar to Gyuk *et al.* (1999) ($\gamma_{\mathrm{warp}} = 0.18$). Preliminary results obtained from the analysis of the DENIS survey (Derrière & Robin 2001; Derrière 2001) indicate the best value for a starting galactocentric radius of the warp to be at $R_{\mathrm{warp}} = 8.4$ kpc, close to $R_\odot$.

It appears from Fig. 2 that the slope of the warp is not correct in model simulations. Apparently, the adopted slope is too high. Using a value of 0.09 gives very good agreement between the data and simulations at positive longitudes (see Fig. 3), while at negative longitudes that is still unsatisfactory. This effect seems to be independent of the assumed slope, or the disk scale length (assumed to be 2.2 kpc), or to the sun-center distance (assumed to be 7.9 kpc). Alternative models of the warp could explain this structure. It is well known that the HI warp is not symmetrical : the gas warp bends back to the Galactic plane in the southern hemisphere at R> 15 kpc (Burton 1988). The stellar warp may also follow this feature. Tweaking the model may allow for the adjustment of this feature. However, a dynamical approach is needed to constrain the origin of this warp and its link to the gaseous warp.

4. Kinematics and dynamics of the Galactic stellar populations

The objective of the construction of this model of stellar population is to have a totally self-consistent and realistic model of the Galaxy, consistent with our knowledge

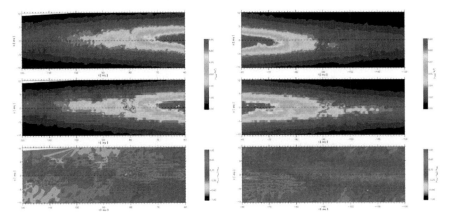

Figure 3. Comparison of 2MASS star counts (top panel) with modeled ones (middle panel), and relative difference between the two (bottom panel) for the warped region at positive longitudes (on the left) and at negative longitudes (on the right), with a warp slope of 0.09.

of Galactic evolution, including chemical and dynamical aspects. However, Galactic dynamics studies are complex and require the modelling of some kinematical features of stellar populations such as invisible matter, non-axisymmetries, inhomogeneities. We do not discuss here the kinematical and dynamical features in the halo, nor the thick disk, which are detailed in Majewski (2008) and Haywood (2008) at this conference. Instead, we concentrate on the disk kinematical features and their relation with dynamics.

4.1. *Disk non-axisymmetries*

The stellar disk shows a number of kinematical features related to its axisymmetries : the stars rotate in the disk with a nearly flat rotation curve in a large part of the disk. The stellar kinematics can be described as a velocity ellipsoid (σ_U, σ_V, σ_W). It has been shown that the values of velocity dispersion follows the stellar age : old stars have larger velocity dispersions than the young ones. That is generally explained by the orbit diffusion due to interactions with the gaseous disk (spiral structure and giant molecular clouds). The vertical velocity dispersion runs from about 5 km/s for young stars (this is about the velocity dispersion from the gas they were born) to a range of 15 to 30 km/s for the oldest disk stars (about 10 Gyr). There are large discrepancies in the literature for the latter values. Gómez *et al.* (1997) find a maximum value of 17 km/s, like Famaey *et al.* (2005), while Rocha-Pinto *et al.* (2004) argue for 30 km/s and Nordstrom *et al.* (2004) find 25 km/s. This disagreement may be explained by contamination of the oldest disk star samples by thick disk stars, which are not easily distinguished from the old thin disk neither by their abundances nor by their rotational velocity (Haywood 2008). Hence, the efficiency of the orbit diffusion in the disk is still a matter of debate, as well as the physical processes at work.

On the top of these axisymmetric features, non-axisymmetries are distinguishable even at the solar position, coming from the effects of bar (Fux 2001; Dehnen 2000) or from the spiral structure (Famaey *et al.* 2005). These non-axisymmetries translate into observables in the form of non zero mean motion of the Local Standard of Rest towards the Galactic center, or away from the Galactic Plane, or by a vertex deviation of the velocity ellipsoid. This last feature, long known from local kinematical studies, may also depend on the ages of a star sample considered.

4.2. *Disk inhomogeneities*

Disk inhomogeneities in positions are also well known. Clusters and associations are numerous in the disk but they participate to the mass distribution to a small level and do not perturb much the whole dynamics of the system. On the contrary, there exist inhomogeneities in the phase space which are very interesting to study as they inform on the overall dynamics. Famaey *et al.* (2005), following an earlier study of Eggen (1958) have identified a number of streams in the local neighborhood from the Hipparcos catalogue. From the analysis of 3D velocities of K and M giants they identify up to 6 kinematical groups and among them, the Hercules stream, Hyades supercluster, Sirius Moving group and the Pleiades supercluster. They show that these groups are not homogeneous in age. Their common motions cannot originate from a common birth history. Hence, the most probable explanation is that they are due to dynamical perturbation by transient spiral waves or related to the bar's outer Lindblad resonance.

The vertex deviation gives another way to investigate the non-axisymmetry in the disk. Soubiran et al. (2003) investigated the vertex deviation from a non local sample of stars towards the North Galactic Pole. They found no evidence of vertex deviation for old disk stars at different distances from the Galactic plane, showing that this population looks well relaxed and non perturbed. This is clear disagreement with Famaey *et al.* (2005) conclusions which argue that the vertex deviation is noticeable even for old populations and probably related to the Hercules stream. The complex dynamics of the thin disk is yet to be understood.

5. Implications of large scale astrometric surveys

Present and future large scale astrometric surveys will allow us to take a big step towards the understanding of the Galaxy formation and evolution. The Galactic potential and Galactic dynamical features are difficult to constrain when one has only 1 or 2 velocity vectors (radial velocities or proper motions) and 2 out of 3 accurate positions, the distances being very inaccurate in the absence of trigonometric parallaxes. There have been significant attempts to produce dynamical models since the Hipparcos mission, and several surveys of radial velocities and proper motions have been completed. However, the determination of a unique dynamical model is nearly impossible and the solutions remain degenerate in the absence of good distances. This is particularly true for the bulge, where different models exist with quite different dynamical properties. This will change when true 6D phase space will be available, for example from the Gaia space mission. Gaia will supply us with very accurate parallactic distances (with an accuracy of about 10% at the Galactic center and much better in the solar neighbourhood), very accurate proper motions and reasonably good radial velocities. It could be complemented by the ground-based radial velocity measurements. RAVE (RAdial Velocity Experiment) (Steinmetz *et al.* 2006) has already started a survey of a significant part of the sky with 1 million stars to be measured from medium resolution spectroscopy giving the radial velocity with an accuracy of about 2 km/s together with stellar atmosphere parameters (effective temperature, metallicity and gravity). This survey is cross-identified with ground based proper motions surveys giving access to at least 5D phase space data (accurate distances are still missing). This will furnish very good data towards the construction of a realistic dynamical model of the Galaxy, but still be limited to a region of a few kiloparsecs around the sun. They are some attempts to deduce such a model by inversion of the parameter space (Veltz *et al.* 2007). Even if this study has not given firm conclusions because it is limited by the sample size, it shows that this approach is feasible and could lead to a unique solution if the data are accurate enough and the sample is large enough. This kind of analysis is expected to be much more useful during the analysis of multivariate data from the Gaia mission.

6. Conclusions

Population synthesis models are important tools for interpretation of data coming from large scale surveys, either photometric, astrometric or both. Although imperfect, they allow to examinate the distribution functions corresponding to a given scenario, by a way of producing simulations to be directly compared with the data sample. Self consistent dynamical models coupled with population synthesis (which reproduce different aspects of the data and can account for the photometric selection and observational bias) are somehow lacking. Large efforts to produce complete chemo-dynamical models have already been started by different groups and need to be continued. Future astrometry missions will need these models to ease their interpretation and to help to construct more realistic scenarios for Galactic formation and evolution.

References

Bailin, J., 2003, ApJ, 583, L79.
Burton, W. B.: 1988, Galactic and Extragalactic Radio Astronomy, 2nd version, p. 295, Springer-Verlag
Bienaymé, O., Robin, A. C., & Crézé, M. 1987, A&A, 180, 94
Cutri, R. M., Skrutskie, M. F., van Dyk, S. *et al.*, 2003, Explanatory Supplement to the 2MASS All Sky Data Release
Dehnen W., 2000, AJ, 119, 800
Derrière, S. & Robin, A. C. 2001, ASP Conf. Ser. 232: The New Era of Wide Field Astronomy, p. 229
Derrière, S. 2001, Thèse de doctorat, Université Louis Pasteur, Strasbourg.
Eggen O. J., 1958, MNRAS, 118, 65
Epchtein, N. *et al.* 1997, The Messenger, 87, 27
Famaey B., Jorissen A., Luri X., *et al.*, 2005, A&A, 430, 165
Fux R., 2001, A&A, 373, 511
Garcia-Ruiz, I., Kuijken, K., Dubinski, J., 2002, MNRAS, 337, 459
Gómez, A. E., Grenier, S., Udry, S., Haywood, M., Meillon, L., Sabas, V., Sellier, A., & Morin, D. 1997, ESA SP-402: Hipparcos - Venice '97, 402, 621
Guillout, P., Haywood, M., Motch, C., & Robin, A. C. 1996, A&A, 316, 89
Gyuk, G., Flynn, C., & Evans, N. W. 1999, ApJ, 521, 190
Haywood, M., Robin, A. C., & Crézé, M. 1997, A&A, 320, 440
Haywood M., 2008, in this volume p.458
López-Corredoira, M., Betancort-Rijo, J., Beckman, J. E., 2002, A&A, 386, 169
Majewski S., 2008, in this volume p.450
Marshall, D. J., Robin, A. C., Reylé, C., Schultheis, M., & Picaud, S. 2006, A&A 453, 635
Nordstrom B., Mayor M., Andersen J., *et al.*, 2004, A&A, 418, 989
Picaud, S., & Robin, A. C. 2004, A&A, 428, 891
Porcel, C., Battaner, E., & Jimenez-Vicente, J.: 1997, A&A 322, 103
Reylé, C. & Robin, A. C. 2001, A&A, 373, 886
Robin, A. C., Reylé, C., Crézé, M. 2000, A&A 359, 103
Robin, A. C., Reylé, C., Derrière, S., & Picaud S. 2003, A&A, 409, 523.
Rocha-Pinto, H. J., *et al.*, 2004, A&A, 423, 517
Soubiran, C., Bienaymé, O., & Siebert, A., 2003, A&A, 398, 141
Steinmetz, M. *et al.*, 2006, AJ 132, 1645
Veltz, L. *et al.*, 2007, submitted to A&A

A Giant Step: from Milli- to Micro-arcsecond Astrometry
Proceedings IAU Symposium No. 248, 2007
W. J. Jin, I. Platais & M. A. C. Perryman, eds.

Precision astrometry, galactic mergers, halo substructure and local dark matter

S. R. Majewski

Department of Astronomy, University of Virginia
P.O. Box 400325, Charlottesville, VA, 22904-4325, U.S.A.
email: srm4n@virginia.edu

Abstract. The concordance Cold Dark Matter model for the formation of structure in the Universe, while remarkably successful at describing observations on large scales, has a number of problems on galactic scales. The Milky Way and its satellite system provide a key laboratory for exploring dark matter (DM) in this regime, but some of the most definitive tests of local DM await microarcsecond astrometry, such as will be delivered by the Space Interferometry Mission (SIM Planetquest). I discuss several tests of Galactic DM enabled by future microarcsecond astrometry.

Keywords. astrometry, stars: kinematics, Galaxy: halo, Galaxy: kinematics and dynamics, galaxies: interactions, dark matter

1. Introduction

Since the seminal study of Searle & Zinn (1978) the notion of accretion of "subgalactic units", including "late infall", has been a central concept of stellar populations studies. N-body simulations of the formation of structure in the Universe in the presence of dark matter (and dark energy) also show galaxies (and all large structures) building up hierarchically. However, while the active merging history on all size scales demonstrated by high resolution, Cold Dark Matter (CDM) numerical simulations have had remarkable success in matching the observed properties of the largest structures in the Universe (like galaxy clusters), they are a challenge to reconcile with the observed properties of structures on galactic scales. The Milky Way (MW) and its satellite system are a particularly important laboratory for testing specific predictions of the CDM models. The era of microarcsecond astrometry such as will be delivered by SIM Planetquest, will enable a number of definitive dynamical tests of local CDM by way of determining the distribution of Galactic dark matter (DM). We discuss several of these tests here.

2. Measuring $\Theta_{\rm LSR}$ with the Sagittarius stream

An understanding of the distribution and amount of DM in the Galaxy — indeed, virtually every problem in Galactic dynamics — depends on establishing the benchmark parameters R_0, the solar Galactocentric distance, and $\Theta_{\rm LSR}$, the Local Standard of Rest (LSR) velocity. A 3% error in both R_0 and $\Theta_{\rm LSR}$ leads to a 5% error in the determination of the Galactic mass scale using, e.g., traditional Jean's equation methods.

Despite decades of effort, the rate of Galactic rotation at the Sun's position remains uncertain, with measurements varying by 20% or more. Hipparcos proper motions have been used to determine that $\Theta_{\rm LSR} = (217.5 \pm 7.0)(R_0/8)$ km s^{-1} using disk Cepheid variables (Feast & Whitelock 1997) and $(240.5 \pm 8.3)(R_0/8)$ km s^{-1} from OB stars (Uemura *et al.* 2000), while Hipparcos proper motions of open clusters have yielded

$207(R_0/8)$ km s^{-1} (Dias & Lépine 2005) [though see the alternative analysis of these open cluster by Frinchaboy 2006 and Frinchaboy & Majewski 2006, which yields (221 ± 3) km s^{-1} for any R_0 in the range $7 \leqslant R_0 \leqslant 9$ kpc]. Meanwhile, Hubble Space Telescope measurements of the proper motions of bulge stars against background galaxies in the field of the globular cluster M4 yield $\Theta_{\mathrm{LSR}} = (202.4 \pm 20.8)(R_0/8)$ km s^{-1} (Kalirai *et al.* 2004) and $(220.8 \pm 13.6)(R_0/8)$ km s^{-1} (Bedin *et al.* 2003). Radio measures of the proper motion of Sgr A* (Reid & Brunthaler 2004) yield a transverse motion of $(235.6 \pm 1.2)(R_0/8)$ km s^{-1}, but, when corrected for the solar peculiar motion, yield $\Theta_{\mathrm{LSR}} = 220(R_0/8)$ km s^{-1} (M. Reid, this proceedings). Of course, as may be seen, most of these measures depend on an accurate measure of R_0 as well as knowledge of the solar motion (which is apparently now more controversial than previously suspected; cf. the significantly smaller solar motion derived by Dehnen & Binney 1998 compared to that from M. Reid, this proceedings). Moreover, considerations of non-axisymmetry of the disk yield corrections to the measurements that suggest Θ_{LSR} may be as low as 184 ± 8 km s^{-1} (Olling & Merrifield 1998) or lower (Kuijken & Tremaine 1994).

Based on recent estimates of the luminosity of the MW (Flynn et al. 2006), any Θ_{LSR} exceeding 220 km s^{-1} puts our Galaxy more than 1σ away from the Tully-Fisher relation. Clearly, additional independent methods to ascertain Θ_{LSR} are valuable for establishing with certainty the MW mass scale and whether our home galaxy is unusual.

An ideal method for ascertaining Θ_{LSR} would be to measure the solar motion with respect to a *nearby* reference known to be at rest in the Galactic reference frame (at least in the disk rotation direction), since this would not only overcome traditional difficulties with working in the highly dust-obscured and crowded Galactic center, but be independent of R_0 and any assumptions that the reference lies in the center of the MW potential. The debris stream from the tidally disrupting Sagittarius (Sgr) dSph galaxy provides almost nearly this ideal situation. This dSph and its extended tidal debris system orbit the MW in nearly a polar orbit, with a Galactic plane line-of-nodes almost exactly along the Galactic X-axis. The remarkable coincidence that the Sun presently lies within a kiloparsec of the Sgr debris plane, which has a pole at $(l_p, b_p) = (272, -12)°$ (Majewski *et al.* 2003), means that the motions of Sgr stars *within* this plane are almost entirely contained in their Galactic U and W velocity components, whereas their V velocities almost entirely reflect *solar motion*. Thus, in principle, the solar motion can be derived directly by the $\mu_l \cos(b)$ dimension of the Sgr stream proper motion — though refined results come from comparing to Sgr models that can account for deviations from the ideal case (Majewski *et al.* 2006). An additional advantage is that stars in the Sgr stream, particularly its substantial M giant population (Majewski *et al.* 2003), are ideally placed for uncrowded field astrometry at high MW latitudes, and at relatively bright magnitudes.

This combination of factors means that our experiment requires only modest precisions from SIM, and is well within the capabilities of Gaia, and may even be within the grasp of future high quality, ground-based astrometric studies. The latter was demonstrated by ground-based astrometry obtained by Casetti-Dinescu *et al.* (2006) in two Kapteyn's Selected Area fields positioned on the Sgr trailing arm, where dozens of 1-3 mas year^{-1} proper motions for $V = 19 - 21$ main sequence stars in the Sgr stream were averaged to derive a mean motion. These data showed that with more stars and more fields at this level of astrometric precision one might hope to distinguish MW models at the level of 10s of km s^{-1}. To achieve ~ 1 km s^{-1} precision in the solar motion with this method will require ~ 0.01 mas year^{-1} *mean* proper motions for sections of the Sgr trailing arm – a result that is within the reach of microarcsecond level space astrometry.

3. Probing the Galactic potential with tidal streams

It is now well known that the Galactic halo is inhomogeneous and coursed by streams of debris tidally pulled out of accreted satellites. Standard methods of measuring the MW's gravitational potential using a tracer population whose orbits are assumed to be random and well-mixed are systematically biased under these circumstances (Yencho *et al.* 2006). However, these streams themselves provide uniquely sensitive probes of the MW potential. If we could measure the distances, angular positions, radial velocities and proper motions of debris stars, we could integrate their orbits backwards in some assumed Galactic potential. Only in the correct potential will the path of the stream stars ever coincide in time, position and velocity with that of the parent satellite (see Fig. 1).

Microarcsecond proper motions and parallaxes combined with ground-based radial velocities provide everything needed to undertake the experiment. If we find a coherent stream without an associated satellite, the same techniques apply, but with the parent satellite's position and velocity as additional free parameters. The usefulness of Galactic tidal streams for probing the MW potential has long been recognized, and the promise of

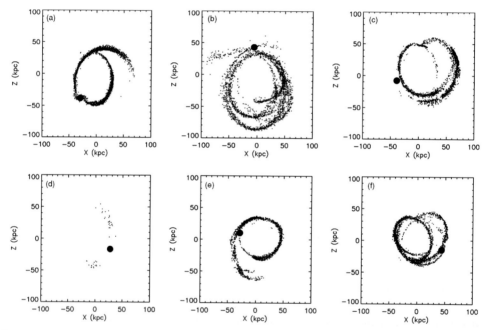

Figure 1. A demonstration of the use of tidal streams to measure the shape and strength of the Galactic potential if precise 6-D information is available for stream stars. A Sgr-like tidal stream was created by the disruption of a dwarf satellite in a rigid Galactic potential through a semi-analytical N-body simulation (e.g., as in Law *et al.* 2005; panel a). With complete 6-D phase space information in hand, guesses may be made on the strength and shape of the Galactic potential, and the orbits of the individual stars in the tidal streams run backwards under these assumed potentials. Panels (b) and (c) demonstrate what happens when the strength of the Galactic potential is underestimated by varying degrees: When the orbits are run backwards, the tidal stream stars orbit at too large a radius and do not converge to a common phase space position. In panels (e) and (f) the strength of the Galactic potential has been overestimated, and when the clock is run backwards the tidal stream stars assume orbits that are too small and once again do not converge on a common phase space position. In panel (d) a Galactic potential of the correct strength was guessed, and, when the stream star orbits are run backwards, the tidal stream stars collect back into the core of the parent satellite (shown as the dot in all panels).

astrometric space missions to provide the last two dimensions of phase space information for stream stars eagerly anticipated (e.g., Johnston *et al.* 1999, Peñarrubia *et al.* 2006).

Applying this method to simulated data observed with the μas yr^{-1} precision proper-motions possible with SIM and km s^{-1} radial velocities suggests that 1% accuracies on Galactic parameters (such as the flattening of the potential and circular speed at the Solar Circle) can be achieved with tidal tail samples as small as 100 stars (Johnston *et al.* 1999, Majewski *et al.* 2006). Dynamical friction is not an important additional consideration if the change in the energy of the satellite's orbit in N_{orb} orbits is less than the range in the energies of debris particles. For $N_{\text{orb}} = 3$, this condition is met for all satellites except the Magellanic Clouds and Sgr. Evolution of the Galactic potential does not affect the current positions of tidal debris, which respond adiabatically to changes in the potential and therefore yield direct information on the *present* Galactic mass distribution independent of how it grew (Peñarrubia *et al.* 2006).

It is important to sample tidal tails to points where stars were torn from the satellite at least one radial orbit ago and have thereby experienced the full range of Galactic potential along the orbit (Johnston 2001). The stars also need to have proper motions measured sufficiently accurately that the difference between their own and their parent satellite's orbits are detectable — this translates to requiring proper motions of order 100 μas yr^{-1} for the Sgr stream but 10 μas yr^{-1} for satellites farther away. Ideally, we would like to probe stars from different tidal tails at a variety of Galactocentric distances and orientations with respect to the Galactic disk to derive the shape and density profile of the DM. In the last few years evidence for dSph galaxy tidal tails has been discovered around systems at large Galactic radii (Muñoz *et al.* 2006a, Sohn *et al.* 2007) and, with SIM, extended versions of such distant streams can be used to trace the Galactic mass distribution as far out as the virial radius with an unprecedented level of detail and accuracy.

A key problem that is currently debated but that will be resolved with precision astrometry of tidal tails is the shape of the Galactic halo. CDM predicts that host DM halos should be oblate (minor to major axis ratio, $q \sim 0.7$) with flattening increasing with radius, whereas warm and self-interacting DM models tend to find rounder halos (Bullock *et al.* 2002). Studies of the halos of external galaxies — whose shapes can be measured via flaring of HI disks, the shapes of diffuse x-ray isophotes (for ellipticals), or with polar rings — typically find rather oblate shapes (see summaries in Combes 2002, Merrifield 2005). Meanwhile a controversy has emerged regarding the true shape of the halo as measured by the Milky Way's own polar ring system — the tidal stream of the Sgr dwarf spheroidal galaxy. Depending on how the extant data are analyzed and modeled, the debris of the Sgr dwarf has been argued to be consistent with a rather prolate ($q \sim 1.4$; Helmi 2004), near spherical (Ibata *et al.* 2001, Fellhauer *et al.* 2006), slightly oblate ($q \sim 0.93$; Johnston *et al.* 2005, Law *et al.* 2005) or oblate ($q = 0.85$; Martinez-Delgado *et al.* 2004) Galactic halo. The definitive measurement of the shape of the Galactic halo within the Sgr orbit will come from SIM proper motions of stars in its tails, and the variation of the halo shape with radius will come by repeating the technique with other tidal streams at other distances. Interestingly, recent measures of the mass profile of the MW using stellar tracers suggest that it gets rounder with radius (Kuijken 2003), in contradiction to the predictions of CDM.

The possible existence of a significant fraction of the halo in the form of dark satellites has been discussed in recent years (Moore *et al.* 1999, Klypin *et al.* 1999). These putative dark subhalos could scatter stars in tidal tails, possibly compromising their use as large-scale potential probes, but astrometric measurements of stars in these tails could, on the other hand, be used to assess the importance of dark substructure (Ibata *et al.* 2002,

Johnston *et al.* 2002). Early tests of such scattering using only radial velocities of the Sgr stream suggest a MW halo smoother than predicted (Majewski *et al.* 2004), but this represents debris from a satellite with an already sizable intrinsic velocity dispersion. Because scattering from subhalos should be most obvious on the narrowest, coldest tails (such as those seen from globular clusters, like Palomar 5 – e.g., Grillmair & Dionatos 2006) these could be used to probe the DM substructures, whereas the stars in tails of satellite galaxies such as Sgr, with larger dispersions initially and so less obviously affected, can still be used as global probes of the Galactic potential.

Gaia will allow these tests to be attempted for nearby streams, whereas with SIM, it will be possible for the first time to probe the full three-dimensional shape, density profile, extent of and substructure within our closest large DM halo.

4. Probing the Galactic potential with hypervelocity stars

A complementary method for sensing the shape of the Galactic potential can come from SIM observations of hypervelocity stars (Gnedin *et al.* 2005). Hills (1988) postulated that such stars would be ejected at speeds exceeding 1000 km s^{-1} after the disruption of a close binary star system deep in the potential well of a massive black hole, but HVSs can also be produced by the interaction of a single star with a binary black hole (Yu & Tremaine 2003). Recently Brown *et al.* (2006) report on five stars with Galactocentric velocities of 550 to 720 km s^{-1}, and argued persuasively that these must be "unbound stars with an extreme velocity that can be explained only by dynamical ejection associated with a massive black hole". After the success of these initial surveys, it is likely that many more HVSs will be discovered in the next few years.

If these stars indeed come from the Galactic center, the orbits are tightly constrained by knowing their point of origin. In this case, as Gnedin *et al.* (2005) demonstrate, the non-spherical shape of the Galactic potential — due in part to the flattened disk and in part to the triaxial dark halo — will induce non-radial inflections (which will be primarily in transverse direction at large radii) in the velocities of the HVSs of order 5–10 km s^{-1}, which corresponds to 10–100 μas yr^{-1}. Each HVS thus provides an independent constraint on the potential, as well as on the solar circular speed and distance from the Galactic center. The magnitudes of the known HVSs range from 16 to 20, so their proper motions will be measurable by SIM with an accuracy of a few μas yr^{-1}, which should define the orientation of their velocity vectors to better than 1%. With a precision of 20 μas yr^{-1} the orientation of the triaxial halo could be well-constrained and at 10 μas yr^{-1} the axial ratios will be well-constrained (Gnedin *et al.* 2005).

5. Dark matter within dwarf galaxies

Dwarf galaxies, and particularly dSph galaxies, are the most DM-dominated systems known to exist. Relatively nearby Galactic dSph satellites provide the opportunity to study the structure of DM halos on the smallest scales, and, with microarcsecond astrometry, make possible a new approach to determining the physical nature of DM (Strigari *et al.* 2007, 2008). CDM particles have negligible velocity dispersion and very large central phase-space density, which results in cuspy density profiles over observable scales (Navarro *et al.* 1997, Moore *et al.* 1998), whereas, in contrast, Warm Dark Matter (WDM) has smaller central phase-space density so that constant central cores develop in the density profiles. Because of their small size, if dSph cores are a result of DM physics then the cores occupy a large fraction of the virial radii, which makes these particular cores more observationally accessible than those in any other galaxy type. Using dSph central velocity dispersions, earlier constraints on dSph cores have excluded extremely warm DM, such as standard massive neutrinos (Lin & Faber 1983, Gerhard & Spergel 1992). More recent studies of the Fornax dSph provide strong constraints on the properties of sterile neutrino DM (Goerdt *et al.* 2006, Strigari *et al.* 2006).

The past decade has seen substantial progress in measuring radial velocities for large numbers of stars in nearby dSph galaxies, and the projected radial velocity dispersion profiles are generally found to be roughly flat as far out as they can be followed (Muñoz *et al.* 2005, 2006a, Walker *et al.* 2006a,b, 2007, Sohn *et al.* 2007, Koch *et al.* 2007a,b, Mateo *et al.* 2007). By combining such radial velocity profiles with the surface density distributions of dSph stars (which are well-fitted by King profiles, modulo slight variations, especially at large radii), it is typical to derive DM density profiles by assuming dynamical equilibrium and solving the Jeans equation (e.g, Richstone & Tremaine 1986). The results of such equilibrium analyses typically imply at least an order of magnitude more dSph mass in dark matter than in stars as well as mass-luminosity ratios that increase with radius — in some cases quite substantially (e.g., Kleyna *et al.* 2002) — though at large radii tidal effects probably complicate this picture (Kuhn 1993, Kroupa 1997, Muñoz *et al.* 2006a, 2007, Sohn *et al.* 2007, Mateo *et al.* 2007). In some cases the mass-follows-light models incorporating tidal disruption describe the observations quite well (Sohn *et al.* 2007, Muñoz *et al.* 2007). Knowing whether mass follows light in dSphs or if the luminous components lie within large extended halos is critical to establishing the regulatory mechanisms that inhibit the formation of galaxies in all subhalos.

Unfortunately, as shown by Strigari *et al.* (2007), equilibrium model solutions to the Jeans equation are degenerate and satisfied by DM density profiles with either cores or cusps. This is because there is a strong degeneracy between the inner slope of the DM density profile and the velocity anisotropy, β, of the stellar orbits, which leads to a strong dependency of the derived dSph masses on β. Radial velocities alone cannot break this degeneracy (Fig. 2) even if the radial velocity samples are increased to several thousand stars (Strigari *et al.* 2007). The problem is further compounded if we add triaxiality, Galactic substructure, and dSph orbital shapes to the allowable range of parameters. The only way to break the mass-anisotropy degeneracy is to measure more phase space coordinates per star, and in particular to acquire transverse velocities for the stars, because the Jeans equation solved in the transverse dimension probes the anisotropy differently than in the radial velocity dimension. Thus, combining high precision proper motions with even the present samples of radial velocities holds the prospect to break the anisotropy-inner slope degeneracy (Fig. 2).

The most promising dSphs to obtain proper motions for will be the nearby (60-90 kpc distant) systems Sculptor, Draco, Ursa Minor, Sextans and Bootes, which include the most DM-dominated systems known (Mateo 1998, Muñoz *et al.* 2006b, Martin *et al.* 2007) as well as a system with a more modest M/L (Sculptor). To sample the velocity dispersions properly will require proper motions of >100 stars per galaxy with accuracies of 7 km s^{-1} or better (less than 15 μas yr^{-1}). Strigari *et al.* (2007) show that with about 200 radial velocities and 200 transverse velocities of this precision it will be possible to reduce the error on the log-slope of the dark matter density profile to about 0.1 — which is an order of magnitude smaller than the errors attainable from a sample of 1000 radial velocities alone, and sensitive enough to rule out nearly all WDM models (Fig. 2). Figure 3 shows that exploring even these nearby dSph systems requires precision proper motions of stars to $V \gtrsim 19$, a task well beyond the capabilities of Gaia, but well-matched to the projected performance of SIM, though requiring a Key Project level of observing time (e.g., 100 days of SIM observing for 200 Draco stars). Note that while the Sgr dSph is several times closer than these other dSphs, it is obviously a system in dynamical

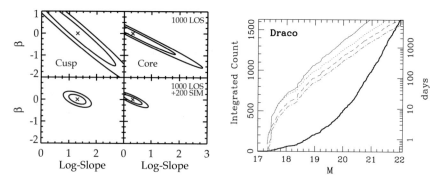

Figure 2. Left panels: A demonstration of the ability to recover information on the nature of DM using observations of dSph stars, from analytical modeling by Strigari *et al.* (2007). Ellipses indicate the 68% and 95% confidence regions for the errors in the measured dark halo density profile slope (measured at twice the King core radius) and velocity anisotropy parameter β in the case where only radial velocities are available for 1000 stars in a particular dSph (top panels). A significant improvement is derived from the addition of 200 SIM proper motions providing 5 km s^{-1} precision transverse velocities (bottom panels). The left (right) panels correspond to a cusp (core) halo model for dSphs and the small x's indicate the fiducial, input model values.

Figure 3. Right panel: Potential SIM PlanetQuest exploration of the Draco dSph would need to probe to $V \sim M = 19$ to derive a sample of 200 red giants as seen by its Washington M-band luminosity function (*thick line and left axis*). The thin lines represent the number of days (*right axis*) necessary to observe all the stars to a given magnitude limit with SIM and for a given transverse velocity uncertainty: 3 km s^{-1} (*thin solid line*), 5 km s^{-1} (*dotted line*), 7 km s^{-1} (*dashed line*) and 10 km s^{-1} (*dotted-dashed line*). From Strigari *et al.* (2008).

non-equilibrium; exploring this system will be of interest as a case study for establishing the internal dynamical effects of tidal interaction on dSphs.

Acknowledgements

I appreciate assistance from James Bullock, Jeffrey Carlin, Peter Frinchaboy, Kathryn Johnston, Ricardo Muñoz, Richard Patterson, Louie Strigari, and Scott Tremaine in preparation of this presentation and contribution.

References

Bedin, L. R., Piotto, G., King, I. R., & Anderson, J. 2003, *AJ*, 126, 247
Bullock, J. S. 2002, "The Shapes of Galaxies and Their Dark Matter Halos", ed. P. Natarajan, (Singapore: World Scientific), 109
Casetti-Dinescu, D. I., Majewski, S. R., Girard, T. M., Carlin, J. L., van Altena, W. F., Patterson, R. J., & Law, D. R. 2006, *AJ*, 132, 2082
Combes, F. 2002, *New Astronomy Review*, 46, 755
Dehnen, W., & Binney, J. J. 1998, *MNRAS*, 298, 387
Dias, W. S., & Lépine, J. R. D. 2005, *ApJ*, 629, 825
Feast, M. W. & Whitelock, P. 1997, *MNRAS*, 291, 683
Fellhauer, M., *et al.* 2006, *ApJ*, 651, 167
Flynn, C., Holmberg, J., Portinari, L., Fuchs, B., & Jahreiß, H. 2006, *MNRAS*, 372, 1149
Frinchaboy, P. M. 2006, Ph.D. Thesis, University of Virginia
Frinchaboy, P. M., & Majewski, S. R. 2006, *BAAS*, 38, 1132
Gerhard, O. E., & Spergel, D. N. 1992, *ApJL*, 389, L9
Gnedin, O. Y., Gould, A., Miralda-Escudé, J., & Zentner, A. R. 2005, *ApJ*, 634, 344
Goerdt, T., Moore, B., Read, J. I., Stadel, J., & Zemp, M. 2006, *MNRAS*, 368, 1073
Grillmair, C. J. & Dionatos, O. 2006, *ApJL*, 641, L37

Helmi, A. 2004, *ApJL*, 610, L97

Ibata, R., Lewis, G. F., Irwin, M., Totten, E., & Quinn, T. 2001, *ApJ*, 551, 294

Ibata, R. A., Lewis, G. F., Irwin, M. J., & Quinn, T. 2002, *MNRAS*, 332, 915

Johnston, K. V. 2001, ASP Conf. Ser. 230: Galaxy Disks and Disk Galaxies, 417

Johnston, K. V., Law, D. R., & Majewski, S. R. 2005, *ApJ*, 619, 800

Johnston, K. V., Spergel, D. N., & Haydn, C. 2002, *ApJ*, 570, 656

Johnston, K. V., Zhao, H., Spergel, D. N., & Hernquist, L. 1999, *ApJL*, 512, L109

Kalirai, J. S., *et al.* 2004, *ApJ*, 601, 277

Koch, A., Wilkinson, M. I., Kleyna, *et al.* 2007, *ApJ*, 657, 241

Koch, A., Kleyna, J. T., Wilkinson, *et al.* 2007, *AJ*, 134, 566

Klypin, A., Kravtsov, A. V., Valenzuela, O. & Prada, F. 1999, *ApJ*, 522, 82

Kuijken, K. 2003, The Mass of Galaxies at Low and High Redshift, eds. R. Bender & A. Renzini, (Berlin: Springer-Verlag), 1

Kuijken, K., & Tremaine, S. 1994, *ApJ*, 421, 178

Law, D. R., Johnston, K. V., & Majewski, S. R. 2005, *ApJ*, 619, 807

Lin, D. N. C., & Faber, S. M. 1983, *ApJL*, 266, L21

Majewski, S. R., *et al.* 2004, *AJ*, 128, 245

Majewski, S. R., Law, D. R., Polak, A. A., & Patterson, R. J. 2006, *ApJL*, 637, L25

Majewski, S. R., Skrutskie, M. F., Weinberg, M. D., & Ostheimer, J. C. 2003, *ApJ*, 599, 1082

Martínez-Delgado, D., Gómez-Flechoso, M. Á., Aparicio, A., & Carrera, R. 2004, *ApJ*, 601, 242

Martin, N. F., Ibata, R. A., Chapman, S. C., Irwin, M., & Lewis, G. F. 2007, *MNRAS*, 380, 281

Mateo, M. L. 1998, *ARAA*, 36, 435, 170

Mateo, M., Olszewski, E. W., & Walker, M. G. 2007, *ApJ*, in press

Merrifield, M. R. 2005, "The Identification of Dark Matter", eds. N. J. C. Spooner & V. Kudryavtsev, (Singapore: World Scientific), 49

Moore, B., Ghigna, S., Governato, F., Lake, G., Quinn, T., Stadel, J. & Tozzi, P. 1999, *ApJL*, 524, L19

Moore, B., Governato, F., Quinn, T., Stadel, J., & Lake, G. 1998, *ApJL*, 499, L5

Muñoz, R. R., *et al.* 2005, *ApJL*, 631, L137

Muñoz, R. R., *et al.* 2006a, *ApJ*, 649, 201

Muñoz, R. R., Carlin, J. C., Frinchaboy, P. M., Nidever, D. L., Majewski, S. R., & Patterson, R. J. 2006b, *ApJL*, 650, L51

Muñoz, R. R., Majewski, S. R., & Johnston, K. V. 2007, *ApJ*, submitted

Navarro, J. F., Frenk, C. S., & White, S. D. M. 1997, *ApJ*, 490, 493

Olling, R. P. & Merrifield, M. R. 1998, *MNRAS*, 297, 943

Peñarrubia, J., Benson, A. J., Martínez-Delgado, D., & Rix, H. W. 2006, *ApJ*, 645, 240

Reid, M. J., & Brunthaler, A. 2004, *ApJ*, 616, 872

Richstone, D. O. & Tremaine, S. 1986, *AJ*, 92, 72

Sohn, S. T., *et al.* 2007, *ApJ*, 663, 960

Strigari, L. E., Bullock, J. S., Kaplinghat, M., Kravtsov, A. V., Gnedin, O. Y., Abazajian, K., & Klypin, A. A. 2006, *ApJ*, 652, 306

Strigari, L. E., Bullock, J. S., & Kaplinghat, M. 2007, *ApJL*, 657, L1

Strigari, L. E., Muñoz, R. R., Bullock, J. S., Kaplinghat, M., Majewski, S. R., Kazantzidiz, S., & Kaufmann, T. 2008, in preparation

Uemura, M., Ohashi, H., Hayakawa, T., Ishida, E., Kato, T., & Hirata, R. 2000, *PASJ*, 52, 143

Walker, M. G., Mateo, M., Olszewski, E. W., Pal, J. K., Sen, B., & Woodroofe, M. 2006a, *ApJL*, 642, L41

Walker, M. G., Mateo, M., Olszewski, E. W., Bernstein, R., Wang, X., & Woodroofe, M. 2006b, *AJ*, 131, 2114

Walker, M. G., Mateo, M., Olszewski, E. W., Gnedin, O. Y., Wang, X., Sen, B., & Woodroofe, M. 2007, *ApJL*, 667, L53

Yencho, B. M., Johnston, K. V., Bullock, J. S., & Rhode, K. L. 2006, *ApJ*, 643, 154

Yu, Q., & Tremaine, S. 2003, *ApJ*, 599 1129

A Giant Step: from Milli- to Micro-arcsecond Astrometry
Proceedings IAU Symposium No. 248, 2007
W. J. Jin, I. Platais & M. A. C. Perryman, eds.

© 2008 International Astronomical Union
doi:10.1017/S1743921308019807

The transition between the thick and thin Galactic disks

M. Haywood†

GEPI, Observatoire de Paris, CNRS, Université Paris Diderot; 92190 Meudon,France
email: Misha.Haywood@obspm.fr

Abstract. We study the transition between the thick and thick disks using solar neighbourhood data, focusing in particular on the status of local metal-poor thin disk stars ([Fe/H]<-0.3 dex, [α/Fe]<0.1 dex). The orbital properties of these stars, which are responsible for the hiatus in metallicity between the two disks, suggest that they most likely originate from the outer Galactic thin disk. It implies that the transition between the two stellar populations at a solar galactocentric distance must have occurred at a metallicity of about -0.3 dex. Transition stars at this metallicity are in fact present in local samples and fill the gap in α-element between the thick and thin disks. These results imply that, at least from the local data, there is a clear evolutionary link between the thick and thin disks.

Keywords. Galaxy: abundances, Galaxy: disk, Galaxy: evolution, solar neighborhood

1. Introduction

The Galactic thick disk is, according to the local census, a population with characteristics neighbouring those of the thin disk : it is rotationally supported, with a somewhat higher internal kinematic dispersion, slightly metal-poor with an overlap in metallicity with this population. It is however possibly much older, and shows clear discontinuities with the thin disk, as visible in particular from metallicities. Several scenarios of the origin of this population have been put forward to explain these characteristics. The following three scenarious are often evoked: (1) the thick disk is an accreted population, (2) it is the first phase of the thin disk, heated up by an interaction episode with another galaxy (3) it could be the result of multiple, early mergers of gas rich subsystems from which the stars of thick disk would have formed. This last suggestion has been made by Brook *et al.* (2007) from numerical simulations where they apparently succeeded to produce a population with the main properties similar to those of the thick disk. In the first scenario, the properties of the thick disk are those of the accreted population. This is a difficulty, because while the range of satellite properties, encompassing the Milky Way is rather large, none of them approach those of the thick disk (in particular on the level of α elements). Furthermore, we have no direct clue to what these satellites may have been 10 Gyr ago, and there is no evidence that a dwarf galaxy with properties similar to those of the thick disk may have formed at that epoch. In the second scenario, it is not entirely clear what exactly occurs in the transition phase (which may have lasted several Gyr) between the interaction with the satellite and the beginning of the thin disk phase, and, in particular, what discontinuities it may give rise to. However, while it is not clear what may differentiate the resulting thick disk scenario (2) and (3), a kind of parenthood is certainly expected with the thin disk, whereas no link is expected in scenario (1).

In the recent years, detailed spectroscopic data have demonstrated that the two disks

† Present address: GEPI, Observatoire de Paris-Meudon; 92190 Meudon,France

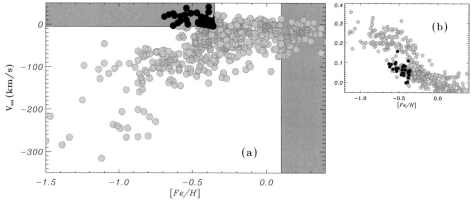

Figure 1. Galactic rotational velocity (relative to the Sun) as a function of metallicity and [α/Fe] *vs* [Fe/H] for the samples of Reddy *et al.* (2003, 2006) and Gilli *et al.* (2006). Black dots for stars with [Fe/H] <-0.35 dex and V_{rot} >-5km.s^{-1}. Grey areas illustrate the two selections made on the catalogue of Nordström *et al.* (2004) and displayed in the next figure.

are apparently distinct in their chemical properties with two main differences : first, at a given metallicity, the α-element content of the thick disk is offset by about +0.1 dex from the thin disk. Second, there is a hiatus of about 0.4 dex between the metal-rich stars of the thick disk (at [Fe/H]≈-0.2 dex) and metal-poor stars of the thin disk (at [Fe/H]≈-0.6 dex). The present study focuses on these two points, showing that these discontinuities are only apparent, and that a real continuity between the two disks does exist. The present results are obtained from the solar neighbourhood stars.

2. Radial mixing in the Galactic thin disk

Figure 1a shows the rotational velocity relative to the Sun of stars in the samples of Reddy *et al.* (2003, 2006) and Gilli *et al.* (2006) as a function of metallicity, while Fig. 1b shows [α/Fe] *vs* [Fe/H] for the same objects. On these plots, the thick disk stands out as a group enriched in α elements and lagging the LSR by about 50 to 150 km/s. In contrast, the metal-poor thin disk stars, (selected with [Fe/H]<-0.35 dex and V_{rot} >-5km.s^{-1}) rotate faster than the Sun. In order to understand the possible origin of this behaviour, we compare the orbital characteristics of the metal-poor and the metal-rich thin disk by selecting two subsamples from the catalogue of Nordström *et al.* (2004). The first includes stars with [Fe/H]<-0.35 dex and V_{rot} >-5 kms^{-1}, the other stars with [Fe/H]>+0.1 dex. Figure 2ab shows the distribution of apocentres *vs* the pericentres for the two subsamples. Panel (a) shows that the metal-poor thin disk subsample is confined to the upper part of the plot, which is a consequence of the selection made on the rotational lag (V_{rot} >-5 kms^{-1}). Panel (b) shows the metal-rich samples (selected only by metallicity), and clearly illustrates that this subsample populates mainly the inner orbits, or those orbits for which the mean orbital radius (defined as (R_a+R_p)/2) is less than 8kpc, in contrast to the metal-poor sample.

It has been suggested long ago that the stars are probably submitted to radial wandering caused by dynamical processes in the disk. A description of one such processes has been given by Sellwood & Binney (2002), who found that the passage of spiral waves could result in a significant modification of the angular momentum of stars around corotation, with individual stars radially drifting by a few kpc in a few Gyr. Because this is a secular effect, we may expect to detect stars that are in the process of approaching

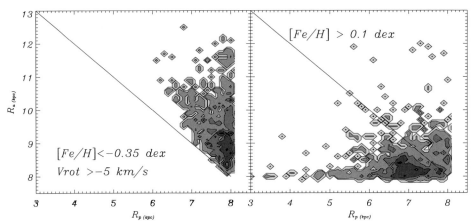

Figure 2. Distribution of apo- *vs* pericentres for the two subsamples selected in the catalogue of Nordström *et al.* (2004). Left, metal-poor thin disk stars with V_{rot} >-5km/s and [Fe/H]<-0.35 dex. Right panel is for metal-rich stars with [Fe/H]>+0.1 dex.

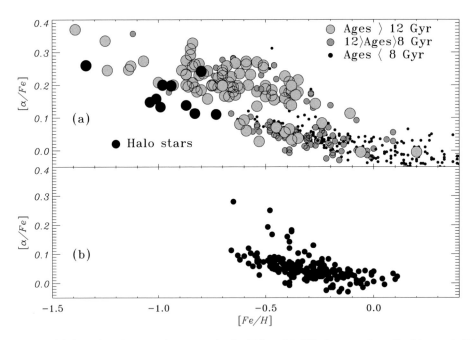

Figure 3. (a) Age distribution of stars in the [α/Fe] *vs* [Fe/H] diagram from Reddy *et al.* (2003, 2006) and Gilli *et al.* (2006). (b) Distribution of stars kinematically selected as thin disk stars in Reddy *et al.* (2006, their Fig. 12). Two branches are clearly visible, one made of metal-poor thin disk stars, the other of transition objects.

the solar orbit by moving outward or inward the Galactic disk, having orbital characteristics slightly off the main population. Combined with the fact that there exist a radial metallicity gradient of about 0.05-0.1 dex/kpc in the disk (the mean metallicity decreasing towards the outer disk), it is expected that metal-rich stars will be prevalent in the group of stars approaching the solar orbit from the inner disk, and metal-poor stars from the outer disk, as seen on Fig. 2.

3. The transition between the thick and thin disks

If the above inferences are correct, then metal-rich stars (at [Fe/H]$\gtrsim$0.15 dex) *and* metal-poor stars of the thin disk (at [Fe/H]$\lesssim$-0.3 dex) in the solar vicinity must be intruders from the inner and outer Galactic disks. It implies that the most metal-poor stars formed in the thin disk at solar galactocentric distance must have [Fe/H]$\approx$-0.3 dex. At what metallicity does the transition between the two disks occur? This is evidenced by the age-metallicity relation of solar neighbourhood stars (Haywood, 2006), which shows that the mean metallicity of the oldest thin disk stars is within [-0.2, -0.3] dex – not -0.6 dex. Figure 3a shows an age-range of stars in the [α/Fe] *vs* [Fe/H] diagram. If we discard the metal-poor thin disk stars on the basis that they are not genuine solar galactocentric objects, there seems to be a clear age discontinuity between the two disks, with the thin disk being essentially younger than 8 Gyr and the thick disk older than 12 Gyr. In spite of the discontinuities visible in age and [α/Fe], there is a real evolutionary link between the two populations. First, the thick disk terminates at the metallicity where the thin disk starts, at [Fe/H]$\approx$-0.3 dex. Most importantly, transition stars between the two disks do exist. In Fig. 12c of Reddy *et al.* (2006) (see Fig. 3b), kinematically selected stars of the thin disk in the [α/Fe] *vs* [Fe/H] diagram clearly show two branches, the first one made of metal-poor thin disk stars discussed above, the other towards the thick disk, starting at a metallicity of about -0.3 dex and with a higher level of α-elements. In any case, statistics of solar neighbourhood stars show that transition objects are relatively rare, confirming that there is a gap in the star formation rate at that epoch in the Galaxy. A few of them are also found in Bensby *et al.* (2005) and Fuhrmann (2004), having [Fe/H] in the range (-0.2, -0.4) dex.

4. Microarsecond astrometry

Virtually every aspect of the present study could be challenged by the expected returns of the space astrometric missions. Gaia in particular should provide samples of FGK stars to more than 1 kpc from the Sun with astrometric accuracy sufficient for accurate age determination. The age-scale itself should also benefit from a drastic improvement from stellar physics. It is therefore expected that several million stars will be available for the kind of studies that are achievable today only with a handful of objects in the solar vicinity. It implies that, with the help of high resolution, high SN spectroscopic data, a detailed map of the interface between the thick and thin disk populations should be obtainable.

References

Bensby T., Feltzing S., Lundström I., & Ilyin I., 2005, *A&A* 433, 185

Brook C., Richard S., Kawata D., Martel H., & Gibson B. K., 2007, *ApJ* 658, 60

Demarque P., Woo J.-H., Kim Y.-C., & Yi S. K., 2004, *ApJS* 155, 667

Fuhrmann K., 2004, AN, 325, 3

Gilli G., Israelian G., Ecuvillon A., Santos N. C., & Mayor M., 2006, *A&A* 449, 723

Haywood M., 2001, *MNRAS* 325, 1365

Haywood M., 2006, *MNRAS* 371, 1760

Nordström B., *et al.*, 2004, *A&A* 418, 989

Reddy, B. E., Tomkin, J., Lambert, D. L., & Allende Prieto, C., 2003, *MNRAS* 340, 304

Reddy, B. E., Lambert, D. L., & Allende Prieto, C., 2006, *MNRAS* 367, 1329

Sellwood, J. A., & Binney, J. J., 2002, *MNRAS* 336, 785

A Giant Step: from Milli- to Micro-arcsecond Astrometry
Proceedings IAU Symposium No. 248, 2007
W. J. Jin, I. Platais & M. A. C. Perryman, eds.

© 2008 International Astronomical Union
doi:10.1017/S1743921308019819

Structure and kinematics of the Milky Way spirals traced by open clusters†

Z. Zhu

Department of Astronomy, Nanjing University, China
email: zhuzi@nju.edu.cn

Abstract. Selecting 301 open clusters with complete spatial velocity measurements and ages, we are able to estimate the disk structure and kinematics of the Milky Way. Our analysis incorporates the disk scale height, the circular velocity of the Galactic rotation, the Galactocentric distance of the Sun and the ellipticity of the weak elliptical potential of the disk. We have derived the distance of the Sun to the Galactic center $R_0 = 8.03 \pm 0.70$ kpc, that is in excellent agreement with the literature. From kinematic analysis, we found an age-dependent rotation of the Milky Way. The mean rotation velocity of the Milky Way is obtained as 235 ± 10 km s^{-1}. Using a dynamic model for an assumed elliptical disk, a clear weak elliptical potential of the disk with ellipticity of $\epsilon(R_0) = 0.060 \pm 0.012$ is detected, the Sun is found to be near the minor axis with a displacement of $30° \pm 3°$. The motion of clusters is suggested to be on elliptical orbits other than the circular rotation.

Keywords. Galaxy: disk, astrometry, Galaxy: kinematics and dynamics, Galaxy: structure, open clusters and associations: general

1. Introduction

The young open clusters have an unusual superiority over other tracers of the disk structure in the Milky Way, providing crucial information and constraints for understanding of the Galactic kinematics and dynamics. Due to the increase of available data on open clusters in the recent years, we are able to refine our study of the Galactic structure. In this paper, we concentrate on a kinematical analysis based on proper motions, distances, radial velocities, and ages of open clusters. Considering the systematic consistency of kinematical data, we decide to use the internally homogeneous Catalogue of Open Cluster Data compiled by Kharchenko *et al.* (2005). In order to increase the number of clusters, we used the New Catalog of Optically Visible Open Clusters and Candidates, that includes 1689 clusters collected from the literature (Dias *et al.* 2002). Finally, a total amount of 301 clusters with complete spatial velocity measurements is selected for this study.

The velocity data are used to inspect the peculiar motions for the individual clusters. The components of peculiar motion are derived from the first order expansion of an asymmetric rotation of the systematic velocity field. Rejecting 22 clusters with peculiar velocities over 50 km s^{-1} (roughly 2.6σ of the total velocity dispersion), 269 clusters are finally used for the kinematic analysis. Figure 1 shows the kinematical structures and distributions of 269 clusters on the Galactic plane.

† Supported by the National Natural Science Foundation of China(Grant Nos. 10333050 and 10673005)

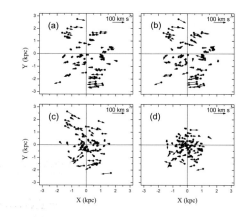

Figure 1. Observed space velocities of clusters projected on the Galactic plane. The solar motion components u_0 and v_0 are removed from the observational data. Panel (a): Simulation of 90 clusters younger than 18 Myr, assuming $R_0 = 8$ kpc, $V_0 = 220$ km s^{-1}, and $(\partial V / \partial R)_{R=R_0} = 4$ km s^{-1} kpc^{-1}. Panel (b): 90 clusters younger than 18 Myr. Panel (c): 88 clusters with ages in the range of 18–120 Myr. Panel (d): 91 clusters older than 120 Myr.

Table 1. Solution of the parameters of the vertical distribution.

	10 pc bins		20 pc bins	
	z_0 [pc]	z_h [pc]	z_0 [pc]	z_h [pc]
all ages	-16.1 ± 4.0	58.3 ± 4.0	-15.6 ± 3.5	56.8 ±3.5
⩽ 50 Myr	-12.3 ± 4.2	51.8 ± 4.6	-12.4 ± 4.2	49.6 ±4.4
> 50 Myr	-20.5 ± 5.6	67.8 ± 5.8	-19.5 ± 5.7	67.1 ±5.7
all ages (PKRSS 2006)			-22 ± 4	56 ± 3
all ages (BKBS 2006)			-14.8 ± 2.4	57.2 ± 2.8
⩽ 200 Myr (BKBS 2006)				47.9 ± 2.8
200 ∼ 1000 Myr (BKBS 2006)				149.8 ± 26.3

2. The disk scale height and kinematics

The recent work concerning the structure and distribution of the Galactic open clusters is detailed by Piskunov *et al.*(2006, PKRSS), and by Bonatto et al.(2006, BKBS). In this Section, we first extend the study on the vertical structure based on open cluster data. In order to study the properties of clusters in different age range, we divide the sample into two subsets: young clusters with ages less than 50 Myr, and old clusters with ages larger than 50 Myr. The results are given in Table 1, where z_0 is the position about the Galactic symmetry plane, z_h is the scale height of the distribution. Note that 17 clusters associated with the Gould's Belt are removed from the samples. Comparing our results with those given by PKRSS and BKBS, we found that most of the solutions are matching very well. However, the value $z_h \simeq 150$ pc for clusters with $200 \sim 1000$ Myr obtained by BKBS significantly deviates from our determination for older clusters. We found that the oldest clusters in BKBS have mostly erroneous distance estimate, explaining the extremal value of $z_h \simeq 150$ pc for clusters older than 200 Myr as derived by BKBS.

Because the objects in our sample are confined to the Galactic plane with a small scale height of ∼60 pc, a two dimensional asymmetric model is sufficient to describe their motion. In this case, we define the Oort's constants A, C (denote the azimuthal and

Table 2. Kinematic parameters derived from proper motions and radial velocities. The unit is in km s^{-1} for components of the solar motion. The Oort's constants are measured in km s^{-1} kpc^{-1}.

	u_0	v_0	w_0	A	B	C	K
all ages (269)	11.8±0.6	13.3±0.6	8.1±0.5	16.4±0.9	-12.9±0.9	0.5±1.0	-2.6±0.9
⩽ 50 Myr (137)	10.8±0.7	14.6±0.7	7.9±0.8	16.5±0.9	-14.6±0.9	2.9±0.8	-1.1±0.9
> 50 Myr (132)	12.3±0.9	12.8±0.8	8.2±0.8	16.4±1.7	-10.9±1.6	-2.7±1.7	-4.6±1.8

radial directions of the velocity field), B (characterizes the vorticity), and K (implies an overall contraction or expansion measured at the Sun). Based on proper motions, radial velocities and heliocentric distances of open clusters, these parameters including solar motions are given in (Table 2).

In an axisymmetric and stationary disk, the Oort's constants A and B describe a differential circular rotation of the Milky Way at the place of the Sun. Feast & Whitelock (1997) found a low angular velocity ($A = 14.8 \pm 0.8$, $B = -12.4 \pm 0.6$ km s^{-1} kpc^{-1}) from the Hipparcos proper motions of the Galactic Cepheids. But the majority of measurements in the recent year have shown a more or less enhanced angular velocity, including our present determination. From the Hipparcos proper motions of the Galactic O-B5 stars, Miyamoto & Zhu (1998) derived $A = 16.1 \pm 1.1$ and $B = -15.6 \pm 0.8$ in km s^{-1} kpc^{-1}. From proper motions of the old red giants from ACT/Tycho-2 catalogues, Olling & Dehnen (2003) found $A - B \simeq 32.8$ km s^{-1} kpc^{-1}. According to observations of the massive compact radio source Sgr A* at the Galactic center by VLBA with respect to the extragalactic sources, Reid & Brunthaler (2004) reported the apparent proper motion of Sgr A* $\mu_\ell = -6.379 \pm 0.026$ and $\mu_b = -0.202 \pm 0.019$ mas yr^{-1}. This apparent motion should fully reflect the Galactic rotation at the Sun, assuming Sgr A* is in rest. Then we have $A - B = -\kappa\mu_\ell - v_0/R_0$, where $v_0 = 5.25 \pm 0.62$km s^{-1} is the component of the solar motion in the direction of Galactic rotation given by Dehnen & Binney (1998). Adopting $R_0 = 8.0$ kpc, $A - B = 29.58 \pm 0.14$km s^{-1} kpc^{-1} was calculated. This is in excellent agreement with our determination. The present determination gives a rotation velocity $V_0 = 235 \pm 10$ km s^{-1} from the complete sample, while $V_0 = 248 \pm 9$ km s^{-1} is for the young and $V_0 = 218 \pm 19$ km s^{-1} for the older sub-sample.

The precision of R_0 is directly related to many astronomical quantities, measurements and theory. According to the statistical analysis from the individual determinations by Reid (1993), $R_0 = 8.0 \pm 0.5$ kpc is currently considered as the best value, whereas the 1985 IAU standard value is 8.5 kpc. Considerable bias and uncertainties may still exist in the determinations of R_0, even if researchers have employed various efforts to improve it, e.g. from the latitude proper motion of the Sgr A* ($\mu_b = -0.202 \pm 0.019$ mas yr^{-1}), we get $R_0 = w_0/\mu_b/\kappa = 7.49 \pm 0.81$ kpc, assuming Sgr A* is at rest. Here $w_0 = 7.17 \pm 0.38$km s^{-1} is the component of the solar motion.

Encouraged by the kinematic analysis, in which we found that the Oort's constant A is independent on the cluster age, we decided to derive the Galactocentric distance of the Sun R_0. Because only small values for C and K of the Oort's constants are found, we are able to simply use an axisymmetric rotation model. The Oort's constant A is independently derived from the proper motions of clusters. We apply this constant to constrain a kinematical model from the radial velocities

$$v_r = 2AR_0\left(\frac{R_0}{R} - 1\right)\sin\ell\cos b - u_0\cos\ell\cos b - v_0\sin\ell\cos b - w_0\sin b - \delta v_r, \quad (2.1)$$

where, δv_r is a possible offset of the radial velocity zero-point. The parameter $2AR_0$ can be calculated from the radial velocities. Using the constant A from the proper-motion solution, the Galactocentric distance R_0 is derived in an iterative way.

The present determination of R_0, based on independent observations for proper motions and radial velocities of clusters, gives $R_0 = 8.03 \pm 0.70\,\mathrm{kpc}$ that is consistent with the current "best estimate" of $R_0 = 8.0 \pm 0.5\,\mathrm{kpc}$ proposed by Reid (1993).

3. The weak elliptical distortion of the disk potential

The persistence of a significant K-term or δv_r for the radial velocities of Galactic young objects has been recognized before. It can be explained either as an overall kinematic contraction or expansion on the Galactic plane, or as a systematic error of the measured radial velocities. On the other hand, if the axisymmetric model is not sufficient to describe the rotation defined by the young disk stars on non-circular orbits in the Galactic plane, a non-axisymmetric model should be introduced to describe such kinematic behavior. Considering rotation velocities of clusters as a function of azimuthal angle, we find that the circular speed gradually decreases in the direction of the Galactic rotation. This fact might be an evidence for the open clusters moving on non-circular orbits. Based on the model of an elliptical disk given by Kuijken & Tremaine (1994), the potential is expressed by

$$\Phi(R, \phi) = \Phi_0(R) + \Phi_1(R) \cos 2(\phi - \phi_{\mathrm b}), \tag{3.1}$$

where $\Phi_0(R)$ is the circular velocity depended axisymmetric part of the potential. In this potential, the mean tangential velocity is expressed by

$$V_\phi(R, \phi) = V_c(R)(1 - c(R)\cos 2\phi - s(R)\sin 2\phi), \tag{3.2}$$

with two components of the ellipticity $c(R) = \epsilon(R)\cos 2\phi_{\mathrm b}$, $s(R) = \epsilon(R)\sin 2\phi_{\mathrm b}$. Here, $\epsilon(R)$ is the potential ellipticity with its minor axis in the direction $\phi_{\mathrm b}$. In order to have a more rigorous evaluation for the ellipticity of the gravitational potential, we have test various solutions by using the observed spatial velocities of clusters. Finally, we obtain the components $c(R_0)$ and $s(R_0)$. The present solution suggest

$$\epsilon(R_0) = 0.060 \pm 0.012, \qquad \phi_b = 30° \pm 3°. \tag{3.3}$$

The present work is the first to succeed in quantifying the two elliptical components of the Milky Way potential via a consistent data set of the disk population of open clusters and based on a simple dynamic model by Kuijken & Tremaine. Using our solution from the open clusters, the motion of objects near the Sun is suggested to be on elliptical orbits.

References

Bonatto, C., Kerber, L. O., Bica, E., & Santiago, B. X. 1997, *A&A* 446,121 (BKBS)

Dehnen, W. & Binney, J. J. 1998, *MNRAS* 298, 387

Dias, W. S., Alessi, B. S., Moitinho, A., & Lépine, J. R. D. 2002, *A&A* 389, 871

Feast, M. W. & Whitelock, P. 1997, *MNRAS* 291, 683

Kharchenko, N. V., Piskunov, A. E., Röser, S., Schilbach, E., & Scholz, R. D. 2005, *A&A* 438, 1163

Kuijken, K. & Tremaine, S. 1994, *ApJ* 421, 178

Miyamoto, M. & Zhu, Z. 1998, *AJ* 115, 1483

Olling, R. P. & Dehnen, W. 2003, *ApJ* 955, 275

Piskunov, A. E., Kharchenko, N. V., Röser, S., *et al.* 2006, *A&A* 445, 545 (PKRSS)

Reid, M. J. 1993, *ARA&A* 31, 345

Reid, M. J. & Brunthaler, A. 2004, *ApJ* 616, 872

A Giant Step: from Milli- to Micro-arcsecond Astrometry
Proceedings IAU Symposium No. 248, 2007
W. J. Jin, I. Platais & M. A. C. Perryman, eds.
© 2008 International Astronomical Union
doi:10.1017/S1743921308019820

15 years of high precision astrometry in the Galactic Center

S. Gillessen[1], **R. Genzel**[1,2], **F. Eisenhauer**[1], **T. Ott**[1], **S. Trippe**[1] **and F. Martins**[1]

[1] Max-Planck-Institute for extraterrestial physics
email: ste@mpe.mpg.de

[2] Physics Department, University of California, Berkeley, CA 94720, USA

Abstract. In 1992, we obtained the first observations of S2 a star close to the supermassive black hole at the Galactic Center. In 2002, S2 passed its periastron and in 2007, it completed a first fully observed revolution. This orbit allowed us to determine the mass of and the distance to the supermassive black hole with unprecedented accuracy. Here we present a re-analysis of the data set, enhancing the astrometric accuracy to 0.5 mas and increasing the number of well-determined stellar orbits to roughly 15. This allows to constrain the extended mass distribution around the massive black hole and will lead in the near future to the detection of post-Newtonian effects. We will also give an outlook on the potential of interferometric near-infrared astrometry with 10 microarcsecond accuracy from the VLTI.

Keywords. Galaxy: center, astrometry, infrared: stars, instrumentation: interferometers

1. Introduction

The Galactic Center (GC) is a stunning example of the power of high precision astrometry. Progress in GC research has been driven by technology, in particular by high resolution techniques in the near infrared (NIR). Only the NIR allows to overcome the interstellar extinction towards the GC at intrinsically high resolution and only with sufficient angular resolution it is possible to beat the confusion of the dense stellar field. The main concern with respect to resolution from an observer's point of view is Earth's atmosphere, which implies the use of special techniques. In 1992 our group obtained the first Speckle observation of the innermost region of our Galaxy, using the custom-built camera SHARP on ESO's NTT, a 3.6m telescope. In 2002 we started using adaptive optics (AO) with the system NACO on ESO's VLT 8.2m telescope.

The most exciting result of these observations is the detection of individual Keplerian stellar orbits (Schödel *et al.* 2002, Ghez *et al.* 2005, Eisenhauer *et al.* 2005), allowing to probe the gravitational potential in which the so-called S-stars move. The data are consistent with a point mass of 4×10^6 solar masses. Spatially it coincides with the compact non-thermal radio source Sgr A* and a spurious NIR counterpart. Taken together, these observations constitute the best proof for the existence of an astrophysical black hole.

We have determined 15 full 3D stellar orbits from our data set of the central arcsecond covering 16 years. We are currently monitoring $\sim$100 stars both in astrometry as well as spectroscopy. The astrometric long-term accuracy of our AO data is $\sim$350 μ as for the stars in the central arcsecond. Given the exquisit quality of the orbital data, it is natural to ask the question, whether the post-Keplerian effects can be detected.

General relativity predicts that the pericenter of the orbit of star S2 (which has the shortest orbital period of our sample and is the brightest one) should precess by 0.2° per revolution of 15 years. While the Schwarzschild metric leads to a prograde precession,

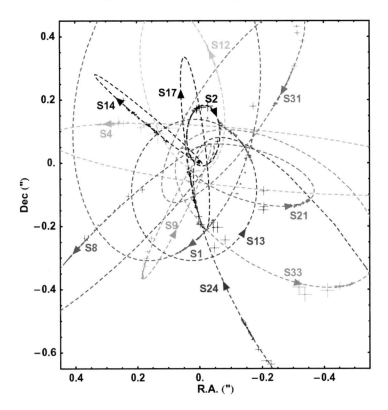

Figure 1. Stellar orbits around Sgr A*. The dashed lines are the best fitting orbits in a common Kepler potential, given the data shown as crosses. The arrows indicate the directions of motion.

there are astrophysical reasons to assume the existence of counteracting effects – a so-far unseen, extended mass component (Morris 1993, Muno *et al.* 2005) would lead to a retrograde precession. Since the two effects depend differently on the distance to the central mass, the system of S-stars offers the opportunity to measure both. Thus one can hope to test the Schwarzschild metric for the heaviest object for which it was ever tested and to determine the profile of a possible extended mass component.

There is a two-fold way to measure these post-Newtonian effects. First, the existing and future data from the same instruments and understanding the sources of systematic errors in the data set should allow us to unambiguously identify these effects. The second way is to increase the angular resolution further and observe objects that are even closer to the event horizon than the S-stars seen nowadays. The latter is only possible with either an increased telescope diameter or with interferometry.

2. Astrometry of S-stars with NIR imaging

A natural choice for the coordinate system in the GC is to put the central mass at rest at the origin of the system. Since Sgr A* is not always detectable in the NIR (either because it is in a faint state or because it is confused with stars) a convenient choice is to refer coordinates to the radio source. The cross-calibration of the two wavelength regimes can be achieved by using nine SiO masers for which the radio coordinates and proper motions can be measured (Reid *et al.* 2007).

The accuracy with which radio Sgr A* can be unrecoved in the NIR images is limited by both the positional errors of the SiO masers in radio as well as the errors in their NIR positions. We have determined the error propagation by means of Monte-Carlo simulations, varying all input coordinates by their respective errors. We find that the radio source can be located to within $(1.3, 1.2)$ mas in both axes. The same sort of simulations also allows to check how well the relative velocity of the two coordinate systems can be matched. We find that the velocity calibration is accurate to the level of $(0.4, 0.2)$ mas/yr (which corresponds to ~ 12 km s^{-1} at the distance of the GC).

In contrast, the large number (60) of available epochs and high intrinsic accuracy of the NIR positions (350μas) allows to locate isolated, bright stars to $\sim 100\mu$as and determine their proper motions to $\sim 20\mu$as/yr. This means that the positional and velocity errors with respect to Sgr A* are dominated by the uncertainty of the coordinate system. When considering the post-Newtonian effects, the velocity calibration is of particular concern. Both precession and incorrect velocity calibration can lead to non-closing orbital figures because the two effects are to some degree degenerate. Therefore, it is important to work in a proper reference frame. The current situation could be improved in a few ways:

- Future measurements of the SiO masers will bring the current errors down.
- The position and velocity of the coordinate system can be left free during the orbital fitting. Apparently this gives up the ability to test whether the mass coincides with radio Sgr A* and due to the degeneracy it decreases the sensitivity to post-Newtonian effects.
- At least for the velocity of the coordinate system one can get an independent calibration. In the NIR a large number of stars are seen in the central star cluster. Dynamically one expects the cluster to be a relaxed system, or being at rest with respect to the central point mass. If this assumption is correct, then one can get the velocity calibration to an accuracy of $\bar{v}/N_{\text{stars}}^{1/2}$ where $\bar{v}$ is the mean velocity per star.

From our data we could use 2700 stars with a mean velocity of 3.8 mas/yr. This yields an accuracy for the velocity calibration of 80 μas/yr, considerably better than that is possible with the SiO masers. We have checked for the such a defined coordinate system, what an orbit fit would yield for a position and velocity of the central point mass. Essentially, both position and velocity agree with the expectations (point mass is at rest at the origin of the system) within 2σ.

Given this coordinate system, we can derive the upper limits on post-Newtonian effects. As an example, we tested a very simple model for an extended mass distribution, namely $\rho(r) = \text{const}$. Fitting a relativistic orbit to the data of S2 then yields the extended mass component inside the orbit of S2 to be $1.7 \pm 0.9|_{\text{stat}} \pm 1.2|_{\text{sys}}$ % of the central mass. This is an upper limit and shows explicitly that the uncertainty is dominated by systematics.

3. The future of GC research: GRAVITY

Given certain orbital elements for an S2-like star, the deviations between a Keplerian orbit and a fully relativistic orbit are on the order of 10 mas. However, the orbital elements are not known a priori but are determined from the same fit. Thus, one has to ask a question how much of the deviation will be there between the models for a certain data set. The result is a smaller number since the orbital elements can absorb much of the changes due to relativity. An explicit simulation shows that the remaining deviations are typically 100 μas, which unfortunately is beyond what is feasible today.

If one could observe test particles closer to the point mass, the detection of post-Newtonian effects would become easier. The density profile of the central star cusp as obtained from number counts (Genzel *et al.* 2003, Schödel *et al.* 2007) predicts that in the central 100 mas there should reside a few (~ 5) stars with K-band magnitudes from

17 to 19. Such stars would have orbital periods of roughly 1 year and the relativistic precession will be a few degrees per revolution.

In order to spatially resolve such stars one needs to increase the angular resolution of the instrument beyond the diffraction limit of existing large telescopes with diameters of $8 - 10$ m. The highest angular resolution will be possible using interferometry, since baselines of ~ 100 m are envisaged, beating even the currently planned extremely large telescopes which will have apertures of 30-40 m. The faintness of the sources aimed at essentially dictates the use of large telescopes for the interferometer. The only observatory that will allow the combination of several 8 m class telescopes is the ESO VLTI. The angular resolution achievable is 1.5 mas, the astrometric accuracy should be around $10 \, \mu$as.

Since both an extended mass distribution and the relativistic effects will influence the shapes of future interferometric orbits in the GC, one actually needs at least two stars to disentangle the two effects. We have simulated the interferometric field of view of VLTI and could show that it is feasible to detect many stars in the field. Since the u-v-plane cannot be sampled very densely due to the limited number of baselines, the effective point spread function of the VLTI is a relatively complicated beam. Therefore one needs deconvolution techniques similar to what was developed for radio interferometry.

Probably the most exciting application of $10 \, \mu$as astrometry in the GC is the observation of flares from Sgr A*, sporadic emission events that last one to two hours and for which it is known that they happen very close to the event horizon that has an apparent size of $10 \, \mu$as. Therefore the ability to measure positions (and thus motions) with a similar accuracy opens a new window for testing gravitational theories in a very strong gravitational field. The apparent path of an assumed plasma hot spot orbiting the black hole is suspect to many relativistic effects, such as gravitational redshift, Doppler boosting and multiple images (Broderick & Loeb 2006). Therefore the GC offers the unique chance to directly observe the effects of strong gravity.

Current instrumentation at the VLTI misses the goal to observe the GC. We have identified the following key features needed for the GC experiment as described:

- Combining the light of four 8 m telescopes
- NIR wavefront sensing for the AO of each telescope.
- Fringe tracking on a nearby field star with $m_K = 10$.

Recently we have finished a feasibility study for an interferometric instrument combining these features, called "GRAVITY". Besides a compelling science case we could show that the instrument is technically feasible and that the GC field offers bright enough field stars for both AO and fringe tracking. The concept for the instrument is presented in poster 2.2.3 of this IAU workshop (Eisenhauer *et al.* 2008).

References

Broderick, A. & Loeb, A. 2006, *ApJ* 636, 109
Eisenhauer, F. *et al.* 2005, *ApJ* 628, 246
Eisenhauer, F. *et al.* 2008, *Proc. IAU 248*, in this volume p.100
Genzel, R. *et al.* 2003, *ApJ* 594, 812
Ghez, A. *et al.* 2005, *ApJ* 620, 744
Morris, M. 2003, *ApJ* 408, 496
Muno, M. 2005, *ApJ* 622, L113
Reid, M. *et al.* 2007, *ApJ* 659, 378
Schödel, R. *et al.* 2002, *Nature* 419, 694
Schödel, R. *et al.* 2007, *A&A* 465, 125

A Giant Step: from Milli- to Micro-arcsecond Astrometry
Proceedings IAU Symposium No. 248, 2007
W. J. Jin, I. Platais & M. A. C. Perryman, eds.

© 2008 International Astronomical Union
doi:10.1017/S1743921308019832

Understanding our Galaxy: from the center to outskirts

Z. Q. Shen[1], Y. Xu[2], J. L. Han[3] and X. W. Zheng[4]

[1]Shanghai Astronomical Observatory, Shanghai 200030, China
email: zshen@shao.ac.cn

[2]Purple Mountain Observatory, Nanjing 210008, China
email: xuye@mpifr-bonn.mpg.de

[3]National Astronomical Observatories, Beijing 100012, China
email: hjl@bao.ac.cn

[4]Department of Astronomy, Nanjing University, Nanjing 210093, China
email: xwzheng@nju.edu.cn

Abstract. We describe the efforts to understand our Milky Way Galaxy, from its center to outskirts, including (1) the measurements of the intrinsic size of the galactic center compact radio source Sgr A*; (2) the determination of the distance from the Sun to the Perseus spiral arm; and (3) the revealing of large scale global magnetic fields of the Galaxy.

With high-resolution millimeter-VLBI observations, Shen *et al.* (2005) have measured the intrinsic size of the radio-emitting region of the galactic center compact radio source Sgr A* to be only 1 AU in diameter at 3.5 mm. When combined with the lower limit on the mass of Sgr A*, this provides strong evidence for Sgr A* being a super-massive black hole. Comparison with the intrinsic size detection at 7 mm indicates a frequency-dependent source size, posing a tight constraint on various theoretical models.

With VLBI phase referencing observations, Xu *et al.* (2006) have measured the trigonometric parallax of W3OH in the Perseus spiral arm with an accuracy of 10 μas and also its absolute velocity with an accuracy of 1 km s^{-1}. This demonstrates the capability of probing the structure and kinematics of the Milky Way by determining distances to 12 GHz methanol (CH3OH) masers in star forming regions of distant spiral arms and Milky Way's outskirts.

With pulsar dispersion measures and rotation measures, Han *et al.* (2006) can directly measure the magnetic fields in a very large region of the Galactic disk. The results show that the large-scale magnetic fields are aligned with the spiral arms but reverse their directions many times from the most inner Norma arm to the outer Perseus arm.

Keywords. Galaxy: center, Galaxy: structure, black hole physics, masers, pulsars: general, techniques: high angular resolution, astrometry, stars: distances, stars: kinematics, ISM: magnetic fields, ISM: structure

1. Introduction

The rapid development of astronomical instruments and techniques has greatly enlarged our knowledge of the structure of Milky Way galaxy, from its center to outskirts. Here, we report some progress made over the past decade, including the measurement of the intrinsic size of the compact radio source Sgr A* with the millimeter-VLBI observations, the high-accuracy distance measurement to a massive star forming region W3OH in the Perseus spiral arm using the trigonometric parallax from the phase-referencing VLBI observations, and the direct measurement of magnetic field in a large region of the Galactic disk using the dispersion measures and rotation measures of pulsars in the Galaxy.

2. The size of Sgr A* at the Galactic center

Sgr A*, an extremely compact non-thermal radio source at the Galactic Center, has been widely recognized as the best and closest candidate for a super massive black hole (SMBH) ever since its first identification in 1974 (Balick & Brown 1974). The accurate determination of its mass and size is of great importance in testing the SMBH hypothesis. The precise determination of the orbital motions of about a dozen stars at the immediate neighborhood of Sgr A* has provided compelling evidence for the existence of a dark mass of 4×10^6 solar masses within a radius of 45 AU (Ghez *et al.* 2005; Schödel *et al.* 2002).

However, past attempts to measure Sgr A* structure with VLBI observations suffered, at long centimeter wavelengths, from the angular broadening caused by the diffractive scattering by the turbulent ionized interstellar medium and, at short millimeter wavelengths, from the large uncertainty in the data calibration due to the low elevation angles at which Sgr A* is visible at the telescopes in the northern hemisphere. To improve the accuracy of the observed source structure measurements, we developed a model fitting method by implicitly using the amplitude closure relation (Shen *et al.* 2003). We then applied this model fitting procedure to various VLBA observations of Sgr A* made at wavelengths from 6 cm to 3.5 mm over the time range from 1994 to 2004.

By fitting the apparent sizes measured at five wavelengths quasi-simultaneously, we were able to revise the 2-dimensional scattering structure (Shen *et al.* 2005). Then, by subtracting in quadrature the extrapolated scattering size, we successfully detected the intrinsic size of Sgr A* at 3.5 mm to be ≈ 1 AU in diameter. Together with a lower limit on the mass of Sgr A* of 0.4×10^6 solar masses from the study of the proper motion of Sgr A* itself (Reid & Brunthaler 2004), a size of 1 AU sets an extraordinarily high mass density for Sgr A*, strongly arguing for the SMBH nature of Sgr A* (Shen *et al.* 2005).

Intriguingly, a size of 1 AU for Sgr A* is about twice the diameter of the shadow caused by the strong gravitational bending of light at the very vicinity of a 4 million solar masses SMBH. Simulation of the black hole shadow image and the corresponding visibility analysis indicate that future sub-millimeter VLBI experiments are critical and promising to probe the Galactic center black hole (Huang *et al.* 2007). We are on the verge of resolving the Galactic Center SMBH.

3. The distance to W3OH in the Perseus spiral arm

Revealing the true plan-view of the Milky Way has been a challenge for nearly a century. We know less about the structure of our own galaxy than many nearby galaxies like Andromeda galaxy simply because we are embedded inside our galaxy and interstellar dust partially blocks our view.

Observations toward HI clouds and HII regions as well as millimeter-wave observations of molecules (e.g. CO) have revealed coherent, large-scale structures, which are probably spiral arms in the Milky Way. Outlines of spiral arms were then drawn by the locations of HI clouds, HII regions and molecular clouds with their kinematic distances. These distances are based on a model of the rotation of the Milky Way (Georgelin & Georgelin 1976). However, such a model is affected by some factors, such as the existence of a non-circular rotation and the streaming motions of objects. This can result in a large uncertainty of kinematic distance estimates.

To improve the distance determination, we adopted a most direct and reliable method for measuring astronomical distance known as trigonometric parallax measurement. In practice, we have performed the phase-referencing VLBI observations of the 12 GHz methanol masers in the massive star-forming region W3OH in the Perseus arm, the nearest arm outward from the Sun. We observed strong and compact maser spots as the reference sources and switched rapidly between masers and three distant quasars. The atmospheric and ionospheric delay errors in the VLBA correlator (about 10 cm in the zenith) were measured during the observations and corrected in the post data processing. Relative position uncertainties were typically about 50 μas for the quasars and 10 μas for the masers. Parallax curves of 9 masers relative to 3 quasars were obtained and fitted by the five required parameters, one for the parallax and two parameters for the proper motion in each of the coordinates. This gives a parallax of W3OH of 0.512 ± 0.010 mas (Xu *et al.* 2006). The corresponding distance is 1.95 ± 0.04 kpc. The distance uncertainty is only 2%, a factor of 100 better than by the Hipparcos satellite. The Perseus arm has a kinematic distance of about 4 kpc and a luminosity distance of 2.2 kpc (Humphreys 1978). Our parallax measurement of W3OH has now resolved this long-standing discrepancy.

Moreover, the demonstration observations of W3OH proved the possibility of reconstructing of our own Galaxy within the size of about 20 kpc. Currently, a large project of measuring parallaxes and proper motions of both 12 GHz methanol and 22 GHz water maser sources in other regions of the Galaxy is ongoing (Reid, this proceedings). This will enable the mapping of the spiral structure with a great accuracy and further help to determine the distribution of dark matter in the Milky Way.

4. The Galactic magnetic fields on large scales

Magnetic fields in a large part of the Galactic disk have been delineated by Faraday rotation data of pulsars, which give a measure of the line-of-sight component of the magnetic field. Pulsars have the advantage of being spread throughout the Galaxy at known distances, allowing a direct mapping of 3-dimensional magnetic field.

For a pulsar at a distance D (in pc), the rotation measure (RM in radians m^{-2}) is given by RM $= 0.810 \int_0^D n_e \mathbf{B} \cdot d\mathbf{l}$, where n_e is the electron density in cm^{-3}, $\mathbf{B}$ is the vector magnetic field in μG and $d\mathbf{l}$ is an elemental vector along the line of sight (positive RMs correspond to fields directed toward us) in pc; and the dispersion measure is DM $= \int_0^D n_e dl$. Thus, pulsars give a direct estimate of the strength of the field through normalization by DM. Furthermore, we can directly estimate the field strength in a given region along the similar lines of sight from the gradient of RM and DM, i.e.

$$\langle B_{||} \rangle_{d1-d0} = 1.232 \frac{\Delta \text{RM}}{\Delta \text{DM}} \qquad (4.1)$$

where $\langle B_{||} \rangle_{d1-d0}$ is the mean line-of-sight field component in μG for the region between distances $d0$ and $d1$, ΔRM $= $ RM$_{d1} - RM_{d0}$ and ΔDM $= $ DM$_{d1} - DM_{d0}$.

Up to now, RMs of 550 pulsars have been observed (e.g., Han *et al.* 1999, Han *et al.* 2006). Most of the new measurements lie in the fourth and first Galactic quadrants and are relatively distant, enabling us to investigate the structure of the Galactic magnetic field over a much larger region than it was previously possible. We detected counterclockwise magnetic fields in the inner Norma arm (Han *et al.* 2002). A more complete analysis for the fields near the tangential regions of most probable spiral of our Galaxy (Han *et al.* 2006) gives such a picture for the coherent large-scale fields aligned with the spiral arm structure in the Galactic disk: magnetic fields in all inner spiral arms are counterclockwise when viewed from the North Galactic pole. On the other hand, at least in the local region and in the inner Galaxy of the fourth quadrant, there is good evidence that the fields in inter-arm regions are coherent, but clockwise in orientation. Thus, there are at least two or three reversals in the inner Galaxy, probably occurring near the boundary of the spiral arms. The magnetic field in the Perseus arm can not be well determined. The negative RMs for distant pulsars and extra-galactic sources in fact suggest that the inter-arm fields both between the Sagittarius and Perseus arms and beyond the Perseus arm are predominantly clockwise.

With much more pulsar RM data available now, Han *et al.* (2006) were able to *measure*, rather than *model*, the regular field strength near the tangential regions in the 1st and 4th Galactic quadrants, and then plot the dependence of regular field strength on the Galactic-radii. Although the "uncertainties", which in fact reflect the random fields, are large, the tendency is clear that fields get stronger at smaller Galactocentric radii and weaker in inter-arm regions. To parameterize the radial variation, an exponential function was used as follows, which not only gives the smallest χ^2 value but also avoids a singularity at $R = 0$ (for $1/R$) and non-physical values at large R (for the linear gradient). That is, $B_{\rm reg}(R) = B_0 \, \exp\left[\frac{-(R-R_\odot)}{R_{\rm B}}\right]$, with the strength of the large-scale or regular field at the Sun, $B_0 = 2.1 \pm 0.3$ μG and the scale radius $R_{\rm B} = 8.5 \pm 4.7$ kpc.

From the RM distribution in the sky, Han *et al.* (1997, 1999) identified the striking antisymmetry in the inner Galaxy with respect to the Galactic coordinates after removing the RM "outliers" compared to their neighborhoods. Such an antisymmetry should be a result from the azimuthal magnetic fields in the Galactic halo with reversed field directions below and above the Galactic plane. Such a field can be produced by an A0 mode of dynamo. The observed filaments near the Galactic center should result from the dipole field in this dynamo scenario. The local vertical field component of 0.2 μG (Han & Qiao 1994, Han *et al.* 1999) may be part of this dipole field in the solar vicinity. At present, we have observed another 1700 radio sources in the Northern sky by the Effelsberg 100 m telescope (Han, Reich *et al.* 2007, in preparation), and we wish to do more in the Southern sky at Parkes, so that the RM sky can be described quantitatively.

Acknowledgements

Chinese astronomers have dedicated to the reported study of the structure of our Galaxy using numerous radio telescopes over the globe. Many results would not be possible without kind help from our international collaborators including K. Y. Lo, R. N. Manchester, M. Reid. We appreciate very much for their long term support and cooperation. Research of JLH is supported by NNSFC under grant numbers 10473015, 10521001 and 10773016. YX is supported by NNSFC under grant numbers 10673024 and 10621303, and NBRPC (973 Program) under grant 2007CB815403. ZQS is supported in part by grants 10573029, 10625314, 10633010, and 06XD14024. ZQS acknowledges the support by the One-Hundred-Talent Program of the Chinese Academy of Sciences.

References

Balick, B. & Brown, R. L. 1974, *ApJ*, 194, 265
Georgelin, Y. M. & Georgelin, Y. P. 1976, *A&A*, 49, 57
Ghez, A. M. *et al.* 2005, *ApJ*, 620, 744
Han, J. L. & Qiao, G. J. 1994, *A&A*, 288, 759
Han, J. L., Manchester, R. N., Berkhuijsen, E. M., & Beck, R. 1997, *A&A*, 322, 98
Han, J. L., Manchester, R. N., & Qiao, G. J. 1999, *MNRAS*, 306, 371
Han, J. L., Manchester, R. N., Lyne, A. G., & Qiao, G. J. 2002, *ApJ*, 570, L17
Han, J. L., Manchester, R. N., Lyne, A. G., Qiao, G. J., & van Straten, W. 2006, *ApJ*, 642, 868.
Huang, L., Cai, M., Shen, Z.-Q., & Yuan, F. 2007, *MNRAS*, 379, 833
Humphreys, R. M. 1978, *ApJS*, 38, 309
Reid, M. J. & Brunthaler, A. 2004, *ApJ*, 616, 872
Schödel, R. *et al.* 2002, *Nature* , 419, 694
Shen, Z.-Q., Liang, M. C., Lo, K. Y., & Miyoshi, M. 2003, *Astron. Nachr.* 324, S1, 383
Shen, Z.-Q., Lo, K. Y., Liang M.-C., Ho, P. T. P., & Zhao J.-H. 2005, *Nature* , 438, 62
Xu, Y., Reid, M. J., Zheng, X. W., & Menten, K. M. 2006, *Science* , 311, 54

A Giant Step: from Milli- to Micro-arcsecond Astrometry
Proceedings IAU Symposium No. 248, 2007
W. J. Jin, I. Platais & M. A. C. Perryman, eds.

Microarcsecond astrometry in the Local Group

A. Brunthaler[1]†, M. J. Reid[2], H. Falcke[3], C. Henkel[1] and K. M. Menten[1]

[1] Max-Planck-Institut für Radioastronomie, Bonn, Germany
[2] Harvard-Smithsonian Center for Astrophysics, Cambridge, USA
[3] Radboud Universiteit Nijmegen, the Netherlands

Abstract. Measuring the proper motions and geometric distances of galaxies within the Local Group is very important for our understanding of its history, present state and future. Currently, proper motion measurements using optical methods are limited only to the closest companions of the Milky Way. However, given that VLBI provides the best angular resolution in astronomy and phase-referencing techniques yield astrometric accuracies of ≈ 10 micro-arcseconds, measurements of proper motions and angular rotation rates of galaxies out to a distance of $\sim$ 1 Mpc are feasible. This paper presents results of VLBI observations in regions of H_2O maser activity of the Local Group galaxies M33 and IC 10. Two masing regions in M33 are on opposite sides of the galaxy. This allows a comparison of the angular rotation rate (as measured by the VLBI observations) with the known inclination and rotation speed of the HI gas disk leading to a determination of a geometric distance of $730 \pm 100 \pm 135$ kpc. The first error indicates the statistical error of the proper-motion measurements, while the second error is the systematic error of the rotation model. Within the errors, this distance is consistent with the most recent Cepheid distance to M33. Since all position measurements were made relative to an extragalactic background source, the proper motion of M33 has also been measured. This provides a three dimensional velocity vector of M33, showing that this galaxy is moving with a velocity of 190 ± 59 km s^{-1} relative to the Milky Way. For IC 10, we obtain a motion of 215 ± 42 km s^{-1} relative to the Milky Way. These measurements promise a new handle on dynamical models for the Local Group and the mass and dark matter halo of Andromeda and the Milky Way.

Keywords. astrometry, galaxies: kinematics and dynamics, Local Group

1. Introduction

The nature of spiral nebulae like M33 was debated in the 1920's. While some astronomers favoured a short distance and Galactic origin, others were convinced of its extragalactic nature. In 1923, van Maanen claimed to have measured a large proper motion and angular rotation of M33 from photographic plates separated by ≈ 12 years (van Maanen 1923). These measurements yielded rotational motions of ≈ 10–30 mas yr^{-1}, indicating a short distance to M33. However, a few years later, Hubble discovered Cepheids in M33, providing evidence for a large distance to M33 and confirming that M33 is indeed an extragalactic object (Hubble 1926). The expected proper motions from the rotation of M33 are then only ≈ 30 μas yr^{-1}, 3 orders of magnitude smaller than the motions claimed by van Maanen. After more than 80 years, the idea behind the experiment to measure the rotation and proper motions of galaxies remains interesting for our understanding of the dynamics and geometry of the Local of the dynamics and geometry of the Local Group. Hence, they are an important science goal of astrometric missions (e.g. SIM and Gaia).

† email: brunthal@mpifr-bonn.mpg.de

The problem when trying to derive the gravitational potential of Local Group is that usually only radial velocities are known and hence and hence statistical approaches have to be used. Kulessa and Lynden-Bell (1992) introduced a maximum-likelihood method that requires only the line-of-sight velocities, but it is also based on some assumptions (eccentricities, equipartition).

Clearly, the most reliable way of deriving masses is using orbits, which requires the knowledge of three-dimensional velocity vectors obtained from measurements of proper motions. However, measuring proper motions of members of the Local Group to determine its mass is difficult. For the LMC a proper motion of 1.2 ± 0.3 mas yr^{-1} was obtained by comparing positions on photographic plates over a time-span of 14 years (Jones *et al.* 1994). The Sculptor dwarf spheroidal galaxy moves with 0.56 ± 0.25 mas yr^{-1} obtained from plates spanning 50 years (Schweitzer *et al.* 1995). It was shown that the inclusion of these marginal proper motions can already significantly improve the estimate for the mass of the Milky Way, since it reduces the strong ambiguity caused by Leo I, which can be treated as either bound or unbound to the Milky Way (Kochanek 1996). In recent years, the proper motions of a number of Galactic satellites were measured using the HST (e.g. Piatek *et al.* 2005, and these proceedings; Kallivayalil *et al.* 2006). These galaxies are all closer than 150 kpc and show motions between 0.2 and a few milliarcseconds (mas) per year. More distant galaxies, such as galaxies in the Andromeda subgroup at distances of ~ 800 kpc, have smaller angular motions, which are currently not measurable with optical telescopes.

With the accuracy obtainable with VLBI, one can in principle measure proper motions for most Local Group members very accurately within less than a decade. The main problem so far is finding appropriate radio sources. Useful sources would be either compact radio cores or strong maser lines associated with star forming regions. The most suitable candidates for such a VLBI phase-referencing experiment are the strong H_2O masers in IC 10 (~ 10 Jy peak flux in 0.5 km s^{-1} line) and IC 133 in M33 (~ 2 Jy, the first ever extragalactic maser discovered). The two galaxies belong to the brightest members of the Local Group and are thought to be associated with M31. In both cases, a relatively bright phase-referencing source is known to exist within a degree. In addition, their galactic rotation is well known from H I observations. Consequently, M33 and IC 10 seem to be the best-known targets for attempting to measure Local Group proper motions with the VLBA.

2. VLBI observations of M33 and IC 10

We observed two regions of H_2O maser activity in M33 (M33/19 and IC 133) eight times with the NRAO Very Long Baseline Array (VLBA) between March 2001 and June 2005 (Brunthaler *et al.* 2005). M33/19 is located in the south-eastern part of M33, while IC 133 is located in the north-east of M33. We observed the usually brightest maser in IC 10-SE with the VLBA thirteen times between February 2001 and June 2005 (Brunthaler *et al.* 2007).

The observations involved rapid switching between the phase calibrator and the target sources. With source changes every 30 seconds, an integration time of 22 seconds per scan was achieved. From the second epoch on, we included *geodetic-like* observations where we observed for 45 minutes 10–15 strong radio sources (> 200 mJy) with accurate positions (< 1 mas) at different elevations to estimate an atmospheric zenith delay error in the VLBA calibrator model (see Reid & Brunthaler 2004 for a discussion). In the second and third epoch, we used two blocks of these *geodetic observations* before and after the phase-referencing observations. From the fourth epoch on, we included a third *geodetic block* in the middle of the observation.

2.1. *Proper motions of M33/19 and IC 133*

The maser emission in M33/19 and IC 133 is variable on timescales of less than one year. Between the epochs, new maser features appeared while others disappeared. However, the motions of four components in M33/19 and six components in IC 133 could be followed over all epochs. The component identification was based on the positions and radial velocities of the maser emission. Each component was usually detected in several frequency channels. A rectilinear motion was fit to each maser feature in each velocity channel separately. Fits with a reduced χ^2 larger than 3 were discarded as they are likely affected by blending. Then, the variance-weighted average of all motions was calculated. This yields an average motion of the maser components in M33/19 of 35.5 ± 2.7 μas yr^{-1} in right ascension and -12.5 ± 6.3 μas yr^{-1} in declination relative to the background source J0137+312. For IC 133, one gets an average motion of 4.7 ± 3.2 μas yr^{-1} in right ascension and -14.1 ± 6.4 μas yr^{-1} in declination.

2.2. *Geometric distance of M33*

The relative motions between M33/19 and IC 133 are independent of the proper motion of M33 and any contribution from the motion of the Sun. Since the rotation curve and inclination of the galaxy disk are known, one can predict the expected relative angular motion of the two masing regions. The rotation of the HI gas in M33 has been measured (Corbelli & Schneider 1997) and one can calculate the expected transverse velocities of M33/19 and IC 133. This gives a relative motion of 106.4 km s^{-1} in right ascension and 35 km s^{-1} in declination between the two regions of maser activity.

The radial velocities of the H2O masers in M33/19 and IC 133 and the nearby HI gas are in very good agreement (< 10 km s^{-1}). This strongly suggests that the maser sources are co-rotating with the HI gas in the galaxy. However, while agreement between the rotation model and the radial velocity of the HI gas at the position of IC 133 is also very good (< 5 km s^{-1}), there is a difference of ~ 15 km s^{-1} at the position of M33/19. Hence, we conservatively assume a systematic error of 20 km s^{-1} in each velocity component for the relative velocity of the two maser components. Comparing the measured angular motion of 30.8 ± 4 μas yr^{-1} in right ascension with the expected linear motion of 106 ± 20 km s^{-1}, one gets a geometric distance of

$$D = 730 \pm 135 \pm 100 \text{ kpc},$$

where the first error indicates the systematic error from the rotation model while the second error is the statistical error from the VLBI proper-motion measurements.

After less than three years of observations, the uncertainty in the distance estimate is already dominated by the uncertainty of the rotation model of M33. However, this can be improved in the near future by determining a better rotation model using higher-resolution (e.g., Very Large Array or Westerbork Synthesis Radio Telescope) data of HI gas in the inner parts of the disk. Also, the precision of the proper-motion measurements will increase with time as $t^{3/2}$ for evenly spaced observations.

Within the current errors, the geometric distance of $730 \pm 100 \pm 135$ kpc is in good agreement with recent Cepheid and the Tip of the Red Giant Branch (TRGB) distances of 802 ± 51 kpc (Lee *et al.* 2002) and 794 ± 23 kpc (McConnachie *et al.* 2005), respectively.

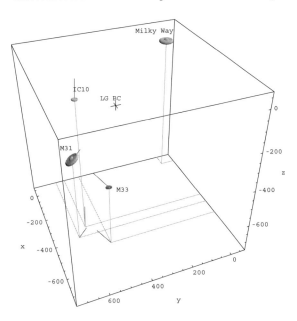

Figure 1. Schematic view of the Local Group with the space velocity of M33 and the radial velocity of Andromeda. The blue cross marks the position of the Local Group Barycenter (LG BC). Taken from Brunthaler *et al.* (2007).

2.3. *Proper motion of M33*

The observed proper motion $\tilde{\vec{v}}_{prop}$ of a maser (M33/19 or IC 133) in M33 can be decomposed into three components $\tilde{\vec{v}}_{prop} = \vec{v}_{rot} + \vec{v}_\odot + \vec{v}_{M33}$. Here, $\vec{v}_{rot}$ is the motion of the masers due to the internal galactic rotation in M33 and $\vec{v}_\odot$ is the apparent motion of M33 caused by the rotation of the Sun around the Galactic Center. The last contribution, $\vec{v}_{M33}$, is the true proper motion of M33 relative to the Galaxy.

Since the motion of the Sun (Reid & Brunthaler 2004; Dehnen & Binney 1998) and the rotation of M33 are known, one can calculate the two contributions $\vec{v}_{rot}$ and $\vec{v}_\odot$. Combining these velocity vectors, one gets the true proper motion of M33:

$$\dot{\alpha}_{M33} = \tilde{\dot{\alpha}}_{prop} - \dot{\alpha}_{rot} - \dot{\alpha}_\odot$$
$$= -29.3 \pm 7.6 \tfrac{\mu as}{yr} = -101 \pm 35 \tfrac{km}{s}$$

and

$$\dot{\delta}_{M33} = \tilde{\dot{\delta}}_{prop} - \dot{\delta}_{rot} - \dot{\delta}_\odot$$
$$= 45.2 \pm 9.1 \tfrac{\mu as}{yr} = 156 \pm 47 \tfrac{km}{s}.$$

Finally, the systemic radial velocity of M33 is -179 km s^{-1}. The radial component of the rotation of the Milky Way toward M33 is -140 ± 9 km s^{-1}. Hence, M33 is moving with -39 ± 9 km s^{-1} toward the Milky Way. This gives now the three-dimensional velocity vector of M33, which is plotted in Fig. 1. The total velocity of M33 relative to the Milky Way is 190 ± 59 km s^{-1}.

2.4. *Proper motion of IC 10*

In IC 10, only the strongest maser component was detected at all epochs. The uncertainties in the observations of the first epoch are larger than the others, because no geodetic-like observations were made to compensate the zenith delay errors. A rectilinear motion was fit to the data and yielded a value of 6 ± 5 μas yr^{-1} toward the East and 23 ± 5 μas yr^{-1} toward the North.

Once again, the contributions $\vec{v}_{rot}$ and $\vec{v}_{\odot}$ can be calculated from the known motion of the Sun and the known rotation of IC 10 (Wilcots & Miller 1998; Shostak & Skillman 1989). The true proper motion of IC 10 is then given by:

$$\dot{\alpha}_{IC\,10} = \dot{\tilde{\alpha}}_{prop} - \dot{\alpha}_{rot} - \dot{\alpha}_{\odot}$$
$$= -39 \pm 9 \; \mu\text{as yr}^{-1} = -122 \pm 31 \text{ km s}^{-1}$$

and

$$\dot{\delta}_{IC\,10} = \dot{\tilde{\delta}}_{prop} - \dot{\delta}_{rot} - \dot{\delta}_{\odot}$$
$$= 31 \pm 8 \; \mu\text{as yr}^{-1} = 97 \pm 27 \text{ km s}^{-1}$$

The measured systematic heliocentric velocity of IC 10 (-344 ± 3 km s^{-1}) is the sum of the radial motion of IC 10 toward the Sun and the component of the solar motion about the Galactic Center toward IC 10, which is -196 ± 10 km s^{-1}. Hence, IC 10 is moving with 148 ± 10 km s^{-1} toward the Sun. The proper motion and the radial velocity combined give the three-dimensional space velocity of IC 10. This velocity vector is shown in the schematic view of the Local Group in Fig. 1. The total velocity is 215 ± 42 km s^{-1} relative to the Milky Way.

3. Local Group dynamics and mass of M31

If IC 10 or M33 are bound to M31, then the velocity of the two galaxies relative to M31 must be smaller than the escape velocity and one can deduce a lower limit on the mass of M31:

$$M_{M31} > \frac{v_{rel}^2 R}{2G}.$$

A relative velocity of 147 km s^{-1} – for a zero tangential motion of M31 – and a distance of 262 kpc between IC 10 and M31 gives a lower limit of $6.6 \times 10^{11} M_{\odot}$. One can carry out this calculation for any tangential motion of M31. The results are shown in Fig. 2 (top). The lowest value of $0.7 \times 10^{11} M_{\odot}$ is found for a tangential motion of M31 of -130 km s^{-1} toward the East and 35 km s^{-1} toward the North.

For a relative motion of 230 km s^{-1} between M33 and M31 – again for a zero tangential motion of M31 – and a distance of 202 kpc, one gets a lower limit of $1.2 \times 10^{12} M_{\odot}$. Fig. 2 (top) shows also the lower limit of the mass of M31 for different tangential motions of M31 if M33 is bound to M31. The lowest value is $4 \times 10^{11} M_{\odot}$ for a tangential motion of M31 of -115 km s^{-1} toward the East and 160 km s^{-1} toward the North. Loeb *et al.* (2005) showed that proper motions of M31 in negative right ascension and positive declination would have lead to close interactions between M31 and M33 in the past. Such proper motions of M31 can be ruled out, since the stellar disk of M33 does not show any signs of strong interactions. Thus, we can rule out certain regions in Fig. 2. This yields a lower limit of $7.5 \times 10^{11} M_{\odot}$ for M31 and agrees with a recent estimate of $12.3^{+18}_{-6} \times 10^{11}$ $M_{\odot}$

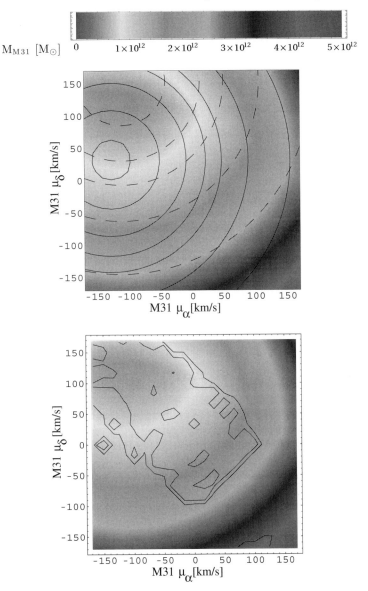

Figure 2. Top: Lower limit on the mass of M31 for different tangential motions of M31 assuming that M33 (dashed) or IC 10 (solid) are bound to M31. The lower limits are $(4, 5, 7.5, 10, 15, 25) \times 10^{11} M_\odot$ for M33, and $(0.7, 1, 2.5, 5, 7.5, 10, 15, 25) \times 10^{11} M_\odot$ for IC 10, rising from inside. The color scale indicates the maximum of both values. **Bottom:** The color scale is the same as above and gives the lower limit on the mass of M31. The contours show ranges of proper motions that would have lead to a large amount of stars stripped from the disk of M33 through interactions with M31 or the Milky Way in the past. The contours delineate 20% and 50% of the total number of stars stripped (Loeb *et al.* 2005). These regions can be excluded, since the stellar disk of M33 shows no signs of such interactions. Taken from Brunthaler *et al.* (2007).

derived from the three-dimensional positions and radial velocities of its satellite galaxies (Evans & Wilkinson 2000).

4. Summary

More than 80 years after van Maanen's observation, we have succeeded in measuring the rotation and proper motion of M33 as well as the proper motion of IC 10. These measurements provide a new handle on dynamical models for the Local Group and the mass and dark matter halo of Andromeda and the Milky Way.

We have presented astrometric VLBA observations of the H_2O masers in the Local Group galaxies M33 and IC 10. We have measured the proper motion of the masers relative to background quasars. Correcting for the internal rotation of M33, IC 10, and the rotation of the Milky Way, these measurements yield proper motions of the two galaxies. The total space velocities relative to the Milky Way of M33 and IC 10 are 190 ± 59 km s^{-1} and 215 ± 42 km s^{-1}, respectively. If IC 10 and M33 are bound to M31, one can calculate the lower mass limit for M31 to be $7.5 \times 10^{11} M_\odot$.

Further VLBI observations within the next few years and an improved rotation model have the potential to improve the accuracy of the distance estimate to less than 10%. At least one additional maser source exists in M33 (Brunthaler *et al.* 2006) that will be used in the future to increase the accuracy of the measurements. A third region of maser activity will also help to check for non-circular velocities of the masers.

Today, we are only able to study the extreme (bright) tip of the maser luminosity distribution for interstellar masers. The Square Kilometer Array (SKA), with substantial collecting area on intercontinental baselines and a frequency coverage up to 22 GHz (Fomalont & Reid 2004), will provide the necessary sensitivity to detect and measure the proper motions of a much greater number of masers in active star-forming regions in the Local Group.

References

Brunthaler, *et al.*, Science 2005, 307, 1440
Brunthaler, *et al.*, A&A 2006, 457, 109
Brunthaler, *et al.*, A&A 2007, 462, 101
Corbelli & Schneider, ApJ 1997, 479, 244
Dehnen & Binney, MNRAS 1998, 298, 387
Evans & Wilkinson, MNRAS 2000, 316, 929
Fomalont & Reid, New Astronomy Review 2004, 48, 1473
Hubble, ApJ 1926, 63, 236
Jones, *et al.*, AJ 1994, 107, 1333
Kallivayalil, *et al.*, ApJ 2006, 638, 772
Kochanek, ApJ 1996, 457, 228
Kulessa & Lynden-Bell, MNRAS 1992, 255, 105
Lee, *et al.*, ApJ 2002, 565, 959
Loeb, *et al.*, ApJ 2005, 633, 894
McConnachie, *et al.*, MNRAS 2005, 356, 979
Piatek, *et al.*, AJ 2006, 131, 1445
Reid & Brunthaler, ApJ 2004, 616, 872
Schweitzer, *et al.*, AJ 1995, 110, 2747
Shostak & Skillman, A&A 1989, 214, 33
van Maanen, ApJ 1923, 57, 264
Wilcots & Miller, AJ 1998, 116, 2363

A Giant Step: from Milli- to Micro-arcsecond Astrometry
Proceedings IAU Symposium No. 248, 2007
W. J. Jin, I. Platais & M. A. C. Perryman, eds.

© 2008 International Astronomical Union
doi:10.1017/S1743921308019856

Mass segregation effects in very young open clusters

L. Chen[1], R. de Grijs[2] and J. L. Zhao[1]

[1]Shanghai Astronomical Observatory, Chinese Academy of Sciences, 80 Nandan Road,
Shanghai 200030, China
email: `chenli@shao.ac.cn, jlzhao@shao.ac.cn`

[2]Department of Physics and Astronomy, University of Sheffield, Hicks Building, Hounsfield
Road, Sheffield S3 7RH, UK; and National Astronomical Observatories, Chinese Academy of
Sciences, 20A Datun Road, Chaoyang District, Beijing 100012, China
email: `r.degrijs@sheffield.ac.uk`

Abstract. We derived proper motions and membership probabilities of stars in the regions of
two very young ($\sim 2 - 4$ Myr-old) open clusters NGC 2244 and NGC 6530. Both clusters show
clear evidence of mass segregation, which provides strong support for the suggestion that the
observed mass segregation is – at least partially – due to the way in which star formation has
proceeded in these complex star-forming regions ("primordial" mass segregation).

Keywords. open clusters and associations: general, open clusters and associations: individual
(NGC2244), open clusters and associations: individual (NGC6530)

1. Introduction

Open clusters are important stellar systems for understanding star formation processes.
One of the important aspects in open cluster studies is related to the effect of mass
segregation. In some previous work on young open clusters, including the Orion Nebula
Cluster (Hillenbrand 1997) and R136 (e.g. Campbell *et al.*1992, Brandl *et al.* 1996) the
authors found evidences for mass segregation. However, the question is, for such young
star clusters, whether this effect is mainly the result of dynamical evolution or possibly
primordial in nature.

In the present work (see Chen *et al.* (2007) for details), based on photographic plate
material with longer time baselines than the published previously, we determined proper
motions and membership probabilities of stars in the regions of two young open clusters
NGC 2244 and NGC 6530, both with ages younger than 5 Myr. Furthermore, we investigate the possible effects of mass segregation in these very young open clusters. The
observed spatial mass segregation might be due to a combined effect of initial conditions
and relaxation process.

2. Mass segregation in NGC 2244 and NGC 6530

The observational data are from the historical photographic plates taken by the 40cm
double astrograph with focal length of 6.9m at Sheshan Station of Shanghai Astronomical
Observatory, Chinese Academy of Sciences. The observational time baseline is 34 and 87
years for open cluster NGC 2244 and NGC 6530, respectively. From this photographic
data, for each cluster, we determined the proper motion of stars (495 stars in NGC 2244
and 364 stars in NGC 6530) in a one square degree area centered on the cluster, and
estimated their membership probabilities using a maximum likelihood principle(Wang
1997, Zhao & He 1987).

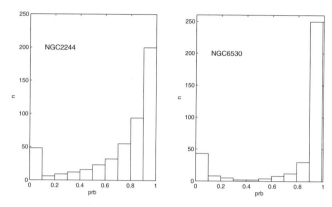

Figure 1. Histogram of the stellar membership probabilities. *Left panel:* 495 stars in the region of NGC 2244; *Right panel:* 364 stars in region of NGC 6530.

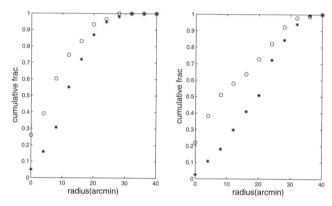

Figure 2. Normalized cumulative radial number density profile for members of NGC 2244 (left) with $m_B \leqslant 13$ (o) and $m_B > 13$ (*) and NGC 6530 (right) with $m_B \leqslant 12$ (o) and $m_B > 12$ (*). Results from Chen *et al.* (2007)

Figure 1 shows the histogram of membership probabilities for NGC 2244 (left) and NGC 6530 (right). From this figure one can deduce that the discrimination of membership for NGC 2244 is quite effective. Of the total of 495 stars, about 16%, or 78 stars, have probabilities in the range of $0.3 < p < 0.7$. For NGC 6530, the separation of members from the field stars is even more effective, with only a few stars having probabilities around $p = 0.5$, i.e., only 4.4% (or 16 stars) have $0.3 < p < 0.7$; 451 stars have $p \geqslant 0.9$.

Since the cluster stars are located at the same distance, and because most of them are main sequence stars, we can use luminosity instead of mass to study the (mass/luminosity) segregation effects.

To examine the mass segregation effects, we investigated the radial density distributions of member stars ($p \geqslant 0.9$) in different luminosity ranges which are shown in Fig. 2 (left for NGC 2244 and right for NGC 6530). In each cluster, a position-dependent LF is found. In essence, brighter (or massive) stars are more concentrated towards the inner part in both of the clusters. This indicates an evident luminosity or mass segregation effect.

We conclude that both open clusters, NGC 6530 and NGC 2244, show clear evidence of mass (luminosity) segregation. These two clusters were too young for this to have only been caused by dynamical relaxation processes, and the observed segregation might be due to a combination of both initial conditions and dynamical evolutions. Therefore, our results are supporting the "primordial mass segregation" scenarios, such as the "competitive accretion" model. (see Larson 1991, Bonnell *et al.* 2001a, Bonnell *et al.* 2001b)

Acknowledgement

LC, and JLZ are supported by the NSFC (Grant Nos. 10773021, 10333050, 10333020,10573022) and NKBRSFG 2007CB815402 and 403. RdG was partially supported by an "International Joint Project" grant, jointly funded by the Royal Society in the UK and the Chinese National Science Foundation (NSFC).

References

Bonnell, I. A., Bate, M. R., Clarke, C. J., & Pringle, J. E. 2001a, *MNRAS*, 323, 785

Bonnell, I. A., Clarke, C. J., Bate, M. R., & Pringle,J. E. 2001b, *MNRAS*, 324, 573

Brandl, B., Sams, B. J., Bertoldi, F., Eckart, A.,Genzel, R., Drapatz, S., Hofmann, R., Löwe, M., & Quirrenbach, A. 1996, *ApJ*, 466, 254

Campbell, B., *et al.* 1992, *AJ*, 104, 1721

Chen, L., de Grijs, R., & Zhao, J. L. 2007, *AJ*, 134, 1368

Hillenbrand, L. A. 1997, *AJ*, 113, 1733

Larson, R. B. 1991, in: Falgarone, E., Boulanger, F., & Duvert, G., (eds.), *Fragmentation of Molecular Clouds and Star Formation, IAU Symp. 147* (Dordrecht: Kluwer), p. 261

Wang, J. J. 1997, *Annals of Shanghai Observatory, Acad. Sinica*, 18, 45

Zhao, J. L. & He, Y. P. 1987, *Acta Astr. Sin.*, 28, 374

A Giant Step: from Milli- to Micro-arcsecond Astrometry
Proceedings IAU Symposium No. 248, 2007
W. J. Jin, I. Platais & M. A. C. Perryman, eds.
© 2008 International Astronomical Union
doi:10.1017/S1743921308019868

The luminosity function of nearby thick-disk sub-dwarfs

M. I. Arifyanto

Astronomy Research Division, Faculty of Mathematics and Natural Sciences,
Institut Teknologi Bandung,
Jl. Ganesha No. 10, Bandung, 40132, Indonesia
email: ikbal@as.itb.ac.d

Abstract. We derived the luminosity function of thick disk using V/V_{max} method for nearby sub-dwarf stars based on the sample stars of Carney *et al.* (1994). Hipparcos parallaxes and proper motions and Tycho2 proper motions were combined with radial velocities and metallicities from CLLA. We found that the luminosity function in the absolute magnitude range $M_V = 4-6$ mag agree well with the luminosity function derived from the initial mass function (Reyle & Robin 2001).

Keywords. Galaxy: general, Galaxy: disk, stars: fundamental parameters (luminosities, masses), subdwarfs

1. Introduction

The thick disk population has a mean metallicity of $-0.7 \leqslant [Fe/H] \leqslant -0.4$ (e.g., Buser *et al.* 1999) which is similar to the disk globular cluster 47 Tuc (Carney *et al.* 1989). The thick-disk LF is thought to have the same shape as the metal rich globular cluster 47 Tuc (Buser *et al.* 1999). Until now no direct measurement of the thick disk LF has been done. Reyle and Robin (2001) derived the LF from their thick disk initial mass function (IMF) based on the star counts at high and intermediate galactic latitudes.

2. Selection of thick disk stars

The data set constructed by Arifyanto *et al.* (2005) was based on the sample of F and G sub-dwarfs of Carney et al. (1994). The sample of AFJW, which forms the basis of our analysis, contains 740 sub-dwarfs with greatly improved parallax and proper motion data. The photometric distances were corrected by a factor of about 10%.

We selected for the thick disk all stars with $-1.0 \leqslant [Fe/H] \leqslant -0.4$. They are all brighter than $m_V = 12.5$ mag and with proper motion larger than $\mu = 155$ mas yr^{-1}. We defined the general restriction of the sample, in which our sample is complete, with $m_V \sim 9.2$ mag. and $\mu \geqslant 180$ mas yr^{-1}. The contamination of thin-disk stars in our proper motion selected sample is minimized by setting up the minimum proper motion cut. The proper motion selection magnifies the contribution from the higher-velocity old populations, since they are effectively sampled over larger volume than the lower-velocity disk stars.

3. The luminosity function

The luminosity function is derived by V/V_{max} method for the 89 thick disk sub-dwarf stars. Each star represents a single sampling over the maximum volume. Therefore each will contribute to the LF $1/V_{max}$ and sum of all the sample stars. We performed our Monte Carlo simulations and derive the correction factor (χ_{TD}) and simulated LF. We ran 200 simulations, where for each simulation 3×10^5 stars were generated and then, after applying our selection criteria, chose 89 surviving stars, equal to the number of thick disk stars in our sample. We scale the thick disk LF following the method used by Digby *et al.* (2003). We consider any possible contamination by the thin disk stars.

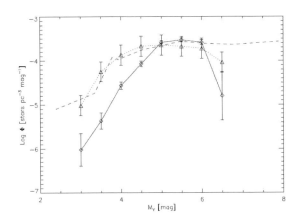

Figure 1. Simulated luminosity function taken from Bergbusch & Vandenberg (1992) for metallicity [Fe/H]= -0.65 with ages 12 Gyrs (dotted line). We plot also the luminosity function of thick disk from our calculation (solid line) and the LF derived from the initial mass function of Reyle & Robin, 2001 (dashed line).

The LF has a steep slope in the range $M_V = +3$ to $+5$ and a constant value for $M_V = 5 - 6$ mag. At $M_V > 6.5$ mag the luminosity function decreases. The reason of this decrease could be the Wielen Dip at $M_V \sim 7$ mag and incompleteness in our sample. We performed Monte Carlo simulations to understand the selection bias in our sample. We use the LFs of Bergbusch & Vandenberg (1992) for metallicity [Fe/H] $= -0.65$ with ages of 12 Gyrs. Bergbusch & Vandenberg (1992) used their LF to fit with the observed luminosity function of globular cluster 47 Tuc. The simulated LFs for metallicity [Fe/H] $= -0.65$ with ages of 12 Gyrs agree well with the luminosity function derived by Reyle & Robin (2001) for $M_V = 3.0 - 6.0$. The differences in the slope between the simulations and observations in the range $M_V = 3 - 5$ mag could be due to the lack of bright stars in the sub-dwarf sample. Gilmore & Reid (1983) found that the LF at $z > 1$ kpc, steepen rapidly for $M_V < 4$ mag.

References

Arifyanto, M. I., Fuchs, B., Jahreiss, H., & Wielen, R. 2005, *A&A*, 433, 911
Bergbusch, P. A. & Vandenberg, D. A. 1992, *ApJS*, 81, 163
Buser, R., Rong, J., & Karaali, S. 1999, *A&A*, 348, 98
Carney, B. W., Latham, D. W., Laird, J. B., & Aguilar, L. A. 1994, *AJ*, 107, 2240
Carney, B. W., Latham, D. W., Laird, J. B., & John, B. 1989, *AJ*, 97, 423
Digby, A. P., Hambly, N. C., Cooke, J. A., Reid, I. N., & Cannon, R.D. 2003, *MNRAS*, 344, 583
Gilmore, G. & Reid, I. N. 1983, *MNRAS*, 202, 1025
Hog, E., Fabricius, C., Makarov, V. V., *et al.* 2000, *A&A*, 355, 367
Perryman, M. A. C., Lindegren, L., Kovalevsky, J. *et al.* 1997, *A&A*, 323, 49
Reyle, C. & Robin, A. C. 2001, *A&A*, 373, 886

A Giant Step: from Milli- to Micro-arcsecond Astrometry
Proceedings IAU Symposium No. 248, 2007
W. J. Jin, I. Platais & M. A. C. Perryman, eds.

Are parallaxes of long-period variable stars and red supergiants reliable?

C. Babusiaux[1] and A. Jorissen[2]

[1] Observatoire de Paris, Place Jules Janssen, 92195 Meudon, France.
email: carine.babusiaux@obspm.fr

[2] IAA, Université Libre de Bruxelles CP 226, 1050 Bruxelles, Belgique.
email: ajorisse@astro.ulb.ac.be

Abstract. We have studied the impact of long-term variability and surface brightness asymmetries on the parallaxes of long-period variable stars and red supergiants.

Keywords. astrometry, stars: AGB and post-AGB, supergiants

1. Introduction

Long-period variables (LPVs) are frequent among highly evolved stars on the asymptotic giant branch (AGB). Period-luminosity relationships have been derived for LPVs in the Large Magellanic Cloud by Wood (2000). It is hoped that Gaia will deliver such relationships for LPVs in the Galaxy (Barthès *et al.* 1999). The astrometric observations of LPVs and red supergiants face, however, several specific problems that may lower the accuracy of the parallax determination. Their very red colours makes it necessary to apply an adopted chromatic correction. This correction is, however, difficult to calibrate because of the small number of well-behaved stars with such red colours. A good illustration of this difficulty may be obtained from the analysis of the Hipparcos Transit Data (Quist & Lindegren 1999), which give the parameter fitting of the stellar signal modulated by the grid atop the detector. For a point source, the ratio R of the modulation coefficients of the second over first harmonics should be 0.35. We observe, however, a clear decrease of this ratio with increasing $V - I$ color. This may explain the large fraction of red stars for which the Hipparcos solution quality flag has been set to S ('suspected non single').

Moreover, the LPVs variability leads to colour variations over the light cycle which calls for the use of *epoch* colour indices to compute the epoch chromaticity corrections (Platais *et al.* 2003). Finally, the very large radii of LPVs and supergiant stars make them extended sources and time-varying brightness asymmetries (spots) on their stellar disc could strongly degrade their parallax accuracy (Eriksson & Lindegren 2007); indeed the angular radius of LPVs can be larger than their parallax as soon as their linear radius exceeds 1 AU.

2. Hipparcos and the LPVs

The reprocessing of the Intermediate Astrometric Data (IAD) of LPVs has been made in Brussels using the epoch $V - I$ colour indices (Knapp *et al.* 2003), leading to much improved parallaxes and also allowed us to reject the Variability-Induced Movers (VIMs) classification for 86% of the LPVs in the Hipparcos catalogue (Pourbaix *et al.* 2003). The quality of these parallaxes may be assessed by confronting the parallax standard deviations σ_{ϖ} with the Hp magnitude at *minimum brightness* (from field H50 of the Hipparcos Catalogue) instead of with the *median Hp*. Only a small number of LPVs

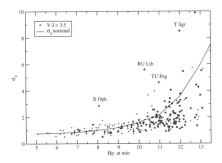

Figure 1. Standard error of the revised parallaxes for LPVs (Knapp *et al.* 2003) as a function of the Hp magnitude at minimum brightness (95th percentile; field H50 of the Hipparcos catalogue). The solid line is the nominal relation derived from the whole Hipparcos catalogue. Stars redder than $V - I = 3.5$ are noted as large dots.

then present too large standard deviations σ_ϖ with respect to the fiducial relationship (σ_ϖ, Hp) for the whole Hipparcos Catalogue (see Fig. 3.2.39 of Volume 1), as seen on Fig. 1. At least two of these remaining outliers are known binaries (X Oph, T Sgr). In a forthcoming paper, we show that the goodness-of-fit parameter provides an even more stringent diagnostic than the parallax standard deviation to assess the parallax reliability.

3. Hipparcos and Betelgeuse

Supergiants do not show such large variability as LPVs, so that there is no need to re-process their astrometry using epoch colours. Three very bright supergiants show a large excess in their parallax standard error: Betelgeuse, Antares and α Her. The two first have been classified DMSA/X and α Her shows a very poor goodness-of-fit. Interestingly, these three stars are also known to have a surface structure (e.g., Tuthill *et al.* 1997). During the Hipparcos mission, two interferometric images of the disc of Betelgeuse have been obtained by Wilson *et al.* (1992) and Tuthill *et al.* (1997). They reveal a very clear change in the surface brightness which could indeed have affected the Hipparcos measurements. We show in a forthcoming paper that it could explain Betelgeuse's classification as DMSA/X. Depending on the data set used (NDAC, FAST, outliers removal), our reprocessing leads to parallaxes of Betelgeuse between 6.8 and 10.5 mas. The knowledge provided by those interferometric data combined to the new reduction should allow us to provide a better estimate of the Betelgeuse parallax, crucial for the theoretical modelling of this well studied red supergiant.

At the same time, this study provides guidance for designing the reduction pipeline for Gaia in such a way that reduces the possible impact of surface brightness asymmetries on the derived parallaxes.

References

Barthès, D., Luri, X., Alvarez, R., & Mennessier, M. O. 1999, *A&AS* 140, 55

Eriksson, U. & Lindegren, L. 2007, *A&A*, in press (astro-ph/0706.1646)

Knapp, G. R., Pourbaix, D., Platais, I., & Jorissen, A. 2003, *A&A* 403, 993

Platais, I., Pourbaix, D., Jorissen, A. *et al.* 2003, *A&A* 397, 997

Pourbaix, D., Platais, I., Detournay, S. *et al.* 2003, *A&A* 399, 1167

Quist C. F. & Lindegren L. 1999, *A&AS* 138, 327

Tuthill, P. G., Haniff, C. A., & Baldwin, J. E. 1997, *MNRAS* 285, 529

Wilson, R. W., Baldwin, J. E., Buscher, D. F., & Warner, P. J. 1992, *MNRAS* 257, 369

Wood P. R. 2000, *PASA* 17, 18

A Giant Step: from Milli- to Micro-arcsecond Astrometry
Proceedings IAU Symposium No. 248, 2007
W. J. Jin, I. Platais & M. A. C. Perryman, eds.

© 2008 International Astronomical Union
doi:10.1017/S1743921308019881

Calibration of stellar parameters using high-precision parallaxes

A. G. Butkevich[1]†, A. V. Berdyugin[2] and P. Teerikorpi[2]

[1]Lohrmann Observatory, Dresden Technical University, D-01062 Dresden, Germany
email: alexey.butkevich@tu-dresden.de

[2]Tuorla Observatory, Turku University, FIN-21500 Piikkiö, Finland
email: andber@utu.fi, pekkatee@utu.fi

Abstract. Utilization of sub-milliarcsecond trigonometric parallaxes shifts the classical problem of calibration of stellar parameters to a new level of complexity. Derivation of stellar luminosity from the parallaxes is not a straightforward task with a number of statistical effects, such as Malmquist bias, to be taken into account. Different methods are to be used in order to derive parameters of luminosity function depending on the nature of underlying stellar sample. It is emphasized that any combination of astrometric parameters (i.e. parallaxes) and astrophysical ones must be handled carefully to avoid or reduce statistical effects, which otherwise may seriously affect the astrophysical applications.

Keywords. stars: distances, stars: fundamental parameters, stars: statistics, methods: statistical

The high-quality parallaxes provided by Hipparcos have allowed astronomers to improve the luminosity calibration for various stellar types. It has also given rise to numerous studies in this field, mainly focusing on the statistical handling of trigonometric parallax data, which appeared to be more complicated than anticipated. One of statistical effects is related to fine details of the Malmquist bias, such as its dependence on distance. This effect had been known for a long time in extragalactic astronomy where redshifts are used as non-photometric distance indicators. One distinction between the stars and galaxies, from the viewpoint of statistics has been the different relative accuracy of distances. Redshifts give a rather good relative distance precision, typically 5-10%, while accurate parallaxes, until Hipparcos were known for a small number of nearby stars. Therefore this type of bias simply was not relevant prior to Hipparcos. With Gaia its analysis will be increasingly important.

The well-known formula for the Malmquist bias states that the mean absolute magnitude calculated for a magnitude-limited sample from a stellar population having Gaussian luminosity function with intrinsic mean M_0 and scatter σ is

$$\overline{M} = M_0 - 1.38\sigma^2 \,.$$

This formula, has to be applied with caution. Malmquist (1922) derived it in his classical study, assuming that the star distribution is spatially uniform.

The Malmquist bias originates from a simple but subtle selection effect: the more distant objects we consider, the brighter objects we get, but at larger distances there is the higher spatial volume. In a magnitude-limited sample this bias at a fixed distance cuts off the stars from the faint side of the luminosity function (Teerikorpi 1975; Sandage 1994). The difference between the mean absolute magnitude calculated for the remaining

---

† On leave from Pulkovo Observatory, 196140 Saint-Petersburg, Russia.

part of the luminosity function and the true intrinsic mean M_0 represents the distance-dependent Malmquist bias.

A quantitative treatment of the distance-dependent Malmquist bias developed by Teerikorpi (1975) and generalized by Butkevich, Berdyugin & Teerikorpi (2005a) implies that, under the following assumptions: (a) there is no interstellar absorption; (b) the luminosity function obeys a Gaussian law with a mean M_0 and a scatter σ and does not depend on distance; (c) the considered stellar sample is complete up to a cutoff apparent magnitude $m_{\rm lim}$; the mean absolute magnitude at a fixed parallax π is given by

$$\overline{M}\left(\pi\right) = -\sigma\sqrt{\frac{2}{\pi}}\frac{\exp\left[-\left(M_{\rm lim}\left(\pi\right) - M_0\right)^2 / \left(2\sigma^2\right)\right]}{{\rm erfc}\left[-\left(M_{\rm lim}\left(\pi\right) - M_0\right) / \left(\sigma\sqrt{2}\right)\right]},$$

where $M_{\rm lim}\left(\pi\right) = m_{\rm lim} + 5 + 5\lg\pi$.

The first empirical demonstration of the distant-dependent Malmquist bias using the Hipparcos parallaxes was done by Oudmaijer, Groenewegen & Schrijver (1999) who pointed out a correlation between the derived absolute magnitude and parallax for a sample of K0V stars with high precision parallaxes. Butkevich, Berdyugin & Teerikorpi (2005b) confirmed this result and showed that the bias behaves in a manner consistent with theoretical predictions.

The bias appears to start from a certain distance, which divides the entire distance range into the regions affected and not affected by the bias. The unbiased region, which is also called unbiased plateau, represents a volume-limited subset of a total sample. Detection of the boundary of the unbiased plateau is not a straightforward task because its position depends on the luminosity function which usually is unknown. Even if we assume some functional form of the luminosity function, say, a Gaussian, one should know its scatter σ in order to calculate where the boundary lies. Fortunately, the unbiased region may be recognised by visual inspection of the Spaenhauer diagram. Moreover, one may simply estimate where the bias is important. Let parallax π_0 and distance r_0 correspond to the condition $M_{\rm lim}\left(\pi_0\right) = M_0$:

$$\pi_0 = 1/r_0 = 10^{(M_0 - m_{\rm lim} - 5)/5}.$$

The bias may be neglected if $\pi > \pi_0$, and should be taken into account if $\pi < \pi_0$.

This criterion and assumed astrometric accuracy allow to predict which part of the HR diagram would be affected by the bias at a given distance. It is worthwhile mentioning that the interstellar extinction can seriously influence the bias at large distances, say, beyond 1 kpc.

Acknowledgements

This study has been supported by the Academy of Finland (the projects 'Fundamental questions of observational cosmology' and 'Study of circumstellar and interstellar medium with polarimetry').

References

Butkevich, A. G., Berdyugin, A. V., & Teerikorpi, P. 2005a, *MNRAS* 362, 321
Butkevich, A. G., Berdyugin, A. V., & Teerikorpi, P. 2005b, *A&A* 435, 949
Malmquist, K. G. 1922, *Lund Medd. Ser. I* 100, 1
Oudmaijer, R. D., Groenewegen, M. A. T., & Schrijver, H. 1999, *A&A* 341, L55
Sandage, A. 1994, *ApJ* 430, 1
Teerikorpi, P. 1975, *A&A* 45, 117

A Giant Step: from Milli- to Micro-arcsecond Astrometry
Proceedings IAU Symposium No. 248, 2007
W. J. Jin, I. Platais & M. A. C. Perryman, eds.

© 2008 International Astronomical Union
doi:10.1017/S1743921308019893

Exploiting kinematics and UBVI*c* photometry to establish high fidelity membership of the open cluster Blanco 1

P. A. Cargile[1], D. J. James[1], I. Platais[2], J.-C. Mermilliod[3], C. Deliyannis[4] and A. Steinhauer[5]

[1] Department of Physics and Astronomy, Vanderbilt University
Nashville, TN 37235, USA
email: `p.cargile@vanderbilt.edu`

[2] Johns-Hopkins University, Baltimore, MD, USA

[3] EPFL Lausanne, Lausanne, Switzerland

[4] Universtiy of Indiana, Bloomington, IN, USA

[5] SUNY-Geneseo, Geneseo, NY, USA

Abstract. We present the results of a wide-field, high-precision $UBVI_c$ CCD photometric survey of the Galactic open cluster Blanco 1. Standardized photometry was acquired using the Y4Kcam on the SMARTS 1m telescope at CTIO. We have also determined new high-precision proper motions ($\sigma_\mu = 0.3\ mas\ yr^{-1}$) over an eight square degree area down to $V = 16.5$. Combined with $1D$ kinematic data, our survey yields a complete list of cluster members down to $\sim 0.5 M_\odot$ and new high-fidelity color-magnitude diagrams are presented for Blanco 1. Having established a bona fide membership catalog, astrophysical characteristics of solar-type cluster members such as X-ray activity and lithium abundance have been studied to gain more insights in the process of internal mixing and convection. Our new results should also help to better understand its peculiar location in the Milky Way and to unravel its dynamical history.

Keywords. astrometry, techniques: photometric, open clusters and associations: individual (Blanco 1)

1. Introduction

Blanco 1 is a rather poor, young ($50 - 100\ Myr$), and nearby ($240\ pc$) open cluster. It is unique among open clusters because of its unusually high Galactic latitude ($b = -79°$), especially for its age. Blanco 1 also shows strong evidence, due to its Galactic latitude and kinematics, that it has passed through, and even interacted with, the Galactic disk. We present here the results of a wide-field, high-precision $UBVI_c$ CCD photometric survey of Blanco 1, including new high-fidelity color-magnitude diagrams for the cluster. Moreover, we have determined new high-precision proper motions ($\sigma_\mu = 0.5\ mas\ yr^{-1}$) over an eight square degree area down to $V = 16.5$.

2. Observations and results

The astrometric reductions are based on 32 sets of photographic plates and CCD frames almost all of them obtained with the 51 cm double astrograph (scale $= 55'mm^{-1}$) at Cesco Observatory in El Leoncito, Argentina. These observations span 40 years ending in September 2007. Proper motions and positions are calculated using a variant of the central plate-overlap method (e.g, Herbig & Jones 1981) and the UCAC2 catalog (Zacharias *et al.* 2004) as a reference frame. The precision of our proper motions, for stars

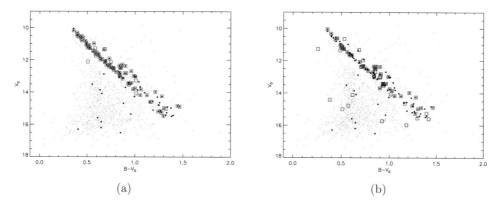

(a) (b)

Figure 1. V vs. $B-V$ CMD for Blanco 1 with proper motion members identified (*bold points*). (a) Objects with radial velocities consistent with the systemic velocity of the cluster are identified (*red boxes*). (b) Optical counterparts to x-ray sources are identified (*blue boxes*).

with optimal image properties, is 0.3 *mas* yr^{-1}. The final catalog contains 6300 objects, among which, more than 3600 have proper-motion precisions better than 2 *mas* yr^{-1}.

The formal proper-motion membership probabilities P_μ were calculated using the probability definition formulated by Vasilevskis *et al.* (1958): $P_\mu = \Phi_c/\Phi_c + \Phi_f$ where Φ_c is the distribution of cluster stars, and Φ_f is the distribution of field stars in the vector-point diagram. We used the so-called local sample method to derive the field and cluster distributions, which works very well in sparse open clusters (Platais *et al.* 2007). The separation between the cluster and field is convincing for the entire magnitude range and a total of 247 stars have their P_μ greater than 0%. $UBVI_c$ CCD photometric data were taken using the SMARTS 1.0m telescope at the CTIO, equipped with the 19.3 arcmin^2 Y4K camera. The central 1.6x1.3 sq. degrees of Blanco 1 were surveyed in a 5x4 grid mosaic (see James *et al.* 2008, in prep).

In Fig. 1 we plot the V vs. $B-V$ CMD for Blanco 1 with proper motion members identified. The majority of astrometric members of the cluster lie along a well defined main-sequence locus. In Fig. 1a, it is clearly seen that stars identified with radial velocities consistent with the systemic velocity of the cluster trace the same main-sequence as the proper motion members (see James *et al.* 2008; Mermilliod et al. 2008). In Fig. 1b, we now identify the optical counterparts to objects with X-ray emission, which can be used as a membership criterion in young clusters (Cargile & James 2008). The vast majority of the X-ray sources in Blanco 1 trace the same cluster sequence as defined by both the proper motion and radial velocity data.

References

Cargile, P. A. & James, D. J. 2008, *AJ*, in prep
Herbig, G. H. & Jones, B. F. 1981, *AJ*, 86, 1232
James, D. J., *et al.* 2008, *AJ*, in print
Mermilliod, J-C., *et al.* 2008, *A&A*, in prep
Platais, I., *et al.* 2007, *A&A*, 461, 509
Vasilevskis, S., Klemola, A., & Preston, G. 1958, *AJ*, 63, 387
Zacharias, N., Urban, S. E., Zacharias, M. I., Wycoff, G. L., Hall, D. M., Monet, D. G., & Rafferty, T. J. 2004, *AJ*, 127, 3043

A Giant Step: from Milli- to Micro-arcsecond Astrometry
Proceedings IAU Symposium No. 248, 2007
W. J. Jin, I. Platais & M. A. C. Perryman, eds.

Preliminary proper motion analysis of the Carina dwarf spheroidal

J. L. Carlin[1], S. R. Majewski, D.I. Casetti-Dinescu, and T. M. Girard

[1]Dept. of Astronomy, Univ. of Virginia, PO Box 400325, Charlottesville, VA 22904-4325 USA

Abstract. We present preliminary results from a proper motion study of the Carina dwarf spheroidal galaxy. Our proper motions show a scatter of $\sim$1.1 mas yr^{-1} per Carina member star, and we determinate the mean ensemble motion to an accuracy of $\sim$7 mas century^{-1}. While this is a precise measurement of the *relative* proper motions of Carina members, our correction to an absolute frame is limited by the small number of measured QSOs in the field.

Keywords. galaxies: dwarf, galaxies: individual (Carina dSph), astrometry

1. Introduction

The discovery by Muñoz *et al.* 2006 (M06) of a population of extratidal stars associated with the Carina dwarf spheroidal (dSph) bolsters the case for the importance of tidal disruption in dSph evolution and the formation of the Milky Way halo. Muñoz *et al.* 2007 explained many observed properties of Carina, including the large inferred M/L ratio, by modeling it as a tidally disrupting dSph. However, the absolute proper motion (PM) is necessary to derive an orbit and assess the importance of tidal forces for Carina. The only existing measurement of the space motion of Carina (Piatek et al. 2003) utilized only two HST fields centered on known QSOs. We will measure the absolute PM for Carina using photographic plates covering a much larger area than the HST observations.

2. Data and measurements

Our dataset consists of 41 CTIO Blanco 4m Prime Focus Camera photographic plates (plate scale 18.60″/mm), spanning the period 1984 - 1998. The plates were digitized in April 2007 using the USNO StarScan system, which produces typical measurement errors of $\sim$0.3 μm, with better than 0.2 μm repeatability (Zacharias *et al.* 2004). Proper motions were measured using similar techniques as in, e.g., Girard et al. 1989, using polynomial geometric terms, and linear color and magnitude terms in the plate solutions.

Carina stellar populations are easily discernible in the CMD (from the photometric catalog of M06) shown in Figure 1(a), from which we have selected a sample of candidate Carina member stars (grey diamonds). Confirmed Carina radial velocity (RV) members (M06) are shown as large filled black circles, and QSOs from Véron-Cetty & Véron (2006) as asterisks. The QSOs will be used as our tie to an absolute, "fixed" reference frame, and include the two QSOs used in the Piatek *et al.* 2003 HST study.

3. Measured proper motions

Proper motions along the two spatial dimensions (x,y $\sim \alpha,\delta$) show (Figure 1(b)) that Carina stars clearly separate from the field (mainly foreground thin/thick disk dwarfs), exhibiting extremely small motion (similar to the "fixed" QSOs) expected for a dwarf galaxy at $\sim$100 kpc (Mateo (1998)). Table 1 shows sigma-clipped mean PMs for each

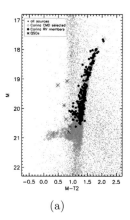

(a)

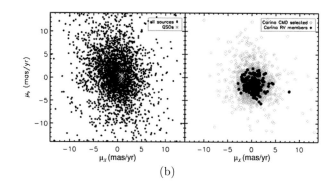

(b)

Figure 1. (a) Washington M, T2 (~V, I) CMD of the Muñoz *et al.* 2006 sample, with known QSOs as large asterisks. Large filled circles represent confirmed Carina RV members (Muñoz *et al.* 2006), and grey diamonds show our CMD selection. **(b)** Relative proper motions (in mas/yr). *Left*: all measured stars, with QSOs as grey asterisks. *Right*: CMD-selected Carina candidates (grey diamonds), with confirmed Carina RV members (black circles). Note the tight clumping of the RV sample, indicative of a well-measured mean motion for Carina.

Table 1. *Relative proper motion measurements (all units mas/yr).*

Sample	#	$\bar{\mu}_x$	$\bar{\mu}_y$	σ_{μ_x}	σ_{μ_y}	$\sigma_{\bar{\mu}_x}$	$\sigma_{\bar{\mu}_y}$
RV members	269	0.344	-1.634	1.003	1.211	0.061	0.073
CMD-selected	1416	0.379	-1.424	1.640	1.862	0.043	0.049
QSOs	7	0.351	-0.724	0.941	0.918	0.356	0.347

of the three samples. The purest Carina sample (the RV members), has per-star errors of $\sigma_{\mu_{x,y}} \sim 1.1$ mas/yr. From the RV members, we calculate a mean bulk motion of the Carina dSph accurate to ~6.5 mas/century. The CMD-selected sample shows larger standard deviations in individual measurements, as many of these stars are near the faint limit of the survey. Clearly the accuracy of our absolute PM for Carina will be dominated by the correction to an absolute frame (using QSOs and galaxies), *not* the measurement precision of the Carina PM center. Further work at identifying well-measured (point-like) galaxies in the field will improve our correction to absolute proper motions. In addition, we are working to address remaining magnitude and color trends in the preliminary PMs, which should reduce the scatter in the reference QSO measurements.

References

Girard, T. M., Grundy, W. M., Lopez, C. E., & van Altena, W. F. 1989, *AJ*, 98, 227
Mateo, M. L. 1998, *ARAA*, 36, 435
Muñoz, R. R., *et al.* 2006, *ApJ*, 649, 201 (M06)
Muñoz, R. R., Majewski, S. R., & Johnston, K. V. 2007, *ApJ*, submitted
Piatek, S., Pryor, C., Olszewski, E. W., *et al.* 2003*AJ*, 126, 2346
Véron-Cetty, M.-P., & Véron, P. 2006, *A&A*, 455, 773
Zacharias, N., Urban, S. E., Zacharias, M. I., *et al.* 2004, *AJ*, 127, 3043

A Giant Step: from Milli- to Micro-arcsecond Astrometry
Proceedings IAU Symposium No. 248, 2007
W. J. Jin, I. Platais & M. A. C. Perryman, eds.

© 2008 International Astronomical Union
doi:10.1017/S1743921308019911

Simultaneous Bayesian estimation of distances and ages from isochrones: SDSS and solar neighborhood FGK stars

M. Franchini[1], C. Morossi[1], P. Di Marcantonio[1] and M. L. Malagnini[1,2]

[1]INAF – Osservatorio Astronomico di Trieste,
Via G.B. Tiepolo, 11, I–34143 Trieste, Italy
email: `franchini@oats.inaf.it`

[2]Dipartimento di Astronomia, Università degli Studi di Trieste, Italy
email: `malagnini@oats.inaf.it`

Abstract. By using a procedure based on the Bayesian probability theory we computed reliable and self–consistent estimates of absolute magnitude and age for about 2000 FKG spectral-type stars from SDSS–DR5, ELODIE, and INDO–US surveys, with effective temperature, surface gravity, and metallicity values homogeneously derived.

Keywords. stars: distances, stars: fundamental parameters, Hertzsprung-Russell diagram

---

Age determination of individual stars from isochrones is a typical inverse problem: usually, stellar age is obtained by selecting (or interpolating) the isochrone nearest to the observed data. A different and more robust method to obtain unbiased estimates of the physical parameters is based on the Bayesian probability theory; in our case, Bayes's theorem relates the posterior probability distribution of age and absolute magnitude given the observed data, effective temperature, surface gravity and metallicity, to the prior probability distribution of age and metallicity, and to the likelihood function of the observed data, assuming for them a Gaussian distribution. By integrating over mass and metallicity, in fixed intervals of age and absolute magnitude, we can find the bi–dimensional posterior probability of age and M_v, $P(\tau, M_v)$. Note that this approach differs from those published up to now since no a priori estimate of M_v is assumed. The importance of deriving simultaneously τ and M_v is illustrated in Fig. 1 where two different cases are illustrated: the maximum of $P(\tau, M_v)$ provides, simultaneously, the best estimate of age and absolute magnitude for the given star.

Our observational data-sets comprise about 2000 FGK stars extracted from the Sloan Digital Sky Survey Data Release (SDSS–DR5; Adelman–McCarthy *et al.*, 2007), the ELODIE collection (Moultaka *et al.*, 2004) and the INDO–US catalogue (Valdes *et al.*, 2004). For all the stars, the homogeneously derived log g, $T_{\rm eff}$ and [Fe/H] values (see Morossi *et al.*, 2008) are used as the input observational parameters.

The theoretical values needed for estimating the likelihood function were taken from two different databases. The first one (Pont & Eyer, 2004) is a Monte Carlo realization, kindly provided to us by Pont, built using the IAC–star stellar population synthesis code. The second one is realized by numerical interpolation of the Padova evolutionary models using a code kindly provided to us by Jørgensen (see for details Jørgensen & Lindegren, 2005).

According to our results, most SDSS stars are old while solar neighbourhood stars are young and span a quite large age interval. A check on the reliability of our method

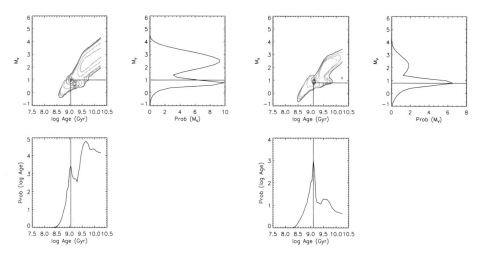

Figure 1. Examples of bidimensional $P(\tau, M_v)$: the cases of the SDSS star spSpec–52025–0529–330 (left) and of HD 112030 (right) are illustrated.

was performed by comparing the derived M_v values with those of the stars with known distance modulus, i.e. the solar neighbourhood dwarf stars with parallactic distances. It shows that the distribution of the difference in absolute magnitudes peaks at $\sim$0.3 mag, with a dispersion of 0.25 mag. As far as τ values are concerned, an indirect estimate of reliability was obtained by applying our procedure to stars with known age. We used the data from de Bruijne *et al.* (2001) for the Hyades main sequence stars. The distribution of the derived ages peaks at 0.7 Gyr, with a mean value of 0.6 ± 0.12 Gyr. Our result is in very good agreement with Hyades age as derived by several authors by comparing observed and theoretical colour–magnitude diagrams. We conclude that our method provides accurate estimates of age and absolute visual magnitude for individual stars. Our technique provides ages and distances used in the contribution to this Symposium by Morossi *et al.* to investigate the Age–Metallicity relation and the chemical vertical gradients in the Galaxy.

Acknowledgements

This work received partial financial support from PRIN-INAF 2005 (P.I. M. Bellazzini). The authors thank B. R. Jørgensen, L. Lindegren and F. Pont for providing the original Bayesian codes to compute input theoretical databases and age estimates. Funding for the SDSS and SDSS–II has been provided by the Alfred P. Sloan Foundation, the Participating Institutions, the National Science Foundation, the U.S. Department of Energy, the National Aeronautics and Space Administration, the Japanese Monbukagakusho, the Max Planck Society, and the Higher Education Funding Council of UK.

References

Adelman–McCarthy, J. K., *et al.* 2007, ApJS, 172, 634 (SDSS-DR5)
de Bruijne, J. H. J., Hoogerwerf, R., & de Zeeuw, P. T. 2001, A&A, 367, 111
Jørgensen, B. R. & Lindegren, L. 2005, A&A, 436, 127
Moultaka, J., Ilovaisky, S. A., Prugniel, P., & Soubiran, C. 2004, PASP, 116, 693 (ELODIE)
Morossi, C., Franchini, M., Di Marcantonio, P., & Malagnini, M. L. 2008, this volume p. 504
Pont, F. & Eyer, L. 2004, MNRAS, 351, 487
Valdes, F., Gupta, R., Rose, J. A., Singh, H. P., & Bell, D. J. 2004, ApJS, 152, 251 (INDO–US)

A Giant Step: from Milli- to Micro-arcsecond Astrometry
Proceedings IAU Symposium No. 248, 2007
W. J. Jin, I. Platais & M. A. C. Perryman, eds.

© 2008 International Astronomical Union
doi:10.1017/S1743921308019923

Precise measurement of the dynamical masses in AB Doradus

J. C. Guirado, I. Martí-Vidal and J. M. Marcaide

Departamento de Astronomía y Astrofísica, Universidad de Valencia
Dr. Moliner 50, 46100 Burjassot, Valencia, Spain
email: Jose.C.Guirado@uv.es

Abstract. The radio stars AB Dor A and AB Dor B (=Rst137B) are the main components of the AB Doradus system. Both stars are double (AB Dor A/AB Dor C and AB Dor Ba/AB Dor Bb) and usual targets of astrometric instruments at optical (Hipparcos), infrared (VLT), and radio (VLBI) wavelengths. From a combination of all astrometric data available, we have obtained precise limits to the dynamical mass of both binaries in AB Doradus. The determination of the mass of AB Dor C ($0.090\pm0.003\,M_\odot$) is important, since this object constitutes one of the few calibration points used to test theoretical evolutionary models of low-mass young stars. Follow-up observations both in radio (VLBI) and optical wavelengths (VLT) should determine with high precision the dynamical mass of the four components of this system.

Keywords. astrometry, binaries: close, stars: individual (AB Doradus)

1. Introduction

Precise astrometry of the orbital motion of stars in binary systems provides precise, model-independent estimates of the masses of the individual components. The case of the PMS star AB Doradus is of particular interest. Hipparcos/VLBI astrometry of the main star AB Dor A (Guirado *et al.* 1997) revealed the presence of a low-mass companion, AB Dor C, which was later imaged by VLT infrared observations (Close *et al.* 2005). AB Dor A has also a wide separation companion 9″ away, AB Dor B, which is actually a close binary (AB Dor Ba/AB Dor Bb, at 0.070″ separation; Close *et al.* 2005; Janson *et al.* 2007). The combination of all astrometric observations of AB Doradus (Hipparcos, VLBI and VLT) provides precise limits to the dynamical masses of the components of this system. In this contribution, we report the status of astrometry for different pairs in AB Doradus and the plans to further constrain the dynamical mass of its four stars.

2. Dynamical masses in AB Doradus

The pair AB Dor A/AB Dor C. The infrared image of AB Dor C (Close *et al.* 2005) provided a valuable astrometric information on the separation of this pair. The relative astrometry data were sufficient to constrain the elements of the reflex orbit of AB Dor A reported in Guirado *et al.* (1997). Recently, we have re-estimated the mass of AB Dor C using an improved method that calculates the reflex orbit of AB Dor A using the existing Hipparcos/VLBI astrometric data of AB Dor A and the near-infrared relative positions of AB Dor C (Guirado *et al.* 2006). In essence, we redefined the χ^2 of our fit to include both types of data in the orbit calculation. Despite the wide range of parameter space investigated, the only acceptable period is 11.76 ± 0.15 yr. With the reflex orbit determined, and using the mass of AB Dor A of $0.865 \pm 0.034 M_\odot$ (extrapolated from empirical PMS tracks; Close *et al.* 2005), we obtained a precise estimate of the mass of the companion AB Dor C to be 0.090 ± 0.003 M⊙.

The pair ABDor A/ABDor B. The $9''$-separation binary ABDor B is physically associated to ABDor A and, consequently, it shows a long-term orbital motion. We attempted to constrain this orbit using the relative positions of ABDor A/ABDor B available in the literature (Guirado *et al.* 2006). We sampled all periods up to 5000 years and eccentricities from 0 to 1. The poor coverage of the orbit results in correlations between the period and semimajor axis, which imposes a constraint on the total mass. According to this, the total mass of the four components in AB Doradus is in the range 0.95–1.35 $M_\odot$.

The pair ABDor Ba/ABDor Bb. Preliminary fits that combine VLBI absolute positions of ABDor Ba and VLT relative positions of ABDor Ba/ABDor Bb do not yield useful limits to the orbital parameters of this pair (probably these observations do not sample properly the expected short period of this binary). However, from the combination of 1) the constraint to the mass of the $9''$ pair ABDor A/ABDor B (0.95–1.35 $M_\odot$), and 2) the dynamical mass of the pair ABDor A/ABDor C (0.955 ± 0.035 $M_\odot$), we can derive an upper limit to the mass of the pair Ba/Bb of 0.4 $M_\odot$. This is in good agreement with the 0.38 $M_\odot$ reported in Janson *et al.* (2007). The period of this close binary is yet to be determined; however, the combination of the 0.4 $M_\odot$ upper limit of the Ba/Bb mass, with the lower limit to the relative semimajor axis (0.9 AU; Janson *et al.* 2007) suggests that the period of this binary should be longer than 1.5 yr. On the other hand, a series of VLT infrared images (Close *et al.* 2007) favor a period of 0.9 yr. Clearly, a better sampling of the relatively fast orbital motion in this pair is needed.

3. Future work and conclusions

Among the four components of AB Doradus, the precise determination of the dynamical mass of ABDor C is perhaps the most relevant. With independent measurements of both the mass (0.090 ± 0.003 $M_\odot$) and the infrared photometry, ABDor C constitutes one of the few calibration points that can be used to test theoretical evolutionary models of low-mass young objects (i.e. those used to calibrate planets and brown dwarf candidates; Close *et al.* 2005, 2007; Luhman *et al.* 2005). Ongoing VLBI and VLT observations of AB Doradus should refine the estimates of the dynamical masses of this system. In particular, further monitoring of the reflex motion of ABDor A, via VLBI observations with the antennas of the Australian Long Baseline Array, will provide a model-independent estimate of the mass function of this pair ($f_{ac} = m_c^3/(m_a + m_c)^2$, with m_a and m_c the mass of ABDor A and ABDor C, respectively). Once combined with the infrared relative astrometry (which in turn provides $m_a + m_c$) it will result in independent estimates of the mass of ABDor A and ABDor C. Likewise, with adequate monitoring, other components of AB Doradus (ABDor Ba/ABDor Bb) may provide new calibration points for models of slightly larger masses.

References

Close, L. M., Lenzen, R., Guirado, J. M. *et al.* 2005, *Nature*, 433, 286
Close, L. M., Thatte, N., Nielsen, E. L. *et al.* 2007, *ApJ*, 665, 736
Guirado, J. C., Reynolds, J. E., Lestrade, J.-F. *et al.* 1997, *ApJ*, 490, 835
Guirado, J. C., Martí-Vidal, I., Marcaide, J. M. *et al.* 2006, *A&A*, 446, 733
Janson, M., Brandner, W., Lenzen, R. *et al.* 2007, *A&A*, 462, 615
Luhman, K. L., Stauffer, J. R., & Mamajek, E. E. 2005, *ApJ*, 628, 69

A Giant Step: from Milli- to Micro-arcsecond Astrometry
Proceedings IAU Symposium No. 248, 2007
W. J. Jin, I. Platais & M. A. C. Perryman, eds.

© 2008 International Astronomical Union
doi:10.1017/S1743921308019935

Dynamics of particles in slowly rotating black holes with dipolar halos

W. B. Han[1,2]

[1] Shanghai Astronomical Observatory, Shanghai
email: wbhan@shao.ac.cn

[2] Graduate School of the Chinese Academy of Sciences, Beijing

Abstract. In general, the model of galaxy assumes a central huge black hole surrounded by a massive halo, disk or ring. In this paper, we investigate the gravitational field structure of a slowly rotating black hole with a dipolar halo, and the dynamics and chaos of test particles moving in it. Using Poincaré sections and fast Lyapunov indicator (FLI) in general relativity, we investigate chaos under different dynamical parameters, and find that the FLI is suitable for detecting chaos and even resonant orbits.

Keywords. Galaxy: kinematics and dynamics, black hole physics, galaxies: halos

1. Introduction

The main motive of our research on the dynamics characteristics of test particles in the relativistic core-shell models is their realistic significance in astrophysics, such as the study of motion of stars in a galaxy with a model of a central bulge surrounded by a halo (Binney and Tremain, 1987), and other astrophysics objects involving active galactic nuclei, black holes, or neutron stars with axisymmetric surrounding sources (for examples accretion disks, massive halos, shells and rings).

In this paper, following Letelier and Vieira (1997), we briefly discuss the spacetime structure of superposition of a slowly rotating Kerr black hole with a dipole along its rotating axis. Furthermore, the dynamics and chaos of test particles in the Weyl field is also studied.

2. Gravity field structure and the stability of test particles

The metric representing the superposition of a Kerr black hole and a dipole along the rotation axis is a stationary axially symmetric spacetime. We only consider slowly rotating black holes. The metric can be simplified by keeping the first order terms in the rotation parameter a (Letelier and Vieira, 1997). The particles moving in it can be written as a four dimensional dynamical system. There are three integration constants: the energy E, the z angular momentum L and the 4-velocity conservation. The last constant is used to check numerical errors, with a precision reaching 10^{-13}.

First we emit five particles along the radial direction with different polar angles. We study the motion of these particles at three different dipole strengths D. Fig. 1 shows the case with $D=0.005$. The gravity field does not possess reflection symmetry with respect to the black hole's equatorial plane. It can be viewed as if another gravity source is possibly located upon the black hole.

We find that with a larger D, the orbits of particles are prone to instabilities (Fig. 2). For every energy and angular momentum, there exists a critical value of D, beyond which we cannot find a stable area.

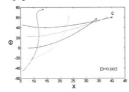

Figure 1. $E = 0.965$, $L = 3.75$ and five particles have different initial inclination: $45°$, $22.5°$, $0°$, $-22.5°$, $-45°$, respectively. The superposition field is not symmetrical about the equator.

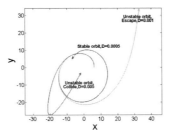

Figure 2. There different orbits of particle with same initial values at varying halo parameter.

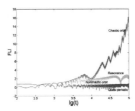

Figure 3. FLIs of three orbits. FLI not only can distinguish chaos, but also resonant orbit.

3. The chaos in varying parameters

Obviously, Poincaré sections are suitable for qualitatively characterizing dynamical systems. With a fixed angular momentum larger energy means emergence of chaos. But if the energy is constant, a bigger angular momentum can enlarge the regular degree of the dynamical system (corresponding Poincaré sections being omitted here).

In Fig. 3, we use FLI (Wu *et al.*, 2006) to study three different aspects: hyperbolic point, resonant trajectory, and quasi-periodic orbits. We find that the first one is chaotic as compared to the last two that are non-chaotic. The FLI also can distinguish between resonant and quasi-periodic orbits.

References

J., Binney & S., Tremaine 1987, *Galactic dynamics*, Princeton U. Press.
P. S., Letelier & W. M., Vieira 1997, *Phys. Rev. D*, 56, 12
X., Wu, T-Y., Huang & H., Zhang 2006, *Phys. Rev. D*, 74, 083001.

A Giant Step: from Milli- to Micro-arcsecond Astrometry
Proceedings IAU Symposium No. 248, 2007
W. J. Jin, I. Platais & M. A. C. Perryman, eds.

Stellar parameters through high precision parallaxes

C. Jordi, C. Fabricius, J. M. Carrasco, F. Figueras, E. Masana, H. Voss and X. Luri

Departament d'Astronomia i Meteorologia. Universitat de Barcelona-IEEC
Martí Franqués 1, E-08028 Barcelona, Spain, email: `carme@am.ub.es`

Abstract. Before Hipparcos, the determination of absolute luminosity was usually done through calibrations based on a few stellar parallaxes measured at the highest precision. The Hipparcos mission meant a giant step on the knowledge of luminosities and fine structure of the HR diagram. The Gaia mission will go an enormous step further. Besides luminosities, Gaia will allow us to derive other stellar parameters like temperature and extinction, gravity, chemical composition, age and mass by the combination of astrometric and spectrometric data. Through simulations during the mission preparation, it has been shown that the astrometric parallax information is essential to deal with the degeneracy between gravity and chemical composition ([Fe/H] and [α/Fe]), that cannot be treated using only spectrophotometry. We show the expected HR diagram for the Gaia domain and the accuracies of stellar parameters.

Keywords. stars: fundamental parameters, stars: distances, stars: variables, Hertzsprung-Russell diagram, techniques: photometric, space vehicles

1. Introduction

Gaia is an all-sky survey satellite (Perryman *et al.* (2001), Lindegren *et al.* (this volume)), to be launched by ESA around 2012, to obtain parallaxes and proper motions to microarcsecond precision, radial velocities and astrophysical parameters for about 10^9 objects down to a limiting magnitude of $\sim$20 mag. A 3D reconstruction of the Galactic velocity field, coupled with the physical properties of the stars, will provide answers to long-standing questions about the Galaxy such as its structure, formation, dynamical and chemical evolution, the halo stream properties, and others.

The derivation of the stellar parameters (luminosity, $T_{\rm eff}$, $\log g$, [Fe/H], etc.) will mainly be done by combining the low resolution spectra (LR), the white light photometry (G band) and the trigonometric parallax. Figure 1 shows two examples of the variation of the 1D low resolution red spectra with stellar parameters. For bright stars, the high resolution spectra will allow chemical composition determination as well.

2. Stellar parameters

From the measurements of unfiltered (white) light, *Gaia* will provide magnitudes in a very broad band (G, 330–1000 nm), with estimated uncertainties are shown in Fig. 2 (left). Figure 2 (right) shows the stars with parallaxes better than 10% in a simulated HR diagram given the instrument model and the Besançon Galaxy model (Robin *et al.* 2004). The 1 mmag uncertainty at $V \sim 18.5 - 19.0$ coupled with high precision parallax will allow to calibrate the absolute luminosities of all types of stars across the HR diagram.

The stellar parameters yield the spectral shape and features of the blue and red LR spectra. However, the variation of the shape and features may be caused by variations in

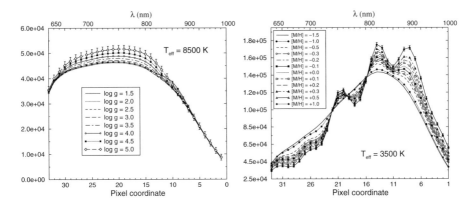

Figure 1. 1D-Red low resolution spectra for several $\log g$ (left) and for several [Fe/H] (right) for $G = 15$ stars. The bars show the end-of-mission errors for $G = 18$.

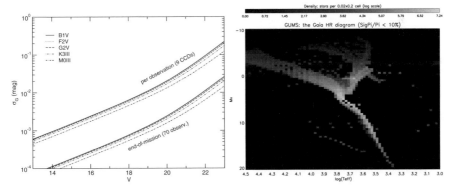

Figure 2. Left: Per transit and the end-of-mission estimated uncertainties of the white light G measurements (calibration errors of about 1 mmag not included). Right: Simulated HR diagram showing the stars with trigonometric parallaxes better than 10% (about 10^8 stars). The Besançon Galaxy model has been used to represent the stellar content of the Galaxy.

more than one stellar parameter. In the case of *Gaia*, the extremely precise parallaxes will allow to handle such degeneracies in many cases (for instance, [Fe/H], [α/Fe] and $\log g$ degeneracy). The uncertainties of stellar parameters taking into account the parallax information have been estimated. The results show that individual ages for halo stars can be derived with a precision of 1.5 and 2.2 Gyr at 2.5 and 5 kpc, respectively. For halo G and K giants, the [Fe/H] values will be precise to 0.07, 0.15 and 0.20 dex at 5, 12 and 17 kpc, respectively. For disk stars, the ages will have precision of 1.2 and 4.7 Gyr at 2 and 3 kpc, respectively towards the galactic center, and 1.1 and 2.6 Gyr at the same distances in the anticenter direction. The [Fe/H] precisions will be of 0.05 and 0.20 dex at 2 and 5 kpc, respectively towards the center and 0.05, 0.10 and 0.15 at 2, 5, and 8 kpc, towards the anticenter.

Acknowledgements: This project is supported by the Spanish MEC under contract PNE2006-13855-C02-01.

References

Perryman, M. A. C., de Boer, K. S., Gilmore, G., *et al.*, 2001, A&A, 369, 339
Robin, A. C., Reylé, C., Derriére, S., & Picaud, S. 2004, A&A, 409, 523

A Giant Step: from Milli- to Micro-arcsecond Astrometry
Proceedings IAU Symposium No. 248, 2007
W. J. Jin, I. Platais & M. A. C. Perryman, eds.

Galactic kinematics near the sun from open clusters†

M. Shen

Department of Astronomy, Nanjing University, China
email: shenming@nju.org.cn

Abstract. A catalogue of open clusters is used to analyze the Galactic kinematics near the Sun. The Galactic open clusters, which are components of Galactic thin disk, were selected for our analysis. Based on kinematical data for around 270 Galactic open clusters, we found the Galactic rotation curve remains flat near the Sun. We also found $V_e = 6.40 + 3.39$ km s^{-1} inside the solar circle, which shows a weak trend of stars moving toward the Galactic anti-center.

Keywords. Galaxy: kinematics and dynamics, open clusters and associations: general

1. Observation data

Kharchenko *et al.* (2005) identified 520 Galactic open clusters and compact associations on the basis of the All-Sky Compiled Catalogue which contains 2.5 million stars (ASCC-2.5) (Kharchenko 2001). Dias *et al.* (2002, 2004) collected all published open clusters from the available literature, resulting in 1637 clusters in their DAML02 catalogue.

In this work, we used the data of 520 open clusters from Kharchenko *et al.* (2005), in which 253 open clusters have heliocentric distances, mean proper motions, mean radial velocities, and ages. For the remaining clusters, 29 mean radial velocity measurements are taken from the DAML02. In addition, another 19 clusters with complete kinematical data in DAML02 were also used. We found spatial velocities for 301 clusters from our sample, in which 10 clusters were excluded because of their suspect peculiar velocity.

Figure 1 displays 137 clusters younger than 50 Myr, projected on the Galactic plane and clearly showing 3 arms. The clusters older that than 120 Myr do not match well the location of spiral arms, and thus were not used in this study.

2. Models & results

Consider a rotating expanding ring around the Galactic center, in which the velocity of an arbitrary point can be written as $\Omega \times \mathbf{R} + (v_e/R)\mathbf{R}$, where Ω is the angular velocity, $\mathbf{R}$ is the Galactocentric position vector of the point, and v_e stands for the velocity to the Galactic center. When observed from the Sun, the line-of-sight velocity can be written as follows,

$$v_{los}(l) = ([\Omega(R) - \Omega(R_0)]R_0 \sin\ell + \frac{d^2 + R^2 - R_0^2}{2Rd}v_e - u_0\cos\ell - v_0\sin\ell)\cos b - w_0\sin b, \quad (2.1)$$

In this equation, $\Omega(R_0)$ is angular velocity at the Sun, and d is the heliocentric distance to the point in the ring.

When analyzing Hipparcos main sequence stars, Dehnen & Binney (1998) found

† Supported by the National Natural Science Foundation of China(Grant Nos. 10333050 and 10673005)

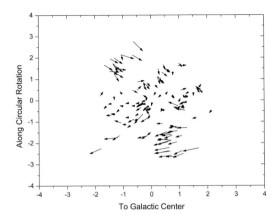

Figure 1. Distribution of 137 young open clusters in the Galactic plane. The vectors show the peculiar motion of clusters.

$u_0 = 10.0 \pm 0.4$ km s^{-1}, $v_0 = 5.2 \pm 0.6$ km s^{-1}, and $w_0 = 7.2 \pm 0.4$ km s^{-1} for the solar motion. In the following analysis, u_0, v_0, w_0 are fixed.

The Galactocentric radius R of a star is given by

$$R = (R_0^2 + d^2\cos^2 b - 2R_0 d\cos\ell\cos b)^{\frac{1}{2}}. \tag{2.2}$$

As mentioned, 137 young open clusters are distribute over the three arms. A few clusters belong to the outer arm, thus we mainly studied the inner and the local arm. We adopted a simple model, in which by changing the arm's radius we tried to minimize the mean distance of our clusters to that arm. The clusters farther than 0.5 kpc from the arm were not used in our analysis.

Equations (2.1) for radial velocities are to be solved via least squares. Then we have,

$$\Omega(R) - \Omega(R_0) = 0.67 \pm 0.35 \ \text{km s}^{-1} \ \text{kpc}^{-1}, \tag{2.3}$$
$$v_e = 6.40 \pm 3.39 \ \text{km s}^{-1} \ \text{for inner arm and,}$$
$$\Omega(R) - \Omega(R_0) = -2.16 \pm 0.59 \ \text{km s}^{-1} \ \text{kpc}^{-1}, \tag{2.4}$$
$$v_e = 3.68 \pm 5.78 \ \text{km s}^{-1} \ \text{for local arm.}$$

This result shows a flat rotation curve near the Sun, and a weak trend of stars moving toward the Galactic anti-center. Due to the large error in v_e, this trend is not definite.

References

Dehnen, W. & Binney, J. 1998, *MNRAS*, 298, 387

Dias, W. S., Alessi, B. S., Motinho, A., & Lépine J. R. D 2002, *A&A*, 389, 871

Dias, W. S, Lépine, J. R. D, Alessi, B. S., & Moitinho, A. 2004, Catalog of Optically Visible Open Clusters and Candidates (DAML02), http://www.astro.iag.usp.br/~wilton/

Kharchenko, N. V. 2001, Kinematics and Physics of Celestial Bodies, 17, 409

Kharchenko, N. V., Piskunov, A. E., Röser S., *et al.*, 2005, *A&A*, 438, 1163

A Giant Step: from Milli- to Micro-arcsecond Astrometry
Proceedings IAU Symposium No. 248, 2007
W. J. Jin, I. Platais & M. A. C. Perryman, eds.

© 2008 International Astronomical Union
doi:10.1017/S1743921308019960

Age-Metallicity Relation, [Fe/H] and [α/Fe] vertical gradients in the Milky Way from the SDSS–DR5 spectroscopic database

C. Morossi[1], M. Franchini[1], P. Di Marcantonio[1] and M. L. Malagnini[1,2]

[1]INAF – Osservatorio Astronomico di Trieste,
Via G.B. Tiepolo, 11, I–34143 Trieste, Italy
email: morossi@oats.inaf.it

[2]Dipartimento di Astronomia, Università degli Studi di Trieste, Italy
email: malagnini@oats.inaf.it

Abstract. Spectra of FGK stars were selected from the SDSS–DR5 spectroscopic database to investigate the Age-Metallicity relation and the [Fe/H] and [α/Fe] vertical gradients in the Milky Way. Atmospheric parameters and [α/Fe] were derived by comparing synthetic and measured Lick/SDSS spectral indices. Results were checked and complemented by analyzing solar neighbourhood stars. Spectroscopic distances and ages were obtained for a subsample of ∼2000 stars using theoretical isochrones via a Bayesian approach. The resulting Age-Metallicity diagram and the [Fe/H] and [α/Fe] vertical gradients are presented.

Keywords. stars: abundances, stars: fundamental parameters, Galaxy: stellar content

---

In recent years, our understanding of the formation and structure of the Milky Way has been gaining importance due to the increasing availability of observational data especially for FGK stars. The Age–Metallicity Relation (AMR) is one of the most popular diagnostic tools for comparing the actual Milky Way with model predictions, since the knowledge about the epochs when stars became population members is essential, but, up to now, there is no general consensus on its shape and interpretation. Recently, Holmberg *et al.* (2007) re-analyzed the Geneva–Copenhagen photometric survey of the Solar neighbourhood (Nordström *et al.* 2004) presenting improved calibrations for temperature and metallicity and an updated AMR. The analysis of FGK stars beyond the solar neighbourhood can now be accomplished using the Sloan Digital Sky Survey (SDSS–DR5, Adelman-McCarthy *et al.* 2007). Its spectroscopic stellar database represents a more extended, both in space coverage and in numbers, collection of mid–resolution spectra. We use this collection to derive the AMR and the chemical gradients at different heights above the Galactic plane; atmospheric and kinematical parameters, distances and ages are homogeneously derived. We adopt a methodology based on Lick/SDSS indices (Franchini *et al.* 2007a,b), calibrated via solar neighbourhood (SN) stars taken from ELODIE (Moultaka *et al.* 2004) and INDO–U.S. (Valdes *et al.* 2004) catalogues, for deriving T_{eff}, log g, [Fe/H] and [α/Fe] estimates. Here we present a preliminary Age-Metallicity Diagram (AMD) and abundance vertical gradients resulting from the analysis of a sub-sample of stars whose atmospheric parameters fall in the range of calibrations.

The program data-sets comprise ∼4500 FGK stars extracted from SDSS–DR5, ELODIE and INDO–US catalogues. For about 2000 stars with parameters in the calibration range, we estimate absolute visual magnitude and age via a Bayesian approach (Franchini *et al.* 2008c). The distributions of atmospheric parameters peak at $T_{eff} = 5800$ K, log $g = 4.2$ dex and [Fe/H]$= -0.7$ dex. The running average trend of [α/Fe] versus [Fe/H]

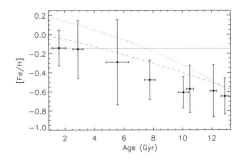

Figure 1. Age–Metallicity Diagram. Three parametric AMR's from Rocha–Pinto *et al.* (2006) are represented by solid (no AMR), dashed ("loose") and dotted ("tight") lines.

shows that our sample is a mixture of stars of different ages and birthplaces. The age distribution for the SDSS and SN samples is illustrated in Franchini *et al.* (2008c): the SDSS sample, with stellar distances in the range $370 \div 15{,}000$ pc and $|Z_{\rm Gal}| > 250$ pc, is older than the SN sample with stars concentrated around the Sun and at $|Z_{\rm Gal}| < 200$ pc.

Figure 1 presents our Age–Metallicity diagram and 3 parametric AMR's from Rocha–Pinto *et al.* (2006). Our data, computed via a running average, are distributed with a slope similar to the "loose" AMR case (dashed line). This result should be taken with caution since AMD's are, in general, affected by sample selection effects (Holmberg *et al.* 2007). As far as the vertical gradients are concerned, the trends of [Fe/H] and [α/H] versus $|Z_{\rm Gal}|$ show a clear decrease of [Fe/H] for $|Z_{\rm Gal}| > 100$ pc while the lowest [α/H] values seem to be confined to $|Z_{\rm Gal}| > 1.0$ kpc.

A work is in progress to disentangle stars of different galactic components by computing stellar orbits and kinematical properties.

Acknowledgements

This work received partial financial support from PRIN-INAF 2005 (P.I. M. Bellazzini). Funding for the SDSS and SDSS–II has been provided by the Alfred P. Sloan Foundation, the Participating Institutions, the National Science Foundation, the U.S. Department of Energy, the National Aeronautics and Space Administration, the Japanese Monbukagakusho, the Max Planck Society, and the Higher Education Funding Council of UK.

References

Adelman–McCarthy, J. K., *et al.* 2007, ApJS, 172, 634 (SDSS–DR5)

Moultaka, J., Ilovaisky, S. A., Prugniel, P., & Soubiran, C. 2004, PASP, 116, 693 (ELODIE)

Franchini, M., Morossi, C., Di Marcantonio, P., Malagnini, M. L., & Chavez, M. 2007a, IAU Symp., 241, p. 245

Franchini, M., Morossi, C., Di Marcantonio, P., Malagnini, M. L., *et al.* , 2007b (in preparation)

Franchini, M., Morossi, C., Di Marcantonio, P., & Malagnini, M. L., 2008c, in this volume p. 494

Holmberg, J., Nordström, B., & Andersen, J. 2007, A&A, in press

Nordström, B., Mayor, M., Andersen, J., Holmberg, J., *et al.* 2004 A&A, 418, 989

Rocha–Pinto, H. J., Rangel, R. H. O., Porto de Mello, G. F., Bragança, G. A., & Maciel, W. J. 2006, A&A 453, L9

Valdes, F., Gupta, R., Rose, J. A., Singh, H. P., & Bell, D. J. 2004, ApJS, 152, 251 (INDO–US)

A Giant Step: from Milli- to Micro-arcsecond Astrometry
Proceedings IAU Symposium No. 248, 2007
W. J. Jin, I. Platais & M. A. C. Perryman, eds.
© 2008 International Astronomical Union
doi:10.1017/S1743921308019972

A spectroscopic survey of late-type giants in the Milky Way disk and local halo substructure

A. A. Sheffield[1,2], S. R. Majewski[2], A. M. Cheung[2], C. M. Hampton[2], J. D. Crane[3], and R. J. Patterson[2]

[1]Department of Physics and Astronomy, Vassar College,
124 Raymond Ave., Poughkeepsie, NY 12604-0075
email: alsheffield@vassar.edu

[2]Department of Astronomy, University of Virginia,
Box 400325, Charlottesville, VA 22904-4325
email: aap5u, srm4n, amc4e, cmh4z, rjp0i@virginia.edu

[3]Carnegie Observatories, 813 Santa Barbara St., Pasadena, CA 91101, USA
email: crane@ociw.edu

Abstract. We report the results of a survey of late-type giants aimed at understanding the nature of the disk and nearby halo Galactic stellar populations. We have obtained medium resolution (2–4 Å) spectra for 749 late K and early M giants at mid-latitudes selected from the 2MASS catalog with the FOBOS system at Fan Mountain Observatory. These spectra provide radial velocities (RVs) at the 5 km s^{-1} level, spectroscopic [Fe/H] good to $\sigma_{[Fe/H]} = 0.25$ dex, and information on the relative abundances of Mg/Fe and Na/Fe in these stars. Proper motions from UCAC2 are used to search for local substructures, in particular the leading arm of the Sagittarius tidal streamer passing through the solar neighborhood. The combined proper motions and RVs yield full 6D stellar space motions. We have, by way of kinematics, relatively cleanly isolated the thick disk from the typically high velocity substructures that compose the nearby halo.

We find evidence for substructure in the kinematics and metallicities of local halo stars.

Keywords. Galaxy: structure, Galaxy: kinematics and dynamics, Galaxy: formation, astrometry

1. Motivation

The formation and evolution of the Galaxy can be explored through the kinematics and abundance patterns of the present Galactic stellar populations. Key to discriminating formation models are an understanding of the existence of smooth or disjointed transitions in properties and the relation of the thick disk to the halo and thin disk. A further constraint is provided by studies of the kinematics of stars within a stellar stream. For example, tidal debris from the accreted Sagittarius (Sgr) dwarf spheroidal (dSph) galaxy provides an excellent means for probing Galactic structure.

We have collected radial velocities (RVs) that, when combined with proper motions and distances, help to unravel the events that led to the current state of the Galaxy.

2. Data and analysis

The 2MASS catalog (Skrutskie *et al.* 2006) was used to select late-type giants for our study of the thick disk and local halo. It has been established (Bessell & Brett 1988) that M giants can be discriminated from M dwarfs based on their $J - H$ and $H - K$

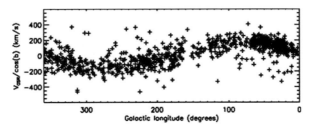

Figure 1. Radial velocity with respect to the Galactic Standard of Rest (GSR).

colors. This technique was used by Majewski *et al.* (2003) to identify streams of M giants from the accreted Sgr dSph galaxy. Photometric parallaxes are estimated using the color-magnitude relations of Ivanov & Borissova (2002). The proper motions used in this survey come from the UCAC2 (Zacharias *et al.* 2004).

Stars in the Galactic disk generally move on nearly circular orbits around the Galactic center and show RVs that fall along a sinusoidal distribution as a function of Galactic longitude. Figure 1 shows the distribution of RVs, with respect to the Galactic Standard of Rest, for the survey stars, where the RVs have been divided by $\cos b$ (this tends to accentuate differences between stars having circular motion from stars on more eccentric orbits). While the majority of the stars appear to be members of the disk based on the RV pattern, there are several notable groupings of outliers. These RV outliers may be associated with tidal debris from an accreted satellite.

Although the resolution of our spectra is not high enough to measure equivalent widths, relative chemical abundances can still be extracted from our spectra by using the Lick spectral indices (Worthey *et al.* 1994). We have combined three Lick indices sensitive to Fe-peak elements for use in calibrating [Fe/H] for our program stars. We use the Lick Mg b and Na D indices to look for relative trends in α/Fe.

3. Results

The distribution of RVs reveals the presence of several groups of stars that deviate from disk-like orbits. These stars also show distinct chemical signatures and may have formed from a gas that underwent a different star formation history than the Galactic disk gas. We find that stars on high-energy orbits are generally more metal poor. This is consistent with the notion of a clumpy halo made up mainly of accreted debris.

Our sample of thick disk stars reveals an asymmetric drift velocity of the thick disk with a vertical gradient of -26 ± 4 km s^{-1} kpc^{-1}. Based on this modest vertical gradient in rotational velocity and the lack of a vertical gradient in the α/Fe-[Fe/H] patterns for our survey, we conclude that our findings support a most likely formation scenario in which the thick disk was formed by the heating of the early disk due to a merger.

References

Bessell, M. S. & Brett, J. M. 1988, *PASP*, 100, 1134
Ivanov, V. D. & Borissova, J. 2002, *A&A*, 390, 937
Majewski, S. R., Skrutskie, M. F., Weinberg, M. D., & Ostheimer, J. C. 2003, *ApJ*, 599, 1082
Skrutskie, M. F., *et al.* 2006, *AJ*, 131, 1163
Worthey, G., Faber, S. M., González, J., & Burstein, D. 1994, *ApJS*, 94, 687
Zacharias, N., Urban, S. E., Zacharias, M. I., Wycoff, G. L., Hall, D. M., Monet, D. G., & Rafferty, T. J. 2004, *AJ*, 127, 3043

A Giant Step: from Milli- to Micro-arcsecond Astrometry
Proceedings IAU Symposium No. 248, 2007
W. J. Jin, I. Platais & M. A. C. Perryman, eds.

TW Hydrae astrometric parameters measurement

R. Teixeira[1], C. Ducourant[2], G. Chauvin[3], A. G. O. Krone-Martins[2,1], J.-F. le Campion[2], I. Song[4] and B. Zuckerman[5]

[1]IAG/São Paulo University, Brazil
email: teixeira@astro.iag.usp.br

[2]Bordeaux Observatory LAB, France
[3]Grenoble Laboratory of Astrophysics, France
[4]Spitzer Science Center, Caltech, USA
[5]UCLA, USA

Abstract. The primary goal of this study was measurement of trigonometric parallaxes and proper motions of a dozen members of the TW Hydrae Association (TWA) that are not present in the Hipparcos catalogue.

Keywords. astrometry, stars: distances, planetary systems, stars: kinematics, solar neighborhood, open clusters and associations: individual (TWA)

1. Introduction

The TW Hydrae Association (TWA) is a young and nearby association composed of objects of particular scientific interest. Belonging to this association is the young brown dwarf 2MASSW J1207334-393254 (2M1207A) hosting the first ever imaged planet (2M1207b) with the estimated mass of about $5 M_{jup}$ and separation from 2M1207A of about 55 AU (Chauvin *et al.* 2004). Estimated distances from photometric and kinematic data for this planetary system are discrepant: 70 pc (Chauvin *et al.* 2004), 53 ± 6pc (Mamajek 2005), 59 ± 7pc (Song *et al.* 2006) and 66 ± 5pc (Mamajek and Meyer 2007). Another interesting object of the TWA Association is SSSPM J1017-5354 (TWA22). This object, although being an accepted TWA member, has a proper motion that is double of that of the other association members and its estimated photometric distance is 22 pc (Song *et al.* 2003). If confirmed as a member of TWA, it will be the nearest known low mass object of TWA (Scholz *et al.* 2005).

Two main reasons motivated this trigonometric parallax program of TW Hydrae members. First of all, the physical characterization of the association and its individual members is strongly dependent of distances. Therefore, precise measurements of trigonometric parallaxes as done in this work are essential. Another point concerns the fact that due to its youth and proximity, the TWA is one of the best targets to test the pre-main sequence models. It is possible to derive dynamical age for this association using astrometric and spectroscopic measurements in a trace-back strategy (Ortega *et al.* 2002, Song *et al.* 2003, de La Reza *et al.* 2006) that is independent of stellar evolutionary models.

At present, about 25–30 TWA members are known and, except for 5 objects, their trigonometric parallaxes have not yet been measured. In this context we started an observational program with the ESO-La Silla NTT/SUSI2 telescope aiming to measure trigonometric parallaxes and proper motions for a dozen of TWA members with no Hipparcos parallax, as described in Ducourant *et al.* 2007.

2. Targets and first results

Table 1. TW Hydrae targets (Mamajek 2005) for our observational program.

target	RA(hms)	DEC(dms)	target	RA(hms)	DEC(dms)	target	RA(hms)	DEC(dms)
TWA22	10 17 27	-53 54 27	TWA02	11 09 14	-30 01 39	TWA23	12 07 27	-32 47 00
TWA06	10 18 28	-31 50 02	TWA03	11 10 28	-37 31 53	TWA27	12 07 33	-39 32 54
TWA07	10 42 30	-33 40 17	TWA12	11 21 06	-38 45 16	TWA25	12 15 31	-39 48 42
TWA28	11 02 10	-34 30 36	TWA05	11 31 55	-34 36 27	TWA10	12 35 04	-41 36 39

Table 2. Proper motions and trigonometric parallaxes for three of our targets.

	$TWA22$	$TWA27$	$TWA28$
$\mu_\alpha \cos\delta (mas/yr)$	-154.9 ± 1.8	-57.5 ± 0.4	-62.1 ± 0.5
$\mu_\delta (mas/yr)$	-27.4 ± 1.8	-22.6 ± 0.4	-12.8 ± 0.5
$\pi (mas)$	53.1 ± 0.8	19.1 ± 0.4	17.7 ± 0.5
$d(pc)$	18.8 ± 0.3	52.4 ± 1.1	56.4 ± 1.6

Until now observations are completed for TWA22, TWA27(2M1207A) and TWA28 (SSSPM1102) and the preliminary results are presented in Table 2.

3. Conclusions

Our results confirm the planetary nature of the sub-stellar companion 2M1207b (Ducourant *et al.* 2007). In this case the proper motion and trigonometric paralax are consistent with those from the two other recently published studies (Gizis *et al.* 2007, Biller and Close 2007). In the case of TWA28 (SSSPM1102), our results reinforce the Scholz hypothesis (Scholz *et al.* 2005) that this object is a sub-stellar companion of the eponymous T Tauri TW Hya. For TWA22, the late spectral-type ($\approx M7$) from the grating spectrometer at Siding Spring Observatory (Song *et al.* 2007), a close binary companion from VLT AO ($\rho \approx 0.1$arcsec, $\Delta(k) \approx 0.4$mag, Chauvin *et al.* 2007), and the small trigonometric distance (18.8pc, this work) will make TWA22 the closest, young, low-mass, brow dwarf binary system known to date.

References

Biller, B. A. & Close, M. L. 2007, *ApJ*, 669, L41

Chauvin, G., Lagrange, A. M., Dumas, C., *et al.*, 2004,

Chauvin, G., *et al.*, 2007, *in preparation*

de la Reza, R., Jilinski, E., & Ortega, V. G. 2006, *AJ*, 131, 2609

Ducourant, C., Teixeira, R., Chauvin, G., *et al.*, 2007, *in press in A&A*

Gizis, J., Jao, W. C., Subsavage, J. P., & Henry, T. J. 2007, *ApJ*, 669, L45

Mamajeck, E., E. 2005, *ApJ*, 634, 1385

Mamajeck, E. E. & Meyer, M. 2007, *ApJ*, 668, L175

Ortega, V., G., de la Reza, R., Jilinski, E., & Bazzanella, B., 2002, *AJ*, 575, L75

Scholz, R. D., McCaughrean, M. J., Zinneker, H., & Lodieu N. 2005, *A&A*,430, L49

Song, I., Zuckerman, B., & Bessel, M. S. 2003, *ApJ*, 599, 342

Song, I., Schneider, G., Zuckerman, B., *et al.* 2006, *ApJ*, 652, 724

Song, I., Zuckerman, B., & Bessel, M. S. 2007, *in preparation*

A Giant Step: from Milli- to Micro-arcsecond Astrometry
Proceedings IAU Symposium No. 248, 2007
W. J. Jin, I. Platais & M. A. C. Perryman, eds.

© 2008 International Astronomical Union
doi:10.1017/S1743921308019996

Development of a near-infrared high-resolution spectrograph (WINERED) for a survey of bulge stars

T. Tsujimoto[1], N. Kobayashi[2], C. Yasui[2], S. Kondo[2], A. Minami[2], K. Motohara[2], Y. Ikeda[3] and N. Gouda[1]

[1]National Astronomical Observatory, Mitaka, Tokyo 181-8588, Japan
email: taku.tsujimoto@nao.ac.jp
[2]Institute of Astronomy, University of Tokyo, Mitaka, Tokyo 181-0015, Japan
[3]Photocoding, Sagamihara, Kanagawa 229-1104, Japan

Abstract. We are developing a new near-infrared high-resolution (R[max] = 100,000) and high-sensitive spectrograph WINERED, which is specifically customized for short NIR bands at 0.9–$1.35\,\mu m$. WINERED employs an innovative optical system; a portable design and a warm optics without any cold stops. The planned astrometric space mission JASMINE will provide precise positions, distances, and proper motions of the bulge stars. The missing components, the radial velocity and chemical composition will be measured by WINERED. These combined data brought by JASMINE and WINERED will certainly reveal the nature of the Galactic bulge. We plan to complete this instrument for observations of single objects by the end of 2008 and to attach it to various 4–10m telescopes as a PI-type instrument. We hope to upgrade WINERED with a multi-object feed in the future for efficient survey of the JASMINE bulge stars.

Keywords. Galaxy: bulge, infrared: stars, stars: abundances, stars: atmospheres, instrumentation: spectrographs, techniques: spectroscopic, techniques: radial velocities

We are developing a new NIR high-resolution spectrograph WINERED (=Warm INfrared Echelle spectrograph to Realize Extreme Dispersion) by employing two novel approaches (Ikeda *et al.* 2006). First, WINERED employs warm optics with no cold stop, which can be realized by limiting the wavelength range to short NIR bands of 0.9–$1.35\,\mu m$. Second, WINERED employs an immersion grating of ZnSe or ZnS, resulting in a high resolving power of $R_{\max} = 100,000$, despite its compactness ($1,500 \times 500 \times 500$ mm). These two approaches make WINERED portable and easy to build, align, and maintain. Therefore, the total cost and time can be significantly reduced as compared with an entirely cooled classical echelle spectrograph.

Warm optics

In the short NIR region ($\lambda \leqslant 1.35\ \mu m$), the ambient thermal and sky backgrounds are negligible if compared to the readout noise (Ikeda *et al.* 2006, Kondo *et al.* 2006). It means that the cooling of entire optics is not necessary as long as we remain in this short NIR region. WINERED employs warm (room temperature) optics except for the camera system that includes the detector array (Fig. 1). The narrow wavelength range significantly increases the performance of the AR coating on the optical elements, e.g., $R < 1\%$ per surface is possible, while $R > 5$ % per surface is inevitable at some wavelengths for 1.0–$5.5\ \mu m$ broad band AR. This leads to a significant throughput improvement for WINERED which uses a large number of refractive optics for the aberration-free design.

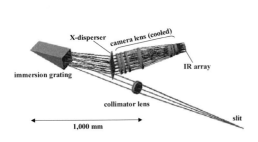

Figure 1. Optical layout of WINERED

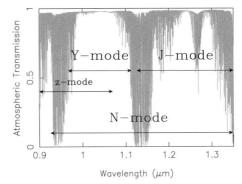

Figure 2. Wavelength coverage of WINERED

Immersion grating

An immersion grating of an infrared material with a high refractive index ($n > 2$) can provide a higher resolution and reduce the size of the collimated beam to $1/n$ for the same resolving power. The main disperser, WINERED uses an immersion grating made of ZnSe ($n \sim 2.4$) or multi-spectral grade ZnS ($n \sim 2.3$), which has a high refractive index and little absorption at NIR wavelengths. By adopting the immersion grating, WINERED can achieve a high resolving power within a very small instrumental volume and compact optics. Our current trial of groove processing on ZnSe blocks uses a nano precision fly-cutting technique at the Lawrence Livermore National Laboratory (Kuzmenko 2006), which was proved to be quite successful.

Optical system

WINERED is a cross-dispersed-type echelle spectrograph that has a refractive collimator, a ZnSe immersion grating, two (blue and red) Volume Phase Holographic gratings (VPHs) for the cross-disperser, and a cryogenic refractive lens system for the camera optics (Fig. 1). We employed this simplest classical optical configuration because it provides the best balance between the system throughput, alignment tolerances, and image quality (high spectral resolution). See Yasui *et al.* (2006) for details of the optical design.

Because we anticipate a long lead time for the development of the ZnSe/ZnS immersion grating, we plan to install a classical reflection echelle grating with a groove density of 31.6 lines mm^{-1} and a blaze angle of 63.9 deg, which will be made switchable to the immersion grating. This normal echelle mode can have a spectrum between 0.9–1.35 μm *simultaneously* with the resolving power of $R_{\max} = 28{,}300$ in one exposure (N-mode in Fig. 2). This mode serves as a full-science mode as well as a practice mode prior to the installation of the immersion grating. We plan to use mostly this mode to cover stellar metal absorption lines in this wavelength range for the JASMINE pre-study of bulge stars.

References

Ikeda, Y., *et al.* 2006, *SPIE*, 6269, 62693T
Kondo, S., *et al.* 2006, *SPIE*, 6269, 626940
Kuzmenko, P. 2006, *SPIE*, 6273, 62733S
Yasui, C., *et al.* 2006, *SPIE*, 6269, 62694P

A Giant Step: from Milli- to Micro-arcsecond Astrometry
Proceedings IAU Symposium No. 248, 2007
W. J. Jin, I. Platais & M. A. C. Perryman, eds.

© 2008 International Astronomical Union
doi:10.1017/S1743921308020000

The globular cluster tidal radii†

Z. Y. Wu

National Astronomical Observatories, Chinese Academy of Sciences, Beijing 100012, China
email: zywu@bao.ac.cn

Abstract. The orbits and theoretical tidal radii for a sample of 45 Galactic globular clusters are calculated. The orbital phase dependence between the theoretical and observed tidal radii is noted.

Keywords. Galaxy: kinematics and dynamics, globular clusters: general

1. Introduction

It has long been recognized that globular clusters must have a finite edge imposed by the Galactic tidal field. Most observed surface density profiles of globular clusters can be fitted by the King model (King 1966), while the edge of a globular cluster has traditionally been estimated using the model-predicted value of the King's tidal radius, r_t. The classical theory has pointed out that the observed tidal radii of globular clusters were obtained when the Galactic tidal force on the clusters was the strongest, that is at the perigalactic position (Hoerner 1957 and King 1962). The classical tidal radius theory does not include the internal dynamical processes: the two-body relaxation, and the energy input caused by shocks when clusters cross through the disk or the central bulge (Wu *et al.* 2003). If the tidal radius of a globular cluster is imposed at the perigalactic postion, then owing to internal and external dynamical effects, this tidal radius will expand along the orbit of a cluster, i.e. there will be an orbital phase dependence between the theoretical tidal radii at the perigalactic position and the presently observed tidal radii.

2. Orbits and tidal radii of globular clusters

Wu, *et al.* (2004) calculated the orbits of 45 globular clusters with a simple Galactic mass model: $M_g(R_g) = V_c^2 R_g / G$, where G is the constant of gravitation, $V_c = 220.0$ km s^{-1} the circle velocity, and R_g the Galactocentric distance. The theoretical tidal radius of a globular cluster r_{te} is determined using the formula of Innanen (1983):

$$r_{te} = \frac{2}{3}[1 - \ln(R_p/A)]^{-1/3} \left(\frac{m}{2M_p}\right)^{1/3} R_p$$

where R_p is the perigalactic distance of the cluster, m is the mass of the cluster, $M_p = M_g(R_p)$ is the mass of the Galaxy within R_p, and A is a distance taken as the radius of a circular orbit which has the same total orbital energy for any given cluster orbit. We use the last perigalactic crossing to compute the theoretical tidal radius r_{te}. The present observed tidal radii of the clusters mostly are taken from the compilation of Trager *et al.* (1993). Panel a of Fig. 1 shows that, for most of the clusters, the theoretical values are small compared to the observed ones. Thus, the theoretical tidal radius calculated at the perigalactic distance gives an underestimated value of the observed one. The theoretical

† This work has been supported in part by the project No.10778720 of National Natural Science Foundation of China.

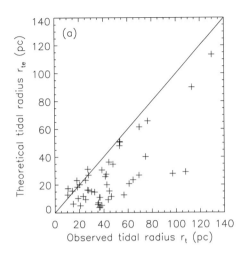 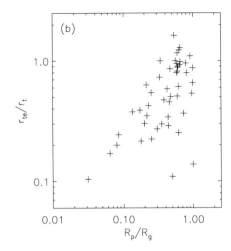

Figure 1. Panel a: Theoretical tidal radii r_{te} versus observed tidal radii r_t. Panel b: Theoretical and observed tidal radii ratio r_{te}/r_t versus perigalactic and present distance ratio R_p/R_g.

tidal radii calculated at the present positions R_g of the clusters are also greater than the observed ones. At the apogalactic distances, computed values are consequently larger. Thus the tidal radius of a cluster is not only determined by the extent tidal field of the Galaxy when it orbits away from the perigalactic position, but is also affected by other dynamical processes.

3. Orbital phase dependance of globular cluster's tidal radii

Following Meziane & Colin (1996), we also define the difference between the theoretical and observed tidal radii by introducing r_{te}/r_t, and show this ratio between the perigalactic and present distances R_p/R_g in panel b of Fig.1. A general trend appears to be in this figure, indicating the dependence of tidal radius on the orbital phase. When a cluster is near its perigalactic position, the theoretical tidal radius gives a good estimation of its observed value. The linear Pearson correlation coefficient of the logarithms between these two ratios is 0.74, corresponding to a probability of correlation of 99.99999%.

References

Hoerner, S. Von. 1957, *ApJ*, 125, 451
Innanen, K. A. 1983 *AJ*, 88, 338
King, I. R. 1962, *AJ*, 67, 471
King, I. R. 1966, *AJ*, 71, 64
Meziane, K. & Colin, J. 1996, *A&A*, 306, 747
Trager, S. C., *et al.* 1993, *ASP-CS*, 50, 347
Wu, Z.-Y., Shu, C.-G., & Chen, W.-P. 2003, *Chin. Phys. Lett.*, 20, 1648
Wu, Z.-Y., Zhou, X., & Ma, J. 2004, *Chin. Phys. Lett.*, 21, 418

A Giant Step: from Milli- to Micro-arcsecond Astrometry
Proceedings IAU Symposium No. 248, 2007
W. J. Jin, I. Platais & M. A. C. Perryman, eds.

© 2008 International Astronomical Union
doi:10.1017/S1743921308020012

Educating astrometry and celestial mechanics students for the 21st century

W. F. van Altena[1] and M. Stavinschi[2]

[1] Yale University, P.O. Box 208101, New Haven, CT 06520, USA
email: vanalten@astro.yale.edu

[2] Astronomical Institute of the Romanian Academy, Cutitul de Argint 5, Bucharest,
RO-040557, Romania
email: magda_stavinschi@yahoo.fr

Abstract. Astrometry and Celestial Mechanics have entered a new era with the advent of Micro-arcsecond positions, parallaxes and proper motions. Cutting-edge science topics will be addressed that were far beyond our grasp only a few years ago. It will be possible to determine definitive distances to Cepheid variables, the center of our Galaxy, the Magellanic Clouds and other Local Group members. We will measure the orbital parameters of dwarf galaxies that are merging with the Milky Way, define the kinematics, dynamics and structure of our Galaxy and search for evidence of the Dark Matter that makes up most of the mass in the universe. Stellar masses will be determined routinely to 1% accuracy and we will be able to make full orbit solutions and mass determinations for Extrasolar planetary systems. If we are to take advantage of Micro-arcsecond astrometry, we need to reformulate our study of reference frames, systems and the equations of motion in the context of special and general relativity. Methods also need to be developed to statistically analyze our data and calibrate our instruments to levels beyond current standards. As a consequence, our curricula must be drastically revised to meet the needs of students in the 21st Century. With the above considerations in mind, we developed a syllabus for an introductory one-semester course in Astrometry and Celestial Mechanics. This course gives broad introductions to most topics in our fields and a base of knowledge from which a student can elect areas for self-study or attendance at centers where advanced courses, workshops or internships are available.

Keywords. astrometry, sociology of astronomy, education

1. Introduction

Astrometry and Celestial Mechanics have entered into a new era with the advent of Micro-arcsecond positions, parallaxes and proper motions. Cutting-edge science topics will be addressed that were far beyond our grasp only a few years ago. It will be possible to determine definitive distances to Cepheid variables, the center of our Galaxy, the Magellanic Clouds and other Local Group members. We will measure the orbital parameters of dwarf galaxies that are merging with the Milky Way, define the kinematics, dynamics and structure of our Galaxy and search for evidence of the Dark Matter that makes up most of the mass in the universe. We have already measured the mass of the Black Hole at the center of the Milky Way and will soon be able to determine the detailed mass distribution of the galactic center. Stellar masses will be determined routinely to 1% accuracy and we will be able to make full orbit solutions and mass determinations for Extrasolar planetary systems.

If we are to take advantage of Micro-arcsecond astrometry, we need to reformulate our study of reference frames, systems and the equations of motion in the context of special and general relativity. The reference frames and systems established by Gaia, an orbiting satellite, must be related to those of the Solar System and the Earth, hence we must consider Celestial Mechanics as an integral part of our students educational program. Finally, methods need to be developed to calibrate our instruments to levels beyond commonly expected standards. In addition to the numerous space-based missions that are planned to provide high accuracy positions, parallaxes and proper motions, several large-scale ground-based telescopes and instruments such as Pan-STARRS, LSST and large-scale CCD mosaic camera surveys will dramatically increase the quantity of data available and require that students be educated in techniques to analyze massive amounts of data to cull out trends and special objects. As a consequence, our curricula must be drastically revised to meet the needs of students in the 21st Century.

A full range of courses is available to students at only a few centers in Europe and Asia, so it will be necessary to utilize a variety of teaching methods and opportunities, such as lecture courses, internships, workshops and summer schools. The 2005 Yale Summer School on Astrometry is an example of an intensive introduction to Astrometry. The School web site (http://www.astro.yale.edu/workshop) contains the lectures, which may be freely downloaded. The IAU Division I Working Group on Astrometry by small ground-based telescopes can provide numerous educational and training opportunities for students and a means to accomplish different international campaigns, such as PHEURA, PHESAT and PHEMU.

2. An introductory course on astrometry and celestial mechanics(ACM1)

With the above considerations in mind, we developed a syllabus for an introductory 40-hour one-semester course in Astrometry and Celestial Mechanics, which we refer to here as ACM 1. A course length may vary at different institutes, but it should be easy to revise this to a 20-hour course by reducing the depth covered in subjects or by selectively omitting some areas. In addition, the relative stress on subject areas should be adjusted to meet the special needs of different institutions. As noted above, there are very few centers where a full range of courses can be offered, therefore we developed the syllabus for a one-semester course that can be offered at many institutions and provide a maximum number of students with an introduction to Astrometry and Celestial Mechanics. This course gives broad introductions to most topics in our fields and a base of knowledge from which a student can elect areas for self-study or attendance at centers where advanced courses, workshops or internships are available.

Our fields are small and specialized so it is vital that ACM 1 be integrated into a general astronomical curriculum with offerings in fields that are significantly impacted by the application of modern astrometric and celestial mechanical techniques. Such areas include: galactic structure and dynamics, stellar astrophysics, the formation and evolution of planets and stars, as well as introductions to extragalactic astronomy and cosmology. With such a diverse background our students will be able to integrate into the astronomical community and optimally apply astrometric and celestial mechanical techniques to the solution of problems relevant to modern astrophysics. It is important that each subject area be introduced with science applications that will help to motivate students to study the topic.

ACM 1 was developed to meet the first level of educational needs for a Ph.D. in our fields. It will be necessary for each institute to revise the subject areas included in ACM 1 for Masters degree programs where the students are normally expected to be more specialized and to eventually play supporting roles in large scale research programs.

With the above factors taken into consideration we present here our proposed syllabus for an introduction to the study of Astrometry and Celestial Mechanics and hope that it will be of use to our colleagues in preparing the next generation of scientists in our fields. Each subject area is preceded by the recommended number of hours to be spent on the topic in the context of a 40-hour course.

The Syllabus for ACM 1

- (1 hr) **Introduction and background:**
  - Hipparcos, Gaia, SIM and other astrometric projects as motivations to study astrometry

- (2 hr) **Trigonometric Parallaxes:**
  - Highlights: Gaia and SIM give parallaxes of Cepheids, RR Lyraes, the Galactic Center and the Magellanic Clouds
  - Absolute *versus* relative parallax
  - Calibration of luminosities
  - Mass-Luminosity relation

- (2.5 hrs) **Binary and multiple stars:**
  - Highlights: Mass of the Milky Way's Black Hole, ...
  - Determining orbital parameters from observations
  - Speckle interferometric observations
  - Space-based methods and the principal results
  - Long baseline optical interferometry (CHARA, VLTI, Keck, PRIMA, ...)
  - Surveys of binary parameters

- (0.75 hr) **Star clusters:**
  - Highlights: Gaia and SIM give definitive distances, kinematical masses and membership
  - Membership determination
  - Internal motion determination

- (1.0 hr) **Solar System astrometry:**
  - Highlights: Finding extra-solar planets, ...
  - Moving object astrometry
    - NEOs, Minor planets and comets
    - Algorithms
  - Observing in the glare of a bright planet
  - Lunar occultations
  - Extra-solar planets
    - Optical astrometry
    - Radial velocity observations
    - Space-based discovery (SIM, Gaia, Kepler, ...)

- (2 hrs) **Galactic Structure astrometry:**
  - Highlights: Gaia and SIM define the structure and dynamics of our Milky Way
  - The role that astrometry plays in defining the spatial, kinematical and dynamical structure of the Galaxy and the Local Group of galaxies
  - Modifications to the equations required by Micro-arcsecond proper motions

- (1.5 hr) **Statistical astronomy:**
  - Highlights: Modeling and correcting for systematic effects are critical
  - Statistical and secular parallaxes

– The Malmquist and Lutz-Kelker corrections
– Monte-Carlo modeling

– (0.5 hr) **Cosmology astrometry:**
  – Highlights: How astrometry can help even at the edge of the universe
  – Constraining critical cosmological parameters.
  – Search for primordial gravitational waves
  – Astrometry of gamma-ray bursters

– (1 hr) **The Earth's atmosphere (troposphere):**
  – Highlights: How we model the atmosphere
  – Models of the atmosphere and turbulence
  – Refraction through a turbulent atmosphere
  – Refraction and absorption as a function of wavelength
  – The effect of refraction on positions
  – Special cases for Radio Astronomy

– (1 hr) **Atmospheric limits to positional precision (optical and radio):**
  – Highlights: Beating atmospheric turbulence in large fields of view
  – Observational determination of the limits
  – Theoretical interpretation of the observations
  – Tip-tilt and adaptive optics correction systems
  – Image reconstruction
  – Phase referencing

– (0.5 hr) **Diffraction-limited imaging:**
  – Highlights: Getting the ultimate resolution from your optics
  – HST, Gaia, and Hipparcos
  – Speckle imaging
  – Ground-based observations in the infra-red

– (2 hrs) **Interferometry (Optical):**
  – Highlights: Beating atmospheric turbulence in small fields of view
  – Theory
    – Sparse vs filled apertures
    – Imaging
    – Coronagraphy
  – Optical
    – Michelson interferometers: theory and calibration
      – Existing systems: VLTI, Keck, CHARA, NPOI, ...
      – Planned systems: PRIMA, SIM
    – Intensity interferometers
    – White-light interferometers, such as the HST/FGS
    – Speckle Interferometers

– (2 hrs) **Interferometry (Radio):**
  – Highlights: The VLBI micro-arcsecond parallaxes
    – http://veraserver.mtk.nao.ac.jp/hilight/pub070711/pub070711-e.html
  – Theory and calibration of existing systems
    – Connected-element, VLA, VLBI, VLBA, VERA
  – Phase referencing
  – Physics of water masers
  – Gravitational deflection of light

– Future instruments
  – ALMA, SKA

– (3 hrs) **Geometrical optics for Astrometry:**
  – Highlights: The imaging properties of optical telescopes
  – Refracting optics
    – Filters & CCD dewar windows
      – Beam shift, focus and image position errors
    – Seidel aberrations
      – Theory
      – Aberrations relevant to optical astrometry
      – Observational examples
  – Reflecting optics
    – Conic sections
    – Seidel aberrations
  – Two-mirror telescopes
    – Aberrations
    – Alignment
  – Astrometric compensation for optical aberrations

– (1.5 hrs) **Imaging detectors:**
  – Highlights: The complexities of our modern detectors
  – Characteristics of astronomical CCD and CMOS detectors
  – Calibrating
    – The signal sources
    – Noise sources (flat, bias, dark and read-out)
    – Charge-transfer efficiency (CTE)
    – Linearity and saturation
    – Fringing
  – Optimizing the signal-to-noise
  – Astrometry
    – Image centroids and PSF
  – Photometry
    – Surface, Aperture and PSF photometry
  – Statistical techniques useful for imaging analyzes

– (0.25 hr) **Time-delayed integration:**
  – Highlights: Letting the sky drift by; observing efficiently
  – Gaia, CTI, SDSS, and QuEST.

– (1 hr) **Image deconvolution:**
  – Highlights: Increase your telescope aperture by 40%, ...
  – Theory
  – Noise sources
  – Aberrated PSFs
  – Good PSFs but correcting for atmospheric degradation
  – Potential gain in precision
  – Potential gain in limiting magnitude

– (1 hr) **Plate measuring machines:**
  – Highlights: Old plates can improve our proper motions
  – Scanning machines
    – SuperCOSMOS, PDS, STScI GAMMA

- Calibrating for thermo-mechanical drifts and non-linearities
- Step-and-repeat imaging of a plate
  - PMM, StarScan, next generation scanners
  - Stitching the "footprints" together
- Algorithms for photographic images
  - Astrometry
  - Photometry
- Calibration
  - Orthogonality of axes, periodic errors, detector time lags, and others

- (3 hrs) **From measures to celestial coordinates:**
  - Highlights: Projecting the sky onto the focal plane
  - Primer on spherical astronomy
  - Telescope-modeling techniques.
    - Gnomonic to focal plane calibrations
    - Focal-plane arrays of CCDs
    - Looking for systematic errors
  - Extending astrometric calibration regions to fainter objects and to different passbands.

- (1 hr) **Calibration of complex instrumental systems:**
  - Highlights: Calibration makes your instrument and data valuable
  - Mosaic cameras
  - HST Fine Guidance Sensors
  - Hipparcos and Gaia
  - Time-delayed integration systems
    - SDSS, QuEST

- (3 hrs) **Relativistic foundations of astrometry and celestial mechanics:**
  - Basics of special and general relativity
  - Post-Newtonian approximation scheme
  - Relativistic astronomical reference systems and frames
  - Coordinate-dependent and measurable quantities
  - Relativistic astronomical time scales and their realizations
  - Relativistic data reduction modeling
  - Relativistic equations of motion (test body around a spherically symmetric body and N-body problem)
  - Relativistic effects needed for micro-arcsecond astrometry

- (3 hrs) **Stellar Coordinate Systems and Positions:**
  - Glossary of terms used and definitions
  - Transformation from ICRS to Observed Places of Stars
  - Flow charts and formulae for the transformations
  - IAU Nomenclature for Fundamental Astronomy — "Explanatory Document B, July 2006" — http://syrte.obspm.fr/iauWGnfa/NFA_B.pdf

- (4 hrs) **Celestial Mechanics of the N-body problem:**
  - The Sun, solar oblateness
  - Major planets, planetary rings
  - Minor planets, asteroid belts
  - The role of analytic techniques and numerical methods
  - The role of pulsar timing, laser and radio ranging
  - High-precision radial velocities

- Analyzing binary and multiple systems with micro-arcsecond precision
- (2 hrs) **Catalogs:**
  - Highlights: Making the observations usable and accessible
  - Fundamental
    - ICRF and radio astrometry
  - Optical reference system
    - ICRS, Hipparcos (HCRF) and Tycho
    - Looking to Gaia
    - Extending to fainter magnitudes
      - UCAC, NPM & SPM
    - Value of "non-astrometric" catalogs: 2MASS, DENIS, SDSS
  - Schmidt-based catalogs
    - GSC, Digital Sky Survey, USNO A2.0 & B-1, Cosmos, SRC
  - The Virtual Observatory
    - National versions of the VO
    - NOMAD — the "astrometric" VO
- (0.5 hr) **Analyzing poorly-sampled images:**
  - Highlights: Increasing accuracy and compensating for systematic effects
  - HST imaging astrometry and photometry

3. Conclusion

The evolution of Astrometry from ground-based to space-based has resulted in greatly increased accuracy that requires us to reformulate our reference frames, systems and equations of motion in the context of special and general relativity. Our old curricula are outdated and need to be completely revised especially in countries that have direct access to space-based observations. A range of teaching methods and opportunities are needed, such as lecture courses, workshops, summer schools and internships. The IAU Division I Working Group on Astrometry by small ground-based telescopes can provide numerous educational and training opportunities for students and a means to accomplish different international campaigns, such as PHEURA, PHESAT and PHEMU.

Acknowledgements

We are grateful to colleagues who provided us with many suggestions for the one-semester syllabus in Astrometry and Celestial Mechanics: Sergei Kopeikin, Sergei Klioner, Nicole Capitaine, Anita Gomez, Wenjing Jin, Stephen Unwin, Elliott Horch, Stephen Majewski, Michael Efroimsky and others.

We acknowledge NSF, NASA, ESA, ESO and our national observatories all of whom provide us with outstanding facilities for astrometric research.

A Giant Step: from Milli- to Micro-arcsecond Astrometry
Proceedings IAU Symposium No. 248, 2007
W. J. Jin, I. Platais & M. A. C. Perryman, eds.

© 2008 International Astronomical Union
doi:10.1017/S1743921308020024

A project of teaching ground-based astrometry

J.-E. Arlot[1], W. J. Jin[2], J. Zhu[3], Q. Y. Peng[4], F. Colas[1], K. X. Shen[5], Z. H. Tang[2], Z. Zhu[6], V. Lainey[1], W. Thuillot[1] and A. Vienne[1]

[1]IMCCE, UMR8028 du CNRS, Paris Observatory, UPMC, USTL
77 avenue Denfert-Rochereau, F-75014 Paris, France
email: arlot@imcce.fr

[2]Shanghai Astronomical Observatory, Shanghai, China email: jwj@shao.ac.cn
email: zhtang@shao.ac.cn

[3]Beijing Planetarium, Beijing, China email: jinzhu@bjp.org.cn

[4]Jinnan University, Guangzhou, China email: pengqy@jnu.edu.cn

[5]National Time Service Center, China email: shenkx@ntsc.ac.cn

[6]Department of Astronomy, Nanjing University, Nanjing, China email: zhuzi@nju.edu.cn

Abstract. The optical ground-based astrometry of solar system objects may have its accuracy strongly improved by using new methods for making observations and reductions of them. New photometric methods of observating the mutual phenomena occurring in the solar system, may provide astrometric data with a higher precision than the classical direct imaging. In order to help preparing observers for the future campaigns of observations (2008–2010) and to promote this kind of high-accuracy astrometry, we plan to organize a spring school in 2008 in Beijing, China, for PhD and post-doctoral students, and for interested young astronomers.

Keywords. astrometry, eclipses, ephemerides, occultations, solar system: general

1. Introduction

The astrometric observations of the solar system bodies are fundamental for several purposes:
– space navigation (ephemerides);
– dynamics, stability, evolution, scale of the solar system;
– impact hazard assessment (ephemerides);
– space and ground based observations (ephemerides);
– physics of the surfaces and interiors of solar system objects;
– gravitational and relativistic studies;
– reference systems.

Hence, we might develop campaigns of observations in order to gather high accuracy-observations allowing theoretical developments. The goal of the proposed spring school is to help students and young astronomers to organize or to participate in campaigns of astrometric observations. The program of the Spring School is as follows.

2. The courses of the spring school

2.1. Fundamental astrometry

– definition of a measurement on the celestial sphere, absolute, relative or direct measurement;
– focal plane observations, the methods for their reduction;

– astrometric reduction of optical ground-based observations, the gnomonic
   projection, biases from telescopes, from the sky;
– the link methods, star catalogues;
– the methods of reduction without reference stars.

2.2. *Detectors, telescopes and images*

– making of images, diffraction, refraction, atmospheric effects;
– the case of extended objects: solar system bodies ; the reflection laws on the
   surface of the bodies, the centre of mass and the photocenter;
– electro-magnetic signal, the optical wavelengths;
– the astrometric and photometric detectors, CCD, the reduction of a CCD image;
– scanning of photographic plates;
– the objects to be observed, planets, inner satellites, outer satellites, large objects,
   asteroids, fast objects, comets·

2.3. *Astrometry through photometry*

– astrometric observation without an angular measure, elements of photometry;
– occultations (occultations by the Moon, occultations of stars);
– the mutual occultations and eclipses, occurrence during the planetary equinoxes;
– the observation and reduction of the mutual events;
– the phase defect, albedo;
– analysis of the photometric light curves;
– preparation for the observational campaigns in 2008-2010.

2.4. *Practical astrometry at night*

– CCD observation and reduction;
– observation of different objects: asteroids, inner and outer natural planetary satellites;
– the case of giant planets.

3. Conclusion

It seems interesting to start a spring school on astrometry by focusing on making observations for a specific campaign of observations using the opportunity of the equinox on
Jupiter and Saturn allowing the occurrence of mutual occultations and eclipses. Tutorials
and recommendations will be provided mostly for these observations, their reduction and
their astrometric interest but general information will be given in order to have a link
with classical astrometry.

A Giant Step: from Milli- to Micro-arcsecond Astrometry
Proceedings IAU Symposium No. 248, 2007
W. J. Jin, I. Platais & M. A. C. Perryman, eds.

© 2008 International Astronomical Union
doi:10.1017/S1743921308020036

Astrometry course at University of Tokyo

T. Fukushima

National Astronomical Observatory,
2-21-1, Ohsawa, Mitaka, Tokyo 181-8588, Japan
e-mail: Toshio.Fukushima@nao.ac.jp

Abstract. The astrometry course at Department of Astronomy, University of Tokyo, is reviewed as an example of educational efforts for top-class students, the possible candidates of professional astronomers, in Japan. The method of teaching is unique in the sense that it gives lectures by using *incomplete* text books both as MS Powerpoint slides posted at a web site, http://chiron.mtk.nao.ac.jp/ toshio/education.html, and as printed materials in the form of self-study notebooks. Also there are self-study notebooks on the related issues; the courses of relativistic astrometry, of rotational motions, of numerical astronomy, and of orbital motions, the last of which is under development.

Keywords. astrometry, education

1. Introduction

We report by this short article our efforts to teach the very basics of astronomy, astrometry, to students in astronomy course. In Japan, a very limited number of universities with Department of Astronomy offer a complete program of teaching astronomy. One rare exception is the University of Tokyo (UoT).

This university is one of the universities with the emphasis on graduate-course education. As a result, the number of graduate students are many more than that of undergraduates. In the case of Department of Astronomy, which is a branch of Faculty of Science, we welcome around two dozen new graduate students and 10 undergraduates each school year. This means that more than a half of UoT graduate students are not educated in the manner the UoT undergraduate students are.

To mitigate this situation, we select three basic courses to teach both the undergraduate students and graduate students without dedicated astronomy course in their curricula. The astrometry is one of the three. The other two are the optical observational astronomy and the galaxy dynamics. This is an exception in the days of lacking astrometry courses worldwide.

At any event, the astrometry or the positional astronomy in its old name has been taught at UoT by famous scholars such as Profs Yoshihide Kozai in 70's, Gen-Ichiro Hori in 80's, and Masanori Miyamoto in 90's. Since 1998, the author has taken on this important task until now.

2. Contents of course

The current style of course is to teach basic concepts of astrometry in one semester. Actually the course is scheduled for the first semester, which is April to July in Japanese academic year, of the fourth year undergraduate students and freshman graduate students. The form of teaching is ordinary. Namely, we give 90 minute lecture in 15 consecutive weeks. Since it is rare to have all 15 classes as scheduled, we design the course to consist of one introduction and 11 topics. The selected topics are; (1) Observation, (2) Time, (3) Space, (4) Coordinate System, (5) Motion of Celestial Bodies, (6) Rotation,

(7) Earth Rotation, (8) Keplerian Motion, (9) Signal Propagation, (10) Least-Squares Method, and (11) General Relativity Effects.

Each topic, which is aimed to be taught in 90 min class, is composed from around a dozen items. For example, the topic of "Signal Propagation" is divided into the following subtopics; (a) One-Way Propagation, (b) Passive Observables, (c) Equation of Light Time, (d) Light Direction, (e) Aberration, (f) Parallax, (g) Doppler Shift, (h) Propagation Delay, (i) Refraction, (j) Multi-Way Propagation, (k) Round-Trip Propagation, (l) Interferometric Observation Equation, and (m) Arrival-Time Observation Equation. Finally, each subtopic consists of a few MS Powerpoint slides. This number would be the maximum if considering the time for each subtopic is several minutes at most. The detailed content is available at the author's home page shown below.

3. Self-study note

None of the previous lecturers of astrometry at UoT adopted standard text books on astrometry. Rather they gave lectures based on their own original manuscripts. Unfortunately none of them are published or available in any form. To change this situation, in 2002, the author started to publish the materials in an easy way, i.e. posting them in downloadable files in the PDF format, MS Powerpoint file, or plain ASCII text, at the author's web site;

$$\mathrm{http://chiron.mtk.nao.ac.jp/\,toshio/education.html}$$

For the educational purposes, especially to encourage students to think by themselves, we intended to provide not every detail of the topic but only selected key descriptions. Besides the astrometry course, we created similar materials on the related fields such as the general relativistic astrometry, the rotation dynamics, and the numerical astronomy in the chronological order of preparation. Also we prepared an introductory course of astronomy, the contents of which is basic and general but with an emphasis on astrometry.

Currently available are the following materials; (i) Theoretical Astrometry, in English, self-study notes, 214 pages, 3.9MB, `Astrometry(Note-BW).pdf`, (ii) Rotational Dynamics, in English, self-study notes, (Vol.1) 183 pages, 0.6 MB, `rotation/Rotation1_e.pdf`, (Vol.2) 178 pages, 0.7 MB, `rotation/Rotation2_e.pdf`, (iii) General Relativistic Astrometry, in Japanese, full text, 75 pages, 0.5MB, `astrometry.pdf`, (iv) Numerical Astronomy, in Japanese, MS Powerpoint slides, 519 slides, 6.4MB, `numerical2006.ppt`, and (v) Introduction to Astronomy, in Japanese, for college students/general public, MS Powerpoint slides, (Vol.1) 111 slides, 8.7 MB, `showa2005-1.ppt`, (Vol.2) 210 slides, 13.4 MB, `showa2005-2.ppt`. The first two are available both in English and in Japanese. Meanwhile the remaining three are in Japanese only. Under development is the preparation of self-study notes on orbital dynamics, which will complement that of rotational dynamics. Also, the English translation of Numerical Astronomy is planned in the near future.

Based on about 10 years experience of teaching the astrometry class by using these downloadable materials, we have come to the conclusion that the Powerpoint slides is not sufficient for educational purposes. Then, we initiated freely providing printed materials, or the so-called self-study notebooks. Each page of the notebook consists of one Powerpoint slide in the top half and blank space in the bottom half. The space is for students to take notes or make exercises. Each blank space is covered by lightly-colored graphs of a 4mm×4mm bin size. This is for students to draw diagrams, lists, equations, and/or comparison tables.

A limited number of copies of printed materials are available from the author.

A Giant Step: from Milli- to Micro-arcsecond Astrometry
Proceedings IAU Symposium No. 248, 2007
W. J. Jin, I. Platais & M. A. C. Perryman, eds.
© 2008 International Astronomical Union
doi:10.1017/S1743921308020048

Astrometric education in China†

W. J. Jin[1] and Z. Zhu[2]

[1]Shanghai Astronomical Observatory, Shanghai 200030, China
email: jwj@shao.ac.cn
[2]Department of Astronomy, University of Nanjing, Nanjing 210093, China

Abstract. With measuring precision on the order of milli-arcseconds for ground-based survey facilities or even micro-arcseconds for space astrometric satellites, the importance of astrometric education continues to be important. The content of astrometric courses in China during the past fifteen years is reviewed and the current astrometric courses for undergraduate and graduate students at universities and observatories in China are presented. Finally the improvements of astrometric education in content and teaching methods are suggested.

Keywords. astrometry, sociology of astronomy, education

1. Introduction

Astrometry is a branch of astronomy dealing with the precise measurements of the positions of celestial objects. With measuring precision on the order of milli-arcseconds for Hipparcos and ground-based survey facilities such as LSST, Pan-STARRS, or even micro-arcseconds for space astrometric satellites such as Gaia, SIM a new era of astrometry has entered. For example, photographic plate is replaced by the CCD detectors; transit circles and astrolabes are replaced by optical interferometers, and the optical reference frame, which was used for almost a hundred years, is replaced by the extragalactic reference frame. Data must be processed in the general relativistic framework and new definition of the ecliptic, models of precession and nutation are used. Finally, the study of the structure and kinematics of the Galaxy by means of stellar proper motions and positions is also included in astrometry. As a consequence of these changes, astrometric education faces new challenges.

2. Current astrometric courses

Astrometric education is implemented at 4 Universities: Nanjing University, Beijing Normal University, University of Science and Technology of China, as well as Beijing University where there are under graduate and graduate students learning astrometry. The astrometric courses for graduate students are set up at 4 observatories: Shanghai Astronomical Observatory(SHAO), Purple Mountain Astronomical Observatory, National Astronomical Observatories and National Time Service Center(Bai, 2006). Due to historical reasons the research field of astrometry in China includes two aspects: (a) the mechanism of Earth Rotation, plate motions etc., and (b) the determination of stellar positions and proper motions. In the past fifteen years the former has become an independent discipline, named Astro-geodynamics and the latter includes a wider range of astrometry, such as observations and studies of the solar system objects, the structure and kinematics of our Galaxy.

† Supported by the Science & Technology Commission of Shanghai Municipality (Grant No. 06DZ22101) and the National Natural Science Foundation of China(Grant No.10333050)

86% and 35% of graduate students at SHAO and Nanjing University(2003-2007) have not had a prior astronomical education, so the required courses: Introduction of astronomy, Introduction of astrometry which is similar Spherical Astronomy(at Nanjing University the content of these is included in the course of Astronomical reference frame), Measuring data analysis and processing. The elective courses are galactic astronomy, general relativity, and the theory of artificial satellite orbits. 82% and 40% of PhD students at SHAO and Nanjing University have a good background in astronomy and usually there is no special courses for them. The supplementary education for graduate students are seminars which are given by foreign and domestic experts each week. Summer schools have been held abroad and at the domestic locations, since it is the best way for PhD students to receive a systematic astrometric education in short time. For example, an astrometric summer school at Yale university was held in 2005, VLBI techniques and applications was held at Shanghai in 2007, a Chinese–French spring school will be held in 2008. In addition, the individual courses such as atmospheric refraction, radio astronomical method and technique etc. are given to individual students.

3. Suggestion

Due to the retirement of lecturers and the development of new techniques since the 1990s, it is important to update our teachers' knowledge and to produce new textbooks. After the publication of vectorial astrometry written by C.A. Murray, good textbooks or reference books have not appeared for more than 15 years. This has encouraged some professional researchers who are retired to write textbooks or reference books, such as Modern Astrometry (Kovalevsky, 1995), Astrometry of Fundamental Catalogues (Walter & Sovers, 2001), Introduction of Fundamental Astrometry (Kovalevsky & Seidelmann, 2003), Astrometric method - past, present and future(Li et al. 2006).

We feel that teaching methods should be changed to compensate the lack of good textbooks and professional courses. Several methods are suggested such as: (a) Share the teaching materials. We recommend these to be written in English since it is then possible to share the materials over the internet, as done by Drs. Fukushima, Tom Loredo and Phil Gregory who have put their materials of statistical mathematics on web sites. (b) Summer school. The lectures should be given by outstanding teachers and experts in the field. The teaching materials would then serve as a textbook or reference book to be published in the future. It has been proven that this is an effective and fast method for attaining results in short time. (c) Education over the network. This method is often used in amateur education such as television programs, education demonstrations but it needs to be run by an organization to obtain adequate support, however, the latter is difficult to obtain. (d) Seminars. International cultural and scientific exchange and collaboration are often carried out between the countries. It is a good method to expand the knowledge and learn new techniques. This method has been used many years at various institutes and proved successful.

References

Bai, C. L. 2006, *Chinese Academy of Sciences: Mathematics and Astronomy*

Kovalevsky, J. 1995, Modern Astrometry, Springer, New York

Kovalevsky, J. & Seidekmann, P. K. 2003, Introduction of fundamental astrometry, Springer, New York

Li, D. M., Jin W. J., Xia Y. F. *et al.* 2006, Astrometric method —past, present and future, Scientific publisher, Beijing

Walter, H. G. & Sovers, O. J. 2001, Astrometry of fundamental catalogues, Springer, New York

A Giant Step: from Milli- to Micro-arcsecond Astrometry
Proceedings IAU Symposium No. 248, 2007
W. J. Jin, I. Platais & M. A. C. Perryman, eds.

Astrometric education in Saint-Petersburg State University, Russia

I. I. Kumkova and V. V. Vityazev

Sobolev Astronimical Instituute, Saint-Petersburg State University, 198504, Petrodvorets,
Universitetsky pr., 28., S. Petersburg, Russia
email: `vityazev@list.ru`

Abstract. The present system of astrometric education in the Saint-Petersburg State University is presented. The general courses, specific programs, seminars and observations that a student takes during a 5-year educational program are described.

Keywords. astrometry, education

1. Introduction

The Astronomy Chair at Saint-Petersburg University has existed since 1819. The names of the world wide known scientists, such as C.P. Glasenapp, A.A. Belopolsky, G.A. Tikhonov, V.V. Sharonov, K.F. Ogorodnikov, A.T. Agekyan, V.A. Ambartsumyan, V.V. Sobolev are connected with the University. The subject of special pride of the Astronomy Department is the fact that two former students who graduated from the Department, were later elected as presidents of International Astronomical Union (IAU). They are academicians V.A. Ambartsumian and A.A. Boyarchuck The current astronomical research in Saint-Petersburg University are focused on the following topics: fractal structure of the Universe, active galactic nuclei, spiral structure of our Galaxy, dark matter in galaxies, interaction of radiation and matter in different galactic objects, synthesis of elements in stars, stars with protoplanetary systems, radio emission of the Sun, dynamics of interplanetary matter, evolution of orbits in planetary and satellite systems, astronomical data analysis, stellar kinematics based on modern astrometric data, Earth Orientation Parameters from VLBI observations, and astrometric applications of the GPS/GLONASS techniques.

2. Educational programme

Annually, 20 students begin a 5-year program to become professional astronomers. During this time they study three principal sets of disciplines: (a) Mathematics, Numerical methods, Computer Science, Physics and Mechanics – 2000 h; (b) Astronomy – 2230 h; (c) Languages, History Philosophy, and others – 900 h.

The first part of the astronomical set includes the following general courses: (a) General astronomy – 118 h; (b) Spherical astronomy – 72 h; (c) Celestial mechanics – 82 h; (d) Astrometry – 74 h; (e) General astrophysics – 104 h; (f) Theoretical astrophysics – 60 h; (g) Galactic astronomy – 56 h; (h) History of Astronomy – 42 h. The aim of general courses is to present the problems in various branches of modern astronomy. It is necessary to mention that the background of all subjects is a high level mathematics, programming and computer sciences, since the Astronomy Department is a part of Mathematics and Mechanics Faculty of the University. At the beginning of the 4-th year of study each student of the Astronomical Department should choose the specific area of the astronomy to continue his/her education. The offered areas are: Astrometry, Celestial Mechanics, Radio astronomy, Observational astrophysics, Theoretical astrophysics, and Galactic astronomy. The student also has additional courses connected with the chosen field of astronomy. For example, a student who is specializing in astrometry has special courses, such as: (a) VLBI astrometry – 32 h; (b) GPS/GLONASS astrometry – 32 h; (c) Space astrometry – 32 h; (d) Small field astrometry – 98 h; (e) The instruments of optical ground-based astrometry – 32 h; (f) Rotation of the Earth – 32 h; (g) Geodynamics – 64 h; (h) Relativity in astrometry and Celestial mechanics – 32 h. Each year, up to 5 students choose Astrometry, while the rest choose Astrophysics and Celestial mechanics where they have their own sets of special courses. The faculty of the Astronomical Department is in charge of providing the general astronomical courses, whereas the special courses are provided mainly by the leading scientists from Pulkovo Observatory and Institute of Applied Astronomy. 720 h of practical astrometric work is added to the theoretical courses mentioned above since research is an indispensable part of the graduate program. For example, students in astrometry have opportunities to work with faculty and staff of Pulkovo Observatory and of Institute of Applied Astronomy on a broad range of astrometric problems. At the end of the 5-year educational program, a student presents the results of a research project that is evaluated by independent scientists. If the student is successful, the student is awarded the first degree – Specialist – corresponding to a Master of Science degree in the USA. The second degree, Candidate of Science, (the counterpart of PhD) is awarded to a person after completing a Postgraduate course (up to 3 years) and presenting a thesis based on the results of his/her scientific activity. The third and highest degree is Doctor of Science, which is awarded for large contribution to science.

3. Conclusion

Where can a former graduate student with a degree from Saint-Petersburg University find a job? There are a few institutions in Saint-Petersburg interested in hiring astrometrists, such as Pulkovo observatory of RAS, the Institute of Applied Astronomy of RAS, and several institutions connected with space geodesy research. If this is not successful, then our astrometrists may go into computer science or industry where the acquired math and programming skills are highly valued.

A Giant Step: from Milli- to Micro-arcsecond Astrometry
Proceedings IAU Symposium No. 248, 2007
W. J. Jin, I. Platais & M. A. C. Perryman, eds.

© 2008 International Astronomical Union
doi:10.1017/S1743921308020061

ELSA – training the next generation of space astrometrists

L. Lindegren[1], A. Bijaoui[2], A. G. A. Brown[3], R. Drimmel[4], L. Eyer[5], S. Jordan[6], M. Kontizas[7], F. van Leeuwen[8], K. Muinonen[9], D. Pourbaix[10], J. Torra[11], C. Turon[12], J. de Vries[13] and T. Zwitter[14]

[1] Lund Observatory, Lund University, Sweden, email: lennart@astro.lu.se

[2] Observatoire de la Côte d'Azur, Nice, France, email: albert.bijaoui@obs-nice.fr

[3] Leiden Observatory, Leiden University, The Netherlands, email: brown@strw.leidenuniv.nl

[4] Istituto Nazionale di Astrofisica, Torino, Italy, email: drimmel@oato.inaf.it

[5] Geneva Observatory, Switzerland, email: Laurent.Eyer@obs.unige.ch

[6] Astronomische Rechen-Institut, Heidelberg, Germany, email: jordan@ari.uni-heidelberg.de

[7] Dpt. of Astrophysics, Astronomy and Mechanics, Athens, Greece, email: mkontiza@cc.uoa.gr

[8] Institute of Astronomy, Cambridge, United Kingdom, email: fvl@ast.cam.ac.uk

[9] Observatory, Univ. Helsinki, Finland, email: karri.muinonen@helsinki.fi

[10] Institute of Astronomy and Astrophysics, Brussels, Belgium, email: pourbaix@astro.ulb.ac.be

[11] Departament d'Astronomia i Meteorologia, Barcelona, Spain, email: jordi@am.ub.es

[12] Observatoire de Paris, Meudon, France, email: catherine.turon@obspm.fr

[13] Dutch Space B. V., Leiden, The Netherlands, email: j.de.vries@dutchspace.nl

[14] Faculty of Mathematics and Physics, Ljubljana, Slovenia, email: tomaz.zwitter@fmf.uni-lj.si

Abstract. ELSA (European Leadership in Space Astrometry) is an EU-funded research project 2006–2010, contributing to the scientific preparations for the Gaia mission while training young researchers in space astrometry and related subjects. Nine postgraduate (PhD) students and five postdocs have been recruited to the network. Their research focuses on the principles of global astrometric, photometric, and spectroscopic measurements from space, instrument modelling and calibration, and numerical analysis tools and data processing methods relevant for Gaia.

Keywords. astrometry, instrumentation: detectors, methods: data analysis, methods: numerical, space vehicles, techniques: photometric, techniques: radial velocities

1. Background

The experience and success of the Hipparcos mission, launched in 1989 by the European Space Agency (ESA), have made space astrometry into a specialty of Europe. This was emphasised by the selection, in 2000, of the Gaia project as one of the major future missions within ESA's scientific programme. Gaia builds directly on the specific know-how accumulated in Europe during the Hipparcos era. Many of the scientists involved in defining Gaia gained their experience with Hipparcos 20–30 years ago, and are now approaching retirement, or are already retired. A new generation of scientists must therefore take the lead into the future. Their expertise is needed both to formulate new projects and to make proper use of the vast output of Gaia and other large-scale surveys. ELSA (European Leadership in Space Astrometry) was formed partly in response to the need to foster the development of a new generation of scientists in this research area.

ELSA is a Marie Curie Research Training Network (RTN) supported by the European Community's Sixth Framework Programme (FP6). The purpose of an RTN is to allow research teams to implement a structured training programme for young researchers in the context of a well-defined collaborative research project. In our case this project is part of the scientific preparations for the Gaia mission, and therefore strongly linked to – but not formally part of – the Gaia Data Processing and Analysis Consortium. The ELSA contract covers four years (2006–2010), has a budget of 2.8 MEuro and involves 14 partners with Lund University as coordinator (see list of authors).

2. Research project

The scientific objective of ELSA is to develop theoretical understanding and practical analysis tools of importance for the Gaia mission. The title emphasizes astrometry, but this term should be understood in a wide sense including the measurement of the complementary (and strongly interrelated) photometric, spectroscopic, orbital, and structural properties of the objects possible with a scanning space observatory such as Gaia.

The project focuses on certain aspects of the mission where a number of common methods are applied to address the problems and develop new tools:

• Space astrometry: Global methods of space astrometry in the framework of General Relativity; sources of systematic errors in Gaia and their effects on the final results; interdependence of astrometric, photometric, and radial velocity information for space astrometry missions, including requirements for ground-based standards.

• Instrument modelling: CCD radiation damage effects on astrometry, photometry, and radial velocity; chromaticity effects; attitude modelling.

• Numerical analysis: Scaled-down model solutions; methods for detection and management of outliers; impact of weighting schemes on accuracy and convergence; alternative solution methods.

• Data processing: Implementation and optimisation of parallel super-computing and GRID technology for Gaia data processing and instrument simulation; data processing methods for the analysis of complex sources and irregular time series observations.

3. Implementation

The network has appointed nine postgraduate (PhD) students and five postdocs. To take up their positions the fellows have to move to a new country. The PhD students are enrolled in the regular postgraduate programme at their respective host institute, but also participate in a number of network activities designed to promote collaboration, mobility, and training of project-related and general skills. The role of the postdocs is to carry out research, add specific expertise and transfer knowledge within the network.

As part of these activities, a series of network-wide meetings are organised with participation from the science community. The first one is the *ELSA School on the Science of Gaia* held in Leiden, 19–28 November 2007, and the final one will be the *ELSA Conference on the Simulation and Analysis of Space Astrometry Data*, to be held in Paris (planned for May 2010). In between, there will be two schools or workshops on more technical matters. For more and updated information, see http://www.astro.lu.se/ELSA.

Acknowledgements

The Marie Curie Research Training Network ELSA is supported by the European Community's Sixth Framework Programme under contract MRTN-CT-2006-033481.

A Giant Step: from Milli- to Micro-arcsecond Astrometry
Proceedings IAU Symposium No. 248, 2007
W. J. Jin, I. Platais & M. A. C. Perryman, eds.

Acoustic astrometry
with a VLBI-like interferometer

I. Martí-Vidal and J. M. Marcaide

Dpt. Astronomia i Astrofísica, Universitat de València, C/ Dr. Moliner 50, 46100 Burjassot,
Valencia (SPAIN)
email: i.marti-vidal@uv.es

Abstract. We show how loud-speakers, home digital recorders, and a common personal computer can be used to emulate VLBI observations on a small scale. These audio-VLBI observations allow for single-field astrometry (sources within the same interferometric field), differential group-delay astrometry, etc. These experiments can be set up very easily and in many possible configurations. Students may find these experiments very useful to learn about the innermost details of the interferometric technique.

Keywords. techniques: interferometric

1. Experimental setup

We performed VLBI-like observations (see, e.g., Thomson, Moran & Swenson 1986, and references therein, for details on the VLBI technique) using, as signal, sound waves generated by two stereo loud-speakers separated by a distance d (upper side of Fig. 1 (a)). We used home digital audio recorders, located at a distance D from the speakers, as receiving antennas (lower side of Fig. 1 (a)). Our interferometer consisted of 10 antennas, ordered by us in the East-West direction at random distances.

The recorded audio signals were treated following all the usual steps of real VLBI data: down-conversion from the observing frequency using a local oscillator, Nyquist sampling, amplitude digitization, phase rotation, correlation (we performed both, XF and FX correlations), Van Vleck correction, amplitude calibration, and Global Fringe Fitting. All these steps were performed using the software *Mathematica* (Wolfram 2003).

2. Single-field astrometry

When the signal bandwidth is narrow enough (in our case, $10\,\text{Hz}$ with the central frequency at $3\,\text{KHz}$), for an appropriate geometry ($d = 0.5\,\text{m}$ and $D = 2\,\text{m}$) the signals from both speakers will fall within the same fringe. Thus, the resulting visibilities will correspond to the Fourier transform of a double source. In Fig. 1 (b) we show the fringe, in delay space, resulting from the correlation of the data corresponding to one of our shortest baselines. The plot of correlated amplitudes vs. baseline length (Fig. 1 (c)) shows a clear modulation produced by the double source.

Due to the relatively high central frequency and to our low visual precision for the alignment of the signals in delay space, we need to fringe-fit the visibility phases prior to Fourier inverting into the "sky" plane. As shown in Fig. 1 (d), after performing Global Fringe-Fitting and one Hybrid Mapping iteration to the data, we recover a "sky" map with a clear double source (notice that we have also synthesized resolution in the North-South direction, just reproducing the same baselines at different vertical coordinates).

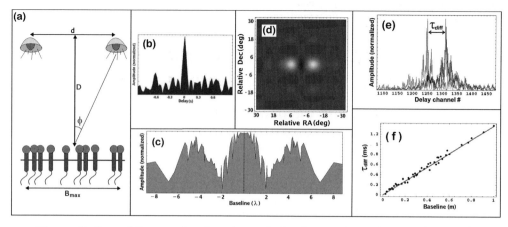

Figure 1. Several figures related to the analysis of our audio-VLBI data (see text).

3. Group-delay astrometry

For a wide signal bandwidth (22 KHz centered at 11KHz in our case), the signal from each speaker generates its own fringe (i.e., the interferometer sees two single sources instead of one double source), provided measurements are taken for an appropriate geometry ($d = 6.5$ m and $D = 12$ m) with D large enough to grant a fraction of a phase cycle precision in the approximation $D \rightarrow \infty$. In that case, we can separately perform Global Fringe Fitting to the signals coming from each speaker and compute, for each baseline, the differential (group) delays. The latter are the time separations between the fringe peaks generated by both signals (see Fig. 1 (e)).

Figure 1 (e) shows example fringes, in delay space, corresponding to our longest baseline. Fig. 1 (f) shows the results of representing the measured differential group-delays (τ_{diff}) as a function of baseline length for all baselines of our interferometer. The line corresponds to the delay model $\tau_{\mathrm{diff}} = 2(B/c)sin(\phi)$, where c is the speed of sound.

4. Conclusions

We have shown that it is possible to perform realistic VLBI emulations using sound waves as signal, home digital recorders as antennas, and a common personal computer as correlator. The observations can be carried out under a large number of different configurations: using the same or different clocks for the recorders, changing the sound pitch, adding spectral lines (i.e., tones) to the continuum, moving the sources during the data acquisition, etc. The educational and training capabilities of these relatively simple hands-on experiments is very promising.

Acknowledgements

I. M. V. thanks Sergio Jiménez and Diana Dunca for their help setting up these experiments, Loli González and Edén Sorolla for their logistic support, and grant AYA2006-14986-C02 of the Spanish DGCYT for financial support.

References

Thomson, A. R., Moran, J. M., & Swenson, G. W. 1986, *Interferometry and Synthesis in Radio Astronomy*, Wiley, New York.

Wolfram, S., *The Mathematica Book*, 2003, 5th Ed., Wolfram Media

A Giant Step: from Milli- to Micro-arcsecond Astrometry
Proceedings IAU Symposium No. 248, 2007
W. J. Jin, I. Platais & M. A. C. Perryman, eds.

© 2008 International Astronomical Union
doi:10.1017/S1743921308020085

Hipparcos data supporting
the IB school curriculum

K. S. O'Flaherty[1], A. Brumfitt[2] and C. Lawton[3]

[1] Research & Scientific Support Dept., Directorate of the Scientific Programme,
European Space Agency
email: koflaher@rssd.esa.int

[2] Directorate Research and Education Support, Space Qualified Ltd., Australia
email: anne.brumfitt@spacequalified.org

[3] Scientific Projects Dept., Directorate of the Scientific Programme, European Space Agency
email: Christopher.Lawton@esa.int

Abstract. The Hipparcos and Tycho catalogues are ideal data sources for classwork since they provide real data illustrating fundamental astronomical concepts in a simple and easy to manipulate format. In this poster we describe how some of this data is used as support material within the International Baccalaureate (IB) science curriculum. The preparation and deployment of a series of Teacher Notes, specifically constructed to support the IB but applicable to other curricula, is described. Specific attention is given to the Teacher Notes on Stellar Distances which employs data from the Hipparcos and Tycho catalogues.

Keywords. astrometry, sociology of astronomy, education, catalogs

1. Introduction

Europe is facing a crisis in science education with a marked decline in the numbers of students opting to study science, particularly physical sciences (OECD, 2006). The resulting deficit in the number of highly qualified science and technology professionals will have important consequences for Europe's prospects as a knowledge-based society.

A key factor governing the decline in the number of students following science courses arises from the way science is, or has been, taught (Eurobarometer, 2005). Within the traditional classroom-based teaching approach methods supporting Inquiry-Based Science Education (IBSE), in which observation and experimentation play a key role, are now acknowledged as one way of rekindling the interest of a new generation of students. Teachers also need to be inspired and supported in their delivery of the science curriculum since they are the cornerstone of any renewal of science education.

The Teacher Notes project (ESA, 2005) was initiated, after consultation with the International Baccalaureate Organisation, with the aim of creating support material for teachers of the IB Astrophysics option using real data from ESA's science missions, in particular from the Hipparcos mission.

2. The International Baccalaureate science curriculum

The International Baccalaureate school system (http://www.ibo.org) offers a programme of international education, covering the age range 3 to 19. The programmes are taught in English, French, Spanish or Chinese in 2,125 IB schools in 125 countries.

For the Teacher Notes project we focussed on the Diploma Programme for students aged 16 to 19. This is a demanding two-year curriculum leading to final examinations and a qualification that is recognised by leading universities around the world. Science subjects form part of the mandatory curriculum of this programme.

3. Teacher notes - support material for the IB astrophysics option

3.1. *Structure of the teacher notes*

The Teacher Notes closely follow the structure of the IB Astrophysics option. Each topic is introduced and clearly explained using familiar examples, whether from day-to-day life or from other areas of the programme curriculum. Where appropriate, tables of values are given for familiar or standard objects. Clear and simple diagrams are used to illustrate concepts and ideas. Worked examples are provided to familiarise the teacher with using the ESA data, and extended data sets are made available for use in the classroom and for further experimentation by the students. The entire package provides a teacher with the basic information and tools needed to teach the topic. The notes are written in a clear and concise manner, and careful attention has been paid to the language used.

3.2. *Content of the teacher notes*

Within the IB Diploma Programme 15 teaching hours are assigned for the four topics delivered at Standard Level: *Introduction to the Universe, Stellar radiation and stellar types, Stellar distances,* and *Cosmology.* An additional 7 hours are assigned for the two extra topics which are taught at Higher Level: *Stellar processes and stellar evolution,* and *Galaxies and the expanding universe.*

3.3. *Hipparcos data: use and access*

The Hipparcos and Tycho Catalogues (ESA, 1997) are the primary data product of the European Space Agency's Hipparcos mission. They contain fundamental astronomical parameters such as the positions, proper motions, parallaxes, magnitude, variability, multiplicity, etc., for a large number of stars. The data are of immediate application to the astrophysics option of the IB, in particular to the section dealing with stellar distances. The fact that the data are easy to manipulate is an important consideration for the end users. Available as simple ASCII tables, and requiring no special software programs to extract the fields that are needed, makes them appealing to teachers and students alike. The topic of stellar distances is notable for the emphasis on problem solving. Students must (a) determine distances using parallax, (b) solve problems involving ratios of apparent brightness and of apparent magnitudes, (c) solve problems involving distances, apparent brightness and luminosity, and (d) determine distances to Cepheids using the Period-Luminosity relationship. Extracts of Hipparcos data are provided as data sources for these problems allowing teachers to illustrate the concepts, and students to explore them.

Acknowledgements

We acknowledge the key contribution to this project of Jo Turner, science writer.

References

ESA, 1997, *The Hipparcos and Tycho Catalogues,* ESA SP-1200
ESA, 2005, *Teacher Notes,* available online at http://sci.esa.int/teachernotes
Eurobarometer, 2005, *Europeans, Science & Technology,* Special Eurobarometer 224
OECD, 2006, *Evolution of Student Interest in Science and Technology Studies - Policy Report*

A Giant Step: from Milli- to Micro-arcsecond Astrometry
Proceedings IAU Symposium No. 248, 2007
W. J. Jin, I. Platais & M. A. C. Perryman, eds.

© 2008 International Astronomical Union
doi:10.1017/S1743921308020097

The Gaia mission
– a rich resource for outreach activities

K. S. O'Flaherty, J. Douglas and T. Prusti

Research & Scientific Support Department, Directorate of the Scientific Programme,
European Space Agency
email: koflaher@rssd.esa.int; jdouglas@rssd.esa.int; tprusti@rssd.esa.int

Abstract. Space science missions, and astronomy missions in particular, capture the public imagination at all levels. ESA's Gaia mission is no exception to this. In addition to its key scientific goal of providing new insight into the origin, formation, and evolution of the Milky Way, Gaia also touches on many other scientific topics of broad appeal, for example, solar system objects, stars (including rare and exotic ones), dark matter, gravitational light bending. The mission naturally provides a rich resource for outreach possibilities whether it be to the general public, or to specific interest groups, such as scientists from other fields or educators. We present some examples of possible outreach activities for Gaia.

Keywords. astrometry, sociology of astronomy, education, space vehicles

1. Introduction

A number of key papers describing the science of Gaia are included in these proceedings and we will not elaborate on this area except to note that the key science questions addressed by Gaia are of immediate and apparent appeal to many diverse audiences. Gaia is one of Europe's flagship astronomy missions of the next decade. All parties involved have a responsibility to contribute to, or support, outreach activities aimed at promoting the mission, and its discoveries, to the widest audience possible.

What is meant by outreach? Here we consider outreach activities in the broadest of terms: *those which go beyond our normal scope of activity to provide information, resources or services to audiences not in our immediate service area.* The goals of astronomy outreach include: connecting the public to the wonder and excitement of astronomy, cultivating an interest and appreciation of science in students (of all ages), sharing results and discoveries with colleagues, peers and interested parties. The actors who participate, and their roles, are varied. ESA, national funding agencies, scientific institutes, and individual astronomers will all participate in Gaia outreach. Each of these groups has their own constituency, motivation, speciality and body of resources. This will determine the activities in which they participate, and the timing and content of their outreach contributions. To be effective, outreach activities need to be focussed to the interests, expectations and delivery means appropriate for each audience.

One of the immediate challenges for Gaia will be how to represent the results. The problem is encapsulated in this observation: *"Beautiful colour images of the sky are both a blessing and a curse for the communication of astronomy to the public. While undoubtably attractive they can obscure the fact that discoveries are often made in astrophysics using techniques and measurements that are much more difficult to grasp and certainly less appealing to view."* (Fosbury, 2005). The Hubble era has led audiences to expect stunning 'photographs' of the Universe. Careful consideration of how to represent and visualise the Gaia results will be required to respond to audience expectations.

2. Outreach activities for Gaia: some examples

2.1. *Outreach to school children*

The school-going population of today, students aged 5 to 18 years, are the scientists and software engineers of the Gaia catalogue era. In Europe, the decline in interest in science amongst this age group is a concern as it forecasts a deficit of skilled professionals in the decades ahead. Outreach activities which bring the real experience of science to students play a role in efforts to reverse the declining number of scientists. An example of this is exemplified by the *proefstuderen* programme at Leiden University. Pre-university students are invited to attend a half day of lectures with the aim of giving them a sense of what is involved in university study. Anthony Brown (Leiden Observatory) uses material from the Hipparcos and Gaia missions in customized lectures. Descriptions of the missions, explanations of the science, and problem-solving exercises using real data are employed to expose the students to realistic scientific work. The students have the chance to be scientists for a day and at the same time they learn about the mission.

2.2. *A Gaia presence in every European planetarium*

More than 100 million visits to planetaria are registered each year (Petersen, 2005). These visitors are the "motivated, interested public" – a key audience for outreach professionals. Planetaria shows are delivered by talented presenters, planetarians, whose professional role is that of astronomy information disseminator. These are the people who decide what is relevant, interesting and appropriate for their audiences. Much of the material produced by outreach offices on behalf of science missions can be used by planetarians for their theatre presentations (Petersen, 2005). Press releases, images, animations, and background information may influence the content of live or pre-recorded planetarium shows. Ensuring, as a minimum, that outreach material on Gaia is made available to every planetarium in Europe will facilitate the inclusion of the mission in some of the awe-inspiring and enthralling shows that are created. Forging links with local planetarians, and making oneself available to discuss the relevance of recent discoveries could ensure that a significant fraction of visitors to European planetaria are aware of the Gaia mission.

2.3. *The virtual world of Gaia*

Since its inception the internet has evolved from being a means of information exchange to one facilitating the interaction of web users and promoting aspects of social networking. One of the growth areas of the internet is that of the virtual world, for example, Second Life (http://secondlife.com). Outreach professionals have been quick to explore this new outlet: several museums have a virtual presence there, and a number of educational institutions are already running virtual classrooms. Although the number of users is presently small (a few million members) virtual worlds are expected to become a mainstay of internet interaction in the future. Outreach within a virtual world could take many forms: a virtual astronomer interacting with other avatars and giving night-sky tours, the virtual planetarium in which trips into the Gaia 3-d galaxy are one of the featured shows, virtual astronomy courses, ... the possibilities are vast. Those interested in outreach for Gaia should pay attention to developments in the virtual online world and be ready to take up the challenge offered by this new forum for communication.

References

Petersen, C. C., 2005, *The Unique role of the planetarium/ science centre in science*, Communicating Astronomy with the Public 2005
Fosbury, B., 2005, *Difficult Concepts*, Communicating Astronomy with the Public 2005

A Giant Step: from Milli- to Micro-arcsecond Astrometry
Proceedings IAU Symposium No. 248, 2007
W. J. Jin, I. Platais & M. A. C. Perryman, eds.

© 2008 International Astronomical Union
doi:10.1017/S1743921308020103

Astrometry with digital sky surveys: from SDSS to LSST

Ž. Ivezić[1], D. G. Monet[2], N. Bond[3], M. Jurić[4], B. Sesar[1], J. A. Munn[2], R. H. Lupton[3], J. E. Gunn[3], G. R. Knapp[3], A. J. Tyson[5], P. Pinto[6], K. Cook[7,8], SDSS Collaboration and LSST Collaboration

[1] Department of Astronomy, University of Washington, Box 351580, Seattle, WA 98195
email: ivezic@astro.washington.edu

[2] U.S. Naval Observatory, Flagstaff Station, P.O. Box 1149, Flagstaff, AZ 86002 [3] Princeton University Observatory, Princeton, NJ 08544 [4] Institute for Advanced Study, 1 Einstein Drive, Princeton, NJ 08540 [5] Physics Department, University of California, One Shields Avenue, Davis, CA 95616 [6] Steward Observatory, University of Arizona, 933 N Cherry Ave., Tucson, AZ 85721 [7] Lawrence Livermore National Laboratory, 7000 East Avenue, Livermore, CA 94550 [8] KHC's work was performed under the auspices of the U.S. D.O.E. by LLNL under contract DE-AC52-07NA27344

Abstract. Major advances in our understanding of the Universe have historically come from dramatic improvements in our ability to accurately measure astronomical quantities. The astrometric observations obtained by modern digital sky surveys are enabling unprecedentedly massive and robust studies of the kinematics of the Milky Way. For example, the astrometric data from the Sloan Digital Sky Survey (SDSS), together with half a century old astrometry from the Palomar Observatory Sky Survey (POSS), have enabled the construction of a catalog that includes absolute proper motions as accurate as 3 mas/year for about 20 million stars brighter than V=20, and for 80,000 spectroscopically confirmed quasars which provide exquisite error assessment. We discuss here several ongoing studies of Milky Way kinematics based on this catalog. The upcoming next-generation surveys will maintain this revolutionary progress. For example, we show using realistic simulations that the Large Synoptic Survey Telescope (LSST) will measure proper motions accurate to 1 mas/year to a limit 4 magnitude fainter than possible with SDSS and POSS catalogs, or with the Gaia survey. LSST will also obtain geometric parallaxes with accuracy similar to Gaia's at its faint end (0.3 mas at V=20), and extend them to V=24 with an accuracy of 3 mas. We discuss the impact that these LSST measurements will have on studies of the Milky Way kinematics, and potential synergies with the Gaia survey.

Keywords. astrometry, catalogs, surveys

1. Introduction

The astrometric obervations obtained by modern massive digital sky surveys, such as SDSS, 2MASS, and FIRST, are enabling numerous unprecedented studies extending from solar system to extragalactic astronomy. In this contribution we describe several results on the kinematics of Milky Way stars based on proper motions obtained by comparing POSS and SDSS astrometric measurements. The main advantages of this data set are the large sample size and distance limits, as well as excellent photometry that enables accurate metallicity estimates for F/G main sequence stars. All these data characteristics will be signficantly enhanced by LSST thanks to its 2 magnitude fainter imaging limit, and close to a thousand observations in six bandpasses over about half of the sky.

2. The analysis of the Milky Way stellar kinematics based on SDSS

2.1. *An overview of the Sloan Digital Sky Survey*

The SDSS is a digital photometric and spectroscopic survey which will cover about one quarter of the Celestial Sphere in the North Galactic cap, and produce a smaller area ($\sim$300 deg^2) but much deeper survey in the Southern Galactic hemisphere (Adelman-McCarthy *et al.* 2006, and references therein). SDSS provides deep ($r < 22.5$) photometry in five *ugriz* bandpasses with errors of order 0.02 mag for sources not limited by photon statistics. The recent Data Release 6 catalogs 287 million unique objects detected in 9583 deg^2 of sky. A compendium of technical details about SDSS can be found on the SDSS web site (http://www.sdss.org), which also provides the interface for public data access.

The SDSS astrometric reductions are described in detail by Pier *et al.* (2003). Briefly, SDSS absolute astrometric accuracy is better than 100 mas, with relative (band-to-band) accuracy of about 20-30 mas (rms, for sources not limited by photon statistics). In addition to providing positions for a large number of objects with remarkable accuracy (and thus enabling recalibration of other less accurate surveys, as described below), an important characteristic of SDSS astrometric observations is that measurements in five photometric bands are obtained over a five minute long period (with 54 sec per exposure). The multi-color nature allows the discovery of the so-called Color Induced Displacement binary stars (for details see Pourbaix *et al.* 2004), and the time delay allows the recognition of asteroids (Ivezić *et al.* 2001). Here we focus on studies of stellar proper motions enabled by SDSS astrometric measurements.

2.2. *Proper motions determined from SDSS and POSS astrometric observations*

We use proper motion measurements from the Munn *et al.* (2004) catalog, which is based on astrometric measurements from SDSS and a collection of Schmidt photographic surveys. Despite the sizable random and systematic astrometric errors in the Schmidt surveys, the combination of a long baseline ($\sim$50 years for POSS-I survey), and a recalibration of the photographic data using positions of SDSS galaxies, results in median random errors for proper motions of only $\sim$3 mas/year for $g < 19.5$ (per component). Systematic errors are typically an order of magnitude smaller, as robustly determined using $\sim$80,000 spectroscopically confirmed SDSS quasars from Schneider *et al.* (2007). At a distance of 1 kpc, a random error of 3 mas/year corresponds to a velocity error of $\sim$15 km/s, which is comparable to the radial velocity accuracy delivered by the SDSS stellar spectroscopic survey.

The kinematics of the SDSS stellar sample, including mutual consistency of kinematics based on radial velocity and proper motion measurements, are discussed in detail by Bond *et al.* (2008, in prep.). Here we briefly present a few results that are directly related to the conclusions of this paper.

2.3. *Analysis of SDSS-POSS proper motion database*

Most available studies of stellar motions are confined to the solar neighborhood. For example, Hipparcos has enabled detailed mapping of velocity distribution for $\sim$12,000 stars within $\sim$100 pc (Dehnen & Binney 1998). With the SDSS-POSS proper motion measurements, analogous analysis can be extended to a distance limit of 10 kpc, with a sample including close to 20 million main-sequence stars.

In this contribution we limit our analysis to $\sim$330,000 stars observed towards the north galactic pole ($b > 80°$), with $14.5 < r < 20$, and colors consistent with main-sequence stars (Jurić *et al.* 2008). We further split this sample into three subsamples, as follows. The main-sequence stars with F/G spectral types, selected by $0.2 < g - r < 0.4$, are

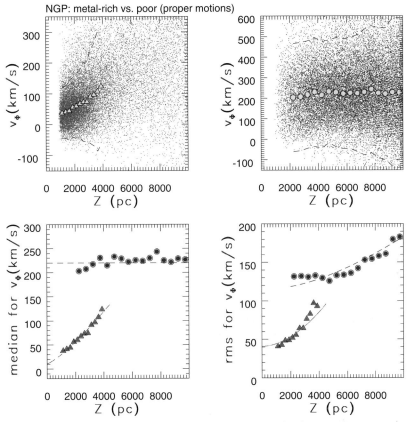

Figure 1. A comparison of the dependence of the rotational velocity component (v_Φ, relative to the Sun, towards $l = 270$) on distance from the galactic plane (Z) for $\sim$21,000 high-metallicity (top left) and $\sim$43,000 low-metallicity (top right) F/G stars. In the top two panels, the small dots show individual stars, and the large symbols show the medians for Z bins. The dashed lines show the 2σ envelope around these medians. The bottom two panels compare the medians (left) and velocity dispersions (right) for the two subsamples. The slightly curved dashed line in the bottom left panel describes velocity shear for disk stars: $v_\Phi = (11 + 19.2(Z/\mathrm{kpc})^{1.25})$ km/s. The dashed lines in the bottom right panel are predictions for the velocity dispersion based on known proper motion errors and an assumption of constant intrinsic dispersions (41 km/s for disk and 115 km/s for halo stars).

separated into 43,000 low-metallicity stars ($[Fe/H] < -1$, approximately $u - g < 1$), and 21,000 high-metallicity stars. We use photometric metallicity estimates from Ivezić *et al.* (2008). The remaining 260,000 main-sequence stars with GKM spectral types, selected by $g - r > 0.4$, represent the third subsample. For all three subsamples, we estimate distance using the "bright" photometric parallax relation from Jurić *et al.* (2008). These distances have random errors of $\sim$10%, with comparable systematic errors for stars with the same metallicity.

The kinematic behavior of high-metallicity and low-metallicity F/G stars is remakably different, as illustrated in Fig. 1. While this difference has been known since the seminal paper by Eggen, Lynden-Bell & Sandage (1962), here it is extended far from the solar neighborhood with $\sim$100 times larger samples. The disk stars display a strong non-linear velocity shear that can be reproduced with all color-selected subsamples. Although such subsamples can have vastly different apparent magnitudes and colors (and thus may be susceptible to different measurement errors), they all produce consistent results for

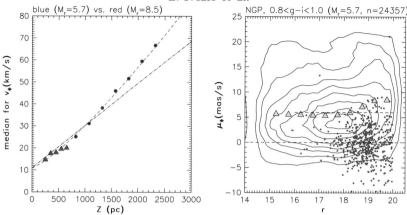

Figure 2. The left panel shows the dependence of the rotational velocity component on distance from the galactic plane for two color-selected samples of disk stars whose M_r differ by ~ 3 mag. The two lines show $v_\Phi = (11 + 19.2(Z/\mathrm{kpc})^k)$ km/s, for $k = 1.0$ (lower, dot-dashed) and $k = 1.25$ (upper, dashed). The right panel shows the rotational proper motion component as a function of apparent magnitude for the bluer sample as contours, with the medians shown by the upper set of large symbols. The non-linearity of the v_Φ vs. Z dependence shown in the left panel is seen here as an upturn of median proper motion for $r > 18$. This upturn is robust, as demonstrated by the quasar sample (dots) whose medians are consistent with zero (the lower set of large symbols).

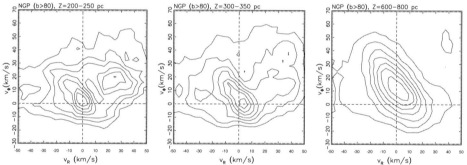

Figure 3. The velocity distribution for GKM dwarfs towards the north galactic pole for three narrow ranges of the distance from the galactic plane (Z, with 3600, 5600 and 29000 stars), as indicated on top of each panel. Note the significant changes of the distribution as Z increases.

the velocity shear, as illustrated in Fig. 2. As the color selection is extended to the more numerous and nearby M dwarfs, the velocity distribution function in the 0.1–1 kpc range can be studied in great detail, as illustrated in Fig. 3. Very close to the plane, the distribution resembles the rich structure seen in the solar neighborhood using Hipparcos data (Dehnen 1998), but we detect significant variation with Z.

3.   The proper motion and parallax accuracy expected from LSST

The revolutionary progress in studies of the Milky Way kinematics enabled by the SDSS-POSS proper motion database will be maintained by the upcoming next-generation surveys, such as Pan-STARRS (Kaiser 2002) and LSST (Tyson 2002). The expected performance by Pan-STARRS is described by E. Magnier in these Proceedings. Here we focus on estimating the accuracy of proper motion and parallax measurements based on anticipated LSST astrometric data.

3.1. *An overview of LSST*

The LSST is currently the most ambitious proposed ground-based optical survey. The main science themes that drive the LSST system design are

(*a*) Constraining Dark Energy and Dark Matter
(*b*) Taking an Inventory of the Solar System
(*c*) Exploring the Transient Optical Sky
(*d*) Mapping the Milky Way

Driven by these science themes, the LSST will be a large, wide-field ground based telescope designed to obtain sequential images covering the entire visible sky from Cerro Pachon in Chile. The current baseline design (details are available at http://www.lsst.org), with an 8.4m (6.5m effective) primary mirror and a 9.6 sq.deg. field of view, will allow about 15,000 sq.deg. of sky to be covered in two photometric bands every three nights (assuming two 15-second exposures per field). The system is designed to yield high image quality as well as superb astrometric and photometric accuracy. The survey area will include 30,000 sq.deg. south from $\delta = +34.5°$, and will be imaged multiple times in six bands covering the wavelength range 0.32–1.05 μm (SDSS-like *ugriz* bands and the *y* band centered on 1 μm). The vast majority (about 90%) of the observing time will be devoted to a deep-wide-fast survey mode which will observe a 20,000 sq.deg. region close to 1000 times (including all bands) during the 10-year survey. The deep-wide-fast survey data will serve the majority of science programs. The remaining 10% of observing time will be allocated to special programs such as Very Deep and Very Fast time domain surveys.

3.2. *Estimates of the proper motion and parallax accuracy*

To estimate the proper motion and parallax accuracy, we use a sophisticated Operations Simulator which creates detailed realizations of the survey sky coverage and depth. The Simulator contains detailed models of site conditions, hardware and software performance, and an algorithm for scheduling observations which will, eventually, drive the robotic observatory.

Observing conditions include a model for seeing derived from an extensive body of on-site MASS/DIMM measurements obtained during site selection and characterization. Weather data are taken from ten years of nightly measurements at nearby Cerro Tololo. The signal to noise ratio of each observation is determined using a sky background model which includes the dark sky brightness in each filter, the effects of seeing and atmospheric transparency, and a model for scattered light from the moon and/or twilight at each observation. The time taken to move from one observation to the next is given by a detailed model of the camera, telescope, and dome. It includes such effects as the acceleration/deceleration profiles employed in moving in altitude, azimuth, camera rotator, dome azimuth, and wind/stray light screen altitude, the time taken to damp vibrations excited by each slew, cable wrap, and time taken for active optics lock and correction as a function of slew distance, filter change, and focal plane readout.

The result of a simulator run is a detailed history over a many-year period of which locations on the sky were observed when, in what filter, and with what sky background, seeing and other observing conditions. The median expected number of visits (two 15-second back-to-back exposures) for the baseline main survey ($\sim$20,000 sq.deg.) is listed in Table 1.

Given the observing sequence for each sky position in the main survey, we generate a time sequence of mock astrometric measurements. The assumed astrometric accuracy is a function of signal-to-noise, SNR. Random astrometric errors are modeled as θ/SNR, with $\theta = 700$ mas. The estimated proper motion and parallax accuracy at the bright end

Filter	u	g	r	i	z	y
Single visit depths (5σ)	23.9	25.0	24.7	24.0	23.3	22.1
Mean number of visits	56	80	184	184	160	160
Final (coadded) depths (5σ)	26.2	27.4	27.6	26.9	26.1	24.8

Table 1. The expected median 5σ depths for point sources per single visit (two 15-second exposures), the median number of visits over a 10-year long LSST baseline survey, and expected depths when all observations are co-added.

($r < 20$, note that $V \sim r$) is driven by systematic errors. Systematic errors of 10 mas are added in quadrature, and are assumed to be *uncorrelated* between different observations. Systematic and random errors become similar at about $r = 22$, and there are about 100 stars per LSST CCD (0.05 sq. deg) to this depth (and fainter than LSST saturation limit) even at the galactic poles.

Pre-cursor data from the Subaru telescope indicate that systematic errors of 10 mas are not too optimistic. Even a drift-scanning survey such as SDSS delivers uncorrelated systematic errors (dominated by seeing effects) at the level 20-30 mas (measured from repeated scans), and the expected image quality for LSST will be twice as good as for SDSS. Furthermore, there are about 1000 galaxies per CCD with $r < 22$, which will provide exquisite control of systematic astrometric errors as a function of magnitude, color and other parameters.

The astrometric transformations for a given CCD and exposure, and proper motion and parallax for all the stars from a given CCD, are simultaneously solved for using an iterative algorithm. The astrometric transformations from pixel to sky coordinates are modeled using low-order polynomials and standard techniques developed at the US Naval Observatory. The expected proper motion and parallax errors for a 10-year long baseline survey, as a function of apparent magnitude, are summarized in Table 2. Blue stars (e.g. F/G) fainter than $r \sim 23$ will have about 50% larger proper motion and parallax errors due to decreased number of the z and y band detections. The impact on red stars is smaller due to relatively small number of observations in the u and g bands, but extremely red objects, such as L and T dwarfs, will definitely have larger errors, depending on details of their spectral energy distribution. As a function of time, the proper motion errors are about five times as large, and parallax errors are about twice as large after the first three years of the survey.

r mag	σ_{xy}^a mas	σ_π^b mas	σ_μ^c mas/yr	σ_1^d mag	σ_C^e mag
21	11	0.6	0.2	0.01	0.005
22	15	0.8	0.3	0.02	0.005
23	31	1.3	0.5	0.04	0.006
24	74	2.9	1.0	0.10	0.009

Notes:
[a] Typical astrometric accuracy (per coordinate); [b] Parallax accuracy; [c] Proper motion accuracy; [d] Photometric error for a single visit (two 15-second exposures); [e] Photometric error for coadded observations (see Table 1)

Table 2. The expected proper motion, parallax and photometric accuracy, as a function of apparent magnitude r, for a 10-year long LSST baseline survey. For comparison, the SDSS-POSS proper motion measurements have an accuracy of ~5 mas/yr per coordinate at $r = 20$. Gaia is expected to deliver parallax errors of 0.3 mas, and proper motion errors of 0.2 mas/yr at its faint end at $r \sim 20$. LSST will smoothly extend Gaia's error vs. magnitude curve to a 4 magnitude fainter level.

4. Discussion and conclusions

The SDSS-POSS proper motion database developed by Munn *et al.* (2004) is a powerful tool to study the Milky Way kinematics. The accuracy of ~ 3 mas/yr for tens of millions of stars with $r < 20$ is not only a quantitative, but also a major qualitative step forward. This accuracy to such a depth enables the extension of kinematic studies beyond the solar neighborhood, and the availability of a large number of stars (essentially a complete flux-limited sample in a given region of the sky) offers an unprecedented spatial resolution. We presented here only a small selection of kinematic results that are based on this dataset, and discussed in detail by Bond *et al.* (2008). These results vividly demonstrate the power of massive and accurate proper motion measurements to faint flux levels for understanding kinematics, dynamics and the assembly history of the Galaxy.

As suggested by its expected proper motion and parallax accuracy listed in Table 2, LSST will enable major breakthroughs in the mapping of stellar motions by delivering the Hipparcos accuracy to the SDSS depth for hundreds of millions of stars. For example, numerous main sequence F/G stars with $0.2 < g - r < 0.4$ have $M_r \sim 4$. LSST will detect about 100 million such stars with $r < 22$ (Ivezić *et al.* 2008). At $r = 22$, they probe the outer halo at distances of ~ 40 kpc, and LSST will measure their tangential velocity with errors as small as ~ 50 km/s. Neither SDSS-POSS proper motions, nor expected Gaia data, can be used to perform such massive and accurate kinematic measurements in the outer halo. At the red end of the main sequence, M dwarfs with $M_r = 12$ (i.e. the peak of the main-sequence luminosity function) will have tangential velocity errors of 1 km/s at 1 kpc ($r = 22$), and 12 km/s at 2.5 kpc ($r = 24$). For sub-stellar objects with $r = 22$, geometric parallax measurements will yield 10% accurate distances for *all* objects brighter than $M_r = 17$ within a 100 pc half-sphere. We estimate that this sample will include about 10,000 objects with $16 < M_r < 17$ ($r < 22$), that will provide unprecedentedly robust and accurate measurement of the stellar luminosity function at its faint end.

LSST and Gaia will be highly complementary surveys. Gaia will obtain much more accurate measurements for $r < 20$, but LSST will extend them 4 magnitudes deeper. Thanks to the 3-4 magnitudes of overlap between the surveys, Gaia's parallax measurements will be used to train photometric parallax estimators as a function of effective temperature and metallicity, and perhaps gravity, and Gaia's proper motion measurements will be used to verify the accuracy of measurements by LSST.

Ž.I. and B.S. acknowledge support by NSF grants AST-0551161 and AST-0707901.

References

Adelman-McCarthy, J. K., Agüeros, M. A., Allam, S. S., *et al.* 2006, ApJS, 162, 38
Dehnen, W. 1998, AJ, 115, 2384
Dehnen, W. & Binney, J. J. 1998, MNRAS, 298, 387
Eggen, O. J., Lynden-Bell, D., & Sandage, A. R. 1962, ApJ, 136, 748
Ivezić, Ž., Tabachnik, S., Rafikov, R., *et al.* 2001, AJ, 122, 2749
Ivezić, Ž., Sesar, B., Jurić, M., *et al.* 2008, submitted to ApJ
Jurić, M., Ivezić, Ž., Brooks, A., *et al.* 2008, ApJ, in press
Kaiser, N., Aussel, It., Burke, B. E., *et al.* 2002, in "Survey and Other Telescope Technologies and Discoveries", Tyson, J. A. & Wolff, S., eds. Proceedings of the SPIE, 4836, 154
Munn, J. A., Monet, D. G., Levine, S. E., *et al.* 2004, AJ, 127, 3034
Pier, J. R., Munn, J. A., Hindsley, R. B., *et al.* 2003, AJ, 125, 1559
Pourbaix, D., Knapp, G. R., Szkody, P., *et al.* 2005, A&A, 444, 643
Schneider, D. P., Hall, P. B., Richards, G. T. *et al.* 2007, AJ, 134, 102
Tyson, J. A. 2002, in *Survey and Other Telescope Technologies and Discoveries*, Tyson, J. A. & Wolff, S., eds. Proceedings of the SPIE, 4836, 10

A Giant Step: from Milli- to Micro-arcsecond Astrometry
Proceedings IAU Symposium No. 248, 2007
W. J. Jin, I. Platais & M. A. C. Perryman, eds.
© 2008 International Astronomical Union
doi:10.1017/S1743921308020115

Astrometric data for NEAs extracted from the infrared DENIS survey

W. Thuillot[1], J. Berthier[1], J. Iglésias[1], G. Simon[2] and V. Lainey[1]

[1]Institut de mécanique céleste et de calcul des éphémérides, IMCCE-Paris Observatory
77, av. Denfert Rochereau, 75014, Paris, France
email: thuillot@imcce.fr

[2]GEPI, Paris Observatory
61 av. de l'Observatoire , Paris, France
email: Guy.Simon@obspm.fr

Abstract. When astrometric data can be extracted from archives, this generally allows us to get very strong constraints for the orbital modeling of Solar System objects. This is particularly important for Near-Earth Asteroids. We have developed tools in the Virtual Observatory framework in order to carry out such a task. We have applied them to the DENIS survey. This survey has been performed from 1995 to 2001 in the I, J, K' spectral bands with a 1m telescope at ESO La Silla. Many sources associated to Solar System Objects have been identified and we present our preliminary results.

Keywords. astrometry, surveys, asteroids

1. Introduction

The orbital models of Near Earth Objects (NEO) can be drastically improved when a long enough time of astrometric observations is available for a fit. In case of detection of new objects, we know that any old observation can be very important for this improvement and consequently for estimating the risks of collision. Nowadays, the numerous sky surveys available provide with a huge amount of astronomical observational data for which their mining requires fast and reliable processing tools in order to detect and identify Solar System objects, in particular to seek for precovery observations. With this goal, we have developed some specific tools within the Virtual Observatory (VO) framework: SkyBoT, and its data-mining pipeline counterpart AstroId wich performs an automatic identification of the Solar System objects present in astronomical archives. We have applied this VO workflow on the sources catalogue of the infrared survey DENIS, obtaining valuable astrometric results related to many asteroids.

2. The DENIS survey

The DENIS infrared survey was operating between 1995 and 2001 from the 1 meter telescope at ESO La Silla (Epchtein *et al.* 1999). Three bands were used -I, J K'- and a specific observation strategy was established in order to get a wide scan of the southern sky: 5206 strips (a column-like scan of the sky at a fixed right ascension), each containing 180 12x12 arcmin frames. The limiting magnitudes were 18.5 (I), 16.5 (J) and 14 (K'). As a result of the survey, a large catalogue of sources (around 355 mill. point sources) was generated and made available. The initial goal of the DENIS survey was to produce one of the first large infrared catalogues directly digitized, allowing the study of several astrophysical objects, in particular the distribution of various stellar populations in our Galaxy and the research on the "missing mass". Initially, it was not addressed for the study of Solar System Objects and much less for the astrometry or the dynamics of such objects. Nevertheless, a preliminary study done in 2001 (Baudrand *et al.* 2001) and 2004 (Baudrand *et al.* 2004) showed that valuable asteroids observations could be extracted from it despite its low astrometric accuracy at about 1 arcsec. However, this work was very limited due to the lack of specific, automatic tools and to the low number of asteroids known at that time.

3. SkyBoT, a VO tool for Solar System objects identification

The fast and accurate identification of Solar System objects on a given stellar field at a certain epoch is not a trivial problem. In order to give an answer to that question, we have developed a powerful tool -SkyBoT (Sky Body Tracker)- within the Virtual Observatory (VO) framework. This tool is based on the pre-computation of a large database of asteroid ephemerides. To this date, more than 380 000 asteroids have been identified. Their orbital elements are regularly computed and stored on the publicly-accessed "Astorb" database from the Lowell Observatory (Bowel 2007). SkyBoT computes the asteroids ephemerides in advance for a period of time within 1949-2009, and then a PHP/MySQL software deals with these data. First, it divides the celestial sphere in boxes of about 20 arcmin in size and determines in advance all the asteroids present in each box on a 10-days interval basis. The equatorial and heliocentric coordinates are associated to each box and asteroid for further interpolations. This computation is based on an accurate numerical integration weekly updated in sync with the update of the Astorb database of orbital elements. Furthermore, we also provide ephemerides for the planets and for 33 of their natural satellites by means of an on-line connection to their dynamical models. Therefore, a query providing the date (on the 1949-2009 interval) and the coordinates of the FOV center, the characteristics of the Solar System objects identified in that field are returned after a few seconds of time (depending on the FOV size). SkyBoT implementation as a web service permits any external software using standard VO protocols to include it as part of its VO workflow. It has already been implemented in Aladin -the CDS (Centre de Donnes de Strasbourg) Sky Atlas (Bonnarel *et al.* 2000)- and is publicly available through this software since 2006, as well as through the address skybot.imcce.fr

4. AstroId, a VO Workflow for data mining

One of the key concepts behind the Virtual Observatory framework is interoperability, which in fact makes this platform well adapted for instance to cross correlations of data between multiple VO users or VO data centers. In this context, we have developed a pipeline based on the use of the SkyBoT tool. This VO workflow, called AstroId, furnishes the users with the capability of data-mining extensive archives with the purpose of identifying the Solar System objects (SSO) contained in them. AstroId's input comprises, on one hand and for SkyBoT identification purposes, information on the date, coordinates of the centre and FOV of the different images from the archive; and, on the other hand, the extracted sources catalogue for correlation purposes. This correlation part consists of a cross-matching of "SSO candidates" with point sources by means of a Chi-square statistical test (based on angular distance as well as astrometric accuracies) and a comparison of magnitudes (although this latter will require a more extensive development in the future). At this point the user disposes of a certain amount of choice to tune up the correlation based on his particular needs. The "SSO candidates" come from

Table 1. Number of asteroids detected in the analysis of the DENIS survey.

	Number of associations	Number of objects
Correlated objects	16 993	15181
Correlated NEO	308	273
Precovery objects	9975	9221
Precovery NEO	198	176

the catalogued celestial bodies that our VO web service SkyBoT furnishes as an output for each FOV (ephemeris computation). Moreover, on this correlation module we also do a pre-matching of sources with "non-SSO" objects in order to eliminate some stars from the cross-matching pipeline. For these "non-SSO" objects we have made use of the NOMAD catalogue (a compilation of the Hipparcos, Tycho-2, UCAC2 and USNO-B1.0 catalogues) accessed through the VizieR VO service from the CDS. Finally, as for the output, AstroId provides a set of lists (VOTable documents) containing the identified and unidentified SSO with respect to their reference sources.

5. Mining the DENIS survey: first results

We have applied a first prototype of AstroId to explore the DENIS survey in search for Solar System objects. In this sense, the previous work done by Baudrand (Baudrand *et al.* 2004) provided some promising results by detecting 1931 asteroid positions from the DENIS sources catalogue thanks to conventional means and a limited list of 9000 known objects. In comparison, our workflow has made use of 40 times more catalogued SSO as well as of several stellar catalogues. This work has allowed us to detect numerous asteroids and, in particular, various Near-Earth asteroids. Our results are given in Table 1 where we show the number of observations (point sources) associated to SSO (among those catalogued in SkyBoT) and the resulting number of SSO identified (once we take into account the multiple observations).

These results are giving us the opportunity to improve the dynamical models of several asteroids, specially those that have been recently discovered and have not been observed for a long period of time. While this application is still in progress, we have already made multiple tests on several detected objects. Let us consider for the sake of illustration, the main belt asteroid 2005 ED93. On one hand we have identified 3 DENIS observations of this body (all dating back from 1998) and, on the other hand, we have had at our disposal about 51 other astrometric observations made between 2003 and 2007. Our dynamical model fitted on these 51 old observations gave a preliminary orbit in large discrepancy with the DENIS observations, with the O-C being around 5-6 arcsec in right ascension and declination at the time of the DENIS observations (1998). Therefore these new observations furnished by DENIS appear to be a strong constraint for the orbit as the comparison between the preliminary orbit and the orbit fitted on the whole set of observations exhibits differences in R.A. and declination as large as 8 arcsec in 1999 and 2002. These differences correspond to a shift of around 12000 km in space in 10 years of time. We note that the discrepancy is small in the time period where observations were already available and so used in the model's fit. Therefore, this example perfectly illustrates the usefulness of data mining on improving orbits of asteroids and in particular for NEOs.

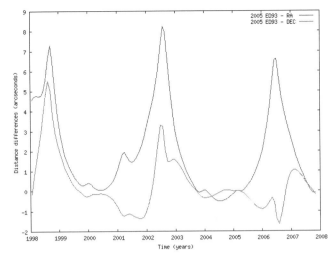

Figure 1. Asteroid 2005ED93: Shift in right ascension and declination between the orbital model fitted on the observations available before this analysis and the orbital model obtained by including the DENIS observations. This shift corresponds to about 12,000 km over a 10-year period

6. Conclusion

Thanks to a set of Virtual Observatory compliant tools, we have explored the DENIS sources catalogue, performing an identification of the Solar System objects present in that survey. The first of this tools -SkyBoT- is a VO web service which can be implemented on a user's software and which is also publicly available either through our web site or through the CDS service Aladin. The second tool -AstroId- which is still in the prototype phase, is also aimed to become a publicly available service. AstroId is a VO workflow dedicated to the data mining of large surveys in search for Solar System Objects (and so it makes use of SkyBoT). In this case, we have applied AstroId to exploit the DENIS infrared survey which was done on the 1995-2001 period at ESO-La Silla. As a result, we have detected 15181 Solar System objects, including 273 Near-Earth Objects including about around 170 NEO precoveries which must be now confirmed. The first test that we have made to test the applications of this concept has proved to be very encouraging as the new data discovered has provided some strong constraints that can be applied to the orbital modeling of these objects.

References

Baudrand, A., Bec-Borsenberger, A., Borsenberger, J., & Barucci, M. A. 2001, *AA*, 375, 275

Baudrand, A., Bec-Borsenberger, A., & Borsenberger, J. 2004, *AA*, 423, 381

Berthier, J., Vachier, F., Thuillot, W., Fernique, P, Ochsenbein, F., Genova, F., Lainey, V., & Arlot, J.-E. 2006, Astronomical Data Analysis Software and Systems XV 2005, Madrid, El Escorial, *ASP Conference Series* 351, 363

Bonnarel, F., Fernique, P., Bienayme, O., Egret, D., Genova, F., Louys, M., Ochsenbein, F., Wenger, M., & Bartlett, J. G. 2000, *AAS* 143, 33

Bowell, E. 2007, Astorb database on-line at the Centre de donnes de Strasbourg *Vizier, CDS*

Epchtein, N., Deul, E., Derriére, S., *et al.* 1999, *AA*, 349, 236

Thuillot, W., Vaubaillon, J., Scholl, H., Colas, F., Rocher, P., Birlan, M., & Arlot, J.-E. 2005 *Comptes Rendus Acad. des Science Paris, Ser. B Sciences Physiques* 6, 327

Thuillot, W., Berthier, J., Vachier, F., *et al.* 2005, *AAS, DPS meeting* 37, 14.10

A Giant Step: from Milli- to Micro-arcsecond Astrometry
Proceedings IAU Symposium No. 248, 2007
W. J. Jin, I. Platais & M. A. C. Perryman, eds.

Astronomical databases of Nikolaev Observatory

Y. Protsyuk and A. Mazhaev

Nikolaev Astronomical Observatory, Ukraine
email: yuri@mao.nikolaev.ua

Abstract. Several astronomical databases were created at Nikolaev Observatory during the last years. The databases are built by using MySQL search engine and PHP scripts. They are available on NAO web-site http://www.mao.nikolaev.ua.

Keywords. astronomical data bases: miscellaneous, catalogs

1. Development of the astronomical databank

Development of the astronomical databank of NAO has begun since 1995, the year of the first CCD observations with the Axial Meridian Circle. Some astrometric catalogues and data of the CCD observations were included into this databank in 1995 (Kovalchuk *et al.* 1997).

A local area network has considerably grown since 1995, and today it includes about 50 PC in four separate buildings. Two dedicated servers manage the network. Control computers of three CCD telescopes, such as the Axial Meridian Circle (AMC), the Multi Channel Telescope (MCT), the Fast Robotic Telescope (FRT) are also connected to the network.

All obtained data and results of data processing are added to the common databank of NAO (Protsyuk *et al.* 2005, Pinigin *et al.* 2005). The daily average volume of the new astronomical information obtained from the CCD instruments makes from 300MB up to 7GB, depending on the purposes and conditions of observations. The total volume of information was about 340GB in the middle of 2007 (Fig. 1). The total volume of information obtained from other sources such as astronomical catalogues makes about 50GB.

The data of CCD observations obtained from the telescopes of NAO and other observatories (Fig. 2) has been handled in automatic and semi-automatic mode for determination of coordinates of observed stars, minor planets and other objects since 1996 (Kovalchuk *et al.* 1997, Protsyuk 2000).

All received data are stored on servers in two copies, namely, working and backup. Also we make two copies of observational data as archives on DVD. One copy is available for users. Another copy is stored in a central storage of NAO. All raw observational data have been stored in FITS format since 1998. Also, the databank of NAO includes the results of data processing as pixel coordinates for all star-shaped objects captured on the CCD frames.

The network has been connected to the Internet via the first dedicated line since the end of 2004 and the second one since the middle of 2007 (Fig. 3). It allows us to carry out fast transmission of raw observational data through the Internet.

The glass library of NAO contains more than 8000 photographic plates. We have started a program of digitizing of photographic plates with resolution of 2400DPI for

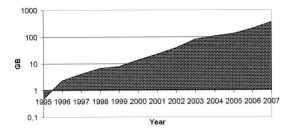

Figure 1. Growth of data volume obtained during CCD observations, GB

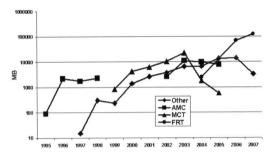

Figure 2. Volume of data obtained from different telescopes, MB

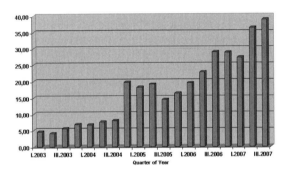

Figure 3. Volume of the Internet traffic, GB

astrometry and 600DPI for preview since the beginning of 2007 (Fig. 4). Now we have about 1000 images for preview and 150 for astrometry.

We have also started digitizing and optical character recognition of printed catalogues and scientific publications (Fig. 5). 27 stellar catalogues were digitized and included in the database of catalogues.

2. Development of the astronomical databases

Development of astronomical databases has started since 2004. Several astronomical databases were created during last years. They are available on NAO web site *http : //www.mao.nikolaev.ua* (Fig. 7):
- The databases of photographic and CCD observations;
- The database of astrometric catalogues;
- The catalogue of artificial satellites and space debris;
- The database of ionosphere sounding.

Figure 4. Sample of digitized photoplate

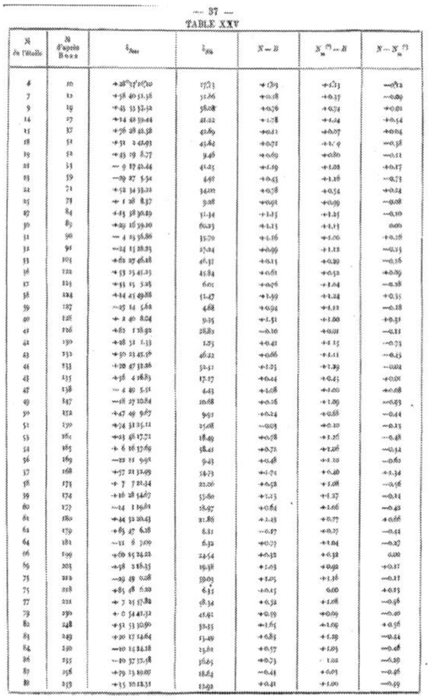

Figure 5. Sample of digitized catalogue

Overall size of databases is about 200MB. Development of new databases is continuing. The databases allow users to search information about observations by using seven different graphical interfaces. The databases are built by using MySQL search engine and PHP scripts.

Now these databases contain information about CCD observations obtained in $1996 - 2006$ and photographic observations obtained in $1929 - 1931$ and $1961 - 1999$ (Fig. 6). The database of catalogues can be connected to interactive client side application such as Aladin allowing the user to visualize data in the field of search. The catalogues can also be visualized in the VOTable or ASCII formats by using such application as TopCat, which allows the user to carry out wide range of data processing.

The databases of photographic and CCD observations allow users to search information by using five different graphical interfaces. The search is possible by coordinates or by

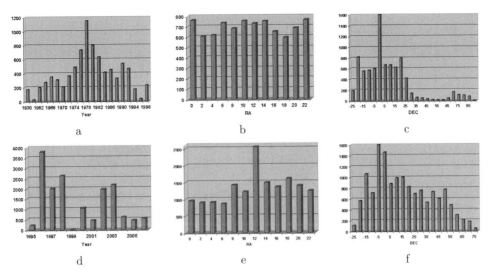

Figure 6. Distribution of photoplates (top) and CCD frames (bottom) by year of observation (a,d), RA (b,e) and DEC (c,f) in databases of photographic and CCD observations.

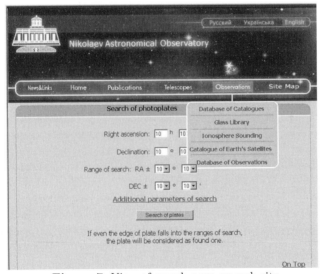

Figure 7. View of search page on web site

object. The databases of photographic and CCD observations allow users to search plates by using the center of search and ranges of search in right ascension and declination. The database of photographic observations allows users to search plates by using the period of observations and observational campaign. The database of CCD observations allows users to search CCD images by using the period and objects of observations. The database of observations also allows users to search plates and CCD frames simultaneously.

The database of astrometric catalogues allows users to search information by using the coordinate center and radius of search.

Database of ionosphere sounding has been recorded since 2002. The graphical data is updated every 5 minutes and is available for online search (Fig. 8).

The Earth satellites in catalogue are separated in two groups: Geostationary Earth Orbit (GEO) satellites and space debris; Low Earth Orbit (LEO) satellites. GEO

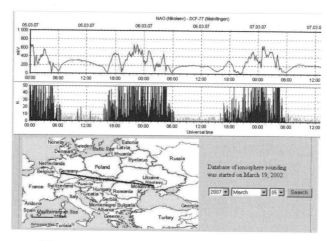

Figure 8. View of ionosphere sounding record

satellites and space debris were observed with the Multi-Channel Telescope in 2002−2004 and with the Fast Robotic Telescope in 2004 − 2007. LEO satellites were observed with the Fast Robotic Telescope in 2004 − 2007.

Several our catalogues were included into the international databases, such as the CDS in Strasbourg, the IMCCE in Paris, the WFPDB in Sofia. The database of astrometric catalogues was recorded in the registry of the NVO in the USA. We hope, that in the near future all our databases will be integrated to the registries of the International Virtual Observatory.

References

Kovalchuk, A. N., Pinigin, G. I., Protsyuk, Yu. I., & Shulga, A. V. 1997, *Journees 1997 Systemes de Reference Spatio-Temporels. Prague.*, 14-17

Protsyuk, Yu. 2000, *Baltic Astronomy*, 9, 554-556

Protsyuk, Yu., Pinigin, G., & Shulga, A. 2005, *Kinematics and Physics of Celestial Bodies, Suppl. Ser.*, 5, 580-584

Pinigin, G., Protsyuk, Yu., & Shulga, A. 2005, *Romanian Astronomical Journal, Suppl.* , 15, 51-56

A Giant Step: from Milli- to Micro-arcsecond Astrometry
Proceedings IAU Symposium No. 248, 2007
W. J. Jin, I. Platais & M. A. C. Perryman, eds.
© 2008 International Astronomical Union
doi:10.1017/S1743921308020139

The extended solar neighborhood: precision astrometry from the Pan-STARRS 1 3π Survey

E. A. Magnier[1], M. Liu[1], D. G. Monet[2], K. C. Chambers[1]

[1]Insitute for Astronomy, University of Hawaii, 2680 Woodlawn Dr., Honolulu, HI 96822, USA
email: eugene@ifa.hawaii.edu
email: mliu@ifa.hawaii.edu
email: chambers@ifa.hawaii.edu

[2]USNO, Flagstaff Station, NM, USA
email: dgm@usno.navy.mil

Abstract. The Pan-STARRS pathfinding telescope PS1 will begin a major set of surveys starting in 2008, and lasting for 3.5 years. One of these, the PS1 3π Survey, will repeatedly observe the entire sky north of -30 degrees, visiting every position 12 times in each of 5 filters. With single-epoch astrometry of 10 milliarcseconds, these observations will yield parallaxes for stars within 100 pc and proper motions out to several hundred pc. The result will be an unprecedented view on nearby stellar populations and insight into the dynamical structure of the local portions of the Galaxy. One exciting science product will be a volume-limited sample of nearby low-mass objects including thousands of L dwarfs, hundreds of T dwarfs, and perhaps even cooler sub-stellar objects. Another project will use proper-motion measurements to improve the membership of nearby star forming regions.

Keywords. astrometry, surveys, stars: low-mass, brown dwarfs, stars: formation

1. The Pan-STARRS project & PS1

The Pan-STARRS project, led by the University of Hawaii, is developing a unique optical survey instrument (PS4) consisting of four co-aligned wide-field telescopes. Each telescope has a 1.8 m diameter primary mirror and a 7 degree^2 field-of-view, imaged with a well-sampled 1.4 Gigapixel camera. This instrument is planned to be completed on Mauna Kea in the 2010 timeframe. As a pathfinder for the PS4 system, the Pan-STARRS project has built a single telescope system (PS1) on the summit of Haleakala on the Hawaiian island of Maui, consisting of a full-scale version of one of the PS4 telescope systems.

PS1 was originally envisioned as a test-bed for the commissioning, testing, and calibration of the Pan-STARRS hardware and software in anticipation of the full Pan-STARRS 4 telescope array. Recognizing the excellent survey capabilities of the PS1 Telescope, the Pan-STARRS project has made the science survey from PS1 a priority, and has formed the PS1 Science Consortium to guide the science observations and extract the scientific results from this telescope. Joining the University of Hawaii Institute for Astronomy in the PS1SC are the Max Planck Institute for Astronomy, the Max Plank Institute for Extraterrestrial Physics, the Johns Hopkins University, University of Durham, University of Edinburgh, Queen's University Belfast, the Harvard-Smithsonian Center for Astrophysics, and the Las Cumbres Observatory Global Telescope.

2. The surveys & the PS1 3π survey

The baseline PS1 survey plan consists of several science surveys, each with a trade off between coverage area and frequency (and/or depth) of observations. Three PS1 survey programs will perform repeated observations of a total of 14 fields, with science goals of detecting supernovae, planetary transits, and for one field, microlensing and variables in M31. One survey program will observe the sky in the direction of the Earth's orbit to detect near Earth asteroids. The majority of the available time (56%) will be dedicated to the PS1 3π Survey, a multi-epoch, multi-filter survey of the observable sky from Haleakala, roughly the 3π steradians north of -30 degrees Declination. Many science projects will make use of this survey, and the survey parameters have been carefully chosen to optimize between these different goals. These science goals include a survey of the Solar System (which motivates much of the timing), an examination of the local solar neighborhood (which motivates the large time-baseline observations and use of the reddest filters), and a large-scale galaxy and weak-lensing survey (which motivates the depth and color choices). The PS1 Survey Mission is expected to run for 3.5 years, starting soon after the telescope is commissioned in early-2008.

The PS1 3π Survey will be a major advance in several respects over existing large-scale surveys. The depth in a single visit will equal or exceed the Sloan Digital Sky Survey with the added dimension of multi-epoch observations. Each field will be observed 12 times in each of the 5 stellar filters ($grizy$), with a 5σ detection limit in the range 20.6–23.2 depending on the filter (see Table 1). The survey cadence is designed to enhance the detection of high-proper motion stars and the measurement of the trignometric parallaxes. A given field is observed 20 times within the course of a year, with observations distributed to enhance the detection of the parallax. The gri filters, which are most sensitive to the solar system objects, will be observed within the 3 months surrounding opposition, i.e., observed near the meridian within 2 hours of midnight. The reddest filters (zy), which are most sensitive to the low-mass stars and substellar objects of primary interest in the local solar neighborhood, will be used to observe fields between 70 and 90 degrees from opposition. These are observed near the meridian in the hours after sunrise and before sunset, and thus maximize the parallax factor for these fields.

3. Astrometry expectations

The goals of the PS1 survey include relative astrometry with an error floor of 10 milliarcseconds. These values are consistent with our measurements from data taken at other sites, combined with the seeing expectations for Haleakala. We have used a variety of test data to evaluate the analysis process and to test the accuracy achieved in similar systems and circumstances. A major component of the test data consists of observations from the CFHT MegaPrime 1 square degree camera, which approaches the PS1 Gigapixel Camera in scale.

Table 1. Predicted PS1 zero-points, sky brightness, magnitude limits.

filter	λ (nm)	$\Delta\lambda$ (nm)	m_0 (mag)	μ (mag asec^{-2})	Magnitude at S/N 100	25	5
g	480	128	24.9	21.9	20.0	21.4	23.2
r	620	137	25.1	20.9	19.4	20.9	22.7
i	750	126	25.0	20.1	19.4	20.9	22.6
z	870	102	24.6	19.3	18.4	19.9	21.6
y	990	57	23.0	18.0	17.0	18.4	20.1

The astrometric analysis of data from a large-scale mosaic camera offers interesting challenges, as well as possibilities. With a mosaic behind a single optical system, we have the opportunity to use the large number of sources in the wide field to constrain the single optical distortion pattern. The challenge in this analysis is to converge on a solution for the detector positions, rotations, and possible tips, tilts, or higher order shape terms, such as warps, while also solving for the optical distortion. It is difficult to solve such a minimization problem directly because of the degeneracy between the per-chip terms and the optical distortion. We have approached this problem by measuring instead the local gradient of the optical distortion term, which is substantially less sensitive to the chip positions.

At the current time (Nov 2007), the PS1 telescope is in the process of being commissioned. The Gigapixel Camera obtained the first-light images on Aug 23, 2007. The most crucial and challenging aspect of the PS1 commissioning is the determination of the proper collimation and alignment of the telescope optics and camera. The wide-field optical design is very sensitive to the collimation, and precise mechanical alignment of the components *a priori* is not possible. The PS1 telescope has 5 axis control over the secondary mirror and 4 axis control over the primary mirror (displacement along the altitude axis is not active). The collimation process employs the observation of star fields with the M1 and M2 positions dithered about a local optimum. These observations are then compared with ray tracing models of the optical system in an attempt to predict the observed stellar shape variations, and thus predict the required motions of the other, uncontrolled elements, such as the corrector lens placement. Once confidence is obtained for a particular predicted move of one of the elements, the mechanical work needed shift or tilt or change the lens spacing may be performed. This work may require removing the camera or machining small components, and so requires 1 or 2 weeks of effort. It is expected to require several months to complete this somewhat laborious iterative process.

At the time of this writing, the image quality from PS1 was still substantially affected by aberrations as well as mirror seeing due to the unfinished cooling system. For example, a typical 'good' set of observations of the Pelican Nebula have stellar images near the field center with roughly Gaussian profiles and FWHM of 1.5–2.0 arcseconds, but which degrade to distended, irregular, multi-peaked shapes in the outer portions of the camera. After collimation and alignment is complete, the optical performance is expected to be dominated by the atmosphere, with an expected median seeing of ~ 0.8 arcseconds. Nonetheless, with the existing data, we can already demonstrate relative astrometry residuals between pairs of images which approach our error goals. In sample tests, we have obtained relative astrometry residuals between pairs of PS1 images across 50 of the 60 chips ranging from 35–50 milliarcsec. The remaining devices were either excessively aberrated for a solution or were contaminated by excessive dark current (also being tuned in commissioning). Given the poor image quality (and the factor of $\sqrt{2}$ increase in the noise due to the comparison of two images), these numbers are very much in line with our goal of 10 mas per observation.

We have performed simulations of our ability to measure parallax and proper motion for sources observed in the PS1 3π Survey, given the 10 mas error floor and the planned survey observing cadence. With the full set of 60 observations available at the end of the 3.5 year survey, we expect to measure proper motion with an accuracy of 1.2 mas/year, and parallaxes of 1.5–2.2 mas, depending on the ecliptic latitude. We have also examined the behavior of the proper-motion and parallax errors as we modify the available observations. For example, for those objects which are only detected in the y-band observations (e.g., ultracool brown dwarfs), the parallax accuracy is reduced because of the smaller number of detections to 2.9 mas.

4. The extended Solar neighborhood from PS1 astrometry

The PS1 3π Survey will be a transformational data set for a wide range of science areas. This paper addresses only the benefits for the study of the extended solar neighborhood, the region within a few hundred parsecs of the Sun, and in particular the low-mass and substellar populations. Some of the other topics, beyond the scope of this article, that will benefit from the improved astrometry resource include the studies of the Galactic dynamical structure, the large-scale streams and the Galaxy formation and evolution, and the impact on improved orbits for objects within our own Solar System. Other important topics within the context of the extended solar neighborhood include the discovery of nearby white dwarfs and the membership of moving groups. In this article, we will focus on two particular science topics: the parallax-selected field dwarf sample and the study of star forming regions via proper-motion selections.

4.1. *Field dwarf parallax sample*

The first generation of digital wide-field surveys (SDSS, 2MASS, DENIS) have contributed most of the discoveries of brown dwarfs and very low-mass stars to date (Delfosse *et al.*, 1999; Kirkpatrick *et al.*, 2000; Hawley *et al.*, 2002; Burgasser *et al.*, 2004). Although these ground-breaking surveys were highly sensitive and covered very large areas on the sky, the low luminosity and very red colors of the coolest dwarfs limit the total number of detected objects. The degeneracy of their optical and/or IR colors with reddened background giants and/or ordinary higher mass stars is also a major hindrance, introducing substantial contamination. In order to find an interesting number of confirmed very low-mass objects, color-selected samples must be further filtered, with proper motions, photometric distance indicators, or spectroscopy. As a result, the efficiency of these searches is limited, and require large amounts of follow-up time. Follow-up parallax measurements for these objects have been particularly time consuming.

The astrometric capabilities of the PS1 3π Survey will allow us to minimize the biases in the observational selection process, and to construct a very large sample of very low-mass stars and brown dwarfs. With such a sample, we will be able to address several key areas in our understanding of these ultracool objects. The multiple epochs and high astrometric precision will make the PS1 3π Survey the most sensitive single large-scale survey for parallax and proper-motions. The PS1 3π Survey will cover a much larger area than the SDSS, and will have a higher sensitivity to very low-mass stars and brown dwarfs as a result of its deeper magnitude limits and very red y-band filter. At 1μm, the PS1 system is roughly a factor of 10 more sensitive than SDSS. PS1 will be orders of magnitude more sensitive than the USNO-B survey for these red objects, and will be significantly more sensitive than the 2MASS survey for all but the reddest T dwarfs. Finally, the PS1 3π Survey will have very stringent photometric calibration requirements, yielding extremely high-quality magnitudes and colors.

By combining parallax and proper-motion measurements with precise photometric selections, the PS1 3π Survey will yield a wide-area sample of very-low-mass stars and brown dwarfs with an unprecedented level of characterization and reliability. Not only will this sample increase the number of known L and T stars by an order of magnitude, yielding thousands of L dwarfs and hundreds of T dwarfs with measured parallaxes (compared with the current total number of less than a 100), but their identifications will be extremely reliable and their basic properties will be well-determined largely without additional follow-up. The PS1 3π Survey will thus yield a gold-mine of an unbiased, volume-limited sample of very low-mass stars and brown dwarfs for a huge range of studies into the properties and evolution of ultracool objects.

4.2. *Star formation regions*

The nearby star forming regions represent an important keystone in the study of the star formation process. By studying the nearby star forming regions we observe a collection of objects formed within a common context, as with any cluster study. More importantly, the nearest clusters are sufficiently close that we can examine the faintest, lowest-mass objects in relation to the higher mass populations and apply detailed spectroscopic follow-up observations to these low-mass sources. Our current understanding of these regions is hampered by a variety of observational selection effects and biases. In particular, the completeness of the low-mass stars and substellar objects is often poor, spatially limited, and difficult to assess because of the heterogeneous selection processes.

The improvement from PS1 will come from two important features: first, the combination of the depth and large-area of the PS1 3π Survey will allow us to search for low-mass members associated with the star forming regions over a very large area. The full extent of nearby star forming regions such as Taurus, Ophichus, Upper and Lower Sco, cover many hundreds of square degrees on the sky. Surveys to date have either been shallow, using photographic plates or small telescopes with wide field, undersampled CCDs, to observe the entire regions; or else they have focused on the cores of these clouds with deep observations on large telescopes. The PS1 3π Survey will span these two regimes, pushing deep across the widest possible regions (north of −30 deg in Dec).

In addition to the wide, deep photometric coverage, the PS1 3π Survey will also use its proper-motion sensitivity to make a quantum leap in the study of the young star formation regions. With high-quality proper-motion measurements, we will be able to select the members of the these star forming regions with substantially higher reliability than prior single-epoch surveys. The photometric selection process is severely hampered by the degeneracy of spectral classes and extinction in the color-color planes, and the degeneracy between the combination of extinction and luminosity classes and distances. Using proper motions to pre-filter potential candidates before performing a photometric selection will allow us to greatly improve the completeness of the cluster membership lists. Even at the start of the survey, we will be able to use the 2MASS observations from the late 1990s as a first epoch for the initial proper motion measurements. These studies will be similar to recent results on the Pleiades based on proper-motion studies (e.g., Lodieu *et al.*, 2007), but will extend the spatial coverage substantially and increase the proper-motion accuracy.

We have been using the CFHT MegaPrime camera to perform a demonstration of this process, using archival CFHT data as a first epoch for the proper motion measurements. To date, we have observed 25 square degrees in Taurus and 10 square degrees in Rho Ophiuchus. Observations of several other nearby clusters are scheduled. The initial analysis of these data show the power of the technique. By applying our precision astrometry analysis to the MegaPrime observations separated by a baseline of 2.5 years, we are able to achieve proper motion measurements with an accuracy of 2.5 milliarcsecond/year. Selecting sources within 2σ of the Taurus proper motion vector reduces the number of potential candidates which would be selected photometrically from ∼2200 to 73 (for an initial 10 square degree subset). We are currently scheduled to observe these targets with the IRTF in Nov 2007 to identify the young dwarfs and confirm their membership.

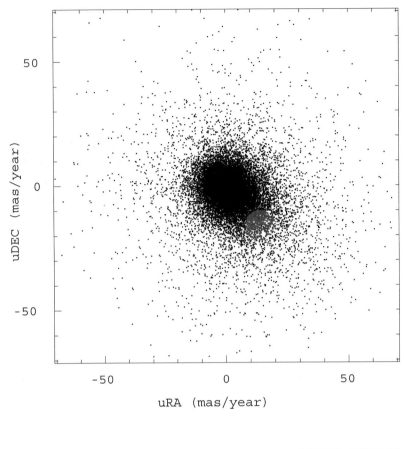

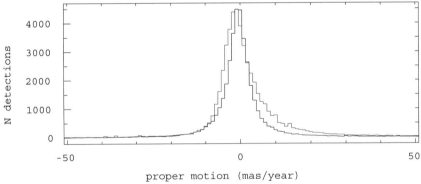

Figure 1. Our proper motion measurements for faint objects in the field of Taurus, based on CFHT *i*-band. Sources within 6 mas/year of the Taurus velocity are indicated in red. The histogram above shows the cuts along (red) and perpendicular (black) to this velocity vector.

5. Conclusions

The PS1 Project will soon begin the PS1 3π Survey, covering the 3/4 of the sky visible from Hawaii with many repeated observations over 3.5 years. The high-quality astrometric and photometric measurements expected from this survey will be a treasure trove for research covering a wide range of Galactic astronomy. The parallax accuracy (∼1.5 mas) will enable the creation of volume-limited samples of low-mass stars and brown dwarfs in the extended solar neighborhood out to roughly 100pc. The proper-motion accuracy of 1.2 mas/year will allow for the identification of the currently missing members of the nearby star formation regions. The PS1 telescope is in the process of being commissioned, and preliminary astrometric tests show the accuracy goals should be achievable. Survey operations are expected to begin in mid-2008.

References

Burgasser, A. J., McElwain, M. W., Kirkpatrick, J. D., Cruz, K. L., Tinney, C. G., & Reid, I. N. 2004, *AJ*, 127, 2856

Delfosse, X., Tinney, C. G., Forveille, T., Epchtein, N., Borsenberger, J., Fouqué, P., Kimeswenger, S., & Tiphène, D. 1999, *A&A*, 135, 41

Hawley, S. L., Covey, K. R., Knapp, G. R., Golimowski, D. A., Fan, X., Anderson, S. F., Gunn, J. E., Harris, H. C., Ivezić, Z., Long, G. M., *et al.* 2002, *AJ*, 123, 3409

Kirkpatrick, J. D., Reid, I. N., Liebert, J., Gizis, J. E., Burgasser, A. J., Monet, D. G., Dahn, C. C., Nelson, B., & Williams, R. J. 2000, *AJ*, 120, 447

Lodieu, N., Dobbie, P. D., Deacon, N. R., Hodgkin, S. T., Hambly, N. C., & Jameson, R. F. 2007, *MNRAS*, 380, 712

A Giant Step: from Milli- to Micro-arcsecond Astrometry
Proceedings IAU Symposium No. 248, 2007
W. J. Jin, I. Platais & M. A. C. Perryman, eds.

© 2008 International Astronomical Union
doi:10.1017/S1743921308020140

PMOE planetary/lunar ephemeris framework†

G. Y. Li[1], H. B. Zhao[1,2], Y. Xia[1,2], F. Zeng[1,2] and Y. J. Luo[1,2]

[1]Purple Mountain Observatory, Chinese Academy of Sciences, Nanjing 210008, China
email: gyl@pmo.ac.cn

[2]Graduate University of Chinese Academy of Sciences, Beijing 100049, China

Abstract. The PMOE planetary/lunar ephemeris framework was established in 2003, and has been improved in recent years. In the framework of the post-Newtonian effects, the figure perturbation effects arising from the a finite size of the Sun, Moon and the Earth, and the effect of the Earth tide were taken into account. The accuracy of using the PMOE ephemeris to predict the positions of the planets in the solar system are the same as that of JPL DE 405. Based on this framework, the orbit optimization for the LISA, ASTROD and ASTROD I missions, and the computation of celestial phenomena and lunar phases in the Xia Shang and Zhou period of ancient China have been completed.

Keywords. astrometry, ephemerides, methods: data analysis, methods: numerical

1. Introduction

E.M. Standish and J.G. Williams *et al.* in the NASA JPL published the DE/LE planetary/lunar ephemerides based on observations of the radar ranging, lunar laser raging (LLR) and others in the solar system after 1960s. The important DE/LE ephemerides are DE/LE 96 in 1975, DE/LE 102 in 1977 (Newhall, Standish & Williams 1983), DE/LE 200 in 1982 (Standish 1990), DE/LE 403 in 1995, and DE/LE 405 in 1998 (Standish 1998). Based upon the resolutions of IAU, DE/LE 200 and DE/LE 405 were acknowledged generally as the standards of the ephemerides. P. Bretagnon and G. Francou *et al.* in the IMCCE and Bureau International de l'Heure published the VSOP 87 (Bretagnon & Francou1988), VSOP 2000 (Bretagnon, 2001), PS-1996 solar/planetary ephemerides and ELP 2000 lunar ephemeris (Chapront-Touz & J. Chapront, 1983) based on the DE 200 and DE 403 ephemerides. G.A. Krasinsky and E.V. Pitjeva at the Institute of Applied Astronomy of RAS published the EPM 1998, EPM 2000 and EPM 2002 ephemerides (Pitjeva, 2001).

In China, Jiaxiang Zhang studied the orbital evalution of the comets 60P/Tsuchinshan 2, 62P/Tsuchinshan 1, 1P/Halley and asteroid Icarus with his model and improved Cowell integrator in 1970s. Zhang made a series of successful predictions of the impact of Comet Shoemaker-Levy 9 on Jupiter in 1994 which compared with Galileo's observation yielded the RMS of $8\rlap{.}''46$ (Zhang 1996). The predictions given by Zhang and those of JPL are about the same accuracy. Using the same ephemeris, Guangyu Li and Haibin Zhao found the origin of particles in the second peak of the Leonids in 1998 (Li & Zhao 2002). At the same time Weitou Ni *et al.* from the Center for Gravitation and Cosmology of Tsing Hua University published the CGC 1 and CGC 2 ephemeris frame, and worked on the orbit simulation and the parameter determination for the ASTROD and ASTROD I mission concepts (Ni 2006).

† Supported by Foundation of Minor Planets of Purple Mountain Observatory and the National Science Foundation (Grant No. 10503013).

According to the requirement of the solar system exploration, especially for the study of the ASTROD I mission, in 2001 Guangyu Li cooperated with Weitou Ni to build a more accurate ephemeris. By the end of the 2002 the PMOE 2003 ephemeris framework had been published (Li 2003) and was further improved in 2005.

2. PMOE ephemeris framework

Referred to the Zhang's Model and CGC 2, the dynamical model of the solar system and the integrator of the PMOE 2003 ephemeris framework was proposed. The following effects on the bodies in the solar system have been taken into consideration: (1)the post-Newtonian effects; (2) the figure perturbation effects arising from the finite size of the Sun, the Moon and the Earth; (3) the effect of the Earth tide on the motion of the Moon; (4) effects of gravitational forces of the Big 3 and 314 other large asteroids. With the identical astronomical constants, initial position and velocities as those in DE 405 ephemeris, the orbits of the Earth-Moon barycenter, 7 planets, Pluto and the Moon around the Earth, are integrated for 1,200 days and 36,000 days (about 98 years), the differences between PMOE and DE 405 ephemeris are shown in Table 1, Table 2 and Figure 1, respectively.

Table 1. Differences between PMOE and DE 405 after integrating 1,200 days

Bodies	Heliocentric distance Δ r(m)	Heliocentric longitude $\Delta\lambda$(mas)	Heliocentric latitude δ(mas)
Mercury	-3.1−3.1	-0.3−0.2	-0.0−0.01
Venus	-9.9−9.6	-0.03−0.04	-0.0−0.02
Earth	-3.3−2.9	0.0−0.04	-0.1−0.0
Moon*	-0.13−0.13	-1.4−0.0	-0.6−0.5
Earth−Moon	-3.3−2.8	0.0−0.04	-0.1−0.01
Mars	-25.6−14.7	-0.02−0.08	-0.01−0.03
Jupiter	-512.2−0.0	0.0−0.19	0.0−0.02
Saturn	-178.6−0.0	0.0−0.02	0.0−0.002
Uranus	-62.8−0.0	< 0.0004	< 0.0002
Neptune	-30.1−0.0	< 0.0004	< -0.00007
Pluto	-26.8−0.0	< -0.0002	< 0.0002

*Geocentric distance, longitude and latitude for the Moon.

Table 2. Differences between PMOE and DE 405 after integrating 36,000 days

Bodies	Heliocentric distance Δ r(m)	Heliocentric longitude $\Delta\lambda$(mas)	Heliocentric latitude δ(mas)
Mercury	-67.6−69.1	-1.6−0.1	-0.5−0.4
Venus	-20.9−20.5	-0.3−0.0	-0.1−0.1
Earth	-26.5−19.6	0−0.7	-0.2−0.2
Moon*	-32.2−32.3	-336.3−0.0	-121.3−138.2
Earth−Moon	-23.6−17.7	0.0−0.7	-0.2−0.2
Mars	-1302−1347	0−10.2	-2.5−3.9
Jupiter	-3765−2274	0−17.4	-5.6−6.3
Saturn	-1637−419	0−5.0	-1.7−1.6
Uranus	-3797−549	0−1.7	-0.7−0.3
Neptune	-5023−0	0−1.1	0.0−0.2
Pluto	-5628−0	0−0.5	0.0−0.2

*Geocentric distance, longitude and latitude for the Moon.

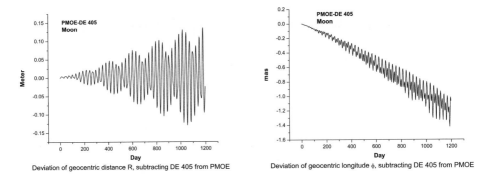

Figure 1. Differences between PMOE and DE 405.

3. Preliminary applications

Based on the PMOE 2003 ephemeris framework, Li *et al.* have calculated the orbits of the LISA, ASTROD I and ASTROD spacecrafts (Ni 2006), and proposed the methods for the orbital optimization of the LISA spacecrafts, cooperating with H. Gerhard *et al.* at ESA (Li, Yi & Heinzel 2008). Also the celestial phenomena and lunar phases in Xia Shang and Zhou Period in China (BC 2100-BC 771) have been calculated and published. After further improvements of the lunar dynamical model and fitting the latest ranging data, lander ranging data, VLBI data, and LLR data, an independent Chinese ephemeris will be completed.

Along with the development of the exploration of the solar system and laser ranging between the bodies in the solar system, the precision of the measurements of the bodies in the solar system will be greatly improved. The present planetary/lunar ephemerides based on the first post-Newtonian (1PN) theory, including the DE ephemerides, will not satisfy the requirement of the space exploration. The planetary/lunar ephemeris should be to establish on the basis of the second post-Newtonian (2PN) theory, which is studied by Weitou Ni's group at the Purple Mountain Observatory (Ni 2006).

References

Bretagnon, P. & Francou, G. 1988, *A&A*, 202, 309.
Bretagnon, P. 2001, *Celest. Mech. & Dyn. Ast.*, 80, 3, 205.
Chapront-Touzé, M. & Chapront, J. 1983, *A&A*, 124, 50.
Li, G. & Zhao, H. 2002, *Int. J. Mod. Phys. D*, 11, 7, 1021.
Li, G. & Ni, W. T. 2003, *Publications of Purple Mountain Observatory*, 22, 3-4, in Chinese .
Li, G., Yi, Z., Heinzel, G., *et al.*, 2008, *Int. J. Mod. Phys. D*, 17, 6.
Newhall, X. X., Standish, Jr., E. M., & Williams, J. G. 1983, *A&A*, 125, 150.
Ni, W. T., *et al.*, 2006, *Gen. Rel. Grav.*, 39.
Pitjeva, E. V. 2001, *A&A*, 371, 760.
Standish, Jr., E .M. 1990, *A&A*, 233, 252.
Standish, Jr., E. M. 1998, *JPL IOM 312, F-98-048*.
Zhang, J. X., *et al.*, 1996, *Science in China (Series A)*, 39, 2, 207.

A Giant Step: from Milli- to Micro-arcsecond Astrometry
Proceedings IAU Symposium No. 248, 2007
W. J. Jin, I. Platais & M. A. C. Perryman, eds.

Worldwide R&D of Virtual Observatory

C. Z. Cui[1] and Y. H. Zhao[1]

[1]National Astronomical Observatories, Chinese Academy of Sciences, Beijing 100012, China
email: ccz@bao.ac.cn

Abstract. Virtual Observatory (VO) is a data intensive online astronomical research and education environment, taking advantages of advanced information technologies to achieve seamless and uniform access to astronomical information. The concept of VO was introduced in the late 1990s to meet the challenges brought up with data avalanche in astronomy. In the paper, current status of International Virtual Observatory Alliance, technical highlights from world wide VO projects are reviewed, a brief introduction of Chinese Virtual Observatory is given.

Keywords. miscellaneous, methods: miscellaneous, astronomical data bases: miscellaneous

1. Introduction

During the last decade, advances in technologies have been changing the abilities and ambitions of astronomers. New technologies on telescope design and fabrication bring more powerful telescopes then ever to astronomers. Large scale digital sky surveys are prospering with the appearance of CCD mosaic camera. The scale of numerical simulations is also increasing rapidly with the development of hardware and software. As a result of these advances, data avalanche is occurring in astronomy. Furthermore, driven by the multi-waveband sky surveys and observations, new astronomical fields appear and are becoming more and more popular such as multi-waveband research, multi-archive data mining, time domain analysis, precise cosmology (Lawrence, 2006). To meet the challenges brought up by the above changes, a Virtual Observatory (VO) concept was initiated (Szalay & Gray, 2000). Virtual Observatory is a data intensive online astronomical research and education environment, taking advantages of advanced information technologies to achieve seamless and uniform access to astronomical information. The power of the World Wide Web is its transparency. It is as if all the documents in the world are inside your PC. The idea of the VO is to achieve the same transparency for astronomical data and information (Quinn *et al.*, 2004). VO is a science driven cyber-infrastructure for the 21st century astronomy.

2. International VO activities

After National Virtual Observatory (US-VO), the first funded VO project, VO projects were initiated in different countries and regions. International Virtual Observatory Alliance (IVOA) was formed in June 2002 with a mission to "facilitate the international coordination and collaboration necessary for the development and deployment of the tools, systems and organizational structures necessary to enable the international utilization of astronomical archives as an integrated and interoperating virtual observatory." At present, the IVOA consists of 16 projects. VO framework includes agreed standards, interoperable data collections, interoperable services and applications. IVOA focuses its work on development of standards. Now, more then 20 specifications have been published by IVOA working groups.

Following the concept of VO and IVOA specifications, a new prosperous era is coming for astronomical softwares and services. New applications, VO-enhanced legacy services are supporting astronomer's research more and more strongly. Core services from US-VO provide functions including data discover, service registry, catalog coverage, object cross match, source extraction and identification. UK VO project, AstroGrid, deploying the world's only unified VO operational service "AstroGrid", gives astronomers a "one-stop" access to all the world's astronomy data from a desktop. Many other new applications are released by world-wide VO projects and contributors such as VOPlot, VOSpec, VisIVO, SAADA. At the same time, existing applications and systems are upgraded and enhanced to support VO including CDS services (SIMBAD, VizieR, Aladin), NED, SciSoft, IRAF, TOPCAT, Montage, SExtractor.

3. VO in China

Chinese Virtual Observatory (China-VO) is the national VO project in China initiated in 2002 (Cui & Zhao, 2004). The China-VO aims to provide VO environments for Chinese astronomers. It focuses its research and development on applications and VO science in the following five fields: (1) China-VO Platform, providing VO environments for Chinese astronomical community; (2) Uniform Access to Global Astronomical Resources and Services, importing international resources to China and sharing Chinese resources to international community; (3) VO-ready Projects and Facilities, collaborating with astronomical projects to ensure they are VO-compliant; (4) VO-based Astronomical Research Activities, guiding and training astronomers to use VO; (5) VO-based Public Education, developing non-professional services for the public.

During the last several years, several VO applications and services have been initiated and developed by the China-VO. For example, VOFilter, an XML filter for OpenOffice.org Calc to load VOTable files; SkyMouse, a smart interface for astronomical on-line resources and services; FitHAS (FITS Header Archiving System), a toolkit for FITS file providers; VO-DAS, an OGSA-DAI based service system to provide unified access to astronomy data. Furthermore, the China-VO is collaborating with LAMOST, a Chinese ambitious spectral survey project, to make it VO-enabled, sharing its archives and software to the VO.

VO is trying to link on-line resources and services together at a higher level. More and more VO supported resources and services are available. How to do a better science using these tools? VO is not a simple thing – it is a new kind of research environment astronomers need to learn about. Astronomers need learn how to survive in the VO era.

Acknowledgements

The China-VO is supported by NSFC under contract No. 60603057, 90412016 and 10778623, and Beijing Science and Technology Nova Project under contract No. 2007A085.

References

Cui, C. & Zhao, Y. 2004, *PNAOC* 1(3), 203
Lawrence, A. 2008, in: K. A. van der Hucht (eds.), *Highlights of Astronomy*, The Virtual Observatory in action : new science, new technology, and next generation facilities. vol. 14, in print
Quinn, P. J., Barnes, D. G., Csabai, I., *et al.* 2004, in: P. J. Quinn & A. Bridger (eds.), *Proceedings of the SPIE*, Optimizing Scientific Return for Astronomy through Information Technologies, vol. 5493, p. 137
Szalay, A. & Gray, J. 2000, *Science* 293, 2037

A Giant Step: from Milli- to Micro-arcsecond Astrometry
Proceedings IAU Symposium No. 248, 2007
W. J. Jin, I. Platais & M. A. C. Perryman, eds.

© 2008 International Astronomical Union
doi:10.1017/S1743921308020164

China NEO Survey Telescope and its preliminary achievement

H. B. Zhao[1,2]†, J. S. Yao[1,2] and H. Lu[1,2]

[1]Purple Mountain Observatory, Chinese Academy of Sciences, Nanjing 210008, China
email: meteorzh@pmo.ac.cn

[2]National Astronomical Observatories, Chinese Academy of Sciences, Beijing 100012, China

Abstract. In recent years, there has been an increasing appreciation for the hazards posed by Near Earth Objects (NEOs), those asteroids and periodic comets whose motions can bring them into the Earth's neighborhood. An NEO Survey Telescope (NEOST) was built in China to be taken part in the international NEO joint survey. This telescope is a 1.0/1.2m Schmidt telescope, equipped with a 4K by 4K CCD detector with a drift-scanning function. After adjusting the telescope and test observations, in December 2006 the NEOST began its NEO survey program. We have found 188 new asteroids including an NEO – 2007 JW2 and one periodic comet – P/2007 S1 (Zhao).

Keywords. telescopes, asteroids, surveys, astrometry

1. Introduction

The rate of asteroid discoveries shows an exponential growth. After astronomer Guiseppe Piazzi of Palermo, Sicily, discovered the first asteroid on January 1, 1801, the number of new finds per year increased to five by 1865, 15 per year by 1895, 25 by 1910 and up to about 40 by 1930. By the end of September of 2007, the number of numbered asteroids was more than 160,000 including about 800 Potential Hazardous Asteroids (PHA). (see http://neo.jpl.nasa.gov/stats/)

According to NASA's report (NASA 2007), the further objectives of NEO Survey Program are to detect, track, catalogue, and characterize the physical characteristics of NEO equal to or larger than 140 meters in diameter with a perihelion distance of less than 1.3 AU from the Sun, achieving 90% completion of the survey within 15 years after enactment of the NASA Authorization Act of 2005.

Chinese scientists have contributed substantially to the field of asteroid survey and related aspects. In the early 1960s, Purple Mountain Observatory began observations of asteroids and found over 130 new numbered asteroids during the following decades. The Schmidt CCD Asteroid Program (SCAP) of Beijing Astronomical Observatory started in 1995 and found 575 asteroids in several years (Ma, Zhao & Yao 2007).

2. NEOST and observation

In October 2006, the 1.0/1.2 m NEOST equipped with a 4096×4096 SI CCD detector was installed and began the test observations. Due to the small focal ratio and the high quantum efficiency (QE) of the CCD detector, the observational system can reach $B=22.5$ with a 40 s exposure, which makes the asteroid survey very efficient. About 22 Gb of raw image data, corresponding to the sky coverage of 2700 deg^2, are produced each good

---

† Present address: Purple Mountain Observatory, 2# West Beijing Road, Nanjing 210008, China.

observing night providing on average more than 2000 asteroid positions. To reduce the observational data and to report the asteroid positions to Minor Planet Center (MPC) in a timely fashion is a challenge to us. We have established a set processing software to reduce the data with good precision (Table1 where D29 is the station code for NEOST).

Table 1. Residuals statistics of asteroid position for top-ten observational program (up to 2007-09-30).(see http://www.cfa.harvard.edu/iau/special/residuals.txt)

COD	Tot	$< 1''$	$< 2''$	$< 3''$	$< 4''$	$\geqslant 4''$	R.A.	Decl.
G96	4417	4307	101	7	2	0	-0.01 ± 0.33	-0.09 ± 0.31
D29	57486	52095	4678	624	89	0	-0.04 ± 0.45	-0.08 ± 0.46
644	288266	286027	2127	94	18	0	$+0.02 \pm 0.26$	$+0.11 \pm 0.23$
E12	27058	24408	2524	116	10	0	-0.05 ± 0.42	$+0.24 \pm 0.43$
691	374098	366226	6582	1012	278	0	-0.09 ± 0.28	$+0.07 \pm 0.28$
D35	21884	20233	1586	49	16	0	$+0.09 \pm 0.42$	$+0.26 \pm 0.37$
704	177620	119287	52736	5147	450	0	-0.09 ± 0.61	$+0.45 \pm 0.64$
291	21476	20559	819	83	15	0	-0.13 ± 0.38	$+0.13 \pm 0.30$
699	28249	20993	6616	599	41	0	-0.13 ± 0.57	$+0.44 \pm 0.55$
703	3532	2568	867	93	4	0	-0.23 ± 0.63	$+0.12 \pm 0.63$

3. Preliminary achievement

From December 2006 to September 2007, we carried out the test observations of asteroids and accumulated more than 600 Gb of raw image data with the sky coverage more than $20000 \deg^2$ near the opposite position. Our observations ranked among the top-ten observational programs in the world. Until September of 2007, we have found 188 new asteroids including an Apollo-type NEO - 2007 JW2. In the middle of September, a new Jupiter-family comet, P/2007 S1, was found.

We used 25 observations from May 7, 2007 to May 10, 2007 to obtain the 2007 JW2's orbital elements and their uncertainties. Dynamical evolution shows that 2007 JW2 has no chance to impact the Earth within 200 years and its MOID is 0.22 AU. For the periodic comet, it is more fortunate that there are many more observations to determinate the orbit. According to its orbital elements, P/2007 S1 is a typical Jupiter-family comet. The astrometric errors of this comet are much larger than that of an asteroid, resulting in higher orbital element uncertainties.

Acknowledgements

We thank Yuezhen Wu, Getu Zhaori, Ming Wang, Renquan Hong and Longfei Hu for their valuable observations. We would like to acknowledge the support of the National Natural Science Foundation of China (Nos. 10503013, 10778637), the Minor Planet Foundation of Purple Mountain Observatory and the exchange program between the Finnish Academy and NSFC.

References

Ma, Y., Zhao, H., & Yao, D., 2007, *Proceedings IAU Symposium No. 236*, 381-384.
NASA, 2007, *Near-Earth Object Survey and Deflection Analysis of Alternatives (Report to Congress)*.

A Giant Step: from Milli- to Micro-arcsecond Astrometry
Proceedings IAU Symposium No. 248, 2007
W. J. Jin, I. Platais & M. A. C. Perryman, eds.

Getting ready for the micro-arcsecond era

A. G. A. Brown

Sterrewacht Leiden, Leiden University,
P.O. Box 9513, 2300 RA Leiden, The Netherlands
email: brown@strw.leidenuniv.nl

Abstract. As the title of this symposium implies, one of the aims is to examine the future of astrometry as we move from an era in which thanks to the Hipparcos Catalogue everyone has become familiar with milliarcsecond astrometry to an era in which microarcsecond astrometry will become the norm. I will take this look into the future by first providing an overview of present astrometric programmes and how they fit together and then I will attempt to identify the most promising future directions. In addition I discuss the important conditions for the maximization of the scientific return of future large and highly accurate astrometric catalogues; catalogue access and analysis tools, the availability of sufficient auxiliary data and theoretical knowledge, and the education of the future generation of astrometrists.

Keywords. astrometry, catalogs, surveys

1. Overview of astrometric programmes

Figure 1 presents a graphical overview of past, present, and future astrometric programmes and the relationships between them. I have grouped a number of the programmes together. For example, 'wide-field CCD surveys' stands for astrometry obtained from observing programmes such as the Sloan Digital Sky Survey (SDSS) or the future Large Synoptic Survey Telescope (LSST, see Ivezić 2008). Similarly 'radio interferometry' stands for a number of astrometric programmes conducted with different VLBI networks. The overview in figure 1 is not intended to be complete but serves to illustrate the variety of astrometric programmes and how they fit together.

All astrometric programmes must ultimately be linked to the International Celestial Reference System (ICRS, the idealized barycentric coordinate system to which celestial position are referred) through procedures which involve a number of steps that depend on the details of the astrometric observing programme. For ground-based programmes where the observations are typically done over a small field of view the obtained relative parallaxes and proper motions are converted to an absolute scale through the observations of astrometric standards. In the case of missions like Hipparcos although the measured parallaxes and proper motions are absolute by design, the reference frame still has six degrees of freedom with respect to the ICRS (three orientation and three spin parameters) which have to be taken out through observations of extra-galactic reference sources. The ICRS thus ties all astrometric programmes together and this is symbolized by the background image in figure 1.

The practical realization of the ICRS is the International Celestial Reference Frame (ICRF) consisting of a few hundred extra-galactic radio sources with adopted ICRS positions. The ICRF sources are constantly monitored in dedicated VLBI observing programmes, which is why the radio interferometry observations are portrayed separately in figure 1 as they underpin the practical realization of the reference system. In the optical, the ICRF is materialized by the Hipparcos Catalogue. For more details see the chapters on reference frames in this volume. For Solar system investigations radar astrometric

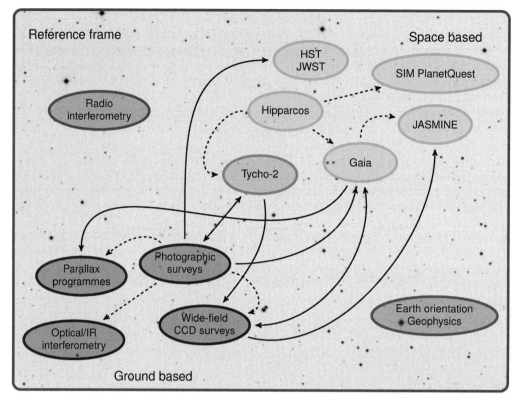

Figure 1. Graphical overview of astrometric programmes and the relations between them. There is no ordering in terms of timing or accuracy of the programmes. Refer to the text for further explanations of this figure. The background image, symbolizing the reference frame, is a 20×15 arcmin^2 region of the sky centred on the Hyades cluster and was taken from the Digitized Sky Survey (POSS-II plates).

observations (ranging and Doppler) are instrumental in providing the reference for accurate ephemerides of the inner planets and the determination of various astronomical constants (see Pitjeva in this volume).

A well established reference system can be used to interpret accurate observations of celestial sources from earth in terms of the orientation of the earth's spin axis and its precession and nutation (bottom right of figure 1). Two particularly interesting examples are given in the contributions by Vondrák and Huang in this volume. The former concerns a study of earth rotation over the past 100 years and the latter a study of the precession and nutation of the earth's rotation axis in terms of a non-rigid model of our planet.

The other astrometric programmes in figure 1 have been lumped together into the categories 'ground based' and 'space based', with the Tycho-2 catalogue located in between as that was constructed from a combination of photographic plate material and observations made with the Hipparcos satellite. The arrows illustrate some of the relations within the ground based or space based programmes (dashed lines) and between ground based and space based astrometry (solid lines).

The photographic surveys that were carried out during the 20th century have formed the basis for the optical reference frame until the Hipparcos catalogue was published and the astrometric catalogues derived from them, now reduced to the Hipparcos reference frame, still serve as basic input for modern ground based astrometric programmes. Recently the photographic material has been combined with CCD astrometric observations such as in the UCAC catalogues (Zacharias *et al.* 2004). Eventually the wide field CCD surveys will provide a much denser optical reference frame, although they will still have to be linked to more absolute frames, e.g. Hipparcos or Gaia.

A good example of how ground based astrometric surveys support the space missions is the Guide Star Catalogue (GSC, Bucciarelli 2008). It serves as the basic input for HST pointing and guidance and will also form the basis of the input catalogue for JWST. For Gaia the GSC will be one of the inputs for the construction of an initial source list to support the identification of sources in the early phases of the mission. The latest version of the GSC makes use of Tycho-2 data as a reference for the astrometric calibration and as a supplement to its bright end.

The ground based CCD surveys go much deeper than missions like Gaia or JASMINE will and the corresponding astrometric catalogues can serve to provide information on sources beyond the survey limits of these missions. This is important in regions on the sky where the density is high enough for faint sources to disturb the astrometry and photometry of the target sources. Conversely, the ground based CCD surveys can make use of the reference frame provided by space based missions. The Tycho-2 catalogue is already being used for this purpose and in the future the Gaia and SIM PlanetQuest catalogues will provide an excellent reference, in particular for the ground based parallax programmes (Smart 2008).

Finally, the space astrometry missions are also interrelated. The Hipparcos catalogue has been used to select candidate grid stars for SIM PlanetQuest (see e.g. Hekker *et al.* 2006) and will be used to predict the positions of sources on the sky that are too bright to be observed by Gaia. The latter will affect the measurements of nearby fainter sources and will influence the state of Gaia's CCDs with respect to charge transfer inefficiency (see Lindegren 2008). The JASMINE mission will rely on a plate overlap technique and observe only a relatively small part of the sky. Thus this mission will benefit greatly from the reference frame which will be provided by Gaia.

In summary all the astrometric programmes schematically indicated in figure 1 form a very tightly interlocked set of observational programmes, none of which be obsolete for some time to come. In the next section I will look at how these astrometric programmes cover the space of survey parameters and what science is consequently not well covered.

2. Science covered by astrometric surveys: are we missing something?

Astrometric programmes can be characterized by the following parameters:

Precision and accuracy: These are the most obvious selling points of any astrometric survey and much effort goes into maximizing these parameters. The precision reflects the quality of the relative measurements that can be made with a particular instrument and optimizing this parameter is enough if one is only interested in differential measurements, such as when hunting for exoplanets. The accuracy of a survey is limited by the external calibration of the measurements and by the design of the experiment. A well known example is the case of ground-based parallax programmes which are done over small fields of view and thus require corrections of precisely measured relative parallaxes to absolute ones. The accuracy in this case is limited by our knowledge of the spatial distribution of the reference sources.

Time baseline: Parallax and proper motion measurements benefit from a long time baseline, which is needed to disentangle the parallaxes from the proper motions and to ensure that proper motions of binary systems reflect the centre of mass motion as opposed to the orbital motions of the components. The sampling in time also plays an important role, especially for parallax measurements carried out over only a few years.

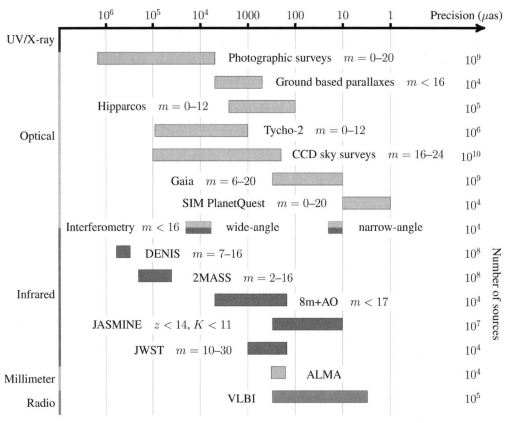

Figure 2. Graphical overview of astrometric surveys. The shading of the boxes and the right vertical axis serves to aid the eye in segregating the wavelength domains. Refer to the text for further explanation of the diagram. The survey parameters are based on the following references: Photographic surveys — Morrison *et al.* (2001), Monet *et al.* (2003); Ground based parallaxes — Jao (2008), Smart (2008); Hipparcos — ESA (1997), van Leeuwen (2007); Tycho-2 — Høg *et al.* (2000); CCD sky surveys — Ivezić (2008), Magnier (2008); Gaia — Lindegren (2008); SIM PlanetQuest — Shao (2008); Optical/IR interferometry — Boden (2008), Richichi (2008); DENIS — Borsenberger *et al.* (2006), G. Simon priv. comm.; 2MASS — Skrutskie *et al.* (2006); 8m+AO Röll *et al.* (2008), Ghez (2008); JASMINE — Gouda (2008); JWST — Beichman (2008); ALMA — Lestrade (2008); VLBI — Reid (2008), Ma (2008).

Depth and dynamic range: The faint limit of the survey determines the kinds of sources that can be observed (faint nearby low-mass objects, distant standard candles, extra-galactic reference sources, etc). Future astrometric surveys will push this parameter as much as possible but it has to be traded off against the dynamic range of the survey. Instruments capable of observing very faint sources often cannot deal with very bright sources. For example the Hipparcos astrometry for the brightest ($V < 6$) stars will not be superseded by any of the planned space missions. On the other hand optical/infrared interferometry programmes can typically only deal with bright sources.

Number of sources: For sources of arbitrary brightness achieving high precision always comes at the expense of more observing time and so a trade-off has to be made between the number of sources observed and the astrometric precision achieved. Targeted high precision programmes (such as SIM or the ground-based parallaxes) are very good at addressing specific scientific questions whereas less precise 'all-sky' surveys are scientifically important because of the advantages of statistical studies of large samples and they come with the added bonus of serendipity.

Density of sources: The density of sources is important if the astrometric catalogue is to be used as a reference frame for other astrometric programmes. This is especially needed for astrometric surveys targeting faint sources over small fields of view.

Wavelength range: This parameter sets the 'window on the universe' for an astrometric survey and is a very important driver for the science that one can do.

Narrow- vs. wide-angle: In ground-based astrometry programmes sub-milliarcsec to micro-arcsec precision can be achieved but only through differential measurements over small fields. This makes it difficult to construct reference frames free from zonal systematic errors and necessitates the correction from relative to absolute parallaxes. On the contrary astrometry over wide angles by design gives absolute parallaxes and a rigid reference frame (e.g. Lindegren 2005). The problem of converting relative to absolute astrometry can in principle be overcome for narrow angle measurements if the observations go deep enough to find enough suitable extra-galactic reference sources in the field of view.

Ground based vs. space based: This parameter is closely related to the previous one. The choice of going to space offers the advantages of a stable thermal environment, freedom from gravity and the atmosphere, and full-sky visibility (Lindegren 2005). These factors in fact enable wide-angle astrometry as implemented on missions such as Hipparcos, Gaia, and SIM PlanetQuest.

Figure 2 is my attempt at summarizing the coverage of the parameter space described above by the present and planned astrometric catalogues and surveys. The main goal of this diagram is to identify major gaps in the parameter space and not to provide a complete and precise overview of all astrometric surveys or programmes. Figure 2 shows the distribution of astrometric programmes over wavelength, precision, depth and dynamic range, and the number of sources observed. The precision figures encompass positions, parallaxes and proper motions and in most cases also reflect the accuracy of the survey. Precision numbers for proper motions are generally numerically smaller than those for positions, especially for photographic surveys. The magnitude limits and numbers of objects are crude indicators. Within each wavelength range the surveys have been ordered roughly in time. From figure 2 a number of observations can be extracted which can help in identifying interesting future directions for astrometry.

In the optical we can expect an enormous amount of astrometric data, with accuracies in the 10 microarcsec to 10 milliarcsec range, coming in over the next 10 to 15 years from the CCD surveys of the sky (LSST, Pan-Starrs) and Gaia. A preview of what we can expect from this data is given in the contribution of Ivezić in this volume in which he discusses the power of using SDSS Data Release 6 for studies of the structure of the Milky Way. The latest SDSS data release also contains proper motions with ~ 3–6 mas/yr errors out to $r = 20$ for 30×10^6 stars (Munn *et al.* 2004). This offers photometry for a huge number of stars over a large volume in our galaxy combined with kinematic data at a level of accuracy which enables both a quantitative and qualitative breakthrough. This dataset will be dwarfed by the combined Gaia/LSST/Pan-Starrs data which will constitute Hipparcos quality astrometry over the magnitude range 20–24!

The ground-based parallax programmes and the optical/IR interferometric observations will fill very important niches at the bright end in the characterization of the Sun's nearest neighbours and in the hunt for exoplanets.

From figure 2 it is very clear that the SIM PlanetQuest mission is unique in its capability of providing astrometry in the 1–10μas range all the way out to 20th magnitude. It will thus be the only instrument capable of directly addressing detailed questions about the outer reaches of our galaxy or nearby external galaxies.

All the optical programmes obviously suffer from the extinction barrier due to dust, especially in the Milky Way disk and towards the Galactic centre. Infrared surveys are thus essential in achieving a complete understanding of our galaxy and for studying star formation in all its aspects. In the infrared the astrometric data is currently of relatively poor quality compared to what is available in the optical even though the DENIS and 2MASS surveys represent a tremendous step forward compared to the situation 10 years ago. The only ongoing astrometric programmes in the infrared are the interferometric and adaptive optics programmes which mainly deliver relative astrometry over small fields of view (although at high precision). The only planned future infrared astrometric survey is the JASMINE mission which will cover the Bulge and inner Milky Way disk regions.

A very exciting prospect in the near future is the possibility of obtaining milliarcsecond astrometry in the millimetre wavelength regime. Combined with the already available radio astrometry at this accuracy this will enable the direct determination of the distances and kinematics of deeply embedded star clusters or even proto-stellar clusters (see contributions by Loinard and Lestrade).

3. Future directions

From the discussion of parameter space coverage by current and planned astrometric programmes a number of gaps can be identified from which proposals for future astrometric observing programmes can be extracted. I briefly discuss here a number of possibilities, some of which are already covered by planned instruments.

All-sky infrared astrometry at high accuracy: JASMINE covers the inner Milky Way disk only and does not go very deep. I think a deep all sky astrometric survey in the infrared provides the biggest 'discovery space', offering a tremendous improvement in our knowledge of the Milky Way disk, spiral arms and star forming regions. In addition low mass stars, brown dwarfs and 'free-floating planets' can be studied over a much larger volume. Such a mission is probably most efficiently carried out as a scanning spacecraft using the principles of Hipparcos and Gaia. The question is whether it will be technically feasible for a reasonable cost. Can a large cryogenic instrument be realised which can operate over the length of time required for high precision astrometry? The experience from the design of missions like ESA's DARWIN will be very useful in this respect.

High-precision narrow angle astrometry: This is an area which is already under development through the interferometric and adaptive optics programmes at the current generation of large (8 m class) telescopes. Pushing the precision of small fields of view as much as possible towards the μas mark will increase the sensitivity of exoplanet searches to the level where earth-sized planets around nearby stars can be discovered; allow us to probe space-time near the massive black hole in the centre of the Milky Way; and enable the study of stellar orbits around the central black holes in external galaxies. This increase in astrometric accuracy can be achieved with new instruments on 8 m class telescopes used as interferometers and with instruments for the future extremely large telescopes (see the contributions by Ghez and Richichi in this volume).

Geometric distances and motions of galaxies in the local volume: The study of the dynamics of galaxies in the local volume is currently limited by the lack of knowledge of their proper motions. As discussed in the contribution by Bruntahler proper motions can be obtained as far out as the M31 subgroup through VLBI measurements of masers with respect to background quasars. He also discusses the possibility to go beyond the local group. An important goal for the future is to extend these studies by improving the precision and the enlarging the sample.

Micro-arcsecond astrometry to 25th magnitude: An obvious future instrument to consider would be a 'super-Gaia', capable of delivering microarcsecond astrometry to 25th magnitude. This would require careful consideration of a number of issues. What will be gained scientifically? Is Hipparcos quality data to 24th magnitude enough? The answer to this question will depend on what we learn from Gaia and the CCD surveys. The technical difficulties of simply scaling up Gaia will be driven by the need to collect 2 to 4 orders of magnitude more photons over the magnitude range 20–25. This cannot be achieved through more sensitive detectors (the Gaia CCDs already having $\sim 80\%$ quantum efficiency) but will have to come from larger optics and a larger field of view. This will make the requirements on instrument stability very difficult to achieve (thermal and mechanical design) if not impossible. It is probably better to think about alternative ways of efficiently doing all-sky absolute astrometry at the microarcsecond level.

Astrometry at wavelengths below 300 nm: I included the UV/X-ray wavelength regime in figure 2 to highlight that currently there is no high precision astrometry available shortward of optical wavelengths. It is not clear that efforts to acquire high accuracy astrometry in UV/X-ray domain will pay off. It is technically difficult and most UV/X-ray sources may have optical counterparts which can be measured astrometrically.

4. Maximizing the scientific return

Although thinking about future directions for astrometric surveys is interesting it is much more important to ensure that we optimize the scientific exploitation of the large amounts of high accuracy astrometric data we can expect over the next two decades. This concerns the following issues:

Complementary data: All the large surveys (Gaia, LSST, Pan-Starrs) will deliver multi-colour photometry which is essential in the scientific exploitation of the astrometric data. Photometry provides the astrophysical characterization of the stars observed and allows the derivation of photometric distances. The latter can be calibrated using Gaia data for bright stars and will provide more reliable distances for the faint stars than can be derived from parallaxes. In the case of Gaia, at the faint end of the survey the low resolution prism spectra will not have much discriminating power with respect to surface gravity and metallicity of the stars. This can be remedied (for most of the sky) by supplementing Gaia photometry with the data from LSST and Pan-Starrs. Hence collaboration between these projects is essential.

Phase space for stars and galaxies consists of six dimensions but in the surveys targeting the stars in our galaxy the 6th dimension, radial velocity, will not be measured for the majority of the 10^{10} stars for which astrometric data will be collected. The RAVE project (Steinmetz *et al.* 2006) will collect radial velocities for $\sim 10^6$ stars and Gaia will do so for $\sim 10^8$ stars. The rewards for gathering radial velocities for the remaining stars are substantial and we should think about efficient ways of doing so. The accuracies of the radial velocities should be matched to the astrometric accuracies (translated to linear velocities) which for the faint end of the surveys lead to rather modest requirements and which can be met by low resolution spectrographs optimized for faint stars. From the ground a radial velocity survey, even with large multi-object instruments, may be very difficult to complete for the whole sky to 20–24th magnitude and dedicated follow-up programmes to pursue specific questions appear more viable. From space an all-sky survey may be possible using a scanning satellite optimized for radial velocities.

An important issue concerning radial velocities is the accuracy of the zero-point of the velocity scale. Given the huge efforts to acquire highly accurate absolute astrometry, the radial velocities should be absolute to the same level of accuracy. This point has not been addressed over the last decade as most of the effort has gone into refining the relative precision of radial velocity surveys in order to optimize them for planet detection. A dedicated effort is needed now to improve the zero-point accuracy across the Hertzsprung-Russell diagram. Astrometric data from Gaia can be used for this through the astrometric radial velocity technique (Dravins, Lindegren, & Madsen 1999) which provides radial velocities free of effects intrinsic to the star.

The compromise between resolution and depth of the spectroscopic surveys carried out by the RAVE project and Gaia is driven by the need for high resolution spectroscopy for the detailed determination of the astrophysical parameters of stars, including rotational velocities. Gathering complementary high resolution spectra is a very expensive undertaking and is probably best done with well designed follow-up projects that serve, for example, to better calibrate photometric indicators or answer specific questions.

The last remark on complementary data concerns the determination of accurate reddening laws and extinction towards the stars and galaxies observed. Especially for the subset of stars with very high accuracy parallaxes the knowledge of reddening and extinction will become a limiting factor in the determination of luminosities and stellar parameters and hence in the improvement of our understanding of stellar structure and evolution. There is a dedicated effort within the Gaia data processing consortium to ensure a proper determination of reddening and extinction. This effort will benefit greatly from collaboration with other survey projects and from investigating the use of complementary data such as infrared photometry and direct measurements of the gas and dust components of the Milky Way.

Preparing for the analysis of large astrometric/photometric catalogues: Optimizing the science return depends critically on having analysis tools and theoretical models that enable a proper interpretation of the vast quantities of high accuracy astrometric and photometric data. For stars the determination of the fundamental parameters luminosity, age, mass, and chemical composition depends heavily on the quality of stellar structure and atmosphere models. As discussed in the contributions by Lebreton and Chaboyer many uncertainties still remain in stellar models. Although addressing these problems will depend on the availability of highly accurate astrometric data, steps can already be taken now to improve our understanding of stars through better theoretical models, the development of 3D model atmospheres and laboratory experiments aimed at obtaining better opacities, thermonuclear reaction rates and equations of state.

For the interpretation of the astrometric data in terms of Galactic structure and dynamics analysis tools should be developed that can deal with data for 10^9 stars. Specific consideration should be given to the way parallax data is treated in order to avoid the biases that can be introduced when converting relatively imprecise parallaxes to distances. The best way forward may be to design analysis methods that rely on predicting the observations of Galactic phase space rather than the phase space variables themselves.

Providing optimized access to large catalogues: If the data for 10^9–10^{10} stars were to land on one's desk today it would be very difficult to decide how to begin exploring these data and extracting interesting science. The accessibility of these large catalogues will be essential in determining their success. The catalogue producers should aim at making tools available that allow easy access trough both high-level, easy to use, and low-level, more complex, interfaces. This includes good visualization tools. The catalogues should not be static finished products but should allow users to re-process data based on newly acquired knowledge about certain objects or classes of objects (see for example the discussion by Pourbaix in this volume). Improved knowledge should be

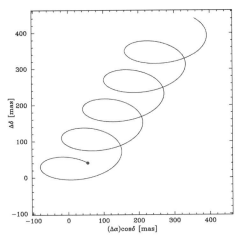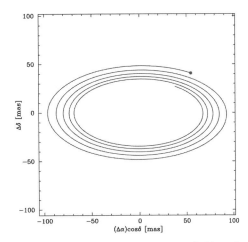

Figure 3. The paths on the sky for a star with a given proper motion and parallax (left) and for a star moving radially away from the Sun (right). This illustrates that from astrometry alone the 3D positions and motions of stars (including radial velocities) can be inferred. The right hand figure is not relevant in practice but as discussed in the paper by Dravins *et al.* (1999) there are a number of other astrometric effects from which radial velocities can be inferred. A practical application using Hipparcos data is given in Madsen *et al.* (2002). This figure was produced with the aid of the IAU's SOFA library (http://www.iau-sofa.rl.ac.uk/).

added to these databases when it becomes available. This could be, for example, radial velocities from follow-up programmes or a better astrophysical parametrization of stars based on improvements in stellar atmosphere models.

Users of the catalogues should be educated through documentation which has to be designed to be easy to understand for beginners while at the same time allowing experts to quickly navigate to the relevant details.

Training the next generation of astrometrists: The design and operation of astrometric programmes is a highly specialized undertaking and it is essential to ensure continuity of the available expertise in the astronomical community. This expertise should cover all aspects of carrying out astrometric programmes, whether they are scanning or interferometric space missions or ground based astrometry programmes, ranging from the establishment of the reference frame with VLBI observations to 'classical' parallax programmes. In addition astrometry at the microarcsecond level requires orders of magnitude better control over instrument calibration and the routine inclusion of effects such as general relativity in the data analysis. Without dedicated astrometry experts the whole edifice illustrated in figure 1 will fall apart. The experts are also needed to ensure a thorough understanding of the contents and the interpretation of astrometric catalogues. At this symposium a session was therefore dedicated to astrometry education, and I refer to the corresponding chapter in this volume for a summary.

5. Final remarks

I will conclude with a little 'pep-talk'. Astrometry is unfortunately considered to be a rather unglamourous discipline which is often even put aside as not being 'science'. I disagree with this; designing and carrying out high accuracy astrometric experiments is a highly scientific undertaking in itself and the results are of fundamental importance to astronomy, astrophysics and geodynamics. Astrometry tells us about our place in the universe, provides the only direct distance measurements to the stars, and can even be used to look inside our own planet as discussed by Huang in this volume. If we seek to motivate people to pursue a career in astrometry we should advertise that the beauty of this discipline, illustrated in figure 3, lies in that fact that:

> By carefully watching the positions of stars on the sky over the course of time you can find out (almost) all there is to know about the universe.

Acknowledgements

I thank Michael Perryman for his comments and Andreas Quirrenbach for reminding me that astrometric radial velocities are interesting in practice, not just in principle.

The background image in figure 1 was obtained from the Digitized Sky Surveys, which were produced at the Space Telescope Science Institute under U.S. Government grant NAG W-2166. The Second Palomar Observatory Sky Survey (POSS-II) was made by the California Institute of Technology with funds from the National Science Foundation, the National Geographic Society, the Sloan Foundation, the Samuel Oschin Foundation, and the Eastman Kodak Corporation. The Oschin Schmidt Telescope is operated by the California Institute of Technology and Palomar Observatory.

References

Beichman, Ch. 2008, *this volume*, p.238
Boden, A. F. 2008, *this volume*, p.36
Borsenberger, J., *et al.* 2006, in *Visions for Infrared Astronomy*, eds. V. Coudé du Foresto, D. Rouan, & G. Rousset, Paris, RS série I2M vol. 6, Lavoisier
Bucciarelli, B. 2008, *this volume*, p.316
Dravins, D., Lindegren, L., & Madsen, S. 1999, *A&A*, 348, 1040
ESA 1997, *The Hipparcos and Tycho Catalogues*, ESA SP-1200
Ghez, A. M. 2008, *this volume*, p.52
Gouda, N. 2008, *this volume*, p.248
Hekker, S., Reffert, S., Quirrenbach, A., Mitchell, D. S., Fischer, D. A., Marcy, G. W., Butler, R. P. 2006, *A&A*, 454, 943
Høg, E., Fabricius, C., Makarov, V. V., *et al.* 2000, *A&A*, 355, L27
Ivezić, Ž. 2008, *this volume*, p.537
Jao, W. 2008, *this volume*, p.421
Lestrade, J.-F. 2008, *this volume*, p.170
Lindegren, L. 2005, in *The Three-Dimensional Universe with Gaia*, eds. C. Turon, K. S. O'Flaherty & M. A. C. Perryman, ESA SP-576, p.29
Lindegren, L. 2008, *this volume*, p.217
Ma, C. 2008, *this volume*, p.337
Madsen, S., Dravins, D., & Lindegren, L. 2002, *A&A*, 381, 446
Magnier, E. A. 2008, *this volume*, p.552
Monet, D. G., Levine, S. E., Canzian, B., *et al.* 2003, *AJ*, 125, 984
Morrison, J. E., Röser, S., McLean, B., *et al.* 2001, *AJ*, 121, 1752
Reid, M. 2008, *this volume*, p.141
Richichi, A. 2008, *this volume*, p.44
Röll, T., Seifahrt, T., Neuhauser, R. 2008, *this volume*, p.48
Skrutskie, M. F., Cutri, R. M., Stiening M. D., *et al.* 2006, *AJ*, 131, 1163
Shao, M. 2008, *this volume*, p.231
Smart, R. 2008, *this volume*, p. 429
Steinmetz, M., Zwitter, T., Siebert, A., *et al.* 2006, *AJ*, 132, 1645
van Leeuwen, F. 2007, *Hipparcos, the New Reduction of the Raw Data*, Springer
Zacharias, N., Urban, S. E., Zacharias, M. I., *et al.* 2004, *AJ*, 127, 3043

Position Formulated
by the Scientific Organizing Committee

IAU Symposium No. 248, 'A Giant Step: From Milli- to Micro-Arcsecond Astrometry', was held in Shanghai between 15–19 October 2007. It represented a gathering of 200 international scientists involved in all aspects of astrometry, ranging from involvement in completed and planned ground and space instrumentation in the optical, infrared, and radio, as well as those involved in the modelling and astrophysical interpretation of advanced astrometric data, now and in the future.

Over the coming years, funding and support will be required for all aspects of this fundamental and rapidly developing field, including maintenance and improvement of the International Celestial Reference Frame in the optical and radio, continued provision of state-of-the-art Solar System ephemerides, relativistic modelling, plate scanning machines, advanced data processing and archiving tools, theoretical and numerical modelling, provision of basic infrastructure, education, and other related areas. Approved facilities, and those in an advanced stage of development, such as Gaia, astrometric VLBI, VLTI (and other ground-based interferometers), and the wide-field CCD surveys such as Pan-STARRS and LSST, will all play central roles in these much-anticipated developments.

During the course of the symposium, the Scientific Organizing Committee took note of four particularly substantial scientific areas with current funding problems, considered of great importance for the continued development of the field. In each case, 'non-overlapping' plans exist across the international scene. However, current funding constraints might inhibit their timely development, to the potential detriment of the field as a whole. Not representing any particular priority order, these are:

(a) microarcsec-level infrared astrometry of the Galactic bulge, from which large-scale distances and dynamics of the central region of the Galaxy would become uniquely accessible; this scientific domain is exemplified by the Japanese-led JASMINE space mission concept;

(b) microarcsec-level optical astrometry focused on tens of thousands of selected targets over the 0-20 mag range (including individual objects in external galaxies, Earth-like planets around nearby stars, primary distance calibrators, etc); this scientific domain is exemplified by the NASA/JPL-led SIM PlanetQuest space mission concept;

(c) microarcsec-level radio astrometry on hundreds of objects, crucial for the maintenance and extension of the International Celestial Reference Frame, and other projects for the calibration of fundamental distance indicators in the Milky Way and for galaxies in the Hubble flow; this scientific domain is exemplified by the existing and highly-successful VLBA radio telescope;

(d) densification and consistency of the Optical Reference Frame towards fainter magnitudes, driven by the needs of the near-future wide-field surveys such as Pan-STARRS, VST, and LSST. The US Naval Observatory has been undertaking this effort embodied by the all-sky UCAC and URAT programmes. The special instrumental calibration needs will be ensured by the Deep Astrometric Standards initiative. These efforts will assume even greater importance and complexity in the future, as more and higher accuracy surveys will become routine tools applied across many fields of astronomy.

Relevant funding and decision-making bodies are invited to take note of the central importance of these programmes, as assessed by the international Scientific Organizing Committee during the course of the symposium.

Michael Perryman, Mark Reid, and Francois Mignard
on behalf of the Scientific Organising Committee

Author Index

Object Index

Subject Index

asteroids – 130, 266, 286, 328, 363, 544, 564

astrometry – 1, 8, 14, 18, 20, 23, 30, 36, 44, 48, 59, 66, 78, 82, 89, 93, 96, 98, 104, 108, 110, 112, 114, 116, 120, 122, 124, 128, 132, 134, 136, 138, 141, 156, 170, 182, 186, 192, 194, 196, 198, 202, 206, 208, 210, 212, 214, 217, 224, 231, 238, 244, 248, 252, 256, 260, 262, 266, 268, 272, 278, 282, 284, 286, 288, 296, 298, 300, 303, 310, 316, 320, 324, 326, 328, 330, 332, 334, 337, 344, 348, 352, 356, 363, 367, 379, 383, 387, 391, 395, 397, 401, 405, 417, 421, 450, 462, 466, 470, 474, 486, 490, 492, 496, 506, 508, 514, 521, 523, 525, 527, 529, 533, 535, 537, 544, 553, 560, 565, 567

astronomical data bases: miscellaneous – 260, 282, 548, 563

atmospheric effects – 210, 212

binaries: close – 116, 126, 130, 496

binaries: eclipsing – 118

binaries: general – 16, 30, 36, 48, 59, 104, 186, 208

BL Lacertae objects: general – 352

black hole physics – 141, 470, 498

catalogs – 14, 74, 89, 108, 276, 303, 310, 316, 320, 326, 330, 332, 334, 337, 395, 397, 533, 537, 548, 567

celestial mechanics – 356, 379

Cepheids – 23

circumstellar matter – 170

cosmological parameters – 141

dark matter – 450

distance scale – 23

Earth – 182, 194, 367, 374, 403, 409

eclipses – 114, 521

education – 514, 523, 525, 527, 533, 535

ephemerides – 20, 66, 93, 96, 182, 252, 367, 521, 560

Galactic: structure – 141

galaxies: active – 348

galaxies: dwarf – 244, 492

galaxies: fundamental parameters – 276

galaxies: halos – 498

galaxies: individual (Carina dSph) – 492

galaxies: interactions – 450

galaxies: jets – 348

galaxies: kinematics and dynamics – 474

galaxies: luminosity function – 276

Galaxy: abundances – 458

Galaxy: bulge – 248, 296, 510

Galaxy: center – 52, 100, 204, 466, 470

Galaxy: disk – 198, 433, 458, 462, 484

Galaxy: evolution – 443, 458

Galaxy: formation – 506

Galaxy: fundamental parameters – 1, 18, 52

Galaxy: general – 8, 484

Galaxy: halo – 30, 450

Galaxy: kinematics and dynamics – 52, 217, 443, 450, 462, 498, 502, 506, 512

Galaxy: solar neighbourhood – 1

Galaxy: stellar content – 74, 443, 504

Galaxy: structure – 196, 214, 462, 470, 506

general relativity – 397

globular clusters: general – 440, 512

globular clusters: individual (47 Tuc) – 48

gravitation – 264, 356, 379, 383, 391

gravitational lensing – 274, 387, 391

gravitational waves – 178

Hertzsprung-Russell diagram – 494, 500

infrared: general – 248, 326

infrared: stars – 132, 417, 466, 510

instrumentation: adaptive optics – 48, 100

instrumentation: detectors – 262, 270, 272, 310, 529

instrumentation: high angular resolution – 44, 170, 387

instrumentation: interferometers – 36, 100, 106, 124, 132, 141, 148, 164, 170, 274, 280, 288, 387, 417, 466

instrumentation: miscellaneous – 82

instrumentation: photometers – 224, 262, 300

instrumentation: spectrographs – 262, 510

ISM: clouds – 186

ISM: individual (G34.4+0.23) – 202